Table of Measurement Abbreviations ———

U.S. Customary System

Length		Capacity		Weight		Area	
in.	inch	oz	ounce	oz	ounce	in.^{2}	square inches
ft	feet	c	cup	lb	pound	ft^2	square feet
yd	yard	qt	quart				
mi	mile	gal	gallon				

Metric System

Length		Capacity		Weight/Mass		Area	
mm	millimeter (0.001 m)	ml	milliliter (0.001 L)	mg	milligram (0.001 g)	cm^2	square centimeters
cm	centimeter (0.01 m)	cl	centiliter (0.01 L)	cg	centigram (0.01 g)	m^2	square meters
dm	decimeter (0.1 m)	dl	deciliter (0.1 L)	dg	decigram (0.1 g)		
m	meter	L	liter	g	gram		
dam	decameter (10 m)	dal	decaliter (10 L)	dag	decagram (10 g)		
hm	hectometer (100 m)	hl	hectoliter (100 L)	hg	hectogram (100 g)		
km	kilometer (1000 m)	kl	kiloliter (1000 L)	kg	kilogram (1000 g)		

Time

h	hours	min	minutes	s	seconds

QA 154.2 .A93 1989

Aufmann, Richard N.

Intermediate algebra, with applications

BRADNER LIBRARY
SCHOOLCRAFT COLLEGE
LIVONIA, MICHIGAN 48152

Intermediate Algebra

WITH APPLICATIONS

Second Edition

Richard N. Aufmann
Palomar College, California

Vernon C. Barker
Palomar College, California

Joanne S. Lockwood
Plymouth State College, New Hampshire

HOUGHTON MIFFLIN COMPANY   **Boston**

Dallas   Geneva, Illinois   Palo Alto   Princeton, New Jersey

QA
154.2
.A93
1989

Cover Photograph by Tony Wong/ T. C. Patterson Design, Inc.

IBM is a registered trademark of International Business Machines Corporation. Apple is a registered trademark of Apple Computer, Inc.

Copyright © 1989 by Houghton Mifflin Company. All rights reserved.

No part of this work may be reproduced or transmitted in any form or by any means, electronic or mechanical, including photocopying and recording, or by any information storage or retrieval system without the prior written permission of Houghton Mifflin Company unless such copying is expressly permitted by federal copyright law. Address inquiries to College Permissions, Houghton Mifflin Company, One Beacon Street, Boston, MA 02108.

Printed in the U.S.A.
Library of Congress Catalog Card Number: 88-81320

ISBN Numbers:
Text: 0-395-43192-1
Instructor's Annotated Edition: 0-395-50501-1
ABCDEFGHIJ-DOH-9543210/898

Contents

Preface

The second edition of *Intermediate Algebra With Applications* provides mathematically sound and comprehensive coverage of the topics considered essential in an intermediate algebra course. Our strategy in preparing this revision has been to build on the successful features of the first edition, features designed to enhance the student's mastery of math skills. In addition, we have expanded the ancillary package by adding new teaching aids for the instructor and new computerized aids for the student.

Features

Immediate Reinforcement of Skills

Instructors have long recognized the need for a text that requires the student to use a skill as it is being taught. *Intermediate Algebra With Applications* uses an interactive technique that meets this need. Each section is divided into objectives, and every objective contains one or more sets of matched-pair examples. The first example in each set is worked out; the second example is not. By solving this second problem, the student interacts with the text. The complete worked-out solutions to these examples are provided in the answer section at the end of the book, so the student can obtain immediate feedback on and reinforcement of the skill being learned.

Emphasis on Problem-Solving Strategies

Intermediate Algebra With Applications features a carefully developed approach to problem solving that emphasizes developing strategies to solve problems. For each type of word problem contained in the text, the student is prompted to use a "strategy step" before performing the actual manipulation of numbers and variables. By developing problem-solving strategies, the student will know better how to analyze and solve those word problems encountered in an intermediate algebra course.

Applications

The traditional approach to teaching or reviewing algebra covers only the straightforward manipulation of numbers and variables and thereby fails to teach students the practical value of algebra. By contrast, *Intermediate Algebra With Applications* emphasizes applications. Wherever appropriate, the last objective in each section presents applications that require the student to use the skills covered in that section to solve practical problems. Also, almost all of Chapter 2 and portions of several other chapters are devoted to certain standard types of applications. This carefully integrated applied approach generates awareness on the student's part of the value of algebra as a real-life tool.

Keyed Objectives

Each chapter begins with a list of the learning objectives included within the sections of that chapter. Each of the objectives in the chapter is keyed by a numbered bullet (in color).

Exercises

In this edition, we have dramatically increased the number of exercises. There are now more than 6,000 exercises, grouped in the following categories:

- **End-of-section exercise sets,** which are keyed to the corresponding learning objectives to organize the material for easy reference, provide ample practice and review of each skill.
- **Supplemental exercise sets,** designed to increase the student's ability to solve problems requiring a combination of skills, have been added at the end of each section.
- **Cumulative review exercises,** which appear at the end of each chapter (beginning with Chapter 2), help the student retain math skills learned earlier.
- **Final Exam,** which follows the last chapter, can be used as a review item or practice final.

New to This Edition

Topical Coverage

In this edition of the text, Chapter 1, a review of the real numbers, has been condensed in order to shorten the material devoted to reviewing first-year algebra. The chapter on linear equations now precedes Chapter 7, *Quadratic Equations.* In the first edition, all the lessons on inequalities were contained within a single chapter; in this edition, the material on inequalities has been placed appropriately within other chapters of the text.

The topical coverage in this second edition has been expanded to include an entire chapter on functions and relations (Chapter 8). The chapter on systems of equations has been expanded to include solving systems of equations by using matrices.

Chapter Openers

Each chapter begins with a mathematically oriented topic of interest. These range from such topics as the origins of equal signs to the four-color theorem.

Calculator Exercises

Selected calculator exercises are included throughout the exercise sets. These are marked by the symbol ▦ and are designed to provide the student with the opportunity to practice using a hand-held calculator.

Calculator and Computer Enrichment Topics

Each chapter also contains optional calculator or computer enrichment topics. Calculator topics (indicated by a ▦) provide the student with valuable key-stroking instructions and practice in using a hand-held calculator. Computer topics (marked by a ▣) correspond directly to the programs on the Student Disk. These topics range from solving first-degree equations to graphing linear equations in two variables.

Supplements For The Student

In addition to the student Study Guide, two computerized study aids, the Computer Tutor and the Student Disk, accompany *Intermediate Algebra With Applications*.

Study Guide

Each chapter of the Study Guide begins by providing insight into the practical value of the material in the chapter. Complete worked-out solutions to all odd-numbered exercises contained in the text are also included. A test concludes each chapter, with solutions to the test provided in the answer section at the back of the Study Guide.

The COMPUTER TUTOR™

The Computer Tutor™ is an interactive instructional microcomputer program for student use. Each learning objective in the text is supported by a lesson on the Computer Tutor™. As a reminder of this, the symbol ▯ appears to the right of each objective title in the text. Lessons on the tutor provide additional instruction and practice and can be used in several ways: (1) to cover material the student missed because of absence from class; (2) to repeat instruction on the skill or concept that the student has not yet mastered; or (3) to review material in preparation for examinations. This tutorial program is available for both the Apple II family of computers and the IBM PC and compatible computers.

Student Disk

The Student Disk contains a number of computational and drill-and-practice programs that correspond to selected Calculator and Computer Enrichment Topics in the text. The logo ▪ next to a computer topic indicates that a corresponding program can be found on the Student Disk. These programs are available for both the Apple II family of computers and the IBM PC or compatible computers.

Supplements For The Instructor

Intermediate Algebra With Applications has an unusually complete set of teaching aids for the instructor.

Instructor's Annotated Edition

The Instructor's Annotated Edition is an exact replica of the student text except that the answers to all of the exercises are printed in color next to the problems.

Solutions Manual

The Solutions Manual contains worked-out solutions for all end-of-section and supplemental exercise sets, cumulative reviews, and the final exam.

Instructor's Manual with Testing Program

The Instructor's Manual contains the printed testing program, which is the first of three sources of testing material available to users of *Intermediate Algebra With Applications*. Four printed tests (in two formats — free response and multiple choice) are provided for each chapter, as are cumulative and final exams. In addition, the Instructor's Manual includes documentation for all the software ancillaries — the Student Disk, the Computer Tutor™, and the Instructor's Computerized Test Generator.

Instructor's Computerized Test Generator

The Instructor's Computerized Test Generator is the second source of testing material for use with *Intermediate Algebra With Applications*. The database contains more than 1,500 test items. These questions are unique to the test generator and do not repeat items provided in the Instructor's Manual testing program. Organized according to the keyed objectives in the text, the Test Generator is designed to produce an unlimited number of tests for each chapter of the text, including cumulative tests and final exams. It is available for both the Apple II family of computers and the IBM PC or compatible computers.

Printed Test Bank

The Printed Test Bank, the third component of the testing material, is a printout of all items in the Instructor's Computerized Test Generator. Instructors using the Test Generator can use the test bank to select specific items from the database. Instructors who do not have access to a computer can use the test bank to select items to be included on a test being prepared by hand.

GPA: Grade Performance Analyzer

GPA is a microcomputer grading and record-keeping program available for use on the IBM PC and Apple II.

Videotapes

Approximately 25 half-hour videotape lessons accompany *Intermediate Algebra With Applications*. These lessons follow the format and style of the text and are closely tied to specific sections of the text.

Acknowledgments

The authors would like to thank the people who have reviewed this manuscript and provided many valuable suggestions:

Glenn Adamson
Emporia State University, KS

Joseph Buggan
Community College of Allegheny County, PA

Arthur P. Dull
Diablo Valley College, CA

Aparna B. Ganguli
University of Minnesota, MI

Henry Graves
Trident Technical College, SC

Steve E. Green
Tyler Junior College, TX

Joel Greenstein
New York City Technical, NY

Miriam Keesey
San Diego State University, CA

Jamie King
Orange Coast College, CA

Norman Ladd
College of the Redwoods, CA

Randy Leifson
Pierce College, WA

David Longshore
Victor Valley College, CA

Carol Jean Martin
Dodge City Community College, KS

Micheal J. Mears
Manatee Community College, FL

Judith Miller
Delta College, MI

Lillian Netlitzky
California State Polytechnic University, CA

Carol C. Oelkers
Fullerton College, CA

Martha Sklar
Los Angelos City College, CA

Sally Stevens
Midway College, KY

Hubert C. Voltz
Slippery Rock University of Pennsylvania, PA

Ray Westergard
Chabot College, CA

Loyd Wilcox
Golden West College, CA

Ray Wilson
Central Piedmont Community College, NC

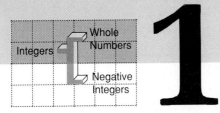

1

Review of Real Numbers

OBJECTIVES

- Absolute value and additive inverse
- Operations on rational numbers
- Exponential expressions
- The Order of Operations Agreement
- Evaluate variable expressions
- The Properties of the Real Numbers
- Simplify variable expressions
- Translate a verbal expression into a variable expression and simplify the resulting expression
- The union and intersection of sets
- Graph the solution set of an inequality of one variable

Early Egyptian Numeration System

The early Egyptian type of picture writing shown at the right is known as hieroglyphics.

The Egyptian hieroglyphic method of representing numbers differs from our modern version in an important way. For the hieroglyphic number, the symbol indicated the value. For example, the symbol ∩ meant 10. The symbol was repeated to get larger values.

The markings at the right represent the number 743. Each vertical stroke represents 1, each ∩ represents 10, and each 9 represents 100.

There are 3 ones, 4 tens, and 7 hundreds representing the number 743.

$$3 + 40 + 700$$
$$743$$

In our system, the position of a number is important. The number 5 in 356 means 5 tens, but the number 5 in 3517 means 5 hundreds. Our system is called a positional number system.

Consider the hieroglyphic number at the right. Notice that the ∩ is missing. For the early Egyptians, when a certain group of ten was not needed, it was just omitted.

There are 4 ones and 5 hundreds. Thus the number 504 is represented by this group of markings.

$$4 + 500$$
$$504$$

In a positional system of notation like ours, a zero is used to show that a certain group of ten is not needed. This may seem like a fairly simple idea, but it was not until nearly the end of the seventh century that a zero was introduced to the number system.

1.1

Operations on the Real Numbers

1 ## Absolute value and additive inverse

The **integers** are $\ldots, -4, -3, -2, -1, 0, 1, 2, 3, 4, \ldots$

The three dots before and after the list of integers mean the list continues without end, and that there is no smallest integer and there is no largest integer.

The integers can be shown on the number line. The integers to the left of zero on the number line are **negative integers.** The integers to the right of zero are **positive integers.** Zero is neither a positive nor a negative integer.

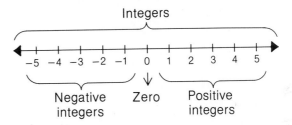

The positive integers are also called the **natural numbers.** $1, 2, 3, 4, \ldots$

The positive integers and zero are called the **whole numbers.** $0, 1, 2, 3, \ldots$

Just as the word *it* is used in language to stand for an object, a letter of the alphabet can be used in mathematics to stand for a number. Such a letter is called a **variable.**

A **rational number** is the quotient of two integers. Therefore, a rational number is a number which can be written in the form $\frac{a}{b}$, where a and b are integers, and $b \neq 0$ (b is not equal to zero). A rational number written in this way is commonly called a fraction.

Every integer is a rational number since an integer can be written as the quotient of the integer and 1; for example, $6 = \frac{6}{1}$. A number written in decimal notation is also a rational number; for example, $0.7 = \frac{7}{10}$.

3

Every rational number can be written as a terminating or repeating decimal.

Some numbers, for example $\sqrt{7}$ and π, have decimal representations which never terminate or repeat. These numbers are called **irrational numbers.**

$$\sqrt{7} = 2.6457513\ldots \qquad\qquad \pi = 3.141592604\ldots$$

The rational numbers and the irrational numbers taken together are called the **real numbers.**

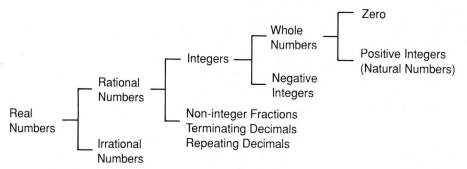

A number line can be used to show the relative order of two numbers a and b. If a is to the left of b on the number line, then a is less than b ($a < b$). If a is to the right of b on the number line, then a is greater than b ($a > b$).

Negative 2 is greater than negative 4.

$$-2 > -4$$

Negative 1 is less than 2.

$$-1 < 2$$

Two numbers that are the same distance from zero on the number line but on opposite sides of zero are **opposite numbers,** or **opposites.** The opposite of a number is called its **additive inverse.**

-3 is the additive inverse of 3.

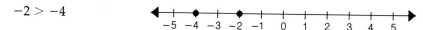

The additive inverse of -4 is 4.

Example 1   Find the additive inverse.

A. -16 B. $\dfrac{3}{4}$

Solution   A. 16   B. $-\dfrac{3}{4}$

Problem 1 Find the additive inverse.

A. 18   B. −5.2

Solution See page A5.

A. −18   B. 5.2

The **absolute value** of a number is a measure of its distance from zero on the number line. Therefore, the absolute value of a number is a positive number or zero. The symbol for absolute value is $|\ |$.

The absolute value of a positive number is the number itself. $|7| = 7$

The absolute value of a negative number is its additive inverse. $|-7| = 7$

The absolute value of zero is zero. $|0| = 0$

Example 2 Evaluate.

A. $|-8|$ B. $-|-2.9|$

Solution A. 8   B. −2.9   • The absolute value sign does not affect the negative sign in front of the absolute value sign.

Problem 2 Evaluate.

A. $|-11|$ B. $-\left|-\dfrac{4}{5}\right|$

Solution See page A5.

A. 11   B. $-\dfrac{4}{5}$

2 Operations on rational numbers

To add numbers with like signs, add the absolute values of the numbers. Then attach the sign of the addends.

$$-15 + (-27) = -42$$

To add numbers with unlike signs, find the difference between the absolute values of the numbers. Then attach the sign of the number with the greater absolute value.

$$-18.42 + 6.3 = -12.12$$

Addition of Rational Numbers

Addition of rational numbers is defined as

$$\frac{a}{b} + \frac{c}{b} = \frac{a + c}{b}.$$

To add fractions with unlike denominators, first rewrite the fractions as equivalent fractions using the least common multiple (LCM) of the denominators as the common denominator. Then add the fractions.

$$\frac{5}{8} + \frac{7}{12} = \frac{15}{24} + \frac{14}{24} = \frac{29}{24}$$

Subtraction is defined as the addition of the additive inverse.

If a and b are integers, then $a - b = a + (-b)$.

$$-\frac{3}{8} - \frac{1}{4} = -\frac{3}{8} + \left(-\frac{1}{4}\right) = -\frac{3}{8} + \left(-\frac{2}{8}\right) = -\frac{5}{8}$$

To multiply numbers with like signs, multiply the absolute values of the factors. The product is positive.

$$-31.7 \times (-0.05) = 1.585$$

To multiply numbers with unlike signs, multiply the absolute values of the factors. The product is negative.

$$-24.8 \times 0.02 = -0.496$$

Multiplication of Rational Numbers

The product of two fractions is the product of the numerators over the product of the denominators.

$$\frac{a}{b} \cdot \frac{c}{d} = \frac{ac}{bd}$$

$$-\frac{5}{12} \cdot \frac{8}{15} = -\frac{5 \cdot 8}{12 \cdot 15} = -\frac{\overset{1}{\cancel{5}} \cdot \overset{1}{\cancel{2}} \cdot \overset{1}{\cancel{2}} \cdot 2}{\underset{1}{\cancel{2}} \cdot \underset{1}{\cancel{2}} \cdot 3 \cdot 3 \cdot \underset{1}{\cancel{5}}} = -\frac{2}{9}$$

The quotient of two numbers with like signs is positive.

$$-1.5 \div (-0.05) = 30$$

The quotient of two numbers with unlike signs is negative.

$$3.8 \div (-0.19) = -20$$

The **reciprocal** of a fraction is the fraction with the numerator and denominator interchanged. The process of interchanging the numerator and denominator of a fraction is called **inverting**.

Division of Rational Numbers

To divide two fractions, multiply by the reciprocal of the divisor.

$$\frac{a}{b} \div \frac{c}{d} = \frac{a}{b} \cdot \frac{d}{c}$$

$$-\frac{3}{8} \div \frac{9}{16} = -\frac{3}{8} \cdot \frac{16}{9} = -\frac{\overset{1}{3} \cdot \overset{1}{2} \cdot \overset{1}{2} \cdot \overset{1}{2} \cdot 2}{\underset{1}{2} \cdot \underset{1}{2} \cdot \underset{1}{2} \cdot \underset{1}{3} \cdot 3} = -\frac{2}{3}$$

Example 3 Simplify.

A. $\dfrac{3}{8} + \dfrac{5}{12} - \dfrac{9}{16}$ B. $6.329 - 12.49$ C. $-\dfrac{5}{12} \cdot \dfrac{4}{25}$

Solution A. $\dfrac{3}{8} + \dfrac{5}{12} - \dfrac{9}{16} = \dfrac{18}{48} + \dfrac{20}{48} - \dfrac{27}{48} = \dfrac{18 + 20 - 27}{48} = \dfrac{11}{48}$

B. $6.329 - 12.49 = 6.329 + (-12.49)$

$$\begin{array}{r} 12.490 \\ -\ 6.329 \\ \hline 6.161 \end{array} \qquad 6.329 - 12.49 = -6.161$$

C. $-\dfrac{5}{12} \cdot \dfrac{4}{25} = -\dfrac{5 \cdot 4}{12 \cdot 25} = -\dfrac{\overset{1}{5} \cdot \overset{1}{2} \cdot \overset{1}{2}}{\underset{1}{2} \cdot \underset{1}{2} \cdot 3 \cdot \underset{1}{5} \cdot 5} = -\dfrac{1}{15}$

Problem 3   Simplify.

A. $\dfrac{5}{6} - \dfrac{3}{8} + \dfrac{7}{9}$ B. $-8.729 + 12.094$ C. $\dfrac{5}{8} \div \left(-\dfrac{15}{40}\right)$

Solution   See page A5.

A. $\dfrac{89}{72}$ B. 3.365 C. $-\dfrac{5}{3}$

Example 4   Simplify: $0.0527 \div (-0.27)$ Round to the nearest hundredth.

Solution

$$
\begin{array}{r}
0.195 \approx 0.20 \\
0.27\,)\overline{0.05\,270} \\
-2\,7 \\
\hline
2\,57 \\
-2\,43 \\
\hline
140 \\
-135 \\
\hline
5
\end{array}
$$

• The symbol $\approx$ is used to indicate that the quotient is an approximate value after being rounded off.

$0.0527 \div (-0.27) \approx -0.20$

Problem 4   Simplify: $-4.027(0.49)$ Round to the nearest hundredth.

Solution   See page A5.
-1.97

3 Exponential expressions

Repeated multiplication of the same factor can be written using an exponent.

$$2 \cdot 2 \cdot 2 \cdot 2 \cdot 2 \cdot 2 = 2^6 \longleftarrow \textbf{Exponent}$$
$$\uparrow \rule{0pt}{0pt}\,\underline{}\ \textbf{Base}$$

$$b \cdot b \cdot b \cdot b \cdot b = b^5 \longleftarrow \textbf{Exponent}$$
$$\uparrow \underline{}\ \textbf{Base}$$

The **exponent** indicates how many times the factor, called the **base**, occurs in the multiplication. The multiplication $2 \cdot 2 \cdot 2 \cdot 2 \cdot 2 \cdot 2$ is in **factored form**. The exponential expression 2^6 is in **exponential form.**

2^1 is read "the first power of two" or just "two." $\longrightarrow$ Usually the exponent 1 is not written.

2^2 is read "the second power of two" or "two squared."

2^3 is read "the third power of two" or "two cubed."

2^4 is read "the fourth power of two."

2^5 is read "the fifth power of two."

b^5 is read "the fifth power of b."

If b is an integer and n is a positive integer, the nth power of b is defined as the product of n factors of b.

$$b^n = \underbrace{b \cdot b \cdot b \ldots b}_{n \text{ factors}}$$

To evaluate an exponential expression, write each factor as many times as indicated by the exponent. Then multiply.

$$5^4 = 5 \cdot 5 \cdot 5 \cdot 5 = 625$$

$$3^2 \cdot 5^3 = (3 \cdot 3)(5 \cdot 5 \cdot 5) = 9 \cdot 125 = 1125$$

Example 5   Evaluate $(-3)^4$ and -3^4.

Solution $(-3)^4 = (-3)(-3)(-3)(-3) = 81$
$-3^4 = -(3 \cdot 3 \cdot 3 \cdot 3) = -81$

• The negative of a number is taken to a power only when the negative sign is *inside* the parentheses.

Problem 5   Evaluate $(-2)^4$ and -2^4.

Solution   See page A5.
16 and -16

Example 6   Evaluate.
A. $(-2)^3 \cdot 3^2$ B. $\left(-\dfrac{2}{3}\right)^2 \cdot 3^3$

Solution   A. $(-2)^3 \cdot 3^2 = (-2)(-2)(-2) \cdot (3)(3) = -8 \cdot 9 = -72$
B. $\left(-\dfrac{2}{3}\right)^2 \cdot 3^3 = \left(-\dfrac{2}{3}\right)\left(-\dfrac{2}{3}\right) \cdot (3)(3)(3) = 12$

Problem 6   Evaluate.
A. $-3^3 \cdot 2^2$ B. $-\left(\dfrac{2}{5}\right)^3 \cdot 5^2$

Solution   See page A5.
A. -108 B. $-\dfrac{8}{5}$

4 The Order of Operations Agreement

In order to prevent more than one answer to the same problem, an Order of Operations Agreement is followed.

The Order of Operations Agreement

Step 1 Perform operations inside grouping symbols. Grouping symbols include parentheses ( ), brackets [ ], and the fraction bar.

Step 2 Simplify exponential expressions.

Step 3 Do multiplication and division as they occur from left to right.

Step 4 Do addition and subtraction as they occur from left to right.

Simplify: $8 - \dfrac{12 - 2}{4 + 1} \div 2^2$

$$8 - \dfrac{12 - 2}{4 + 1} \div 2^2$$

Perform operations above and below the fraction bar.

$$8 - \dfrac{10}{5} \div 2^2$$

Simplify exponential expressions.

$$8 - \dfrac{10}{5} \div 4$$

Do multiplication and division as they occur from left to right. Note that a fraction bar can be read "÷".

$$8 - 2 \div 4$$

$$8 - \dfrac{1}{2}$$

Do addition and subtraction as they occur from left to right.

$$\dfrac{15}{2}$$

One or more of the above steps may not be needed to simplify an expression. In that case, proceed to the next step in the Order of Operations Agreement.

When an expression has grouping symbols inside grouping symbols, perform the operations inside the inner grouping symbols first.

Simplify: $4.3 - [(25 - 9) \div 2]^2$

Perform operations inside grouping symbols.

$$4.3 - [(25 - 9) \div 2]^2$$
$$4.3 - [16 \div 2]^2$$
$$4.3 - [8]^2$$

Simplify exponential expressions.

$$4.3 - 64$$

Perform addition and subtraction as they occur from left to right.

$$-59.7$$

A **complex fraction** is a fraction whose numerator or denominator contains one or more fractions. Examples of complex fractions are shown below.

$$\dfrac{\dfrac{2}{3}}{\dfrac{1}{3}} \qquad\qquad \dfrac{\dfrac{2}{5}+1}{\dfrac{7}{8}} \quad \longleftarrow \text{Main Fraction Bar}$$

To simplify a complex fraction, perform operations above and below the main fraction bar as the first step in the Order of Operations Agreement.

Simplify: $\dfrac{\dfrac{3}{4}-1}{\dfrac{5}{8}}$

Perform operations above and below the main fraction bar.

Multiply the numerator of the complex fraction by the reciprocal of the denominator of the complex fraction.

$$\dfrac{\dfrac{3}{4}-1}{\dfrac{5}{8}}$$

$$\dfrac{-\dfrac{1}{4}}{\dfrac{5}{8}}$$

$$-\dfrac{1}{4}\cdot\dfrac{8}{5}$$

$$-\dfrac{1\cdot 8}{4\cdot 5}$$

$$-\dfrac{1\cdot\overset{1}{\cancel{2}}\cdot\overset{1}{\cancel{2}}\cdot 2}{\underset{1}{\cancel{2}}\cdot\underset{1}{\cancel{2}}\cdot 5}$$

$$-\dfrac{2}{5}$$

Example 7 Simplify: $(-1.2)^3 - 8.4 \div 2.1$

Solution $(-1.2)^3 - 8.4 \div 2.1$
$-1.728 - 8.4 \div 2.1$ • Simplify exponential expressions.
$-1.728 - 4$ • Do multiplication and division.
-5.728 • Do addition and subtraction.

Problem 7 Simplify: $(3.81 - 1.41)^2 \div 0.036 - 1.89$

Solution See page A5.
158.11

Example 8   Simplify: $\left(\frac{1}{2}\right)^3 - \left[\left(\frac{2}{3} + \frac{1}{4}\right) \div \frac{5}{6}\right]$

Solution $\left(\frac{1}{2}\right)^3 - \left[\left(\frac{2}{3} + \frac{1}{4}\right) \div \frac{5}{6}\right]$

$\left(\frac{1}{2}\right)^3 - \left[\frac{11}{12} \div \frac{5}{6}\right]$ • Perform operations inside the inner grouping symbols.

$\left(\frac{1}{2}\right)^3 - \frac{11}{10}$ • Perform operations inside the grouping symbols.

$\frac{1}{8} - \frac{11}{10}$ • Simplify exponential expressions.

$-\frac{39}{40}$ • Do addition and subtraction.

Problem 8   Simplify: $\frac{1}{3} + \frac{5}{8} \div \frac{15}{16} - \frac{7}{12}$

Solution   See page A6. $\frac{5}{12}$

Example 9   Simplify: $9 \cdot \dfrac{\frac{5}{6} - 2}{\frac{3}{8}} \div \frac{7}{6}$

Solution $9 \cdot \dfrac{\frac{5}{6} - 2}{\frac{3}{8}} \div \frac{7}{6}$

$9 \cdot \dfrac{-\frac{7}{6}}{\frac{3}{8}} \div \frac{7}{6}$ • Do operations above the main fraction bar.

$9 \cdot \left(-\frac{7}{6} \cdot \frac{8}{3}\right) \div \frac{7}{6}$ • Multiply the numerator of the complex fraction by the reciprocal of the denominator of the complex fraction.

$9 \cdot \left(-\frac{28}{9}\right) \div \frac{7}{6}$

$-28 \div \frac{7}{6}$ • Do multiplication and division as they occur from left to right.

$-28 \cdot \frac{6}{7}$

-24

Problem 9   Simplify: $\dfrac{11}{12} - \dfrac{\dfrac{5}{4}}{2 - \dfrac{7}{2}} \cdot \dfrac{3}{4}$

Solution   See page A6. $\dfrac{37}{24}$

EXERCISES 1.1

1   Find the additive inverse.

1. 83
   -83

2. 51
   -51

3. -75
   75

4. -126
   126

5. 9.3
   -9.3

6. 2.7

   -2.7

7. -6.4

   6.4

8. -43.9

   43.9

9. $-\dfrac{11}{12}$

   $\dfrac{11}{12}$

10. $-\dfrac{8}{9}$

   $\dfrac{8}{9}$

Evaluate.

11. $-|126|$
   -126

12. $-|89|$
   -89

13. $|-436|$
   436

14. $|-502|$
   502

15. $-|-16|$
   -16

16. $-|-22|$

   -22

17. $|-4.93|$

   4.93

18. $-|72.1|$

   -72.1

19. $-\left|-\dfrac{7}{8}\right|$

   $-\dfrac{7}{8}$

20. $\left|-\dfrac{15}{16}\right|$

   $\dfrac{15}{16}$

2   Simplify.

21. $\dfrac{7}{12} + \dfrac{5}{16}$ $\dfrac{43}{48}$

22. $\dfrac{3}{8} - \dfrac{5}{12}$ $-\dfrac{1}{24}$

23. $-\dfrac{5}{9} - \dfrac{14}{15}$ $-\dfrac{67}{45}$

24. $\dfrac{1}{2} + \dfrac{1}{7} - \dfrac{5}{8}$ $\dfrac{1}{56}$

25. $-\dfrac{1}{3} + \dfrac{5}{9} - \dfrac{7}{12}$ $-\dfrac{13}{36}$

26. $\dfrac{1}{3} + \dfrac{19}{24} - \dfrac{7}{8}$ $\dfrac{1}{4}$

27. $\dfrac{2}{3} - \dfrac{5}{12} + \dfrac{5}{24}$ $\dfrac{11}{24}$

28. $-\dfrac{7}{10} + \dfrac{4}{5} + \dfrac{5}{6}$ $\dfrac{14}{15}$

29. $\dfrac{5}{8} - \dfrac{7}{12} + \dfrac{1}{2}$ $\dfrac{13}{24}$

30. $-\dfrac{1}{3} \cdot \dfrac{5}{8}$ $-\dfrac{5}{24}$

31. $\left(\dfrac{6}{35}\right)\left(-\dfrac{5}{16}\right)$ $-\dfrac{3}{56}$

32. $\dfrac{2}{3}\left(-\dfrac{9}{20}\right) \cdot \dfrac{5}{12}$ $-\dfrac{1}{8}$

33. $-\dfrac{8}{15} \div \dfrac{4}{5}$ $-\dfrac{2}{3}$

34. $-\dfrac{2}{3} \div \left(-\dfrac{6}{7}\right)$ $-\dfrac{7}{9}$

35. $-\dfrac{11}{24} \div \dfrac{7}{12}$ $-\dfrac{11}{14}$

36. $\dfrac{7}{9} \div \left(-\dfrac{14}{27}\right)$ $-\dfrac{3}{2}$

37. $\left(-\dfrac{5}{12}\right)\left(\dfrac{4}{35}\right)\left(\dfrac{7}{8}\right)$ $-\dfrac{1}{24}$

38. $\dfrac{6}{35}\left(-\dfrac{7}{40}\right)\left(-\dfrac{8}{21}\right)$ $\dfrac{2}{175}$

39. $-14.27 + 1.296$ -12.974

40. $-0.4355 + 172.5$ 172.0645

41. $1.832 - 7.84$ -6.008

42. $(3.52)(4.7)$ 16.544

43. $(0.03)(10.5)(6.1)$ 1.9215

44. $(1.2)(3.1)(-6.4)$ -23.808

45. $5.418 \div (-0.9)$ -6.02

46. $-0.2645 \div (-0.023)$ 11.5

47. $-0.4355 \div 0.065$ -6.7

48. $-6.58 - 3.97 + 0.875$ -9.675

49. $8.8 + 3.072 - 56.457$ -44.585

Simplify. Round to the nearest hundredth.

50. $38.241 - (-6.027) - 7.453$ 36.82

51. $-9.0508 - (-3.177) + 24.77$ 18.90

52. $-287.3069 \div 0.1415$ -2030.44

53. $6472.3018 \div (-3.59)$ -1802.87

3   Simplify.

54. -2^3
-8

55. -4^3
-64

56. $(-5)^3$
-125

57. $(-8)^2$
64

58. $2^2 \cdot 3^4$
324

59. $4^2 \cdot 3^3$
432

60. $-2^2 \cdot 3^2$
-36

61. $-3^2 \cdot 5^3$
-1125

62. $(-2)^3(-3)^2$
-72

63. $(-4)^3(-2)^3$
512

64. $-4(-3)^2(4^2)$
-576

65. $2^2(-10)(-2)^2$
-160

66. $\left(-\dfrac{3}{4}\right)^2(2)^4$
9

67. $\left(\dfrac{2}{3}\right)^3(-9)^2$
24

68. $-\left(\dfrac{4}{5}\right)^2(5)^3$
-80

69. $-\left(\dfrac{3}{5}\right)^2(10)^2$
-36

4   Simplify.

70. $5 - 3(8 \div 4)^2$ -7

71. $4^2 - (5 - 2)^2 \cdot 3$ -11

72. $16 - \dfrac{2^2 - 5}{3^2 + 2}$ $\dfrac{177}{11}$

73. $\dfrac{4(5 - 2)}{4^2 - 2^2} \div 4$ $\dfrac{1}{4}$

74. $\dfrac{3 + \dfrac{2}{3}}{\dfrac{11}{16}}$ $\dfrac{16}{3}$

75. $\dfrac{\dfrac{11}{14}}{4 - \dfrac{6}{7}}$ $\dfrac{1}{4}$

76. $5[(2 - 4) \cdot 3 - 2]$ -40

77. $2[(16 \div 8) - (-2)] + 4$ 12

78. $16 - 4\left(\dfrac{8 - 2}{3 - 6}\right) \div \dfrac{1}{2}$ 32

79. $25 \div 5\left(\dfrac{16 + 8}{-2^2 + 8}\right) - 5$ 25

80. $6[3 - (-4 + 2) \div 2]$ 24

81. $12 - 4[2 - (-3 + 5) - 8]$ 44

82. $\dfrac{1}{2} - \left(\dfrac{2}{3} \div \dfrac{5}{9}\right) + \dfrac{5}{6}$ $\dfrac{2}{15}$

83. $\left(-\dfrac{3}{5}\right)^2 - \dfrac{3}{5} \cdot \dfrac{5}{9} + \dfrac{7}{10}$ $\dfrac{109}{150}$

84. $\dfrac{1}{2} - \dfrac{\frac{17}{25}}{4 - \frac{3}{5}} \div \dfrac{1}{5}$ $-\dfrac{1}{2}$

85. $\dfrac{3}{4} + \dfrac{3 - \frac{7}{9}}{\frac{5}{6}} \cdot \dfrac{2}{3}$ $\dfrac{91}{36}$

86. $0.4(1.2 - 2.3)^2 + 5.8$ 6.284

87. $5.4 - (0.3)^2 \div 0.09$ 4.4

⌨ Use your calculator for the following exercises.

88. $1.75 \div 0.25 - (1.25)^2$ 5.4375

89. $(3.5 - 4.2)^2 - 3.50 \div 2.5$ -0.91

SUPPLEMENTAL EXERCISES 1.1

Classify each of the following numbers as a natural number, an integer, a
positive integer, a negative integer, a rational number, an irrational number,
and a real number.

90. -1
   integer, neg. integer, rational no., real no.

91. 0
   integer, rational no., real no.

92. 65
   natural no., integer, pos. integer, rational no., real no.

93. $\dfrac{17}{32}$
   rational no., real no.

94. $-\dfrac{19}{21}$
   rational no., real no.

95. -8.43
   rational no., real no.

96. $\sqrt{5}$
   irrational no., real no.

97. $-\sqrt{3}$
   irrational no., real no.

98. $-6.\overline{3}$
   rational no., real no.

99. $0.232332333\ldots$
   irrational no., real no.

100. 3.14
   rational no., real no.

101. π
   irrational no., real no.

Complete.

102. If a is a positive number, then $-a$ is a __neg__ number.

103. If a is a negative number, then $-a$ is a __pos__ number.

104. The product of an even number of negative factors is a __pos__ number.

105. The product of an odd number of negative factors is a __neg__ number.

106. A number that is its own additive inverse is ___0___ .

107. A number that is its own reciprocal is __1, −1__ .

Solve.

108. What is the sum of the three largest prime numbers less than 100?
269

109. How many positive integers less than 10 have an odd number of positive integral divisors?
3 integers; 1, 4, and 9

110. What is the tens' digit in 11^{22}?
2

111. In 1963, mathematicians used a computer to show that $2^{11213} - 1$ is a prime number. When written in standard notation, this number contains 3376 digits. What is the one's digit?
1

112. For any rational number $\dfrac{m}{n}$, what is the maximum number of digits in a repeating cycle of the repeating digits?
$n - 1$

SECTION 1.2 _____

Variable Expressions

1 Evaluate variable expressions

An expression that contains one or more variables is called a **variable expression.**

A variable expression is shown at the right. The expression has 4 addends, which are called **terms** of the expression. The variable expression has 3 variable terms and 1 constant term.

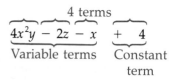

4 terms

$$4x^2y - 2z - x + 4$$

Variable terms   Constant term

Each variable term is composed of a **numerical coefficient** and a **variable part.** When the numerical coefficient is 1 or −1, the 1 is usually not written.

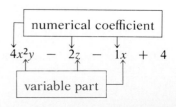

numerical coefficient

$$4x^2y \quad - \quad 2z \quad - \quad 1x \quad + \quad 4$$

variable part

Replacing the variable in a variable expression by a numerical value and then simplifying the resulting expression is called **evaluating the variable expression.**

Example 1   Evaluate $a^2 - (ab - c)$ when $a = -2$, $b = 3$, and $c = -4$.

Solution $a^2 - (ab - c)$
$(-2)^2 - [(-2)(3) - (-4)]$ • Replace each variable in the expression with the number it stands for.

$(-2)^2 - [-6 - (-4)]$ • Use the Order of Operations Agreement to simplify the resulting numerical expression.

$(-2)^2 - [-2]$
$4 - [-2]$
6

Problem 1   Evaluate $(b - c)^2 \div ab$ when $a = -3$, $b = 2$, and $c = -4$.

Solution   See page A6.
-24

2   The Properties of the Real Numbers

The Properties of the Real Numbers describe the way operations on numbers can be performed. Here is a list of some of the real number properties and an example of each property.

PROPERTIES OF THE REAL NUMBERS

> **The Commutative Property of Addition**
> $a + b = b + a$ $3 + 2 = 2 + 3$
> $5 = 5$

> **The Commutative Property of Multiplication**
> $a \cdot b = b \cdot a$ $(3)(-2) = (-2)(3)$
> $-6 = -6$

> **The Associative Property of Addition**
> $(a + b) + c = a + (b + c)$ $(3 + 4) + 5 = 3 + (4 + 5)$
> $7 + 5 = 3 + 9$
> $12 = 12$

The Associative Property of Multiplication

$(a \cdot b) \cdot c = a \cdot (b \cdot c)$

$$(3 \cdot 4) \cdot 5 = 3 \cdot (4 \cdot 5)$$
$$12 \cdot 5 = 3 \cdot 20$$
$$60 = 60$$

The Addition Property of Zero

$a + 0 = 0 + a = a$

$3 + 0 = 0 + 3 = 3$

The Multiplication Property of Zero

$a \cdot 0 = 0 \cdot a = 0$

$3 \cdot 0 = 0 \cdot 3 = 0$

The Multiplication Property of One

$a \cdot 1 = 1 \cdot a = a$

$5 \cdot 1 = 1 \cdot 5 = 5$

The Inverse Property of Addition

$a + (-a) = (-a) + a = 0$

$4 + (-4) = (-4) + 4 = 0$

a is called the additive inverse of $-a$.

$-a$ is called the additive inverse of a.

The sum of a number and its additive inverse is 0.

The Inverse Property of Multiplication

For $a \neq 0$, $a \cdot \dfrac{1}{a} = \dfrac{1}{a} \cdot a = 1$.

$(4)\left(\dfrac{1}{4}\right) = \left(\dfrac{1}{4}\right)(4) = 1$

$\dfrac{1}{a}$ is called the reciprocal of a.

$\dfrac{1}{a}$ is also called the multiplicative inverse of a.

The product of a number and its multiplicative inverse is 1.

> **The Distributive Property**
>
> $a(b + c) = ab + ac$ $3(4 + 5) = 3 \cdot 4 + 3 \cdot 5$
> $3 \cdot 9 = 12 + 15$
> $27 = 27$
>
> $(b + c)a = ba + ca$ $(4 + 5)2 = 4 \cdot 2 + 5 \cdot 2$
> $9 \cdot 2 = 8 + 10$
> $18 = 18$

> **Division Properties of Zero and One**
>
> Zero divided by any number other than zero is zero.
>
> For $a \neq 0$, $\dfrac{0}{a} = 0$. $\dfrac{0}{4} = 0$
>
> Division by zero is not defined.
>
> $\dfrac{a}{0}$ is undefined. $\dfrac{4}{0}$ is undefined
>
> Any number other than zero divided by itself is 1.
>
> For $a \neq 0$, $\dfrac{a}{a} = 1$. $\dfrac{-7}{-7} = 1$
>
> $\dfrac{0}{0}$ is not a real number.

Example 2   Complete the statement by using the Inverse Property of Addition.
$3x + ? = 0$

Solution $3x + (-3x) = 0$

Problem 2   Complete the statement by using the Commutative Property of Multiplication.
$(x)\left(\dfrac{1}{4}\right) = (?)(x)$

Solution   See page A7.
$(x)\left(\dfrac{1}{4}\right) = \left(\dfrac{1}{4}\right)(x)$

Example 3   Identify the property that justifies the statement.
$3(x + 4) = 3x + 12$

Solution   The Distributive Property

Problem 3   Identify the property that justifies the statement.
$(a + 3b) + c = a + (3b + c)$

Solution   See page A7.
The Associative Property of Addition

3   Simplify variable expressions

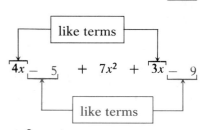

Like terms of a variable expression are the terms with the same variable part.

Constant terms are like terms.

like terms

$4x - 5 + 7x^2 + 3x - 9$

like terms

To **combine** like terms, use the Distributive Property $ba + ca = (b + c)a$ to add the coefficients.

$3x + 2x$
$(3 + 2)x$
$5x$

Example 4   Simplify: $2(x + y) + 3(y - 3x)$

Solution $2(x + y) + 3(y - 3x)$
$2x + 2y + 3y - 9x$ • Use the Distributive Property to remove parentheses.
$(2x - 9x) + (2y + 3y)$ • Use the Commutative and Associative Properties of Addition to rearrange and group like terms. (Do this step mentally.)

$-7x + 5y$ • Combine like terms.

Problem 4   Simplify: $(2x + xy - y) - (5x - 7xy + y)$

Solution   See page A7.
$-3x + 8xy - 2y$

Example 5   Simplify: $4y - 2[x - 3(x + y) - 5y]$

Solution $4y - 2[x - 3(x + y) - 5y]$
$4y - 2[x - 3x - 3y - 5y]$ • Use the Distributive Property to remove parentheses.

$4y - 2[-2x - 8y]$ • Combine like terms.
$4y + 4x + 16y$ • Use the Distributive Property to remove brackets.
$20y + 4x$ • Combine like terms.

Problem 5   Simplify: $2x - 3[y - 3(x - 2y + 4)]$

Solution   See page A7.
$11x - 21y + 36$

4   Translate a verbal expression into a variable expression and simplify the resulting expression

One of the major skills required in applied mathematics is to translate a verbal expression into a variable expression. This requires recognizing the verbal phrases that translate into mathematical operations. Some of the verbal phrases used to indicate the different mathematical operations are given below.

Addition	added to	5 added to x	$x + 5$
	more than	2 more than t	$t + 2$
	the sum of	the sum of s and r	$s + r$
	increased by	z increased by 7	$z + 7$
	the total of	the total of 6 and b	$6 + b$
Subtraction	minus	y minus 8	$y - 8$
	less than	5 less than p	$p - 5$
	decreased by	n decreased by 1	$n - 1$
	the difference between	the difference between t and 4	$t - 4$
Multiplication	times	9 times y	$9y$
	of	one third of p	$\frac{1}{3}p$
	the product of	the product of x and y	xy
	multiplied by	t multiplied by 43	$43t$
	twice	twice n	$2n$
Division	divided by	a divided by 6	$\frac{a}{6}$
	the quotient of	the quotient of s and t	$\frac{s}{t}$
	the ratio of	the ratio of r to 5	$\frac{r}{5}$
Power	the square of	the square of b	b^2
	the cube of	the cube of x	x^3

In most applications that involve translating phrases into variable expressions, the variable to be used is not given. To translate these phrases, a variable must be assigned to an unknown quantity before the variable expression can be written. After translating a verbal expression into a variable expression, simplify the variable expression by using the Addition, Multiplication, and Distributive Properties.

Example 6   Translate and simplify "the total of five times a number and twice the difference between the number and three."

Solution   the unknown number: n

- Assign a variable to one of the unknown quantities.

five times the number: $5n$
the difference between the number and three: $n - 3$
twice the difference between the number and three: $2(n - 3)$

- Use the assigned variable to write an expression for any other unknown quantity.

$5n + 2(n - 3)$

- Use the assigned variable to write the variable expression.

$5n + 2n - 6$
$7n - 6$

- Simplify the variable expression.

Problem 6   Translate and simplify "the sum of three consecutive integers."

Solution   See page A7.
$n + (n + 1) + (n + 2)$; $3n + 3$

Example 7   Translate and simplify "fifteen minus one half the sum of a number and ten."

Solution   the unknown number: n

- Assign a variable to one of the unknown quantities.

the sum of the number and ten: $n + 10$

- Use the assigned variable to write an expression for any other unknown quantity.

one half the sum of the number and ten: $\dfrac{1}{2}(n + 10)$

$15 - \dfrac{1}{2}(n + 10)$

- Use the assigned variable to write the variable expression.

$15 - \dfrac{1}{2}n - 5$

- Simplify the variable expression.

$10 - \dfrac{1}{2}n$

Problem 7   Translate and simplify "the sum of three eighths of a number and five twelfths of the number."

Solution   See page A7.
$\dfrac{3}{8}n + \dfrac{5}{12}n$; $\dfrac{19}{24}n$

EXERCISES 1.2

1   Evaluate the variable expression when $a = 2$, $b = 3$, $c = -1$, and $d = -4$.

1. $ab + dc$ 10

2. $2ab - 3dc$ 0

3. $4cd \div a^2$ 4

4. $b^2 - (d - c)^2$ 0

5. $(b - 2a)^2 + c$ 0

6. $(b - d)^2 \div (b - d)$ 7

7. $(bc + a)^2 \div (d - b)$ $-\dfrac{1}{7}$

8. $\dfrac{1}{3}b^3 - \dfrac{1}{4}d^3$ 25

9. $\dfrac{1}{4}a^4 - \dfrac{1}{6}bc$ $\dfrac{9}{2}$

10. $2b^2 \div \dfrac{ad}{2}$ $-\dfrac{9}{2}$

11. $\dfrac{3ac}{-4} - c^2$ $\dfrac{1}{2}$

12. $\dfrac{2d - 2a}{2bc}$ 2

13. $\dfrac{3b - 5c}{3a - c}$ 2

14. $\dfrac{2d - a}{b - 2c}$ -2

15. $\dfrac{a - d}{b + c}$ 3

16. $|a^2 + d|$ 0

17. $-a|a + 2d|$ -12

18. $d|b - 2d|$ -44

19. $\dfrac{2a - 4d}{3b - c}$ 2

20. $\dfrac{3d - b}{b - 2c}$ -3

21. $-3d \div \left| \dfrac{ab - 4c}{2b + c} \right|$ 6

22. $-2bc + \left| \dfrac{bc + d}{ab - c} \right|$ 7

23. $2(d - b) \div (3a - c)$ -2

24. $(d - 4a)^2 \div c^3$ -144

Evaluate the variable expression when $a = 1.5$, $b = -2.4$, and $c = -0.5$.

25. $b^2 - ac$

6.51

26. $(a - b)^2 \div c^2$

60.84

27. $\dfrac{a^2}{c} - (b - a)^2$

-19.71

2   Use the given Property of the Real Numbers to complete the statement.

28. The Commutative Property of Multiplication
$3 \cdot 4 = 4 \cdot ?$
$3 \cdot 4 = 4 \cdot 3$

29. The Commutative Property of Addition
$7 + 15 = ? + 7$
$7 + 15 = 15 + 7$

30. The Associative Property of Addition
$(3 + 4) + 5 = ? + (4 + 5)$
$(3 + 4) + 5 = 3 + (4 + 5)$

31. The Associative Property of Multiplication
$(3 \cdot 4) \cdot 5 = 3 \cdot (? \cdot 5)$
$(3 \cdot 4) \cdot 5 = 3 \cdot (4 \cdot 5)$

32. The Division Property of Zero

$\dfrac{5}{?}$ is undefined.

$\dfrac{5}{0}$

33. The Multiplication Property of Zero

$5 \cdot ? = 0$

$5 \cdot 0 = 0$

34. The Distributive Property

$3(x + 2) = 3x + ?$

$3(x + 2) = 3x + 6$

35. The Distributive Property

$5(y + 4) = ? \cdot y + 20$

$5(y + 4) = 5y + 20$

36. The Division Property of Zero

$\dfrac{?}{-6} = 0$

$\dfrac{0}{-6} = 0$

37. The Inverse Property of Addition

$(x + y) + ? = 0$

$x + y + [-(x + y)] = 0$

38. The Inverse Property of Multiplication

$\dfrac{1}{mn}(mn) = ?$

$\dfrac{1}{mn} \cdot mn = 1$

39. The Multiplication Property of One

$? \cdot 1 = x$

$x \cdot 1 = x$

40. The Associative Property of Multiplication

$2(3x) = ? \cdot x$

$2(3x) = (2 \cdot 3) \cdot x$

41. The Commutative Property of Addition

$ab + bc = bc + ?$

$ab + bc = bc + ab$

Identify the property that justifies the statement.

42. $\dfrac{0}{-5} = 0$

The Division Property of Zero

43. $-8 + 8 = 0$

The Inverse Property of Addition

44. $(-12)\left(-\dfrac{1}{12}\right) = 1$

The Inverse Property of Multiplication

45. $(3 \cdot 4) \cdot 2 = 2 \cdot (3 \cdot 4)$

The Commutative Property of Multiplication

46. $y + 0 = y$

The Addition Property of Zero

47. $2x + (5y + 8) = (2x + 5y) + 8$

The Associative Property of Addition

48. $\dfrac{-9}{0}$ is undefined.

The Division Property of Zero

49. $(x + y)z = xz + yz$

The Distributive Property

50. $6(x + y) = 6x + 6y$

The Distributive Property

51. $(-12y)(0) = 0$

The Multiplication Property of Zero

52. $(ab)c = a(bc)$

The Associative Property of Multiplication

53. $(x + y) + z = (y + x) + z$

The Commutative Property of Addition

3   Simplify.

54. $5x + 7x$

$12x$

55. $3x + 10x$

$13x$

56. $-8ab - 5ab$

$-13ab$

57. $-2x + 5x - 7x$
$-4x$

58. $3x - 5x + 9x$
$7x$

59. $-2a + 7b + 9a$
$7a + 7b$

60. $5b - 8a - 12b$
$-8a - 7b$

61. $12\left(\dfrac{1}{12}x\right)$
x

62. $\dfrac{1}{3}(3y)$
y

63. $-3(x - 2)$
$-3x + 6$

64. $-5(x - 9)$
$-5x + 45$

65. $(x + 2)5$
$5x + 10$

66. $-(x + y)$
$-x - y$

67. $-(-x - y)$
$x + y$

68. $3(-2a + 3a - 5)$
$3a - 15$

69. $3(x - 2y) - 5$
$3x - 6y - 5$

70. $4x - 3(2y - 5)$
$4x - 6y + 15$

71. $-2a - 3(3a - 7)$
$-11a + 21$

72. $3x - 2(5x - 7)$
$-7x + 14$

73. $2x - 3(x - 2y)$
$-x + 6y$

74. $3[a - 5(5 - 3a)]$
$48a - 75$

75. $5[-2 - 6(a - 5)]$
$-30a + 140$

76. $3[x - 2(x + 2y)]$
$-3x - 12y$

77. $5[y - 3(y - 2x)]$
$30x - 10y$

78. $-2(x - 3y) + 2(3y - 5x)$
$-12x + 12y$

79. $4(-a - 2b) - 2(3a - 5b)$
$-10a + 2b$

80. $5(3a - 2b) - 3(-6a + 5b)$
$33a - 25b$

81. $-7(2a - b) + 2(-3b + a)$
$-12a + b$

82. $3x - 2[y - 2(x + 3[2x - y])]$
$31x - 14y$

83. $2x - 4[x - 4(y - 2[5y + 3])]$
$-2x - 144y - 96$

84. $4 - 2(7x - 2y) - 3(-2x + 3y)$
$-8x - 5y + 4$

85. $3x + 8(x - 4) - 3(2x - y)$
$5x - 32 + 3y$

86. $\dfrac{1}{3}[8x - 2(x - 12) + 3]$ $2x + 9$

87. $\dfrac{1}{4}[14x - 3(x - 8) - 7x]$ $x + 6$

4   Translate into a variable expression. Then simplify.

88. a number minus the sum of the number and two
$n - (n + 2); -2$

89. a number decreased by the difference between five and the number
$n - (5 - n); 2n - 5$

90. the sum of one third of a number and four fifths of the number
$\dfrac{1}{3}n + \dfrac{4}{5}n; \dfrac{17}{15}n$

91. the difference between three eighths of a number and one sixth of the number
$\dfrac{3}{8}n - \dfrac{1}{6}n; \dfrac{5}{24}n$

92. five times the product of eight and a number
$5(8n); 40n$

93. a number increased by two thirds of the number
$n + \dfrac{2}{3}n; \dfrac{5}{3}n$

94. the difference between the product of seventeen and a number and twice the number
$17n - 2n$; $15n$

95. one half the total of six times a number and twenty-two
$\frac{1}{2}(6n + 22)$; $3n + 11$

96. three times the total of two consecutive integers
$3[n + (n + 1)]$; $6n + 3$

97. the sum of the second and third of three consecutive integers
$(n + 1) + (n + 2)$; $2n + 3$

98. thirty less than the product of six and the sum of the first and third of three consecutive integers
$6[n + (n + 2)] - 30$; $12n - 18$

99. twelve more than the product of four and the second of three consecutive integers
$4(n + 1) + 12$; $4n + 16$

100. three more than the sum of the first and third of three consecutive integers
$[n + (n + 2)] + 3$; $2n + 5$

101. three times the quotient of twice a number and six
$3\left(\frac{2n}{6}\right)$; n

102. the sum of five times a number and twelve added to the product of fifteen and the number
$(5n + 12) + 15n$; $20n + 12$

103. four less than twice the sum of a number and eleven
$2(n + 11) - 4$; $2n + 18$

104. twice a number plus the product of two more than the number and eight
$2n + (n + 2)8$; $10n + 16$

105. twenty minus the product of four more than a number and twelve
$20 - (4 + n)12$; $-28 - 12n$

106. a number added to the product of five plus the number and four
$n + (5 + n)4$; $5n + 20$

107. a number plus the product of the number minus twelve and three
$n + (n - 12)3$; $4n - 36$

SUPPLEMENTAL EXERCISES 1.2

Name the property that justifies each lettered step used in simplifying the expression.

108. $3(x + y) + 2x$
   a. $(3x + 3y) + 2x$
      The Distributive Property
   b. $(3y + 3x) + 2x$
      The Commutative Property of Addition
   c. $3y + (3x + 2x)$
      The Associative Property of Addition
   d. $3y + (3 + 2)x$
      The Distributive Property
      $3y + 5x$

109. $3a + 4(b + a)$
   a. $3a + (4b + 4a)$
      The Distributive Property
   b. $3a + (4a + 4b)$
      The Commutative Property of Addition
   c. $(3a + 4a) + 4b$
      The Associative Property of Addition
   d. $(3 + 4)a + 4b$
      The Distributive Property
      $7a + 4b$

110. $y + (3 + y)$
  a. $y + (y + 3)$
    The Commutative Property of Addition
  b. $(y + y) + 3$
    The Associative Property of Addition
  c. $(1y + 1y) + 3$
    The Multiplication Property of One
  d. $(1 + 1)y + 3$
    $2y + 3$
    The Distributive Property

111. $8(b \cdot 4)$
  a. $8(4b)$
    The Commutative Property of Mult.
  b. $(8 \cdot 4)b$
    $32b$
    The Associative Property of Mult.

112. $(9x)(3x)$
  a. $(9x \cdot 3)x$
    The Associative Property of Mult.
  b. $(3 \cdot 9x)x$
    The Commutative Property of Mult.
  c. $[(3 \cdot 9)x]x$
    $(27x)x$
    The Associative Property of Mult.
  d. $27(x \cdot x)$
    $27x^2$
    The Associative Property of Mult.

113. $5(3a + 1)$
  a. $5(3a) + 5(1)$
    The Distributive Property
  b. $(5 \cdot 3)a + 5(1)$
    $15a + 5(1)$
    The Associative Property of Mult.
  c. $15a + 5$
    The Mult. Property of One

Write a variable expression.

114. The length of a rectangle is 5 m more than the width. Write a variable expression for the length of the rectangle in terms of the width.
$w + 5$

115. A mixture contains three times as many peanuts as cashews. Write a variable expression for the amount of peanuts in terms of the amount of cashews.
$3c$

116. One cyclist rode 4 mph faster than a second cyclist. Write a variable expression for the speed of the first cyclist in terms of the speed of the second.
$x + 4$

117. In a triangle, the measure of one angle is one third the measure of the largest angle. Write a variable expression for the measure of the smaller angle in terms of the largest angle.
$\frac{1}{3}x$

118. In a coin bank, the number of nickels is 3 more than twice the number of dimes. Write a variable expression for the number of nickels in terms of the number of dimes.
$2d + 3$

119. The age of a gold coin is 15 years less than twice the age of a silver coin. Write a variable expression for the age of the gold coin in terms of the age of the silver coin.
$2s - 15$

Sets

1 The union and intersection of sets

A **set** is a collection of objects. The objects in a set are the **elements** of the set.

A set can be written in various ways. The **roster method** of writing a set encloses the list of the elements of the set in braces.

The set of the three planets nearest the sun is written {Mercury, Venus, Earth}.

The set of even natural numbers less than 10 is written {2, 4, 6, 8}. This is an example of a **finite set.** All the elements of the set can be listed.

The set of natural numbers greater than 5 is written {6, 7, 8, 9, 10, 11, . . .}. The three dots mean that the pattern of numbers continues without end. This is an example of an **infinite set.** It is impossible to list all of the elements of the set.

Example 1   Use the roster method to write the set of positive prime numbers less than 15.

Solution $A = \{2, 3, 5, 7, 11, 13\}$ • A set is usually designated by a capital letter.

Problem 1   Use the roster method to write the set of even natural numbers less than 12.

Solution   See page A8.
$\{2, 4, 6, 8, 10\}$

The symbol $\in$ means *is an element of.*

$9 \in B$ is read "9 is an element of set B."

Given $A = \{-1, 5, 9\}$, then $-1 \in A$, $5 \in A$, and $9 \in A$. $12 \notin A$ is read "12 is not an element of A."

The **empty set,** or **null set,** is the set that contains no elements. The symbol $\varnothing$ or { } is used to represent the empty set. (Note: It is incorrect to write $\{\varnothing\}$ as the empty set.)

The set of trees over 1000 ft tall is the empty set.

A second method of representing a set is **set builder notation.** The set of all integers greater than −3 would be written

$$\{x \mid x > -3, \, x \text{ is an integer}\}$$

and is read "the set of all x such that x is greater than −3 and x is an integer."

Set builder notation can be used to describe almost any set, but it is especially useful when writing infinite sets.

Using set builder notation, the set of real numbers less than 5 is written

$$\{x \mid x < 5, \, x \in \text{real numbers}\}$$

and is read "the set of all x such that x is less than 5 and x is an element of the real numbers." This is an infinite set. It is impossible to list all the elements in this set.

Example 2   Write $\{x \mid x < 10, \, x \text{ is a natural number}\}$ by using the roster method.

Solution $A = \{1, 2, 3, 4, 5, 6, 7, 8, 9\}$

Problem 2   Write $\{x \mid x < 10, \, x \text{ is a positive odd integer}\}$ by using the roster method.

Solution   See page A8.
$\{1, 3, 5, 7, 9\}$

Example 3   Is $-7 \in \{x \mid x > 5, \, x \in \text{real numbers}\}$?

Solution   −7 is not greater than 5.
No, −7 is not an element of the set.

Problem 3   Is $0.35 \in \{x \mid x < 2, \, x \in \text{real numbers}\}$?

Solution   See page A8.
Yes

Just as operations such as addition and multiplication are performed on real numbers, operations are performed on sets.

The **union** of two sets, written $A \cup B$, is the set of all elements which belong to either A **or** B. In set builder notation, this is written

$$A \cup B = \{x \mid x \in A \text{ or } x \in B\}$$

Given $A = \{2, 3, 5, 7\}$ and $B = \{0, 1, 2, 3, 4\}$, the union of A and B contains all the elements which belong to either A or B. The elements which belong to both sets are listed only once.

$A \cup B = \{0, 1, 2, 3, 4, 5, 7\}$

The **intersection** of two sets, written $A \cap B$, is the set of all elements which are common to both A **and** B. In set builder notation, this is written

$$A \cap B = \{x \mid x \in A \text{ and } x \in B\}$$

Given $A = \{2, 3, 5, 7\}$ and $B = \{0, 1, 2, 3, 4\}$, the intersection of A and B contains all the elements which are common to both A and B.

$A \cap B = \{2, 3\}$

Example 4   Find $C \cup D$ given $C = \{1, 5, 9, 13, 17\}$ and $D = \{3, 5, 7, 9, 11\}$.

Solution $C \cup D = \{1, 3, 5, 7, 9, 11, 13, 17\}$

Problem 4   Find $A \cup C$ given $A = \{-2, -1, 0, 1, 2\}$ and $C = \{-5, -1, 0, 1, 5\}$.

Solution   See page A8.
$A \cup C = \{-5, -2, -1, 0, 1, 2, 5\}$

Example 5   Find $A \cap B$ given $A = \{x \mid x \text{ is a natural number}\}$ and $B = \{x \mid x \text{ is a negative integer}\}$.

Solution   There are no natural numbers that are also negative numbers.
$A \cap B = \varnothing$

Problem 5   Find $E \cap F$ given $E = \{x \mid x \text{ is an odd integer}\}$ and $F = \{x \mid x \text{ is an even integer}\}$.

Solution   See page A8.
$E \cap F = \varnothing$

2   Graph the solution set of an inequality in one variable

An **inequality** expresses the relative order of two mathematical expressions. The symbols $>$, $<$, $\leq$, and $\geq$ are used to write inequalities. The symbol $\leq$ means is less than or equal to. The symbol $\geq$ means is greater than or equal to.

$7 > -2$
$2x < 4$
$2x - 3y \geq 6$
$x^2 - 2x - 3 \leq 0$

The **solution set of an inequality** is the set of real numbers that makes the inequality true. The solution set can be graphed on the number line.

The graph of the solution set of $x > -2$ is shown at the right. The solution set is the real numbers greater than -2. The circle on the graph indicates that -2 is not included in the solution set.

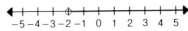

The solution set is written $\{x \mid x > -2,\ x \in \text{real numbers}\}$.

The graph of the solution set of $x \geq -2$ is shown at the right. The dot at -2 indicates that -2 is included in the solution set.

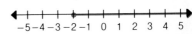

The solution set is written $\{x \mid x \geq -2,\ x \in \text{real numbers}\}$.

For the remainder of this section, all variables will represent real numbers. Using this convention, the solution set above would be written $\{x \mid x \geq -2\}$.

Example 6   Graph the solution set of $x \leq 3$.

Solution   The solution set is $\{x \mid x \leq 3\}$.

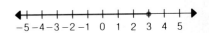

Problem 6   Graph the solution set of $x > -3$.

Solution   See page A8.

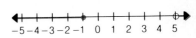

The union of two sets is the set of all elements belonging to either one or the other of the two sets.

The solution set of $\{x \mid x \leq -1\} \cup \{x \mid x > 5\}$ is the set of all numbers that are either less than or equal to -1 or greater than 5.

The solution set is written $\{x \mid x \leq -1 \text{ or } x > 5\}$.

The solution set of $\{x \mid x > 2\} \cup \{x \mid x > 4\}$ is the set of all numbers that are either greater than 2 or greater than 4.

The solution set is written $\{x \mid x > 2\}$

The intersection of two sets is the set that contains the elements common to both sets.

The solution set of $\{x \mid x > -2\} \cap \{x \mid x < 5\}$ is the set of numbers that are greater than -2 and less than 5.

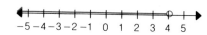

The solution set can be written $\{x \mid x > -2 \text{ and } x < 5\}$. However, it is more commonly written $\{x \mid -2 < x < 5\}$.

The solution set of $\{x \mid x < 4\} \cap \{x \mid x < 5\}$ is the set of numbers that are less than 4 and less than 5.

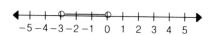

The solution set is written $\{x \mid x < 4\}$.

Example 7 Graph the solution set of $\{x \mid x < 0\} \cap \{x \mid x > -3\}$.

Solution The solution set is $\{x \mid -3 < x < 0\}$.

Problem 7 Graph the solution set of $\{x \mid x \geq 1\} \cup \{x \mid x \leq -3\}$.

Solution See page A8.

EXERCISES 1.3

1 Use the roster method to write the set.

1. the integers between -3 and 5
   $\{-2, -1, 0, 1, 2, 3, 4\}$

2. the integers between -4 and 0.
   $\{-3, -2, -1\}$

3. the even natural numbers less than 13
   $\{2, 4, 6, 8, 10, 12\}$

4. the odd natural numbers less than 13
   $\{1, 3, 5, 7, 9, 11\}$

5. the prime numbers between 2 and 5
   $\{3\}$

6. the prime numbers between 30 and 40
   $\{31, 37\}$

7. the positive integers less than 20 that are divisible by 3
   $\{3, 6, 9, 12, 15, 18\}$

8. the perfect square integers less than 100
   $\{1, 4, 9, 16, 25, 36, 49, 64, 81\}$

Use set builder notation to write the set.

9. the integers greater than 4
   $\{x \mid x > 4, x \text{ is an integer}\}$

10. the integers less than -2
    $\{x \mid x < -2, x \text{ is an integer}\}$

11. the real numbers greater than or equal to 1
    $\{x \mid x \geq 1, x \in \text{real numbers}\}$

12. the real numbers less than or equal to -3
    $\{x \mid x \leq -3, x \in \text{real numbers}\}$

13. the integers between -2 and 5
$\{x \mid -2 < x < 5,\ x \text{ is an integer}\}$

14. the integers between -5 and 2
$\{x \mid -5 < x < 2,\ x \text{ is an integer}\}$

15. the real numbers between 0 and 1
$\{x \mid 0 < x < 1,\ x \in \text{real numbers}\}$

16. the real numbers between -2 and 4
$\{x \mid -2 < x < 4,\ x \in \text{real numbers}\}$

Find $A \cup B$.

17. $A = \{1, 4, 9\},\ B = \{2, 4, 6\}$
$A \cup B = \{1, 2, 4, 6, 9\}$

18. $A = \{-1, 0, 1\},\ B = \{0, 1, 2\}$
$A \cup B = \{-1, 0, 1, 2\}$

19. $A = \{2, 3, 5, 8\},\ B = \{9, 10\}$
$A \cup B = \{2, 3, 5, 8, 9, 10\}$

20. $A = \{1, 3, 5, 7\},\ B = \{2, 4, 6, 8\}$
$A \cup B = \{1, 2, 3, 4, 5, 6, 7, 8\}$

21. $A = \{-4, -2, 0, 2, 4\},\ B = \{0, 4, 8\}$
$A \cup B = \{-4, -2, 0, 2, 4, 8\}$

22. $A = \{-3, -2, -1\},\ B = \{-2, -1, 0, 1\}$
$A \cup B = \{-3, -2, -1, 0, 1\}$

23. $A = \{1, 2, 3, 4, 5\},\ B = \{3, 4, 5\}$
$A \cup B = \{1, 2, 3, 4, 5\}$

24. $A = \{2, 4\},\ B = \{0, 1, 2, 3, 4, 5\}$
$A \cup B = \{0, 1, 2, 3, 4, 5\}$

Find $A \cap B$.

25. $A = \{6, 12, 18\},\ B = \{3, 6, 9\}$
$A \cap B = \{6\}$

26. $A = \{-4, 0, 4\},\ B = \{-2, 0, 2\}$
$A \cap B = \{0\}$

27. $A = \{1, 5, 10, 20\},\ B = \{5, 10, 15, 20\}$
$A \cap B = \{5, 10, 20\}$

28. $A = \{-9, -5, 0, 7\},\ B = \{-7, -5, 0, 5, 7\}$
$A \cap B = \{-5, 0, 7\}$

29. $A = \{1, 2, 4, 8\},\ B = \{3, 5, 6, 7\}$
$A \cap B = \varnothing$

30. $A = \{-3, -2, -1, 0\},\ B = \{1, 2, 3, 4\}$
$A \cap B = \varnothing$

31. $A = \{2, 4, 6, 8, 10\},\ B = \{4, 6\}$
$A \cap B = \{4, 6\}$

32. $A = \{1, 3, 5, 7, 9\},\ B = \{1, 9\}$
$A \cap B = \{1, 9\}$

2   Graph the solution set.

33. $\{x \mid x \geq 3\}$

34. $\{x \mid x \leq -2\}$

35. $\{x \mid x > 0\}$

36. $\{x \mid x < 4\}$

37. $\{x \mid x > 1\} \cap \{x \mid x < -1\}$

38. $\{x \mid x \leq 2\} \cup \{x \mid x > 4\}$

39. $\{x \mid x \leq 2\} \cap \{x \mid x \geq 0\}$

40. $\{x \mid x > -1\} \cap \{x \mid x \leq 4\}$

41. $\{x \mid x > 1\} \cap \{x \mid x \geq -2\}$

42. $\{x \mid x < 4\} \cap \{x \mid x \leq 0\}$

43. $\{x \mid x > 2\} \cup \{x \mid x > 1\}$

44. $\{x \mid x < -2\} \cup \{x \mid x < -4\}$

SUPPLEMENTAL EXERCISES 1.3

Graph the solution set.

45. $\left\{x \mid x > \dfrac{3}{2}\right\} \cup \left\{x \mid x < -\dfrac{1}{2}\right\}$

46. $\{x \mid x \leq -2.5\} \cup \{x \mid x > 1.5\}$

47. $\left\{x \mid x > -\dfrac{5}{2}\right\} \cap \left\{x \mid x \leq \dfrac{7}{3}\right\}$

48. $\{x \mid x > -0.5\} \cap \{x \mid x > 3.5\}$

49. $|x| < 2$

50. $|x| < 5$

51. $|x| > 3$

52. $|x| > 4$

Use set builder notation to write $A \cup B$.

53. $A = \{1, 3, 5, 7, \ldots\}$
$B = \{2, 4, 6, 8, \ldots\}$
$\{x \mid x > 0, x \text{ is an integer}\}$

54. $A = \{\ldots, -6, -4, -2\}$
$B = \{\ldots, -5, -3, -1\}$
$\{x \mid x < 0, x \text{ is an integer}\}$

Use set builder notation to write $A \cap B$.

55. $A = \{15, 17, 19, 21, \ldots\}$
$B = \{11, 13, 15, 17, \ldots\}$
$\{x \mid x \geq 15, x \text{ is an odd integer}\}$

56. $A = \{-12, -10, -8, -6, \ldots\}$
$B = \{-4, -2, 0, 2, \ldots\}$
$\{x \mid x \geq -4, x \text{ is an even integer}\}$

Solve.

57. Given that a, b, c, and d are real numbers, which of the following will ensure that $a + c < b + d$?

   a. $a < b$ and $c < d$

   c. $a < b$ and $c > d$

   b. $a > b$ and $c > d$

   d. $a > b$ and $c < d$

   a

58. Given that a and b are real numbers, which of the following will ensure that $a^2 < b^2$?

   a. $a < b$

   c. $0 < a < b$

   b. $a > b$

   d. $a < b < 0$

   c

59. Given that a, b, c, and d are positive real numbers, which of the following will ensure that $\dfrac{a - b}{c - d} \le 0$?

   a. $a \ge b$ and $c > d$

   c. $a \ge b$ and $c < d$

   b. $a \le b$ and $c > d$

   d. $a \le b$ and $c < d$

   b and c

A set is "closed" under an operation if it is not possible to get an answer that is not in the set by performing that operation on the numbers in the set. For example, the set $\{0, 1\}$ is closed under the operation of multiplication because $0 \times 0 = 0$, $0 \times 1 = 0$, and $1 \times 1 = 1$. All the possible answers are in the set $\{0, 1\}$. The set is not closed under addition because $1 + 1 = 2$ and $2 \notin \{0, 1\}$. For each set given, state whether or not it is closed under the given operation.

60. $\{0, 1\}$; subtraction

   not closed

61. $\{-1, 1\}$; multiplication

   closed

62. $\{-1, 0, 1\}$; multiplication

   closed

63. the set of whole numbers; addition

   closed

64. the set of integers; subtraction

   closed

65. the set of natural numbers; multiplication

   closed

66. the set of positive integers; division

   not closed

67. the set of negative numbers; multiplication

   not closed

68. the set of real numbers; addition

   closed

69. the set of positive numbers; subtraction

   not closed

CALCULATORS AND COMPUTERS

Euclidean Algorithm

An algorithm is a method or procedure that repeats the same sequence of steps over and over. For example, to divide two whole numbers, the division algorithm is used.

$$
\begin{array}{r}
2 \\
3\overline{)747} \\
-6 \\
\hline
14
\end{array}
$$

1. Estimate the quotient. $\left(3\overline{)7}^{\,2}\right)$
2. Multiply the quotient times the divisor. $(2 \times 3 = 6)$
3. Subtract. $(7 - 6)$
4. "Bring down" the next digit. (4)

$$
\begin{array}{r}
249 \\
3\overline{)747} \\
-6 \\
\hline
14 \\
-12 \\
\hline
27 \\
-27 \\
\hline
0
\end{array}
$$

The division problem is completed by repeating these four steps until there are no numbers to bring down. The algorithm is the four steps: estimate, multiply, subtract, and bring down.

The Euclidean Algorithm is a procedure to find the greatest common divisor (GCD) of two numbers.

Find the GCD of 15 and 25 using the Euclidean Algorithm.

$$
\begin{array}{r}
1 \\
15\overline{)25} \\
-15 \\
\hline
10
\end{array}
$$

1. Divide the smaller number into the larger number.

$$
\begin{array}{r}
1 \\
10\overline{)15} \\
-10 \\
\hline
5
\end{array}
$$

2. Divide the remainder (10) into the divisor (15).

$$
\begin{array}{r}
2 \\
5\overline{)10} \\
-10 \\
\hline
0
\end{array}
$$

3. Repeat Step 2 until the remainder is zero.

4. The divisor in the last division is the GCD.

The GCD is 5.

The Euclidean Algorithm is the four listed steps. A computer program can be written to carry out this procedure for any two positive integers. One such program is on the Student Disk.

Find the GCD of 5 and 8 using the Euclidean Algorithm. You may use the program on the Student Disk to check your answer.

CHAPTER SUMMARY

Key Words

The **integers** are $\ldots, -4, -3, -2, -1, 0, 1, 2, 3, 4, \ldots$

The **negative integers** are the integers $\ldots, -4, -3, -2, -1$.

The **positive integers** are the integers $1, 2, 3, 4, \ldots$ (The positive integers are also called the **natural numbers.**)

The positive integers and zero are called the **whole numbers.**

A **rational number** is a number of the form $\dfrac{a}{b}$, where a and b are integers, and b is not equal to zero.

An **irrational number** is a number whose decimal representation never terminates or repeats.

The rational numbers and the irrational numbers taken together are called the **real numbers.**

The **absolute value** of a number is a measure of its distance from zero on the number line.

The expression a^n is in **exponential form,** where a is the base, and n is the exponent.

A **complex fraction** is a fraction whose numerator or denominator contains one or more fractions.

The Order of Operations Agreement is used to simplify numerical expressions.

A **variable expression** is an expression that contains one or more variables.

The **terms** of a variable expression are the addends of the expression.

A **variable term** is composed of a numerical coefficient and a variable part.

The **additive inverse** of a number is the opposite of the number.

The **multiplicative inverse** of a number is the reciprocal of the number.

A **set** is a collection of objects. The objects of the set are the **elements** of the set.

The **roster method** of writing a set encloses a list of the elements of the set in braces.

The **finite set** is a set in which the elements can be counted.

An **infinite set** is a set in which it is impossible to list all the elements.

The **empty set,** or **null set,** written $\varnothing$ or { }, is the set that contains no elements.

The **union** of two sets, written $A \cup B$, is the set that contains all the elements of A and all the elements of B. (The elements in both set A and set B are listed only once.)

The **intersection** of two sets, written $A \cap B$, is the set that contains the elements which are common to both A and B.

An **inequality** is an expression that contains the symbol $<$, $>$, $\leq$, or $\geq$.

The **solution set of an inequality** is the set of real numbers that makes the inequality true.

Essential Rules

The Commutative Property of Addition

If a and b are real numbers, then $a + b = b + a$.

The Associative Property of Addition

If a, b, and c are real numbers, then $a + (b + c) = (a + b) + c$.

The Commutative Property of Multiplication

If a and b are real numbers, then $ab = ba$.

The Associative Property of Multiplication

If a, b, and c are real numbers, then $a(bc) = (ab)c$.

The Addition Property of Zero	If a is a real number, then $a + 0 = 0 + a = a.$
The Multiplication Property of One	If a is a real number, then $a \cdot 1 = 1 \cdot a = a.$
The Inverse Property of Addition	If a is a real number, then $a + (-a) = (-a) + a = 0.$
The Inverse Property of Multiplication	If a is a nonzero real number, then $a \cdot \dfrac{1}{a} = \dfrac{1}{a} \cdot a = 1.$
The Distributive Property	If a, b, and c are real numbers, then $a(b + c) = ab + ac.$

CHAPTER REVIEW

1. Find the additive inverse of -12.
12

2. Simplify: $|-3 - (-5)|$
2

3. Simplify: $4.27 - 6.98 + 1.3$
-1.41

4. Simplify: $2x - 4[2 - 3(x + 4y) - 2]$
$14x + 48y$

5. Simplify: $8 - 4(2 - 3)^2 \div 2$
6

6. Evaluate $(a - b)^2 \div (2b + 1)$ when $a = 2$ and $b = -3$.
-5

7. Identify the property that justifies the statement.
$-2(x + y) = -2x - 2y$
The Distributive Property

8. Use the Commutative Property of Addition to complete the statement.
$(3 + 4) + 2 = (? + 3) + 2$
$(3 + 4) + 2 = (4 + 3) + 2$

9. Evaluate: $-|-2|$
-2

10. Simplify: $-5^2 \cdot 4$
-100

11. Simplify: $-15.092 \div 3.08$
-4.9

12. Find $A \cup B$ given $A = \{2, 3, 5, 8\}$ and $B = \{2, 4, 6, 8\}$.
$A \cup B = \{2, 3, 4, 5, 6, 8\}$

13. Simplify: $3x - 2(x - y) - 3(y - 4x)$
$13x - y$

14. Graph the solution set of $\{x \mid x \le -1\}$.

15. Simplify: $12 - 4\left(\dfrac{5^2 - 1}{3}\right) \div 16$

10

16. Simplify: $(-2)^3(-3)^2$

-72

17. Find $A \cup B$ given $A = \{-3, -1, 0, 1, 3\}$ and $B = \{0, 1, 3, 5\}$.

$A \cup B = \{-3, -1, 0, 1, 3, 5\}$

18. Evaluate $\dfrac{b^2 - c^2}{a - 2c}$ when $a = 2$, $b = 3$, and $c = -1$.

2

19. Simplify: $\dfrac{2}{3} - \dfrac{5}{12} + \dfrac{4}{9}$

$\dfrac{25}{36}$

20. Find $A \cap B$ given $A = \{-3, -2, -1, 0, 1, 2\}$ and $B = \{-1, 0, 1\}$.

$A \cap B = \{-1, 0, 1\}$

21. Graph the solution set of $\{x \mid x \geq 2\} \cup \{x \mid x < -1\}$.

22. Simplify: $\left(-\dfrac{2}{3}\right)\left(\dfrac{9}{15}\right)\left(\dfrac{10}{27}\right)$

$-\dfrac{4}{27}$

23. Simplify: $15 \div \dfrac{\frac{2}{3} - 2}{4} \cdot 6$

-270

24. Find the additive inverse of 3.

-3

25. Find $A \cap B$ given $A = \{-4, -2, 0, 2\}$ and $B = \{-4, 0, 4, 8\}$.

$A \cap B = \{-4, 0\}$

26. Simplify: $-3^2 \cdot (-2)^2$

-36

27. Graph the solution set of $\{x \mid x < 3\}$.

28. Evaluate $-2b - \left|\dfrac{b - a}{b^3 + 2a}\right|$ when $a = 2$ and $b = -1$.

1

29. Simplify: $3x - [2x - 3(y - 2x) + 5y]$

$-5x - 2y$

30. Graph the solution set of $\{x \mid x \leq 3\} \cap \{x \mid x > -1\}$.

31. Translate and simplify "thirteen decreased by the product of three less than a number and nine."

$13 - 9(n - 3)$; $40 - 9n$

32. Translate and simplify "five times the sum of two consecutive integers."

$5[n + (n + 1)]$; $10n + 5$

33. Translate and simplify: "eight minus the difference between four times a number and twelve."

$8 - (4n - 12)$; $-4n + 20$

34. Translate and simplify "six plus the sum of two consecutive integers."

$6 + [n + (n + 1)]$; $2n + 7$

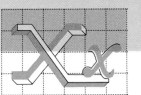

2

First-Degree Equations and Inequalities

OBJECTIVES

- Solve equations using the Addition and the Multiplication Properties of Equality
- Solve equations using the Distributive Property
- Integer problems
- Solve inequalities in one variable
- Solve compound inequalities
- Application problems
- Absolute value equations
- Absolute value inequalities
- Application problems
- Coin and stamp problems
- Value mixture problems
- Uniform motion problems
- Investment problems
- Percent mixture problems

Moscow Papyrus

Most of the early history of mathematics can be traced to the Egyptians and Babylonians. This early work began around 3000 B.C. The Babylonians used clay tablets and a type of writing called *cuneiform* (wedge-shape) to record their thoughts and discoveries. Here are the symbols for 10, 1, and subtraction.

Simple groupings of these symbols were used to represent numbers less than 60. The numbers 32 and 28 are shown below.

The Egyptians used papyrus, a plant reed dried and pounded thin, and hieratic writing that was derived from hieroglyphics to make records. Several papyrus documents have been discovered over the years. One particularly famous one is called the Moscow Papyrus, which dates from 1850 B.C. It is approximately 18 feet long and three inches wide and contains 25 problems.

The Moscow Papyrus dealt with solving practical problems related to geometry, food preparation, and grain allotments. Here is one of the problems:

The width of a rectangle is $\frac{3}{4}$ the length, and the area is 12. Find the dimensions of the rectangle.

You might recognize this as a type of problem you have solved before. Word problems are very old indeed. This one is around 3900 years old.

Equations in One Variable

1 Solve equations using the Addition and Multiplication Properties of Equality

An **equation** expresses the equality of two mathematical expressions. The expressions can be either numerical or variable expressions.

$$2 + 8 = 10$$
$$x + 8 = 11$$ } Equations
$$x^2 + 2y = 7$$

The equation at the right is a **conditional equation.** The equation is true if the variable is replaced by 3. The equation is false if the variable is replaced by 4.

$x + 2 = 5$	Conditional Equation
$3 + 2 = 5$	A true equation
$4 + 2 = 5$	A false equation

The replacement values of the variable that will make an equation true are called the **roots**, or **solutions**, of the equation.

The solution of the equation $x + 2 = 5$ is 3.

The equation at the right is an **identity**. Any replacement for x will result in a true equation.

$$x + 2 = x + 2 \qquad \text{Identity}$$

The equation at the right has **no solution** since there is no number that equals itself plus 1. Any replacement value for x will result in a false equation.

$$x = x + 1 \qquad \text{No solution}$$

Each of the equations at the right is a **first-degree equation in one variable.** All variables have an exponent of 1.

$$x + 2 = 12$$
$$3y - 2 = 5y$$ First-Degree Equations
$$3(a + 2) = 14a$$

To **solve** an equation means to find a root or solution of the equation. The simplest equation to solve is an equation of the form

$$\textbf{variable} = \textbf{constant}$$

since the constant is the solution.

If $x = 3$, then 3 is the solution of the equation since $3 = 3$ is a true equation.

In solving an equation, the goal is to rewrite the given equation in the form *variable = constant*. The Addition Property of Equality can be used to rewrite an equation in this form.

The Addition Property of Equality

The same number can be added to each side of an equation without changing the solution of the equation.

$$\text{If } a = b, \text{ then } a + c = b + c.$$

The Addition Property of Equality is used to **remove a term** from one side of an equation **by adding the additive inverse of that term** to each side of the equation.

Solve: $x + 3 = 7$

$$x + 3 = 7$$

Add the additive inverse of the constant term 3 to each side of the equation.

$$x + 3 + (-3) = 7 + (-3)$$

Simplify using the Inverse Property of Addition and the Addition Property of Zero.

$$x + 0 = 4$$
$$x = 4$$

Now the equation is in the form variable = constant.

Write the solution.

The solution is 4.

Check: $x + 3 = 7$
$$\frac{4 + 3 \mid 7}{7 = 7}$$

Example 1   Solve: $x - 7 = -12$

Solution $x - 7 = -12$
$x - 7 + 7 = -12 + 7$ • Add the additive inverse of the constant term to each side of the equation.

$x + 0 = -5$ • Simplify using the Inverse Property of Addition. (Do this step mentally.)

$x = -5$ • Simplify using the Addition Property of Zero.
• Write the solution.

The solution is -5.

Problem 1   Solve: $x + 4 = -3$

Solution   See page A9.
-7

The Multiplication Property of Equality can be used to rewrite an equation in the form *variable = constant*.

> ### The Multiplication Property of Equality
> Each side of an equation can be multiplied by the same nonzero number without changing the solution of the equation.
> $$\text{If } a = b \text{ and } c \neq 0, \text{ then } ac = bc.$$

The Multiplication Property of Equality is used to **remove a coefficient** from one side of an equation **by multiplying** each side of the equation **by the reciprocal of the coefficient.**

Solve: $-\dfrac{3}{4}x = 12$

$$-\frac{3}{4}x = 12$$

Multiply each side of the equation by the reciprocal of the coefficient $-\dfrac{3}{4}$.

$$\left(-\frac{4}{3}\right)\left(-\frac{3}{4}x\right) = \left(-\frac{4}{3}\right)(12)$$

Simplify using the Inverse Property of Multiplication and the Multiplication Property of One. Now the equation is in the form *variable = constant*.

$$1x = -16$$
$$x = -16$$

Write the solution.

The solution is -16.

Check:
$$-\frac{3}{4}x = 12$$
$$\begin{array}{c|c} -\dfrac{3}{4}(-16) & 12 \\ \hline & \\ 12 & = 12 \end{array}$$

Example 2   Solve: $-\dfrac{2}{7}x = \dfrac{4}{21}$

Solution
$$-\frac{2}{7}x = \frac{4}{21}$$

$$\left(-\frac{7}{2}\right)\left(-\frac{2}{7}x\right) = \left(-\frac{7}{2}\right)\left(\frac{4}{21}\right)$$

$$1x = -\frac{2}{3}$$

$$x = -\frac{2}{3}$$

The solution is $-\dfrac{2}{3}$.

• Multiply each side of the equation by the reciprocal of the coefficient $-\dfrac{2}{7}$.

• Simplify using the Inverse Property of Multiplication. (Do this step mentally.)

• Simplify using the Multiplication Property of One.

• Write the solution.

Problem 2 Solve: $-3x = 18$

Solution See page A9.
-6

In solving an equation, the application of both the Addition and the Multiplication Properties of Equality are frequently required.

Solve: $4x - 3 + 2x = 8 + 9x - 12$

Simplify each side of the equation by combining like terms.

$$4x - 3 + 2x = 8 + 9x - 12$$
$$6x - 3 = -4 + 9x$$

Add the additive inverse of the variable term $9x$ to each side of the equation. Simplify.

$$6x - 3 + (-9x) = -4 + 9x + (-9x)$$
$$-3x - 3 = -4$$

Add the additive inverse of the constant term -3 to each side of the equation. Simplify.

$$-3x - 3 + 3 = -4 + 3$$
$$-3x = -1$$

Multiply each side of the equation by the reciprocal of the coefficient -3. Simplify.

$$-\frac{1}{3}(-3x) = -\frac{1}{3}(-1)$$
$$x = \frac{1}{3}$$

Write the solution.

The solution is $\frac{1}{3}$.

Check:

$$4x - 3 + 2x = 8 + 9x - 12$$

$4\left(\frac{1}{3}\right) - 3 + 2\left(\frac{1}{3}\right)$	$8 + 9\left(\frac{1}{3}\right) - 12$
$\frac{4}{3} - 3 + \frac{2}{3}$	$8 + 3 - 12$
$-\frac{5}{3} + \frac{2}{3}$	$11 - 12$

$$-1 = -1$$

Example 3 Solve: $3x - 5 = x + 2 - 7x$

Solution
$$3x - 5 = x + 2 - 7x$$
$$3x - 5 = -6x + 2$$

$$3x + 6x - 5 = -6x + 6x + 2$$

$$9x - 5 = 2$$

• Simplify the right-hand side of the equation.
• Add the additive inverse of the variable term $-6x$ to each side of the equation. (Do this step mentally.)
• Simplify.

$$9x - 5 + 5 = 2 + 5$$

• Add the additive inverse of the constant term -5 to each side of the equation. (Do this step mentally.)

$$9x = 7$$

$$\frac{1}{9}(9x) = \frac{1}{9}(7)$$

• Simplify.
• Multiply each side of the equation by the reciprocal of the coefficient 9. (Do this step mentally.)

$$x = \frac{7}{9}$$

The solution is $\frac{7}{9}$.

Problem 3   Solve: $6x - 5 - 3x = 14 - 5x$

Solution   See page A10. $\frac{19}{8}$

2   ## Solve equations using the Distributive Property

When an equation contains parentheses, one of the steps required in solving the equation involves the use of the Distributive Property.

Solve: $3(x - 2) + 3 = 2(6 - x)$

Use the Distributive Property to remove parentheses. Simplify.

$$3(x - 2) + 3 = 2(6 - x)$$
$$3x - 6 + 3 = 12 - 2x$$
$$3x - 3 = 12 - 2x$$

Add the additive inverse of the variable term $-2x$ to each side of the equation.

$$5x - 3 = 12$$

Add the additive inverse of the constant term -3 to each side of the equation.

$$5x = 15$$

Multiply each side of the equation by the reciprocal of the coefficient 5.

$$x = 3$$

Write the solution.

The solution is 3.

Check:
$$3(x - 2) + 3 = 2(6 - x)$$

$3(3 - 2) + 3$	$2(6 - 3)$
$3(1) + 3$	$2(3)$
$3 + 3$	6
$6 = 6$	

Example 4   Solve: $5(2x - 7) + 2 = 3(4 - x) - 12$

Solution $5(2x - 7) + 2 = 3(4 - x) - 12$

$10x - 35 + 2 = 12 - 3x - 12$ • Use the Distributive Property.

$10x - 33 = -3x$ • Simplify.

$-33 = -13x$ • Add $-10x$ to each side of the equation.

$\dfrac{33}{13} = x$ • Multiply each side of the equation by $-\dfrac{1}{13}$.

The solution is $\dfrac{33}{13}$.

Problem 4   Solve: $6(5 - x) - 12 = 2x - 3(4 + x)$

Solution   See page A10.

6

To solve an equation containing fractions, first clear denominators by multi-plying each side of the equation by the least common multiple (LCM) of the denominators.

Solve: $\dfrac{x}{2} - \dfrac{7}{9} = \dfrac{x}{6} + \dfrac{2}{3}$

$$\dfrac{x}{2} - \dfrac{7}{9} = \dfrac{x}{6} + \dfrac{2}{3}$$

Multiply each side of the equation by the LCM. The LCM of 2, 9, 6, and 3 is 18.

$$18\left(\dfrac{x}{2} - \dfrac{7}{9}\right) = 18\left(\dfrac{x}{6} + \dfrac{2}{3}\right)$$

Use the Distributive Property to remove parentheses.

$$\dfrac{18x}{2} - \dfrac{18 \cdot 7}{9} = \dfrac{18x}{6} + \dfrac{18 \cdot 2}{3}$$

Simplify.

$$9x - 14 = 3x + 12$$

Add the additive inverse of the variable term $3x$ to each side of the equation.

$$6x - 14 = 12$$

Add the additive inverse of the constant term -14 to each side of the equation.

$$6x = 26$$

Multiply each side of the equation by the re-ciprocal of the coefficient 6. Simplify.

$$x = \dfrac{13}{3}$$

Write the solution.   The solution is $\dfrac{13}{3}$.

$$\text{Check: } \frac{x}{2} - \frac{7}{9} = \frac{x}{6} + \frac{2}{3}$$

$$
\begin{array}{c|c}
\dfrac{\frac{13}{3}}{2} - \dfrac{7}{9} & \dfrac{\frac{13}{3}}{6} + \dfrac{2}{3} \\[4mm]
\dfrac{13}{6} - \dfrac{7}{9} & \dfrac{13}{18} + \dfrac{2}{3} \\[3mm]
\dfrac{39}{18} - \dfrac{14}{18} & \dfrac{13}{18} + \dfrac{12}{18} \\[3mm]
\dfrac{25}{18} & = \dfrac{25}{18}
\end{array}
$$

Example 5   Solve: $\dfrac{3x - 2}{12} - \dfrac{x}{9} = \dfrac{2x}{4}$

Solution
$$\frac{3x - 2}{12} - \frac{x}{9} = \frac{2x}{4}$$ • The LCM of 12, 9, and 4 is 36.

$$36\!\left(\frac{3x - 2}{12} - \frac{x}{9}\right) = 36\!\left(\frac{2x}{4}\right)$$ • Multiply each side of the equation by the LCM.

$$\frac{36(3x - 2)}{12} - \frac{36x}{9} = \frac{36(2x)}{4}$$ • Use the Distributive Property.

$$3(3x - 2) - 4x = 9(2x)$$ • Simplify.
$$9x - 6 - 4x = 18x$$
$$5x - 6 = 18x$$
$$-6 = 13x$$
$$-\frac{6}{13} = x$$

The solution is $-\dfrac{6}{13}$.

Problem 5   Solve: $\dfrac{2x - 7}{3} - \dfrac{5x + 4}{5} = \dfrac{-x - 4}{30}$

Solution   See page A10.
-10

3   ## Integer problems

An equation states that two mathematical expressions are equal. Therefore, to translate a sentence into an equation requires recognizing the words or phrases that mean equals. These phrases include "is," "is equal to," "amounts to," and "represents."

Once the sentence is translated into an equation, the equation can be solved by rewriting the equation in the form *variable = constant.*

Recall that an **even integer** is an integer that is divisible by 2. An **odd integer** is an integer that is not divisible by 2.

Consecutive integers are integers that follow one another in order. Examples of consecutive integers are shown at the right.	$8, 9, 10$ $-3, -2, -1$ $n, n + 1, n + 2,$ where n is an integer
Examples of **consecutive even integers** are shown at the right.	$16, 18, 20$ $-6, -4, -2$ $n, n + 2, n + 4,$ where n is an even integer
Examples of **consecutive odd integers** are shown at the right.	$11, 13, 15$ $-23, -21, -19$ $n, n + 2, n + 4,$ where n is an odd integer

There is a basic strategy for solving word problems that involve integers.

The sum of three consecutive even integers is seventy-eight. Find the integers.

STRATEGY *for solving an integer problem*

▶ **Let a variable represent one of the integers. Express each of the other integers in terms of that variable. Remember that for consecutive integer problems, consecutive integers will differ by 1. Consecutive even or consecutive odd integers will differ by 2.**

Represent three consecutive even integers.

First even integer: n
Second even integer: $n + 2$
Third even integer: $n + 4$

▶ **Determine the relationship among the integers.**

The sum of the three even integers is 78.

$$n + (n + 2) + (n + 4) = 78$$
$$3n + 6 = 78$$
$$3n = 72$$
$$n = 24$$

$n + 2 = 24 + 2 = 26$
$n + 4 = 24 + 4 = 28$

The three consecutive even integers are 24, 26, and 28.

Example 6 One number is four more than another number. The sum of the two numbers is sixty-six. Find the two numbers.

Strategy ▶ The smaller number: n
The larger number: $n + 4$
▶ The sum of the numbers is 66.

Solution
$$n + (n + 4) = 66$$
$$2n + 4 = 66$$
$$2n = 62$$
$$n = 31$$ • The smaller number is 31.

$$n + 4 = 31 + 4 = 35$$ • Substitute the value of n into the variable expression for the larger number.

The numbers are 31 and 35.

Problem 6 The sum of three numbers is eighty-one. The second number is twice the first number, and the third number is three less than four times the first number. Find the numbers.

Solution See page A10.
12, 24, 45

Example 7 Five times the first of three consecutive even integers is five more than the product of four and the third integer. Find the integers.

Strategy ▶ First even integer: n
Second even integer: $n + 2$
Third even integer: $n + 4$
▶ Five times the first integer equals five more than the product of four and the third integer.

Solution
$$5n = 4(n + 4) + 5$$
$$5n = 4n + 16 + 5$$
$$5n = 4n + 21$$
$$n = 21$$

Since 21 is not an even integer, there is no solution.

Problem 7 Find three consecutive odd integers such that three times the sum of the first two integers is ten more than the product of the third integer and four.

Solution See page A11.
no solution

EXERCISES 2.1

1 Solve and check.

1. $x - 2 = 7$ 9
2. $x - 8 = 4$ 12
3. $a + 3 = -7$ -10

4. $a + 5 = -2$ -7
5. $b - 3 = -5$ -2
6. $c - 8 = -2$ 6

7. $10 + m = 2$ -8
8. $8 + m = -5$ -13
9. $-7 = x + 8$ -15

10. $-12 = x - 3$ 9
11. $3x = 12$ 4
12. $8x = 4$ $\dfrac{1}{2}$

13. $-3x = 2$ $-\dfrac{2}{3}$
14. $-5a = 7$ $-\dfrac{7}{5}$
15. $-\dfrac{3}{2} + x = \dfrac{4}{3}$ $\dfrac{17}{6}$

16. $\dfrac{2}{7} + x = \dfrac{17}{21}$ $\dfrac{11}{21}$
17. $x + \dfrac{2}{3} = \dfrac{5}{6}$ $\dfrac{1}{6}$
18. $\dfrac{5}{8} - y = \dfrac{3}{4}$ $-\dfrac{1}{8}$

19. $\dfrac{2}{3}y = 5$ $\dfrac{15}{2}$
20. $\dfrac{3}{5}y = 12$ 20
21. $\dfrac{3t}{8} = -15$ -40

22. $\dfrac{3a}{7} = -21$ -49
23. $-\dfrac{5}{8}x = \dfrac{4}{5}$ $-\dfrac{32}{25}$
24. $-\dfrac{5}{12}y = \dfrac{7}{16}$ $-\dfrac{21}{20}$

25. $-\dfrac{3b}{5} = -\dfrac{3}{5}$ 1
26. $-\dfrac{7b}{12} = \dfrac{7}{8}$ $-\dfrac{3}{2}$
27. $-\dfrac{2}{3}x = -\dfrac{5}{8}$ $\dfrac{15}{16}$

28. $-\dfrac{3}{4}x = -\dfrac{4}{7}$
    $\dfrac{16}{21}$
29. $b - 14.72 = -18.45$

    -3.73
30. $b + 3.87 = -2.19$

    -6.06

Solve and check.

31. $3x + 5x = 12$
    $\dfrac{3}{2}$
32. $2x - 7x = 15$

    -3
33. $2x - 4 = 12$

    8

34. $5x + 9 = 6$
    $-\dfrac{3}{5}$
35. $3y - 5y = 0$

    0
36. $2y - 9 = -9$

    0

37. $4x - 6 = 3x$

    6
38. $2a - 7 = 5a$

    $-\dfrac{7}{3}$
39. $7x + 12 = 9x$

    6

40. $3x - 12 = 5x$
    -6
41. $4x + 2 = 4x$
    no solution
42. $3m - 7 = 3m$
    no solution

43. $2x + 2 = 3x + 5$

    -3
44. $7x - 9 = 3 - 4x$

    $\dfrac{12}{11}$
45. $2 - 3t = 3t - 4$

    1

46. $7 - 5t = 2t - 9$

$\dfrac{16}{7}$

47. $2a - 3a = 7 - 5a$

$\dfrac{7}{4}$

48. $3a - 5a = 8a + 4$

$-\dfrac{2}{5}$

49. $3b - 2b = 4 - 2b$

$\dfrac{4}{3}$

50. $3x - 5 = 6 - 9x$

$\dfrac{11}{12}$

51. $3x + 7 = 3 + 7x$

1

52. $\dfrac{5}{8}b - 3 = 12$

24

53. $\dfrac{1}{3} - 2b = 3$

$-\dfrac{4}{3}$

54. $b + \dfrac{1}{5}b = 2$

$\dfrac{5}{3}$

55. $\dfrac{2}{3}y + y = \dfrac{2}{3}$

$\dfrac{2}{5}$

56. $\dfrac{2}{3}y - 5 = 7$

18

57. $\dfrac{7}{12}x - 3 = 4$

12

58. $3x - 2x + 7 = 12 - 4x$

1

59. $2x - 9x + 3 = 6 - 5x$

$-\dfrac{3}{2}$

60. $2t - 7 + 5t = 16 - 8t$

$\dfrac{23}{15}$

61. $-2t + 8 - 4t = -8t + 2$

-3

62. $7 + 8y - 12 = 3y - 8 + 5y$
no solution

63. $2y - 4 + 8y = 7y - 8 + 3y$
no solution

64. $2x - 5 + 7x = 11 - 3x + 4x$

2

65. $9 + 4x - 12 = -3x + 5x + 8$

$\dfrac{11}{2}$

Use your calculator for the following exercises.

66. $3.24a + 7.14 = 5.34a$
3.4

67. $5.3y + 0.35 = 5.02y$
-1.25

68. $1.2b - 3.8b = 1.6b + 4.494$
-1.07

69. $6.7a - 9.2a = 6.55a - 3.91865$
0.433

2   Solve and check.

70. $2x + 2(x + 1) = 10$
2

71. $2x + 3(x - 5) = 15$
6

72. $2(a - 3) = 2(4 - 2a)$ $\dfrac{7}{3}$

73. $5(2 - b) = -3(b - 3)$ $\dfrac{1}{2}$

74. $-8(3w - 4) = 16w$

$\dfrac{4}{5}$

75. $6(2n + 3) = 10n$

-9

76. $-4(c + 2) = 3c + 6$

-2

77. $-5(t - 4) = 10t + 10$

$\dfrac{2}{3}$

78. $8(5 + m) = 10m + 4$

18

79. $3(-4b + 1) = 6b + 9$

$-\dfrac{1}{3}$

80. $3 - 2(y - 3) = 4y - 7$

$\dfrac{8}{3}$

81. $3(y - 5) - 5y = 2y + 9$

-6

82. $2(3x - 2) - 5x = 3 - 2x$

$\dfrac{7}{3}$

83. $4 - 3x = 7x - 2(3 - x)$

$\dfrac{5}{6}$

84. $4(x - 2) + 2 = 4x - 2(2 - x)$

-1

85. $2x - 3(x - 4) = 2(3 - 2x) + 2$

$-\dfrac{4}{3}$

86. $-2c + 4(3c + 8) = -5c - 2(6 - 3c)$

$-\dfrac{44}{9}$

87. $2n - 5(2 - 3n) = -4n + 3(n + 2)$

$\dfrac{8}{9}$

88. $3(p + 2) + 4p = 6(3p - 1) + 5p$

$\dfrac{3}{4}$

89. $10(2x - 3) + 5x = 6(x + 1) + 3x$

$\dfrac{9}{4}$

90. $2(2d + 1) - 3d = 5(3d - 2) + 4d$

$\dfrac{2}{3}$

91. $-4(7y - 1) + 5y = -2(3y + 4) - 3y$

$\dfrac{6}{7}$

92. $3[2 - 3(y - 2)] = 12$

$\dfrac{4}{3}$

93. $3y = 2[5 - 3(2 - y)]$

$\dfrac{2}{3}$

94. $4[3 + 5(3 - x) + 2x] = 6 - 2x$

$\dfrac{33}{5}$

95. $2[4 + 2(5 - x) - 2x] = 4x - 7$

$\dfrac{35}{12}$

96. $3[4 - 2(a - 2)] = 3(2 - 4a)$

-3

97. $2[3 - 2(z + 4)] = 3(4 - z)$

-22

98. $2[b - (4b - 5)] = 3b + 4$

$\dfrac{2}{3}$

99. $-3[x + 4(x + 1)] = x + 4$

-1

100. $4[a - (3a - 5)] = a - 7$

3

101. $5 - 6[2t - 2(t + 3)] = 8 - t$

-33

102. $m + 2[3m + 3(m - 7)] = 7$

$\dfrac{49}{13}$

103. $2p - 3 = 1 - 2[p + 3(p - 6)]$

4

104. $-3(x - 2) = 2[x - 4(x - 2) + x]$
10

105. $3[x - (2 - x) - 2x] = 3(4 - x)$
6

106. $\dfrac{2}{9}t - \dfrac{5}{6} = \dfrac{1}{12}t$
6

107. $\dfrac{3}{4}t - \dfrac{7}{12}t = 1$
6

108. $\dfrac{2}{3}x - \dfrac{5}{6}x - 3 = \dfrac{1}{2}x - 5$
3

109. $\dfrac{1}{2}x - \dfrac{3}{4}x + \dfrac{5}{8} = \dfrac{3}{2}x - \dfrac{5}{2}$
$\dfrac{25}{14}$

110. $\dfrac{3x - 2}{4} - 3x = 12$
$-\dfrac{50}{9}$

111. $\dfrac{2a - 9}{5} + 3 = 2a$
$\dfrac{3}{4}$

112. $\dfrac{5 - 2x}{5} + \dfrac{x - 4}{10} = \dfrac{3}{10}$
1

113. $\dfrac{2x - 5}{12} - \dfrac{3 - x}{6} = \dfrac{11}{12}$
$\dfrac{11}{2}$

114. $\dfrac{x - 2}{4} - \dfrac{x + 5}{6} = \dfrac{5x - 2}{9}$
$-\dfrac{40}{17}$

115. $\dfrac{2x - 1}{4} + \dfrac{3x + 4}{8} = \dfrac{1 - 4x}{12}$
$-\dfrac{4}{29}$

116. $\dfrac{2}{3}(15 - 6a) = \dfrac{5}{6}(12a + 18)$
$-\dfrac{5}{14}$

117. $\dfrac{1}{5}(20x + 30) = \dfrac{1}{3}(6x + 36)$
3

118. $\dfrac{1}{3}(x - 7) + 5 = 6x + 4$
$-\dfrac{4}{17}$

119. $2(y - 4) + 8 = \dfrac{1}{2}(6y + 20)$
-10

120. $\dfrac{1}{4}(2b + 50) = \dfrac{5}{2}\left(15 - \dfrac{1}{5}b\right)$
25

121. $\dfrac{1}{4}(7 - x) = \dfrac{2}{3}(x + 2)$
$\dfrac{5}{11}$

Use your calculator for the following exercises.

122. $2.3(x - 1.8) = 3.91$
3.5

123. $0.4(m + 6.5) = 3.14$
1.35

124. $-4.2(p + 3.4) = 11.13$
-6.05

125. $-1.6(b - 2.35) = -11.28$
9.4

126. $0.1y + 6 = 0.12(y + 40)$
60

127. $0.35(n + 40) = 0.1n + 20$
24

128. $2.1x + 18 = 2.16(x + 8)$
12

129. $0.85d + 0.91(84 - d) = 73.08$
56

130. $0.492 - 0.34(y - 2.5) = 3.2(4.2y - 5.8)$
1.4442671

131. $2.9(x - 3.4) = -1.432 - 6.7(x - 4.4)$
3.94875

3 Solve.

132. One integer is three more than another integer. The sum of the two integers is sixty-one. Find the integers.
29, 32

133. One integer is eight less than another integer. The sum of the two integers is forty. Find the integers.
24, 16

134. The sum of two integers is twenty. The difference between three times the smaller integer and four is one more than twice the larger integer. Find the integers.
9, 11

135. The sum of two integers is eight. Three times the smaller integer is one less than twice the larger integer. Find the integers.
3, 5

136. Find two consecutive even integers such that six more than three times the first integer is ten more than twice the second integer. Find the integers.
8, 10

137. Find two consecutive odd integers such that six more than nine times the first integer is five times the second integer. Find the integers.
1, 3

138. The sum of three numbers is forty-two. The second number is twice the first number, and the third number is three less than the second number. Find the three numbers.
9, 18, 15

139. The sum of three numbers is one hundred twenty-three. The second number is two more than twice the first number. The third number is five less than the product of three and the first number. Find the three numbers.
21, 44, 58

140. The sum of three consecutive integers is seventy-eight. Find the integers.
25, 26, 27

141. The sum of three consecutive integers is negative eighty-seven. Find the integers.
−30, −29, −28

142. The sum of 10 and three consecutive integers is equal to four times the smallest of the integers. Find the smallest integer.
13

143. The sum of three consecutive integers and one is equal to twenty more than the smallest of the integers. Find the smallest integer.
8

144. Find three consecutive even integers such that twice the sum of the first and third integers is twenty-one more than the second integer.
no solution

145. Five times the smallest of three consecutive odd integers is ten more than twice the largest. Find the integers.
no solution

146. Find three consecutive even integers such that eight times the sum of the first two integers is eight more than twelve times the third integer.
10, 12, 14

147. Find three consecutive odd integers such that five times the middle integer is fifteen more than the sum of the first and third integers.
3, 5, 7

SUPPLEMENTAL EXERCISES 2.1

Solve.

148. $8 \div \dfrac{1}{x} = -3$ $-\dfrac{3}{8}$

149. $5 \div \dfrac{1}{y} = -2$ $-\dfrac{2}{5}$

150. $\dfrac{1}{\frac{1}{x}} = -4$ -4

151. $\dfrac{1}{\frac{1}{y}} = -9$ -9

152. $\dfrac{1}{\frac{1}{x}} = -14$ -14

153. $\dfrac{1}{\frac{1}{y}} + 18 = 17$ -1

154. $\dfrac{4}{\frac{5}{b}} = 8$ 10

155. $\dfrac{6}{\frac{7}{a}} = -18$ -21

156. $\dfrac{10}{\frac{3}{x}} - 5 = 4x$ $-\dfrac{15}{2}$

Solve. If the equation has no solution, write "No solution."

157. $3[4(y + 2) - (y + 5)] = 3(3y + 1)$
no solution

158. $2[3(x + 4) - 2(x + 1)] = 5x + 3(1 - x)$
no solution

159. $\dfrac{2(3x - 4) - (x + 1)}{3} = x - 7$

-6

160. $\dfrac{3(2x - 1) - (x + 2)}{5} = x - 4$

no solution

161. $\dfrac{4(3a + 2) - 3(a + 1)}{9} = a + 5$

no solution

162. $\dfrac{4[(x - 3) + 2(1 - x)]}{5} = x + 1$

-1

Solve for x.

163. $2584 \div x = 54\dfrac{46}{x}$

47

164. $3479 \div x = 66\dfrac{47}{x}$

52

165. $(2^4 - 2^3)(2^3 - 2^2) = 2^{x+1}$

4

166. $(2^5 - 2^4)(2^4 - 2^3) = 2^{x-2}$

9

Solve.

167. The sum of four consecutive even integers is 100. What is the sum of the smallest and the largest of the four integers?
50

168. The sum of four consecutive odd integers is -64. What is the sum of the smallest and the largest of the four integers?
-32

169. A library charges a fine for each overdue book. The fine is 15¢ the first day plus 7¢ a day for each additional day the book is overdue. If the fine for a book is 78¢, how many days overdue is the book?
10 days

170. At a museum, the admission charge for a family is $1.75 for the first person and $.75 for each additional member of the family. How many people are in a family that is charged $5.50 for admission?
6 people

171. The charges for a long-distance telephone call are $1.63 for the first three minutes and $.88 for each additional minute or fraction of a minute. If the charges for a call were $7.79, how many minutes did the phone call last?
10 min

172. Show why the solution of an equation of the form $ax + b = cx + d$ is
$x = \dfrac{d - b}{a - c}$, $a - c \ne 0$.
Complete solution available in Solutions Manual.

SECTION 2.2 _____

Inequalities in One Variable

1 Solve inequalities in one variable

The **solution set of an inequality** is a set of numbers, each element of which, when substituted for the variable, results in a true inequality.

The inequality at the right is true if the variable is replaced by 3, -1.98, or $\frac{2}{3}$.

$$x - 1 < 4$$
$$3 - 1 < 4$$
$$-1.98 - 1 < 4$$
$$\frac{2}{3} - 1 < 4$$

There are many values of the variable x that will make the inequality $x - 1 < 4$ true. The solution set of the inequality is any number less than 5. The solution set can be written in set builder notation as $\{x \mid x < 5\}$.

The graph of the solution set of $x - 1 < 4$ is shown below.

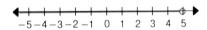

In solving an inequality, the Addition and Multiplication Properties of Inequalities are used to rewrite the inequality in the form _variable_ < _constant_ or _variable_ > _constant_.

The Addition Property of Inequalities

The same number can be added to each side of an inequality without changing the solution set of the inequality.

$$\text{If } a > b, \text{ then } a + c > b + c.$$
$$\text{If } a < b, \text{ then } a + c < b + c.$$

This property is also true for an inequality that contains the symbol $\leq$ or $\geq$.

The Addition Property of Inequalities is used to remove a term from one side of an inequality by adding the inverse of that term to each side of the inequality.

Solve: $3x - 4 < 2x - 1$

Add the additive inverse of the variable term $2x$ to each side of the inequality.

$$3x - 4 < 2x - 1$$
$$x - 4 < -1$$

Add the additive inverse of the constant term to each side of the inequality.

$$x < 3$$

Write the solution set.

$$\{x \mid x < 3\}$$

The Multiplication Property of Inequalities is used to remove a coefficient from one side of an inequality by multiplying each side of the inequality by the reciprocal of the coefficient.

The Multiplication Property of Inequalities

Rule 1   Each side of an inequality can be multiplied by the same positive number without changing the solution set of the inequality.

If $a > b$ and $c > 0$, then $ac > bc$.
If $a < b$ and $c > 0$, then $ac < bc$.

Rule 2   If each side of an inequality is multiplied by the same negative number and the inequality symbol is reversed, then the solution set of the inequality is not changed.

If $a > b$ and $c < 0$, then $ac < bc$.
If $a < b$ and $c < 0$, then $ac > bc$.

The Multiplication Property of Inequalities is also true for an inequality that contains the symbol $\leq$ or $\geq$.

The two examples shown below illustrate Rule 1 of the Multiplication Property of Inequalities.

$$3 > 2 \qquad\qquad 2 < 5$$
$$3(4) > 2(4) \qquad\qquad 2(4) < 5(4)$$
$$12 > 8 \qquad\qquad 8 < 20$$

Rule 2 states that when each side of an inequality is multiplied by a negative number, the inequality symbol must be reversed. The two examples shown below illustrate Rule 2 of the Multiplication Property of Inequalities.

$$3 > 2 \qquad\qquad 2 < 5$$
$$3(-2) < 2(-2) \qquad\qquad 2(-4) > 5(-4)$$
$$-6 < -4 \qquad\qquad -8 > -20$$

Solve: $-3x > 9$

$$-3x > 9$$

Multiply each side of the inequality by the reciprocal of the coefficient -3. Since $-\dfrac{1}{3}$ is a negative number, the inequality symbol must be reversed.

$$-\frac{1}{3}(-3x) < -\frac{1}{3}(9)$$

$$x < -3$$

Write the solution set.

$$\{x \mid x < -3\}$$

Solve: $2x - 9 > 4x + 5$

$$2x - 9 > 4x + 5$$

Add the additive inverse of the variable term $4x$ to each side of the inequality.

$$-2x - 9 > 5$$

Add the additive inverse of the constant term -9 to each side of the inequality.

$$-2x > 14$$

Multiply each side of the inequality by the reciprocal of the coefficient -2. Reverse the inequality symbol.

$$-\frac{1}{2}(-2x) < -\frac{1}{2}(14)$$

$$x < -7$$

Write the solution set.

$$\{x \mid x < -7\}$$

Example 1   Solve: $x + 3 > 4x + 6$

Solution
$$x + 3 > 4x + 6$$
$$-3x + 3 > 6$$ • Add $-4x$ to each side of the inequality.
$$-3x > 3$$ • Add -3 to each side of the inequality.

$$x < -1$$ • Multiply each side of the inequality by $-\dfrac{1}{3}$.
Reverse the inequality symbol.

$$\{x \mid x < -1\}$$ • Write the solution set.

Problem 1   Solve: $2x - 1 < 6x + 7$

Solution   See page A12.
$$\{x \mid x > -2\}$$

When an inequality contains parentheses, the first step in solving the inequality is to use the Distributive Property to remove the parentheses.

Example 2   Solve: $5(x - 2) \geq 9x - 3(2x - 4)$

Solution $5(x - 2) \geq 9x - 3(2x - 4)$

$5x - 10 \geq 9x - 6x + 12$ • Use the Distributive Property to remove parentheses.

$5x - 10 \geq 3x + 12$ • Simplify.

$2x - 10 \geq 12$ • Add $-3x$ to each side of the inequality.

$2x \geq 22$ • Add 10 to each side of the inequality.

$x \geq 11$ • Multiply each side of the inequality by $\frac{1}{2}$.

$\{x \,|\, x \geq 11\}$

Problem 2   Solve: $5x - 2 \leq 4 - 3(x - 2)$

Solution   See page A12.
$\left\{ x \,\middle|\, x \leq \dfrac{3}{2} \right\}$

2   # Solve compound inequalities

A **compound inequality** is formed by joining two inequalities with a connective word such as "and" or "or." The inequalities shown below are compound inequalities.

$$2x < 4 \quad \text{and} \quad 3x - 2 > -8$$

$$2x + 3 > 5 \quad \text{or} \quad x + 2 < 5$$

The **solution set of a compound inequality** with the connective word **and** is
the set of all elements common to the solution sets of each inequality. Therefore, it is the intersection of the solution sets of the two inequalities.

Solve: $2x < 6$ and $3x + 2 > -4$

$$2x < 6 \quad \text{and} \quad 3x + 2 > -4$$

Solve each inequality.

$x < 3$ $3x > -6$
$x > -2$
$\{x \,|\, x < 3\}$ $\{x \,|\, x > -2\}$

Find the intersection of the $\{x \,|\, x < 3\} \cap \{x \,|\, x > -2\} = \{x \,|\, -2 < x < 3\}$
solution sets.

Solve: $-3 < 2x + 1 < 5$

This inequality is equivalent to the compound inequality $-3 < 2x + 1$ and $2x + 1 < 5$.

$$-3 < 2x + 1 < 5$$

$-3 < 2x + 1 \quad$ and $\quad 2x + 1 < 5$
$-4 < 2x \qquad\qquad\qquad 2x < 4$
$-2 < x \qquad\qquad\qquad\quad x < 2$

$\{x | x > -2\} \qquad\qquad \{x | x < 2\}$

The solution set of $-3 < 2x + 1 < 5$ is the intersection of the solution sets of $-3 < 2x + 1$ and $2x + 1 < 5$.

$\{x | x > -2\} \cap \{x | x < 2\} = \{x | -2 < x < 2\}$

There is an alternative method for solving the inequality in the last example.

Add the additive inverse of the constant term 1 to each of the three parts of the inequality.

$$-3 < 2x + 1 < 5$$
$$-3 + (-1) < 2x + 1 + (-1) < 5 + (-1)$$
$$-4 < 2x < 4$$

Multiply each of the three parts of the inequality by the reciprocal of the coefficient 2.

$$\frac{1}{2}(-4) < \frac{1}{2}(2x) < \frac{1}{2}(4)$$
$$-2 < x < 2$$

$$\{x | -2 < x < 2\}$$

The **solution set of a compound inequality** with the connective word **or** is found by taking the union of the solution sets of the two inequalities.

Solve: $2x + 3 > 7$ or $4x - 1 < 3$

Solve each inequality.

$2x + 3 > 7 \quad$ or $\quad 4x - 1 < 3$
$2x > 4 \qquad\qquad\qquad 4x < 4$
$x > 2 \qquad\qquad\qquad\quad x < 1$

$\{x | x > 2\} \qquad\qquad \{x | x < 1\}$

Find the union of the solution sets.

$\{x | x > 2\} \cup \{x | x < 1\} =$
$\{x | x > 2 \text{ or } x < 1\}$

Example 3 Solve: $1 < 3x - 5 < 4$

Solution
$$1 < 3x - 5 < 4$$
$$1 + 5 < 3x - 5 + 5 < 4 + 5$$

• Add the additive inverse of -5 to each of the three parts of the inequality.

$$6 < 3x < 9$$

• Simplify.

$$\frac{1}{3}(6) < \frac{1}{3}(3x) < \frac{1}{3}(9)$$

• Multiply each of the three parts of the inequality by the reciprocal of 3.

$$2 < x < 3$$

$$\{x | 2 < x < 3\}$$

• Write the solution set.

Problem 3   Solve: $-2 \le 5x + 3 \le 13$

Solution   See page A12.
$\{x \mid -1 \le x \le 2\}$

Example 4   Solve: $11 - 2x > -3$ and $7 - 3x < 4$

Solution

$$\begin{array}{lll} 11 - 2x > -3 & \text{and} & 7 - 3x < 4 \\ -2x > -14 & & -3x < -3 \\ x < 7 & & x > 1 \end{array}$$ • Solve each inequality.

$\{x \mid x < 7\}$ $\qquad$ $\{x \mid x > 1\}$

$\{x \mid x < 7\} \cap \{x \mid x > 1\} =$ $\qquad$ • Find the intersection of the solution sets.
$\{x \mid 1 < x < 7\}$

Problem 4   Solve: $5 - 4x > 1$ and $6 - 5x < 11$

Solution   See page A12.
$\{x \mid -1 < x < 1\}$

Example 5   Solve: $3 - 4x > 7$ or $4x + 5 < 9$

Solution

$$\begin{array}{ll} 3 - 4x > 7 & 4x + 5 < 9 \\ -4x > 4 & 4x < 4 \\ x < -1 & x < 1 \end{array}$$ • Solve each inequality.

$\{x \mid x < -1\}$ $\qquad$ $\{x \mid x < 1\}$

$\{x \mid x < -1\} \cup \{x \mid x < 1\} =$ $\qquad$ • Find the union of the solution sets.
$\{x \mid x < 1\}$

Problem 5   Solve: $2 - 3x > 11$ or $5 + 2x > 7$

Solution   See page A12.
$\{x \mid x < -3 \text{ or } x > 1\}$

3   Application problems

Example 6   Company A rents cars for $6 a day and 14¢ for every mile driven. Company B rents cars for $12 a day and 8¢ for every mile driven. You want to rent a car for 5 days. How many miles can you drive a Company A car during the 5 days if it is to cost less than a Company B car?

Strategy   ▶ To find the number of miles, write and solve an inequality using N to represent the number of miles.

Solution   Cost of Company A car $<$ cost of Company B car
$$6(5) + 0.14N < 12(5) + 0.08N$$
$$30 + 0.14N < 60 + 0.08N$$
$$30 + 0.06N < 60$$
$$0.06N < 30$$
$$N < 500$$

It is less expensive to rent from Company A if the car is driven less than 500 mi.

Problem 6   The base of a triangle is 12 in., and the height is $(x + 2)$ in. Express as an integer the maximum height of the triangle when the area is less than 50 in.[2]. (The formula for the area of a triangle is $A = \frac{1}{2}bh$.)

Solution   See page A13.
8 in.

Example 7   Find three consecutive odd integers whose sum is between 27 and 51.

Strategy   ▶ To find the three consecutive odd integers, write and solve a compound inequality using x to represent the first odd integer.

Solution $\begin{array}{c}\text{Lower limit} \\ \text{of the sum}\end{array} < \text{sum} < \begin{array}{c}\text{upper limit} \\ \text{of the sum}\end{array}$

$$27 < x + (x + 2) + (x + 4) < 51$$
$$27 < 3x + 6 < 51$$
$$27 - 6 < 3x + 6 - 6 < 51 - 6$$
$$21 < 3x < 45$$
$$\frac{1}{3}(21) < \frac{1}{3}(3x) < \frac{1}{3}(45)$$
$$7 < x < 15$$

The three integers are 9, 11, and 13; or 11, 13, and 15; or 13, 15, and 17.

Problem 7   An average score of 80 to 89 in a history course receives a B grade. A student has grades of 72, 94, 83, and 70 on four exams. Find the range of scores on the fifth exam that will give the student a B for the course.

Solution   See page A13.
$81 \le N \le 100$

EXERCISES 2.2

1 Solve.

1. $x - 3 < 2$
   $\{x \mid x < 5\}$

2. $x + 4 \geq 2$
   $\{x \mid x \geq -2\}$

3. $4x \leq 8$
   $\{x \mid x \leq 2\}$

4. $6x > 12$
   $\{x \mid x > 2\}$

5. $-2x > 8$
   $\{x \mid x < -4\}$

6. $-3x \leq -9$
   $\{x \mid x \geq 3\}$

7. $3x - 1 > 2x + 2$
   $\{x \mid x > 3\}$

8. $5x + 2 \geq 4x - 1$
   $\{x \mid x \geq -3\}$

9. $2x - 1 > 7$
   $\{x \mid x > 4\}$

10. $3x + 2 < 8$
    $\{x \mid x < 2\}$

11. $5x - 2 \leq 8$
    $\{x \mid x \leq 2\}$

12. $4x + 3 \leq -1$
    $\{x \mid x \leq -1\}$

13. $6x + 3 > 4x - 1$
    $\{x \mid x > -2\}$

14. $7x + 4 < 2x - 6$
    $\{x \mid x < -2\}$

15. $8x + 1 \geq 2x + 13$
    $\{x \mid x \geq 2\}$

16. $5x - 4 < 2x + 5$
    $\{x \mid x < 3\}$

17. $4 - 3x < 10$
    $\{x \mid x > -2\}$

18. $2 - 5x > 7$
    $\{x \mid x < -1\}$

19. $7 - 2x \geq 1$
    $\{x \mid x \leq 3\}$

20. $3 - 5x \leq 18$
    $\{x \mid x \geq -3\}$

21. $-3 - 4x > -11$
    $\{x \mid x < 2\}$

22. $-2 - x < 7$
    $\{x \mid x > -9\}$

23. $4x - 2 < x - 11$
    $\{x \mid x < -3\}$

24. $6x + 5 \geq x - 10$
    $\{x \mid x \geq -3\}$

25. $x + 7 \geq 4x - 8$
    $\{x \mid x \leq 5\}$

26. $3x + 1 \leq 7x - 15$
    $\{x \mid x \geq 4\}$

27. $6 - 2(x - 4) \leq 2x + 10$
    $\{x \mid x \geq 1\}$

28. $4(2x - 1) > 3x - 2(3x - 5)$
    $\left\{x \mid x > \dfrac{14}{11}\right\}$

29. $2(1 - 3x) - 4 > 10 + 3(1 - x)$
    $\{x \mid x < -5\}$

30. $2 - 5(x + 1) \geq 3(x - 1) - 8$
    $\{x \mid x \leq 1\}$

31. $7 + 2(4 - x) < 9 - 3(6 + x)$
    $\{x \mid x < -24\}$

32. $3(4x + 3) \leq 7 - 4(x - 2)$
    $\left\{x \mid x \leq \dfrac{3}{8}\right\}$

33. $2 - 2(7 - 2x) < 3(3 - x)$

$\{x \mid x < 3\}$

34. $3 + 2(x + 5) \geq x + 5(x + 1) + 1$

$\left\{x \mid x \leq \dfrac{7}{4}\right\}$

35. $10 - 13(2 - x) < 5(3x - 2)$

$\{x \mid x > -3\}$

36. $3 - 4(x + 2) \leq 6 + 4(2x + 1)$

$\left\{x \mid x \geq -\dfrac{5}{4}\right\}$

2   Solve.

37. $3x < 6$ and $x + 2 > 1$
$\{x \mid -1 < x < 2\}$

38. $x - 3 \leq 1$ and $2x \geq -4$
$\{x \mid -2 \leq x \leq 4\}$

39. $x + 2 \geq 5$ or $3x \leq 3$
$\{x \mid x \geq 3 \text{ or } x \leq 1\}$

40. $2x < 6$ or $x - 4 > 1$
$\{x \mid x < 3 \text{ or } x > 5\}$

41. $-2x > -8$ and $-3x < 6$
$\{x \mid -2 < x < 4\}$

42. $\dfrac{1}{2}x > -2$ and $5x < 10$
$\{x \mid -4 < x < 2\}$

43. $\dfrac{1}{3}x < -1$ or $2x > 0$
$\{x \mid x < -3 \text{ or } x > 0\}$

44. $\dfrac{2}{3}x > 4$ or $2x < -8$
$\{x \mid x > 6 \text{ or } x < -4\}$

45. $x + 4 \geq 5$ and $2x \geq 6$
$\{x \mid x \geq 3\}$

46. $3x < -9$ and $x - 2 < 2$
$\{x \mid x < -3\}$

47. $-5x > 10$ and $x + 1 > 6$
$\varnothing$

48. $7x < 14$ and $1 - x < 4$
$\{x \mid -3 < x < 2\}$

49. $2x - 3 > 1$ and $3x - 1 < 2$
$\varnothing$

50. $4x + 1 < 5$ and $4x + 7 > -1$
$\{x \mid -2 < x < 1\}$

51. $3x + 7 < 10$ or $2x - 1 > 5$
$\{x \mid x < 1 \text{ or } x > 3\}$

52. $6x - 2 < -14$ or $5x + 1 > 11$
$\{x \mid x < -2 \text{ or } x > 2\}$

53. $-5 < 3x + 4 < 16$
$\{x \mid -3 < x < 4\}$

54. $5 < 4x - 3 < 21$
$\{x \mid 2 < x < 6\}$

55. $0 < 2x - 6 < 4$
$\{x \mid 3 < x < 5\}$

56. $-2 < 3x + 7 < 1$
$\{x \mid -3 < x < -2\}$

57. $4x - 1 > 11$ or $4x - 1 \leq -11$
$\left\{x \mid x > 3 \text{ or } x \leq -\dfrac{5}{2}\right\}$

58. $3x - 5 > 10$ or $3x - 5 < -10$
$\left\{x \mid x > 5 \text{ or } x < -\dfrac{5}{3}\right\}$

59. $2x + 3 \geq 5$ and $3x - 1 > 11$

$\{x \mid x > 4\}$

60. $6x - 2 < 5$ or $7x - 5 < 16$
$\left\{x \mid x < \dfrac{7}{6}\right\}$

61. $9x - 2 < 7$ and $3x - 5 > 10$
$\varnothing$

62. $8x + 2 \leq -14$ and $4x - 2 > 10$
$\varnothing$

63. $3x - 11 < 4$ or $4x + 9 \geq 1$
the set of real numbers

64. $5x + 12 \geq 2$ or $7x - 1 \leq 13$
the set of real numbers

65. $-6 \leq 5x + 14 \leq 24$

$\{x \mid -4 \leq x \leq 2\}$

66. $3 \leq 7x - 14 \leq 31$

$\left\{x \mid \dfrac{17}{7} \leq x \leq \dfrac{45}{7}\right\}$

67. $3 - 2x > 7$ and $5x + 2 > -18$

$\{x \mid -4 < x < -2\}$

68. $1 - 3x < 16$ and $1 - 3x > -16$

$\left\{x \mid -5 < x < \dfrac{17}{3}\right\}$

69. $5 - 4x > 21$ or $7x - 2 > 19$
$\{x \mid x < -4 \text{ or } x > 3\}$

70. $6x + 5 < -1$ or $1 - 2x < 7$
the set of real numbers

71. $3 - 7x \leq 31$ and $5 - 4x > 1$
$\{x \mid -4 \leq x < 1\}$

72. $9 - x \geq 7$ and $9 - 2x < 3$
$\varnothing$

73. $\dfrac{2}{3}x - 4 > 5$ or $x + \dfrac{1}{2} < 3$

$\left\{x \mid x > \dfrac{27}{2} \text{ or } x < \dfrac{5}{2}\right\}$

74. $\dfrac{5}{8}x + 2 < -3$ or $2 - \dfrac{3}{5}x < -7$

$\{x \mid x < -8 \text{ or } x > 15\}$

75. $\dfrac{5}{7}x - 2 > 3$ and $3 - \dfrac{2}{3}x > -2$

$\left\{x \mid 7 < x < \dfrac{15}{2}\right\}$

76. $5 - \dfrac{2}{7}x < 4$ and $\dfrac{3}{4}x - 2 < 7$

$\left\{x \mid \dfrac{7}{2} < x < 12\right\}$

77. $-\dfrac{3}{8} \leq 1 - \dfrac{1}{4}x \leq \dfrac{7}{2}$

$\left\{x \mid -10 \leq x \leq \dfrac{11}{2}\right\}$

78. $-2 \leq \dfrac{2}{3}x - 1 \leq 3$

$\left\{x \mid -\dfrac{3}{2} \leq x \leq 6\right\}$

3   Solve.

79. Three times the difference between a number and ten is less than or equal to five times the sum of the number and two. Find the smallest number that will satisfy the inequality.
−20

80. Four times the difference between a number and five is greater than the quotient of two times the number and three. Find the smallest integer that will satisfy the inequality.
7

81. The length of a rectangle is one foot longer than twice the width. Express as an integer the maximum width of the rectangle when the perimeter is less than 80 in.
12 in.

82. The length of a rectangle is 2 m less than three times the width. Express as an integer the maximum width of the rectangle when the perimeter is less than 60 m.

7 m

83. Company A rents cars for $8 a day and 12¢ for every mile driven. Company B rents cars for $21.80 per day with unlimited mileage. How many miles per day can you drive a Company A car if it is to cost you less than a Company B car?

less than 115 mi

84. Company A rents cars for $10 a day and 10¢ for every mile driven. Company B rents cars for $24.30 per day with unlimited mileage. How many miles per day can you drive a Company A car if it is to cost you less than a Company B car?

less than 142 mi

85. Company A rents cars for $11 a day and 12¢ for every mile driven. Company B rents cars for $8 per day and 18¢ for every mile driven. You want to rent a car for one week. How many miles can you drive a Company B car during the week if it is to cost you less than a Company A car?

less than 350 mi

86. Company A rents cars for $15 a day and 10¢ for every mile driven. Company B rents cars for $10 per day and 24¢ for every mile driven. You want to rent a car for one week. How many miles can you drive a Company B car during the week if it is to cost you less than a Company A car?

less than 250 mi

87. An account executive earns $1000 per month plus 5% commission on the amount of sales. The executive's goal is to earn a minimum of $3200 per month. What amount of sales will enable the executive to earn $3200 or more per month?

$44,000 or more

88. An account executive earns $1200 per month plus 6% commission on the amount of sales. The executive's goal is to earn a minimum of $6000 per month. What amount of sales will enable the executive to earn $6000 or more per month?

$80,000 or more

89. A bank offers two types of checking accounts. One account has a charge of $4 per month plus 4¢ per check. The second account has a charge of $1 per month and 10¢ per check. How many checks can a customer who has the second type of account write if it is to cost the customer less than the first type of account?

less than 50 checks

90.   A bank offers two types of checking accounts. One account has a charge of $6 per month plus 2¢ per check. The second account has a charge of $2 per month and 7¢ per check. How many checks can a customer who has the second type of account write if it is to cost the customer less than the first type of account?
less than 80 checks

91.   An average score of 70 to 79 in a mathematics class receives a C grade. A student has grades of 56, 91, 83, and 62 on four tests. Find the range of scores on the fifth test that will give the student a C for the course.
$58 \leq N \leq 100$

92.   An average score of 90 or above in a physics course receives an A grade. A student has grades of 97, 86, and 82 on three exams. Find the range of scores on the fourth exam that will give the student an A grade for the course.
$95 \leq N \leq 100$

93.   Find three consecutive even integers whose sum is between 30 and 51.
10, 12, 14; 12, 14, 16; or 14, 16, 18

94.   Find four consecutive integers whose sum is between 66 and 82.
16, 17, 18, 19; 17, 18, 19, 20; or 18, 19, 20, 21

SUPPLEMENTAL EXERCISES 2.2

Use the roster method to list the set of positive integers that are solutions of the inequality.

95. $3x - 4 \leq 7$
{1, 2, 3}

96. $4 - 5x \geq -16$
{1, 2, 3, 4}

97. $8x - 7 < 2x + 9$
• {1, 2}

98. $2x + 9 \geq 5x - 4$
{1, 2, 3, 4}

99. $5 + 3(2 + x) > 8 + 4(x - 1)$
{1, 2, 3, 4, 5, 6}

100. $6 + 4(2 - x) > 7 + 3(x + 5)$
∅

101. $-3x < 15$ and $x + 2 < 7$
{1, 2, 3, 4}

102. $3x - 2 > 1$ and $2x - 3 < 5$
{2, 3}

103. $-4 \leq 3x + 8 < 16$
{1, 2}

104. $5 < 7x - 3 \leq 24$
{2, 3}

105. $2x - 9 \leq 3$ and $4x + 8 \geq 1$
{1, 2, 3, 4, 5, 6}

106. $6x + 11 \geq 3$ and $8x - 1 \leq 12$
{1}

Graph the solution set.

107. $-5 \leq 2x + 3 \leq 7$ or $-7 < 3x + 2 < 11$

108. $-4 < 3x + 2 < 8$ or $-9 \leq 4x - 1 \leq 11$

109. $8 < 5x + 3 < 18$ and $-4 < 3x + 2 < 8$

110. $-3 \leq 2x - 5 \leq 7$ and $6 < 5x + 1 < 16$

Solve.

111. The relationship between Celsius temperature and Fahrenheit temperature is given by the formula $F = \dfrac{9}{5}C + 32$. If the temperature is between 68°F and 77°F, what is the temperature range in degrees Celsius?
20°C to 25°C

112. The relationship between Celsius temperature and Fahrenheit temperature is given by the formula $F = \dfrac{9}{5}C + 32$. If the temperature is between −22°F and −13°F, what is the temperature range in degrees Celsius?
−30°C to −25°C

113. The average of two positive integers is less than or equal to 40. The larger integer is 8 more than the smaller integer. Find the greatest possible value for the larger integer.
44

114. The average of two negative integers is less than or equal to −19. The smaller integer is 14 less than the larger integer. Find the greatest possible value for the smaller integer.
−26

115. The charges for a long-distance telephone call are $1.42 for the first three minutes and $.56 for each additional minute or fraction of a minute. What is the largest whole number of minutes a call could last if it is to cost you less than $6?
11 min

116. Your class decides to publish a calendar to raise money. The initial cost, regardless of the number of calendars printed, is $650. After the initial cost, each calendar costs $1.25 to produce. What is the minimum number of calendars your class must sell at $7 per calendar to make a profit of at least $1250?
331 calenders

SECTION 2.3 _____

Absolute Value Equations and Inequalities

1 Absolute value equations

The **absolute value** of a number is its distance from zero on the number line. Distance is always a positive number or zero. Therefore, the absolute value of a number is always a positive number or zero.

The distance from 0 to 3 or from 0 to −3 is 3 units.

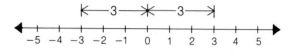

$|3| = 3$ $|-3| = 3$

Absolute value can also be used to represent the distance between any two points on the number line. The **distance between two points** on the number line is the absolute value of the difference between the coordinates of the two points.

The distance between point a and point b is given by $|b - a|$.

The distance between 4 and −3 on the number line is 7 units. Note that the order in which the coordinates are subtracted does not affect the distance.

$$\text{Distance} = |-3 - 4| \qquad \text{Distance} = |4 - (-3)|$$
$$= |-7| \qquad\qquad\qquad = |7|$$
$$= 7 \qquad\qquad\qquad\quad = 7$$

For any two numbers a and b, $|b - a| = |a - b|$.

An equation containing an absolute value symbol is called an **absolute value equation.**

$|x| = 3$ Absolute
$|x + 2| = 8$ Value
$|3x - 4| = 5x - 9$ Equations

Solving Absolute Value Equations

Solving an absolute value equation is based on the following property:

If $a \geq 0$ and $|x| = a$, then $x = a$ or $x = -a$ since $|a| = a$ and $|-a| = a$.

Given $|x| = 3$, then $x = 3$ or $x = -3$ since $|3| = 3$ and $|-3| = 3$.

Solve: $|x + 2| = 8$

Remove the absolute value sign, and rewrite as two equations.

$$|x + 2| = 8$$
$$x + 2 = 8 \qquad x + 2 = -8$$

Solve each equation.

Write the solution.

$x = 6$ $\quad\quad\quad\quad$ $x = -10$

The solutions are 6 and -10.

Check:
$$
\begin{array}{r|r}
|x + 2| = 8 & |x + 2| = 8 \\ \hline
|6 + 2| \quad\Big|\quad 8 & |-10 + 2| \quad\Big|\quad 8 \\
|8| \quad\Big|\quad 8 & |-8| \quad\Big|\quad 8 \\
8 = 8 & 8 = 8
\end{array}
$$

Example 1 Solve.

A. $|x| = 15$ $\quad$ B. $|2 - x| = 12$ $\quad$ C. $|2x| = -4$ $\quad$ D. $3 - |2x - 4| = -5$

Solution A. $|x| = 15$

$x = 15$ $\quad\quad\quad$ $x = -15$ $\quad\quad\quad$ • Remove the absolute value sign, and rewrite as two equations.

The solutions are 15 and -15.

B. $|2 - x| = 12$

$2 - x = 12$ $\quad\quad\quad$ $2 - x = -12$ $\quad\quad\quad$ • Remove the absolute value sign, and rewrite as two equations.

$-x = 10$ $\quad\quad\quad$ $-x = -14$ $\quad\quad\quad$ • Solve each equation.

$x = -10$ $\quad\quad\quad$ $x = 14$

The solutions are -10 and 14.

C. $|2x| = -4$ $\quad\quad\quad\quad\quad\quad\quad\quad\quad\quad$ • The absolute value of a number must be nonnegative.

There is no solution to the equation.

D. $3 - |2x - 4| = -5$

$-|2x - 4| = -8$ $\quad\quad\quad\quad\quad$ • Solve for the absolute value. Multiply each side of the equation by -1.

$|2x - 4| = 8$

$2x - 4 = 8$ $\quad\quad\quad$ $2x - 4 = -8$ $\quad\quad\quad$ • Remove the absolute value sign, and rewrite as two equations.

$2x = 12$ $\quad\quad\quad$ $2x = -4$ $\quad\quad\quad$ • Solve each equation.

$x = 6$ $\quad\quad\quad$ $x = -2$

The solutions are 6 and -2.

Problem 1 Solve.

A. $|x| = 25$ $\quad$ B. $|2x - 3| = 5$ $\quad$ C. $|x - 3| = -2$ $\quad$ D. $5 - |3x + 5| = 3$

Solution See page A14.

A. 25 and -25 $\quad\quad$ B. 4 and -1 $\quad\quad$ C. no solution $\quad\quad$ D. -1 and $-\dfrac{7}{3}$

2 Absolute value inequalities

Recall that absolute value represents the distance between two points. For example, the solutions of the absolute value equation $|x - 1| = 3$, are the numbers whose distance from 1 is 3. Therefore, the solutions are -2 and 4.

The solutions of the absolute value inequality $|x - 1| < 3$ are the numbers whose distance from 1 is *less than* 3. Therefore, the solutions are the numbers greater than -2 and less than 4. The solution set is $\{x \mid -2 < x < 4\}$

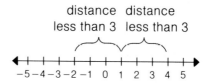

distance   distance
less than 3 less than 3

Absolute Value Inequalities of the Form $|ax + b| < c$

To solve an absolute value inequality of the form $|ax + b| < c$, solve the equivalent compound inequality $-c < ax + b < c$.

Solve: $|3x - 1| < 5$

Solve the equivalent compound inequality.

$$|3x - 1| < 5$$
$$-5 < 3x - 1 < 5$$
$$-5 + 1 < 3x - 1 + 1 < 5 + 1$$
$$-4 < 3x < 6$$
$$\frac{1}{3}(-4) < \frac{1}{3}(3x) < \frac{1}{3}(6)$$
$$-\frac{4}{3} < x < 2$$
$$\left\{ x \mid -\frac{4}{3} < x < 2 \right\}$$

Example 2   Solve: $|4x - 3| < 5$

Solution $|4x - 3| < 5$
$$-5 < 4x - 3 < 5$$ • Solve the equivalent compound inequality.
$$-5 + 3 < 4x - 3 + 3 < 5 + 3$$ • Add 3 to each of the three parts of the inequality. (Do this step mentally.)

$$-2 < 4x < 8$$
$$\frac{1}{4}(-2) < \frac{1}{4}(4x) < \frac{1}{4}(8)$$ • Multiply each of the three parts of the inequality by $\frac{1}{4}$. (Do this step mentally.)

$$-\frac{1}{2} < x < 2$$

$$\left\{ x \mid -\frac{1}{2} < x < 2 \right\}$$ • Write the solution set.

Problem 2   Solve: $|3x + 2| < 8$

Solution   See page A14.
$$\left\{x \,\middle|\, -\frac{10}{3} < x < 2\right\}$$

Example 3   Solve: $|x - 3| < 0$

Solution $|x - 3| < 0$ • The absolute value of a number must be nonnegative.
$\varnothing$ • The solution set is the empty set.

Problem 3   Solve: $|3x - 7| < 0$

Solution   See page A14.
$\varnothing$

The solutions of the absolute value inequality $|x + 1| > 2$ are the numbers whose distance from -1 is *greater than* 2. Therefore, the solutions are the numbers less than -3 or greater than 1. The solution set is $\{x \,|\, x < -3 \text{ or } x > 1\}$.

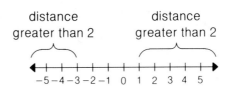

distance   distance
greater than 2   greater than 2

$-5\ -4\ -3\ -2\ -1\ \ \ 0\ \ \ 1\ \ \ 2\ \ \ 3\ \ \ 4\ \ \ 5$

Absolute Value Inequalities of the Form $|ax + b| > c$

To solve an absolute value inequality of the form $|ax + b| > c$, solve the equivalent compound inequality $ax + b < -c$ or $ax + b > c$.

Solve: $|3 - 2x| > 1$

$|3 - 2x| > 1$

Solve the equivalent compound inequality.

$3 - 2x < -1$ or $3 - 2x > 1$
$-2x < -4$ $-2x > -2$
$x > 2$ $x < 1$

$\{x \,|\, x > 2\}$ $\{x \,|\, x < 1\}$

Find the union of the solution sets.

$\{x \,|\, x > 2\} \cup \{x \,|\, x < 1\} = \{x \,|\, x > 2 \text{ or } x < 1\}$

Example 4 Solve: $|2x - 1| > 7$

Solution $|2x - 1| > 7$

$2x - 1 < -7$ or $2x - 1 > 7$ • Solve the equivalent compound inequality.

$\qquad 2x < -6 \qquad\qquad 2x > 8$

$\qquad\quad x < -3 \qquad\qquad\quad x > 4$

$\{x \,|\, x < -3\} \qquad\qquad \{x \,|\, x > 4\}$

$\{x \,|\, x < -3\} \cup \{x \,|\, x > 4\} =$ • Find the union of the solution sets.
$\{x \,|\, x < -3 \text{ or } x > 4\}$

Problem 4 Solve: $|5x + 3| > 8$

Solution See page A15.

$\left\{ x \,\middle|\, x > 1 \text{ or } x < -\dfrac{11}{5} \right\}$

3 Application problems

The **tolerance** of a component, or part, is the acceptable amount by which the component may vary from a given measurement. For example, the diameter of a piston may vary from the given measurement of 9 cm by 0.001 cm. This is written as 9 cm ± 0.001 cm, read "9 centimeters plus or minus 0.001 centimeters." The maximum diameter, or **upper limit,** of the piston is 9 cm + 0.001 cm = 9.001 cm. The minimum diameter, or **lower limit,** is 9 cm − 0.001 cm = 8.999 cm.

The lower and upper limits of the diameter of the piston could also be found by solving the absolute value inequality $|d - 9| < 0.001$, where d is the diameter of the piston.

$$|d - 9| \leq 0.001$$
$$-0.001 \leq d - 9 \leq 0.001$$
$$-0.001 + 9 \leq d - 9 + 9 \leq 0.001 + 9$$
$$8.999 \leq d \leq 9.001$$

The lower and upper limits of the diameter of the piston are 8.999 cm and 9.001 cm.

Example 5 A doctor has prescribed 2 cc of medication for a patient. The tolerance is 0.03 cc. Find the lower and upper limits of the amount of medication to be given.

Strategy ▶ Let p represent the prescribed amount of medication, T the tolerance, and m the given amount of medication. Solve the absolute value inequality $|m - p| \leq T$ for m.

Solution $|m - p| \leq T$
$|m - 2| \leq 0.03$ • Substitute the values of p and T into the inequality.
$-0.03 \leq m - 2 \leq 0.03$ • Solve the equivalent compound inequality.
$1.97 \leq m \leq 2.03$

The lower and upper limits of the amount of medication to be given to the patient are 1.97 cc and 2.03 cc.

Problem 5   A machinist must make a bushing that has a tolerance of 0.003 in. The diameter of the bushing is 2.55 in. Find the lower and upper limits of the diameter of the bushing.

Solution   See page A15.
2.547 in. and 2.553 in.

EXERCISES 2.3

1   Solve.

1. $|x| = 7$
−7 and 7

2. $|a| = 2$
−2 and 2

3. $|b| = 4$
−4 and 4

4. $|c| = 12$
−12 and 12

5. $|-y| = 6$
6 and −6

6. $|-t| = 3$
−3 and 3

7. $|-a| = 7$
−7 and 7

8. $|-x| = 3$
3 and −3

9. $|x| = -4$
no solution

10. $|y| = -3$
no solution

11. $|-t| = -3$
no solution

12. $|-y| = -2$
no solution

13. $|x + 2| = 3$
−5 and 1

14. $|x + 5| = 2$
−7 and −3

15. $|y - 5| = 3$
8 and 2

16. $|y - 8| = 4$
4 and 12

17. $|a - 2| = 0$
2

18. $|a + 7| = 0$
−7

19. $|x - 2| = -4$

no solution

20. $|x + 8| = -2$

no solution

21. $|2x - 5| = 4$
$\frac{1}{2}$ and $\frac{9}{2}$

22. $|3x - 9| = 3$

4 and 2

23. $|3x - 2| = 7$

$-\frac{5}{3}$ and 3

24. $|5x + 8| = 2$

$-\frac{6}{5}$ and −2

25. $|2x + 5| = 8$
$\frac{3}{2}$ and $-\frac{13}{2}$

26. $|2x - 3| = 5$

−1 and 4

27. $|4 - 2x| = 6$

−1 and 5

28. $|7 - 3x| = 5$
$\frac{2}{3}$ and 4

29. $|2 - 3y| = 4$
$-\frac{2}{3}$ and 2

30. $|4 - 9x| = 14$
$-\frac{10}{9}$ and 2

31. $|4 - 3x| = 4$

0 and $\dfrac{8}{3}$

32. $|2 - 5x| = 2$

0 and $\dfrac{4}{5}$

33. $|3 - 4x| = 9$

$-\dfrac{3}{2}$ and 3

34. $|2 - 5x| = 3$

$-\dfrac{1}{5}$ and 1

35. $|2x - 3| = 0$

$\dfrac{3}{2}$

36. $|5x + 5| = 0$

-1

37. $|3x - 2| = -4$

no solution

38. $|2x + 5| = -2$

no solution

39. $|x - 2| - 2 = 3$

-3 and 7

40. $|x - 9| - 3 = 2$

4 and 14

41. $|3a + 2| - 4 = 4$

$-\dfrac{10}{3}$ and 2

42. $|2a + 9| + 4 = 5$

-4 and -5

43. $|2 - y| + 3 = 4$

1 and 3

44. $|8 - y| - 3 = 1$

4 and 12

45. $|2x - 3| + 3 = 3$

$\dfrac{3}{2}$

46. $|4x - 7| - 5 = -5$

$\dfrac{7}{4}$

47. $|2x - 3| + 4 = -4$

no solution

48. $|3x - 2| + 1 = -1$

no solution

49. $|6x - 5| - 2 = 4$

$\dfrac{11}{6}$ and $-\dfrac{1}{6}$

50. $|4b + 3| - 2 = 7$

-3 and $\dfrac{3}{2}$

51. $|3t + 2| + 3 = 4$

$-\dfrac{1}{3}$ and -1

52. $|5x - 2| + 5 = 7$

$\dfrac{4}{5}$ and 0

53. $3 - |x - 4| = 5$

no solution

54. $2 - |x - 5| = 4$

no solution

55. $8 - |2x - 3| = 5$

3 and 0

56. $8 - |3x + 2| = 3$

$-\dfrac{7}{3}$ and 1

57. $|2 - 3x| + 7 = 2$

no solution

58. $|1 - 5a| + 2 = 3$

0 and $\dfrac{2}{5}$

59. $|8 - 3x| - 3 = 2$

1 and $\dfrac{13}{3}$

60. $|6 - 5b| - 4 = 3$

$-\dfrac{1}{5}$ and $\dfrac{13}{5}$

61. $|2x - 8| + 12 = 2$

no solution

62. $|3x - 4| + 8 = 3$

no solution

63. $2 + |3x - 4| = 5$

$\dfrac{7}{3}$ and $\dfrac{1}{3}$

64. $5 + |2x + 1| = 8$

-2 and 1

65. $5 - |2x + 1| = 5$

$-\dfrac{1}{2}$

66. $3 - |5x + 3| = 3$

$-\dfrac{3}{5}$

67. $6 - |2x + 4| = 3$

$-\dfrac{1}{2}$ and $-\dfrac{7}{2}$

68. $8 - |3x - 2| = 5$

$-\dfrac{1}{3}$ and $\dfrac{5}{3}$

69. $8 - |1 - 3x| = -1$

$-\dfrac{8}{3}$ and $\dfrac{10}{3}$

70. $3 - |3 - 5x| = -2$

$-\dfrac{2}{5}$ and $\dfrac{8}{5}$

71. $5 + |2 - x| = 3$

no solution

72. $6 + |3 - 2x| = 2$

no solution   .

2   Solve.

73. $|x| > 3$
$\{x \mid x > 3 \text{ or } x < -3\}$

74. $|x| < 5$
$\{x \mid -5 < x < 5\}$

75. $|x + 1| > 2$
$\{x \mid x > 1 \text{ or } x < -3\}$

76. $|x - 2| > 1$
$\{x \mid x < 1 \text{ or } x > 3\}$

77. $|x - 5| \le 1$
$\{x \mid 4 \le x \le 6\}$

78. $|x - 4| \le 3$
$\{x \mid 1 \le x \le 7\}$

79. $|2 - x| \ge 3$
$\{x \mid x \le -1 \text{ or } x \ge 5\}$

80. $|3 - x| \ge 2$
$\{x \mid x \le 1 \text{ or } x \ge 5\}$

81. $|2x + 1| < 5$

$\{x \mid -3 < x < 2\}$

82. $|3x - 2| < 4$

$\left\{ x \mid -\dfrac{2}{3} < x < 2 \right\}$

83. $|5x + 2| > 12$

$\left\{ x \mid x > 2 \text{ or } x < -\dfrac{14}{5} \right\}$

84. $|7x - 1| > 13$

$\left\{ x \mid x < -\dfrac{12}{7} \text{ or } x > 2 \right\}$

85. $|4x - 3| \le -2$
$\varnothing$

86. $|5x + 1| \le -4$
$\varnothing$

87. $|2x + 7| > -5$
the set of real numbers

88. $|3x - 1| > -4$
the set of real numbers

89. $|4 - 3x| \ge 5$

$\left\{ x \mid x \le -\dfrac{1}{3} \text{ or } x \ge 3 \right\}$

90. $|7 - 2x| > 9$

$\{x \mid x < -1 \text{ or } x > 8\}$

91. $|5 - 4x| \le 13$

$\left\{ x \mid -2 \le x \le \dfrac{9}{2} \right\}$

92. $|3 - 7x| < 17$

$\left\{ x \mid -2 < x < \dfrac{20}{7} \right\}$

93. $|6 - 3x| \le 0$
2

94. $|10 - 5x| \ge 0$
the set of real numbers

95. $|2 - 9x| > 20$

$\left\{ x \mid x < -2 \text{ or } x > \dfrac{22}{9} \right\}$

96. $|5x - 1| < 16$

$\left\{ x \mid -3 < x < \dfrac{17}{5} \right\}$

3   Solve.

97. A doctor has prescribed 4 cc of medication for a patient. The tolerance is 0.05 cc. Find the lower and upper limits of the amount of medication to be given.
3.95 cc, 4.05 cc

98. A doctor has prescribed 3 cc of medication for a patient. The tolerance is 0.04 cc. Find the lower and upper limits of the amount of medication to be given.
2.96 cc, 3.04 cc

99. A machinist must make a bushing that has a tolerance of 0.002 in. The diameter of the bushing is 2.65 in. Find the lower and upper limits of the diameter of the bushing.
2.648 in., 2.652 in.

100. The diameter of a bushing is 2.45 in. The bushing has a tolerance of 0.001 in. Find the lower and upper limits of the diameter of the bushing.
2.449 in., 2.451 in.

101. A piston rod for an automobile is $9\frac{5}{8}$ in. long with a tolerance of $\frac{1}{32}$ in.
Find the lower and upper limits of the length of the piston rod.
$9\frac{19}{32}$ in., $9\frac{21}{32}$ in.

102. A piston rod for an automobile is $9\frac{3}{8}$ in. long with a tolerance of $\frac{1}{64}$ in.
Find the lower and upper limits of the length of the piston rod.
$9\frac{23}{64}$ in., $9\frac{25}{64}$ in.

The tolerance of the resistors used in electronics is given as a percent. Use your calculator for the following exercises.

103. Find the lower and upper limits of a 29,000-ohm resistor with a 2% tolerance.
28,420 ohms, 29,580 ohms

104. Find the lower and upper limits of a 15,000-ohm resistor with a 10% tolerance.
13,500 ohms, 16,500 ohms

105. Find the lower and upper limits of a 25,000-ohm resistor with a 5% tolerance.
23,750 ohms, 26,250 ohms

106. Find the lower and upper limits of a 56-ohm resistor with a 5% tolerance.   53.2 ohms, 58.8 ohms

SUPPLEMENTAL EXERCISES 2.3

Solve.

107. $\left|\dfrac{2x-5}{3}\right| = 7$

$13, -8$

108. $\left|\dfrac{4x-3}{5}\right| = 5$

$7, -\dfrac{11}{2}$

109. $\left|\dfrac{3x-2}{4}\right| + 5 = 6$

$2, -\dfrac{2}{3}$

110. $\left|\dfrac{5x+1}{6}\right| - 1 = 2$

$\dfrac{17}{5}, -\dfrac{19}{5}$

111. $\left|\dfrac{4x-2}{3}\right| > 6$

$\{x \mid x > 5 \text{ or } x < -4\}$

112. $\left|\dfrac{2x-3}{3}\right| \geq 9$

$\{x \mid x \geq 15 \text{ or } x \leq -12\}$

113. $\left|\dfrac{2x-1}{5}\right| \leq 3$

$\{x \mid -7 \leq x \leq 8\}$

114. $\left|\dfrac{3x-5}{2}\right| < 7$

$\left\{x \mid -3 < x < \dfrac{19}{3}\right\}$

Graph the solution set.

115. $|2x - 3| > 5$

116. $|2x + 1| \geq 3$

117. $|4x - 2| \leq 10$

118. $|2x - 3| < 5$

119. $|4x + 2| \geq 6$

120. $|5 - 2x| \geq 3$

121. $|3 - 2x| \leq 5$

122. $|3 - 6x| \leq 9$

For what values of the variable is the equation true? Write the answer in set builder notation.

123. $|x + 3| = x + 3$
$\{x \mid x \geq -3\}$

124. $|y + 6| = y + 6$
$\{y \mid y \geq -6\}$

125. $|a - 4| = a - 4$
$\{a \mid a \geq 4\}$

126. $|b - 7| = b - 7$
$\{b \mid b \geq 7\}$

127. $|c + 5| = -(c + 5)$
$\{c \mid c \leq -5\}$

128. $|x + 9| = -(x + 9)$
$\{x \mid x \leq -9\}$

129. $|y - 8| = -(y - 8)$
$\{y \mid y \leq 8\}$

130. $|a - 1| = -(a - 1)$
$\{a \mid a \leq 1\}$

Solve.

131. Show that the solution set of an absolute value equation of the form

$$|ax + b| < c \text{ is } \left\{ x \left| \frac{-c - b}{a} < x < \frac{c - b}{a} \right. \right\}.$$

Complete solution available in Solutions Manual.

SECTION **2.4**

Applications

1   Coin and stamp problems

In solving problems dealing with coins or stamps of different values, it is necessary to represent the value of the coins or stamps in the same unit of money. The unit of money is frequently cents. For example:

The value of five 8¢ stamps is 5 · 8, or 40 cents.
The value of four 20¢ stamps is 4 · 20, or 80 cents.
The value of n 10¢ stamps is n · 10, or $10n$ cents.

A collection of stamps consists of 5¢, 13¢, and 18¢ stamps. The number of 13¢ stamps is two more than three times the number of 5¢ stamps. The number of 18¢ stamps is five less than the number of 13¢ stamps. The total value of all the stamps is $1.68. Find the number of 18¢ stamps.

STRATEGY *for solving a stamp problem*

▶ For each denomination of stamp, write a numerical or variable expression for the number of stamps, the value of the stamp, and the total value of the stamps in cents. The results can be recorded in a table.

The number of 5¢ stamps: x
The number of 13¢ stamps: $3x + 2$
The number of 18¢ stamps: $(3x + 2) - 5 = 3x - 3$

Stamp	Number of stamps	·	Value of stamp in cents	=	Total value in cents
5¢	x	·	5	=	$5x$
13¢	$3x + 2$	·	13	=	$13(3x + 2)$
18¢	$3x - 3$	·	18	=	$18(3x - 3)$

▶ **Determine the relationship between the total values of the stamps. Use the fact that the sum of the total values of each denomination of stamp is equal to the total value of all the stamps.**

The sum of the total values of each denomination of stamp is equal to the total value of all the stamps (168 cents).

$$5x + 13(3x + 2) + 18(3x - 3) = 168$$
$$5x + 39x + 26 + 54x - 54 = 168$$
$$98x - 28 = 168$$
$$98x = 196$$
$$x = 2$$

The number of 18¢ stamps is $3x - 3$. Replace x by 2 and evaluate.

$$3x - 3 = 3(2) - 3 = 3$$

There are three 18¢ stamps in the collection.

Example 1 A coin bank contains $1.80 in nickels and dimes. In all, there are twenty-two coins in the bank. Find the number of nickels and the number of dimes in the bank.

Strategy ▶ Number of nickels: x
Number of dimes: $22 - x$

Coin	Number	Value	Total Value
Nickel	x	5	$5x$
Dime	$22 - x$	10	$10(22 - x)$

▶ The sum of the total values of each denomination of coin equals the total value of all the coins (180 cents).

Solution
$$5x + 10(22 - x) = 180$$
$$5x + 220 - 10x = 180$$
$$-5x + 220 = 180$$
$$-5x = -40$$
$$x = 8 \qquad \text{• There are 8 nickels in the bank.}$$

$$22 - x = 22 - 8 = 14 \qquad \text{• Substitute the value of } x \text{ into the variable expression}$$
for the number of dimes.

The bank contains 8 nickels and 14 dimes.

Problem 1   A collection of stamps contains 3¢, 10¢, and 15¢ stamps. The number of 10¢ stamps is two more than twice the number of 3¢ stamps. There are three times as many 15¢ stamps as there are 3¢ stamps. The total value of the stamps is $1.56. Find the number of 15¢ stamps.

Solution   See page A16.
6 stamps

2   Value mixture problems

A **value mixture problem** involves combining two ingredients that have different prices into a single blend. For example, a coffee manufacturer may blend two types of coffee into a single blend.

The solution of a value mixture problem is based on the equation $AC = V$, where A is the amount of the ingredient, C is the cost per unit of the ingredient, and V is the value of the ingredient.

The value of 12 lb of coffee costing $5.25 per pound is:

$$V = AC$$
$$V = 12(\$5.25)$$
$$V = \$63$$

How many pounds of peanuts that cost $2.25 per pound must be mixed with 40 lb of cashews that cost $6.00 per pound to make a mixture that costs $3.50 per pound?

STRATEGY *for solving a value mixture problem*

▶ For each ingredient in the mixture, write a numerical or variable expression for the amount of the ingredient used, the unit cost of the ingredient, and the value of the amount used. For the mixture, write a numerical or variable expression for the amount, the unit cost of the mixture, and the value of the amount. The results can be recorded in a table.

Amount of peanuts: x
Amount of cashews: 40

	Amount, A	·	Unit cost, C	=	Value, V
Peanuts	x	·	$2.25	=	2.25x
Cashews	40	·	$6.00	=	6.00(40)
Mixture	40 + x	·	$3.50	=	3.50(40 + x)

▶ **Determine how the values of each ingredient are related. Use the fact that the sum of the values of each ingredient is equal to the value of the mixture.**

The sum of the values of the peanuts and the cashews is equal to the value of the mixture.

$$2.25x + 6.00(40) = 3.50(40 + x)$$
$$2.25x + 240 = 140 + 3.50x$$
$$-1.25x + 240 = 140$$
$$-1.25x = -100$$
$$x = 80$$

The mixture must contain 80 lb of peanuts.

Example 2   How many ounces of a gold alloy that costs $320 an ounce must be mixed with 100 oz of an alloy that costs $100 an ounce to make a mixture that costs $160 an ounce?

Strategy   ▶ Ounces of the $320 gold alloy: x
Ounces of the $100 gold alloy: 100

	Amount	Cost	Value
$320 alloy	x	320	$320x$
$100 alloy	100	100	100(100)
Mixture	$x + 100$	160	$160(x + 100)$

▶ The sum of the values before mixing equals the value after mixing.

Solution $320x + 100(100) = 160(x + 100)$
$320x + 10{,}000 = 160x + 16{,}000$
$160x + 10{,}000 = 16{,}000$
$160x = 6000$
$x = 37.5$

The mixture must contain 37.5 oz of the $320 gold alloy.

Problem 2   A butcher combined hamburger that cost $3.00 per pound with hamburger that cost $1.80 per pound. How many pounds of each were used to make a 75-lb mixture that cost $2.20 per pound?

Solution   See page A17.
$3/lb hamburger: 25 lb; $1.80/lb hamburger: 50 lb

3   Uniform motion problems

A car that travels constantly in a straight line at 55 mph is in uniform motion. **Uniform motion** means the speed of an object does not change.

The solution of a uniform motion problem is based on the equation $rt = d$, where r is the rate of travel, t is the time spent traveling, and d is the distance traveled.

An executive has an appointment 785 mi from the office. The executive takes a helicopter from the office to the airport and a plane from the airport to the business appointment. The helicopter averages 70 mph, and the plane averages 500 mph. The total time spent traveling is 2 h. Find the distance from the executive's office to the airport.

STRATEGY *for solving a uniform motion problem*

▶ For each object, write a numerical or variable expression for the distance, rate, and time. The results can be recorded in a table.

The total time of travel is 2 h.

Unknown time in the helicopter: t
Time in the plane: $2 - t$

	Rate, r	·	Time, t	=	Distance, d
Helicopter	70	·	t	=	$70t$
Plane	500	·	$2 - t$	=	$500(2 - t)$

▶ Determine how the distances traveled by each object are related. For example, the total distance traveled by both objects may be known, or it may be known that the two objects traveled the same distance.

The total distance traveled is 785 mi.

$$70t + 500(2 - t) = 785$$
$$70t + 1000 - 500t = 785$$
$$-430 + 1000 = 785$$
$$-430t = -215$$
$$t = 0.5$$

The time spent traveling from the office to the airport in the helicopter is 0.5 h. To find the distance between these two points, substitute the values of r and t into the equation $rt = d$.

$$rt = d$$
$$70 \cdot 0.5 = d$$
$$35 = d$$

The distance from the office to the airport is 35 mi.

Example 3   A long-distance runner started a course running at an average speed of 6 mph. One and one half hours later, a cyclist traveled the same course at an average speed of 12 mph. How long after the runner started did the cyclist overtake the runner?

Strategy   ▶ Unknown time for the cyclist: t
Time for the runner: $t + 1.5$

	Rate	Time	Distance
Runner	6	$t + 1.5$	$6(t + 1.5)$
Cyclist	12	t	$12t$

▶ The runner and the cyclist traveled the same distance.

Solution
$$6(t + 1.5) = 12t$$
$$6t + 9 = 12t$$
$$9 = 6t$$
$$\frac{3}{2} = t \qquad \text{• The cyclist traveled for 1.5 h.}$$

$t + 1.5 = 1.5 + 1.5 = 3$ • Substitute the value of t into the variable expression for the runner's time.

The cyclist overtook the runner 3 h after the runner started.

Problem 3   Two small planes start from the same point and fly in opposite directions. The first plane is flying 30 mph faster than the second plane. In 4 h the planes are 1160 mi apart. Find the rate of each plane.

Solution   See page A17.
1st plane: 160 mph; 2nd plane: 130 mph

Example 4   A student bicycles to school at an average speed of 15 mph and bicycles from school to home at an average speed of 10 mph. If the entire trip takes 5 h, what is the distance from the student's home to school?

Strategy   ▶ Time bicycling at 15 mph: t
Time bicycling at 10 mph: $5 - t$

	Rate	Time	Distance
At 15 mph	15	t	$15t$
At 10 mph	10	$5 - t$	$10(5 - t)$

▶ The distance bicycling at 15 mph is equal to the distance bicycling at 10 mph.

Solution $15t = 10(5 - t)$
$15t = 50 - 10t$
$25t = 50$
    $t = 2$ • The time bicycling at 15 mph is 2 h.

$d = rt = 15(2) = 30$ • Substitute the values of r and t into the equation $d = rt$, and solve for d.

The distance from the student's home to school is 30 mi.

Problem 4   A bus driver's average speed driving to a mountain resort was 60 mph. The average speed returning from the resort was 40 mph. If the total driving time was 10 h, what is the distance to the mountain resort?

Solution   See page A18.
240 mi

EXERCISES 2.4

1   Solve.

1. A coin bank contains 22 coins in dimes and quarters. The coins have a total value of $4.75. Find the number of dimes and the number of quarters in the bank.
5 dimes and 17 quarters

2. A coin bank contains 53 coins in nickels and dimes. The coins have a total value of $3.70. Find the number of nickels and the number of dimes in the bank.
32 nickels and 21 dimes

3. A child's piggy bank contains nickels, dimes, and quarters. There are twice as many dimes as quarters and seven more nickels than dimes. The total value of all the coins is $2.00. How many nickels are in the bank?   13 nickels

4. A coin purse contains 22 coins in nickels, dimes, and quarters. There are four times as many dimes as quarters. The total value of the coins is $2.30. How many nickels are in the coin purse?
7 nickels

5. A cashier's cash drawer contains a total of $4.60 in dimes, quarters, and nickels. There are twice as many dimes as quarters and four more nickels than dimes. How many dimes are in the cash drawer?
16 dimes

6. A child's piggy bank contains dimes, quarters, and nickels. There are twice as many dimes as quarters and twice as many quarters as nickels in the bank. The total value of all the coins is $4.75. How many dimes are in the bank?
20 dimes

7. A clothing store uses twice as many five-dollar bills in conducting its daily business as ten-dollar bills. $2500 was obtained in five- and ten-dollar bills for the day's business. How many ten-dollar bills were obtained?
125 ten-dollar bills

8. A bank teller cashes a check for $170 using twenty-dollar bills and ten-dollar bills. In all, ten bills were handed to the customer. Find the number of twenty-dollar bills given to the customer.
7 twenty-dollar bills

9. A postal clerk sold some 20¢ stamps and some 28¢ stamps. Altogether 140 stamps were sold for a total cost of $31.20. How many of each type of stamp were sold?
20¢ stamps: 100; 28¢ stamps: 40

10. A drawer contains some 15¢ stamps and some 20¢ stamps. The number of 15¢ stamps is eight less than three times the number of 20¢ stamps. The total value is $4. Find the number of each type of stamp in the drawer.
15¢ stamps: 16; 20¢ stamps: 8

11. A stamp collection consists of 6¢ stamps and 15¢ stamps. There are 80 stamps in the collection, and the value of the collection is $6.60. Find the number of 15¢ stamps in the collection.
20 stamps

12. A stamp collection consists of 18¢ stamps and 12¢ stamps. There are 100 stamps in the collection, and the value of the collection is $15. Find the number of 12¢ stamps in the collection.
50 stamps

13. A business executive bought 330 stamps for $79.50. The purchase included 15¢ stamps, 20¢ stamps, and 40¢ stamps. The number of 20¢ stamps is four times the number of 15¢ stamps. How many 40¢ stamps were purchased?
80 stamps

14. A stamp collection consists of 3¢, 8¢, and 13¢ stamps. The number of 8¢ stamps is three less than twice the number of 3¢ stamps. The number of 13¢ stamps is twice the number of 8¢ stamps. The total value of all the stamps is $2.53. Find the number of 13¢ stamps in the collection.
14 stamps

15. A stamp collection consists of 3¢, 12¢, and 15¢ stamps. The number of 3¢ stamps is five times the number of 12¢ stamps. The number of 15¢ stamps is four less than the number of 12¢ stamps. The total value of the stamps in the collection is $3.18. Find the number of 3¢ stamps in the collection.
45 stamps

16. A stamp collection consists of 8¢, 13¢, and 18¢ stamps. The number of 8¢ stamps is twice the number of 18¢ stamps. There are three more 13¢ stamps than 18¢ stamps. The total value of the stamps in the collection is $3.68. Find the number of 8¢ stamps in the collection.
14 stamps

2   Solve.

17. A coffee merchant wants to make 25 lb of a blend of coffee costing $4.25 per pound. The blend is made by using a $6 grade and a $3.50 grade of coffee. How many pounds of each grade of coffee should be used?
7.5 lb at $6; 17.5 lb at $3.50

18. A butcher combines 100 lb of hamburger that costs $1.61 per pound with 60 lb of hamburger that costs $3.45 per pound. Find the cost of the hamburger mixture.
$2.30/lb

19. Tickets for a school play sold for $2.50 for each adult and $1.00 for each child. The total receipt for 113 tickets sold was $221. Find the number of adult tickets sold.
72 adult tickets

20. For one performance of a play, 420 tickets were sold. Adult tickets sold for $4.00 each, and children's tickets sold for $1.00 each. Receipts from the sale of the tickets totaled $1410. Find the number of adult tickets sold.
330 adult tickets

21. To make a flour mix, a miller combined soybeans that cost $8.50 per bushel with wheat that costs $4.50 per bushel. How many bushels of each were used to make a mixture of 1000 bushels costing $5.50 per bushel?
250 bushels of soybeans; 750 bushels of wheat

22. Fifty liters of pure maple syrup that costs $9.50 per liter is mixed with imitation maple syrup that costs $4.00 per liter. How much imitation maple syrup is needed to make a mixture costing $5.00 per liter?
225 L

23. How many ounces of pure silver that costs $10.40 an ounce must be mixed with 50 oz of a silver alloy that costs $6.48 an ounce to make an alloy that costs $9.00 an ounce?
90 oz

24. A goldsmith combined pure gold that costs $450 per ounce with an alloy of gold costing $200 per ounce. How many ounces of each were used to make 50 oz of gold alloy costing $300 per ounce?
20 oz of pure gold; 30 oz of alloy

25. Find the cost per ounce of a face cream mixture made from 100 oz of face cream that costs $3.46 per ounce and 60 oz of face cream that costs $12.50 per ounce.
$6.85/oz

26. Find the cost per pound of a tea mixture made from 40 lb of tea costing $5.40 per pound and 60 lb of tea costing $3.25 per pound.
$4.11/lb

Use your calculator for the following exercises.

27. A grocer combined walnuts that cost $4.05 per kilogram with cashews that cost $7.25 per kilogram. How many kilograms of each were used to make a 50-kg mixture costing $6.25 per kilogram?
walnuts: 15.625 kg; cashews: 34.375 kg

28. Cranberry juice that costs $4.60 per gallon was mixed with 50 gal of apple juice that costs $2.24 per gallon. How much cranberry juice was used to make cranapple juice costing $3.00 per gallon?
23.75 gal

3 Solve.

29. Two hikers start from the same point and hike in opposite directions around a lake whose shoreline is 13 mi. One hiker walks 0.5 mph faster than the other hiker. How fast did each hiker walk if they meet in 2 h?
3 mph, 3.5 mph

30. Two planes start from the same point and fly in opposite directions. The first plane is flying 50 mph slower than the second plane. In 2.5 h, the planes are 1400 mi apart. Find the rate of each plane.
1st plane: 255 mph; 2nd plane: 305 mph

31. A speeding car going 80 mph has a one-half hour head start on a helicopter trying to overtake the car. The helicopter is traveling at 130 mph. How far from the starting point does the helicopter overtake the car?
104 mi

32. A cyclist traveling at 16 mph overtakes a long distance runner who, running at 8 mph, had a 1.5-h head start. How far from the starting point does the cyclist overtake the runner?
24 mi

33. A commuter plane flew to a small town from a major airport. The average speed flying to the small town was 250 mph, and the average speed returning was 150 mph. The total flying time was 4 h. Find the distance between the two airports.
375 mi

34. A cabin cruiser left a harbor and traveled to a small island at an average speed of 20 mph. On the return trip, the cabin cruiser traveled at an average speed of 12 mph because of fog. The total time for the trip was 5 h. How far was the island from the harbor?
37.5 mi

35. A passenger train leaves a depot 1.5 h after a freight train leaves the same depot. The passenger train is traveling 18 mph faster than the freight train. Find the rate of each train if the passenger train overtakes the freight train in 2.5 h.
freight train: 30 mph; passenger train: 48 mph

36. A jogger ran a distance at a speed of 8 mph and returned the same distance running at a speed of 6 mph. Find the total distance that the jogger ran if the total time running was one hour forty-five minutes.
12 mi

37. Two cars are 310 mi apart and traveling toward each other. One car travels 8 mph faster than the other car. The cars meet in 2.5 h. Find the speed of each car.
1st car: 58 mph; 2nd car: 66 mph

38. Two planes are 1620 mi apart and traveling toward each other. One plane is traveling 120 mph faster than the other plane. The planes meet in 1.5 h. Find the speed of each plane.
480 mph, 600 mph

39. A jogger and a cyclist set out at 9 A.M. from the same point headed in the same direction. The average speed of the cyclist is four times the speed of the jogger. In 2 h, the cyclist is 33 mi ahead of the jogger. How far did the cyclist ride?
    44 mi

40. Two planes set out at 3 P.M. from the same airfield headed in the same direction. The average speed of the first plane is 250 mph slower than twice the speed of the second plane. In 3 h, the first plane is 300 mi ahead of the second plane. How far did the second plane travel?
    1050 mi

SUPPLEMENTAL EXERCISES 2.4

Solve.

41. Find the cost per pound of a mixture made from 16 lb of chocolate that cost $4.00 per pound and 24 lb of chocolate that cost $2.50 per pound.
    $3.10/lb

42. A truck leaves a depot at 10 A.M. and travels at 50 mph. At 10:30 A.M., a van leaves the same place and travels the same route at 65 mph. At what time does the van overtake the truck?
    12:10 P.M.

43. A coin bank contains only nickels and dimes. The number of dimes in the bank is two less than twice the number of nickels. There are 52 coins in the bank. How much money is in the bank?
    $4.30

44. Four times the first of three consecutive odd integers is five less than the product of three and the third integer. Find the integers.
    7, 9, 11

45. A collection of stamps consists of 3¢ stamps, 5¢ stamps, and 7¢ stamps. There are six more 3¢ stamps than 5¢ stamps, and two more 7¢ stamps than 3¢ stamps. The total value of the stamps is $1.94. How many 3¢ stamps are in the collection?
    14 stamps

46. A grocer combines 50 gal of cranberry juice that costs $3.50 per gallon with apple juice that costs $2.50 per gallon. How many gallons of apple juice must be used to make cranapple juice costing no more than $2.75 per gallon?
    150 gal

47. A commuter plane flew to a small town from a major airport at an average speed of 300 mph. The average speed on the return trip was
200 mph. What is the maximum distance between the two airports if
the total flying time was less than 4 h? Round to the nearest whole
number.
479 mi

48. A coin purse contains only dimes and quarters. The value of the coins
is less than $3.50. If there are four times as many dimes as quarters,
what is the maximum number of quarters in the coin purse?
5 quarters

49. Three times the difference between a number and 8 is less than four
times the sum of the number and 5. Find the smallest integer that will
satisfy the inequality.
−43

50. A postal clerk sold some 13¢ stamps and some 22¢ stamps. Altogether,
80 stamps were sold. If the total cost of the stamps was less than $15,
what is the maximum number of 22¢ stamps sold?
51 stamps

SECTION 2.5

Applications: Problems Involving Percent

1 Investment problems

The annual simple interest that an investment earns is given by the equation
$Pr = I$, where P is the principal, or the amount invested, r is the simple interest rate, and I is the simple interest. The solution of an investment problem is based on this equation.

The annual interest rate on a $3000 investment is 9%. The annual simple in-
terest earned on the investment is:

$$I = Pr$$
$$I = \$3000(0.09)$$
$$I = \$270$$

You have a total of $8000 invested in two simple interest accounts. On one account, a money market fund, the annual simple interest rate is 11.5%. On the second account, a bond fund, the annual simple interest rate is 9.75%. The total annual interest earned by the two accounts is $823.75. How much do you have invested in each account?

STRATEGY *for solving a problem involving money deposited in two simple interest accounts*

▶ For each amount invested, use the equation $Pr = I$. Write a numerical or variable expression for the principal, the interest rate, and the interest earned. The results can be recorded in a table.

The total amount invested is $8000.

Amount invested at 11.5%: x
Amount invested at 9.75%: $8000 - x$

	Principal, P	.	Interest rate, r	=	Interest earned, I
Amount at 11.5%	x	.	0.115	=	$0.115x$
Amount at 9.75%	$8000 - x$	.	0.0975	=	$0.0975(8000 - x)$

▶ Determine how the amounts of interest earned on each amount are related. For example, the total interest earned by both accounts may be known, or it may be known that the interest earned on one account is equal to the interest earned on the other account.

The total annual interest earned is $823.75.

$$0.115x + 0.0975(8000 - x) = 823.75$$
$$0.115x + 780 - 0.0975x = 823.75$$
$$0.0175x + 780 = 823.75$$
$$0.0175x = 43.75$$
$$x = 2500$$

The amount invested at 9.75% is $8000 - x$. Replace x by 2500 and evaluate.

$$8000 - x = 8000 - 2500 = 5500$$

The amount invested at 11.5% is $2500.
The amount invested at 9.75% is $5500.

Example 1   An investment of $4000 is made at an annual simple interest rate of 10.9%. How much additional money must be invested at an annual simple interest rate of 14.5% so that the total interest earned is 12% of the total investment?

Strategy   ▶ Additional amount to be invested at 14.5%: x

	Principal	Rate	Interest
Amount at 10.9%	4000	0.109	0.109(4000)
Amount at 14.5%	x	0.145	0.145x
Amount at 12%	4000 + x	0.12	0.12(4000 + x)

▶ The sum of the interest earned by the two investments equals the interest earned by the total investment.

Solution $0.109(4000) + 0.145x = 0.12(4000 + x)$
$$436 + 0.145x = 480 + 0.12x$$
$$436 + 0.025x = 480$$
$$0.025x = 44$$
$$x = 1760$$

An additional $1760 must be invested at an annual simple interest rate of 14.5%.

Problem 1   An investment of $3500 is made at an annual simple interest rate of 13.2%. How much additional money must be invested at an annual simple interest rate of 11.5% so that the total interest earned is $1037?

Solution   See page A18.
$5000

2 Percent mixture problems

The amount of a substance in a solution or alloy can be given as a percent of the total solution or alloy. For example, a 10% hydrogen peroxide solution means 10% of the total solution is hydrogen peroxide. The remaining 90% is water.

The solution of a percent mixture problem is based on the equation $Ar = Q$, where A is the amount of solution or alloy, r is the percent of concentration, and Q is the quantity of a substance in the solution or alloy.

The number of grams of silver in 50 g of a 40% silver alloy is:

$$Q = Ar$$
$$Q = 50(0.40)$$
$$Q = 20$$

A chemist mixes an 11% acid solution with a 4% acid solution. How many milliliters of each solution should the chemist use to make a 700-ml solution that is 6% acid?

STRATEGY *for solving a percent mixture problem*

▶ For each solution, use the equation $Ar = Q$. Write a numerical or variable expression for the amount of solution, percent of concentration, and the quantity of the substance in the solution. The results can be recorded in a table.

The total amount of solution is 700 ml.

Amount of 11% solution: x
Amount of 4% solution: $700 - x$

	Amount of solution, A	·	Percent of concentration, r	=	Quantity of substance, Q
11% solution	x	·	0.11	=	$0.11x$
4% solution	$700 - x$	·	0.04	=	$0.04(700 - x)$
6% solution	700	·	0.06	=	$0.06(700)$

▶ Determine how the quantities of the substance in each solution are related. Use the fact that the sum of the quantities of the substances being mixed is equal to the quantity of the substance after mixing.

The sum of the quantities of the substances in the 11% solution and the 4% solution is equal to the quantity of the substances in the 6% solution.

$$0.11x + 0.04(700 - x) = 0.06(700)$$
$$0.11x + 28 - 0.04x = 42$$
$$0.07x + 28 = 42$$
$$0.07x = 14$$
$$x = 200$$

The amount of 4% solution is $700 - x$. Replace x by 200 and evaluate.

$$700 - x = 700 - 200 = 500$$

The chemist should use 200 ml of the 11% solution and 500 ml of the solution.

Example 2 How many grams of pure acid must be added to 60 g of an 8% acid solution to make a 20% acid solution?

Strategy ▶ Grams of pure acid: x

	Amount	Percent	Quantity
Pure acid (100%)	x	1.00	x
8%	60	0.08	0.08(60)
20%	60 + x	0.20	0.20(60 + x)

▶ The sum of the quantities before mixing equals the quantity after mixing.

Solution $x + 0.08(60) = 0.20(60 + x)$
$$x + 4.8 = 12 + 0.20x$$
$$0.8x + 4.8 = 12$$
$$0.8x = 7.2$$
$$x = 9$$

To make the 20% acid solution, 9 g of pure acid must be used.

Problem 2 A butcher has some hamburger that is 22% fat and some that is 12% fat. How many pounds of each should be mixed to make 80 lb of hamburger that is 18% fat?

Solution See page A19.
48 lb of the hamburger that is 22% fat; 32 lb of the hamburger that is 12% fat

EXERCISES 2.5

1 Solve.

1. A total of $8000 is deposited into two simple interest accounts. On one account, the annual simple interest rate is 10.5%, while on the second account, the annual simple interest rate is 14%. How much should be invested in each account so that the total annual interest earned is $1015?
$3000 at 10.5%; $5000 at 14%

2. An investment club invested a part of $12,500 in a 9.8% tax-free annual simple interest account and the remainder in a 13.5% annual simple interest account. The amount of interest earned for one year was $1465.50. How much was invested in each account?
$6000 at 9.8%; $6500 at 13.5%

3. An investment of $4500 is made at an annual simple interest rate of 9.25%. How much additional money must be invested at an annual simple interest rate of 13.5% so that the total interest earned is 12% of the total investment?
$8250

4. An investment of $5000 is made at an annual simple interest rate of 9.2%. How much additional money must be invested at an annual simple interest rate of 15% so that the total interest earned is 11% of the total investment?
$2250

5. An investment of $10,800 is deposited into two simple interest accounts. On one account, the annual simple interest rate is 11.2%. On the other, the annual simple interest rate is 14%. How much should be invested in each account so that the interest earned by each account is the same?
$6000 at 11.2%; $4800 at 14%

6. An investment advisor deposited $42,000 into two simple interest accounts. On the tax-free account the annual simple interest rate is 8.5%, while on the money market fund, the annual simple interest rate is 15.5%. How much should be invested in each account so that the interest earned by each account is the same?
$27,125 at 8.5%; $14,875 at 15.5%

Use your calculator for the following exercises.

7. The manager of a trust account invests 25% of a client's account in a money market fund that earns 8% annual simple interest, 40% in bonds that earn 10.5% annual simple interest, and the remainder in trust deeds that earn 20% annual simple interest. How much should be invested in each type of investment so that the total interest earned is $2904?
$5500 at 8%; $8800 at 10.5%; $7700 at 20%

8. A financial manager recommended an investment plan in which 30% of a client's investment be placed in an 8% annual simple interest tax-free account, 45% in 12% high-grade bonds, and the remainder in a 20% high-risk investment. The total interest earned from the investments would be $3840. Find the total amount to be invested.
$30,000

9. An investment advisor invested $18,000 in two accounts. One investment earned 17.2% annual simple interest, while the other investment lost 4.7%. The total earnings from both investments were $1782. Find the amount invested at 17.2%.
$12,000

10. An investment advisor invested $12,000 in two accounts. One investment earned 12.6% annual simple interest, while the other investment lost 5%. The total earnings from both investments were $104. Find the amount invested at 12.6%.
$4000

2   Solve.

11. A silversmith mixed 30 g of a 60% silver alloy with 70 g of a 20% silver alloy. What is the percent concentration of the resulting alloy?
32%

12. A goldsmith mixed 10 g of a 50% gold alloy with 40 g of a 15% gold alloy. What is the percent concentration of the resulting alloy?
22%

13. How many milliliters of pure acid must be added to 80 ml of a 32% acid solution to make a 50% acid solution?
28.8 ml

14. How many ounces of pure water must be added to 75 oz of an 8% salt solution to make a 5% salt solution?
45 oz

15. A chemist mixed 10 L of a 25% acid solution with 20 L of a 40% acid solution. Find the percent concentration of the resulting mixture.
35%

16. A silversmith mixed 80 g of a 60% silver alloy with 120 g of a 40% silver alloy. Find the percent concentration of the resulting alloy.
48%

17. A butcher has some hamburger that is 21% fat and some hamburger that is 15% fat. How many pounds of each should be mixed to make 84 lb of hamburger that is 17% fat?
28 lb of 21% fat; 56 lb of 15% fat

18. A hospital staff mixed a 75% disinfectant solution with a 25% disinfectant solution. How many liters of each were used to make 20 L of a 40% disinfectant solution?
6 L of 75%; 14 L of 25%

19. How many grams of a 3% salt solution must be mixed with 40 g of an 8% salt solution to make a 5% salt solution?
60 g

20. How many pounds of a 15% aluminum alloy must be mixed with 500 lb of a 22% aluminum alloy to make a 20% aluminum alloy?
200 lb

21. How many pounds of a 60% copper alloy must be mixed with 20 lb of a 25% copper alloy to make an alloy that is 50% copper?

    50 lb

22. How many kilograms of a 4% solution must be mixed with 10 kg of a 10% solution to make an 8% solution?

    5 kg

23. A chemist mixed 150 L of a 45% hydrogen peroxide solution with 50 L of an 85% solution. Find the percent concentration of the resulting mixture.

    55%

24. A druggist mixed 45 g of a 10% alcohol solution with 15 g of a 6% solution. Find the percent concentration of the resulting mixture.

    9%

25. A chemist mixed a 60% alcohol solution with a 20% alcohol solution to make a 50% alcohol solution. How many liters of each were used to make 100 L of a 50% solution?

    75 L of 60% solution; 25 L of 20% solution

26. An alloy containing 30% tin is mixed with an alloy containing 70% tin. How many pounds of each were used to make 500 lb of an alloy containing 40% tin?

    375 lb of 30% tin; 125 lb of 70% tin

▦ Use your calculator for the following exercises.

27. A student mixed 40 ml of a 4% hydrogen peroxide solution with 25 ml of a 12% solution. Find the percent concentration of the resulting mixture. Round to the nearest tenth of a percent.

    7.1%

28. A car radiator contains 12 qt of a 40% antifreeze solution. How many quarts will have to be replaced with pure antifreeze if the resulting solution is to be 60% antifreeze? Round to the nearest tenth.

    4.0 qt

SUPPLEMENTAL EXERCISES 2.5

Solve.

29. A financial manager invested 25% of a client's money in bonds paying 9% annual simple interest, 30% in an 8% simple interest account, and the remainder in 9.5% corporate bonds. Find the amount invested in each if the total annual interest earned is $1785.

    $5000 at 9%; $6000 at 8%; $9000 at 9.5%

30. A silversmith mixed 90 g of a 40% silver alloy with 120 g of a 60% silver alloy. Find the percent concentration of the resulting alloy. Round to the nearest tenth of a percent.
51.4%

31. Find the cost per pound of a tea mixture made from 50 lb of tea costing $5.50 per pound and 75 lb of tea costing $4.40 per pound.
$4.84/lb

32. Find four consecutive integers such that the difference between the sum of the three largest and the product of two and the smallest is equal to sixteen.
10, 11, 12, and 13

33. Two planes start from the same point and fly in opposite directions. The first plane is flying 40 mph faster than the second plane. In 2 h, the planes are 980 mi apart. Find the rate of each plane.
1st plane: 265 mph; 2nd plane: 225 mph

34. A coin bank contains 35 coins in nickels, dimes, and quarters. There are three times as many dimes as quarters. The total value of the coins is $3.85. How many nickels are in the coin bank?
11 nickels

35. How many kilograms of water must be evaporated from 75 kg of a 15% salt solution to produce a 20% salt solution?
18.75 kg

36. How many grams of pure water must be added to 20 g of pure acid to make a solution that is 25% acid?
60 g

37. An investment of $6000 is made at an annual simple interest rate of 7.8%. How much additional money must be invested at an annual simple interest rate of 9% so that the total interest earned is at least 8.5% of the total investment?
$8400 or more

38. How many pounds of a 70% copper alloy must be mixed with 15 lb of a 20% copper alloy to make an alloy that is at least 50% copper?
22.5 lb or more

CALCULATORS AND COMPUTERS

Solving First-Degree Equations and Absolute Value Inequalities

The program SOLVE A FIRST-DEGREE EQUATION on the Student Disk will allow you to practice solving the following three types of equations:

1. $ax + b = c$
2. $ax + b = cx + d$
3. Equations with parentheses

After you select the type of equation you want to practice, a problem will be displayed on the screen. Using paper and pencil, solve the problem. When you are ready, press the RETURN key, and the complete solution will be displayed. Compare your solution with the displayed solution.

When you finish a problem, you may continue practicing the type of problem you have selected or return to the main menu and select a different type, or quit the program.

The program ABSOLUTE VALUE INEQUALITIES on the Student Disk will allow you to practice solving an absolute value inequality of the form

$$|ax + b| < c \quad \text{or} \quad |ax + b| > c$$

A problem will be displayed on the screen and you, using paper and pencil, are to solve the problem. When you are ready to see the solution, press the RETURN key. The solution will be displayed.

CHAPTER SUMMARY

Key Words

An **equation** expresses the equality of two mathematical expressions.

A **solution** or **root** of an equation is a replacement value for the variable that will make the equation true.

To **solve** an equation means to find its solutions.

Consecutive integers are integers that follow one another in order.

The **solution set of an inequality** is a set of numbers, each element of which, when substituted for the variable, results in a true inequality.

A **compound inequality** is formed by joining two inequalities with a connective word such as "and" or "or."

An **absolute value equation** is an equation containing an absolute value symbol.

Essential Rules

The Addition Property of Equality   The same number can be added to each side of an equation without changing the solution of the equation.

$$\text{If } a = b, \text{ then } a + c = b + c.$$

The Multiplication Property of Equality   Each side of an equation can be multiplied by the same nonzero number without changing the solution of the equation.

$$\text{If } a = b \text{ and } c \neq 0, \text{ then } ac = bc.$$

The Addition Property of Inequalities   The same number can be added to each side of an inequality without changing the solution set of the inequality.

$$\text{If } a > b, \text{ then } a + c > b + c.$$
$$\text{If } a < b, \text{ then } a + c < b + c.$$

The Multiplication Property of Inequalities
Rule 1 Each side of an inequality can be multiplied by the same positive number without changing the solution set of the inequality.

$$\text{If } a > b \text{ and } c > 0, \text{ then } ac > bc.$$
$$\text{If } a < b \text{ and } c > 0, \text{ then } ac < bc.$$

Rule 2 If each side of an inequality is multiplied by the same negative number and the inequality symbol is reversed, then the solution set of the inequality is not changed.

$$\text{If } a > b \text{ and } c < 0, \text{ then } ac < bc.$$
$$\text{If } a < b \text{ and } c < 0, \text{ then } ac > bc.$$

To solve an absolute value inequality of the form $|ax + b| < c$:

solve the equivalent compound inequality
$-c < ax + b < c.$

To solve an absolute value inequality of the form $|ax + b| > c$:

solve the equivalent compound inequality
$ax + b < -c$ or $ax + b > c.$

Value Mixture Equation

$$\text{Value} = \text{amount} \times \text{unit cost}$$
$$V = A \times C$$

Uniform Motion Equation

$$\text{Distance} = \text{rate} \times \text{time}$$
$$d = r \times t$$

Annual Simple Interest Equation

$$\text{Simple interest} = \text{principal} \times \text{simple interest rate}$$
$$I = p \times r$$

Percent Mixture Equation

$$\text{Quantity} = \text{amount} \times \text{percent of concentration}$$
$$Q = A \times r$$

CHAPTER REVIEW

1. Solve: $-\dfrac{3}{4}y = -\dfrac{5}{8}$

   $\dfrac{5}{6}$

2. Solve: $2x - 3 - 5x = 8 + 2x - 10$

   $-\dfrac{1}{5}$

3. Solve: $2x - 1 > 3$ or $1 - 3x > 7$

   $\{x \mid x > 2 \text{ or } x < -2\}$

4. Solve: $\dfrac{2x + 1}{3} - \dfrac{3x + 4}{6} = \dfrac{5x - 9}{9}$

   $\dfrac{12}{7}$

5. Solve: $x - 2 = -4$

   -2

6. Solve: $|3x - 1| > 5$

   $\left\{x \mid x > 2 \text{ or } x < -\dfrac{4}{3}\right\}$

7. Solve: $\frac{3}{4}y - 2 = 6$

$$\frac{32}{3}$$

8. Solve: $|3 - 5x| = 12$

3 and $-\frac{9}{5}$

9. Solve: $4 - 3(x + 2) < 2(2x + 3) - 1$

$\{x \mid x > -1\}$

10. Solve: $\frac{2}{3}a = -8$

-12

11. Solve: $b + \frac{3}{4} = \frac{5}{8}$

$-\frac{1}{8}$

12. Solve: $2[a - 2(2 - 3a) - 4] = a - 5$

$\frac{11}{13}$

13. Solve: $4x + 7 < 6x + 3$

$\{x \mid x > 2\}$

14. Solve: $3a - 5a = 2a + 4$

-1

15. Solve: $|2x - 5| \leq 3$

$\{x \mid 1 \leq x \leq 4\}$

16. Solve: $\frac{2}{3}t - \frac{5}{6}t = 4$

-24

17. Solve: $3x + 2 \leq 5$ and $x + 5 \geq 1$

$\{x \mid -4 \leq x \leq 1\}$

18. Solve: $3x - 5 = 7$

4

19. Solve: $5(2x - 3) - 5 = -2(4 - 3x)$

3

20. Solve: $3 - 2(2x - 1) \geq 3(2x - 2) + 1$

$\{x \mid x \leq 1\}$

21. Solve: $2 - |2x - 5| = -7$

7 and -2

22. Solve: $4 - 3x \geq 7$ and $2x + 3 \geq 7$

$\varnothing$

23. A coin purse contains 45 coins in nickels, dimes, and quarters. There are twice as many nickels as dimes. The total value of the coins is $4.65. How many dimes are in the coin purse?

12 dimes

24. A car traveling at 50 mph overtakes a cyclist who, riding at 15 mph, has a 3.5-h head start. How far from the starting point does the car overtake the cyclist?

75 mi

25. A silversmith mixed 50 g of silver with 200 g of a 25% silver alloy. What is the percent concentration of the resulting alloy?

40%

26. The sum of two integers is eighteen. Five times the smaller integer is six more than twice the larger integer. Find the integers.

12, 6

27. A bank offers two types of checking accounts. One account has a charge of $5 per month plus 3¢ for each check. The second account has a charge of $2 per month plus 10¢ for each check. How many checks can a customer who has the second type of account write if it is to cost the customer less than the first type of account?

42 checks

28. An investment club invested a part of $16,000 in a 9.5% tax-free annual simple interest account and the remainder in a 11.2% annual simple interest account. The amount of interest earned for one year on both accounts was $1707. How much was invested in each account?
$5000 at 9.5%; $11,000 at 11.2%

29. A grocer mixes 20 lb of peanuts that cost $3.50 per pound with cashews that cost $5.00 per pound. How many pounds of cashews must be used so that the mixture costs $4.00 per pound?
10 lb

30. A doctor has prescribed 4 cc of medication for a patient. The tolerance is 0.03 cc. Find the upper and lower limits of the amount of medication to be given.
4.03 cc, 3.97 cc

CUMULATIVE REVIEW

1. Simplify: $-2^2 \cdot 3^3$
-108

2. Simplify: $4 - (2 - 5)^2 \div 3 + 2$
3

3. Simplify: $4 \div \dfrac{\frac{3}{8} - 1}{5} \cdot 2$
-64

4. Evaluate $2a^2 - (b - c)^2$ when $a = 2$, $b = 3$, and $c = -1$.
-8

5. Identify the property that justifies the statement.
$(2x + 3y) + 2 = (3y + 2x) + 2$
The Commutative Property of Addition

6. Find $A \cap B$ given $A = \{3, 5, 7, 9\}$ and $B = \{3, 6, 9\}$.
$A \cap B = \{3, 9\}$

7. Graph the solution set of $\{x \mid x \geq -2\}$.

8. Graph the solution set of $\{x \mid x \geq 1\} \cup \{x \mid x < -2\}$.

9. Simplify: $3x - 2[x - 3(2 - 3x) + 5]$
$-17x + 2$

10. Simplify: $5[y - 2(3 - 2y) + 6]$
$25y$

11. Solve: $4 - 3x = -2$
2

12. Solve: $-\dfrac{5}{6}b = -\dfrac{5}{12}$
$\dfrac{1}{2}$

13. Solve: $2x + 5 = 5x + 2$
1

14. Solve: $\dfrac{5}{12}x - 3 = 7$
24

15. Solve: $2[3 - 2(3 - 2x)] = 2(3 + x)$
2

16. Solve: $3[2x - 3(4 - x)] = 2(1 - 2x)$
2

17. Solve: $\dfrac{3x - 1}{4} - \dfrac{4x - 1}{12} = \dfrac{3 + 5x}{8} - \dfrac{13}{5}$

18. Solve: $3x - 2 \geq 6x + 7$
$\{x \mid x \leq -3\}$

19. Solve: $5 - 2x \geq 6$ and $3x + 2 \geq 5$
$\varnothing$

20. Solve: $4x - 1 > 5$ or $2 - 3x < 8$
$\{x \mid x > -2\}$

21. Solve: $|3 - 2x| = 5$
-1 and 4

22. Solve: $3 - |2x - 3| = -8$
-4 and 7

23. Solve: $|3x - 5| \leq 4$ $\left\{x \mid \dfrac{1}{3} \leq x \leq 3\right\}$

24. Solve: $|4x - 3| > 5$ $\left\{x \mid x > 2 \text{ or } x < -\dfrac{1}{2}\right\}$

25. Translate and simplify "the sum of three times a number and six added to the product of three and the number."
$(3n + 6) + 3n; 6n + 6$

26. Three times the sum of the first and third of three consecutive odd integers is fifteen more than the second integer. Find the first integer.
1

27. A stamp collection consists of 9¢ and 11¢ stamps. The number of 9¢ stamps is five less than twice the number of 11¢ stamps. The total value of the stamps is $1.87. Find the number of 9¢ stamps.
11 stamps

28. Tickets for a school play sold for $2.25 for each adult and $.75 for each child. The total receipts for 75 tickets were $128.25. Find the number of adult tickets sold.
48 adult tickets

29. Two planes were 1400 mi apart and traveling toward each other. One plane is traveling 120 mph faster than the other plane. The planes meet in 2.5 h. Find the speed of the faster plane.
340 mph

30. How many liters of a 12% acid solution must be mixed with 4 L of a 5% acid solution to make an 8% acid solution?
3 L

31. An investment advisor invested $10,000 in two accounts. One investment earned 9.8% annual simple interest, while the other investment earned 12.8% annual simple interest. The amount of interest earned in one year was $1085. How much was invested in the 9.8% account?
$6500

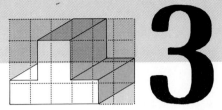

3

Polynomials

OBJECTIVES

- Add and subtract polynomials
- Multiply monomials
- Divide monomials
- Rewrite a monomial as a product of factors
- Multiply a polynomial by a monomial
- Multiply two polynomials
- Multiply polynomials that have special products
- Application problems
- Factor a monomial from a polynomial
- Factor trinomials of the form $x^2 + bx + c$
- Factor trinomials of the form $ax^2 + bx + c$
- Factor trinomials that are quadratic in form
- Factor completely
- Factor the difference of two perfect squares and perfect square trinomials
- Factor the sum or the difference of two cubes
- Factor by grouping
- Factor completely
- Solve equations by factoring
- Solve inequalities by factoring
- Application problems

Origins of the Word Algebra

The word *algebra* has its origins in an Arabic book written around A.D. 825 titled *Hisab al-jabr w' almuqa-balah* by al-Khowarizmi. The word *al-jabr*, which literally translated means reunion, was written as the word *algebra* in Latin translations of al-Khowarizmi's work and became synonymous with equations and the solutions of equations. Interestingly, an early meaning of the Spanish word *algebrista* was bonesetter or reuniter of broken bones.

There is actually a second contribution to our language of mathematics by al-Khowarizmi. One of the translations of his work into Latin shortened his name to Algoritmi. A further modification of this word gives us our present word "algorithm." An algorithm is a procedure or set of instructions that is used to solve different types of problems. Computer scientists use algorithms when writing computer programs.

A further historical note is not about the word *algebra*, but about Omar Khayyam, a Persian who probably read al-Khowarizmi's work. Omar Khayyam is especially noted as a poet and the author of the *Rubiat*. However, he was also an excellent mathematician and astronomer and made many contributions to mathematics.

Operations on Polynomials

1 Add and subtract polynomials

A **monomial** is a number, a variable, or a product of a number and variables.

The examples at the right are monomials.	x	degree 1 $(x = x^1)$
The **degree of a monomial** is the sum of	$3x^2$	degree 2
the exponents of the variables.	$4x^2y$	degree 3
	$6x^3y^4z^2$	degree 9

In this chapter, the letter n is considered a positive integer when used as an exponent.	x^n	degree n

The degree of a nonzero constant term is zero.	6	degree 0

$\dfrac{1}{x}$ is not a monomial because a variable appears in the denominator.

A **polynomial** is a variable expression in which the terms are monomials.

A **monomial** is a polynomial of one term.	$3x$
A **binomial** is a polynomial of two terms.	$5x^2y + 6x$
A **trinomial** is a polynomial of three terms.	$3x^2 + 9xy - 5y$

Polynomials with more than three terms do not have special names.

The **degree of a polynomial** is the greatest of the degrees of any of its terms.	$3x + 2$	degree 1
	$3x^2 + 2x - 4$	degree 2
	$4x^3y^2 + 6x^4 - y$	degree 5
	$3x^{2n} - 5x^n + 2$	degree $2n$

The terms of a polynomial in one variable are usually arranged so that the exponents of the variable decrease from left to right. This is called **descending order**.	$2x^2 - x + 8$
	$3y^5 - 3y^3 + y^2 - 12$

For a polynomial in more than one variable, descending order may refer to any one of the variables.

The polynomial at the right is shown first in descending order of the x variable and then in descending order of the y variable.

$$2x^2 + 3xy + 5y^2$$

$$5y^2 + 3xy + 2x^2$$

Polynomials can be added, using either a vertical or a horizontal format, by combining like terms.

Example 1   Simplify: $(4x^2 + 5x - 3) + (7x^3 - 7x + 1) + (4x^3 - 3x^2 + 2x + 1)$
Use a vertical format.

Solution
$$\begin{array}{r} 4x^2 + 5x - 3 \\ 7x^3 \quad\quad - 7x + 1 \\ +\ \ 4x^3 - 3x^2 + 2x + 1 \\ \hline 11x^3 + x^2 \quad\quad - 1 \end{array}$$

• Arrange the terms of each polynomial in descending order with like terms in the same column.

• Combine like terms in each column.

Problem 1   Simplify: $(5x^2 + 3x - 1) + (2x^2 + 4x - 6) + (x^2 - 7x + 8)$
Use a vertical format.

Solution   See page A20.
$8x^2 + 1$

Example 2   Simplify: $(3x^2 + 2x - 7) + (7x^3 - 3 + 4x^2)$
Use a horizontal format.

Solution $(3x^2 + 2x - 7) + (7x^3 - 3 + 4x^2)$
$7x^3 + (3x^2 + 4x^2) + 2x + (-7 - 3)$

• Use the Commutative and Associative Properties of Addition to rearrange and group like terms. (Do this step mentally.)

$7x^3 + 7x^2 + 2x - 10$

• Combine like terms. Write the polynomial in descending order.

Problem 2   Simplify: $(4x^2 + 3x - 5) + (6x^3 - 2 + x^2)$
Use a horizontal format.

Solution   See page A20.
$6x^3 + 5x^2 + 3x - 7$

The **additive inverse** of the polynomial $x^2 + 5x - 4$ is $-(x^2 + 5x - 4)$.

To find the additive inverse of a polynomial, change the sign of every term inside the parentheses.

$$-(x^2 + 5x - 4) = -x^2 - 5x + 4$$

Polynomials can be subtracted using either a horizontal or a vertical format. To subtract, add the additive inverse of the second polynomial to the first.

To simplify $(6x^2 - 3x + 7) - (3x^2 - 5x + 12)$ using a vertical format, arrange the terms of each polynomial in descending order with like terms in the same column. Then rewrite subtraction as the addition of the additive inverse.

$$
\begin{array}{r}
6x^2 - 3x + 7 \\
- \quad (3x^2 - 5x + 12)
\end{array}
\quad = \quad
\begin{array}{r}
6x^2 - 3x + 7 \\
+ \quad -3x^2 + 5x - 12 \\
\hline
3x^2 + 2x - 5
\end{array}
$$

Combine the terms in each column.

Example 3   Simplify: $(3x^2 - 2x + 4) - (7x^2 + 3x - 12)$
Use a vertical format.

Solution
$$
\begin{array}{r}
3x^2 - 2x + 4 \\
- \quad 7x^2 + 3x - 12
\end{array}
\quad = \quad
\begin{array}{r}
3x^2 - 2x + 4 \\
+ \quad -7x^2 - 3x + 12 \\
\hline
-4x^2 - 5x + 16
\end{array}
$$

• Rewrite subtraction as the addition of the additive inverse.
• Combine like terms in each column.

Problem 3   Simplify: $(-5x^2 + 2x - 3) - (6x^2 + 3x - 7)$
Use a vertical format.

Solution   See page A20.
$-11x^2 - x + 4$

Example 4   Simplify: $(2x^{2n} - 3x^n + 7) - (3x^{2n} + 3x^n + 5)$
Use a horizontal format.

Solution $(2x^{2n} - 3x^n + 7) - (3x^{2n} + 3x^n + 5)$
$(2x^{2n} - 3x^n + 7) + (-3x^{2n} - 3x^n - 5)$

$-x^{2n} - 6x^n + 2$

• Rewrite subtraction as the addition of the additive inverse.
• Combine like terms.

Problem 4   Simplify: $(5x^{2n} - 3x^n - 7) - (-2x^{2n} - 5x^n + 8)$
Use a horizontal format.

Solution   See page A20.
$7x^{2n} + 2x^n - 15$

2   Multiply monomials

In an exponential expression, the exponent indicates the number of times the base occurs as a factor.

The product of exponential expressions with the *same* base can be simplified by adding the exponents.

$$x^3 \cdot x^4 = (x \cdot x \cdot x) \cdot (x \cdot x \cdot x \cdot x) = x^7$$
$$x^3 \cdot x^4 = x^{3+4} = x^7$$

> **Rule for Multiplying Exponential Expressions**
> If m and n are positive integers, then $x^m \cdot x^n = x^{m+n}$.

Example 5   Simplify: $(5a^2b^4)(2ab^5)$

Solution $(5a^2b^4)(2ab^5) =$
$(5 \cdot 2)(a^2 \cdot a)(b^4 \cdot b^5) =$ • Use the Commutative and Associative Properties to rearrange and group factors. (Do this step mentally.)

$10a^{2+1}b^{4+5} =$ • Multiply variables with like bases by adding the exponents. (Do this step mentally.)

$10a^3b^9$

Problem 5   Simplify: $(7xy^3)(-5x^2y^2)(-xy^2)$

Solution   See page A20.
$35x^4y^7$

A power of an exponential expression can be simplified by multiplying the exponents.

$$(x^4)^3 = x^4 \cdot x^4 \cdot x^4 = x^{4+4+4} = x^{12}$$
$$(x^4)^3 = x^{4 \cdot 3} = x^{12}$$

> **Rule for Simplifying Powers of Exponential Expressions**
> If m and n are positive integers, then $(x^m)^n = x^{m \cdot n}$.

Example 6   Simplify.
A. $(x^4)^5$ B. $(x^2)^n$

Solution   A. $(x^4)^5 = x^{4 \cdot 5}$ • Multiply the exponents. (Do this step mentally.)
$= x^{20}$

B. $(x^2)^n = x^{2n}$ • Multiply the exponents.

Problem 6   Simplify.

A. $(y^3)^6$ B. $(x^n)^3$

Solution   See page A20.

A. y^{18} B. x^{3n}

A power of the product of exponential expressions can be simplified by multiplying each exponent inside the parentheses by the exponent outside the parentheses.

$$(x^3 \cdot y^4)^2 = (x^3 \cdot y^4)(x^3 \cdot y^4) = x^3 \cdot x^3 \cdot y^4 \cdot y^4 = x^6 y^8$$

$$(x^3 \cdot y^4)^2 = x^{3 \cdot 2} \cdot y^{4 \cdot 2} = x^6 y^8$$

Rule for Simplifying Powers of Products

If m, n, and p are positive integers, then $(x^m \cdot y^n)^p = x^{m \cdot p} y^{n \cdot p}$.

Example 7   Simplify: $(2a^3 b^4)^3$

Solution $(2a^3 b^4)^3 = 2^{1 \cdot 3} a^{3 \cdot 3} b^{4 \cdot 3}$ • Multiply each exponent inside the parentheses by the exponent outside the parentheses. (Do this step mentally.)

$= 2^3 a^9 b^{12}$

$= 8a^9 b^{12}$

Problem 7   Simplify: $(-2ab^3)^4$

Solution   See page A20.

$16a^4 b^{12}$

Example 8   Simplify: $(2ab)(3a)^2 + 5a(2a^2 b)$

Solution $(2ab)(3a)^2 + 5a(2a^2 b) = (2ab)(3^2 a^2) + 10a^3 b$

$= (2ab)(9a^2) + 10a^3 b$

$= 18a^3 b + 10a^3 b$

$= 28a^3 b$

Problem 8   Simplify: $6a(2a)^2 + 3a(2a^2)$

Solution   See page A20.

$30a^3$

3 Divide monomials

The quotient of two exponential expressions with the same base can be simplified by subtracting the smaller exponent from the larger exponent.

$$\frac{x^6}{x^4} = \frac{\overset{1}{\cancel{x}} \cdot \overset{1}{\cancel{x}} \cdot \overset{1}{\cancel{x}} \cdot \overset{1}{\cancel{x}} \cdot x \cdot x}{\underset{1}{\cancel{x}} \cdot \underset{1}{\cancel{x}} \cdot \underset{1}{\cancel{x}} \cdot \underset{1}{\cancel{x}}} = x^2 \qquad\qquad \frac{x^6}{x^4} = x^{6-4} = x^2$$

$$\frac{x^4}{x^6} = \frac{\overset{1}{\cancel{x}} \cdot \overset{1}{\cancel{x}} \cdot \overset{1}{\cancel{x}} \cdot \overset{1}{\cancel{x}}}{\underset{1}{\cancel{x}} \cdot \underset{1}{\cancel{x}} \cdot \underset{1}{\cancel{x}} \cdot \underset{1}{\cancel{x}} \cdot x \cdot x} = \frac{1}{x^2} \qquad\qquad \frac{x^4}{x^6} = \frac{1}{x^{6-4}} = \frac{1}{x^2}$$

Rule for Dividing Exponential Expressions

If m and n are positive integers and $x \neq 0$, then $\dfrac{x^m}{x^n} = x^{m-n}$ if $m > n$

and $\dfrac{x^m}{x^n} = \dfrac{1}{x^{n-m}}$ if $m < n$.

Example 9 Simplify.

A. $\dfrac{a^8b^2}{-a^3b^5}$ B. $\dfrac{x^{4n-2}}{x^{2n-5}}$

Solution A. $\dfrac{a^8b^2}{-a^3b^5} = -\dfrac{a^8b^2}{a^3b^5} =$

• A negative sign is placed in front of a fraction.

$-\dfrac{a^{8-3}}{b^{5-2}} = -\dfrac{a^5}{b^3}$

• Divide variables with like bases by subtracting the exponents.

B. $\dfrac{x^{4n-2}}{x^{2n-5}} = x^{4n-2-(2n-5)} =$

$x^{4n-2-2n+5} = x^{2n+3}$

• Divide variables with like bases by subtracting the exponents. (Do these steps mentally.)

Problem 9 Simplify.

A. $\dfrac{x^2y^4}{x^3y}$ B. $\dfrac{a^{2n+1}}{a^{n+3}}$

Solution See page A21.

A. $\dfrac{y^3}{x}$ B. a^{n-2}

A power of the quotient of two exponential expressions can be simplified by multiplying each exponent in the quotient by the exponent outside the parentheses.

$$\left(\frac{x^3}{y^4}\right)^2 = \frac{x^3}{y^4} \cdot \frac{x^3}{y^4} = \frac{x^3 \cdot x^3}{y^4 \cdot y^4} = \frac{x^6}{y^8}$$

$$\left(\frac{x^3}{y^4}\right)^2 = \frac{x^{3 \cdot 2}}{y^{4 \cdot 2}} = \frac{x^6}{y^8}$$

Rule for Simplifying Powers of Quotients

If m, n, and p are positive integers and $y \neq 0$, then $\left(\dfrac{x^m}{y^n}\right)^p = \dfrac{x^{mp}}{y^{np}}$.

Example 10   Simplify: $\left(\dfrac{9x^2y^4}{6xy^7}\right)^3$

Solution $\left(\dfrac{9x^2y^4}{6xy^7}\right)^3 = \left(\dfrac{3x}{2y^3}\right)^3$ • Simplify inside the parentheses.

$= \dfrac{3^3x^3}{2^3y^9}$ • Multiply each exponent inside the parentheses by the exponent outside the parentheses.

$= \dfrac{27x^3}{8y^9}$

Problem 10   Simplify: $\left(\dfrac{4a^2}{5b^3}\right)^3\left(\dfrac{-10a^3}{2b^8}\right)^2$

Solution   See page A21. $\dfrac{64a^{12}}{5b^{25}}$

4 Rewrite a monomial as a product of factors

A monomial can be written as the product of two monomial factors.

The monomial a^8 can be written as the product of the monomial factors a^6 and a^2. Note that one factor can be obtained by dividing the monomial by the other factor.

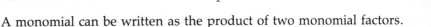

$a^8 = a^6 \cdot a^2 \qquad \dfrac{a^8}{a^6} = a^2 \qquad \dfrac{a^8}{a^2} = a^6$

Write $20a^2b^4c^3$ as a product of factors using $5ab^2c$ as one of the factors.

The other factor can be found by dividing the monomial by the given factor.

$$\frac{20a^2b^4c^3}{5ab^2c} = 4ab^2c^2$$

$$20a^2b^4c^3 = 5ab^2c(4ab^2c^2)$$

Example 11   Write the monomial as a product of factors.

A. $-10x^3y^4 = (2xy)\,(?)$ B. $x^{2n}y^{4n} = (x^ny^{2n})\,(?)$

Solution   A. $-10x^3y^4 = (2xy)(-5x^2y^3)$ $\bullet \dfrac{-10x^3y^4}{2xy} = -5x^2y^3$

B. $x^{2n}y^{4n} = (x^ny^{2n})(x^ny^{2n})$ $\bullet \dfrac{x^{2n}y^{4n}}{x^ny^{2n}} = x^ny^{2n}$

Problem 11   Write the monomial as a product of factors.

A. $12a^{12}b^4 = (6a^5b^3)\,(?)$ B. $a^{4n}b^{6n} = (a^{2n}b^{3n})\,(?)$

Solution   See page A21.

A. $2a^7b$ B. $a^{2n}b^{3n}$

EXERCISES 3.1

1   Simplify. Use a vertical format.

1. $(5x^2 + 2x - 7) + (x^2 - 8x + 12)$
$6x^2 - 6x + 5$

2. $(3x^2 - 2x + 7) + (-3x^2 + 2x - 12)$
-5

3. $(x^2 - 3xy + y^2) + (2x^2 - 3y^2)$
$3x^2 - 3xy - 2y^2$

4. $(3x^2 + 2y^2) + (-5x^2 + 2xy - 3y^2)$
$-2x^2 + 2xy - y^2$

5. $(x^2 - 3x + 8) - (2x^2 - 3x + 7)$
$-x^2 + 1$

6. $(2x^2 + 3x - 7) - (5x^2 - 8x - 2)$
$-3x^2 + 11x - 5$

7. $(x^{2n} + 7x^n - 3) + (-x^{2n} + 2x^n + 8)$
$9x^n + 5$

8. $(2x^{2n} - x^n - 1) + (5x^{2n} + 7x^n + 1)$
$7x^{2n} + 6x^n$

Simplify. Use a horizontal format.

9. $(3y^3 - 7y) + (2y^2 - 8y + 2)$
$3y^3 + 2y^2 - 15y + 2$

10. $(-2y^2 - 4y - 12) + (5y^2 - 5y)$
$3y^2 - 9y - 12$

11. $(2a^2 - 3a - 7) - (-5a^2 - 2a - 9)$
$7a^2 - a + 2$

12. $(3a^2 - 9a) - (-5a^2 + 7a - 6)$
$8a^2 - 16a + 6$

13. $(3x^4 - 3x^3 - x^2) + (3x^3 - 7x^2 + 2x)$
$3x^4 - 8x^2 + 2x$

14. $(3x^4 - 2x + 1) + (3x^3 - 5x - 8)$
$3x^4 + 3x^3 - 7x - 7$

15. $(3a^3 - 5a^2 - 6) + (-3a^3 - 6a + 2)$
$-5a^2 - 6a - 4$

16. $(6a^3 - 2a^2 - 12) + (6a^2 + 3a + 9)$
$6a^3 + 4a^2 + 3a - 3$

17. $(b^{2n} - b^n - 3) - (2b^{2n} - 3b^n + 4)$
$-b^{2n} + 2b^n - 7$

18. $(x^{2n} - x^n + 2) - (3x^{2n} - x^n + 5)$
$-2x^{2n} - 3$

19. $(3x^2 + 5x + 2) + (4x^2 - 3x - 1) + (-5x^2 + 2x - 8)$
$2x^2 + 4x - 7$

20. $(2a^2 + 3a - 6) + (5a^2 - 7a) + (a^3 - 4a + 1)$
$a^3 + 7a^2 - 8a - 5$

21. $(4x^2 - 3x - 9) - (2x^3 - 3x^2 + 6x) - (x^3 - 4x + 1)$
$-3x^3 + 7x^2 - 5x - 10$

22. $(8b^2 + 2b - 5) - (3b^2 - 4b) - (-6b^3 - 5b^2 + 3)$
$6b^3 + 10b^2 + 6b - 8$

23. $(6x^4 - 5x^3 + 2x) - (4x^3 + 3x^2 - 1) + (x^4 - 2x^2 + 7x - 3)$
$7x^4 - 9x^3 - 5x^2 + 9x - 2$

24. $(3x^2 - 4xy + 6y^2) + (5x^2 + 2xy - 4y^2) - (7x^2 + xy - 3y^2)$
$x^2 - 3xy + 5y^2$

25. $(2a^3 - 2a^2b + 2ab^2 + 3b^3) - (4a^2b - 3ab^2 - 5b^3) + (3a^3 - 2a^2b + 5ab^2 - b^3)$
$5a^3 - 8a^2b + 10ab^2 + 7b^3$

26. $(x^{2n} - 2x^n + 3) + (4x^{2n} - 3x^n - 1) - (2x^{2n} - 6x^n - 3)$
$3x^{2n} + x^n + 5$

2   Simplify.

27. $(ab^3)(a^3b)$ a^4b^4

28. $(-2ab^4)(-3a^2b^4)$ $6a^3b^8$

29. $(9xy^2)(-2x^2y^2)$ $-18x^3y^4$

30. $(x^2y)^2$ x^4y^2

31. $(x^2y^4)^4$ x^8y^{16}

32. $(-2ab^2)^3$ $-8a^3b^6$

33. $(-3x^2y^3)^4$ $81x^8y^{12}$

34. $(2^2a^2b^3)^3$ $64a^6b^9$

35. $(3^3a^5b^3)^2$ $729a^{10}b^6$

36. $(xy)(x^2y)^4$ x^9y^5

37. $(x^2y^2)(xy^3)^3$ x^5y^{11}

38. $(2a^2b)^3(-3ab^4)^2$ $72a^8b^{11}$

39. $(-3ab^3)^3(-2^2a^2b)^2$ $-432a^7b^{11}$

40. $(4ab)^2(-2ab^2c^3)^3$ $-128a^5b^8c^9$

41. $(-2ab^2)(-3a^4b^5)^3$ $54a^{13}b^{17}$

42. $y^n \cdot y^{2n}$ y^{3n}

43. $x^n \cdot x^{n+1}$ x^{2n+1}

44. $y^{2n} \cdot y^{4n+1}$ y^{6n+1}

45. $y^{3n} \cdot y^{3n-2}$ y^{6n-2}

46. $(a^n)^{2n}$ a^{2n^2}

47. $(a^{n-3})^{2n}$ a^{2n^2-6n}

48. $(y^{2n-1})^3$ y^{6n-3}

49. $(x^{3n+2})^5$ x^{15n+10}

50. $(b^{2n-1})^n$ b^{2n^2-n}

51. $(2xy)(-3x^2yz)(x^2y^3z^3)$ $-6x^5y^5z^4$

52. $(x^2z^4)(2xyz^4)(-3x^3y^2)$ $-6x^6y^3z^8$

53. $(3b^5)(2ab^2)(-2ab^2c^2)$ $-12a^2b^9c^2$

54. $(-c^3)(-2a^2bc)(3a^2b)$ $6a^4b^2c^4$

55. $(-2x^2y^3z)(3x^2yz^4)$ $\quad -6x^4y^4z^5$

56. $(5a^2b^2c)(-2a^4b^2c^5)$ $\quad -10a^6b^4c^6$

57. $(3x^2yz^3)(-2x^2y^2z)(-2x^3z^4)^3$ $\quad 48x^{13}y^3z^{16}$

58. $(-5xy^2z)^2(-3x^2yz^4)(-2x^2y^4z^2)$ $\quad 150x^6y^9z^8$

59. $(-2xyz^3)^2(3x^2y^2z)^2(-2x^3yz^3)$ $\quad -72x^9y^7z^{11}$

60. $(3xy^2)^2(-2x^2y^2z^2)(-5x^3y^4z)^3$ $\quad 2250x^{13}y^{18}z^5$

61. $(2ab)^2(3a^2b) + (6a^2b)(a^2b^2)$ $\quad 18a^4b^3$

62. $(2a^2)(3ab^2)^3 - (2a^2b)^2(ab^4)$ $\quad 50a^5b^6$

63. $(3x^2y)^2(2xy) - (3xy)(x^2y)^2$ $\quad 15x^5y^3$

64. $(5x^2y^3)(3x^4) + (2x^2y)^2(x^2y)$ $\quad 19x^6y^3$

65. $(3a^nbc^n)(-2ab^nc)(3a^nc^n)^2$ $\quad -54a^{3n+1}b^{n+1}c^{3n+1}$

66. $(2x^ny^n)^3(-3x^ny^n)^2(-2x^ny^n)$ $\quad -144x^{6n}y^{6n}$

67. $(2a^nb)^2(3ab) + (4ab)(a^nb)^2$ $\quad 16a^{2n+1}b^3$

68. $(4x^{2n}y^{2n})(3x^ny^n) + (2x^ny^n)^2(-3x^ny^n)$ $\quad 0$

3 Simplify.

69. $\dfrac{50b^{10}}{70b^5}$ $\quad \dfrac{5b^5}{7}$

70. $\dfrac{x^3y^6}{x^6y^2}$ $\quad \dfrac{y^4}{x^3}$

71. $\dfrac{x^{17}y^5}{-x^7y^{10}}$ $\quad -\dfrac{x^{10}}{y^5}$

72. $\dfrac{-6x^2y}{12x^4y}$ $\quad -\dfrac{1}{2x^2}$

73. $\dfrac{-2x^4y^2}{-6x^4y^4}$ $\quad \dfrac{1}{3y^2}$

74. $\dfrac{25a^2b^{12}}{10a^5b^7}$ $\quad \dfrac{5b^5}{2a^3}$

75. $\dfrac{-48ab^{10}}{32a^4b^3}$ $\quad -\dfrac{3b^7}{2a^3}$

76. $\dfrac{a^2b^3c^7}{a^6bc^5}$ $\quad \dfrac{b^2c^2}{a^4}$

77. $\dfrac{a^5b^6c}{a^9b^2c^4}$ $\quad \dfrac{b^4}{a^4c^3}$

78. $\dfrac{(2x^2y)^2}{8x^2y^3}$ $\quad \dfrac{x^2}{2y}$

79. $\dfrac{(3xy^2)^2}{9x^3y^4}$ $\quad \dfrac{1}{x}$

80. $\dfrac{25x^6y^7}{(5x^2y^3)^2}$ $\quad x^2y$

81. $\dfrac{2x^2y^4}{(3xy^2)^3}$ $\quad \dfrac{2}{27xy^2}$

82. $\dfrac{-3ab^2}{(9a^2b^4)^3}$ $\quad -\dfrac{1}{243a^5b^{10}}$

83. $\left(\dfrac{-12a^2b^3}{9a^5b^9}\right)^3$ $\quad -\dfrac{64}{27a^9b^{18}}$

84. $\left(\dfrac{12x^3y^2z}{18xy^3z^4}\right)^4$ $\quad \dfrac{16x^8}{81y^4z^{12}}$

85. $\dfrac{(4x^2y)^2}{(2xy^3)^3}$ $\quad \dfrac{2x}{y^7}$

86. $\dfrac{(3a^2b)^3}{(-6ab^3)^2}$ $\quad \dfrac{3a^4}{4b^3}$

87. $\dfrac{(-4xy^3)^3}{(-2x^7y)^4}$ $\quad -\dfrac{4y^5}{x^{25}}$

88. $\dfrac{(-8x^2y^2)^4}{(16x^3y^7)^2}$ $\quad \dfrac{16x^2}{y^6}$

89. $\dfrac{a^{5n}}{a^{3n}}$ $\quad a^{2n}$

90. $\dfrac{b^{6n}}{b^{10n}}$ $\quad \dfrac{1}{b^{4n}}$

91. $\dfrac{-x^{5n}}{x^{2n}}$ $-x^{3n}$

92. $\dfrac{y^{2n}}{-y^{8n}}$ $-\dfrac{1}{y^{6n}}$

93. $\dfrac{x^{2n-1}}{x^{n-3}}$ x^{n+2}

94. $\dfrac{y^{3n+2}}{y^{2n+4}}$ y^{n-2}

95. $\left(\dfrac{2a^2b}{3b}\right)^2\left(\dfrac{a^3}{-2ab}\right)^3$ $-\dfrac{a^{10}}{18b^3}$

96. $\left(\dfrac{4x^2y}{-12y^4}\right)^4\left(\dfrac{3xy}{2x^3}\right)^5$ $\dfrac{3}{32x^2y^7}$

97. $\dfrac{3a(a-b)^3}{12a^2(a-b)}$ $\dfrac{(a-b)^2}{4a}$

98. $\dfrac{-6a^2b(a+b)^2}{8ab^4(a+b)}$ $-\dfrac{3a(a+b)}{4b^3}$

99. $\dfrac{a^{3n}b^n}{a^nb^{2n}}$ $\dfrac{a^{2n}}{b^n}$

100. $\dfrac{x^ny^{3n}}{x^ny^{5n}}$ $\dfrac{1}{y^{2n}}$

101. $\dfrac{a^{3n-2}b^{n+1}}{a^{2n+1}b^{2n+2}}$ $\dfrac{a^{n-3}}{b^{n+1}}$

4   Write the monomial as a product of factors.

102. $10b^6 = 5b^2(?)$ $2b^4$

103. $6a^5 = 3a^2(?)$ $2a^3$

104. $8x^2y = 8x(?)$ xy

105. $16ab^4 = 8b(?)$ $2ab^3$

106. $-9ab^2 = -3a(?)$ $3b^2$

107. $-4x^2y = -2y(?)$ $2x^2$

108. $10x^3y = 5xy(?)$ $2x^2$

109. $22x^4y^2 = 11xy(?)$ $2x^3y$

110. $15a^2b^2c = 5abc(?)$ $3ab$

111. $4x^2y^3z^2 = 2xy(?)$ $2xy^2z^2$

112. $16x^2y^3 = 4x^2y(?)$ $4y^2$

113. $24a^2b^5 = 6ab^2(?)$ $4ab^3$

114. $-14a^5b^3 = 7a^2b^2(?)$ $-2a^3b$

115. $-18x^4y^2 = 9x^2y^2(?)$ $-2x^2$

116. $x^{4n} = x^n(?)$ x^{3n}

117. $a^{5n} = a^{3n}(?)$ a^{2n}

118. $a^{n+2} = a^2(?)$ a^n

119. $z^{n+1} = z^n(?)$ z

SUPPLEMENTAL EXERCISES 3.1

State whether or not the expression is a polynomial.

120. $\dfrac{1}{3}x - 1$ yes

121. $\dfrac{3}{x} - 1$ no

122. $\dfrac{1}{2}x^2y^2 + \dfrac{1}{2}$ yes

123. $5\sqrt{x} + 2$ no

124. $\sqrt{5}x + 2$ yes

125. $0y^6$ yes

126. $\dfrac{1}{4y^2} + \dfrac{1}{3y}$ no

127. $x + \sqrt{3}$ yes

128. $|x - 7|$ no

Write the polynomial in descending order.

129. $6x + 3 + 9x^2$
$9x^2 + 6x + 3$

130. $2y^2 + 5y^3 + 4y + 8$
$5y^3 + 2y^2 + 4y + 8$

131. $7 - b^2 + 4b + b^3$
$b^3 - b^2 + 4b + 7$

132. $16r - r^4 + 3r^5 - 2r^3$
$3r^5 - r^4 - 2r^3 + 16r$

133. $5a - 8 + 2a^4 - 3a^3 + a^2$
$2a^4 - 3a^3 + a^2 + 5a - 8$

134. $6a^{n+1} + 4 - 5a^n + 3a^{n+2}$
$3a^{n+2} + 6a^{n+1} - 5a^n + 4$

Simplify.

135. $\left(\dfrac{5y^2}{y}\right) + \left(\dfrac{3xy^4}{xy^3}\right)$ $8y$

136. $\left(\dfrac{4a^3b^2}{ab}\right) - \left(\dfrac{6a^5b^5}{a^3b^4}\right)$ $-2a^2b$

137. $\left(\dfrac{8a^3b^5c^4}{4ab^3c}\right) + \left(\dfrac{5a^6b^7c^3}{a^4b^5}\right)$ $7a^2b^2c^3$

138. $\left(\dfrac{9x^2y^2z}{3xy}\right) - \left(\dfrac{8x^3yz^4}{2x^2y^3}\right)$ $3xyz - \dfrac{4xz^4}{y^2}$

139. $\left(\dfrac{6x^2yz^3}{2xyz^2}\right)^2 - \left(\dfrac{4x^3y^2z^2}{x^2y^2z}\right)^2$ $-7x^2z^2$

140. $\left(\dfrac{12a^{2n}bc^n}{3bc^n}\right)^2 - \left(\dfrac{10a^{3n}b^n}{5a^{2n}b^n}\right)^4$ 0

For what value of k is the given equation an identity?

141. $(2x^3 + 3x^2 + kx + 5) - (x^3 + x^2 - 5x - 2) = x^3 + 2x^2 + 3x + 7$
-2

142. $(6x^3 + kx^2 - 2x - 1) - (4x^3 - 3x^2 + 1) = 2x^3 - x^2 - 2x - 2$
-4

Solve.

143. The width of a rectangle is x^n. The length of the rectangle is $3x^n$. Find the perimeter of the rectangle in terms of x^n.
$8x^n$

144. The base of an isosceles triangle is $4x^n$. The length of each of the equal sides is $3x^n$. Find the perimeter of the triangle in terms of x^n.
$10x^n$

145. The width of a rectangle is $4ab$. The length is $6ab$. Find the area of the rectangle in terms of ab.
$24a^2b^2$

146. The height of a triangle is $5xy$. The length of the base of the triangle is $8xy$. Find the area of the triangle in terms of xy.
$20x^2y^2$

147. The product of a monomial and $3x^3y^4$ is $12x^5y^7$. Find the monomial.
$4x^2y^3$

148. The product of a monomial and $5a^4b^2c$ is $25a^5b^2c^2$. Find the monomial.
$5ac$

149. The area of a rectangle is $16a^2b^3$. The width of the rectangle is $4ab^2$. Find the length.
$4ab$

150. The area of a rectangle is $18x^3y^2$. The length of the rectangle is $9x^2y$. Find the width. $2xy$

SECTION 3.2

Multiplication of Polynomials

1 Multiply a polynomial by a monomial

To multiply a polynomial by a monomial, use the Distributive Property and the Rule for Multiplying Exponential Expressions.

Example 1 Simplify.
A. $-5x(x^2 - 2x + 3)$ B. $x^2 - x[3 - x(x - 2) + 3]$ C. $x^n(x^n - x^2 + 1)$

Solution A. $-5x(x^2 - 2x + 3)$
$-5x(x^2) - (-5x)(2x) + (-5x)(3)$ • Use the Distributive Property. (Do this step mentally.)

$-5x^3 + 10x^2 - 15x$ • Use the Rule for Multiplying Exponential Expressions.

B. $x^2 - x[3 - x(x - 2) + 3]$
$x^2 - x[3 - x^2 + 2x + 3]$ • Use the Distributive Property to remove the inner grouping symbols.

$x^2 - x[6 - x^2 + 2x]$ • Combine like terms.
$x^2 - 6x + x^3 - 2x^2$ • Use the Distributive Property to remove the brackets.

$x^3 - x^2 - 6x$ • Combine like terms, and write the polynomial in descending order.

C. $x^n(x^n - x^2 + 1)$
$x^{2n} - x^{n+2} + x^n$ • Use the Distributive Property and the Rule for Multiplying Exponential Expressions.

Problem 1 Simplify.
A. $-4y(y^2 - 3y + 2)$ $-4y^3 + 12y^2 - 8y$
B. $x^2 - 2x[x - x(4x - 5) + x^2]$ $6x^3 - 11x^2$
C. $y^{n+3}(y^{n-2} - 3y^2 + 2)$ $y^{2n+1} - 3y^{n+5} + 2y^{n+3}$

Solution See page A21.

2 Multiply two polynomials

The product of two polynomials is the polynomial obtained by multiplying each term of one polynomial by each term of the other polynomial and then combining like terms.

Multiply: $(2x^2 - 2x + 1)(3x + 2)$

Use the Distributive Property to multiply the trinomial by each term of the binomial.	$(2x^2 - 2x + 1)(3x + 2)$ $(2x^2 - 2x + 1)(3x) + (2x^2 - 2x + 1)(2)$
Use the Distributive Property.	$(6x^3 - 6x^2 + 3x) + (4x^2 - 4x + 2)$
Combine like terms.	$6x^3 - 2x^2 - x + 2$

A more convenient method of multiplying two polynomials is to use a vertical format similar to that used for multiplication of whole numbers.

$$
\begin{array}{r}
2x^2 - 2x + 1 \\
\times \quad\quad 3x + 2 \\
\hline
4x^2 - 4x + 2 \\
6x^3 - 6x^2 + 3x \quad\quad\\
\hline
6x^3 - 2x^2 - x + 2
\end{array}
$$

Like terms are written in the same column.

$4x^2 - 4x + 2 = 2(2x^2 - 2x + 1)$

$6x^3 - 6x^2 + 3x = 3x(2x^2 - 2x + 1)$

Combine like terms.

Example 2 Simplify: $(4a^3 - 3a + 7)(a - 5)$

Solution

$$
\begin{array}{r}
4a^3 - 3a + 7 \\
\times \quad\quad a - 5 \\
\hline
-20a^3 \quad\quad + 15a - 35 \\
4a^4 \quad\quad - 3a^2 + 7a \quad\quad\quad\\
\hline
4a^4 - 20a^3 - 3a^2 + 22a - 35
\end{array}
$$

Problem 2 Simplify: $(-2b^2 + 5b - 4)(-3b + 2)$

Solution See page A21.

$6b^3 - 19b^2 + 22b - 8$

It is frequently necessary to find the product of two binomials. The product can be found by using a method called **FOIL**, which is based on the Distributive Property. The letters of FOIL stand for First, **O**uter, **I**nner, and **L**ast.

Simplify: $(3x - 2)(2x + 5)$

Multiply the First terms.	$(3x - 2)(2x + 5)$	$3x \cdot 2x = 6x^2$
Multiply the Outer terms.	$(3x - 2)(2x + 5)$	$3x \cdot 5 = 15x$
Multiply the Inner terms.	$(3x - 2)(2x + 5)$	$-2 \cdot 2x = -4x$
Multiply the Last terms.	$(3x - 2)(2x + 5)$	$-2 \cdot 5 = -10$

$$\qquad\qquad \text{F} \qquad \text{O} \qquad \text{I} \qquad \text{L}$$

Add the products.

Combine like terms.

$$(3x - 2)(2x + 5) = 6x^2 + 15x - 4x - 10$$
$$= 6x^2 + 11x - 10$$

Example 3   Simplify: $(6x - 5)(3x - 4)$

Solution $(6x - 5)(3x - 4) =$

$6x(3x) + 6x(-4) + (-5)(3x) + (-5)(-4) =$ • Do this step mentally.

$18x^2 - 24x - 15x + 20 =$

$18x^2 - 39x + 20$

Problem 3   Simplify: $(5a - 3b)(2a + 7b)$

Solution   See page A21.

$10a^2 + 29ab - 21b^2$

3   ## Multiply polynomials that have special products

Using FOIL, a pattern for the product of the sum and the difference of two terms and for the square of a binomial can be found.

The Sum and Difference of Two Terms

$$(a + b)(a - b) = a^2 - ab + ab - b^2$$
$$= a^2 - b^2$$

Square of the first term ⸻⸻⸻⸻⸻

Square of the second term ⸻⸻⸻⸻⸻

The Square of a Binomial

$$(a + b)^2 = (a + b)(a + b) = a^2 + ab + ab + b^2$$
$$= a^2 + 2ab + b^2$$

Square of the first term ⸻⸻⸻⸻⸻

Twice the product of the two terms ⸻⸻⸻⸻⸻

Square of the last term ⸻⸻⸻⸻⸻

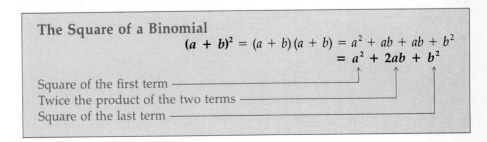

Example 4   Simplify.

A. $(4x + 3)(4x - 3)$

B. $(2x - 3y)^2$

C. $(x^n + 5)(x^n - 5)$

D. $(x^{2n} - 2)^2$

Solution   A. $(4x + 3)(4x - 3) =$ • This is the sum and difference of two terms.

$(4x)^2 - 3^2 =$ • Do this step mentally.
$16x^2 - 9$

B. $(2x - 3y)^2 =$ • This is the square of a binomial.
$(2x)^2 + 2(2x)(-3y) + (-3y)^2 =$ • Do this step mentally.
$4x^2 - 12xy + 9y^2$

C. $(x^n + 5)(x^n - 5) = x^{2n} - 25$

D. $(x^{2n} - 2)^2 = x^{4n} - 4x^{2n} + 4$

Problem 4   Simplify.

A. $(3x - 7)(3x + 7)$ B. $(3x - 4y)^2$

C. $(2x^n + 3)(2x^n - 3)$ D. $(2x^n - 8)^2$

Solution   See page A22.
A. $9x^2 - 49$ B. $9x^2 - 24xy + 16y^2$
C. $4x^{2n} - 9$ D. $4x^{2n} - 32x^n + 64$

4   # Application problems

Example 5   The length of a rectangle is $(2x + 3)$ ft. The width is $(x - 5)$ ft. Find the area of the rectangle in terms of the variable x.

$x - 5$ ┌─────────────┐
        │   │
        └─────────────┘
        $2x + 3$

Strategy   To find the area, replace the variables L and W in the equation $A = LW$ by the given values, and solve for A.

Solution $A = LW$
$A = (2x + 3)(x - 5)$
$A = 2x^2 - 10x + 3x - 15$
$A = 2x^2 - 7x - 15$

The area is $(2x^2 - 7x - 15)$ ft^2.

Problem 5   The base of a triangle is $(2x + 6)$ ft. The height is $(x - 4)$ ft. Find the area of the triangle in terms of the variable x.

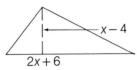

Solution   See page A22.
$(x^2 - x - 12)$ ft^2

Example 6   The corners are cut from a rectangular piece of cardboard measuring 8 in. by 12 in. The sides are folded up to make a box. Find the volume of the box in terms of the variable x, where x is the length of the side of the square cut from each corner of the rectangle.

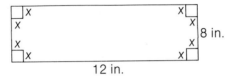

Strategy   Length of the box: $12 - 2x$
Width of the box: $8 - 2x$
Height of the box: x
To find the volume, replace the variables L, W, and H in the equation $V = LWH$, and solve for V.

Solution $V = LWH$
$V = (12 - 2x)(8 - 2x)x$
$V = (96 - 24x - 16x + 4x^2)x$
$V = (96 - 40x + 4x^2)x$
$V = 96x - 40x^2 + 4x^3$
$V = 4x^3 - 40x^2 + 96x$

The volume is $(4x^3 - 40x^2 + 96x)$ in.^{3}.

Problem 6   Find the volume of the rectangular solid shown in the diagram below. All dimensions are given in feet.

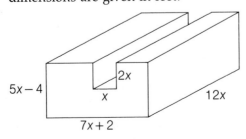

Solution   See page A22.
$(396x^3 - 216x^2 - 96x)$ ft^3

EXERCISES 3.2

1   Simplify.

1. $2x(x - 3)$
$2x^2 - 6x$

2. $2a(2a + 4)$
$4a^2 + 8a$

3. $3x^2(2x^2 - x)$
$6x^4 - 3x^3$

4. $-4y^2(4y - 6y^2)$
$-16y^3 + 24y^4$

5. $3xy(2x - 3y)$
$6x^2y - 9xy^2$

6. $-4ab(5a - 3b)$
$-20a^2b + 12ab^2$

7. $x^n(x + 1)$
$x^{n+1} + x^n$

8. $y^n(y^{2n} - 3)$
$y^{3n} - 3y^n$

9. $x^n(x^n + y^n)$
$x^{2n} + x^ny^n$

10. $x - 2x(x - 2)$
$-2x^2 + 5x$

11. $2b + 4b(2 - b)$
$-4b^2 + 10b$

12. $-2y(3 - y) + 2y^2$
$4y^2 - 6y$

13. $-2a^2(3a^2 - 2a + 3)$
$-6a^4 + 4a^3 - 6a^2$

14. $4b(3b^3 - 12b^2 - 6)$
$12b^4 - 48b^3 - 24b$

15. $3b(3b^4 - 3b^2 + 8)$
$9b^5 - 9b^3 + 24b$

16. $(2x^2 - 3x - 7)(-2x^2)$
$-4x^4 + 6x^3 + 14x^2$

17. $(-3y^2 - 4y + 2)(y^2)$
$-3y^4 - 4y^3 + 2y^2$

18. $(6b^4 - 5b^2 - 3)(-2b^3)$
$-12b^7 + 10b^5 + 6b^3$

19. $-5x^2(4 - 3x + 3x^2 + 4x^3)$
$-20x^5 - 15x^4 + 15x^3 - 20x^2$

20. $-2y^2(3 - 2y - 3y^2 + 2y^3)$
$-4y^5 + 6y^4 + 4y^3 - 6y^2$

21. $-2x^2y(x^2 - 3xy + 2y^2)$
$-2x^4y + 6x^3y^2 - 4x^2y^3$

22. $3ab^2(3a^2 - 2ab + 4b^2)$
$9a^3b^2 - 6a^2b^3 + 12ab^4$

23. $x^n(x^{2n} + x^n + x)$
$x^{3n} + x^{2n} + x^{n+1}$

24. $x^{2n}(x^{2n-2} + x^{2n} + x)$
$x^{4n-2} + x^{4n} + x^{2n+1}$

25. $a^{n+1}(a^n - 3a + 2)$
$a^{2n+1} - 3a^{n+2} + 2a^{n+1}$

26. $a^{n+4}(a^{n-2} + 5a^2 - 3)$
$a^{2n+2} + 5a^{n+6} - 3a^{n+4}$

27. $2y^2 - y[3 - 2(y - 4) - y]$
$5y^2 - 11y$

28. $3x^2 - x[x - 2(3x - 4)]$
$8x^2 - 8x$

29. $2y - 3[y - 2y(y - 3) + 4y]$
$6y^2 - 31y$

30. $4a^2 - 2a[3 - a(2 - a + a^2)]$
$2a^4 - 2a^3 + 8a^2 - 6a$

31. $7n - 4[3 + 2n(1 - 2n - 3n^2)]$
$24n^3 + 16n^2 - n - 12$

2   Simplify.

32. $(x - 2)(x + 7)$
$x^2 + 5x - 14$

33. $(y + 8)(y + 3)$
$y^2 + 11y + 24$

34. $(2y - 3)(4y + 7)$
$8y^2 + 2y - 21$

35. $(5x - 7)(3x - 8)$
$15x^2 - 61x + 56$

36. $(2x - 3y)(2x + 5y)$
$4x^2 + 4xy - 15y^2$

37. $(7x - 3y)(2x - 9y)$
$14x^2 - 69xy + 27y^2$

38. $(2a - 3b)(5a + 4b)$
$10a^2 - 7ab - 12b^2$

39. $(3a - 5b)(a + 7b)$
$3a^2 + 16ab - 35b^2$

40. $(5a + 2b)(3a + 7b)$
$15a^2 + 41ab + 14b^2$

41. $(5x + 9y)(3x + 2y)$
$15x^2 + 37xy + 18y^2$

42. $(3x - 7y)(7x + 2y)$
$21x^2 - 43xy - 14y^2$

43. $(5x - 9y)(6x - 5y)$
$30x^2 - 79xy + 45y^2$

44. $(xy + 4)(xy - 3)$
$x^2y^2 + xy - 12$

45. $(xy - 5)(2xy + 7)$
$2x^2y^2 - 3xy - 35$

46. $(2x^2 - 5)(x^2 - 5)$
$2x^4 - 15x^2 + 25$

47. $(x^2 - 4)(x^2 - 6)$
$x^4 - 10x^2 + 24$

48. $(5x^2 - 5y)(2x^2 - y)$
$10x^4 - 15x^2y + 5y^2$

49. $(x^2 - 2y^2)(x^2 + 4y^2)$
$x^4 + 2x^2y^2 - 8y^4$

50. $(x^n + 2)(x^n - 3)$
$x^{2n} - x^n - 6$

51. $(x^n - 4)(x^n - 5)$
$x^{2n} - 9x^n + 20$

52. $(2a^n - 3)(3a^n + 5)$
$6a^{2n} + a^n - 15$

53. $(5b^n - 1)(2b^n + 4)$
$10b^{2n} + 18b^n - 4$

54. $(2a^n - b^n)(3a^n + 2b^n)$
$6a^{2n} + a^nb^n - 2b^{2n}$

55. $(3x^n + b^n)(x^n + 2b^n)$
$3x^{2n} + 7x^nb^n + 2b^{2n}$

56. $(x - 2)(x^2 - 3x + 7)$
$x^3 - 5x^2 + 13x - 14$

57. $(x + 3)(x^2 + 5x - 8)$
$x^3 + 8x^2 + 7x - 24$

58. $(x + 5)(x^3 - 3x + 4)$
$x^4 + 5x^3 - 3x^2 - 11x + 20$

59. $(a + 2)(a^3 - 3a^2 + 7)$
$a^4 - a^3 - 6a^2 + 7a + 14$

60. $(2a - 3b)(5a^2 - 6ab + 4b^2)$
$10a^3 - 27a^2b + 26ab^2 - 12b^3$

61. $(3a + b)(2a^2 - 5ab - 3b^2)$
$6a^3 - 13a^2b - 14ab^2 - 3b^3$

62. $(2y^2 - 1)(y^3 - 5y^2 - 3)$
$2y^5 - 10y^4 - y^3 - y^2 + 3$

63. $(2b^2 - 3)(3b^2 - 3b + 6)$
$6b^4 - 6b^3 + 3b^2 + 9b - 18$

64. $(2x - 5)(2x^4 - 3x^3 - 2x + 9)$
$4x^5 - 16x^4 + 15x^3 - 4x^2 + 28x - 45$

65. $(2a - 5)(3a^4 - 3a^2 + 2a - 5)$
$6a^5 - 15a^4 - 6a^3 + 19a^2 - 20a + 25$

66. $(x^2 + 2x - 3)(x^2 - 5x + 7)$
$x^4 - 3x^3 - 6x^2 + 29x - 21$

67. $(x^2 - 3x + 1)(x^2 - 2x + 7)$
$x^4 - 5x^3 + 14x^2 - 23x + 7$

68. $(a - 2)(2a - 3)(a + 7)$
$2a^3 + 7a^2 - 43a + 42$

69. $(b - 3)(3b - 2)(b - 1)$
$3b^3 - 14b^2 + 17b - 6$

70. $(x^n + 1)(x^{2n} + x^n + 1)$
$x^{3n} + 2x^{2n} + 2x^n + 1$

71. $(a^{2n} - 3)(a^{5n} - a^{2n} + a^n)$
$a^{7n} - 3a^{5n} - a^{4n} + a^{3n} + 3a^{2n} - 3a^n$

72. $(x^n + y^n)(x^n - 2x^ny^n + 3y^n)$
$x^{2n} - 2x^{2n}y^n + 4x^ny^n - 2x^ny^{2n} + 3y^{2n}$

73. $(x^n - y^n)(x^{2n} - 3x^ny^n - y^{2n})$
$x^{3n} - 4x^{2n}y^n + 2x^ny^{2n} + y^{3n}$

3 Simplify.

74. $(a - 4)(a + 4)$
$a^2 - 16$

75. $(b - 7)(b + 7)$
$b^2 - 49$

76. $(3x - 2)(3x + 2)$
$9x^2 - 4$

77. $(b - 11)(b + 11)$
$b^2 - 121$

78. $(a - 5b)(a + 5b)$
$a^2 - 25b^2$

79. $(x - yz)(x + yz)$
$x^2 - y^2z^2$

80. $(4y + 1)(4y - 1)$
$16y^2 - 1$

81. $(6 - x)(6 + x)$
$36 - x^2$

82. $(10 + b)(10 - b)$
$100 - b^2$

83. $(2a - 3b)(2a + 3b)$
$4a^2 - 9b^2$

84. $(5x - 7y)(5x + 7y)$
$25x^2 - 49y^2$

85. $(x^2 + 1)(x^2 - 1)$
$x^4 - 1$

86. $(x^2 + y^2)(x^2 - y^2)$
$x^4 - y^4$

87. $(x^n + 3)(x^n - 3)$
$x^{2n} - 9$

88. $(x^n + y^n)(x^n - y^n)$
$x^{2n} - y^{2n}$

89. $(5a - 9b)(5a + 9b)$
$25a^2 - 81b^2$

90. $(3x + 7y)(3x - 7y)$
$9x^2 - 49y^2$

91. $(2x^n - 5)(2x^n + 5)$
$4x^{2n} - 25$

92. $(x - 5)^2$
$x^2 - 10x + 25$

93. $(y + 2)^2$
$y^2 + 4y + 4$

94. $(2a - 3)^2$
$4a^2 - 12a + 9$

95. $(2x - y)^2$
$4x^2 - 4xy + y^2$

96. $(3a + 5b)^2$
$9a^3 + 30ab + 25b$

97. $(5x - 4y)^2$
$25x^2 - 40xy + 16y^2$

98. $(x^2 - 3)^2$
$x^4 - 6x^2 + 9$

99. $(x^2 + y^2)^2$
$x^4 + 2x^2y^2 + y^4$

100. $(2x^2 - 3y^2)^2$
$4x^4 - 12x^2y^2 + 9y^4$

101. $(3a - 4b)^2$
$9a^2 - 24ab + 16b^2$

102. $(2x^2 + 5)^2$
$4x^4 + 20x^2 + 25$

103. $(3x^n + 2)^2$
$9x^{2n} + 12x^n + 4$

104. $(4b^n - 3)^2$
$16b^{2n} - 24b^n + 9$

105. $(2x^n + y^n)^2$
$4x^{2n} + 4x^ny^n + y^{2n}$

106. $(a^n + 5b^n)^2$
$a^{2n} + 10a^nb^n + 25b^{2n}$

107. $(x^n - 1)^2$
$x^{2n} - 2x^n + 1$

108. $(a^n - b^n)^2$
$a^{2n} - 2a^nb^n + b^{2n}$

109. $(2x^n + 5y^n)^2$
$4x^{2n} + 20x^ny^n + 25y^{2n}$

4 Solve.

110. The length of a rectangle is $(3x - 2)$ ft. The width is $(x + 4)$ ft. Find the area of the rectangle in terms of the variable x.
$(3x^2 + 10x - 8)$ ft^2

111. The base of a triangle is $(x - 4)$ ft. The height is $(3x + 2)$ ft. Find the area of the triangle in terms of the variable x.
$\left(\dfrac{3}{2}x^2 - 5x - 4\right)$ ft^2

112. Find the area of the figure shown below. All dimensions given are in meters.

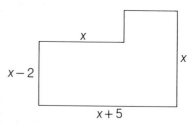

$(x^2 + 3x)$ m^2

113. Find the area of the figure shown below. All dimensions given are in feet.

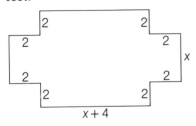

$(x^2 + 12x + 16)$ ft^2

114. The length of the side of a cube is $(x + 3)$ cm. Find the volume of the cube in terms of the variable x.
$(x^3 + 9x^2 + 27x + 27)$ cm^3

115. The length of a box is $(3x + 2)$ cm, the width is $(x - 4)$ cm, and the height is x cm. Find the volume of the box in terms of the variable x.
$(3x^3 - 10x^2 - 8x)$ cm^3

116. Find the volume of the figure shown below. All dimensions are given in inches.

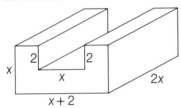

$2x^3$ in.^{3}

117. Find the volume of the figure shown below. All dimensions given are in centimeters.

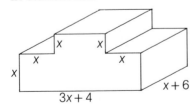

$(4x^2 + 32x^2 + 48x)$ cm^3

Use your calculator for the following exercises.

118. The radius of a circle is $(5x + 4)$ in. Find the area of the circle in terms of the variable x. Use 3.14 for π.
$(78.5x^2 + 125.6x + 50.24)$ in.^{2}

119. The radius of a circle is $(x - 2)$ in. Find the area of the circle in terms of the variable x. Use 3.14 for π.
$(3.14x^2 - 12.56x + 12.56)$ in.^{2}

SUPPLEMENTAL EXERCISES 3.2

Simplify.

120. $\dfrac{(2x + 1)^5}{(2x + 1)^3}$
$4x^2 + 4x + 1$

121. $\dfrac{(3x - 5)^6}{(3x - 5)^4}$
$9x^2 - 30x + 25$

122. $(a - b)^2 - (a + b)^2$
$-4ab$

123. $(x + 2y)^2 + (x + 2y)(x - 2y)$
$2x^2 + 4xy$

124. $(x^n + 1)(x^n - 1) - (x^n + 1)^2$
$-2x^n - 2$

125. $(3a - 2)^2 - 4a(a - 3)$
$5a^2 + 4$

126. $(4y + 3)^2 - (3y + 4)^2$
$7y^2 - 7$

127. $2x^2(3x^3 + 4x - 1) - 5x^2(x^2 - 3)$
$6x^5 - 5x^4 + 8x^3 + 13x^2$

128. $(2b + 3)(b - 4) + (3 + b)(3 - 2b)$
$-8b - 3$

129. $(3x - 2y)^2 - (2x - 3y)^2$
$5x^2 - 5y^2$

130. $[x + (y + 1)][x - (y + 1)]$
$x^2 - y^2 - 2y - 1$

131. $[x^2(2y - 1)]^2$
$4x^4y^2 - 4x^4y + x^4$

For what value of k is the given equation an identity?

132. $(kx + 3)^2 = k^2x^2 + 24x + 9$
4

133. $(kx - 2)^2 = k^2x^2 - 12x + 4$
3

134. $(3x - k)(2x + k) = 6x^2 + 5x - k^2$
5

135. $(4x + k)^2 = 16x^2 + 8x + k^2$
1

136. $(5x - k)(3x + k) = 15x^2 + 4x - k^2$
2

137. $(2x - k)(3x - k) = 6x^2 - 25x + k^2$
5

138. $(kx + 1)(kx - 6) = k^2x^2 - 15x - 6$
3

139. $(kx - 7)(kx + 2) = k^2x^2 + 5x - 14$
-1

Complete.

140. If $m = n + 1$, then $\dfrac{a^m}{a^n} = \underline{\quad a \quad}$.

141. If $m = n + 2$, then $\dfrac{a^m}{a^n} = \underline{\quad a^2 \quad}$.

Solve.

142. Find the square of $4a^2 - 2a + 3$.
$16a^4 - 16a^3 + 30a^2 - 12a + 9$

143. Find the cube of $x - 5$.
$x^3 - 15x^2 + 75x - 125$

144. What polynomial when divided by $x - 4$ has a quotient of $2x + 3$?
$2x^2 - 5x - 12$

145. What polynomial when divided by $2x - 3$ has a quotient of $x + 7$?
$2x^2 + 11x - 21$

146. Subtract the product of $4a + b$ and $2a - b$ from $9a^2 - 2ab$.
$a^2 + b^2$

147. Subtract the product of $5x - y$ and $x + 3y$ from $6x^2 + 12xy - 2y^2$.
$x^2 - 2xy + y^2$

148. Find $(3n^4)^3$ if $5(n - 1) = 2(3n - 2)$.
27

149. Find $(-2n^3)^2$ if $3(2n - 1) = 5(n - 1)$.
256

150. Use the expression $(x^2 + y^2)^2$ to show that $(x^a + y^b)^c \neq x^{ac} + y^{bc}$.
$(x^2 + y^2)^2 \neq x^4 + y^4$

151. Use the expression $(x^2 - y^2)^2$ to show that $(x^a - y^b)^c \neq x^{ac} - y^{bc}$.
$(x^2 - y^2)^2 \neq x^4 - y^4$

SECTION 3.3

Factoring Polynomials

1 Factor a monomial from a polynomial

The GCF of two or more exponential expressions with the same base is the exponential expression with the smallest exponent.

$$2^5$$
$$2^2$$
$$2^9$$
$$\text{GCF} = 2^2 = 4$$

$$x^5$$
$$x^7$$
$$x$$
$$\text{GCF} = x$$

The GCF of two or more monomials is the product of the GCF of each common factor with the smallest exponent.

$$16a^4b = 2^4 \cdot a^4 \cdot b$$
$$40a^2b^5 = 2^3 \cdot 5 \cdot a^2 \cdot b^5$$
$$\text{GCF} = 2^3 \cdot a^2 \cdot b = 8a^2b$$

To **factor a polynomial** means to write the polynomial as a product of other polynomials.

In the example at the right, $3x$ is the GCF of the terms $3x^2$ and $6x$. $3x$ is a **common monomial factor** of the terms of the binomial. $x - 2$ is a **binomial factor** of $3x^2 - 6x$.

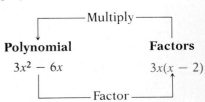

Multiply

| **Polynomial** | **Factors** |
| $3x^2 - 6x$ | $3x(x - 2)$ |

Factor

Example 1   Factor.

A. $4x^3y^2 + 12x^3y + 20xy^2$ B. $x^{2n} + x^{n+1} + x^n$

Solution   A. The GCF of $4x^3y^2$, $12x^3y$, and $20xy^2$ is $4xy$.

• Find the GCF of the terms of the polynomial.
$$4x^3y^2 = 2^2 \cdot \quad x^3 \cdot y^2$$
$$12x^3y = 2^2 \cdot 3 \cdot x^3 \cdot y$$
$$20xy^2 = 2^2 \cdot 5 \cdot x \cdot y^2$$
$$\text{GCF} = 2^2 \cdot \quad x \cdot y$$

$$4x^3y^2 + 12x^3y + 20xy^2 =$$
$$4xy(x^2y) + 4xy(3x^2) + 4xy(5y) =$$

• Rewrite each term of the polynomial as a product with the GCF as one of the factors. (Do this step mentally.)

$$4xy(x^2y + 3x^2 + 5y)$$

• Use the Distributive Property to write the polynomial as a product of factors.

B. The GCF of x^{2n}, x^{n+1}, and x^n is x^n.

$$x^{2n} + x^{n+1} + x^n = x^n(x^n + x + 1)$$

• The GCF is x^n because $n < 2n$ and $n < n + 1$.

Problem 1   Factor.

A. $3x^3y - 6x^2y^2 - 3xy^3$ B. $6t^{2n} - 9t^n$

Solution   See page A23.
A. $3xy(x^2 - 2xy - y^2)$ B. $3t^n(2t^n - 3)$

2   Factor trinomials of the form $x^2 + bx + c$

A **quadratic trinomial** is a trinomial of the form $ax^2 + bx + c$, where a, b, and c are nonzero constants. The degree of a quadratic trinomial is 2. Examples of quadratic trinomials are shown below.

$$3x^2 + 4x + 7 \quad (a = 3, b = 4, \quad c = 7)$$
$$y^2 + 2y - 9 \quad (a = 1, b = 2, \quad c = -9)$$
$$6x^2 - 5x + 1 \quad (a = 6, b = -5, c = 1)$$

To **factor a quadratic trinomial** of the form $x^2 + bx + c$ means to express the trinomial as the product of two binomials.

The method by which factors of a trinomial are found is based on FOIL. Consider the binomial products at the top of the following page, noting the relationship between the constant terms of the binomials and the terms of the trinomials:

$$(x + 4)(x + 7) = x^2 + 7x + 4x + (4)(7) \quad = x^2 + 11x + 28$$
$$(x - 3)(x - 5) = x^2 - 5x - 3x + (-3)(-5) = x^2 - 8x + 15$$
$$(x + 9)(x - 6) = x^2 - 6x + 9x + (9)(-6) \quad = x^2 + 3x - 54$$
$$(x + 2)(x - 8) = x^2 - 8x + 2x + (2)(-8) \quad = x^2 - 6x - 16$$

The coefficient of x is the sum of the _____ constant terms of the binomials.

The constant term is the product of _____ the constant terms of the binomials.

Points to Remember to Factor $x^2 + bx + c$

1. In the trinomial, the coefficient of x is the sum of the constant terms of the binomials.

2. In the trinomial, the constant term is the product of the constant terms of the binomials.

3. When the constant term of the trinomial is positive, the constant terms of the binomials have the same sign as the coefficient of x in the trinomial.

4. When the constant term of the trinomial is negative, the constant terms of the binomials have opposite signs.

Use the four points listed above to factor a trinomial. For example, to factor

$$x^2 - 5x - 24,$$

find two numbers whose sum is -5 and whose product is -24 [Points 1 and 2]. Because the constant term of the trinomial is negative (-24), the numbers will have opposite signs [Point 4].

A systematic method of finding these numbers involves listing the factors of the constant term of the trinomial and the sum of those factors.

Factors of -24	Sum of the Factors
1, −24	$1 + (-24) = -23$
−1, 24	$-1 + 24 = 23$
2, −12	$2 + (-12) = -10$
−2, 12	$-2 + 12 = 10$
3, −8	**$3 + (-8) = -5$**
−3, 8	$-3 + 8 = 5$
4, −6	$4 + (-6) = -2$
−4, 6	$-4 + 6 = 2$

3 and -8 are two numbers whose sum is -5 and whose product is -24. Write the binomial factors of the trinomial.

$$x^2 - 5x - 24 = (x + 3)(x - 8)$$

Check: $(x + 3)(x - 8) = x^2 - 8x + 3x - 24 = x^2 - 5x - 24$

By the Commutative Property of Multiplication, the binomial factors can also be written as

$$x^2 - 5x - 24 = (x - 8)(x + 3)$$

Example 2   Factor.

A. $x^2 + 8x + 12$ B. $x^2 + 5x - 84$ C. $10 - 3x - x^2$

Solution   A. $x^2 + 8x + 12$

Factors of 12	Sum
1, 12	13
2, 6	8
3, 4	7

• Try only positive factors of 12 [Point 3].

• Once the correct pair is found, the other factors need not be tried.

$x^2 + 8x + 12 = (x + 2)(x + 6)$

• Write the factors of the trinomial.

Check: $(x + 2)(x + 6) =$
$\quad x^2 + 6x + 2x + 12 =$
$\quad x^2 + 8x + 12$

B. $x^2 + 5x - 84$

$(-7)(12) = -84$
$-7 + 12 = 5$

• The factors must be of opposite signs [Point 4]. Find two factors of -84 whose sum is 5. (Do this step mentally.)

$x^2 + 5x - 84 = (x + 12)(x - 7)$

Check: $(x + 12)(x - 7) =$
$\quad x^2 - 7x + 12x - 84 =$
$\quad x^2 + 5x - 84$

C. $10 - 3x - x^2 = (5 + x)(2 - x)$

Check: $(5 + x)(2 - x) =$
$\quad 10 - 5x + 2x - x^2 =$
$\quad 10 - 3x - x^2$

Problem 2   Factor.

A. $x^2 + 13x + 42$ B. $x^2 - x - 20$ C. $x^2 + 5xy + 6y^2$

Solution   See page A23.

A. $(x + 6)(x + 7)$ B. $(x + 4)(x - 5)$ C. $(x + 2y)(x + 3y)$

Not all trinomials can be factored when using only integers. Consider $x^2 - 3x - 6$.

Factors of -6	Sum
1, -6	-5
-1, 6	5
2, -3	-1
-2, 3	1

Because none of the pairs of factors of -6 have a sum of -3, the trinomial is not factorable. The trinomial is said to be **nonfactorable** over the integers.

Example 3   Factor: $x^2 + 2x - 4$

Solution   The trinomial is nonfactorable over the integers.   • There are no factors of -4 whose sum is 2.

Problem 3   Factor: $x^2 + 5x - 1$

Solution   See page A23.
Nonfactorable over the integers

3   # Factor trinomials of the form $ax^2 + bx + c$

To factor a trinomial of the form $ax^2 + bx + c$, a trial-and-error method is used. Trial factors are written, using the factors of a and c to write the binomials. Then FOIL is used to check for b, the coefficient of the middle term.

Factoring polynomials of this type by trial and error may require testing many trial factors. To reduce the number of trial factors, remember the following points:

> **Points to Remember to Factor $ax^2 + bx + c$**
>
> **1.** If the terms of the trinomial do not have a common factor, then a binomial factor cannot have a common factor.
>
> **2.** When the constant term of the trinomial is positive, the constant terms of the binomials have the same sign as the coefficient of x in the trinomial.
>
> **3.** When the constant term of the trinomial is negative, the constant terms of the binomials have opposite signs.

Factor: $4x^2 + 31x - 8$

The terms of the trinomial do not have a common factor; therefore, the binomial factors will not have a common factor.

Because the constant term, c, of the trinomial is negative (-8), the constant terms of the binomial factors will have opposite signs.

Find the factors of a (4) and the factors of c (-8).

Factors of 4	Factors of -8
1, 4	1, -8
2, 2	-1, 8
	2, -4
	-2, 4

Using these factors, write trial factors, and use FOIL to check the middle term of the trinomial.

Remember that if the terms of the trinomial do not have a common factor, then a binomial factor cannot have a common factor (Point 1). Such trial factors need not be checked.

Trial Factors	Middle Term
$(x + 1)(4x - 8)$	Common factor
$(x - 1)(4x + 8)$	Common factor
$(x + 2)(4x - 4)$	Common factor
$(x - 2)(4x + 4)$	Common factor
$(2x + 1)(2x - 8)$	Common factor
$(2x - 1)(2x + 8)$	Common factor
$(2x + 2)(2x - 4)$	Common factor
$(2x - 2)(2x + 4)$	Common factor
$(4x + 1)(x - 8)$	$-32x + x = -31x$
$(4x - 1)(x + 8)$	**$32x - x = 31x$**

The correct factors have been found. The remaining trial factors need not be checked.

$$4x^2 + 31x - 8 = (4x - 1)(x + 8)$$

The last example illustrates that many of the trial factors may have common factors and thus need not be tried. For the remainder of this chapter, the trial factors with a common factor will not be listed.

Example 4 Factor.

A. $2x^2 - 21x + 10$ B. $6x^2 + 17x - 10$

Solution A. $2x^2 - 21x + 10$

Factors of 2	Factors of 10
1, 2	-1, -10
	-2, -5

• Use negative factors of 10. [Point 2]

Trial Factors	Middle Term
$(x - 2)(2x - 5)$	$-5x - 4x = -9x$
$(2x - 1)(x - 10)$	$-20x - x = -21x$

• Write trial factors. Use FOIL to check the middle term.

$2x^2 - 21x + 10 = (2x - 1)(x - 10)$

B. $6x^2 + 17x - 10$

Factors of 6	Factors of -10
1, 6	1, -10
2, 3	-1, 10
	2, -5
	-2, 5

• Find the factors of a (6) and the factors of c (-10).

Trial Factors	Middle Term
$(x + 2)(6x - 5)$	$-5x + 12x = 7x$
$(x - 2)(6x + 5)$	$5x - 12x = -7x$
$(2x + 1)(3x - 10)$	$-20x + 3x = -17x$
$(2x - 1)(3x + 10)$	$20x - 3x = 17x$

• Write trial factors. Use FOIL to check the middle term.

$6x^2 + 17x - 10 = (2x - 1)(3x + 10)$

Problem 4   Factor.

A. $4x^2 + 15x - 4$ B. $10x^2 + 39x + 14$

Solution   See page A23.

A. $(4x - 1)(x + 4)$ B. $(2x + 7)(5x + 2)$

4   Factor trinomials that are quadratic in form

Certain trinomials that are not quadratic can be expressed as a quadratic trinomial by making suitable variable substitutions. A trinomial is **quadratic in form** if it can be written as

$$au^2 + bu + c.$$

Each of the trinomials shown below is quadratic in form.

	$x^4 + 5x^2 + 6$		$2x^2y^2 + 3xy - 9$
	$(x^2)^2 + 5(x^2) + 6$		$2(xy)^2 + 3(xy) - 9$
Let $x^2 = u$.	$u^2 + 5u + 6$	Let $xy = u$.	$2u^2 + 3u - 9$

When a trinomial that is quadratic in form is factored, the variable part of the first term in each binomial factor will be u. For example, because $x^4 + 5x^2 + 6$ is quadratic in form when $x^2 = u$, the first term in each binomial factor will be x^2.

$$x^4 + 5x^2 + 6 = (x^2)^2 + 5(x^2) + 6 = (x^2 + 2)(x^2 + 3)$$

The trinomial $x^2y^2 - 2xy - 15$ is quadratic in form when $xy = u$. The first term in each binomial factor will be xy.

Find two factors of -15 whose sum is -2. $3(-5) = -15$ $3 + (-5) = -2$

$$x^2y^2 - 2xy - 15 = (xy + 3)(xy - 5)$$

Example 5   Factor.

A. $6x^2y^2 - xy - 12$ B. $2x^4 + 5x^2 - 12$

Solution   A. $6x^2y^2 - xy - 12 =$ • The trinomial is quadratic in form when $xy = u$.
$(3xy + 4)(2xy - 3)$

B. $2x^4 + 5x^2 - 12 =$ • The trinomial is quadratic in form when $x^2 = u$.
$(x^2 + 4)(2x^2 - 3)$

Problem 5   Factor.

A. $6x^2y^2 - 19xy + 10$ B. $3x^4 + 4x^2 - 4$

Solution   See page A24.
A. $(3xy - 2)(2xy - 5)$ B. $(3x^2 - 2)(x^2 + 2)$

5   # Factor completely

A polynomial is factored completely when it is written as a product of factors that are nonfactorable over the integers.

Factor: $4x^3 + 12x^2 - 160x$

$$4x^3 + 12x^2 - 160x =$$

Factor out the GCF of the terms.
The GCF of $4x^3$, $12x^2$, and $160x$ is $4x$. $4x(x^2 + 3x - 40) =$

Factor the trinomial. $4x(x + 8)(x - 5)$

Check: $4x(x + 8)(x - 5) =$
$(4x^2 + 32x)(x - 5) =$
$4x^3 + 12x^2 - 160x$

Example 6   Factor.

A. $30y + 2xy - 4x^2y$ B. $12x^3y^2 + 14x^2y - 6x$

Solution   A. $30y + 2xy - 4x^2y =$ • The GCF of $30y$, $2xy$, and $4x^2y$ is $2y$.
$2y(15 + x - 2x^2) =$ • Factor out the GCF.
$2y(5 + 2x)(3 - x)$ • Factor the trinomial.

B. $12x^3y^2 + 14x^2y - 6x =$ • The GCF of $12x^3y^2$, $14x^2y$, and $6x$ is $2x$.
$2x(6x^2y^2 + 7xy - 3) =$
$2x(3xy - 1)(2xy + 3)$

Problem 6   Factor.

A. $3a^3b^3 + 3a^2b^2 - 60ab$ B. $40a - 10a^2 - 15a^3$

Solution   See page A24.
A. $3ab(ab + 5)(ab - 4)$ B. $5a(2 + a)(4 - 3a)$

EXERCISES 3.3

1 Factor.

1. $6a^2 - 15a$
$3a(2a - 5)$

2. $32b^2 + 12b$
$4b(8b + 3)$

3. $4x^3 - 3x^2$
$x^2(4x - 3)$

4. $12a^5b^2 + 16a^4b$
$4a^4b(3ab + 4)$

5. $3a^2 - 10b^3$
nonfactorable

6. $9x^2 + 14y^4$
nonfactorable

7. $x^5 - x^3 - x$
$x(x^4 - x^2 - 1)$

8. $y^4 - 3y^2 - 2y$
$y(y^3 - 3y - 2)$

9. $16x^2 - 12x + 24$
$4(4x^2 - 3x + 6)$

10. $2x^5 + 3x^4 - 4x^2$
$x^2(2x^3 + 3x^2 - 4)$

11. $5b^2 - 10b^3 + 25b^4$
$5b^2(1 - 2b + 5b^2)$

12. $x^2y^4 - x^2y - 4x^2$
$x^2(y^4 - y - 4)$

13. $x^{2n} - x^n$
$x^n(x^n - 1)$

14. $a^{5n} + a^{2n}$
$a^{2n}(a^{3n} + 1)$

15. $x^{3n} - x^{2n}$
$x^{2n}(x^n - 1)$

16. $y^{4n} + y^{2n}$
$y^{2n}(y^{2n} + 1)$

17. $a^{2n+2} + a^2$
$a^2(a^{2n} + 1)$

18. $b^{n+5} - b^5$
$b^5(b^n - 1)$

19. $12x^2y^2 - 18x^3y + 24x^2y$
$6x^2y(2y - 3x + 4)$

20. $14a^4b^4 - 42a^3b^3 + 28a^3b^2$
$14a^3b^2(ab^2 - 3b + 2)$

21. $-16a^2b^4 - 4a^2b^2 + 24a^3b^2$
$4a^2b^2(-4b^2 - 1 + 6a)$

22. $10x^2y + 20x^2y^2 + 30x^2y^3$
$10x^2y(1 + 2y + 3y^2)$

23. $y^{2n+2} + y^{n+2} - y^2$
$y^2(y^{2n} + y^n - 1)$

24. $a^{2n+2} + a^{2n+1} + a^n$
$a^n(a^{n+2} + a^{n+1} + 1)$

2 Factor.

25. $x^2 - 8x + 15$
$(x - 5)(x - 3)$

26. $x^2 + 12x + 20$
$(x + 10)(x + 2)$

27. $a^2 + 12a + 11$
$(a + 11)(a + 1)$

28. $a^2 + a - 72$
$(a + 9)(a - 8)$

29. $b^2 + 2b - 35$
$(b + 7)(b - 5)$

30. $a^2 + 7a + 6$
$(a + 6)(a + 1)$

31. $y^2 - 16y + 39$
$(y - 3)(y - 13)$

32. $y^2 - 18y + 72$
$(y - 6)(y - 12)$

33. $b^2 + 4b - 32$
$(b + 8)(b - 4)$

34. $x^2 + x - 132$
$(x + 12)(x - 11)$

35. $a^2 - 15a + 56$
$(a - 7)(a - 8)$

36. $x^2 + 15x + 50$
$(x + 10)(x + 5)$

37. $y^2 + 13y + 12$
$(y + 12)(y + 1)$

38. $b^2 - 6b - 16$
$(b - 8)(b + 2)$

39. $x^2 + 4x - 5$
$(x + 5)(x - 1)$

40. $a^2 - 3ab + 2b^2$
$(a - 2b)(a - b)$

41. $a^2 + 11ab + 30b^2$
$(a + 6b)(a + 5b)$

42. $a^2 + 8ab - 33b^2$
$(a + 11b)(a - 3b)$

43. $x^2 - 14xy + 24y^2$
$(x - 12y)(x - 2y)$

44. $x^2 + 5xy + 6y^2$
$(x + 2y)(x + 3y)$

45. $y^2 + 2xy - 63x^2$
$(y + 9x)(y - 7x)$

46. $2 + x - x^2$
$(2 - x)(1 + x)$

47. $21 - 4x - x^2$
$(7 + x)(3 - x)$

48. $5 + 4x - x^2$
$(5 - x)(1 + x)$

49. $50 + 5a - a^2$
$(5 + a)(10 - a)$

50. $x^2 - 5x + 6$
$(x - 3)(x - 2)$

51. $x^2 - 7x - 12$
nonfactorable

3   Factor.

52. $2x^2 + 7x + 3$
$(2x + 1)(x + 3)$

53. $2x^2 - 11x - 40$
$(2x + 5)(x - 8)$

54. $6y^2 + 5y - 6$
$(2y + 3)(3y - 2)$

55. $4y^2 - 15y + 9$
$(4y - 3)(y - 3)$

56. $6b^2 - b - 35$
$(2b - 5)(3b + 7)$

57. $2a^2 + 13a + 6$
$(2a + 1)(a + 6)$

58. $3y^2 - 22y + 39$
$(3y - 13)(y - 3)$

59. $12y^2 - 13y - 72$
nonfactorable

60. $6a^2 - 26a + 15$
nonfactorable

61. $5x^2 + 26x + 5$
$(5x + 1)(x + 5)$

62. $4a^2 - a - 5$
$(4a - 5)(a + 1)$

63. $11x^2 - 122x + 11$
$(11x - 1)(x - 11)$

64. $10x^2 - 7x - 12$
$(2x - 3)(5x + 4)$

65. $12x^2 + 16x - 3$
$(6x - 1)(2x + 3)$

66. $4a^2 + 7a - 15$
$(4a - 5)(1a + 3)$

67. $12a^2 - 17a - 5$
$(4a + 1)(3a - 5)$

68. $2x^2 + 17x + 35$
$(2x + 7)(x + 5)$

69. $15x^2 + 19x + 6$
$(3x + 2)(5x + 3)$

70. $11y^2 - 47y + 12$
$(11y - 3)(y - 4)$

71. $12x^2 - 17x + 5$
$(12x - 5)(x - 1)$

72. $12x^2 - 40x + 25$
$(6x - 5)(2x - 5)$

73. $8y^2 - 18y + 9$
$(4y - 3)(2y - 3)$

74. $4x^2 + 9x + 10$
nonfactorable

75. $6a^2 - 5a - 2$
nonfactorable

76. $10x^2 - 29x + 10$
$(5x - 2)(2x - 5)$

77. $2x^2 + 5x + 12$
nonfactorable

78. $4x^2 - 6x + 1$
nonfactorable

79. $6x^2 + 5xy - 21y^2$
$(2x - 3y)(3x + 7y)$

80. $6x^2 + 41xy - 7y^2$
$(6x - y)(x + 7y)$

81. $4a^2 + 43ab + 63b^2$
$(4a + 7b)(a + 9b)$

82. $7a^2 + 46ab - 21b^2$
$(7a - 3b)(a + 7b)$

83. $10x^2 - 23xy + 12y^2$
$(5x - 4y)(2x - 3y)$

84. $18x^2 + 27xy + 10y^2$
$(6x + 5y)(3x + 2y)$

85. $24 + 13x - 2x^2$
$(8 - x)(3 + 2x)$

86. $6 - 7x - 5x^2$
$(2 + x)(3 - 5x)$

87. $8 - 13x + 6x^2$
nonfactorable

88. $30 + 17a - 20a^2$
nonfactorable

89. $15 - 14a - 8a^2$
$(3 - 4a)(5 + 2a)$

90. $35 - 6b - 8b^2$
$(7 - 4b)(5 + 2b)$

91. $12 - 5a - 28a^2$
$(4 - 7a)(3 + 4a)$

92. $24 - 2x - 15x^2$
$(6 - 5x)(4 + 3x)$

93. $15 - 44a - 20a^2$
$(3 - 10a)(5 + 2a)$

4   Factor.

94. $x^2y^2 - 8xy + 15$
$(xy - 3)(xy - 5)$

95. $x^2y^2 - 8xy - 33$
$(xy - 11)(xy + 3)$

96. $x^2y^2 - 17xy + 60$
$(xy - 5)(xy - 12)$

97. $a^2b^2 + 10ab + 24$
$(ab + 4)(ab + 6)$

98. $x^4 - 9x^2 + 18$
$(x^2 - 6)(x^2 - 3)$

99. $y^4 - 6y^2 - 16$
$(y^2 - 8)(y^2 + 2)$

100. $b^4 - 13b^2 - 90$
$(b^2 - 18)(b^2 + 5)$

101. $a^4 + 14a^2 + 45$
$(a^2 + 9)(a^2 + 5)$

102. $x^4y^4 - 8x^2y^2 + 12$
$(x^2y^2 - 6)(x^2y^2 - 2)$

103. $a^4b^4 + 11a^2b^2 - 26$
$(a^2b^2 + 13)(a^2b^2 - 2)$

104. $x^{2n} + 3x^n + 2$
$(x^n + 2)(x^n + 1)$

105. $a^{2n} - a^n - 12$
$(a^n - 4)(a^n + 3)$

106. $3x^2y^2 - 14xy + 15$
$(3xy - 5)(xy - 3)$

107. $5x^2y^2 - 59xy + 44$
$(5xy - 4)(xy - 11)$

108. $6a^2b^2 - 23ab + 21$
$(2ab - 3)(3ab - 7)$

109. $10a^2b^2 + 3ab - 7$
$(ab + 1)(10ab - 7)$

110. $2x^4 - 13x^2 - 15$
$(2x^2 - 15)(x^2 + 1)$

111. $3x^4 + 20x^2 + 32$
$(3x^2 + 8)(x^2 + 4)$

112. $2x^{2n} - 7x^n + 3$
$(2x^n - 1)(x^n - 3)$

113. $4x^{2n} + 8x^n - 5$
$(2x^n - 1)(2x^n + 5)$

114. $6a^{2n} + 19a^n + 10$
$(2a^n + 5)(3a^n + 2)$

5 Factor.

115. $4x^3 - 10x^2 + 6x$
$2x(2x - 3)(x - 1)$

116. $9a^3 + 30a^2 - 24a$
$3a(3a - 2)(a + 4)$

117. $12y^3 + 22y^2 - 70y$
$2y(3y - 5)(2y + 7)$

118. $5y^4 - 29y^3 + 20y^2$
$y^2(5y - 4)(y - 5)$

119. $30a^2 + 85ab + 60b^2$
$5(2a + 3b)(3a + 4b)$

120. $4x^3 + 10x^2y - 24xy^2$
$2x(2x - 3y)(x + 4y)$

121. $8a^4 + 37a^3b - 15a^2b^2$
$a^2(8a - 3b)(a + 5b)$

122. $100 - 5x - 5x^2$
$5(5 + x)(4 - x)$

123. $50x^2 + 25x^3 - 12x^4$
$x^2(5 + 4x)(10 - 3x)$

124. $320x - 8x^2 - 4x^3$
$4x(10 + x)(8 - x)$

125. $96y - 16xy - 2x^2y$
$2y(12 + x)(4 - x)$

126. $20x^2 - 38x^3 - 30x^4$
$2x^2(5 + 3x)(2 - 5x)$

127. $4x^2y^2 - 32xy + 60$
$4(xy - 3)(xy - 5)$

128. $a^4b^4 - 3a^3b^3 - 10a^2b^2$
$a^2b^2(ab - 5)(ab + 2)$

129. $2a^2b^4 + 9ab^3 - 18b^2$
$b^2(2ab - 3)(ab + 6)$

130. $90a^2b^2 + 45ab + 10$
$5(18a^2b^2 + 9ab + 2)$

131. $3x^3y^2 + 12x^2y - 96x$
$3x(xy + 8)(xy - 4)$

132. $4x^4 - 45x^2 + 80$
nonfactorable

133. $x^4 + 2x^2 + 15$
nonfactorable

134. $2a^5 + 14a^3 + 20a$
$2a(a^2 + 5)(a^2 + 2)$

135. $3b^6 - 9b^4 - 30b^2$
$3b^2(b^2 - 5)(b^2 + 2)$

136. $3x^4y^2 - 39x^2y^2 + 120y^2$
$3y^2(x^2 - 5)(x^2 - 8)$

137. $2x^4y - 7x^3y - 30x^2y$
$x^2y(2x + 5)(x - 6)$

138. $45a^2b^2 + 6ab^2 - 72b^2$
$3b^2(3a + 4)(5a - 6)$

139. $16x^2y^3 + 36x^2y^2 + 20x^2y$
$4x^2y(4y + 5)(y + 1)$

140. $36x^3y + 24x^2y^2 - 45xy^3$
$3xy(6x - 5y)(2x + 3y)$

141. $12a^3b - 70a^2b^2 - 12ab^3$
$2ab(6a + b)(a - 6b)$

142. $48a^2b^2 - 36ab^3 - 54b^4$
$6b^2(2a - 3b)(4a + 3b)$

143. $x^{3n} + 10x^{2n} + 16x^n$
$x^n(x^n + 8)(x^n + 2)$

144. $10x^{2n} + 25x^n - 60$
$5(2x^n - 3)(x^n + 4)$

SUPPLEMENTAL EXERCISES 3.3

Find all integers k such that the trinomial can be factored.

145. $x^2 + kx + 8$
$6, -6, 9, -9$

146. $x^2 + kx - 6$
$5, -5, 1, -1$

147. $x^2 + kx + 20$
$21, -21, 12, -12, 9, -9$

148. $x^2 + kx - 30$
$29, -29, 7, -7, 13, -13$

149. $2x^2 - kx + 3$
$7, -7, 5, -5$

150. $2x^2 - kx - 5$
$3, -3, 9, -9$

151. $3x^2 + kx + 5$
$16, -16, 8, -8$

152. $2x^2 + kx - 3$
$1, -1, 5, -5$

153. $6x^2 + kx + 1$
$7, -7, 5, -5$

154. $4x^2 + kx - 1$
$0, 3, -3$

155. $7x^4 - kx^2 + 3$
$22, -22, 10, -10$

156. $3x^4 - kx^2 - 2$
$5, -5, 1, -1$

Factor.

157. $3y^2 - 2 - 5y$
$(3y + 1)(y - 2)$

158. $8b^2 - 12 - 29b$
$(8b + 3)(b - 4)$

159. $4a + 3a^3 + 13a^2$
$a(3a + 1)(a + 4)$

160. $4p^2 - 3p + 4p^3$
$p(2p - 1)(2p + 3)$

161. $2a^3b - ab^3 - a^2b^2$
$ab(2a + b)(a - b)$

162. $3x^3y - xy^3 - 2x^2y^2$
$xy(3x + y)(x - y)$

163. $2y^3 + 2y^5 - 24y$
$2y(y^2 + 4)(y^2 - 3)$

164. $9b^3 + 3b^5 - 30b$
$3b(b^2 + 5)(b^2 - 2)$

165. $3(p + 1)^2 - (p + 1) - 2$
$p(3p + 5)$

166. $4(a - 1)^2 + 7(a - 1) - 2$
$(4a - 5)(a + 1)$

167. $6(b - 2)^2 + (b - 2) - 2$
$(2b - 5)(3b - 4)$

168. $12(y + 2)^2 + 5(y + 2) - 3$
$(3y + 5)(4y + 11)$

169. $4y(x - 1)^2 + 5y(x - 1) - 6y$
$y(4x - 7)(x + 1)$

170. $5b(a - 2)^2 - 11b(a - 2) - 12b$
$b(5a - 6)(a - 5)$

SECTION **3.4**

Special Factoring

1 Factor the difference of two perfect squares and perfect square trinomials

The product of a term and itself is called a **perfect square.** The exponents on variables of perfect squares are always even numbers.

Term		Perfect Square
5	$5 \cdot 5 =$	25
x	$x \cdot x =$	x^2
$3y^4$	$3y^4 \cdot 3y^4 =$	$9y^8$
x^n	$x^n \cdot x^n =$	x^{2n}

The **square root** of a perfect square is one of the two equal factors of the perfect square. $\sqrt{}$ is the symbol for square root. To find the exponent of the square root of a perfect square variable term, multiply the exponent by $\frac{1}{2}$.

$$\sqrt{25} = 5$$
$$\sqrt{x^2} = x$$
$$\sqrt{9y^8} = 3y^4$$
$$\sqrt{x^{2n}} = x^n$$

The difference of two perfect squares is the product of the sum and the difference of two terms. The factors of the difference of two squares are the sum and difference of the square roots of the perfect squares.

Sum and Difference of Two Terms   **Difference of Two Perfect Squares**

$$(a + b)(a - b) \qquad = \qquad a^2 - b^2$$

The expression $4x^2 - 81y^2$ is the difference of two perfect squares.

$$4x^2 - 81y^2 =$$
$$(2x)^2 - (9y)^2 =$$

The factors are the sum and difference of the square roots of the perfect squares.

$$(2x + 9y)(2x - 9y)$$

The sum of two perfect squares, $a^2 + b^2$, is nonfactorable over the integers.

Example 1 Factor: $25x^2 - 1$

Solution $25x^2 - 1 = (5x)^2 - 1^2$
   • Write the binomial as the difference of two perfect squares.

$\qquad\qquad = (5x + 1)(5x - 1)$
   • The factors are the sum and difference of the square roots of the perfect squares.

Problem 1 Factor: $x^2 - 36y^4$

Solution   See page A24.
$(x + 6y^2)(x - 6y^2)$

A perfect square trinomial is the square of a binomial.

Square of a Binomial				**Perfect Square Trinomial**
$(a + b)^2$	$=$	$(a + b)(a + b)$	$=$	$a^2 + 2ab + b^2$
$(a - b)^2$	$=$	$(a - b)(a - b)$	$=$	$a^2 - 2ab + b^2$

In factoring a perfect square trinomial, remember that the terms of the binomial are the square roots of the perfect squares of the trinomial. The sign in the binomial is the sign of the middle term of the trinomial.

The trinomial $x^2 - 14x + 49$ is a perfect square. Write the factors as the square of a binomial.

$$x^2 - 14x + 49 =$$
$$(x - 7)^2$$

Example 2   Factor: $4x^2 - 20x + 25$

Solution $4x^2 - 20x + 25 = (2x - 5)^2$

Problem 2   Factor: $9x^2 + 12x + 4$

Solution   See page A24.
$(3x + 2)^2$

2   Factor the sum or the difference of two cubes

The product of the same three factors is called a **perfect cube.** The exponents on variables of perfect cubes are always divisible by 3.

Term		Perfect Cube
2	$2 \cdot 2 \cdot 2 =$	8
$3y$	$3y \cdot 3y \cdot 3y =$	$27y^3$
y^2	$y^2 \cdot y^2 \cdot y^2 =$	y^6

The **cube root** of a perfect cube is one of the three equal factors of the perfect cube. $\sqrt[3]{}$ is the symbol for cube root. To find the exponent of the cube root of a perfect cube variable term, multiply the exponent by $\frac{1}{3}$.

$$\sqrt[3]{8} = 2$$
$$\sqrt[3]{27y^3} = 3y$$
$$\sqrt[3]{y^6} = y^2$$

> **The Sum or Difference of Two Cubes**
>
> The **sum or the difference of two cubes** is the product of a binomial and a trinomial.
>
> $$a^3 + b^3 = (a + b)(a^2 - ab + b^2)$$
> $$a^3 - b^3 = (a - b)(a^2 + ab + b^2)$$

Factor: $8x^3 - 27$

Write the binomial as the difference of two perfect cubes.

$$8x^3 - 27 = (2x)^3 - 3^3$$

The terms of the binomial factor are the cube roots of the perfect cubes. The sign of the binomial factor is the same sign as in the given binomial. The trinomial factor is obtained from the binomial factor.

$$= (2x - 3)(4x^2 + 6x + 9)$$

Square of the first term

Opposite of the product of the two terms

Square of the last term

Factor: $a^3 + 64y^3$

$$a^3 + 64y^3 =$$

Write the binomial as the sum of two perfect cubes.

$$a^3 + (4y)^3 =$$

Write the binomial factor and the trinomial factor.

$$(a + 4y)(a^2 - 4ay + 16y^2)$$

Example 3 Factor.

A. $x^3y^3 - 1$ B. $(x + y)^3 - x^3$

Solution A. $x^3y^3 - 1 =$
$(xy)^3 - 1^3 =$

• Write the binomial as the difference of two perfect cubes.

$(xy - 1)(x^2y^2 + xy + 1)$

B. $(x + y)^3 - x^3 =$

• This is the difference of two perfect cubes.

$[(x + y) - x][(x + y)^2 + x(x + y) + x^2] =$
$y(x^2 + 2xy + y^2 + x^2 + xy + x^2) =$
$y(3x^2 + 3xy + y^2)$

• Simplify.

Problem 3 Factor.

A. $8x^3 + y^3z^3$ B. $(x - y)^3 + (x + y)^3$

Solution See page A24.

A. $(2x + yz)(4x^2 - 2xyz + y^2z^2)$ B. $2x(x^2 + 3y^2)$

3   Factor by grouping

The Distributive Property is used to factor a common binomial factor from a variable expression.

$$3(x - y) - x(x - y) = (x - y)(3 - x)$$

To factor $3x^2(y - 2) - 5(2 - y)$, write the expression as a sum of terms that have a common factor.
Note that $(2 - y) = (-y + 2) = -(y - 2)$.
Thus $-5(2 - y) = +5(y - 2)$.
Write the expression as a product of factors.

$3x^2(y - 2) - 5(2 - y) =$

$3x^2(y - 2) - 5[-(y - 2)] =$
$3x^2(y - 2) + 5(y - 2) =$

$(y - 2)(3x^2 + 5)$

Example 4   Factor: $3x(y - 4) - 2(4 - y)$

Solution $3x(y - 4) - 2(4 - y) =$
$3x(y - 4) + 2(y - 4) =$

• Write the expression as a sum of terms that have a common factor. Note that $4 - y = -y + 4 = -(y - 4)$.

$(y - 4)(3x + 2)$

• Write the expression as a product of factors.

Problem 4   Factor: $6a(2b - 5) + 7(5 - 2b)$

Solution   See page A24.
$(2b - 5)(6a - 7)$

Some polynomials can be factored by grouping the terms so that a common binomial factor is found.

For example, the expression $3x^2 + 2x + 3xy + 2y$ can be factored by grouping the first two terms and the last two terms.

$3x^2 + 2x + 3xy + 2y =$

$(3x^2 + 2x) + (3xy + 2y) =$

Factor out the GCF from each group.

$x(3x + 2) + y(3x + 2) =$

Write the expression as a product of factors.

$(3x + 2)(x + y)$

Example 5   Factor: $xy - 4x - 2y + 8$

Solution $xy - 4x - 2y + 8 =$
$(xy - 4x) + (-2y + 8) =$
$x(y - 4) - 2(y - 4) =$
$(y - 4)(x - 2)$

• Group the first two terms and the last two terms.
• Factor out the GCF from each group.

Problem 5   Factor: $3rs - 2r - 3s + 2$

Solution   See page A24.
$(r - 1)(3s - 2)$

4 Factor completely

When factoring a polynomial completely, ask yourself the following questions about the polynomial:

1. Is there a common factor? If so, factor out the GCF.

2. If the polynomial is a binomial, is it the difference of two perfect squares, the sum of two cubes, or the difference of two cubes? If so, factor.

3. If the polynomial is a trinomial, is it a perfect square trinomial or the product of two binomials? If so, factor.

4. Can the polynomial be factored by grouping? If so, factor.

5. Is each factor nonfactorable over the integers? If not, factor.

Example 6 Factor.
A. $x^2y + 2x^2 - y - 2$ B. $x^{4n} - y^{4n}$

Solution A. $x^2y + 2x^2 - y - 2 =$
$(x^2y + 2x^2) + (-y - 2) =$ • Factor by grouping.
$x^2(y + 2) - (y + 2) =$
$(y + 2)(x^2 - 1) =$
$(y + 2)(x + 1)(x - 1)$ • Factor the difference of two perfect squares.

B. $x^{4n} - y^{4n} =$
$(x^{2n})^2 - (y^{2n})^2 =$
$(x^{2n} + y^{2n})(x^{2n} - y^{2n}) =$ • Factor the difference of two perfect squares.
$(x^{2n} + y^{2n})[(x^n)^2 - (y^n)^2] =$ • Factor the difference of two perfect squares.
$(x^{2n} + y^{2n})(x^n + y^n)(x^n - y^n)$ • Factor the difference of two perfect squares.

Problem 6 Factor.
A. $4x - 4y - x^3 + x^2y$ B. $x^{4n} - x^{2n}y^{2n}$

Solution See page A25.
A. $(2 + x)(2 - x)(x - y)$ B. $x^{2n}(x^n + y^n)(x^n - y^n)$

EXERCISES 3.4

1 Factor.

1. $x^2 - 16$
    $(x + 4)(x - 4)$

2. $y^2 - 49$
    $(y - 7)(y + 7)$

3. $4x^2 - 1$
    $(2x + 1)(2x - 1)$

4. $81x^2 - 4$
    $(9x + 2)(9x - 2)$

5. $16x^2 - 121$
    $(4x - 11)(4x + 11)$

6. $49y^2 - 36$
    $(7y - 6)(7y + 6)$

7. $1 - 9a^2$
    $(1 + 3a)(1 - 3a)$

8. $16 - 81y^2$
    $(4 - 9y)(4 + 9y)$

9. $x^2y^2 - 100$
    $(xy + 10)(xy - 10)$

10. $a^2b^2 - 25$
    $(ab + 5)(ab - 5)$

11. $x^2 + 4$
    nonfactorable

12. $a^2 + 16$
    nonfactorable

13. $25 - a^2b^2$
    $(5 - ab)(5 + ab)$

14. $64 - x^2y^2$
    $(8 - xy)(8 + xy)$

15. $4x^2 - y^2$
    $(2x + y)(2x - y)$

16. $49a^2 - 16b^4$
    $(7a - 4b^2)(7a + 4b^2)$

17. $a^{2n} - 1$
    $(a^n - 1)(a^n + 1)$

18. $b^{2n} - 16$
    $(b^n - 4)(b^n + 4)$

19. $a^2 + 4a + 4$
    $(a + 2)^2$

20. $b^2 - 18b + 81$
    $(b - 9)^2$

21. $x^2 - 12x + 36$
    $(x - 6)^2$

22. $y^2 - 6y + 9$
    $(y - 3)^2$

23. $b^2 - 2b + 1$
    $(b - 1)^2$

24. $a^2 + 14a + 49$
    $(a + 7)^2$

25. $16x^2 - 40x + 25$
    $(4x - 5)^2$

26. $49x^2 + 28x + 4$
    $(7x + 2)^2$

27. $4a^2 + 4a - 1$
    nonfactorable

28. $9x^2 + 12x - 4$
    nonfactorable

29. $b^2 + 7b + 14$
    nonfactorable

30. $y^2 - 5y + 25$
    nonfactorable

31. $x^2 + 6xy + 9y^2$
    $(x + 3y)^2$

32. $4x^2y^2 + 12xy + 9$
    $(2xy + 3)^2$

33. $25a^2 - 40ab + 16b^2$
    $(5a - 4b)^2$

34. $4a^2 - 36ab + 81b^2$
    $(2a - 9b)^2$

35. $x^{2n} + 6x^n + 9$
    $(x^n + 3)^2$

36. $y^{2n} - 16y^n + 64$
    $(y^n - 8)^2$

2   Factor.

37. $x^3 - 27$
    $(x - 3)(x^2 + 3x + 9)$

38. $y^3 + 125$
    $(y + 5)(y^2 - 5y + 25)$

39. $8x^3 - 1$
    $(2x - 1)(4x^2 + 2x + 1)$

40. $64a^3 + 27$
    $(4a + 3)(16a^2 - 12a + 9)$

41. $x^3 - y^3$
    $(x - y)(x^2 + xy + y^2)$

42. $x^3 - 8y^3$
    $(x - 2y)(x^2 + 2xy + 4y^2)$

43. $m^3 + n^3$
    $(m + n)(m^2 - mn + n^2)$

44. $27a^3 + b^3$
    $(3a + b)(9a^2 - 3ab + b^2)$

45. $64x^3 + 1$
    $(4x + 1)(16x^2 - 4x + 1)$

46. $1 - 125b^3$
    $(1 - 5b)(1 + 5b + 25b^2)$

47. $27x^3 - 8y^3$
    $(3x - 2y)(9x^2 + 6xy + 4y^2)$

48. $64x^3 + 27y^3$
    $(4x + 3y)(16x^2 - 12xy + 9y^2)$

49. $x^3y^3 + 64$
    $(xy + 4)(x^2y^2 - 4xy + 16)$

50. $8x^3y^3 + 27$
    $(2xy + 3)(4x^2y^2 - 6xy + 9)$

51. $16x^3 - y^3$
nonfactorable

52. $27x^3 - 8y^2$
$(3x - 2y)(9x^2 + 6xy + 4y^2)$

53. $8x^3 - 9y^3$
nonfactorable

54. $27a^3 - 16$
nonfactorable

55. $(a - b)^3 - b^3$
$(a - 2b)(a^2 - ab + b^2)$

56. $a^3 + (a + b)^3$
$(2a + b)(a^2 + ab + b^2)$

57. $x^{6n} + y^{3n}$
$(x^{2n} + y^n)(x^{4n} - x^{2n}y^n + y^{2n})$

58. $x^{3n} + y^{3n}$
$(x^n + y^n)(x^{2n} - x^n y^n + y^{2n})$

59. $x^{3n} + 8$
$(x^n + 2)(x^{2n} - 2x^n + 4)$

60. $a^{3n} + 64$
$(a^n + 4)(a^{2n} - 4a^n + 16)$

3   Factor.

61. $x(a + 2) - 2(a + 2)$
$(x - 2)(a + 2)$

62. $3(x + y) + a(x + y)$
$(x + y)(3 + a)$

63. $a(x - 2) - b(2 - x)$
$(x - 2)(a + b)$

64. $3(a - 7) - b(7 - a)$
$(3 + b)(a - 7)$

65. $x^2 + 3x + 2x + 6$
$(x + 2)(x + 3)$

66. $x^2 - 5x + 4x - 20$
$(x + 4)(x - 5)$

67. $xy + 4y - 2x - 8$
$(y - 2)(x + 4)$

68. $ab + 7b - 3a - 21$
$(b - 3)(a + 7)$

69. $ax + bx - ay - by$
$(x - y)(a + b)$

70. $2ax - 3ay - 2bx + 3by$
$(a - b)(2x - 3y)$

71. $x^2y - 3x^2 - 2y + 6$
$(x^2 - 2)(y - 3)$

72. $a^2b + 3a^2 + 2b + 6$
$(a^2 + 2)(b + 3)$

73. $6 + 2y + 3x^2 + x^2y$
$(3 + y)(2 + x^2)$

74. $15 + 3b - 5a^2 - a^2b$
$(3 - a^2)(5 + b)$

75. $2ax^2 + bx^2 - 4ay - 2by$
$(x^2 - 2y)(2a + b)$

76. $4a^2x + 2a^2y - 6bx - 3by$
$(2a^2 - 3b)(2x + y)$

77. $x^ny - 5x^n + y - 5$
$(x^n + 1)(y - 5)$

78. $a^nx^n + 2a^n + x^n + 2$
$(a^n + 1)(x^n + 2)$

4   Factor.

79. $5x^2 + 10x + 5$
$5(x + 1)^2$

80. $12x^2 - 36x + 27$
$3(2x - 3)^2$

81. $3x^4 - 81x$
$3x(x - 3)(x^2 + 3x + 9)$

82. $27a^4 - a$
$a(3a - 1)(9a^2 + 3a + 1)$

83. $7x^2 - 28$
$7(x + 2)(x - 2)$

84. $20x^2 - 5$
$5(2x + 1)(2x - 1)$

85. $y^4 - 10y^3 + 21y^2$
$y^2(y - 7)(y - 3)$

86. $y^5 + 6y^4 - 55y^3$
$y^3(y + 11)(y - 5)$

87. $x^4 - 16$
$(x^2 + 4)(x + 2)(x - 2)$

88. $16x^4 - 81$
$(4x^2 + 9)(2x - 3)(2x + 3)$

89. $8x^5 - 98x^3$
$2x^3(2x + 7)(2x - 7)$

90. $16a - 2a^4$
$2a(2 - a)(4 + 2a + a^2)$

91. $x^3y^3 - x^3$
$x^3(y - 1)(y^2 + y + 1)$

92. $a^3b^6 - b^3$
$b^3(ab - 1)(a^2b^2 + ab + 1)$

93. $x^6y^6 - x^3y^3$
$x^3y^3(xy - 1)(x^2y^2 + xy + 1)$

94. $x^4 - 2x^3 - 35x^2$
$x^2(x - 7)(x + 5)$

95. $x^4 + 15x^3 - 56x^2$
$x^2(x^2 + 15x - 56)$

96. $4x^2 + 4x - 1$
nonfactorable

97. $8x^4 - 40x^3 + 50x^2$
$2x^2(2x - 5)^2$

98. $6x^5 + 74x^4 + 24x^3$
$2x^3(3x + 1)(x + 12)$

99. $x^4 - y^4$
$(x^2 + y^2)(x + y)(x - y)$

100. $16a^4 - b^4$
$(4a^2 + b^2)(2a - b)(2a + b)$

101. $x^6 + y^6$
$(x^2 + y^2)(x^4 - x^2y^2 + y^4)$

102. $x^4 - 5x^2 - 4$
nonfactorable

103. $a^4 - 25a^2 - 144$
nonfactorable

104. $3b^5 - 24b^2$
$3b^2(b - 2)(b^2 + 2b + 4)$

105. $16a^4 - 2a$
$2a(2a - 1)(4a^2 + 2a + 1)$

106. $x^4y^2 - 5x^3y^3 + 6x^2y^4$
$x^2y^2(x - 3y)(x - 2y)$

107. $a^4b^2 - 8a^3b^3 - 48a^2b^4$
$a^2b^2(a + 4b)(a - 12b)$

108. $16x^3y + 4x^2y^2 - 42xy^3$
$2xy(4x + 7y)(2x - 3y)$

109. $24a^2b^2 - 14ab^3 - 90b^4$
$2b^2(4a - 9b)(3a + 5b)$

110. $x^3 - 2x^2 - x + 2$
$(x + 1)(x - 1)(x - 2)$

111. $x^3 - 2x^2 - 4x + 8$
$(x - 2)^2(x + 2)$

112. $8xb - 8x - 4b + 4$
$4(b - 1)(2x - 1)$

113. $4xy + 8x + 4y + 8$
$4(x + 1)(y + 2)$

114. $4x^2y^2 - 4x^2 - 9y^2 + 9$
$(y + 1)(y - 1)(2x + 3)(2x - 3)$

115. $4x^4 - x^2 - 4x^2y^2 + y^2$
$(x + y)(x - y)(2x + 1)(2x - 1)$

116. $x^5 - 4x^3 - 8x^2 + 32$
$(x + 2)(x - 2)^2(x^2 + 2x + 4)$

117. $x^6y^3 + x^3 - x^3y^3 - 1$
$(x - 1)(x^2 + x + 1)(xy + 1)(x^2y^2 - xy + 1)$

118. $a^{2n+2} - 6a^{n+2} + 9a^2$
$a^2(a^n - 3)^2$

119. $x^{2n+1} + 2x^{n+1} + x$
$x(x^n + 1)^2$

120. $2x^{n+2} - 7x^{n+1} + 3x^n$
$x^n(2x - 1)(x - 3)$

121. $3b^{n+2} + 4b^{n+1} - 4b^n$
$b^n(3b - 2)(b + 2)$

SUPPLEMENTAL EXERCISES 3.4

Find all integers k such that the trinomial is a perfect square trinomial.

122. $x^2 + kx + 36$
$12, -12$

123. $x^2 - kx + 81$
$18, -18$

124. $4x^2 - kx + 25$
20, −20

125. $9x^2 - kx + 1$
6, −6

126. $16x^2 + kxy + y^2$
8, −8

127. $49x^2 + kxy + 64y^2$
112, −112

Find k such that the trinomial is a perfect square trinomial.

128. $x^2 + 8x + k$
16

129. $x^2 - 12x + k$
36

130. $x^2 + 4xy + ky^2$
4

131. $x^2 - 6xy + ky^2$
9

Factor.

132. $ax^3 + b - bx^3 - a$
$(x - 1)(x^2 + x + 1)(a - b)$

133. $xy^2 - 2b - x + 2by^2$
$(y + 1)(y - 1)(x + 2b)$

134. $(p - 1)^2 - 6(p - 1) + 9$
$(p - 4)^2$

135. $4(r - 1)^2 - 4(r - 1) + 1$
$(2r - 3)^2$

136. $(a - 3)^2 - (a + 2)^2$
$-5(2a - 1)$

137. $(b + 4)^2 - (b - 5)^2$
$9(2b - 1)$

138. $y^{8n} - 2y^{4n} + 1$
$(y^{2n} + 1)^2 (y^n - 1)^2 (y^n + 1)^2$

139. $x^{6n} - 1$
$(x^n - 1)(x^{2n} + x^n + 1)(x^n + 1)(x^{2n} - x^n + 1)$

140. $(y^2 - y - 6)^2 - (y^2 + y - 2)^2$
$-4(y + 2)^2 (y - 2)$

141. $(a^2 + a - 6)^2 - (a^2 + 2a - 8)^2$
$-(a - 2)^2 (2a + 7)$

Solve.

142. The product of two numbers is 63. One of the two numbers is a perfect square. The other is a prime number. Find the sum of the two numbers.
16

143. What is the smallest whole number by which 250 can be multiplied so that the product will be a perfect square?
10

144. Palindromic numbers are natural numbers that remain unchanged when their digits are written in reverse order. Find all perfect squares less than 500 that are palindromic numbers.
1, 4, 9, 121, 484

145. A circular cookie is cut from a square piece of dough. The diameter of the cookie is x cm. The piece of dough is x cm on a side and is 1 cm deep. In terms of x, how many cubic centimeters of dough is left over? Use 3.14 for π.
$0.215x^2$ cm^3

146. The following "proof" shows that $2 = 1$. Find the error.

$$a = b$$
$$a^2 = ab$$
$$a^2 - b^2 = ab - b^2$$
$$(a + b)(a - b) = b(a - b) \qquad \text{If } a = b, \text{ then } a - b = 0$$
$$a + b = b \qquad\qquad\qquad \text{and division by zero}$$
$$b + b = b \qquad\qquad\qquad \text{is undefined.}$$
$$2b = b$$
$$2 = 1$$

SECTION **3.5** _____

Solving Equations and Inequalities by Factoring

1 Solve equations by factoring

The Multiplication Property of Zero states that the product of a number and zero is zero.

If a is a real number, then $a \cdot 0 = 0 \cdot a = 0$.

Consider the equation $x \cdot y = 0$. If this is a true equation, then either $x = 0$ or $y = 0$.

> ### The Principle of Zero Products
>
> If the product of two factors is zero, then at least one of the factors must be zero.
>
> $$\text{If } a \cdot b = 0, \text{ then } a = 0 \text{ or } b = 0.$$

The Principle of Zero Products is used in solving equations.

Solve: $(x - 3)(x - 4) = 0$

If $(x - 3)(x - 4) = 0$, then $(x - 3) = 0$ or $(x - 4) = 0$.

$$(x - 3)(x - 4) = 0$$

Rewrite each equation in the form variable = constant.

$$x - 3 = 0 \qquad\qquad x - 4 = 0$$
$$x = 3 \qquad\qquad\quad x = 4$$

Write the solution.

The solutions are 3 and 4.

Check:

$$\frac{(x - 3)(x - 4) = 0}{\begin{array}{c|c} (3 - 3)(3 - 4) & 0 \\ 0(-1) & 0 \\ & 0 = 0 \end{array}} \qquad \frac{(x - 3)(x - 4) = 0}{\begin{array}{c|c} (4 - 3)(4 - 4) & 0 \\ 1(0) & 0 \\ & 0 = 0 \end{array}}$$

An equation of the form $ax^2 + bx + c = 0$, $a \neq 0$, is a **quadratic equation.** A quadratic equation is in **standard form** when the polynomial is in descending order and equal to zero.

$$2x^2 + 3x + 1 = 0$$

$$5x^2 - 4x + 3 = 0$$

Some quadratic equations can be solved by factoring and using the Principle of Zero Products.

Example 1 Solve.

A. $3x^2 + 5x = 2$ B. $(x + 4)(x - 3) = 8$

Solution A.

$$3x^2 + 5x = 2$$
$$3x^2 + 5x - 2 = 0$$ • Write the equation in standard form.
$$(3x - 1)(x + 2) = 0$$ • Factor the trinomial.
$$3x - 1 = 0 \qquad x + 2 = 0$$ • Let each factor equal zero (the Principle of Zero Products).

$$3x = 1 \qquad\quad x = -2$$ • Solve each equation for x.

$$x = \frac{1}{3}$$

The solutions are $\frac{1}{3}$ and -2.   • Write the solutions.

B.
$$(x + 4)(x - 3) = 8$$
$$x^2 + x - 12 = 8$$ • Write the equation in standard form by first multiplying the binomials.
$$x^2 + x - 20 = 0$$ • Add -8 to each side of the equation.
$$(x + 5)(x - 4) = 0$$ • Factor the trinomial.

$$x + 5 = 0 \qquad x - 4 = 0$$ • Let each factor equal zero.
$$x = -5 \qquad\quad x = 4$$ • Solve each equation for x.

The solutions are -5 and 4.   • Write the solutions.

Problem 1 Solve.

A. $4x^2 + 11x = 3$ B. $(x - 2)(x + 5) = 8$

Solution See page A25.

A. -3 and $\frac{1}{4}$ B. -6 and 3

2 Solve inequalities by factoring

A **quadratic inequality** is one that can be written in the form $ax^2 + bx + c < 0$ or $ax^2 + bx + c > 0$, where $a \neq 0$. The symbols $\leq$ or $\geq$ can also be used. The solution set of a quadratic inequality can be found by solving a compound inequality.

To solve $x^2 - 3x - 10 > 0$, first factor the trinomial.

$$x^2 - 3x - 10 > 0$$
$$(x + 2)(x - 5) > 0$$

There are two cases for which the product of the factors will be positive:

(1) both factors are positive, **or**
(2) both factors are negative.

(1) $x + 2 > 0$ and $x - 5 > 0$, or
(2) $x + 2 < 0$ and $x - 5 < 0$

Solve each pair of compound inequalities.

(1) $x + 2 > 0$ and $x - 5 > 0$
    $x > -2$ $x > 5$

$$\{x \mid x > -2\} \cap \{x \mid x > 5\} = \{x \mid x > 5\}$$

(2) $x + 2 < 0$ and $x - 5 < 0$
    $x < -2$ $x < 5$
$$\{x \mid x < -2\} \cap \{x \mid x < 5\} = \{x \mid x < -2\}$$

Since the two cases for which the product will be positive are connected by **or**, the solution set is the union of the solution sets of each pair of inequalities.

$$\{x \mid x > 5\} \cup \{x \mid x < -2\} =$$
$$\{x \mid x > 5 \text{ or } x < -2\}$$

Although the solution set of any quadratic inequality can be found by using the method outlined above, a graphical method is often easier to use.

Solve and graph the solution set of $x^2 - x - 6 < 0$.

Factor the trinomial.

$$x^2 - x - 6 < 0$$
$$(x - 3)(x + 2) < 0$$

On a number line, draw lines indicating the numbers that make each factor equal to zero.

$x - 3 = 0$ $x + 2 = 0$
    $x = 3$ $x = -2$

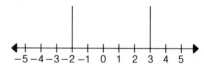

For each factor, place plus signs above the number line for those regions where the factor is positive and negative signs where the factor is negative.

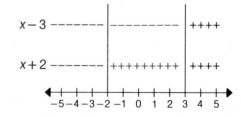

Since $x^2 - x - 6 < 0$, the solution set will be the regions where one factor is positive and the other factor is negative.

Write the solution set.

$\{x\,|-2 < x < 3\}$

The graph of the solution set of
$x^2 - x - 6 < 0$

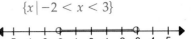

Solve and graph the solution set of $(x - 2)(x + 1)(x - 4) > 0$.

On a number line, identify for each factor the regions where the factor is positive and where the factor is negative.

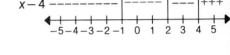

There are two regions where the product of the three factors is positive.

Write the solution set.

$\{x\,|-1 < x < 2 \text{ or } x > 4\}$

The graph of the solution set of
$(x - 2)(x + 1)(x - 4) > 0$

Example 2   Solve and graph the solution set of $2x^2 - x - 3 \ge 0$.

Solution
$$2x^2 - x - 3 \ge 0$$
$$(2x - 3)(x + 1) \ge 0$$

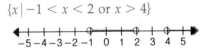

$$\left\{x\,\middle|\,x \le -1 \text{ or } x \ge \frac{3}{2}\right\}$$

Problem 2   Solve and graph the solution set of $2x^2 - x - 10 \le 0$.

Solution   See page A26.
$$\left\{x\,\middle|-2 \le x \le \frac{5}{2}\right\}$$

3 Application problems

Example 3   The sum of the squares of two consecutive positive odd integers is 130. Find the two integers.

Strategy   First positive odd integer: n
Second positive odd integer: $n + 2$

The sum of the square of the first positive odd integer and the square of the second positive odd integer is 130.

Solution

$$n^2 + (n + 2)^2 = 130$$
$$n^2 + n^2 + 4n + 4 = 130$$
$$2n^2 + 4n - 126 = 0$$
$$2(n^2 + 2n - 63) = 0$$
$$2(n + 9)(n - 7) = 0$$

$$n + 9 = 0 \qquad n - 7 = 0$$
$$n = -9 \qquad n = 7$$

• -9 is not a positive odd integer. Therefore, it is not a solution.

$$n + 2 = 7 + 2 = 9$$

• Substitute the value of n into the variable expression for the second positive odd integer and evaluate.

The two integers are 7 and 9.

Problem 3   The length of a rectangle is 5 in. longer than the width. The area of the rectangle is 66 in.^{2}. Find the length and width of the rectangle.

Solution   See page A26.
length: 11 in.; width: 6 in.

EXERCISES 3.5

1   Solve.

1. $(y + 4)(y + 6) = 0$
-4 and -6

2. $(a - 5)(a - 2) = 0$
5 and 2

3. $x(x - 7) = 0$
0 and 7

4. $b(b + 8) = 0$

0 and -8

5. $3z(2z + 5) = 0$

0 and $-\dfrac{5}{2}$

6. $4y(3y - 2) = 0$

0 and $\dfrac{2}{3}$

7. $(2x + 3)(x - 7) = 0$

$-\dfrac{3}{2}$ and 7

8. $(4a - 1)(a + 9) = 0$

$\dfrac{1}{4}$ and -9

9. $b^2 - 49 = 0$

7 and -7

10. $4z^2 - 1 = 0$

$\dfrac{1}{2}$ and $-\dfrac{1}{2}$

11. $9t^2 - 16 = 0$

$\dfrac{4}{3}$ and $-\dfrac{4}{3}$

12. $x^2 + x - 6 = 0$

2 and -3

13. $y^2 + 4y - 5 = 0$

-5 and 1

14. $a^2 - 8a + 16 = 0$

4

15. $2b^2 - 5b - 12 = 0$

$-\dfrac{3}{2}$ and 4

16. $t^2 - 8t = 0$
0 and 8

17. $x^2 - 9x = 0$
0 and 9

18. $2y^2 - 10y = 0$
0 and 5

19. $3a^2 - 12a = 0$
0 and 4

20. $b^2 - 4b = 32$
-4 and 8

21. $z^2 - 3z = 28$
7 and -4

22. $2x^2 - 5x = 12$

   $-\dfrac{3}{2}$ and 4

23. $3t^2 + 13t = 10$

   $\dfrac{2}{3}$ and -5

24. $4y^2 - 19y = 5$

   $-\dfrac{1}{4}$ and 5

25. $5b^2 - 17b = -6$

   $\dfrac{2}{5}$ and 3

26. $6a^2 + a = 2$

   $\dfrac{1}{2}$ and $-\dfrac{2}{3}$

27. $8x^2 - 10x = 3$

   $\dfrac{3}{2}$ and $-\dfrac{1}{4}$

28. $z(z - 1) = 20$

   5 and -4

29. $y(y - 2) = 35$

   7 and -5

30. $t(t + 1) = 42$

   -7 and 6

31. $x(x - 12) = -27$

   9 and 3

32. $x(2x - 5) = 12$

   $-\dfrac{3}{2}$ and 4

33. $y(3y - 2) = 8$

   $-\dfrac{4}{3}$ and 2

34. $2b^2 - 6b = b - 3$

   $\dfrac{1}{2}$ and 3

35. $3a^2 - 4a = 20 - 15a$

   $\dfrac{4}{3}$ and -5

36. $2t^2 + 5t = 6t + 15$

   $-\dfrac{5}{2}$ and 3

37. $(y + 5)(y - 7) = -20$

   5 and -3

38. $(x + 2)(x - 6) = 20$

   8 and -4

39. $(b + 5)(b + 10) = 6$

   -11 and -4

40. $(a - 9)(a - 1) = -7$

   2 and 8

41. $(t - 3)^2 = 1$

   4 and 2

42. $(y - 4)^2 = 4$

   6 and 2

43. $(3 - x)^2 + x^2 = 5$

   2 and 1

44. $(2 - b)^2 + b^2 = 10$

   3 and -1

45. $(a - 1)^2 = 3a - 5$

   2 and 3

2 Solve and graph the solution set.

46. $(x - 4)(x + 2) > 0$

47. $(x + 1)(x - 3) > 0$

48. $x^2 - 3x + 2 \geq 0$

49. $x^2 + 5x + 6 > 0$

50. $x^2 - x - 12 < 0$

51. $x^2 + x - 20 < 0$

52. $(x - 1)(x + 2)(x - 3) < 0$

53. $(x + 4)(x - 2)(x + 1) > 0$

54. $(x + 4)(x - 2)(x - 1) \geq 0$

55. $(x - 1)(x + 5)(x - 2) \leq 0$

Solve.

56. $x^2 - 16 > 0$
$\{x \mid x < -4 \text{ or } x > 4\}$

57. $x^2 - 4 \geq 0$
$\{x \mid x \leq -2 \text{ or } x \geq 2\}$

58. $x^2 - 4x + 4 > 0$
$\{x \mid x > 2 \text{ or } x < 2\}$

59. $x^2 + 6x + 9 > 0$
$\{x \mid x > -3 \text{ or } x < -3\}$

60. $x^2 - 9x \leq 36$
$\{x \mid -3 \leq x \leq 12\}$

61. $x^2 + 4x > 21$
$\{x \mid x < -7 \text{ or } x > 3\}$

62. $2x^2 - 5x + 2 \geq 0$
$\left\{x \mid x \leq \dfrac{1}{2} \text{ or } x \geq 2\right\}$

63. $4x^2 - 9x + 2 < 0$
$\left\{x \mid \dfrac{1}{4} < x < 2\right\}$

64. $4x^2 - 8x + 3 < 0$
$\left\{x \mid \dfrac{1}{2} < x < \dfrac{3}{2}\right\}$

65. $2x^2 + 11x + 12 \geq 0$
$\left\{x \mid x \leq -4 \text{ or } x \geq -\dfrac{3}{2}\right\}$

66. $(x - 6)(x + 3)(x - 2) \leq 0$
$\{x \mid x \leq -3 \text{ or } 2 \leq x \leq 6\}$

67. $(x + 5)(x - 2)(x - 3) > 0$
$\{x \mid -5 < x < 2 \text{ or } x > 3\}$

68. $(2x - 1)(x - 4)(2x + 3) > 0$
$\left\{x \mid -\dfrac{3}{2} < x < \dfrac{1}{2} \text{ or } x > 4\right\}$

69. $(x - 2)(3x - 1)(x + 2) \leq 0$
$\left\{x \mid x \leq -2 \text{ or } \dfrac{1}{3} \leq x \leq 2\right\}$

70. $(x - 5)(2x - 7)(x + 1) < 0$
$\left\{x \mid x < -1 \text{ or } \dfrac{7}{2} < x < 5\right\}$

71. $(x + 4)(2x + 7)(x - 2) > 0$
$\left\{x \mid -4 < x < -\dfrac{7}{2} \text{ or } x > 2\right\}$

3 Solve.

72. The sum of a number and its square is 56. Find the number.
−8 or 7

73. The sum of a number and its square is 72. Find the number.
−9 or 8

74. For what numbers is the sum of a number and its square less than 156?
$\{n \mid -13 < n < 12\}$

75. For what numbers is the sum of a number and its square less than 42?
$\{n \mid -7 < n < 6\}$

76. The square of a number is 80 more than twice the number. Find the number.
−8 or 10

77. The square of a number is 143 more than twice the number. Find the number.
13 or −11

78. The sum of the squares of two consecutive positive integers is equal to 85. Find the two integers.
6 and 7

79. The sum of the squares of two consecutive positive even integers is equal to 244. Find the two integers.
    10 and 12

80. The product of two consecutive even integers is less than 8. Find the integers that will satisfy the inequality.
    −2 and 0 or 0 and 2

81. The product of two consecutive even positive integers is less than 24. Find the integers that will satisfy the inequality.
    2 and 4

82. The sum of the cube of a number and the product of the number and four is equal to five times the square of the number. Find the number.
    0, 1, or 4

83. The sum of the cube of a number and the product of the number and five is equal to six times the square of the number. Find the number.
    0, 1, or 5

84. The length of a rectangle is 5 in. more than twice the width. The area is 75 in.^{2}. Find the width and length of the rectangle.
    length: 15 in.; width: 5 in.

85. The width of a rectangle is 5 ft less than the length. The area of the rectangle is 176 ft^2. Find the length and width of the rectangle.
    length: 16 ft; width: 11 ft

86. The length of the base of a triangle is three times the height. The area of the triangle is 54 cm^2. Find the base and height of the triangle.
    base: 18 cm; height: 6 cm

87. The height of a triangle is 4 cm more than twice the length of the base. The area of the triangle is 80 cm^2. Find the height of the triangle.
    20 cm

88. An object is thrown downward, with an initial speed of 16 ft/s, from the top of a building 480 ft high. How many seconds later will the object hit the ground? Use the equation $d = vt + 16t^2$, where d is the distance in feet, v is the initial speed, and t is the time in seconds.
    5 s

89. A stone is thrown into a well with an initial speed of 8 ft/s. The well is 624 ft deep. How many seconds later will the stone hit the bottom of the well? Use the equation $d = vt + 16t^2$, where d is the distance in feet, v is the initial speed, and t is the time in seconds.
    6 s

90. The length of a rectangle is 6 cm, and the width is 3 cm. If both the length and the width are increased by equal amounts, the area of the rectangle is increased by 70 cm^2. Find the length and width of the larger rectangle. length: 11 cm; width: 8 cm

91. The width of a rectangle is 4 cm, and the length is 8 cm. If both the width and the length are increased by equal amounts, the area of the rectangle is increased by 64 cm². Find the length and width of the larger rectangle.
length: 12 cm; width: 8 cm

SUPPLEMENTAL EXERCISES 3.5

Solve for x in terms of a.

92. $x^2 + 3ax - 10a^2 = 0$
2a and −5a

93. $x^2 + 4ax - 21a^2 = 0$
−7a and 3a

94. $x^2 - 9a^2 = 0$
3a and −3a

95. $x^2 - 16a^2 = 0$
4a and −4a

96. $x^2 + 10ax + 25a^2 = 0$
−5a

97. $x^2 - 12ax + 36a^2 = 0$
6a

98. $2x^2 + ax - 3a^2 = 0$
$-\dfrac{3a}{2}$ and a

99. $3x^2 + 4ax - 4a^2 = 0$
$\dfrac{2a}{3}$ and −2a

100. $x^2 - 5ax = 15a^2 - 3ax$
5a and −3a

101. $x^2 + 3ax = 8ax + 24a^2$
8a and −3a

Graph the solution set.

102. $(x + 2)(x - 3)(x + 1)(x + 4) > 0$

103. $(x - 1)(x + 3)(x - 2)(x - 4) \geq 0$

104. $(x^2 + 2x - 8)(x^2 - 2x - 3) < 0$

105. $(x^2 + 2x - 3)(x^2 + 3x + 2) \geq 0$

Solve.

106. Find $3n^2 + 2n - 1$ if $n(n + 6) = 16$.
175 or 15

107. Find $2n^3 - 3n + 1$ if $n(n - 2) = 15$.
236 or −44

108. A rectangular piece of cardboard is 10 in. longer than it is wide. Squares 2 in. on a side are to be cut from each corner and then the sides folded up to make an open box with a volume of 112 in.³. Find the length and width of the piece of cardboard.
length: 18 in.; width: 8 in.

109. A rectangular piece of cardboard is 12 in. longer than it is wide. Squares 2 in. on a side are to be cut from each corner and then the sides folded up to make an open box with a volume of 216 in.^{3}. Find the length and width of the piece of cardboard.
length: 22 in.; width: 10 in.

110. The perimeter of a rectangular garden is 28 ft. The area of the garden is 40 ft^2. Find the length and width of the garden.
length: 10 ft; width: 4 ft

111. The perimeter of a rectangular garden is 44 m. The area of the garden is 120 m^2. Find the length and width of the garden.
length: 12 m; width: 10 m

112. The sides of a rectangular box have areas of 18 cm^2, 24 cm^2, and 48 cm^2. Find the volume of the box.
144 cm^3

113. The sides of a rectangular box have areas of 16 cm^2, 20 cm^2, and 80 cm^2. Find the volume of the box.
160 cm^3

CALCULATORS AND COMPUTERS

General Factoring

The program GENERAL FACTORING on the Student Disk can be used to practice factoring. You may choose to practice polynomials of the form

$$x^2 + bx + c, \quad ax^2 + bx + c, \quad \text{or} \quad x^3 + a^3.$$

These choices, along with the option of quitting the program, are given on a menu screen. You may practice for as long as you like with any type of problem. At the end of each problem, you may either return to the menu screen or continue practicing.

The program will present you with a polynomial to factor. When you have tried to factor the polynomial using paper and pencil, press the RETURN key on the keyboard. The correct factorization will be displayed.

CHAPTER SUMMARY

Key Words

A **monomial** is a number, a variable, or a product of numbers and variables.

A **polynomial** is a variable expression in which the terms are monomials.

A **monomial** is a polynomial of *one* term.

A **binomial** is a polynomial of *two* terms.

A **trinomial** is a polynomial of *three* terms.

The **degree of a polynomial** is the greatest of the degrees of any of its terms.

To **factor a polynomial** means to write the polynomial as the product of other polynomials.

A **quadratic trinomial** is a polynomial of the form $ax^2 + bx + c$, where a, b, and c are nonzero constants.

A polynomial is **nonfactorable over the integers** if it does not factor using only integers.

To **factor a quadratic trinomial** of the form $ax^2 + bx + c$ means to express the trinomial as the product of two binomials.

The product of a term and itself is a **perfect square.**

The **square root** of a perfect square is one of the two equal factors of the perfect square.

The product of the same three factors is called a **perfect cube.**

The **cube root** of a perfect cube is one of the three equal factors of the perfect cube.

A **quadratic equation** is an equation of the form $ax^2 + bx + c = 0$, where $a \neq 0$. A quadratic equation is in standard form when the polynomial is in descending order and equal to zero.

A **quadratic inequality** is one that can be written in the form $ax^2 + bx + c < 0$ or $ax^2 + bx + c > 0$, where $a \neq 0$. The symbols $\leq$ and $\geq$ can also be used.

Essential Rules

Rule for Multiplying Exponential Expressions	If m and n are integers, then $x^m \cdot x^n = x^{m+n}$.
Rule for Simplifying Powers of Exponential Expressions	If m and n are integers, then $(x^m)^n = x^{m \cdot n}$.
Rule for Simplifying Powers of Products	If m, n, and p are integers, then $(x^m \cdot y^n)^p = x^{mp}y^{np}$.
Rule for Dividing Exponential Expressions	If m and n are integers and $x \neq 0$, then $\dfrac{x^m}{x^n} = x^{m-n}$ if $m > n$ and $\dfrac{x^m}{x^n} = \dfrac{1}{x^{n-m}}$ if $m < n$.
Rule for Simplifying Powers of Quotients	If m, n, and p are positive integers and $y \neq 0$, then $\left(\dfrac{x^m}{y^n}\right)^p = \dfrac{x^{mp}}{y^{np}}$.
The Sum and Difference of Two Terms	$(a + b)(a - b) = a^2 - b^2$
The Square of a Binomial	$(a + b)^2 = a^2 + 2ab + b^2$ $(a - b)^2 = a^2 - 2ab + b^2$
The Sum or the Difference of Two Cubes	$a^3 + b^3 = (a + b)(a^2 - ab + b^2)$ $a^3 - b^3 = (a - b)(a^2 + ab + b^2)$
The Principle of Zero Products	If $ab = 0$, then $a = 0$ or $b = 0$.

CHAPTER REVIEW

1. Simplify: $(-3xy)^2(-2x^2y^4)^3$
   $-72x^8y^{14}$

2. Simplify: $2a[5 - 2a(3 - a) - 3a]$
   $4a^3 - 18a^2 + 10a$

3. Factor: $2x^4 + 5x^2 - 3$
   $(2x^2 - 1)(x^2 + 3)$

4. Factor: $1 - 8a^3$
   $(1 - 2a)(1 + 2a + 4a^2)$

5. Simplify: $(2a - 5b)^2$
   $4a^2 - 20ab + 25b^2$

6. Simplify: $\dfrac{(-3a^2b^3)^3}{(6ab^3)^2}$
   $-\dfrac{3a^4b^3}{4}$

7. Solve: $x(2x - 7) = 0$ 0 and $\dfrac{7}{2}$

8. Factor: $3(x - 2) - y(x - 2)$
   $(3 - y)(x - 2)$

9. Simplify: $(3x - 2y)(3x + 8y)$
   $9x^2 + 18xy - 16y^2$

10. Factor: $4x^3y + 12x^2y - 4xy^3$
    $4xy(x^2 + 3x - y^2)$

11. Factor: $6 - 5x - 6x^2$
    $(2 - 3x)(3 + 2x)$

12. Factor: $9y^2 + 6y + 1$
    $(3y + 1)^2$

13. Solve: $(b - 2)(b + 4) = 16$
    4 and -6

14. Simplify:
    $(3x^3 - 2x^2 + 7x + 1) - (-5x^3 + 2x - 3)$
    $8x^3 - 2x^2 + 5x + 4$

15. Simplify: $-2x(3x^2 - 2x + 4)$
    $-6x^3 + 4x^2 - 8x$

16. Write the monomial as a product of factors.
    $-12a^4b^7 = -3ab^2(?)$
    $4a^3b^5$

17. Factor: $b(a - 4) + 2(a - 4)$
    $(b + 2)(a - 4)$

18. Factor: $4x^2y^2 + 6xy^2 - 4y^2$
    $2y^2(2x - 1)(x + 2)$

19. Solve: $2y^2 - 7y - 4 = 0$ $-\dfrac{1}{2}$ and 4

20. Simplify: $(2a^2 + 2)(2a^3 - 3a - 5)$
    $4a^5 - 2a^3 - 10a^2 - 6a - 10$

21. Factor: $4x^2 - 8x + 3$
    $(2x - 1)(2x - 3)$

22. Factor: $2x^4 - 7x^2 - 4$
    $(2x^2 + 1)(x + 2)(x - 2)$

23. Simplify: $(3x - 1)(3x + 1)$
    $9x^2 - 1$

24. Solve: $3x^2 + 13x = 10$ $\dfrac{2}{3}$ and -5

25. Factor: $x^2 + 2x - 24$
    $(x + 6)(x - 4)$

26. Solve: $x(x - 3) = 40$
    8 and -5

27. Factor: $6ax^2 + 3ax - 10ax - 5a$
    $a(2x + 1)(3x - 5)$

28. Factor: $x^2(y + 1) - y - 1$
    $(y + 1)(x + 1)(x - 1)$

29. Simplify: $\dfrac{(-3x^2y^3)^3}{(-2xy^4)^4}$ $-\dfrac{27x^2}{16y^7}$

30. Solve: $x^2 + 3x - 4 \geq 0$
    $\{x \mid x \leq -4 \text{ or } x \geq 1\}$

31. Factor: $8x^2 - 26x + 15$
    $(4x - 3)(2x - 5)$

32. Factor: $4x^2 - 20xy + 25y^2$
    $(2x - 5y)^2$

33. Solve: $4b^2 - 8b = 0$
    0 and 2

34. Simplify: $(2a - 5b)(2a + 5b)$
    $4a^2 - 25b^2$

35. Factor: $x^3y^3 - 27$
    $(xy - 3)(x^2y^2 + 3xy + 9)$

36. Solve: $3a^2 - 15a = 0$
    0 and 5

37. Solve: $2x^2 - 5x - 3 < 0$
    $\left\{ x \mid -\dfrac{1}{2} < x < 3 \right\}$

38. Simplify: $(2ab^2)^2(-3ab)^4$
    $324a^6b^8$

39. The length of a rectangle is $(3x - 2)$ ft. The width is $(x + 2)$ ft. Find the area of the rectangle in terms of the variable x.
$(3x^2 + 4x - 4)$ ft^2

40. The sum of the squares of two consecutive even integers is 52. Find the two integers.
−6 and −4, or 4 and 6

41. The sum of a number and its square is 90. Find the number.
9 or −10

42. The length of a side of a cube is $(x + 4)$ cm. Find the volume of the cube in terms of the variable x.
$(x^3 + 12x^2 + 48x + 64)$ cm^3

CUMULATIVE REVIEW

1. Simplify: $8 - 2[-3 - (-1)]^2 \div 4$
6

2. Evaluate $\dfrac{2a - b}{b - c}$ when $a = 4$, $b = -2$, and $c = 6$. $\quad -\dfrac{5}{4}$

3. Identify the property that justifies the statement. $2x + (-2x) = 0$
Inverse Property of Addition

4. Simplify: $2x - 4[x - 2(3 - 2x) + 4]$
$-18x + 8$

5. Find $A \cap B$ given $A = \{3, 5, 7, 9\}$ and $B = \{3, 6, 9, 12\}$.
$A \cap B = \{3, 9\}$

6. Graph the solution set of $\{x \mid x \geq 2\}$.

7. Graph the solution set of $\{x \mid x < 0\} \cup \{x \mid x > 3\}$.

8. Solve: $\dfrac{2}{3} - y = \dfrac{5}{6}$ $\quad -\dfrac{1}{6}$

9. Solve: $8x - 3 - x = -6 + 3x - 8$ $\quad -\dfrac{11}{4}$

10. Solve: $\dfrac{3x - 5}{3} - 8 = 2$ $\quad \dfrac{35}{3}$

11. Solve: $3x - 5 \leq 4$
$\{x \mid x \leq 3\}$

12. Solve: $3x - 3 < 7$ or $4x + 3 > 5$
the set of real numbers

13. Solve: $3 - |2 - 3x| = -2$
-1 and $\dfrac{7}{3}$

14. Solve: $|3x + 2| \geq 5$
$\left\{ x \mid x \geq 1 \text{ or } x \leq -\dfrac{7}{3} \right\}$

15. Solve: $|4x - 1| < 11$
$\left\{ x \mid -\dfrac{5}{2} < x < 3 \right\}$

16. Simplify:
$(3x^2 - 7xy + 2y^2) - (-5x^2 - 2xy + 3y^2)$
$8x^2 - 5xy - y^2$

17. Simplify: $(3a^2b)^2(-2ab^2)^4$
$144a^8b^{10}$

18. Simplify: $\dfrac{(-4x^2y^3)^2}{(-2x^4y)^3}$ $-\dfrac{2y^3}{x^8}$

19. Simplify: $3x[8 - 2(3x - 6) + 5x] + 8$
$-3x^2 + 60x + 8$

20. Simplify: $(2x + 3)(2x^2 - 3x + 1)$
$4x^3 - 7x + 3$

21. Simplify: $(x^n + 1)^2$
$x^{2n} + 2x^n + 1$

22. Factor: $6x^2 - 7x - 20$
$(2x - 5)(3x + 4)$

23. Factor: $-4x^3 + 14x^2 - 12x$
$-2x(2x - 3)(x - 2)$

24. Factor: $81x^2 - y^2$
$(9x + y)(9x - y)$

25. Factor: $a(x - y) - b(y - x)$
$(a + b)(x - y)$

26. Factor: $x^4 - 16$
$(x^2 + 4)(x + 2)(x - 2)$

27. Solve: $x^2 + 6x - 72 = 0$
-12 and 6

28. Solve: $3x^2 - 7x = 6$ $-\dfrac{2}{3}$ and 3

29. Solve: $x^2 + 4x - 5 \le 0$
$\{x \,|\, -5 \le x \le 1\}$

30. Solve: $3x^2 - 16x + 5 < 0$
$\left\{ x \,\middle|\, \dfrac{1}{3} < x < 5 \right\}$

31. The sum of two integers is twenty-four. The difference between four times the smaller integer and nine is three less than twice the larger integer. Find the integers.
9 and 15

32. How many ounces of pure gold that costs $360 per ounce must be mixed with 80 oz of an alloy that costs $120 per ounce to make a mixture that costs $200 per ounce?
40 oz

33. Two bicycles are 25 mi apart and traveling toward each other. One cyclist is traveling at $\dfrac{2}{3}$ the rate of the other cyclist. They pass in two hours. Find the rate of each cyclist.
slower cyclist: 5 mph; faster cyclist: 7.5 mph

34. If $3000 is invested at an annual simple interest rate of 7.5%, how much additional money must be invested at an annual simple interest rate of 10% so that the total interest earned is 9% of the total investment?
$4500

35. Find three consecutive even integers whose sum is between 25 and 40.
8, 10, and 12; or 10, 12, and 14

36. The length of a rectangle is 3 in. more than the width. The area of the rectangle is 108 in.^{2}. Find the length of the rectangle.
12 in.

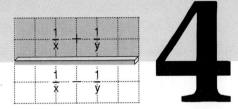

4

Rational Expressions

OBJECTIVES

- Simplify rational expressions
- Divide polynomials
- Divide polynomials by using synthetic division
- Multiply and divide rational expressions
- Add and subtract rational expressions
- Simplify complex fractions
- Solve proportions
- Application problems
- Solve fractional equations
- Solve fractional inequalities
- Literal equations
- Work problems
- Uniform motion problems

The Abacus

The abacus is an ancient device used to add, subtract, multiply, divide, and to calculate square and cube roots. There are many variations of the abacus, but the one shown here has been used in China for hundreds of years.

This abacus consists of thirteen rows of beads. A crossbar separates the beads. Each upper bead represents five units, and each lower bead represents one unit. The place holder of each row is shown above the abacus. The first row of beads represents numbers from one to nine, the second row represents numbers from ten to ninety, and so on.

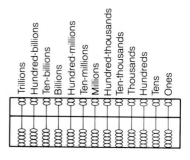

The number shown at the right is 7036.
From the thousands' row—
    one upper bead—5000
    two lower beads—2000
From the hundreds' row—no beads
From the tens' row—three lower beads—30
From the ones' row—one upper bead—5
                one lower bead—1

Add: 1247 + 2516

1247 is shown at the right.

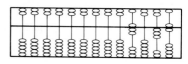

Add 6 to the ones' column.

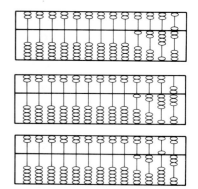

Move the 10 (the upper two beads in the ones' row) to the tens' row (one lower bead). This makes 5 beads in the lower tens' row. Remove 5 beads from the lower row, and place one bead in the upper row.

Continue adding in each row until the problem is completed. Carry if necessary.
1247 + 2516 = 3763

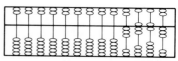

3763

Simplifying Rational Expressions

1   Simplify rational expressions

A fraction in which the numerator and denominator is a polynomial is called a **rational expression.** Examples of rational expressions are shown at the right.

$$\frac{2}{a}, \quad \frac{x^2 + y}{2x + 3}, \quad \frac{x^2 + 2x - 4}{x^4 - 2x}$$

The algebraic expression shown at the right is not a rational expression since $\sqrt{x} + 3$ is not a polynomial.

$$\frac{\sqrt{x} + 3}{2}$$

Rational expressions name a real number for each real number replacement of the variable or variables. Thus all properties of real numbers apply to rational expressions. However, when the variables in a rational expression are replaced with real numbers, the resulting denominator must not equal zero since division by zero is not defined.

For example, the value of x cannot be 3 in the rational expression at the right.

$$\frac{2x - 5}{3x - 9}$$

$$\frac{2 \cdot 3 - 5}{3 \cdot 3 - 9} = \frac{6 - 5}{9 - 9} = \frac{1}{0} \quad \begin{array}{l} \text{Not a real} \\ \text{number} \end{array}$$

A rational expression is in simplest form when the numerator and denominator have no common factors.

Example 1   Simplify.

A. $\dfrac{x^2 - 25}{x^2 + 13x + 40}$ B. $\dfrac{12 + 5x - 2x^2}{2x^2 - 3x - 20}$ C. $\dfrac{x^{2n} + x^n - 2}{x^{2n} - 1}$

Solution   A. $\dfrac{x^2 - 25}{x^2 + 13x + 40} = \dfrac{(x + 5)(x - 5)}{(x + 5)(x + 8)}$ • Factor the numerator and denominator.

$$= \frac{\overset{1}{\cancel{(x + 5)}}(x - 5)}{\underset{1}{\cancel{(x + 5)}}(x + 8)}$$ • Divide by the common factors.

$$= \frac{x - 5}{x + 8}$$ • Write the answer in simplest form.

171

B. $\dfrac{12 + 5x - 2x^2}{2x^2 - 3x - 20} = \dfrac{(4 - x)(3 + 2x)}{(x - 4)(2x + 5)}$

• Factor the numerator and denominator.

$= \dfrac{\overset{-1}{(\cancel{4 - x})}(3 + 2x)}{\underset{1}{(\cancel{x - 4})}(2x + 5)}$

• Divide by the common factors. Remember that $4 - x = -(x - 4)$. Therefore,
$\dfrac{4 - x}{x - 4} = \dfrac{-(x - 4)}{x - 4} = \dfrac{-1}{1} = -1.$

$= -\dfrac{2x + 3}{2x + 5}$

• Write the answer in simplest form.

C. $\dfrac{x^{2n} + x^n - 2}{x^{2n} - 1} = \dfrac{(x^n - 1)(x^n + 2)}{(x^n - 1)(x^n + 1)}$

• Factor the numerator and denominator.

$= \dfrac{\overset{1}{(\cancel{x^n - 1})}(x^n + 2)}{\underset{1}{(\cancel{x^n - 1})}(x^n + 1)}$

• Divide by the common factors.

$= \dfrac{x^n + 2}{x^n + 1}$

Problem 1   Simplify.

A. $\dfrac{6x^4 - 24x^3}{12x^3 - 48x^2}$ B. $\dfrac{20x - 15x^2}{15x^3 - 5x^2 - 20x}$ C. $\dfrac{x^{2n} + x^n - 12}{x^{2n} - 3x^n}$

Solution   See page A28.

A. $\dfrac{x}{2}$ B. $-\dfrac{1}{x + 1}$ C. $\dfrac{x^n + 4}{x^n}$

2 Divide polynomials

To divide two polynomials, use a method similar to that used for division of whole numbers. The same equation used to check division of whole numbers is used to check polynomial division.

Dividend = (Quotient × Divisor) + Remainder

Simplify: $(x^2 + 5x - 7) \div (x + 3)$

Step 1

$$\begin{array}{r} x \\ x + 3\overline{)x^2 + 5x - 7} \\ \underline{x^2 + 3x} \downarrow \\ 2x - 7 \end{array}$$

Think: $x\overline{)x^2} = \dfrac{x^2}{x} = x$

Multiply: $x(x + 3) = x^2 + 3x$

Subtract: $(x^2 + 5x) - (x^2 + 3x) = 2x$

Step 2

$$\begin{array}{r} x + 2 \\ x + 3\overline{)x^2 + 5x - 7} \\ \underline{x^2 + 3x} \\ 2x - 7 \\ \underline{2x + 6} \\ -13 \end{array}$$

Think: $x\overline{)2x} = \dfrac{2x}{x} = 2$

Multiply: $2(x + 3) = 2x + 6$

Subtract: $(2x - 7) - (2x + 6) = -13$

The remainder is -13.

Check: $(x + 2)(x + 3) + (-13) = x^2 + 3x + 2x + 6 - 13 = x^2 + 5x - 7$

$(x^2 + 5x - 7) \div (x + 3) = x + 2 - \dfrac{13}{x + 3}$

Simplify: $\dfrac{6 - 6x^2 + 4x^3}{2x + 3}$

Arrange the terms in descending order. There is no term of x in $4x^3 - 6x^2 + 6$. Insert $0x$ for the missing term so that like terms will be in columns.

$$\begin{array}{r} 2x^2 - 6x + 9 \\ 2x + 3\overline{)4x^3 - 6x^2 + 0x + 6} \\ \underline{4x^3 + 6x^2} \\ -12x^2 + 0x \\ \underline{-12x^2 - 18x} \\ 18x + 6 \\ \underline{18x + 27} \\ -21 \end{array}$$

$$\dfrac{4x^3 - 6x^2 + 6}{2x + 3} = 2x^2 - 6x + 9 - \dfrac{21}{2x + 3}$$

Example 2   Simplify.

A. $\dfrac{12x^2 - 11x + 10}{4x - 5}$

B. $\dfrac{x^3 + 1}{x + 1}$

Solution   **A.**
$$
\begin{array}{r}
3x + 1 \\
4x - 5\overline{)12x^2 - 11x + 10} \\
\underline{12x^2 - 15x} \\
4x + 10 \\
\underline{4x - 5} \\
15
\end{array}
$$

$$\frac{12x^2 - 11x + 10}{4x - 5} = 3x + 1 + \frac{15}{4x - 5}$$

B.
$$
\begin{array}{r}
x^2 - x + 1 \\
x + 1\overline{)x^3 + 0x^2 + 0x + 1} \\
\underline{x^3 + x^2} \\
-x^2 + 0x \\
\underline{-x^2 - x} \\
x + 1 \\
\underline{x + 1} \\
0
\end{array}
$$
   • Insert a $0x^2$ and a $0x$ for the missing terms.

$$\frac{x^3 + 1}{x + 1} = x^2 - x + 1$$

Problem 2   Simplify.

A. $\dfrac{15x^2 + 17x - 20}{3x + 4}$ B. $\dfrac{3x^3 + 8x^2 - 6x + 2}{3x - 1}$

Solution   See page A28.

A. $5x - 1 - \dfrac{16}{3x + 4}$ B. $x^2 + 3x - 1 + \dfrac{1}{3x - 1}$

3   Divide polynomials by using synthetic division

Synthetic division is a shorter method of dividing a polynomial by a binomial of the form $x - a$. This method of dividing uses only the coefficients of the variable terms.

Both long division and synthetic division are used below to simplify $(3x^2 - 4x + 6) \div (x - 2)$.

LONG DIVISION

Compare the coefficients in this problem worked by long division with the coefficients in the same problem worked by synthetic division below.

$$
\begin{array}{r}
3x + 2 \\
x - 2\overline{)3x^2 - 4x + 6} \\
\underline{3x^2 - 6x} \\
2x + 6 \\
\underline{2x - 4} \\
10
\end{array}
$$

$$(3x^2 - 4x + 6) \div (x - 2) = 3x + 2 + \frac{10}{x - 2}$$

SYNTHETIC DIVISION

	Value of a	Coefficients of the dividend

$x - a = x - 2; a = 2$ $\underline{2|}$ 3 -4 6

Bring down the 3.   3

Multiply $2 \cdot 3$ and add $\underline{2|}$ 3 -4 6
the product (6) to -4.   6
                                        3   2

Multiply $2 \cdot 2$ and add $\underline{2|}$ 3 -4 6
the product (4) to 6.   6   4
                                        3   2   10

Coefficients of   Remainder
the quotient

The degree of the first
term of the quotient is
one degree less than
the degree of the first
term of the dividend.

$$(3x^2 - 4x + 6) \div (x - 2) = 3x + 2 + \frac{10}{x - 2}$$

Check:
$$(3x + 2)(x - 2) + 10 = 3x^2 - 6x + 2x - 4 + 10$$
$$= 3x^2 - 4x + 6$$

Simplify: $(2x^3 + 3x^2 - 4x + 8) \div (x + 3)$

Write down the value of a and the $\underline{-3|}$ 2   3 -4 8
coefficients of the dividend.
$x - a = x + 3 = x - (-3); a = -3$ -6 9 -15
Bring down the 2.   2 -3 5 -7
Multiply $-3(2)$. Add the product
to 3.   Coefficients   Remainder
Continue until all the coefficients   of the
have been used.   quotient

Write the quotient. $(2x^3 + 3x^2 - 4x + 8) \div (x + 3) =$
$$2x^2 - 3x + 5 - \frac{7}{x + 3}$$

Example 3   Simplify.

A. $(5x^2 - 3x + 7) \div (x - 1)$

B. $(3x^4 - 8x^2 + 2x + 1) \div (x + 2)$

Solution   A. $\underline{1}|$ 5   −3   7   • $x - a = x - 1; a = 1$

         $\dfrac{5 \quad 2}{5 \quad 2 \quad 9}$

$(5x^2 - 3x + 7) \div (x - 1) =$

$5x + 2 + \dfrac{9}{x - 1}$

B. $\underline{-2}|$ 3   0   −8   2   1   • Insert a zero for the missing term.

         $\dfrac{-6 \quad 12 \quad -8 \quad 12}{3 \quad -6 \quad 4 \quad -6 \quad 13}$ $x - a = x + 2; a = -2$

$(3x^4 - 8x^2 + 2x + 1) \div (x + 2) =$

$3x^3 - 6x^2 + 4x - 6 + \dfrac{13}{x + 2}$

Problem 3   Simplify.
A. $(6x^2 + 8x - 5) \div (x + 2)$ B. $(2x^4 - 3x^3 - 8x^2 - 2) \div (x - 3)$

Solution   See page A29.

A. $6x - 4 + \dfrac{3}{x + 2}$ B. $2x^3 + 3x^2 + x + 3 + \dfrac{7}{x - 3}$

EXERCISES 4.1

1   Simplify.

1. $\dfrac{4 - 8x}{4}$ $1 - 2x$ 2. $\dfrac{8y + 2}{2}$ $4y + 1$ 3. $\dfrac{6x^2 - 2x}{2x}$ $3x - 1$

4. $\dfrac{3y - 12y^2}{3y}$ $1 - 4y$ 5. $\dfrac{8x^2(x - 3)}{4x(x - 3)}$ $2x$ 6. $\dfrac{16y^4(y + 8)}{12y^3(y + 8)}$ $\dfrac{4y}{3}$

7. $\dfrac{2x - 6}{3x - x^2}$ $-\dfrac{2}{x}$ 8. $\dfrac{3a^2 - 6a}{12 - 6a}$ $-\dfrac{a}{2}$ 9. $\dfrac{6x^3 - 15x^2}{12x^2 - 30x}$ $\dfrac{x}{2}$

10. $\dfrac{-36a^2 - 48a}{18a^3 + 24a^2}$ $-\dfrac{2}{a}$ 11. $\dfrac{a^2 + 4a}{4a - 16}$ simplest form   12. $\dfrac{3x - 6}{x^2 + 2x}$ simplest form

13. $\dfrac{16x^3 - 8x^2 + 12x}{4x}$ $4x^2 - 2x + 3$ 14. $\dfrac{3x^3y^3 - 12x^2y^2 + 15xy}{3xy}$ $x^2y^2 - 4xy + 5$

15. $\dfrac{-10a^4 - 20a^3 + 30a^2}{-10a^2}$ $a^2 + 2a - 3$ 16. $\dfrac{-7a^5 - 14a^4 + 21a^3}{-7a^3}$ $a^2 + 2a - 3$

17. $\dfrac{3x^{3n} - 9x^{2n}}{12x^{2n}}$ $\dfrac{x^n - 3}{4}$ 18. $\dfrac{8a^n}{4a^{2n} - 8a^n}$ $\dfrac{2}{a^n - 2}$

19. $\dfrac{x^{2n} + x^n y^n}{x^{2n} - y^{2n}}$ $\dfrac{x^n}{x^n - y^n}$

20. $\dfrac{a^{2n} - b^{2n}}{5a^{3n} + 5a^{2n}b^n}$ $\dfrac{a^n - b^n}{5a^{2n}}$

21. $\dfrac{x^2 - 7x + 12}{x^2 - 9x + 20}$ $\dfrac{x - 3}{x - 5}$

22. $\dfrac{x^2 - x - 20}{x^2 - 2x - 15}$ $\dfrac{x + 4}{x + 3}$

23. $\dfrac{x^2 - xy - 2y^2}{x^2 - 3xy + 2y^2}$ $\dfrac{x + y}{x - y}$

24. $\dfrac{2x^2 + 7xy - 4y^2}{4x^2 - 4xy + y^2}$ $\dfrac{x + 4y}{2x - y}$

25. $\dfrac{6 - x - x^2}{3x^2 - 10x + 8}$ $-\dfrac{x + 3}{3x - 4}$

26. $\dfrac{3x^2 + 10x - 8}{8 - 14x + 3x^2}$ $-\dfrac{x + 4}{4 - x}$

27. $\dfrac{14 - 19x - 3x^2}{3x^2 - 23x + 14}$ $-\dfrac{x + 7}{x - 7}$

28. $\dfrac{x^2 + x - 12}{x^2 - x - 12}$ $\dfrac{(x + 4)(x - 3)}{(x - 4)(x + 3)}$

29. $\dfrac{a^2 - 7a + 10}{a^2 + 9a + 14}$ simplest form

30. $\dfrac{x^2 - 2x}{x^2 + 2x}$ $\dfrac{x - 2}{x + 2}$

31. $\dfrac{a^2 - b^2}{a^3 + b^3}$ $\dfrac{a - b}{a^2 - ab + b^2}$

32. $\dfrac{x^4 - y^4}{x^2 + y^2}$ $(x + y)(x - y)$

33. $\dfrac{8x^3 - y^3}{4x^2 - y^2}$ $\dfrac{4x^2 + 2xy + y^2}{2x + y}$

34. $\dfrac{a^2 - b^2}{a^3 - b^3}$ $\dfrac{a + b}{a^2 + ab + b^2}$

35. $\dfrac{x^3 + y^3}{3x^3 - 3x^2y + 3xy^2}$ $\dfrac{x + y}{3x}$

36. $\dfrac{3x^3 + 3x^2 + 3x}{9x^3 - 9}$ $\dfrac{x}{3(x - 1)}$

37. $\dfrac{3x^3 - 21x^2 + 30x}{6x^4 - 24x^3 - 30x^2}$ $\dfrac{x - 2}{2x(x + 1)}$

38. $\dfrac{3x^2y - 15xy + 18y}{6x^2y + 6xy - 36y}$ $\dfrac{x - 3}{2(x + 3)}$

39. $\dfrac{x^3 - 4xy^2}{3x^3 - 2x^2y - 8xy^2}$ $\dfrac{x + 2y}{3x + 4y}$

40. $\dfrac{4a^2 - 8ab + 4b^2}{4a^2 - 4b^2}$ $\dfrac{a - b}{a + b}$

41. $\dfrac{4x^3 - 14x^2 + 12x}{24x + 4x^2 - 8x^3}$ $-\dfrac{2x - 3}{2(2x + 3)}$

42. $\dfrac{6x^3 - 15x^2 - 75x}{150x + 30x^2 - 12x^3}$ $-\dfrac{1}{2}$

43. $\dfrac{x^2 - 4}{a(x + 2) - b(x + 2)}$ $\dfrac{x - 2}{a - b}$

44. $\dfrac{x^2(a - 2) - a + 2}{ax^2 - ax}$ $\dfrac{(x + 1)(a - 2)}{ax}$

45. $\dfrac{x^4 + 3x^2 + 2}{x^4 - 1}$ $\dfrac{x^2 + 2}{(x + 1)(x - 1)}$

46. $\dfrac{x^4 - 2x^2 - 3}{x^4 + 2x^2 + 1}$ $\dfrac{x^2 - 3}{x^2 + 1}$

47. $\dfrac{x^2y^2 + 4xy - 21}{x^2y^2 - 10xy + 21}$ $\dfrac{xy + 7}{xy - 7}$

48. $\dfrac{6x^2y^2 + 11xy + 4}{9x^2y^2 + 9xy - 4}$ $\dfrac{2xy + 1}{3xy - 1}$

49. $\dfrac{a^{2n} - a^n - 2}{a^{2n} + 3a^n + 2}$ $\dfrac{a^n - 2}{a^n + 2}$

50. $\dfrac{a^{2n} + a^n - 12}{a^{2n} - 2a^n - 3}$ $\dfrac{a^n + 4}{a^n + 1}$

51. $\dfrac{a^{2n} - 1}{a^{2n} - 2a^n + 1}$ $\dfrac{a^n + 1}{a^n - 1}$

52. $\dfrac{a^{2n} + 2a^n b^n + b^{2n}}{a^{2n} - b^{2n}}$ $\dfrac{a^n + b^n}{a^n - b^n}$

53. $\dfrac{(x - 3) - b(x - 3)}{b(x + 3) - x - 3}$ $-\dfrac{x - 3}{x + 3}$

54. $\dfrac{x^2(a + b) + a + b}{x^4 - 1}$ $\dfrac{a + b}{(x + 1)(x - 1)}$

---

2   Divide by using long division.

55. $(x^2 + 3x - 40) \div (x - 5)$
$x + 8$

56. $(x^2 - 14x + 24) \div (x - 2)$
$x - 12$

57. $(x^3 - 3x^2 + 2) \div (x - 3)$
$x^2 + \dfrac{2}{x - 3}$

58. $(x^3 + 4x^2 - 8) \div (x + 4)$
$x^2 - \dfrac{8}{x + 4}$

59. $(6x^2 + 13x + 8) \div (2x + 1)$
$3x + 5 + \dfrac{3}{2x + 1}$

60. $(12x^2 + 13x - 14) \div (3x - 2)$
$4x + 7$

61. $(10x^2 + 9x - 5) \div (2x - 1)$
$5x + 7 + \dfrac{2}{2x - 1}$

62. $(18x^2 - 3x + 2) \div (3x + 2)$
$6x - 5 + \dfrac{12}{3x + 2}$

63. $(8x^3 - 9) \div (2x - 3)$
$4x^2 + 6x + 9 + \dfrac{18}{2x - 3}$

64. $(64x^3 + 4) \div (4x + 2)$
$16x^2 - 8x + 4 - \dfrac{4}{4x + 2}$

65. $(6x^4 - 13x^2 - 4) \div (2x^2 - 5)$
$3x^2 + 1 + \dfrac{1}{2x^2 - 5}$

66. $(12x^4 - 11x^2 + 10) \div (3x^2 + 1)$
$4x^2 - 5 + \dfrac{15}{3x^2 + 1}$

67. $\dfrac{3x^3 - 8x^2 - 33x - 10}{3x + 1}$
$x^2 - 3x - 10$

68. $\dfrac{8x^3 - 38x^2 + 49x - 10}{4x - 1}$
$2x^2 - 9x + 10$

69. $\dfrac{x^3 - 5x^2 + 7x - 4}{x - 3}$
$x^2 - 2x + 1 - \dfrac{1}{x - 3}$

70. $\dfrac{2x^3 - 3x^2 + 6x + 4}{2x + 1}$
$x^2 - 2x + 4$

71. $\dfrac{2x^4 - 13x^3 + 16x^2 - 9x + 20}{x - 5}$
$2x^3 - 3x^2 + x - 4$

72. $\dfrac{3x^4 + 5x^3 - x^2 + x - 2}{x + 2}$
$3x^3 - x^2 + x - 1$

73. $\dfrac{x^3 - 4x^2 + 2x - 1}{x^2 + 1}$
$x - 4 + \dfrac{x + 3}{x^2 + 1}$

74. $\dfrac{3x^3 - 2x^2 - 8}{x^2 + 5}$
$3x - 2 + \dfrac{-15x + 2}{x^2 + 5}$

75. $\dfrac{2x^3 - 3x^2 - x + 4}{x^2 - 1}$ $\quad 2x - 3 + \dfrac{1}{x - 1}$

76. $\dfrac{5x^3 - 3x^2 + 2}{x^2 + 3}$ $\quad 5x - 3 + \dfrac{-15x + 11}{x^2 + 3}$

77. $\dfrac{6x^3 + 2x^2 + x + 4}{2x^2 - 3}$ $\quad 3x + 1 + \dfrac{10x + 7}{2x^2 - 3}$

78. $\dfrac{9x^3 + 6x^2 + 2x + 1}{3x^2 + 2}$ $\quad 3x + 2 - \dfrac{4x + 3}{3x^2 + 2}$

3   Divide by using synthetic division.

79. $(2x^2 - 6x - 8) \div (x + 1)$
$2x - 8$

80. $(3x^2 + 19x + 20) \div (x + 5)$
$3x + 4$

81. $(3x^2 - 14x + 16) \div (x - 2)$
$3x - 8$

82. $(4x^2 - 23x + 28) \div (x - 4)$
$4x - 7$

83. $(3x^2 - 4) \div (x - 1)$
$3x + 3 - \dfrac{1}{x - 1}$

84. $(4x^2 - 8) \div (x - 2)$
$4x + 8 + \dfrac{8}{x - 2}$

85. $(x^2 - 9) \div (x + 4)$
$x - 4 + \dfrac{7}{x + 4}$

86. $(x^2 - 49) \div (x + 5)$
$x - 5 - \dfrac{24}{x + 5}$

87. $(2x^2 + 24) \div (2x + 4)$
$x - 2 + \dfrac{16}{x + 2}$

88. $(3x^2 - 15) \div (x + 3)$
$3x - 9 + \dfrac{12}{x + 3}$

89. $(4x^2 - 8x + 3) \div (x + 1)$
$4x - 12 + \dfrac{15}{x + 1}$

90. $(3x^2 + 7x - 6) \div (x + 4)$
$3x - 5 + \dfrac{14}{x + 4}$

91. $(2x^3 - x^2 + 6x + 9) \div (x + 1)$
$2x^2 - 3x + 9$

92. $(3x^3 + 10x^2 + 6x - 4) \div (x + 2)$
$3x^2 + 4x - 2$

93. $(x^3 - 6x^2 + 11x - 6) \div (x - 3)$
$x^2 - 3x + 2$

94. $(x^3 - 4x^2 + x + 6) \div (x + 1)$
$x^2 - 5x + 6$

95. $(x^3 - 3x^2 + 6x - 9) \div (x + 2)$
$x^2 - 5x + 16 - \dfrac{41}{x + 2}$

96. $(x^3 + 4x^2 - 5x + 5) \div (x - 3)$
$x^2 + 7x + 16 + \dfrac{53}{x - 3}$

97. $(x^3 + x - 2) \div (x + 1)$
$x^2 - x + 2 - \dfrac{4}{x + 1}$

98. $(x^3 + 2x + 5) \div (x - 2)$
$x^2 - 2x + 6 + \dfrac{17}{x - 2}$

99. $(4x^3 - x - 18) \div (x - 2)$
$4x^2 + 8x + 15 + \dfrac{12}{x - 2}$

100. $(x^3 - 3x^2 + 12) \div (x + 3)$
$x^2 - 6x + 18 - \dfrac{42}{x + 3}$

101. $(2x^3 + 5x^2 - 5x + 20) \div (x + 4)$

$2x^2 - 3x + 7 - \dfrac{8}{x + 4}$

102. $(5x^3 + 3x^2 - 17x + 6) \div (x + 2)$

$5x^2 - 7x - 3 + \dfrac{12}{x + 2}$

103. $\dfrac{2x^4 - 13x^3 + 16x^2 - 9x + 20}{x - 5}$

$2x^3 - 3x^2 + x - 4$

104. $\dfrac{x^4 + 2x^3 - x^2 - 10x + 15}{x - 2}$

$x^3 + 4x^2 + 7x + 4 + \dfrac{23}{x - 2}$

105. $\dfrac{3x^4 - 4x^3 + 8x^2 - 5x - 5}{x - 2}$

$3x^3 + 2x^2 + 12x + 19 + \dfrac{33}{x - 2}$

106. $\dfrac{2x^4 - 9x^3 + 5x^2 + 13x - 3}{x - 3}$

$2x^3 - 3x^2 - 4x + 1$

107. $\dfrac{3x^4 + 3x^3 - x^2 + 3x + 2}{x + 1}$

$3x^3 - x + 4 - \dfrac{2}{x + 1}$

108. $\dfrac{4x^4 + 12x^3 - x^2 - x + 2}{x + 3}$

$4x^3 - x + 2 - \dfrac{4}{x + 3}$

109. $\dfrac{2x^4 - x^2 + 2}{x - 3}$

$2x^3 + 6x^2 + 17x + 51 + \dfrac{155}{x - 3}$

110. $\dfrac{x^4 - 3x^3 - 30}{x + 2}$

$x^3 - 5x^2 + 10x - 20 + \dfrac{10}{x + 2}$

111. $\dfrac{x^3 + 125}{x + 5}$

$x^2 - 5x + 25$

112. $\dfrac{x^3 + 343}{x + 7}$

$x^2 - 7x + 49$

SUPPLEMENTAL EXERCISES 4.1

For what values of x is the rational expression undefined?

113. $\dfrac{5x}{x^2 - 9}$

$-3, 3$

114. $\dfrac{4x - 1}{x^2 - 25}$

$-5, 5$

115. $\dfrac{x - 3}{x^2 + 5x - 6}$

$-6, 1$

116. $\dfrac{2x + 3}{x^2 - 2x - 8}$

$4, -2$

117. $\dfrac{6}{x^2 - 3x}$

$0, 3$

118. $\dfrac{9}{x^2 - 7x}$

$0, 7$

119. $\dfrac{4x + 5}{6x^2 + 7x - 3}$

$-\dfrac{3}{2}, \dfrac{1}{3}$

120. $\dfrac{3x + 4}{12x^2 - 5x - 2}$

$-\dfrac{1}{4}, \dfrac{2}{3}$

121. $\dfrac{8x - 1}{3(x + 4) - x(x + 4)}$

$3, -4$

Divide by using long division.

122. $\dfrac{3x^2 - xy - 2y^2}{3x + 2y}$

$x - y$

123. $\dfrac{12x^2 + 11xy + 2y^2}{4x + y}$

$3x + 2y$

124. $\dfrac{4a^2 - 2ab - 5b^2}{2a + b}$

$2a - 2b - \dfrac{3b^2}{2a + b}$

125. $\dfrac{6a^2 - 5ab + 3b^2}{3a - b}$

$2a - b + \dfrac{2b^2}{3a - b}$

126. $\dfrac{2x^2 + 3xy - 5y^2}{2x - y}$

$x + 2y - \dfrac{3y^2}{2x - y}$

127. $\dfrac{6x^2 + 7xy + 3y^2}{2x + y}$

$3x + 2y + \dfrac{y^2}{2x + y}$

128. $\dfrac{a^3 + b^3}{a - b}$

$a^2 + ab + b^2$

130. $\dfrac{x^5 + y^5}{x + y}$

$x^4 - x^3y + x^2y^2 - xy^3 + y^4$

129. $\dfrac{a^4 + b^4}{a + b}$

$a^3 - a^2b + ab^2 - b^3 + \dfrac{2b^4}{a + b}$

131. $\dfrac{x^6 - y^6}{x - y}$

$x^5 + x^4y + x^3y^2 + x^2y^3 + xy^4 + y^5$

For what value of k will the remainder be zero?

132. $(x^3 - 3x^2 - x + k) \div (x - 3)$

3

133. $(x^3 - 2x^2 + x + k) \div (x - 2)$

-2

134. $(x^3 + 4x^2 + 7x + k) \div (x + 2)$

6

135. $(x^3 - x^2 - 3x + k) \div (x + 3)$

27

136. $(2x^3 - x + k) \div (x - 1)$

-1

137. $(3x^3 - 5x + k) \div (x + 1)$

-2

Solve.

138. When $x^2 + x + 2$ is divided by a polynomial, the quotient is $x + 4$, and the remainder is 14. Find the polynomial.

$x - 3$

139. When $x^2 + 3x + 4$ is divided by a polynomial, the quotient is $x + 1$, and the remainder is 2. Find the polynomial.

$x + 2$

SECTION 4.2

Operations on Rational Expressions

1 Multiply and divide rational expressions

The product of two fractions is a fraction whose numerator is the product of the numerators of the two fractions and whose denominator is the product of the denominators of the two fractions.

$$\frac{a}{b} \cdot \frac{c}{d} = \frac{ac}{bd}$$

$$\frac{5}{a + 2} \cdot \frac{b - 3}{3} = \frac{5(b - 3)}{(a + 2)3} = \frac{5b - 15}{3a + 6}$$

The product of two rational expressions can often be simplified by factoring the numerator and the denominator.

Simplify: $\dfrac{x^2 - 2x}{2x^2 + x - 15} \cdot \dfrac{2x^2 - x - 10}{x^2 - 4}$

$$\frac{x^2 - 2x}{2x^2 + x - 15} \cdot \frac{2x^2 - x - 10}{x^2 - 4} =$$

Factor the numerator and denominator of each fraction.

$$\frac{x(x - 2)}{(x + 3)(2x - 5)} \cdot \frac{(x + 2)(2x - 5)}{(x + 2)(x - 2)} =$$

Multiply.

$$\frac{x(x - 2)(x + 2)(2x - 5)}{(x + 3)(2x - 5)(x + 2)(x - 2)} =$$

Divide by the common factors.

$$\frac{x \overset{1}{\cancel{(x - 2)}}\, \overset{1}{\cancel{(x + 2)}}\, \overset{1}{\cancel{(2x - 5)}}}{(x + 3)\underset{1}{\cancel{(2x - 5)}}\, \underset{1}{\cancel{(x + 2)}}\, \underset{1}{\cancel{(x - 2)}}} =$$

Write the answer in simplest form.

$$\frac{x}{x + 3}$$

Example 1 Simplify.

A. $\dfrac{2x^2 - 6x}{3x - 6} \cdot \dfrac{6x - 12}{8x^3 - 12x^2}$

B. $\dfrac{6x^2 + x - 2}{6x^2 + 7x + 2} \cdot \dfrac{2x^2 + 9x + 4}{4 - 7x - 2x^2}$

Solution   A. $\dfrac{2x^2 - 6x}{3x - 6} \cdot \dfrac{6x - 12}{8x^3 - 12x^2} = \dfrac{2x(x - 3)}{3(x - 2)} \cdot \dfrac{6(x - 2)}{4x^2(2x - 3)} =$

$$\dfrac{2x(x - 3) \cdot 6(x - 2)}{3(x - 2) \cdot 4x^2(2x - 3)} = \dfrac{2x(x - 3) \cdot 6\overset{1}{\cancel{(x - 2)}}}{3\underset{1}{\cancel{(x - 2)}} \cdot 4x^2(2x - 3)} = \dfrac{x - 3}{x(2x - 3)}$$

B. $\dfrac{6x^2 + x - 2}{6x^2 + 7x + 2} \cdot \dfrac{2x^2 + 9x + 4}{4 - 7x - 2x^2} = \dfrac{(2x - 1)(3x + 2)}{(3x + 2)(2x + 1)} \cdot \dfrac{(2x + 1)(x + 4)}{(1 - 2x)(4 + x)} =$

$$\dfrac{\overset{-1}{\cancel{(2x - 1)}}\,\overset{1}{\cancel{(3x + 2)}}\,\overset{1}{\cancel{(2x + 1)}}\,\overset{1}{\cancel{(x + 4)}}}{\underset{1}{\cancel{(3x + 2)}}\,\underset{1}{\cancel{(2x + 1)}}\,\underset{1}{\cancel{(1 - 2x)}}\,\underset{1}{\cancel{(x + 4)}}} = -1$$

Problem 1   Simplify.

A. $\dfrac{12 + 5x - 3x^2}{x^2 + 2x - 15} \cdot \dfrac{2x^2 + x - 45}{3x^2 + 4x}$

B. $\dfrac{2x^2 - 13x + 20}{x^2 - 16} \cdot \dfrac{2x^2 + 9x + 4}{6x^2 - 7x - 5}$

Solution   See page A30.

A. $-\dfrac{2x - 9}{x}$ B. $\dfrac{2x - 5}{3x - 5}$

The **reciprocal** of a rational expression is the rational expression with the numerator and denominator interchanged.

$$\begin{array}{c} \text{Rational} \\ \text{Expression} \end{array} \left\{ \begin{array}{cc} \dfrac{a}{b} & \dfrac{b}{a} \\[2mm] \dfrac{a^2 - 2y}{4} & \dfrac{4}{a^2 - 2y} \end{array} \right\} \text{Reciprocal}$$

To divide two rational expressions, multiply by the reciprocal of the divisor.

$$\dfrac{a}{b} \div \dfrac{c}{d} = \dfrac{a}{b} \cdot \dfrac{d}{c}$$

$$\dfrac{2}{a} \div \dfrac{5}{b} = \dfrac{2}{a} \cdot \dfrac{b}{5} = \dfrac{2b}{5a}$$

$$\dfrac{x + y}{2} \div \dfrac{x - y}{5} = \dfrac{x + y}{2} \cdot \dfrac{5}{x - y} = \dfrac{(x + y)5}{2(x - y)} = \dfrac{5x + 5y}{2x - 2y}$$

Example 2   Simplify.

A. $\dfrac{12x^2y^2 - 24xy^2}{5z^2} \div \dfrac{4x^3y - 8x^2y}{3z^4}$ B. $\dfrac{3y^2 - 10y + 8}{3y^2 + 8y - 16} \div \dfrac{2y^2 - 7y + 6}{2y^2 + 5y - 12}$

Solution   A. $\dfrac{12x^2y^2 - 24xy^2}{5z^2} \div \dfrac{4x^3y - 8x^2y}{3z^4} = \dfrac{12x^2y^2 - 24xy^2}{5z^2} \cdot \dfrac{3z^4}{4x^3y - 8x^2y} =$

$$\dfrac{12xy^2(x - 2)}{5z^2} \cdot \dfrac{3z^4}{4x^2y(x - 2)} = \dfrac{12xy^2(\cancel{x - 2})3z^4}{5z^2 \cdot 4x^2y(\cancel{x - 2})} = \dfrac{9yz^2}{5x}$$

B. $\dfrac{3y^2 - 10y + 8}{3y^2 + 8y - 16} \div \dfrac{2y^2 - 7y + 6}{2y^2 + 5y - 12} = \dfrac{3y^2 - 10y + 8}{3y^2 + 8y - 16} \cdot \dfrac{2y^2 + 5y - 12}{2y^2 - 7y + 6} =$

$$\dfrac{(y - 2)(3y - 4)}{(3y - 4)(y + 4)} \cdot \dfrac{(y + 4)(2y - 3)}{(y - 2)(2y - 3)} =$$

$$\dfrac{\cancel{(y - 2)}\,\cancel{(3y - 4)}\,\cancel{(y + 4)}\,\cancel{(2y - 3)}}{\cancel{(3y - 4)}\,\cancel{(y + 4)}\,\cancel{(y - 2)}\,\cancel{(2y - 3)}} = 1$$

Problem 2   Simplify.

A. $\dfrac{6x^2 - 3xy}{10ab^4} \div \dfrac{16x^2y^2 - 8xy^3}{15a^2b^2}$ B. $\dfrac{6x^2 - 7x + 2}{3x^2 + x - 2} \div \dfrac{4x^2 - 8x + 3}{5x^2 + x - 4}$

Solution   See page A30.

A. $\dfrac{9a}{16b^2y^2}$ B. $\dfrac{5x - 4}{2x - 3}$

2   # Add and subtract rational expressions

When adding rational expressions in which the denominators are the same, add the numerators. The denominator of the sum is the common denominator. Write the answer in simplest form.

$$\dfrac{a}{c} + \dfrac{b}{c} = \dfrac{a + b}{c}$$

$$\dfrac{4x}{15} + \dfrac{8x}{15} = \dfrac{4x + 8x}{15} = \dfrac{12x}{15} = \dfrac{4x}{5}$$

$$\dfrac{a}{a^2 - b^2} + \dfrac{b}{a^2 - b^2} = \dfrac{a + b}{a^2 - b^2} = \dfrac{a + b}{(a - b)(a + b)} = \dfrac{\cancel{(a + b)}}{(a - b)\cancel{(a + b)}} = \dfrac{1}{a - b}$$

When subtracting rational expressions with like denominators, subtract the numerators. The denominator of the difference is the common denominator. Write the answer in simplest form.

$$\frac{y}{y-3} - \frac{3}{y-3} = \frac{y-3}{y-3} = \frac{\overset{1}{\cancel{(y-3)}}}{\underset{1}{\cancel{(y-3)}}} = 1$$

$$\frac{7x-12}{2x^2+5x-12} - \frac{3x-6}{2x^2+5x-12} = \frac{(7x-12)-(3x-6)}{2x^2+5x-12} = \frac{4x-6}{2x^2+5x-12}$$

$$= \frac{2(2x-3)}{(2x-3)(x+4)} = \frac{2\overset{1}{\cancel{(2x-3)}}}{\underset{1}{\cancel{(2x-3)}}(x+4)} = \frac{2}{x+4}$$

Before two rational expressions with unlike denominators can be added or subtracted, each rational expression must be expressed in terms of a common denominator. This common denominator is the LCM of the denominators of the rational expressions.

The LCM of two or more polynomials is the simplest polynomial that contains the factors of each polynomial. To find the LCM, first factor each polynomial completely. The LCM is the product of each factor the greatest number of times it occurs in any one factorization.

Find the LCM of $3x^2 + 15x$ and $6x^4 + 24x^3 - 30x^2$.

Factor each polynomial.

$$3x^2 + 15x = 3x(x+5)$$
$$6x^4 + 24x^3 - 30x^2 = 6x^2(x^2+4x-5) = 6x^2(x-1)(x+5)$$

The LCM is the product of the LCM of the numerical coefficients and each variable factor the greatest number of times it occurs in any one factorization.

$$\text{LCM} = 6x^2(x-1)(x+5)$$

Write the fractions $\dfrac{x+2}{x^2-2x}$ and $\dfrac{5x}{3x-6}$ in terms of the LCM of the denominators.

Find the LCM of the denominators.

$$x^2 - 2x = x(x-2)$$
$$3x - 6 = 3(x-2)$$
The LCM is $3x(x-2)$.

For each fraction, multiply the numerator and denominator by the factor whose product with the denominator is the LCM.

$$\frac{x+2}{x^2-2x} = \frac{x+2}{x(x-2)} \cdot \frac{3}{3} = \frac{3x+6}{3x(x-2)}$$
$$\frac{5x}{3x-6} = \frac{5x}{3(x-2)} \cdot \frac{x}{x} = \frac{5x^2}{3x(x-2)}$$

Simplify: $\dfrac{3x}{2x-3} + \dfrac{3x+6}{2x^2+x-6}$

Find the LCM of the denominators.

$\dfrac{3x}{2x-3} + \dfrac{3x+6}{2x^2+x-6} =$

The LCM is $(2x-3)(x+2)$.

Rewrite each fraction in terms of the LCM of the denominators.

$\dfrac{3x}{2x-3} \cdot \dfrac{x+2}{x+2} + \dfrac{3x+6}{(2x-3)(x+2)} =$

$\dfrac{3x^2+6x}{(2x-3)(x+2)} + \dfrac{3x+6}{(2x-3)(x+2)} =$

Add the fractions.

$\dfrac{(3x^2+6x)+(3x+6)}{(2x-3)(x+2)} = \dfrac{3x^2+9x+6}{(2x-3)(x+2)} =$

Factor the numerator to determine whether there are common factors in the numerator and denominator.

$\dfrac{3(x^2+3x+2)}{(2x-3)(x+2)} = \dfrac{3(x+2)(x+1)}{(2x-3)(x+2)} =$

$\dfrac{3(\cancel{x+2})\overset{1}{}(x+1)}{(2x-3)(\cancel{x+2})\underset{1}{}} = \dfrac{3(x+1)}{2x-3}$

Example 3   Simplify: $\dfrac{x}{2x-4} - \dfrac{4-x}{x^2-2x}$

Solution $\dfrac{x}{2x-4} - \dfrac{4-x}{x^2-2x} = \dfrac{x}{2(x-2)} \cdot \dfrac{x}{x} - \dfrac{4-x}{x(x-2)} \cdot \dfrac{2}{2} =$

• Write each fraction in terms of the LCM. The LCM is $2x(x-2)$.

$\dfrac{x^2}{2x(x-2)} - \dfrac{8-2x}{2x(x-2)} = \dfrac{x^2-(8-2x)}{2x(x-2)} =$

• Subtract the fractions.

$\dfrac{x^2+2x-8}{2x(x-2)} = \dfrac{(x+4)(x-2)}{2x(x-2)} = \dfrac{(x+4)(\cancel{x-2})\overset{1}{}}{2x(\cancel{x-2})\underset{1}{}} =$

• Divide by the common factors. (Do this step mentally.)

$\dfrac{x+4}{2x}$

Problem 3   Simplify: $\dfrac{a-3}{a^2-5a} + \dfrac{a-9}{a^2-25}$

Solution   See page A30. $\dfrac{2a+3}{a(a+5)}$

Example 4   Simplify: $\dfrac{6x - 23}{2x^2 + x - 6} + \dfrac{3x}{2x - 3} - \dfrac{5}{x + 2}$

Solution $\dfrac{6x - 23}{2x^2 + x - 6} + \dfrac{3x}{2x - 3} - \dfrac{5}{x + 2} =$

$\dfrac{6x - 23}{(2x - 3)(x + 2)} + \dfrac{3x}{2x - 3} \cdot \dfrac{x + 2}{x + 2} - \dfrac{5}{x + 2} \cdot \dfrac{2x - 3}{2x - 3} =$ • Write each fraction in terms of the LCM, $(2x - 3)(x + 2)$.

$\dfrac{6x - 23}{(2x - 3)(x + 2)} + \dfrac{3x^2 + 6x}{(2x - 3)(x + 2)} - \dfrac{10x - 15}{(2x - 3)(x + 2)} =$

$\dfrac{(6x - 23) + (3x^2 + 6x) - (10x - 15)}{(2x - 3)(x + 2)} =$

$\dfrac{6x - 23 + 3x^2 + 6x - 10x + 15}{(2x - 3)(x + 2)} =$

$\dfrac{3x^2 + 2x - 8}{(2x - 3)(x + 2)} = \dfrac{(3x - 4)(x + 2)}{(2x - 3)(x + 2)} = \dfrac{3x - 4}{2x - 3}$

Problem 4   Simplify: $\dfrac{x - 1}{x - 2} - \dfrac{7 - 6x}{2x^2 - 7x + 6} + \dfrac{4}{2x - 3}$

Solution   See page A31. $\dfrac{x + 4}{x - 2}$

EXERCISES 4.2

1   Simplify.

1. $\dfrac{27a^2b^5}{16xy^2} \cdot \dfrac{20x^2y^3}{9a^2b}$

$\dfrac{15b^4xy}{4}$

2. $\dfrac{15x^2y^4}{24ab^3} \cdot \dfrac{28a^2b^4}{35xy^4}$

$\dfrac{abx}{2}$

3. $\dfrac{3x - 15}{4x^2 - 2x} \cdot \dfrac{20x^2 - 10x}{15x - 75}$

1

4. $\dfrac{2x^2 + 4x}{8x^2 - 40x} \cdot \dfrac{6x^3 - 30x^2}{3x^2 + 6x}$

$\dfrac{x}{2}$

5. $\dfrac{x^2y^3}{x^2 - 4x - 5} \cdot \dfrac{2x^2 - 13x + 15}{x^4y^3}$

$\dfrac{2x - 3}{x^2(x + 1)}$

6. $\dfrac{2x^2 - 5x + 3}{x^6y^3} \cdot \dfrac{x^4y^4}{2x^2 - x - 3}$

$\dfrac{y(x - 1)}{x^2(x + 1)}$

7. $\dfrac{x^2 - 3x + 2}{x^2 - 8x + 15} \cdot \dfrac{x^2 + x - 12}{8 - 2x - x^2}$

$-\dfrac{x - 1}{x - 5}$

8. $\dfrac{x^2 + x - 6}{12 + x - x^2} \cdot \dfrac{x^2 + x - 20}{x^2 - 4x + 4}$

$-\dfrac{x + 5}{x - 2}$

9. $\dfrac{x^{n+1} + 2x^n}{4x^2 - 6x} \cdot \dfrac{8x^2 - 12x}{x^{n+1} - x^n}$

$\dfrac{2(x + 2)}{x - 1}$

10. $\dfrac{x^{2n} + 2x^n}{x^{n+1} + 2x} \cdot \dfrac{x^2 - 3x}{x^{n+1} - 3x^n}$

1

11. $\dfrac{2x^2 - 13x - 7}{3x^2 - 25x + 28} \cdot \dfrac{15x^2 - 17x - 4}{10x^2 + 3x - 1}$

$\dfrac{5x + 1}{5x - 1}$

12. $\dfrac{4x^2 - 9}{6x^2 + 5x - 6} \cdot \dfrac{6x^2 + 5x - 6}{4x^2 - 12x + 9}$

$\dfrac{2x + 3}{2x - 3}$

13. $\dfrac{12 + x - 6x^2}{6x^2 + 29x + 28} \cdot \dfrac{2x^2 + x - 21}{4x^2 - 9}$

$-\dfrac{x - 3}{2x + 3}$

14. $\dfrac{x^2 + 5x + 4}{4 + x - 3x^2} \cdot \dfrac{3x^2 + 2x - 8}{x^2 + 4x}$

$-\dfrac{x + 2}{x}$

15. $\dfrac{x^{2n} - x^n - 6}{x^{2n} + x^n - 2} \cdot \dfrac{x^{2n} - 5x^n - 6}{x^{2n} - 2x^n - 3}$

$\dfrac{x^n - 6}{x^n - 1}$

16. $\dfrac{x^{2n} + 3x^n + 2}{x^{2n} - x^n - 6} \cdot \dfrac{x^{2n} + x^n - 12}{x^{2n} - 1}$

$\dfrac{x^n + 4}{x^n - 1}$

17. $\dfrac{x^3 - y^3}{2x^2 + xy - 3y^2} \cdot \dfrac{2x^2 + 5xy + 3y^2}{x^2 + xy + y^2}$

$x + y$

18. $\dfrac{x^4 - 5x^2 + 4}{3x^2 - 4x - 4} \cdot \dfrac{3x^2 - 10x - 8}{x^2 - 4}$

$\dfrac{(x + 1)(x - 1)(x - 4)}{x - 2}$

19. $\dfrac{6x^2y^4}{35a^2b^5} \div \dfrac{12x^3y^3}{7a^4b^5}$

$\dfrac{a^2y}{10x}$

20. $\dfrac{12a^4b^7}{13x^2y^2} \div \dfrac{18a^5b^6}{26xy^3}$

$\dfrac{4by}{3ax}$

21. $\dfrac{2x - 6}{6x^2 - 15x} \div \dfrac{4x^2 - 12x}{18x^3 - 45x^2}$

$\dfrac{3}{2}$

22. $\dfrac{4x^2 - 4y^2}{6x^2y^2} \div \dfrac{3x^2 + 3xy}{2x^2y - 2xy^2}$

$\dfrac{4(x - y)^2}{9x^2y}$

23. $\dfrac{2x^2 - 2y^2}{14x^2y^4} \div \dfrac{x^2 + 2xy + y^2}{35xy^3}$

$\dfrac{5(x - y)}{xy(x + y)}$

24. $\dfrac{8x^3 + 12x^2y}{4x^2 - 9y^2} \div \dfrac{16x^2y^2}{4x^2 - 12xy + 9y^2}$

$\dfrac{2x - 3y}{4y^2}$

25. $\dfrac{2x^2 - 5x - 3}{2x^2 + 7x + 3} \div \dfrac{2x^2 - 3x - 20}{2x^2 - x - 15}$

$\dfrac{(x - 3)^2}{(x + 3)(x - 4)}$

26. $\dfrac{3x^2 - 10x - 8}{6x^2 + 13x + 6} \div \dfrac{2x^2 - 9x + 10}{4x^2 - 4x - 15}$

$\dfrac{x - 4}{x - 2}$

27. $\dfrac{6x^2 + 23x + 20}{3x^2 - 5x - 12} \div \dfrac{6x^2 - 29x + 35}{3x^2 - 16x + 21}$

$\dfrac{2x + 5}{2x - 5}$

28. $\dfrac{6x^2 + 23x - 4}{6x^2 + 17x - 3} \div \dfrac{4x^2 + 20x + 25}{2x^2 + 11x + 15}$

$\dfrac{x + 4}{2x + 5}$

29. $\dfrac{x^2 - 8x + 15}{x^2 + 2x - 35} \div \dfrac{15 - 2x - x^2}{x^2 + 9x + 14}$

$-\dfrac{x + 2}{x + 5}$

30. $\dfrac{2x^2 + 13x + 20}{8 - 10x - 3x^2} \div \dfrac{6x^2 - 13x - 5}{9x^2 - 3x - 2}$

$-\dfrac{2x + 5}{2x - 5}$

31. $\dfrac{x^{2n} + x^n}{2x - 2} \div \dfrac{4x^n + 4}{x^{n+1} - x^n}$

$\dfrac{x^{2n}}{8}$

32. $\dfrac{x^{2n} - 4}{4x^n + 8} \div \dfrac{x^{n+1} - 2x}{4x^3 - 12x^2}$

$x(x - 3)$

33. $\dfrac{2x^2 - 13x + 21}{2x^2 + 11x + 15} \div \dfrac{2x^2 + x - 28}{3x^2 + 4x - 15}$

$\dfrac{(x - 3)(3x - 5)}{(2x + 5)(x + 4)}$

34. $\dfrac{2x^2 - 13x + 15}{2x^2 - 3x - 35} \div \dfrac{6x^2 + x - 12}{6x^2 + 13x - 28}$

$\dfrac{2x - 3}{2x + 3}$

35. $\dfrac{14 + 17x - 6x^2}{3x^2 + 14x + 8} \div \dfrac{4x^2 - 49}{2x^2 + 15x + 28}$

-1

36. $\dfrac{16x^2 - 9}{6 - 5x - 4x^2} \div \dfrac{16x^2 + 24x + 9}{4x^2 + 11x + 6}$

-1

37. $\dfrac{2x^{2n} - x^n - 6}{x^{2n} - x^n - 2} \div \dfrac{2x^{2n} + x^n - 3}{x^{2n} - 1}$

1

38. $\dfrac{x^{4n} - 1}{x^{2n} + x^n - 2} \div \dfrac{x^{2n} + 1}{x^{2n} + 3x^n + 2}$

$(x^n + 1)^2$

39. $\dfrac{6x^2 + 6x}{3x + 6x^2 + 3x^3} \div \dfrac{x^2 - 1}{1 - x^3}$

$-\dfrac{2(x^2 + x + 1)}{(x + 1)^2}$

40. $\dfrac{x^3 + y^3}{2x^3 + 2x^2y} \div \dfrac{3x^3 - 3x^2y + 3xy^2}{6x^2 - 6y^2}$

$\dfrac{(x + y)(x - y)}{x^3}$

2 Simplify.

41. $\dfrac{3}{2xy} - \dfrac{7}{2xy} - \dfrac{9}{2xy}$

$-\dfrac{13}{2xy}$

42. $-\dfrac{3}{4x^2} + \dfrac{8}{4x^2} - \dfrac{3}{4x^2}$

$\dfrac{1}{2x^2}$

43. $\dfrac{x}{x^2 - 3x + 2} - \dfrac{2}{x^2 - 3x + 2}$

$\dfrac{1}{x - 1}$

44. $\dfrac{3x}{3x^2 + x - 10} - \dfrac{5}{3x^2 + x - 10}$

$\dfrac{1}{x + 2}$

45. $\dfrac{3}{2x^2 y} - \dfrac{8}{5x} - \dfrac{9}{10xy}$

$\dfrac{15 - 16xy - 9x}{10x^2 y}$

46. $\dfrac{2}{5ab} - \dfrac{3}{10a^2 b} + \dfrac{4}{15ab^2}$

$\dfrac{12ab - 9b + 8a}{30a^2 b^2}$

47. $\dfrac{2}{3x} - \dfrac{3}{2xy} + \dfrac{4}{5xy} - \dfrac{5}{6x}$

$\dfrac{-5y + 21}{30xy}$

48. $\dfrac{3}{4ab} - \dfrac{2}{5a} + \dfrac{3}{10b} - \dfrac{5}{8ab}$

$\dfrac{12a - 16b + 5}{40ab}$

49. $\dfrac{2x - 1}{12x} - \dfrac{3x + 4}{9x}$

$-\dfrac{6x + 19}{36x}$

50. $\dfrac{3x - 4}{6x} - \dfrac{2x - 5}{4x}$

$\dfrac{7}{12x}$

51. $\dfrac{3x + 2}{4x^2 y} - \dfrac{y - 5}{6xy^2}$

$\dfrac{10x + 6y + 7xy}{12x^2 y^2}$

52. $\dfrac{2y - 4}{5xy^2} + \dfrac{3 - 2x}{10x^2 y}$

$\dfrac{2xy - 8x + 3y}{10x^2 y^2}$

53. $\dfrac{2x}{x - 3} - \dfrac{3x}{x - 5}$

$-\dfrac{x^2 + x}{(x - 3)(x - 5)}$

54. $\dfrac{3a}{a - 2} - \dfrac{5a}{a + 1}$

$-\dfrac{2a^2 - 13a}{(a - 2)(a + 1)}$

55. $\dfrac{3}{2a - 3} + \dfrac{2a}{3 - 2a}$

-1

56. $\dfrac{x}{2x - 5} - \dfrac{2}{5x - 2}$

$\dfrac{5x^2 - 6x + 10}{(2x - 5)(5x - 2)}$

57. $\dfrac{3}{x + 5} + \dfrac{2x + 7}{x^2 - 25}$

$\dfrac{5x - 8}{(x + 5)(x - 5)}$

58. $\dfrac{x}{4 - x} - \dfrac{4}{x^2 - 16}$

$-\dfrac{(x + 2)^2}{(x + 4)(x - 4)}$

59. $\dfrac{2}{x} - 3 - \dfrac{10}{x - 4}$

$-\dfrac{3x^2 - 4x + 8}{x(x - 4)}$

60. $\dfrac{6a}{a - 3} - 5 + \dfrac{3}{a}$

$\dfrac{a^2 + 18a - 9}{a(a - 3)}$

61. $\dfrac{1}{2x - 3} - \dfrac{5}{2x} + 1$

$\dfrac{4x^2 - 14x + 15}{2x(2x - 3)}$

62. $\dfrac{5}{x} - \dfrac{5x}{5 - 6x} + 2$

$\dfrac{17x^2 + 20x - 25}{x(6x - 5)}$

63. $\dfrac{3}{x^2 - 1} + \dfrac{2x}{x^2 + 2x + 1}$

$\dfrac{2x^2 + x + 3}{(x + 1)^2(x - 1)}$

64. $\dfrac{1}{x^2 - 6x + 9} - \dfrac{1}{x^2 - 9}$

$\dfrac{6}{(x - 3)^2(x + 3)}$

65. $\dfrac{x}{x + 3} - \dfrac{3 - x}{x^2 - 9}$

$\dfrac{x + 1}{x + 3}$

66. $\dfrac{1}{x + 2} - \dfrac{3x}{x^2 + 4x + 4}$

$-\dfrac{2(x - 1)}{(x + 2)^2}$

67. $\dfrac{2x - 3}{x + 5} - \dfrac{x^2 - 4x - 19}{x^2 + 8x + 15}$

$\dfrac{x + 2}{x + 3}$

68. $\dfrac{-3x^2 + 8x + 2}{x^2 + 2x - 8} - \dfrac{2x - 5}{x + 4}$

$-\dfrac{5x^2 - 17x + 8}{(x + 4)(x - 2)}$

69. $\dfrac{x^n}{x^{2n} - 1} - \dfrac{2}{x^n + 1}$

$-\dfrac{x^n - 2}{(x^n + 1)(x^n - 1)}$

70. $\dfrac{2}{x^n - 1} + \dfrac{x^n}{x^{2n} - 1}$

$\dfrac{3x^n + 2}{x^{2n} - 1}$

71. $\dfrac{2}{x^n - 1} - \dfrac{6}{x^{2n} + x^n - 2}$

$\dfrac{2}{x^n + 2}$

72. $\dfrac{2x^n - 6}{x^{2n} - x^n - 6} + \dfrac{x^n}{x^n + 2}$

1

73. $\dfrac{2x - 2}{4x^2 - 9} - \dfrac{5}{3 - 2x}$

$\dfrac{12x + 13}{(2x + 3)(2x - 3)}$

74. $\dfrac{x^2 + 4}{4x^2 - 36} - \dfrac{13}{x + 3}$

$\dfrac{x^2 - 52x + 160}{4(x + 3)(x - 3)}$

75. $\dfrac{x - 2}{x + 1} - \dfrac{3 - 12x}{2x^2 - x - 3}$

$\dfrac{2x + 3}{2x - 3}$

76. $\dfrac{3x - 4}{4x + 1} + \dfrac{3x + 6}{4x^2 + 9x + 2}$

$\dfrac{3x - 1}{4x + 1}$

77. $\dfrac{x + 1}{x^2 + x - 6} - \dfrac{x + 2}{x^2 + 4x + 3}$

$\dfrac{2x + 5}{(x + 3)(x - 2)(x + 1)}$

78. $\dfrac{x + 1}{x^2 + x - 12} - \dfrac{x - 3}{x^2 + 7x + 12}$

$\dfrac{2(5x - 3)}{(x + 4)(x + 3)(x - 3)}$

79. $\dfrac{x - 1}{2x^2 + 11x + 12} + \dfrac{2x}{2x^2 - 3x - 9}$

$\dfrac{3x^2 + 4x + 3}{(2x + 3)(x + 4)(x - 3)}$

80. $\dfrac{x - 2}{4x^2 + 4x - 3} + \dfrac{3 - 2x}{6x^2 + x - 2}$

$-\dfrac{x^2 + 4x - 5}{(2x - 1)(2x + 3)(3x + 2)}$

81. $\dfrac{x}{x - 3} - \dfrac{2}{x + 4} - \dfrac{14}{x^2 + x - 12}$

$\dfrac{x - 2}{x - 3}$

82. $\dfrac{x^2}{x^2 + x - 2} + \dfrac{3}{x - 1} - \dfrac{4}{x + 2}$

$\dfrac{x^2 - x + 10}{(x + 2)(x - 1)}$

83. $\dfrac{x^2 + 6x}{x^2 + 3x - 18} - \dfrac{2x - 1}{x + 6} + \dfrac{x - 2}{3 - x}$

$-\dfrac{2x^2 - 9x - 9}{(x + 6)(x - 3)}$

84. $\dfrac{2x^2 - 2x}{x^2 - 2x - 15} - \dfrac{2}{x + 3} + \dfrac{x}{5 - x}$

$\dfrac{x - 2}{x + 3}$

85. $\dfrac{4 - 20x}{6x^2 + 11x - 10} - \dfrac{4}{2 - 3x} + \dfrac{x}{2x + 5}$

$\dfrac{3x^2 - 14x + 24}{(2x + 5)(3x - 2)}$

86. $\dfrac{x}{4x - 1} + \dfrac{2}{2x + 1} + \dfrac{6}{8x^2 + 2x - 1}$

$\dfrac{x + 4}{4x - 1}$

87. $\dfrac{7 - 4x}{2x^2 - 9x + 10} + \dfrac{x - 3}{x - 2} - \dfrac{x + 1}{2x - 5}$

$\dfrac{x - 12}{2x - 5}$

88. $\dfrac{x}{3x + 4} + \dfrac{3x + 2}{x - 5} - \dfrac{7x^2 + 24x + 28}{3x^2 - 11x - 20}$

1

89. $\dfrac{32x - 9}{2x^2 + 7x - 15} + \dfrac{x - 2}{3 - 2x} + \dfrac{3x + 2}{x + 5}$

$\dfrac{5x - 1}{2x - 3}$

90. $\dfrac{x + 1}{1 - 2x} - \dfrac{x + 3}{4x - 3} + \dfrac{10x^2 + 7x - 9}{8x^2 - 10x + 3}$

$\dfrac{x + 1}{2x - 1}$

91. $\dfrac{x^2}{x^3 - 8} - \dfrac{x + 2}{x^2 + 2x + 4}$

$\dfrac{4}{(x - 2)(x^2 + 2x + 4)}$

92. $\dfrac{2x}{4x^2 + 2x + 1} + \dfrac{4x + 1}{8x^3 - 1}$

$\dfrac{1}{2x - 1}$

93. $\dfrac{2x^2}{x^4 - 1} - \dfrac{1}{x^2 - 1} + \dfrac{1}{x^2 + 1}$

$\dfrac{2}{x^2 + 1}$

94. $\dfrac{x^2 - 12}{x^4 - 16} + \dfrac{1}{x^2 - 4} - \dfrac{1}{x^2 + 4}$

$\dfrac{1}{x^2 + 4}$

SUPPLEMENTAL EXERCISES 4.2

Simplify.

95. $\dfrac{(x + 1)^2}{1 - 2x} \cdot \dfrac{2x - 1}{x + 1}$ $-x - 1$

96. $\dfrac{2y - 3}{a - 6} \cdot \dfrac{(a - 6)^2}{3 - 2y}$ $-a + 6$

97. $\left(\dfrac{3a}{b}\right)^3 \div \left(\dfrac{a}{2b}\right)^2$ $\dfrac{108a}{b}$

98. $\left(\dfrac{2m}{3}\right)^2 \div \left(\dfrac{m^2}{6} + \dfrac{m}{2}\right)$ $\dfrac{8m}{3(m + 3)}$

99. $\left(\dfrac{y - 2}{x^2}\right)^3 \cdot \left(\dfrac{x}{2 - y}\right)^2$ $\dfrac{y - 2}{x^4}$

100. $\dfrac{b + 3}{b - 1} \div \dfrac{b + 3}{b - 2} \cdot \dfrac{b - 1}{b + 4}$ $\dfrac{b - 2}{b + 4}$

101. $\left(\dfrac{y+1}{y-1}\right)^2 - 1$ $\dfrac{4y}{(y-1)^2}$

102. $1 - \left(\dfrac{x-2}{x+2}\right)^2$ $\dfrac{8x}{(x+2)^2}$

103. $\left(\dfrac{1}{3} - \dfrac{2}{a}\right) \div \left(\dfrac{3}{a} - 2 + \dfrac{a}{4}\right)$ $\dfrac{4}{3(a-2)}$

104. $\left(\dfrac{b}{6} - \dfrac{6}{b}\right) \div \left(\dfrac{6}{b} - 4 + \dfrac{b}{2}\right)$ $\dfrac{b+6}{3(b-2)}$

105. $\dfrac{3x^2 + 6x}{4x^2 - 16} \cdot \dfrac{2x+8}{x^2+2x} \div \dfrac{3x-9}{5x-20}$ $\dfrac{5(x+4)(x-4)}{2(x+2)(x-2)(x-3)}$

106. $\dfrac{5y^2 - 20}{3y^2 - 12y} \cdot \dfrac{9y^3 + 6y^2}{2y^2 - 4y} \div \dfrac{y^3 + 2y^2}{2y^2 - 8y}$ $\dfrac{5(3y+2)}{y}$

107. $\dfrac{a^2 + a - 6}{4 + 11a - 3a^2} \cdot \dfrac{15a^2 - a - 2}{4a^2 + 7a - 2} \div \dfrac{6a^2 - 7a - 3}{4 - 17a + 4a^2}$ $-\dfrac{(a+3)(a-2)(5a-2)}{(3a+1)(a+2)(2a-3)}$

108. $\dfrac{25x - x^3}{x^4 - 1} \cdot \dfrac{3 - x - 4x^2}{2x^2 + 7x - 15} \div \dfrac{4x^3 - 23x^2 + 15x}{3 - 5x + 2x^2}$ $\dfrac{1}{x^2 + 1}$

109. $\left(\dfrac{x+1}{2x-1} - \dfrac{x-1}{2x+1}\right) \cdot \left(\dfrac{2x-1}{x} - \dfrac{2x-1}{x^2}\right)$ $\dfrac{6(x-1)}{x(2x+1)}$

110. $\left(\dfrac{y-2}{3y+1} - \dfrac{y+2}{3y-1}\right) \cdot \left(\dfrac{3y+1}{y} - \dfrac{3y-1}{y^2}\right)$ $-\dfrac{14(3y^2 - 2y + 1)}{y(3y+1)(3y-1)}$

Rewrite the expression as the sum of two fractions in simplest form.

111. $\dfrac{3x + 6y}{xy}$ $\dfrac{3}{y} + \dfrac{6}{x}$

112. $\dfrac{5a + 8b}{ab}$ $\dfrac{5}{b} + \dfrac{8}{a}$

113. $\dfrac{4a^2 + 3ab}{a^2 b^2}$ $\dfrac{4}{b^2} + \dfrac{3}{ab}$

114. $\dfrac{3m^2 n + 2mn^2}{12m^3 n^3}$ $\dfrac{1}{4mn^2} + \dfrac{1}{6m^2 n}$

Solve.

115. Use the expressions $\dfrac{1}{3}$ and $\dfrac{1}{5}$ to show that $\dfrac{1}{x} + \dfrac{1}{y} \neq \dfrac{1}{x+y}$.

$\dfrac{8}{15} \neq \dfrac{1}{8}$

116. Use the expressions $\dfrac{1}{3}$ and $\dfrac{1}{5}$ to show that $\dfrac{1}{x} - \dfrac{1}{y} \neq \dfrac{1}{x-y}$. $\dfrac{2}{15} \neq -\dfrac{1}{2}$

SECTION 4.3

Complex Fractions

1 Simplify complex fractions

A **complex fraction** is a fraction whose numerator or denominator contains one or more fractions. Examples of complex fractions are shown below.

$$\frac{5}{2 + \dfrac{1}{2}} \qquad \frac{5 + \dfrac{1}{y}}{5 - \dfrac{1}{y}} \qquad \frac{x + 4 + \dfrac{1}{x + 2}}{x - 2 + \dfrac{1}{x + 2}}$$

Simplify: $\dfrac{\dfrac{1}{x} + \dfrac{1}{y}}{\dfrac{1}{x} - \dfrac{1}{y}}$

Multiply the numerator and denominator of the complex fraction by the LCM of the denominators. The LCM of x and y is xy.

$$\frac{\dfrac{1}{x} + \dfrac{1}{y}}{\dfrac{1}{x} - \dfrac{1}{y}} = \frac{\dfrac{1}{x} + \dfrac{1}{y}}{\dfrac{1}{x} - \dfrac{1}{y}} \cdot \frac{xy}{xy} = \frac{\dfrac{1}{x} \cdot xy + \dfrac{1}{y} \cdot xy}{\dfrac{1}{x} \cdot xy - \dfrac{1}{y} \cdot xy} = \frac{y + x}{y - x}$$

Note that after multiplying the numerator and denominator of the complex fraction by the LCM of the denominators, no fraction remains in the numerator or denominator.

Example 1 Simplify.

A. $\dfrac{2 - \dfrac{11}{x} + \dfrac{15}{x^2}}{3 - \dfrac{5}{x} - \dfrac{12}{x^2}}$

B. $\dfrac{2x - 1 + \dfrac{7}{x + 4}}{3x - 8 + \dfrac{17}{x + 4}}$

Solution   **A.**
$$\dfrac{2 - \dfrac{11}{x} + \dfrac{15}{x^2}}{3 - \dfrac{5}{x} - \dfrac{12}{x^2}} = \dfrac{2 - \dfrac{11}{x} + \dfrac{15}{x^2}}{3 - \dfrac{5}{x} - \dfrac{12}{x^2}} \cdot \dfrac{x^2}{x^2} =$$

• The LCM is x^2.

$$\dfrac{2 \cdot x^2 - \dfrac{11}{x} \cdot x^2 + \dfrac{15}{x^2} \cdot x^2}{3 \cdot x^2 - \dfrac{5}{x} \cdot x^2 - \dfrac{12}{x^2} \cdot x^2} = \dfrac{2x^2 - 11x + 15}{3x^2 - 5x - 12} =$$

$$\dfrac{(2x - 5)(x - 3)}{(3x + 4)(x - 3)} = \dfrac{2x - 5}{3x + 4}$$

B.
$$\dfrac{2x - 1 + \dfrac{7}{x + 4}}{3x - 8 + \dfrac{17}{x + 4}} = \dfrac{2x - 1 + \dfrac{7}{x + 4}}{3x - 8 + \dfrac{17}{x + 4}} \cdot \dfrac{x + 4}{x + 4} =$$

• The LCM is $x + 4$.

$$\dfrac{(2x - 1)(x + 4) + \dfrac{7}{x + 4}(x + 4)}{(3x - 8)(x + 4) + \dfrac{17}{x + 4}(x + 4)} =$$

$$\dfrac{2x^2 + 7x - 4 + 7}{3x^2 + 4x - 32 + 17} = \dfrac{2x^2 + 7x + 3}{3x^2 + 4x - 15} =$$

$$\dfrac{(2x + 1)(x + 3)}{(3x - 5)(x + 3)} = \dfrac{2x + 1}{3x - 5}$$

Problem 1   Simplify.

A.
$$\dfrac{3 + \dfrac{16}{x} + \dfrac{16}{x^2}}{6 + \dfrac{5}{x} - \dfrac{4}{x^2}}$$

B.
$$\dfrac{2x + 5 + \dfrac{14}{x - 3}}{4x + 16 + \dfrac{49}{x - 3}}$$

Solution   See page A32.

A. $\dfrac{x + 4}{2x - 1}$ B. $\dfrac{x - 1}{2x + 1}$

Example 2   Simplify: $2 - \dfrac{2}{2 - \dfrac{2}{2 - x}}$

Solution $2 - \dfrac{2}{2 - \dfrac{2}{2 - x}} = 2 - \dfrac{2}{2 - \dfrac{2}{2 - x}} \cdot \dfrac{2 - x}{2 - x} =$

• Simplify the term that is a complex fraction. The LCM is $2 - x$.

$2 - \dfrac{2(2 - x)}{2(2 - x) - 2} = 2 - \dfrac{4 - 2x}{4 - 2x - 2} =$

$2 - \dfrac{4 - 2x}{2 - 2x} = 2 - \dfrac{2(2 - x)}{2(1 - x)} = 2 - \dfrac{2 - x}{1 - x} =$

$\dfrac{2(1 - x)}{1 - x} - \dfrac{2 - x}{1 - x} = \dfrac{2 - 2x - (2 - x)}{1 - x}$

• Subtract. The LCM of the denominators is $1 - x$.

$\dfrac{2 - 2x - 2 + x}{1 - x} = \dfrac{-x}{1 - x} = \dfrac{-x}{-(x - 1)} = \dfrac{x}{x - 1}$

Problem 2   Simplify: $3 + \dfrac{3}{3 + \dfrac{3}{y}}$

Solution   See page A32.
$\dfrac{4y + 3}{y + 1}$

EXERCISES 4.3

1   Simplify.

1. $\dfrac{1 + \dfrac{1}{x}}{1 - \dfrac{1}{x^2}}$ $\dfrac{x}{x - 1}$

2. $\dfrac{\dfrac{1}{y^2} - 1}{1 + \dfrac{1}{y}}$ $\dfrac{1 - y}{y}$

3. $\dfrac{a - 2}{\dfrac{4}{a} - a}$ $-\dfrac{a}{a + 2}$

4. $\dfrac{\dfrac{25}{a} - a}{5 + a}$ $\dfrac{5 - a}{a}$

5. $\dfrac{\dfrac{1}{a^2} - \dfrac{1}{a}}{\dfrac{1}{a^2} + \dfrac{1}{a}}$ $-\dfrac{a - 1}{a + 1}$

6. $\dfrac{\dfrac{1}{b} + \dfrac{1}{2}}{\dfrac{4}{b^2} - 1}$ $\dfrac{b}{2(2 - b)}$

7. $\dfrac{2 - \dfrac{4}{x + 2}}{5 - \dfrac{10}{x + 2}}$ $\dfrac{2}{5}$

8. $\dfrac{4 + \dfrac{12}{2x - 3}}{5 + \dfrac{15}{2x - 3}}$ $\dfrac{4}{5}$

9. $\dfrac{\dfrac{3}{2a - 3} + 2}{\dfrac{-6}{2a - 3} - 4}$ $-\dfrac{1}{2}$

10. $\dfrac{\dfrac{-5}{b - 5} - 3}{\dfrac{10}{b - 5} + 6}$ $-\dfrac{1}{2}$

11. $\dfrac{\dfrac{x}{x + 1} - \dfrac{1}{x}}{\dfrac{x}{x + 1} + \dfrac{1}{x}}$ $\dfrac{x^2 - x - 1}{x^2 + x + 1}$

12. $\dfrac{\dfrac{2a}{a - 1} - \dfrac{3}{a}}{\dfrac{1}{a - 1} + \dfrac{2}{a}}$ $\dfrac{2a^2 - 3a + 3}{3a - 2}$

13. $\dfrac{1 - \dfrac{1}{x} - \dfrac{6}{x^2}}{1 - \dfrac{4}{x} + \dfrac{3}{x^2}}$ $\dfrac{x + 2}{x - 1}$

14. $\dfrac{1 - \dfrac{3}{x} - \dfrac{10}{x^2}}{1 + \dfrac{11}{x} + \dfrac{18}{x^2}}$ $\dfrac{x - 5}{x + 9}$

15. $\dfrac{1 + \dfrac{1}{x} - \dfrac{12}{x^2}}{\dfrac{9}{x^2} + \dfrac{3}{x} - 2}$ $-\dfrac{x + 4}{2x + 3}$

16. $\dfrac{\dfrac{15}{x^2} - \dfrac{2}{x} - 1}{\dfrac{4}{x^2} - \dfrac{5}{x} + 4}$ $\dfrac{(5 + x)(3 - x)}{4x^2 - 5x + 4}$

17. $\dfrac{6 + \dfrac{2}{x} - \dfrac{20}{x^2}}{3 - \dfrac{17}{x} + \dfrac{20}{x^2}}$ $\dfrac{2x + 4}{x - 4}$

18. $\dfrac{3 + \dfrac{19}{x} + \dfrac{20}{x^2}}{6 + \dfrac{5}{x} - \dfrac{4}{x^2}}$ $\dfrac{x + 5}{2x - 1}$

19. $\dfrac{1 - \dfrac{2x}{3x - 4}}{x - \dfrac{32}{3x - 4}}$ $\dfrac{1}{3x + 8}$

20. $\dfrac{1 - \dfrac{12}{3x + 10}}{x - \dfrac{8}{3x + 10}}$ $\dfrac{1}{x + 4}$

21. $\dfrac{x - 1 + \dfrac{2}{x - 4}}{x + 3 + \dfrac{6}{x - 4}}$ $\dfrac{x - 2}{x + 2}$

22. $\dfrac{x - 5 - \dfrac{18}{x + 2}}{x + 7 + \dfrac{6}{x + 2}}$ $\dfrac{x - 7}{x + 5}$

23. $\dfrac{x - 4 + \dfrac{9}{2x + 3}}{x + 3 - \dfrac{5}{2x + 3}}$ $\dfrac{x - 3}{x + 4}$

24. $\dfrac{2x - 3 - \dfrac{10}{4x - 5}}{3x + 2 + \dfrac{11}{4x - 5}}$ $\dfrac{2x - 5}{3x - 1}$

25. $$\dfrac{3x - 2 - \dfrac{5}{2x - 1}}{x - 6 + \dfrac{9}{2x - 1}} \qquad \dfrac{3x + 1}{x - 5}$$

26. $$\dfrac{x + 4 - \dfrac{7}{2x - 5}}{2x + 7 - \dfrac{28}{2x - 5}} \qquad \dfrac{x - 3}{2x - 7}$$

27. $$\dfrac{\dfrac{1}{a} - \dfrac{3}{a - 2}}{\dfrac{2}{a} + \dfrac{5}{a - 2}} \qquad -\dfrac{2a + 2}{7a - 4}$$

28. $$\dfrac{\dfrac{2}{b} - \dfrac{5}{b + 3}}{\dfrac{3}{b} + \dfrac{3}{b + 3}} \qquad -\dfrac{b - 2}{2b + 3}$$

29. $$\dfrac{\dfrac{1}{y^2} - \dfrac{1}{xy} - \dfrac{2}{x^2}}{\dfrac{1}{y^2} - \dfrac{3}{xy} + \dfrac{2}{x^2}} \qquad \dfrac{x + y}{x - y}$$

30. $$\dfrac{\dfrac{2}{b^2} - \dfrac{5}{ab} - \dfrac{3}{a^2}}{\dfrac{2}{b^2} + \dfrac{7}{ab} + \dfrac{3}{a^2}} \qquad \dfrac{a - 3b}{a + 3b}$$

31. $$\dfrac{\dfrac{x - 1}{x + 1} - \dfrac{x + 1}{x - 1}}{\dfrac{x - 1}{x + 1} + \dfrac{x + 1}{x - 1}} \qquad -\dfrac{2x}{x^2 + 1}$$

32. $$\dfrac{\dfrac{y}{y + 2} - \dfrac{y}{y - 2}}{\dfrac{y}{y + 2} + \dfrac{y}{y - 2}} \qquad -\dfrac{2}{y}$$

33. $$4 - \dfrac{2}{2 - \dfrac{3}{x}} \qquad \dfrac{6x - 12}{2x - 3}$$

34. $$a + \dfrac{a}{a + \dfrac{1}{a}} \qquad \dfrac{a^3 + a^2 + a}{a^2 + 1}$$

35. $$a - \dfrac{a}{1 - \dfrac{a}{1 - a}} \qquad -\dfrac{a^2}{1 - 2a}$$

36. $$3 - \dfrac{3}{3 - \dfrac{3}{3 - x}} \qquad \dfrac{3 - 2x}{2 - x}$$

SUPPLEMENTAL EXERCISES 4.3

Find the reciprocal. Write your answer in simplest form.

37. $$\dfrac{x - \dfrac{1}{x}}{1 + \dfrac{1}{x}} \qquad \dfrac{1}{x - 1}$$

38. $$\dfrac{a - \dfrac{1}{a}}{\dfrac{1}{a} + 1} \qquad \dfrac{1}{a - 1}$$

39. $$1 - \dfrac{1}{1 - \dfrac{1}{b - 2}} \qquad 3 - b$$

40. $$2 - \dfrac{2}{2 - \dfrac{2}{c - 1}} \qquad \dfrac{c - 2}{c - 3}$$

Simplify. (These types of expressions are encountered in the study of calculus.)

41. $$\dfrac{\dfrac{1}{x + h} - \dfrac{1}{x}}{h} \qquad -\dfrac{1}{x(x + h)}$$

42. $$\dfrac{\dfrac{1}{(x + h)^2} - \dfrac{1}{x^2}}{h} \qquad -\dfrac{2x + h}{x^2(x + h)^2}$$

Solve. Write your answer in simplest form.

43. If $a = \dfrac{b^2 + 4b + 4}{b^2 - 4}$ and $b = \dfrac{1}{c}$, express a in terms of c.

$\dfrac{2c + 1}{1 - 2c}$

44. If $x = \dfrac{y^2 - 6y + 9}{y^2 - 9}$ and $y = \dfrac{1}{2z}$, express x in terms of z.

$-\dfrac{6z - 1}{6z + 1}$

45. If $z = \dfrac{3y^2 - 12y}{6y^3 - 96y}$ and $y = \dfrac{1}{x}$, express z in terms of x.

$\dfrac{x}{2 + 8x}$

46. If $c = \dfrac{2b^3 - 8b}{6b^2 + 12b}$ and $b = \dfrac{1}{2a}$, express c in terms of a. $-\dfrac{4a - 1}{6a}$

SECTION 4.4

Ratio and Proportion

1 Solve proportions

Quantities such as 3 feet, 5 liters, and 2 miles are number quantities written with units. In these examples, the units are feet, liters, and miles.

A **ratio** is the quotient of two quantities that have the same unit.

The weekly wages of a painter are $425. The painter spends $50 a week for food. The ratio of wages spent for food to the total weekly wages is written

$\dfrac{\$50}{\$425} = \dfrac{50}{425} = \dfrac{2}{17}$ A ratio is in simplest form when the two numbers do not have a common factor. The units are not written.

A **rate** is the quotient of two quantities that have different units.

A car travels 120 mi on 3 gal of gas. The miles-to-gallons rate is

$\dfrac{120 \text{ mi}}{3 \text{ gal}} = \dfrac{40 \text{ mi}}{1 \text{ gal}}$ A rate is in simplest form when the two numbers do not have a common factor. The units are written as part of the rate.

A **proportion** is an equation that states the equality of two ratios or rates.

For example, $\dfrac{90 \text{ km}}{4 \text{ L}} = \dfrac{45 \text{ km}}{2 \text{ L}}$ and $\dfrac{3}{4} = \dfrac{x + 2}{16}$ are proportions.

Recall that an equation containing fractions can be solved by multiplying each side of the equation by the LCM of the denominators.

Solve: $\dfrac{2}{7} = \dfrac{x}{5}$

$$\dfrac{2}{7} = \dfrac{x}{5}$$

Multiply each side of the proportion by the LCM of the denominators.

$$35 \cdot \dfrac{2}{7} = 35 \cdot \dfrac{x}{5}$$

Solve the equation.

$$10 = 7x$$

$$\dfrac{10}{7} = x$$

The solution is $\dfrac{10}{7}$.

Example 1   Solve.

A. $\dfrac{3}{12} = \dfrac{5}{x + 5}$ B. $\dfrac{x}{x + 2} = \dfrac{6}{x + 5}$

Solution   A. $\dfrac{3}{12} = \dfrac{5}{x + 5}$

$$12(x + 5)\dfrac{3}{12} = 12(x + 5)\dfrac{5}{x + 5}$$ • Multiply each side of the proportion by the LCM.

$$(x + 5)3 = (12)5$$
$$3x + 15 = 60$$
$$3x = 45$$
$$x = 15$$

The solution is 15.

B. $\dfrac{x}{x + 2} = \dfrac{6}{x + 5}$

$$(x + 2)(x + 5)\dfrac{x}{x + 2} = (x + 2)(x + 5)\dfrac{6}{x + 5}$$
$$(x + 5)x = (x + 2)6$$
$$x^2 + 5x = 6x + 12$$
$$x^2 - x - 12 = 0$$ • Write the quadratic equation in standard form.

$$(x - 4)(x + 3) = 0$$ • Factor the trinomial.

$$x - 4 = 0 \qquad x + 3 = 0$$ • Solve for x.
$$x = 4 \qquad\qquad x = -3$$

The solutions are 4 and -3.

Problem 1   Solve.

A. $\dfrac{5}{2x - 3} = \dfrac{-2}{x + 1}$ B. $\dfrac{x}{x + 3} = \dfrac{5}{x + 7}$

Solution   See page A33.

A. $\dfrac{1}{9}$ B. -5 and 3

2 Application problems

Example 2   A stock investment of 50 shares pays a dividend of $106. At this rate, how many additional shares are required to earn a dividend of $424?

Strategy   To find the additional number of shares that are required, write and solve a proportion using x to represent the additional number of shares. Then $50 + x$ is the total number of shares of stock.

Solution

$$\frac{106}{50} = \frac{424}{50 + x}$$

$$\frac{53}{25} = \frac{424}{50 + x}$$

$$25(50 + x)\frac{53}{25} = 25(50 + x)\frac{424}{50 + x}$$

$$(50 + x)53 = (25)424$$
$$2650 + 53x = 10,600$$
$$53x = 7950$$
$$x = 150$$

An additional 150 shares of stock are required.

Problem 2   Two pounds of cashews cost $3.10. At this rate, how much would 15 lb of cashews cost?

Solution   See page A33.
$23.25

EXERCISES 4.4

1   Solve.

1. $\dfrac{x}{30} = \dfrac{3}{10}$ 9

2. $\dfrac{5}{15} = \dfrac{x}{75}$ 25

3. $\dfrac{2}{x} = \dfrac{8}{30}$ $\dfrac{15}{2}$

4. $\dfrac{80}{16} = \dfrac{15}{x}$ 3

5. $\dfrac{x + 1}{10} = \dfrac{2}{5}$ 3

6. $\dfrac{5 - x}{10} = \dfrac{3}{2}$ -10

7. $\dfrac{4}{x + 2} = \dfrac{3}{4}$ $\dfrac{10}{3}$ **8.** $\dfrac{8}{3} = \dfrac{24}{x + 3}$ 6   **9.** $\dfrac{x}{4} = \dfrac{x - 2}{8}$ −2

10. $\dfrac{8}{x - 5} = \dfrac{3}{x}$ −3   **11.** $\dfrac{16}{2 - x} = \dfrac{4}{x}$ $\dfrac{2}{5}$ **12.** $\dfrac{6}{x - 5} = \dfrac{1}{x}$ −1

13. $\dfrac{8}{x - 2} = \dfrac{4}{x + 1}$ −4   **14.** $\dfrac{4}{x - 4} = \dfrac{2}{x - 2}$ 0   **15.** $\dfrac{x}{3} = \dfrac{x + 1}{7}$ $\dfrac{3}{4}$

16. $\dfrac{x - 2}{5} = \dfrac{1}{x + 2}$ 3 and −3   **17.** $\dfrac{x + 4}{10} = \dfrac{6}{x - 3}$ −9 and 8

18. $\dfrac{x}{x - 2} = \dfrac{3}{x - 4}$ 6 and 1   **19.** $\dfrac{x}{x - 1} = \dfrac{10}{x + 3}$ 5 and 2

20. $\dfrac{5}{x + 2} = \dfrac{x}{x + 8}$ 8 and −5   **21.** $\dfrac{6}{x + 5} = \dfrac{2x}{x + 1}$ 1 and −3

2   Solve.

22. The real estate tax for a house that cost $80,000 is $1200. At this rate, what is the value of a house for which the real estate tax is $1725?
$115,000

23. The license fee for a car that cost $6000 was $72. At the same rate, what is the license fee for a car that cost $8200?
$98.40

24. In a wildlife preserve, 50 ducks are captured, tagged, and then released. Later, 120 ducks are examined, and two of the 120 ducks are found to have tags. Estimate the number of ducks in the preserve.
3000 ducks

25. A pre-election survey showed that 5 out of every 8 voters would vote in an election. At this rate, how many people would be expected to vote in a city of 224,000?
140,000 people

26. A quality control inspector found 8 defective transistors in a shipment of 3000 transistors. At this rate, how many transistors would be defective in a shipment of 24,000 transistors?
64 transistors

27. A contractor estimated that 12 ft^2 of window space will be allowed for every 150 ft^2 of floor space. Using this estimate, how much window space will be allowed for 3300 ft^2 of floor space?
264 ft^2

28. The scale on an architectural drawing is $\dfrac{1}{4}$ in. represents one foot. Find the dimensions of a room that measures $4\dfrac{1}{2}$ in. by 6 in. on the drawing.
18 ft by 24 ft

29. One hundred forty-four ceramic tiles are required to tile a 25-ft^2 area. At this rate, how many tiles are required to tile 275 ft^2?
1584 ceramic tiles

30. One and one-half ounces of a medication are required for a 140-lb adult. At the same rate, how many additional ounces of medication are required for a 210-lb adult?
0.75 additional ounces

31. Eight ounces of an insecticide are mixed with 24 gal of water to make a spray for spraying an orange grove. How much additional insecticide is required to be mixed with 60 gal of water?
12 additional ounces

32. A stock investment of 200 shares pays a dividend of $240. At this rate, how many additional shares are required to earn a dividend of $600?
300 additional shares

33. An investment of $4000 earns $460 each year. At the same rate, how much additional money must be invested to earn $805 each year?
$3000

34. A farmer estimates that 8250 bushels of corn can be harvested from 150 acres of land. Using this estimate, how many additional acres are needed to harvest 11,000 bushels of corn?
50 additional acres

35. A caterer estimates that 2 gal of fruit punch will serve 25 people. How much additional punch is necessary to serve 60 people?
2.8 additional gallons

36. A computer printer can print a 600-word document in 25 s. At this rate, how many seconds are required to print a document that contains 780 words?
32.5 s

37. A 200-ft^2 stained glass window contains 1600 individual pieces of glass. At this rate, how many additional pieces of glass would be required for a stained glass window that measures 350 ft^2?
1200 additional pieces

38. It is estimated that a car traveling at 20 mph has an emergency stopping distance of 24 ft. Using this estimate, what is the additional number of feet that would be required for an emergency stop for a car traveling at 24 mph?
4.8 additional feet

39. A contractor estimated that 40 ft^3 of cement is required to make a 120-ft^2 concrete floor. Using this estimate, how many additional cubic feet of cement would be required to make a 150-ft^2 concrete floor?   10 ft^3

40. A magazine pays freelance writers by the word for published articles. If the magazine pays $400 for an article that is 1500 words long, how much will be paid for an article that is 3375 words long?
$900

41. A mechanic's pay for 35 hours of work is $420. At the same rate of pay, how much would the mechanic earn for 50 hours of work?
$600

42. An average jogger burns 100 Calories by jogging 1 mi. How many miles would a jogger need to run in order to burn 475 Calories?
4.75 mi

43. A major league baseball player opened the season by getting 26 base hits in 95 times at bat. At the same rate, how many hits would the player get in 475 times at bat?
130 hits

SUPPLEMENTAL EXERCISES 4.4

Solve.

44. On a map, two cities are $3\frac{3}{8}$ in. apart. If $\frac{3}{8}$ in. on the map represents 25 mi, find the number of miles in the distance between the two cities.
225 mi

45. On a map, two cities are $2\frac{5}{8}$ in. apart. If $\frac{3}{4}$ in. on the map represents 50 mi, find the number of miles in the distance between the two cities.
175 mi

46. An orange paint is created by mixing 5 parts of yellow with every 7 parts of red. How many gallons of yellow paint are needed to make 60 gal of this color?
25 gal

47. A soft drink is made by mixing 3 parts syrup with every 4 parts carbonated water. How many milliliters of syrup are in 350 ml of soft drink?
150 ml

48. A basketball player has made 6 out of every 7 foul shots attempted. If 35 foul shots were missed in the player's career, how many foul shots were made in the player's career?
210 foul shots

49. An advertisement claims that 4 out of every 5 dentists surveyed recommend a particular brand of toothpaste. If 24 of the dentists polled did not recommend this brand of toothpaste, how many dentists were included in the survey?   120 dentists

50. Three people put their money together to buy lottery tickets. The first person put in $20, the second person put in $25, and the third person put in $30. One of their tickets was a winning ticket. If they won $4.5 million, what was the first person's share of the winnings?
$1,200,000

51. Three people pooled their money to buy lottery tickets. The first person contributed $35, the second person contributed $40, and the third person contributed $45. If one of their tickets was a winning ticket and they won $2.4 million, what was the third person's share of the winnings?
$900,000

52. If 5 machines can assemble 10 toy cars in 9 min, how many toy cars can 9 machines assemble in 15 min?
30 toy cars

53. If 6 machines can fill 12 boxes of cereal in 7 min, how many boxes of cereal can be filled by 14 machines in 12 min?   48 boxes

SECTION 4.5
Rational Equations and Inequalities

1 Solve fractional equations

To solve an equation containing fractions, **clear denominators** by multiplying each side of the equation by the LCM of the denominators. Then solve for the variable.

Solve: $\dfrac{3x}{x-5} = 5 - \dfrac{5}{x-5}$

$$\frac{3x}{x-5} = 5 - \frac{5}{x-5}$$

Multiply each side of the equation by the LCM of the denominators.

$$(x-5)\left(\frac{3x}{x-5}\right) = (x-5)\left(5 - \frac{5}{x-5}\right)$$

$$3x = (x-5)5 - (x-5)\left(\frac{5}{x-5}\right)$$

Simplify.

$$3x = 5x - 25 - 5$$
$$3x = 5x - 30$$

Solve the equation for x.

$$-2x = -30$$
$$x = 15$$

15 checks as a solution.
The solution is 15.

Occasionally, a value of the variable that appears to be a solution will make one of the denominators zero. In this case, the equation has no solution for that value of the variable.

Solve: $\dfrac{3x}{x-3} = 2 + \dfrac{9}{x-3}$

$$\dfrac{3x}{x-3} = 2 + \dfrac{9}{x-3}$$

Multiply each side of the equation by the LCM of the denominators.

$$(x-3)\left(\dfrac{3x}{x-3}\right) = (x-3)\left(2 + \dfrac{9}{x-3}\right)$$

Use the Distributive Property.

$$3x = (x-3)2 + (x-3)\left(\dfrac{9}{x-3}\right)$$

$$3x = 2x - 6 + 9$$
$$3x = 2x + 3$$
$$x = 3$$

3 does not check as a solution.
The equation has no solution.

Example 1   Solve: $\dfrac{2x}{x-4} = \dfrac{2}{x-4} + 3$

Solution

$$\dfrac{2x}{x-4} = \dfrac{2}{x-4} + 3$$

$$(x-4)\left(\dfrac{2x}{x-4}\right) = (x-4)\left(\dfrac{2}{x-4} + 3\right)$$

• Multiply each side of the equation by the LCM.

$$2x = (x-4)\left(\dfrac{2}{x-4}\right) + (x-4)3$$

• Use the Distributive Property. (Do this step mentally.)

$$2x = 2 + 3x - 12$$
$$2x = 3x - 10$$
$$-x = -10$$
$$x = 10$$

The solution is 10.

Problem 1   Solve: $\dfrac{4x+1}{2x-1} = 5 + \dfrac{3}{2x-1}$

Solution   See page A34.
No solution

2 ## Solve fractional inequalities

Recall that a graphical method was used to solve quadratic inequalities. The same method is used to solve fractional inequalities.

For example, to solve the fractional inequality $\dfrac{2x - 5}{x - 4} \le 1$,

rewrite the inequality so that 0 appears on the right side of the inequality.
Then simplify.

$$\dfrac{2x - 5}{x - 4} \le 1$$

$$\dfrac{2x - 5}{x - 4} - 1 \le 0$$

$$\dfrac{2x - 5}{x - 4} - \dfrac{x - 4}{x - 4} \le 0$$

$$\dfrac{x - 1}{x - 4} \le 0$$

On a number line, identify for each factor of the numerator and each factor of the denominator the regions where the factor is positive and where the factor is negative.

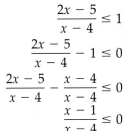

The region where the quotient of the two factors is negative is between 1 and 4.

Write the solution set. $\{x \,|\, 1 \le x < 4\}$

Note that 1 is part of the solution set, but 4 is not part of the solution set since the denominator of the rational expression is zero when $x = 4$.

Example 2   Solve and graph the solution set of $\dfrac{x + 4}{x - 3} \ge 0$.

Solution $\dfrac{x + 4}{x - 3} \ge 0$.

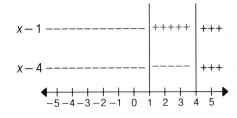

$\{x \,|\, x > 3 \text{ or } x \le -4\}$

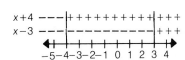

Problem 2   Solve and graph the solution set of $\dfrac{x}{x - 2} \le 0$.

Solution   See page A34.
$\{x \,|\, 0 \le x < 2\}$

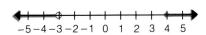

3 Literal equations

A **literal equation** is an equation that contains more than one variable. Examples of literal equations are shown below.

$$3x - 2y = 4$$

$$v^2 = v_0^2 + 2as$$

Formulas are used to express a relationship among physical quantities. A **formula** is a literal equation that states rules about measurement. Examples of formulas are shown below.

$$s = vt + 16t^2 \qquad \text{(Physics)}$$

$$c^2 = a^2 + b^2 \qquad \text{(Geometry)}$$

$$I = P(1 + r)^t \qquad \text{(Business)}$$

The Addition and Multiplication Properties of Equations can be used to solve a literal equation for one of the variables. The goal is to rewrite the equation so that the variable being solved for is alone on one side of the equation, and all the other numbers and variables are on the other side.

Solve $C = \dfrac{5}{9}(F - 32)$ for F.

$$C = \frac{5}{9}(F - 32)$$

Use the Distributive Property to remove parentheses.

$$C = \frac{5}{9}F - \frac{160}{9}$$

Multiply each side of the equation by the LCM of the denominators.

$$9C = 5F - 160$$

Add the additive inverse of the constant term to each side of the equation.

$$9C + 160 = 5F$$

Multiply each side of the equation by the reciprocal of the coefficient 5.

$$\frac{9C + 160}{5} = F$$

Example 3   A. Solve $A = P + Prt$ for P.

B. Solve $\dfrac{S}{S - C} = R$ for C.

Solution   A.
$$A = P + Prt$$
$$A = (1 + rt)P$$ • Factor P from $P + Prt$.

$$\frac{1}{1 + rt} \cdot A = \frac{1}{1 + rt}(1 + rt)P$$ • Multiply each side of the equation by the reciprocal of $1 + rt$.

$$\frac{A}{1 + rt} = P$$

B.
$$\frac{S}{S - C} = R$$

$$(S - C)\frac{S}{S - C} = (S - C)R$$ • Multiply each side of the equation by the LCM. The LCM is $S - C$.

$$S = SR - CR$$
$$CR + S = SR$$ • Add CR to each side of the equation.
$$CR = SR - S$$ • Add $-S$ to each side of the equation.

$$C = \frac{SR - S}{R}$$ • Multiply each side of the equation by the reciprocal of R.

Problem 3   A. Solve $\dfrac{1}{R_1} + \dfrac{1}{R_2} = \dfrac{1}{R}$ for R.   B. Solve $\dfrac{r}{r + 1} = t$ for r.

Solution   See page A34.

A. $R = \dfrac{R_1 R_2}{R_1 + R_2}$ B. $r = \dfrac{t}{1 - t}$

4   Work problems

If a mason can build a retaining wall in 12 h, then in 1 h the mason can build $\dfrac{1}{12}$ of the wall. The mason's rate of work is $\dfrac{1}{12}$ of the wall each hour. The **rate of work** is that part of a task that is completed in one unit of time. If an apprentice can build the wall in x hours, the rate of work for the apprentice is $\dfrac{1}{x}$ of the wall each hour.

In solving a work problem, the goal is to determine the time it takes to complete a task. The basic equation that is used to solve work problems is

Rate of work × Time worked = Part of task completed

For example, if a pipe can fill a tank in 5 h, then in 2 h the pipe will fill $\dfrac{1}{5} \times 2 = \dfrac{2}{5}$ of the tank. In t hours, the pipe will fill $\dfrac{1}{5} \times t = \dfrac{t}{5}$ of the tank.

A mason can build a wall in 10 h. An apprentice can build a wall in 15 h. How long will it take to build a wall when they work together?

STRATEGY *for solving a work problem*

▶ For each person or machine, write a numerical or variable expression for the rate of work, the time worked, and the part of the task completed. The results can be recorded in a table.

Unknown time to build the wall working together: t

	Rate of work	·	Time worked	=	Part of task completed
Mason	$\dfrac{1}{10}$	·	t	=	$\dfrac{t}{10}$
Apprentice	$\dfrac{1}{15}$	·	t	=	$\dfrac{t}{15}$

▶ Determine how the parts of the task completed are related. Use the fact that the sum of the parts of the task completed must equal 1, the complete task.

The sum of the part of the task completed by the mason and the part of the task completed by the apprentice is 1.

$$\frac{t}{10} + \frac{t}{15} = 1$$
$$30\left(\frac{t}{10} + \frac{t}{15}\right) = 30(1)$$
$$3t + 2t = 30$$
$$5t = 30$$
$$t = 6$$

Working together, they will build the wall in 6 h.

Example 4 An electrician requires 12 h to wire a house. The electrician's apprentice can wire a house in 16 h. After working alone on one job for 4 h, the electrician quits, and the apprentice completes the task. How long does it take the apprentice to finish wiring the house?

Strategy ▶ Time required for the apprentice to finish wiring the house: t

	Rate	Time	Part
Electrician	$\dfrac{1}{12}$	4	$\dfrac{4}{12}$
Apprentice	$\dfrac{1}{16}$	t	$\dfrac{t}{16}$

▶ The sum of the part of the task completed by the electrician and the part of the task completed by the apprentice is 1.

Solution

$$\frac{4}{12} + \frac{t}{16} = 1$$

$$\frac{1}{3} + \frac{t}{16} = 1$$

$$48\left(\frac{1}{3} + \frac{t}{16}\right) = 48(1)$$

$$16 + 3t = 48$$

$$3t = 32$$

$$t = \frac{32}{3}$$

It will take the apprentice $10\frac{2}{3}$ h to finish wiring the house.

Problem 4 Two water pipes can fill a tank with water in 6 h. The larger pipe working alone can fill the tank in 9 h. How long will it take the smaller pipe working alone to fill the tank?

Solution See page A35.
18 h

5 Uniform motion problems

A car that travels constantly in a straight line at 55 mph is in uniform motion. **Uniform motion** means that the speed of an object does not change.

The basic equation used to solve uniform motion problems is

Distance = Rate × Time

An alternate form of this equation can be written by solving the equation for time.

$$\frac{\text{Distance}}{\text{Rate}} = \text{Time}$$

This form of the equation is useful when the total time of travel for two objects or the time of travel between two points is known.

A motorist drove 150 mi on country roads before driving 50 mi of mountain roads. The rate of speed on the country roads was three times the rate on the mountain roads. The time spent traveling the 200 mi was 5 h. Find the rate of the motorist on the country roads.

STRATEGY *for solving a uniform motion problem*

▶ For each object, write a numerical or variable expression for the distance, rate, and time. The results can be recorded in a table.

The unknown rate of speed on the mountain roads: r
Rate of speed on the country roads: $3r$

	Distance	÷	Rate	=	Time
Country roads	150	÷	$3r$	=	$\dfrac{150}{3r}$
Mountain roads	50	÷	r	=	$\dfrac{50}{r}$

▶ Determine how the times traveled by each object are related. For example, it may be known that the times are equal, or the total time may be known.

The total time of the trip is 5 h.

$$\frac{150}{3r} + \frac{50}{r} = 5$$
$$3r\left(\frac{150}{3r} + \frac{50}{r}\right) = 3r(5)$$
$$150 + 150 = 15r$$
$$300 = 15r$$
$$20 = r$$

The rate of speed on the country roads was $3r$. Replace r with 20 and evaluate.

$3r = 3(20) = 60$

The rate of speed on the country roads was 60 mph.

Example 5   A marketing executive traveled 810 mi on a corporate jet in the same amount of time that it took to travel an additional 162 mi by helicopter. The rate of the jet was 360 mph faster than the rate of the helicopter. Find the rate of the jet.

Strategy   ▶ Rate of the helicopter: r
Rate of the jet: $r + 360$

	Distance	Rate	Time
Jet	810	$r + 360$	$\dfrac{810}{r + 360}$
Helicopter	162	r	$\dfrac{162}{r}$

▶ The time traveled by jet is equal to the time traveled by helicopter.

Solution
$$\frac{810}{r + 360} = \frac{162}{r}$$

$$r(r + 360)\left(\frac{810}{r + 360}\right) = r(r + 360)\left(\frac{162}{r}\right)$$

$$810r = (r + 360)162$$
$$810r = 162r + 58{,}320$$
$$648r = 58{,}320$$
$$r = 90$$

- The rate of the helicopter is 90 mph.

$$r + 360 = 90 + 360 = 450$$

- Substitute the value of r into the variable expression for the rate of the jet.

The rate of the jet was 450 mph.

Problem 5   A plane can fly at a rate of 150 mph in calm air. Traveling with the wind, the plane flew 700 mi in the same amount of time that it flew 500 mi against the wind. Find the rate of the wind.

Solution   See page A35.

25 mph

EXERCISES 4.5

1   Solve.

1. $\dfrac{x}{2} + \dfrac{5}{6} = \dfrac{x}{3}$ -5

2. $\dfrac{x}{5} - \dfrac{2}{9} = \dfrac{x}{15}$ $\dfrac{5}{3}$

3. $\dfrac{8}{2x - 1} = 2$ $\dfrac{5}{2}$

4. $3 = \dfrac{18}{3x - 4}$ $\dfrac{10}{3}$

5. $1 - \dfrac{3}{y} = 4$ -1

6. $7 + \dfrac{6}{y} = 5$ -3

7. $\dfrac{3}{x - 2} = \dfrac{4}{x}$ 8

8. $\dfrac{5}{x} = \dfrac{2}{x + 3}$ -5

9. $\dfrac{3}{x - 4} + 2 = \dfrac{5}{x - 4}$ 5

10. $\dfrac{5}{y + 3} - 2 = \dfrac{7}{y + 3}$ -4

11. $5 + \dfrac{8}{a - 2} = \dfrac{4a}{a - 2}$ no solution

12. $\dfrac{-4}{a - 4} = 3 - \dfrac{a}{a - 4}$ no solution

13. $-\dfrac{5}{x + 7} + 1 = \dfrac{4}{x + 7}$ 2

14. $5 - \dfrac{2}{2x - 5} = \dfrac{3}{2x - 5}$ 3

15. $\dfrac{2}{4y^2 - 9} + \dfrac{1}{2y - 3} = \dfrac{3}{2y + 3}$ $\dfrac{7}{2}$

16. $\dfrac{5}{x - 2} - \dfrac{2}{x + 2} = \dfrac{3}{x^2 - 4}$ $-\dfrac{11}{3}$

17. $\dfrac{5}{x^2 - 7x + 12} = \dfrac{2}{x - 3} + \dfrac{5}{x - 4}$ no solution

18. $\dfrac{9}{x^2 + 7x + 10} = \dfrac{5}{x + 2} - \dfrac{3}{x + 5}$ no solution

2   Solve and graph the solution set.

19. $\dfrac{x - 4}{x + 2} > 0$

20. $\dfrac{x + 2}{x - 3} > 0$

21. $\dfrac{x - 3}{x + 1} \le 0$

22. $\dfrac{x - 1}{x} > 0$

23. $\dfrac{(x - 1)(x + 2)}{x - 3} \le 0$

24. $\dfrac{(x + 3)(x - 1)}{x - 2} \ge 0$

Solve.

25. $\dfrac{3x}{x - 2} > 1$ $\{x \mid x < -1 \text{ or } x > 2\}$

26. $\dfrac{2x}{x + 1} < 1$ $\{x \mid -1 < x < 1\}$

27. $\dfrac{2}{x + 1} \geq 2$ $\{x \mid -1 < x \leq 0\}$

28. $\dfrac{3}{x - 1} < 2$ $\left\{x \mid x < 1 \text{ or } x > \dfrac{5}{2}\right\}$

29. $\dfrac{x}{(x - 1)(x + 2)} \geq 0$
$\{x \mid -2 < x \leq 0 \text{ or } x > 1\}$

30. $\dfrac{x - 2}{(x + 1)(x - 1)} \leq 0$
$\{x \mid x < -1 \text{ or } 1 < x \leq 2\}$

31. $\dfrac{1}{x} < 2$ $\left\{x \mid x < 0 \text{ or } x > \dfrac{1}{2}\right\}$

32. $\dfrac{x}{2x - 1} \geq 1$ $\left\{x \mid \dfrac{1}{2} < x \leq 1\right\}$

3   Solve the formula for the variable given.

33. $P = 2L + 2w; \; w$ (Geometry)
$\dfrac{P - 2L}{2} = w$

34. $F = \dfrac{9}{5}C + 32; \; C$ (Temperature conversion)
$C = \dfrac{5}{9}(F - 32)$

35. $S = C - rC; \; C$ (Business)
$\dfrac{S}{1 - r} = C$

36. $A = P + Prt; \; t$ (Business)
$\dfrac{A - P}{Pr} - t$

37. $PV = nRT; \; R$ (Chemistry)
$\dfrac{PV}{nT} = R$

38. $A = \dfrac{1}{2}bh; \; h$ (Geometry)
$h = \dfrac{2A}{b}$

39. $F = \dfrac{Gm_1 m_2}{r^2}; \; m_2$ (Physics)
$\dfrac{Fr^2}{Gm_1} = m_2$

40. $\dfrac{P_1 V_1}{T_1} = \dfrac{P_2 V_2}{T_2}; \; P_2$ (Chemistry)
$P_2 = \dfrac{P_1 V_1 T_2}{T_1 V_2}$

41. $I = \dfrac{E}{R + r}; \; R$ (Physics)
$\dfrac{E - Ir}{I} = R$

42. $S = V_0 t - 16t^2; \; V_0$ (Physics)
$\dfrac{S + 16t^2}{t} = V_0$

43. $A = \dfrac{1}{2}h(b_1 + b_2); \; b_2$ (Geometry)
$\dfrac{2A - hb_1}{h} = b_2$

44. $V = \dfrac{1}{3}\pi r^2 h; \; h$ (Geometry)
$\dfrac{3V}{\pi r^2} = h$

45. $\dfrac{1}{R} = \dfrac{1}{R_1} + \dfrac{1}{R_2}$; R_2 (Physics)

$R_2 = \dfrac{RR_1}{R_1 - R}$

46. $\dfrac{1}{f} = \dfrac{1}{a} + \dfrac{1}{b}$; b (Physics)

$b = \dfrac{ab}{a - f}$

47. $a_n = a_1 + (n - 1)d$; d (Mathematics)

$\dfrac{a_n - a_1}{n - 1} = d$

48. $P = \dfrac{R - C}{n}$; R (Business)

$P_n + C = R$

49. $S = 2wh + 2wL + 2Lh$; h (Geometry)

$\dfrac{S - 2wL}{2w + 2L} = h$

50. $S = 2\pi r^2 + 2\pi rh$; h (Geometry)

$\dfrac{S - 2\pi r^2}{2\pi r} = h$

Solve the equation for the variable x.

51. $ax + by + c = 0$

$\dfrac{-by - c}{a} = x$

52. $x = ax + b$

$x = \dfrac{b}{1 - a}$

53. $ax + b = cx + d$

$\dfrac{d - b}{a - c} = x$

54. $y - y_1 = m(x - x_1)$

$\dfrac{y - y_1 + mx_1}{m} = x$

55. $\dfrac{a}{x} = \dfrac{b}{c}$

$x = \dfrac{ac}{b}$

56. $\dfrac{1}{x} + \dfrac{1}{a} = b$

$x = \dfrac{a}{ab - 1}$

57. $\dfrac{1}{a} + \dfrac{1}{b} = \dfrac{1}{x}$ $\dfrac{ab}{a + b} = x$

58. $a(a - x) = b(b - x)$ $a + b = x$

4   Solve.

59. One printer can print the paychecks for the employees of a company in 80 min. A second printer can print the checks in 120 min. How long would it take to print the checks with both printers operating?
48 min

60. A mason can construct a retaining wall in 16 h. The mason's apprentice can do the job in 24 h. How long would it take to construct the wall when they work together?
9.6 h

61. One solar heating panel can raise the temperature of water 1° in 40 min. A second solar heating panel can raise the temperature 1° in 60 min. How long would it take to raise the temperature of the water 1° when both solar panels are operating?
24 min

62. One member of a gardening team can landscape a new lawn in 36 h. The other member of the team can do the job in 45 h. How long would it take to landscape the lawn when both gardeners work together?
20 h

63. One member of a telephone crew can wire new telephone lines in 6 h, while it would take 9 h for the other member of the crew to do the job. How long would it take to wire new telephone lines when both members of the crew are working together?
3.6 h

64. A new printer can print checks four times faster than an old printer. The old printer can print the checks in 40 min. How long would it take to print the checks when both printers are operating?
8 min

65. A new machine can package transistors three times faster than an older machine. Working together, the machines can package the transistors in 6 h. How long would it take the new machine working alone to package the transistors?
8 h

66. An experienced electrician can wire a room twice as fast as an apprentice electrician. Working together, the electricians can wire a room in 3 h. How long would it take the apprentice working alone to wire a room?
9 h

67. One member of a gardening team can mow and clean up a lawn in 6 h. With both members of the team working, the job can be done in 4 h. How long would it take the second member of the team, working alone, to do the job?
12 h

68. A student can type a 42-page term paper in 3 h. With a friend's assistance, the paper can be typed in 2 h. How long would it take the friend working alone to type the paper?
6 h

69. The larger of two printers being used to print the payroll for a major corporation requires 30 min to print the payroll. After both printers have been operating for 10 min, the larger printer malfunctions. The smaller printer requires 40 more minutes to complete the payroll. How long would it take the smaller printer working alone to print the payroll?
75 min

70. An experienced bricklayer can work twice as fast as an apprentice bricklayer. After working together on a job for 6 h, the experienced bricklayer quit. The apprentice required 10 more hours to finish the job. How long would it take the experienced bricklayer working alone to do the job?
14 h

71. A roofer requires 8 h to shingle a roof. After the roofer and an apprentice work on a roof for 2 h, the roofer moves on to another job. The apprentice requires 10 more hours to finish the job. How long would it take the apprentice working alone to do the job?
16 h

72. A welder requires 20 h to do a job. After the welder and an apprentice work on a job for 8 h, the welder quits. The apprentice finishes the job in 13 h. How long would it take the apprentice working alone to do the job?
35 h

73. Three computers can print out a task in 20 min, 30 min, and 60 min, respectively. How long would it take to complete the task when all three computers are working?
10 min

74. Three machines are filling soda bottles. The machines can fill the daily quota of soda bottles in 10 h, 12 h, and 15 h, respectively. How long would it take to fill the daily quota of soda bottles when all three machines are working?
4 h

75. With both hot and cold water running, a bathtub can be filled in 8 min. The drain will empty the tub in 10 min. A child turns both faucets on and leaves the drain open. How long will it be before the bathtub starts to overflow?
40 min

76. The inlet pipe can fill a water tank in 45 min. The outlet pipe can empty the tank in 30 min. How long would it take to empty a full tank when both pipes are open?
90 min

77. An oil tank has two inlet pipes and one outlet pipe. One inlet pipe can fill the tank in 12 h, and the other inlet pipe can fill the tank in 20 h. The outlet pipe can empty the tank in 10 h. How long would it take to fill the tank when all three pipes are open?
30 h

78. Water from a tank is being used for irrigation at the same time as the tank is being filled. The two inlet pipes can fill the tank in 5 h and 10 h, respectively. The outlet pipe can empty the tank in 20 h. How long will it take to fill the tank when all three pipes are open?   4 h

5   Solve.

79. An express bus travels 310 mi in the same amount of time that a car travels 275 mi. The rate of the car is 7 mph less than the rate of the bus. Find the rate of the bus.
62 mph

80. A commercial jet travels 1800 mi in the same amount of time that a corporate jet travels 1350 mi. The rate of the commercial jet is 150 mph faster than the rate of the corporate jet. Find the rate of each jet.
commercial jet: 600 mph; corporate jet: 450 mph

81. A passenger train travels 240 mi in the same amount of time that a freight train travels 168 mi. The rate of the passenger train is 18 mph faster than the rate of the freight train. Find the rate of each train.
passenger train: 60 mph; freight train: 42 mph

82. The rate of a bicyclist is 7 mph faster than the rate of a long-distance runner. The bicyclist travels 30 mi in the same amount of time that the runner travels 16 mi. Find the rate of the runner.
8 mph

83. A cabin cruiser travels 20 mi in the same amount of time that a power boat travels 45 mi. The rate of the cabin cruiser is 10 mph less than the rate of the power boat. Find the rate of the cabin cruiser.
8 mph

84. A motorcycle travels 196 mi in the same amount of time that a car travels 161 mi. The rate of the motorcycle is 10 mph faster than the rate of the car. Find the rate of the motorcycle.
56 mph

85. A cyclist rode 40 mi before having a flat tire and then walking 5 mi to a service station. The cycling rate was four times faster than the walking rate. The time spent cycling and walking was 5 h. Find the rate at which the cyclist was riding.
12 mph

86. A sales executive traveled 55 mi by car and then an additional 990 mi by plane. The rate of the plane was six times faster than the rate of the car. The total time of the trip was 4 h. Find the rate of the plane.
330 mph

87. A motorist drove 72 mi before running out of gas and then walking 4 mi to a gas station. The driving rate of the motorist was twelve times the walking rate. The time spent driving and walking was 2.5 h. Find the rate at which the motorist walks.
4 mph

88. An insurance representative traveled 1200 mi by commercial jet and then an additional 120 mi by helicopter. The rate of the jet was four times the rate of the helicopter. The entire trip took 3.5 h. Find the rate of the jet.
480 mph

89. An express train and a car leave a town at 3 P.M. and head for a town 280 mi away. The rate of the express train is twice the rate of the car. The train arrives 4 h ahead of the car. Find the rate of the train.
70 mph

90. A cyclist and a jogger start from a town at the same time and head for a destination 30 mi away. The rate of the cyclist is twice the rate of the jogger. The cyclist arrives 3 h ahead of the jogger. Find the rate of the cyclist.
10 mph

91. A single-engine plane and a commercial jet leave an airport at 10 A.M. and head for an airport 720 mi away. The rate of the jet is four times the rate of the single-engine plane. The single-engine plane arrives 4.5 h after the jet. Find the rate of each plane.
single-engine plane: 120 mph; commercial jet: 480 mph

92. A single-engine plane and a car start from a town at 6 A.M. and head for a town 450 mi away. The rate of the plane is three times the rate of the car. The plane arrives 6 h ahead of the car. Find the rate of the plane.
150 mph

93. A cabin cruiser and a sailboat start from the same dock and sail to a resort island that is 48 mi away. The average rate of the cruiser is two times the average rate of the sailboat. The cruiser arrives 3 h ahead of the sailboat. Find the rate of each.
sailboat: 8 mph; cabin cuiser: 16 mph

94. A train and a plane leave from a town at 2 P.M. for a destination 900 mi away. The rate of the plane is six times the rate of the train. The plane arrives 15 h before the train. Find the rate of each.
train: 50 mph; plane: 300 mph

95. A motorboat can travel at 15 mph in still water. Traveling with the current of a river, the boat can travel 40 mi in the same amount of time that it takes to go 20 mi against the current. Find the rate of the current.
5 mph

96. An account executive traveled 150 mi by car in the same amount of time that it takes a plane to travel 600 mi. The rate of the plane was 150 mph faster than the rate of the car. Find the rate of the plane.
200 mph

97. A plane can fly at a rate of 175 mph in calm air. Traveling with the wind, the plane flew 615 mi in the same amount of time that it flew 435 mi against the wind. Find the rate of the wind.
30 mph

98. A jet can fly at a rate of 525 mph in calm air. Traveling with the wind, the plane flew 1815 mi in the same amount of time that it flew 1335 mi against the wind. Find the rate of the wind.
80 mph

99. A tour boat used for river excursions can travel 7 mph in calm water. The amount of time it takes to travel 20 mi with the current is the same amount of time that it takes to travel 8 mi against the current. Find the rate of the current.
3 mph

100. A canoe can travel 8 mph in still water. Rowing with the current of a river, the canoe can travel 15 mi in the same amount of time that it takes to travel 9 mi against the current. Find the rate of the current.
2 mph

Use your calculator for the following exercises.

101. A twin-engine plane can travel 180 mph in calm air. Flying with the wind, the plane can travel 900 mi in the same amount of time that it takes to fly 500 mi against the wind. Find the rate of the wind. Round to the nearest hundredth.
51.43 mph

102. A jet can travel 550 mph in calm air. Flying with the wind, the jet can travel 3059 mi in the same amount of time that it takes to fly 2450 mi against the wind. Find the rate of the wind. Round to the nearest hundredth.
60.80 mph

SUPPLEMENTAL EXERCISES 4.5

Solve the proportion for x.

103. $\dfrac{x - y}{y} = \dfrac{x + 5}{2y}$ $5 + 2y$

104. $\dfrac{x - 2}{y} = \dfrac{x + 2}{5y}$ 3

105. $\dfrac{x}{x + y} = \dfrac{2x}{4y}$ $0, y$

106. $\dfrac{2x}{x - 2y} = \dfrac{x}{2y}$ $0, 6y$

107. $\dfrac{x - y}{2x} = \dfrac{x - 3y}{5y}$ $\dfrac{y}{2}, 5y$

108. $\dfrac{x - y}{x} = \dfrac{2x}{9y}$ $3y, \dfrac{3y}{2}$

Solve.

109. The sum of the multiplicative inverses of two consecutive integers is $\frac{17}{72}$. Find the two integers.

8 and 9

110. The sum of the multiplicative inverses of two consecutive integers is $-\frac{13}{42}$. Find the two integers.

−7 and −6

111. The denominator of a fraction is 4 more than the numerator. If the numerator and denominator of the fraction are increased by 3, the new fraction is $\frac{5}{6}$. Find the original fraction. $\frac{17}{21}$

112. The numerator of a fraction is 2 less than the denominator. If the numerator and denominator of the fraction are increased by 5, the new fraction is $\frac{9}{11}$. Find the original fraction. $\frac{4}{6}$

113. The sum of an integer and twice the reciprocal of the integer is greater than 3. What integers satisfy the inequality? Use set builder notation to write your answer.

$\{x \mid x > 2, x \text{ is an integer}\}$

114. The sum of the first and second of three consecutive integers plus three times the reciprocal of the third consecutive integer is greater than 2. What integers satisfy the inequality? Use set builder notation to write your answer. $\left\{x \mid x > -\frac{1}{2}, x \text{ is an integer}\right\}$

115. One pipe can fill a tank in 3 h, a second pipe can fill the tank in 4 h, and a third pipe can fill the tank in 6 h. How long will it take to fill the tank with all three pipes operating? $1\frac{1}{3}$ h

116. One printer can print a company's paychecks in 24 min, a second printer can print the checks in 16 min, and a third printer can complete the job in 12 min. How long would it take to print the checks with all three printers operating? $5\frac{1}{3}$ min

117. By increasing your speed by 10 mph, you can drive the 200-mi trip to your hometown in 40 min less time than it usually takes you to drive the trip. How fast do you usually drive?

50 mph

118. Because of weather conditions, a bus driver reduced the usual speed along a 165-mi bus route by 5 mph. The bus arrived only 15 min later than its usual arrival time. How fast does the bus usually travel?
60 mph

CALCULATORS AND COMPUTERS

Rational Expressions

The program RATIONAL EXPRESSIONS on the Student Disk will give you additional practice in multiplying and dividing rational expressions. There are three levels of difficulty, with the first level the easiest of the problems and the third level the most difficult. You may choose the level you wish to practice.

After you choose a level of difficulty, a problem will be displayed on the screen. Using paper and pencil, simplify the expression. When you are ready, press the RETURN key. The correct solution will be displayed.

After each problem, you may continue the same level of problems, return to the menu to change the level, or quit the program.

CHAPTER SUMMARY

Key Words

A fraction in which the numerator and the denominator are polynomials is a **rational expression.**

A rational expression is in **simplest form** when the numerator and denominator have no common factors.

Synthetic division is a shorter method of dividing a polynomial by a binomial of the form $x - a$. (This method uses only the coefficients of the variable terms.)

The **least common multiple** (LCM) of two or more polynomials is the simplest polynomial that contains the factors of each polynomial.

The **reciprocal** of a rational expression is the rational expression with the numerator and denominator interchanged.

A **complex fraction** is a fraction whose numerator or denominator contains one or more fractions.

A **ratio** is the quotient of two quantities that have the same unit.

A **rate** is the quotient of two quantities that have different units.

A **proportion** is an equation that states the equality of two ratios or rates.

A **literal equation** is an equation that contains more than one variable.

A **formula** is a literal equation that states rules about measurements.

Essential Rules

To add fractions

$$\frac{a}{c} + \frac{b}{c} = \frac{a + b}{c}$$

To multiply fractions

$$\frac{a}{b} \cdot \frac{c}{d} = \frac{ac}{bd}$$

To divide fractions

$$\frac{a}{b} \div \frac{c}{d} = \frac{a}{b} \cdot \frac{d}{c}$$

Equation for Work Problems

$$\text{Rate of work} \times \text{Time worked} = \text{Part of task completed}$$

Uniform Motion Equation

$$\text{Distance} = \text{Rate} \times \text{Time}$$

CHAPTER REVIEW

1. Simplify: $\dfrac{4x^2 + 4x - 3}{4 - 16x^2}$ $-\dfrac{2x + 3}{4(2x + 1)}$

2. Simplify: $\dfrac{2x^2 - 32}{2x^2 + x - 6} \div \dfrac{2x^2 - 4x}{2x^3 - x^2 - 3x}$ $\dfrac{(x + 4)(x - 4)(x + 1)}{(x + 2)(x - 2)}$

3. Simplify: $\dfrac{1 + \dfrac{1}{x} - \dfrac{6}{x^2}}{1 - \dfrac{5}{x} + \dfrac{6}{x^2}}$ $\dfrac{x + 3}{x - 3}$

4. Solve: $3 - \dfrac{4}{x - 4} = \dfrac{-x}{x - 4}$ no solution

5. Solve $\dfrac{1}{f} = \dfrac{1}{a} + \dfrac{1}{b}$ for a.

$a = \dfrac{bf}{b - f}$

6. Simplify: $\dfrac{8x^2 - 12x}{4x^3 - 2x^2 - 6x}$

$\dfrac{2}{x + 1}$

7. Simplify: $\dfrac{6x^2 - 9x}{12x^3 - 30x^2} \cdot \dfrac{4x^2 - 4x - 15}{4x^2 - 9}$

$\dfrac{1}{2x}$

8. Solve: $\dfrac{2}{x - 3} = \dfrac{5}{x}$

5

9. Solve: $\dfrac{5}{x^2 - 25} + \dfrac{1}{x - 5} = \dfrac{3}{x + 5}$

$\dfrac{25}{2}$

10. Simplify: $\dfrac{3x^4 + 2x^3 - 4x^2 + 6x - 8}{x + 2}$

$3x^3 - 4x^2 + 4x - 2 - \dfrac{4}{x + 2}$

11. Simplify: $\dfrac{x - 3}{x^2 + 3x + 2} - \dfrac{2 - x}{x^2 - x - 2}$

$\dfrac{2x - 1}{(x + 2)(x + 1)}$

12. Solve $a - s = sr$ for s.

$\dfrac{a}{r + 1} = s$

13. Solve: $\dfrac{(x + 2)(x - 1)}{x + 3} \le 0$

$\{x \,|\, x < -3 \text{ or } -2 \le x \le 1\}$

14. Simplify: $(4x^3 - x - 10) \div (2x - 3)$

$2x^2 + 3x + 4 + \dfrac{2}{2x - 3}$

15. Simplify: $\dfrac{\dfrac{2}{x} - \dfrac{8}{x - 3}}{\dfrac{1}{x} + \dfrac{5}{x - 3}}$

$-\dfrac{2(x + 1)}{2x - 1}$

16. Simplify: $\dfrac{x^4 + x^3y - 6x^2y^2}{x^3 - 2x^2y}$

$x + 3y$

17. Simplify: $\dfrac{16x^2 - 9y^2}{16x^2y - 12xy^2} \div \dfrac{4x^2 - xy - 3y^2}{12x^2y^2}$

$\dfrac{3xy}{x - y}$

18. Solve: $\dfrac{2}{x - 3} = \dfrac{5}{2x - 3}$

9

19. Solve: $\dfrac{6x}{2x - 3} - \dfrac{1}{x - 3} = 7$

$\dfrac{5}{2}$

20. Simplify: $\dfrac{4a^2 + 8a}{a^3 + a^2 - 2a}$

$\dfrac{4}{a - 1}$

21. Simplify: $\dfrac{x^2 + 2x + 1}{8x^2 + 8x} \cdot \dfrac{4x^3 - 4x^2}{x^2 - 1}$

$\dfrac{x}{2}$

22. Solve $P = \dfrac{R - C}{n}$ for C. $C = R - Pn$

23. Simplify: $\dfrac{4x^3 + 2x^2 - 10x + 1}{x - 2}$

$4x^2 + 10x + 10 + \dfrac{21}{x - 2}$

24. Simplify: $\dfrac{5x}{3x^2 - x - 2} - \dfrac{2x}{x^2 - 1}$

$-\dfrac{x}{(3x + 2)(x + 1)}$

25. Solve: $\dfrac{3}{x^2 - 36} = \dfrac{2}{x - 6} - \dfrac{5}{x + 6}$

13

26. Simplify: $\dfrac{x - 4 + \dfrac{5}{x + 2}}{x + 2 - \dfrac{1}{x + 2}}$ $\dfrac{x - 3}{x + 3}$

27. Solve: $\dfrac{x - 3}{(x + 2)(x - 2)} > 0$

$\{x \mid x > 3 \text{ or } -2 < x < 2\}$

28. Simplify: $(3x^3 - 13x^2 + 10) \div (3x - 4)$

$x^2 - 3x - 4 - \dfrac{6}{3x - 4}$

29. A real estate tax for a house that cost \$75,000 is \$900. At this rate, what is the real estate tax on a house that cost \$125,000?

\$1500

30. A company has two printers for use in printing the employee pay-checks. The smaller printer can print the checks in 40 min. When both printers are operating, the checks can be printed in 15 min. How long would it take the larger printer working alone to print the checks?

24 min

31. A passenger train travels 125 mi in the same amount of time that a freight train travels 95 mi. The rate of the passenger train is 12 mph faster than the rate of the freight train. Find the rate of each train.

passenger train: 50 mph; freight train: 38 mph

CUMULATIVE REVIEW

1. Simplify: $8 - 4[-3 - (-2)]^2 \div 5$

$\dfrac{36}{5}$

2. Evaluate $3a^2 - (b^2 - c)^2$ when $a = 2$, $b = -3$, and $c = 1$.

-52

3. Graph the solution set of $\{x \mid x \ge -3\} \cap \{x \mid x < 4\}$.

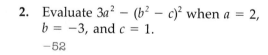

4. Solve: $-\dfrac{2}{3}y = -\dfrac{4}{9}$

$\dfrac{2}{3}$

5. Solve: $\dfrac{2x - 3}{6} - \dfrac{x}{9} = \dfrac{x - 4}{3}$ $\dfrac{15}{2}$

6. Solve: $4x - 9 \le 5x + 7$ $\{x \mid x \ge -16\}$

7. Solve: $4x - 3 < 5$ and $3 - 2x < 7$
$\{x \mid -2 < x < 2\}$

8. Solve $5 - |x - 4| = 2$
7 and 1

9. Solve: $|5x + 6| < 4$ $\left\{x \mid -2 < x < -\dfrac{2}{5}\right\}$

10. Simplify: $(2a^2 - 3a + 1)(-2a^2)$
$-4a^4 + 6a^3 - 2a^2$

11. Factor: $2x^{2n} + 3x^n - 2$
$(2x^n - 1)(x^n + 2)$

12. Factor: $a^3b^3 - 8$
$(ab - 2)(a^2b^2 + 2ab + 4)$

13. Factor: $ax^2 + bx^2 - ay^2 - by^2$
$(a + b)(x + y)(x - y)$

14. Graph the solution set of
$2x^2 + 5x - 3 \geq 0$.

15. Solve: $3x^2 + 5x = 2$
$\dfrac{1}{3}$ and -2

16. Divide: $\dfrac{3x^3 - 4x^2 - 9x + 1}{x - 3}$

$3x^2 + 5x + 6 + \dfrac{19}{x - 3}$

17. Simplify: $\dfrac{25a^2 - 4b^2}{15a^2b - 6ab^2} \div \dfrac{5a^2 - 3ab - 2b^2}{15a^2b^2}$
$\dfrac{5ab}{a - b}$

18. Simplify: $\dfrac{5x}{2x^2 + 5x + 3} - \dfrac{2x}{x^2 - 1}$

$\dfrac{x(x - 11)}{(2x + 3)(x + 1)(x - 1)}$

19. Simplify: $\dfrac{1 - \dfrac{1}{a^2}}{\dfrac{1}{a} + \dfrac{1}{a^2}}$ $a - 1$

20. Solve: $\dfrac{2}{x - 8} = \dfrac{5}{x + 7}$ 18

21. Solve: $\dfrac{2x}{x - 2} = 3 + \dfrac{4}{x - 2}$
no solution

22. Solve: $\dfrac{6}{x^2 - 25} = \dfrac{1}{x - 5} - \dfrac{3}{x + 5}$
7

23. Solve $I = \dfrac{E}{R + r}$ for r. $\dfrac{E - IR}{I} = r$

24. Solve: $\dfrac{3x + 1}{x + 4} \leq 1$ $\left\{x \mid -4 < x \leq \dfrac{3}{2}\right\}$

25. The sum of two integers is 15. Five times the smaller integer is five more than twice the larger integer. Find the integers.
5 and 10

26. Find three consecutive integers whose sum is between 45 and 60.
15, 16, and 17; 16, 17, and 18; 17, 18, and 19; or 18, 19, and 20

27. How many pounds of almonds that cost $5.40 per pound must be mixed with 50 lb of peanuts that cost $2.60 per pound to make a mixture that costs $4.00 per pound?   50 lb

28. A pre-election survey showed that three out of five voters would vote in an election. At this rate, how many people would be expected to vote in a city of 125,000?
75,000 people

29. A new computer can work six times faster than an older computer. Working together, the computers can complete a job in 12 min. How long would it take the new computer working alone to do the job?
14 min

30. A plane can fly at a rate of 300 mph in calm air. Traveling with the wind, the plane flew 900 mi in the same amount of time that it flew 600 mi against the wind. Find the rate of the wind.
60 mph

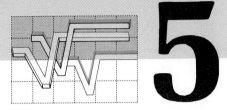

5

Exponents and Radicals

OBJECTIVES

- Simplify expressions with integer exponents
- Scientific notation
- Application problems
- Simplify expressions with rational exponents
- Write exponential expressions as radical expressions and radical expressions as exponential expressions
- Simplify expressions of the form $\sqrt[n]{a^n}$
- Simplify radical expressions
- Add and subtract radical expressions
- Multiply radical expressions
- Divide radical expressions
- Simplify complex numbers
- Add and subtract complex numbers
- Multiply complex numbers
- Divide complex numbers
- Solve equations containing one or more radical expressions
- Application problems

Golden Ratio

The golden rectangle fascinated the early Greeks and appeared in much of their architecture. They considered this particular rectangle the most pleasing to the eye, and consequently, when used in the design of a building, would make the structure pleasant to see.

The golden rectangle is constructed from a square by drawing a line from the midpoint of the base of the square to the opposite vertex. Now extend the base of the square, starting from the midpoint, the length of the line drawn. The resulting rectangle is called the golden rectangle.

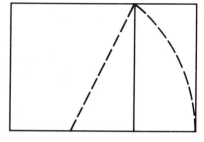

The Parthenon in Athens, Greece, is the classic example of the use of the golden rectangle in Greek architecture. A picture of the Parthenon is shown below.

Alinari/Art Resource

SECTION 5.1

Rules of Exponents

1 Simplify expressions with integer exponents

Recall that the Rule for Dividing Exponential Expressions states that to divide exponential expressions with the same base, subtract the exponents.

$$\frac{a^m}{a^n} = a^{m-n}$$

If $m = n$, and m is substituted for n in the equation, then the result is a^0.

$$\frac{a^m}{a^m} = a^{m-m} = a^0$$

When the same expression is simplified by using the Division Property of One, the result is 1.

$$\frac{a^m}{a^m} = 1$$

To ensure that the two answers are equal, a number or variable to the zero power must be equal to 1.

$$a^0 = 1, a \neq 0$$

The Zero Property of Exponents

For any number a, $a \neq 0$, $a^0 = 1$.

$12^0 = 1$

$(2x)^0 = 1, x \neq 0$

Zero raised to the zero power is not defined.

0^0 is undefined.

Negative integers, as well as positive, can be used as exponents. The rules that have been developed for exponential expressions can be extended to include negative integers.

$$x^m \cdot x^n = x^{m+n} \qquad\qquad a^{-4}b^6 = a^{-4+6} = a^2$$

$$(x^m)^n = x^{m \cdot n} \qquad\qquad (a^2)^{-4} = a^{2(-4)} = a^{-8}$$

$$(x^m \cdot y^n)^p = x^{mp} \cdot y^{np} \qquad\qquad (a^3b^{-2})^{-3} = a^{3(-3)}b^{-2(-3)} = a^{-9}b^6$$

Using integer exponents, the Rule for Dividing Exponential Expressions can be stated as a single rule.

$$\frac{x^m}{x^n} = x^{m-n} \qquad\qquad \frac{a^4}{a^7} = a^{4-7} = a^{-3}$$

The meaning of an expression containing a negative exponent can be shown by using the Rule for Multiplying Exponential Expressions and the Zero Property of Exponents.

The exponential expression at the right is simplified by using the Rule for Multiplying Exponential Expressions. The result is a^0, or 1.

$$a^4 \cdot a^{-4} = a^{4+(-4)} = a^0 = 1$$

Recall that a number times its reciprocal is equal to 1.

$$a^4 \cdot \frac{1}{a^4} = 1$$

Since $a^4 \cdot a^{-4} = 1$ and $a^4 \cdot \frac{1}{a^4} = 1$, $a^4 \cdot a^{-4} - a^4 \cdot \frac{1}{a^4}$.

Thus $a^{-4} = \frac{1}{a^4}$.

The Rule of Negative Exponents

If n is a positive integer and $a \neq 0$, then $a^{-n} = \dfrac{1}{a^n}$ and $\dfrac{1}{a^{-n}} = a^n$.

An exponential expression is considered to be in simplest form when each base occurs only once, no power is raised to a power, and all negative exponents have been rewritten as positive exponents.

Each of the expressions at the right has been rewritten in simplest form.

$$\frac{1}{2^{-3}} = 2^3 = 8 \qquad x^2 y^{-2} = \frac{x^2}{y^2}$$

$$x^{-4} = \frac{1}{x^4} \qquad \frac{1}{4^{-2}x^3} = \frac{4^2}{x^3} = \frac{16}{x^3}$$

Example 1   Simplify.

A. $\dfrac{x^2 y^{-4}}{x^{-5} y^{-2}}$ B. $\left(\dfrac{3a^2 b^{-2} c^{-1}}{27a^{-1}b^2 c^{-4}}\right)^{-2}$ C. $x^{-1}y + xy^{-1}$

Solution   A. $\dfrac{x^2 y^{-4}}{x^{-5} y^{-2}} = x^{2-(-5)}y^{-4-(-2)}$

   $= x^7 y^{-2}$

   $= \dfrac{x^7}{y^2}$

• Use the Rule for Dividing Exponential Expressions. (Do this step mentally.)

• Use the Rule of Negative Exponents to rewrite the expression without negative exponents.

B. $\left(\dfrac{3a^2b^{-2}c^{-1}}{27a^{-1}b^2c^{-4}}\right)^{-2} = \left(\dfrac{a^3b^{-4}c^3}{9}\right)^{-2}$

• Simplify inside the parentheses by using the Rule for Dividing Exponential Expressions.

$= \dfrac{a^{-6}b^8c^{-6}}{9^{-2}}$

• Use the Rule for Simplifying Powers of Quotients.

$= \dfrac{9^2b^8}{a^6c^6}$

• Use the Rule of Negative Exponents to rewrite the expression without negative exponents.

$= \dfrac{81b^8}{a^6c^6}$

• Simplify.

C. $x^{-1}y + xy^{-1} = \dfrac{y}{x} + \dfrac{x}{y}$

• Use the Rule of Negative Exponents.

$= \dfrac{y^2}{xy} + \dfrac{x^2}{xy}$

• Write each fraction in terms of the LCM of the denominators.

$= \dfrac{y^2 + x^2}{xy}$

• Add the two fractions.

Problem 1   Simplify.

A. $\dfrac{a^{-1}b^4}{a^{-2}b^{-2}}$ B. $\left(\dfrac{2^{-1}x^2y^{-3}}{4x^{-2}y^{-5}}\right)^{-2}$ C. $[(a^{-1}b)^{-2}]^3$

Solution   See page A37.   A. ab^6 B. $\dfrac{64}{x^8y^4}$ C. $\dfrac{a^6}{b^6}$

2 Scientific notation

Very large and very small numbers are encountered in the fields of science and engineering. For example, the mass of the electron is 0.00000000000000000000000000009 g. Numbers such as this one are difficult to read and write, so a more convenient system for writing them has been developed. It is called **scientific notation.**

To express a number in scientific notation, write the number as the product of a number between 1 and 10 and a power of 10. The form for scientific notation is $a \times 10^n$, where a is a number between 1 and 10.

For numbers greater than 10, move the decimal point to the right of the first digit. The exponent n is positive and equal to the number of places the decimal point has been moved.

$965,000 = 9.65 \times 10^5$

$3,600,000 = 3.6 \times 10^6$

$92,000,000,000 = 9.2 \times 10^{10}$

For numbers less than 1, move the decimal point to the right of the first nonzero digit. The exponent n is negative. The absolute value of the exponent is equal to the number of places the decimal point has been moved.

$$0.0002 = 2 \times 10^{-4}$$
$$0.0000000974 = 9.74 \times 10^{-8}$$
$$0.000000000086 = 8.6 \times 10^{-11}$$

Example 2 Write 0.000041 in scientific notation.

Solution $0.000041 = 4.1 \times 10^{-5}$ • The decimal point must be moved 5 digits to the right. The exponent is negative.

Problem 2 Write 942,000,000 in scientific notation.

Solution See page A37.
9.42×10^8

Converting a number written in scientific notation to decimal notation requires moving the decimal point.

When the exponent is positive, move the decimal point to the right the same number of places as the exponent.

$$1.32 \times 10^4 = 13,200$$
$$1.4 \times 10^8 = 140,000,000$$

When the exponent is negative, move the decimal point to the left the same number of places as the absolute value of the exponent.

$$1.32 \times 10^{-2} = 0.0132$$
$$1.4 \times 10^{-4} = 0.00014$$

Example 3 Write 3.3×10^7 in decimal notation.

Solution $3.3 \times 10^7 = 33,000,000$ • Move the decimal point 7 places to the right.

Problem 3 Write 2.7×10^{-5} in decimal notation.

Solution See page A37.
0.000027

Numerical calculations involving numbers that have more digits than the hand-held calculator is able to handle can be performed using scientific notation.

For example, to simplify $\dfrac{220,000 \times 0.000000092}{0.0000011}$,

write the numbers in scientific notation.

$$\frac{220,000 \times 0.000000092}{0.0000011} = \frac{2.2 \times 10^5 \times 9.2 \times 10^{-8}}{1.1 \times 10^{-6}}$$

Simplify.

$$= \frac{(2.2)(9.2) \times 10^{5+(-8)-(-6)}}{1.1}$$
$$= 18.4 \times 10^3 = 18,400$$

Example 4   Simplify: $\dfrac{2{,}400{,}000{,}000 \times 0.0000063}{0.00009 \times 480}$

Solution $\dfrac{2{,}400{,}000{,}000 \times 0.0000063}{0.00009 \times 480} = \dfrac{2.4 \times 10^9 \times 6.3 \times 10^{-6}}{9 \times 10^{-5} \times 4.8 \times 10^2}$

$\dfrac{(2.4)(6.3) \times 10^{9+(-6)-(-5)-2}}{(9)(4.8)} = 0.35 \times 10^6 = 350{,}000$

Problem 4   Simplify: $\dfrac{5{,}600{,}000 \times 0.000000081}{900 \times 0.000000028}$

Solution   See page A37.
18,000

3   Application problems

Example 5   How many miles does light travel in one day? The speed of light is 186,000 mi/s. Write the answer in scientific notation.

Strategy   To find the distance traveled:
▶ Write the speed of light in scientific notation.
▶ Write the number of seconds in one day in scientific notation.
▶ Use the equation $d = rt$, where r is the speed of light, and t is the number of seconds in one day.

Solution $186{,}000 = 1.86 \times 10^5$

$24 \cdot 60 \cdot 60 = 86{,}400 = 8.64 \times 10^4$

$d = rt$
$d = (1.86 \times 10^5)(8.64 \times 10^4)$
$d = 1.86 \times 8.64 \times 10^9$
$d = 16.0704 \times 10^9$
$d = 1.60704 \times 10^{10}$

Light travels 1.60704×10^{10} mi in one day.

Problem 5   A computer can do an arithmetic operation in 1×10^{-7} s. How many arithmetic operations can the computer perform in one minute? Write the answer in scientific notation.

Solution   See page A38.
6×10^8 operations

EXERCISES 5.1

1   Simplify.

1. 2^{-3} $\dfrac{1}{8}$

2. $\dfrac{1}{3^{-5}}$ 243

3. $\dfrac{1}{x^{-4}}$ x^4

4. $\dfrac{1}{y^{-3}}$ y^3

5. $\dfrac{2x^{-2}}{y^4}$ $\dfrac{2}{x^2y^4}$

6. $\dfrac{a^3}{4b^{-2}}$ $\dfrac{a^3b^2}{4}$

7. $x^{-3}y$ $\dfrac{y}{x^3}$

8. xy^{-4} $\dfrac{x}{y^4}$

9. $-5x^0$ -5

10. $\dfrac{1}{2x^0}$ $\dfrac{1}{2}$

11. $\dfrac{(2x)^0}{-2^3}$ $-\dfrac{1}{8}$

12. $\dfrac{-3^{-2}}{(2y)^0}$ $-\dfrac{1}{9}$

13. $(3x^{-2})^2$ $\dfrac{9}{x^4}$

14. $(5x^2)^{-3}$ $\dfrac{1}{125x^6}$

15. $x^{-4}x^4$ 1

16. $x^{-3}x^{-5}$ $\dfrac{1}{x^8}$

17. $a^{-2} \cdot a^4$ a^2

18. $a^{-5} \cdot a^7$ a^2

19. $\dfrac{x^{-3}}{x^2}$ $\dfrac{1}{x^5}$

20. $\dfrac{x^4}{x^{-5}}$ x^9

21. $\dfrac{y^{-7}}{y^{-8}}$ y

22. $\dfrac{y^{-2}}{y^6}$ $\dfrac{1}{y^8}$

23. $(x^2y^{-4})^2$ $\dfrac{x^4}{y^8}$

24. $(x^3y^5)^{-2}$ $\dfrac{1}{x^6y^{10}}$

25. $(a^{-2}b^{-3})^{-2}$ a^4b^6

26. $(a^{-1}b^{-2})^{-4}$ a^4b^8

27. $\dfrac{2^2x^{-3}}{2^{-3}x^{-2}}$ $\dfrac{32}{x}$

28. $\dfrac{3^{-2}a^2}{3a^{-3}}$ $\dfrac{a^5}{27}$

29. $\dfrac{x^{-2}y^{-11}}{xy^{-2}}$ $\dfrac{1}{x^3y^9}$

30. $\dfrac{x^4y^3}{x^{-1}y^{-2}}$ x^5y^5

31. $\dfrac{a^{-1}b^{-3}}{a^4b^{-5}}$ $\dfrac{b^2}{a^5}$

32. $\dfrac{a^6b^{-4}}{a^{-2}b^5}$ $\dfrac{a^8}{b^9}$

33. $(3x^{-2}y^4)^2$ $\dfrac{9y^8}{x^4}$

34. $(2^{-1}x^2y^{-2})^3$ $\dfrac{x^6}{8y^6}$

35. $(2a^2)(a^{-2})^2$ $\dfrac{2}{a^2}$

36. $(3b^{-2})(3b)^3$ $81b$

37. $(2a^{-1})^{-2}(2a^{-1})^4$ $\dfrac{4}{a^2}$

38. $(3a)^{-3}(9a^{-1})^{-2}$ $\dfrac{1}{2187a}$

39. $(x^{-2}y)^2(xy)^{-2}$ $\dfrac{1}{x^6}$

40. $(x^{-1}y^2)^{-3}(x^2y^{-4})^{-3}$ $\dfrac{y^6}{x^3}$

41. $(a^{-2}b^{-3})(a^{-1}b^3)^4$ $\dfrac{b^9}{a^6}$

42. $(a^2b^{-5})(a^2b^{-3})^{-3}$ $\dfrac{b^4}{a^4}$

43. $(2^{-1}x^2y^{-3})^2(2^{-2}x^{-3}y^4)$ $\dfrac{x}{16y^2}$

44. $(3^2x^2y^{-3})^{-2}(3^{-1}x^{-3}y^{-2})$ $\dfrac{y^4}{243x^7}$

45. $(x^{-1}y^{-2}z)^{-2}(x^2y^{-4}z)^2$ $\dfrac{x^6}{y^4}$

46. $(x^{-1}y^2z^{-4})^3(x^3y^{-3}z^2)^{-2}$ $\dfrac{y^{12}}{x^7z^{16}}$

47. $\dfrac{4x^2y^{-2}}{2^{-2}x^4y^{-4}}$ $\dfrac{16y^2}{x^2}$

48. $\dfrac{2^4a^{-1}b^{-2}}{2^{-3}a^{-2}b^2}$ $\dfrac{128a}{b^4}$

49. $\dfrac{6^2a^{-2}b^3}{3ab^4}$ $\dfrac{12}{a^3b}$

50. $\dfrac{5^2a^2b^{-3}}{10^{-1}a^{-2}b^3}$ $\dfrac{250a^4}{b^6}$

51. $\dfrac{3^{-3}x^3y^{-3}}{6^{-2}x^4y^{-6}}$ $\dfrac{4y^3}{3x}$

52. $\left(\dfrac{x^2y^{-1}}{xy}\right)^{-4}$ $\dfrac{y^8}{x^4}$

53. $\left(\dfrac{x^{-3}y^{-4}}{x^{-2}y}\right)^{-2}$ x^2y^{10}

54. $\left(\dfrac{a^{-2}b}{a^3b^{-4}}\right)^2$ $\dfrac{b^{10}}{a^{10}}$

55. $\left(\dfrac{a^3b^2}{a^{-5}b^{-2}}\right)^3$ $a^{24}b^{12}$

56. $\dfrac{(2a^{-3}b^{-2})^3}{(a^{-4}b^{-1})^{-2}}$ $\dfrac{8}{a^{17}b^8}$

57. $\dfrac{(3x^{-2}y)^{-2}}{(4xy^{-2})^{-1}}$ $\dfrac{4x^5}{9y^4}$

58. $\left(\dfrac{2x^2y}{x^{-3}y}\right)^{-1}\left(\dfrac{x^{-3}y^2}{4x^{-2}y}\right)^{-2}$ $\dfrac{8}{x^3y^2}$

59. $\left(\dfrac{4^{-2}xy^{-3}}{x^{-3}y}\right)^3\left(\dfrac{8^{-1}x^{-2}y}{x^4y^{-1}}\right)^{-2}$ $\dfrac{x^{24}}{64y^{16}}$

60. $\left(\dfrac{9ab^{-2}}{8a^{-2}b}\right)^{-2}\left(\dfrac{3a^{-2}b}{2a^2b^{-2}}\right)^3$ $\dfrac{8b^{15}}{3a^{18}}$

61. $\left(\dfrac{2ab^{-1}}{ab}\right)^{-1}\left(\dfrac{3a^{-2}b}{a^2b^2}\right)^{-2}$ $\dfrac{a^8b^4}{18}$

62. $\dfrac{x^{-1}+y^{-1}}{x^{-1}-y^{-1}}$ $\dfrac{y+x}{y-x}$

63. $\dfrac{x^{-1}+y}{x^{-1}-y}$ $\dfrac{1+xy}{1-xy}$

64. $(a+b)^{-1}$ $\dfrac{1}{a+b}$

65. $\dfrac{1}{(a-b)^{-1}}$ $a-b$

66. $a+a^{-1}b$ $\dfrac{a^2+b}{a}$

67. $x^{-1}+y^{-1}$ $\dfrac{y+x}{xy}$

68. $\dfrac{x^{-1}}{y^{-1}}+\dfrac{y}{x}$ $\dfrac{2y}{x}$

69. $x^{-1}y^{-1}+xy$ $\dfrac{1+x^2y^2}{xy}$

70. $\dfrac{x^{-1}}{y}+\dfrac{y^{-1}}{x}$ $\dfrac{2}{xy}$

71. $[(xy^{-2})^3]^{-2}$ $\dfrac{y^{12}}{x^6}$

72. $[(a^{-2}b)^2]^3$ $\dfrac{b^6}{a^{12}}$

73. $\left[\left(\dfrac{a^2}{b}\right)^{-1}\right]^2$ $\dfrac{b^2}{a^4}$

74. $\left[\left(\dfrac{x}{y^2}\right)^{-2}\right]^3$ $\dfrac{y^{12}}{x^6}$

75. $[(x^{-2}y^{-1})^2]^{-3}$ $x^{12}y^6$

76. $[(x^{-3}y^{-2})^{-1}]^4$ $x^{12}y^8$

2   Write in scientific notation.

77. 0.00000467 4.67×10^{-6}

78. 0.00000005 5×10^{-8}

79. 0.00000000017 1.7×10^{-10}

80. $4,300,000$ 4.3×10^6

81. $200,000,000,000$ 2×10^{11}

82. $9,800,000,000$ 9.8×10^9

Write in decimal notation.

83. 1.23×10^{-7} 0.000000123

84. 6.2×10^{-12} 0.0000000000062

85. 8.2×10^{15} $8,200,000,000,000,000$

86. 6.34×10^5 $634,000$

87. 3.9×10^{-2} 0.039

88. 4.35×10^9 $4,350,000,000$

Simplify.

89. $(3 \times 10^{-12})(5 \times 10^{16})$
150,000

90. $(8.9 \times 10^{-5})(3.2 \times 10^{-6})$
0.0000000002848

91. $(0.0000065)(3,200,000,000,000)$
20,800,000

92. $(480,000)(0.0000000096)$
0.004608

93. $\dfrac{9 \times 10^{-3}}{6 \times 10^{5}}$ 0.000000015

94. $\dfrac{2.7 \times 10^{4}}{3 \times 10^{-6}}$ 9,000,000,000

95. $\dfrac{0.0089}{500,000,000}$ 0.0000000000178

96. $\dfrac{4,800}{0.00000024}$ 20,000,000,000

97. $\dfrac{0.00056}{0.000000000004}$ 140,000,000

98. $\dfrac{0.000000346}{0.0000005}$ 0.692

Use your calculator for the following exercises.

99. $\dfrac{(3.2 \times 10^{-11})(2.9 \times 10^{15})}{8.1 \times 10^{-3}}$ 11,456,790

100. $\dfrac{(6.9 \times 10^{27})(8.2 \times 10^{-13})}{4.1 \times 10^{15}}$ 1.38

101. $\dfrac{(0.00000004)(84,000)}{(0.0003)(1,400,000)}$ 0.000008

102. $\dfrac{(720)(0.0000000039)}{(26,000,000,000)(0.018)}$ 0.000000000000006

3   Solve. Write the answer in scientific notation.

Use your calculator for the following exercises.

103. How many kilometers does light travel in one day? The speed of light is 300,000 km/s.
2.592×10^{10} km

104. How many meters does light travel in 8 h? The speed of light is 300,000,000 m/s.
8.64×10^{12} m

105. A computer can do an arithmetic operation in 3×10^{-6} s. How many arithmetic operations can the computer perform in one minute?
2×10^{7} operations

106. A computer can do an arithmetic operation in 9×10^{-7} s. How many arithmetic operations can the computer perform in one hour?
4×10^{9} operations

107. The national debt is 2×10^{12} dollars. How much would each American citizen have to pay in order to pay off the national debt? Use a figure of two hundred million for the number of citizens. 1×10^{4} dollars

108. A high speed centrifuge makes 9×10^7 revolutions each minute. Find the time in seconds for the centrifuge to make one revolution.
$6.\overline{6} \times 10^{-7}$ s

109. How long does it take light to travel to the earth from the sun? The sun is 9.3×10^7 mi from the earth, and light travels 1.86×10^5 mi/s.
5×10^2 s

110. The mass of the earth is 5.9×10^{27} g. The mass of the sun is 2×10^{33} g. How many times heavier is the sun than the earth?
3.38983×10^5 times heavier

111. The distance to the sun is 9.3×10^7 mi. A satellite leaves the earth traveling at a constant speed of 1×10^5 mph. How long does it take for the satellite to reach the sun?
9.3×10^2 h

112. The weight of 31 million orchid seeds is one ounce. Find the weight of one orchid seed.
3.225806×10^{-8} oz

113. One light year, an astronomical unit of distance, is the distance that light will travel in one year. Light travels 1.86×10^5 mi/s. Find the measure of one light year in miles. Use a 360-day year.
5.785344×10^{12} mi

114. The light from the star Alpha Centauri takes 4.3 years to reach the earth. Light travels at 1.86×10^5 mi/s. How far is Alpha Centauri from the earth? Use a 360-day year.
2.4876979×10^{13} mi

115. The radius of the earth is approximately 2.1×10^7 ft. Use the formula $SA = 4\pi r^2$ to find the surface area of the earth.
5.54×10^{15} ft^2

116. The radius of a cell is 1.3×10^{-5} cm. Use the formula $V = \frac{4}{3}\pi r^3$ to find the volume of the cell.
9.2×10^{-15} cm^3

117. A computer can perform 12 million computations in one second. How many computations can the computer perform in one hour?
4.32×10^{10} computations

118. One gram of hydrogen contains 6.023×10^{23} atoms. Find the weight of one atom of hydrogen.
1.67×10^{-24} g

119. Our galaxy is estimated to be 6×10^{17} mi across. How long would it take a space ship to cross the galaxy traveling at 20,000 mph? 3×10^{13} h

120. The radius of the moon is 1.7×10^6 m. Use the formula $V = \dfrac{4}{3}\pi r^3$ to find the volume of the moon.
$2.06 \times 10^{19} \text{ m}^3$

SUPPLEMENTAL EXERCISES 5.1

Simplify.

121. $\left(\dfrac{-x^2y^3}{x^{-4}y^{-2}}\right)^3 \qquad -x^{18}y^{15}$

122. $\left(\dfrac{a^4b^{-3}}{-a^{-5}y^2}\right)^{-1} \qquad -\dfrac{b^3y^2}{a^9}$

123. $\left(\dfrac{-2mn^{-2}}{4m^{-2}n}\right)^{-1} \qquad -\dfrac{2n^3}{m^3}$

124. $\left(\dfrac{6xy^{-2}}{-3x^{-1}y^2}\right)^{-3} \qquad -\dfrac{y^{12}}{8x^6}$

125. $\left(\dfrac{2x^2y}{x^{-2}y}\right)^{-1}\left(\dfrac{x^{-3}y}{4x^2y^{-1}}\right)^{-2}\left(\dfrac{3x^{-1}y}{xy^{-2}}\right)^2 \qquad 72x^2y^2$

126. $\left(\dfrac{3a^4b^{-2}}{a^{-1}b^3}\right)^{-1}\left(\dfrac{a^{-2}b}{6a^3b^{-2}}\right)^2\left(\dfrac{a^{-1}b}{9a^3b^{-4}}\right)^{-2} \qquad \dfrac{3b}{4a^7}$

127. $\dfrac{4m^4}{n^{-2}} + \left(\dfrac{n^{-1}}{m^2}\right)^{-2} \qquad 5m^4n^2$

128. $\dfrac{5x^3}{y^{-6}} + \left(\dfrac{x^{-1}}{y^2}\right)^{-3} \qquad 6x^3y^6$

129. $\left(\dfrac{3a^{-2}b}{a^{-4}b^{-1}}\right)^2 \div \left(\dfrac{a^{-1}b}{9a^2b^3}\right)^{-1} \qquad ab^2$

130. $\left(\dfrac{2m^3n^{-2}}{4m^4n}\right)^{-2} \div \left(\dfrac{mn^5}{m^{-1}n^3}\right)^3 \qquad \dfrac{4}{m^4}$

131. $\dfrac{x^{-1} + 5y^{-1}}{x^{-2} - 25y^{-2}} \qquad \dfrac{xy}{y - 5x}$

132. $\dfrac{2a^{-1} + 3b^{-1}}{4a^{-2} - 9b^{-2}} \qquad \dfrac{ab}{2b - 3a}$

133. $\left(\dfrac{x^{n-1}}{x^{n-2}}\right)^{-2} \qquad \dfrac{1}{x^2}$

134. $\left(\dfrac{a^{2n-3}}{a^{2n-4}}\right)^{-1} \qquad \dfrac{1}{a}$

Significant digits are important when working with very large or very small numbers. A **significant digit** of a number is any nonzero digit or a zero that is used not just for the purpose of placing the decimal point. The number 93,000,000 has 2 significant digits, the 9 and the 3; the 6 zeros are used for the purpose of placing the decimal point. The number 0.00354 has 3 significant digits, the 3, 5, and 4; the purpose of the zeros is to place the decimal point. The significant digits in the number 609,000 are underlined. The underlined zero is a significant digit; its purpose is not to place the decimal point. Find the number of significant digits in each of the following numbers.

135. 143,000   3

136. 8,000,000,000   1

137. 0.0001   1

138. 0.0076   2

139. 3,007,000   4

140. 5,020,000   3

141. 0.000902   3

142. 0.07004   4

143. 4.83×10^3 3

144. 5.2×10^{-4} 2

145. 1.987×10^7 4

146. 9×10^{-6} 1

Solve.

147. Use the expressions $(2 + 3)^{-2}$ and $2^{-2} + 3^{-2}$ to show that

$(x + y)^{-2} \neq x^{-2} + y^{-2}$. $\dfrac{1}{25} \neq \dfrac{13}{36}$

148. Use the expressions $(2 - 3)^{-1}$ and $2^{-1} - 3^{-1}$ to show that

$(x - y)^{-1} \neq x^{-1} - y^{-1}$. $-1 \neq \dfrac{1}{6}$

149. Use the expressions $(2 + 3)^{-2}$ and $\dfrac{1}{2^2 + 3^2}$ to show that

$(x + y)^{-2} \neq \dfrac{1}{x^2 + y^2}$. $\dfrac{1}{25} \neq \dfrac{1}{13}$

150. Use the expressions $(2 + 3)^{-2}$ and $\dfrac{1}{2^2} + \dfrac{1}{3^2}$ to show that

$(x + y)^{-2} \neq \dfrac{1}{x^2} + \dfrac{1}{y^2}$. $\dfrac{1}{25} \neq \dfrac{13}{36}$

SECTION 5.2

Rational Exponents and Radical Expressions

1 Simplify expressions with rational exponents

In this section, the definition of a power is extended beyond integers so that any rational number can be used as an exponent. The definition is made so that the Rules of Exponents hold true for rational exponents.

Since the Rules of Exponents will hold true for rational exponents, the expression $\left(a^{\frac{1}{n}}\right)^n$ can be simplified by using the Rule for Simplifying Powers of Exponential Expressions.

$$\left(a^{\frac{1}{n}}\right)^n = a^{\frac{1}{n} \cdot n} = a^1 = a$$

Since $\left(a^{\frac{1}{n}}\right)^n = a$, the number $a^{\frac{1}{n}}$ is the number whose nth power is a. $a^{\frac{1}{n}}$ is called **the nth root of a**.

$$\left(a^{\frac{1}{n}}\right)^n = a \qquad \left(25^{\frac{1}{2}}\right)^2 = 25 \qquad \left(8^{\frac{1}{3}}\right)^3 = 8$$

$25^{\frac{1}{2}}$ is the number whose 2nd power is 25. $25^{\frac{1}{2}} = 5$ because $(5)^2 = 25$.

$8^{\frac{1}{3}}$ is the number whose 3rd power is 8. $8^{\frac{1}{3}} = 2$ because $(2)^3 = 8$.

In the expression $a^{\frac{1}{n}}$, a must be positive when n is a positive even integer. $(-4)^{\frac{1}{2}}$ is not a real number since there is no real number whose 2nd power is -4.

As shown on the previous page, expressions that contain rational exponents do not always represent real numbers when the base of the exponential expression is a negative number. For this reason, all variables in this chapter represent positive numbers unless otherwise stated.

When n is a positive odd integer, a can be a positive or a negative number.

$(-27)^{\frac{1}{3}}$ is the number whose 3rd power is -27.

$(-27)^{\frac{1}{3}} = -3$ because $(-3)^3 = -27$.

Using the definition of $a^{\frac{1}{n}}$ and the Rules of Exponents, it is possible to define any exponential expression that contains a rational exponent.

Definition of $a^{\frac{m}{n}}$

If $a^{\frac{1}{n}}$ is a real number and m and n are positive integers, then

$$a^{\frac{m}{n}} = a^{\frac{1}{n} \cdot m} = \left(a^{\frac{1}{n}}\right)^m$$

$$\text{and} \quad a^{\frac{m}{n}} = a^{m \cdot \frac{1}{n}} = (a^m)^{\frac{1}{n}}.$$

Example 1 Simplify.

A. $27^{\frac{2}{3}}$ B. $32^{-\frac{2}{5}}$ C. $(-49)^{\frac{3}{2}}$

Solution A. $27^{\frac{2}{3}} = (3^3)^{\frac{2}{3}}$ • Rewrite 27 as 3^3.

$= 3^{3\left(\frac{2}{3}\right)}$ • Use the Rule for Simplifying Powers of Exponential Expressions. (Do this step mentally.)

$= 3^2$ • Simplify.

$= 9$

B. $32^{-\frac{2}{5}} = (2^5)^{-\frac{2}{5}}$ • Rewrite 32 as 2^5.

$= 2^{-2}$ • Use the Rule for Simplifying Powers of Exponential Expressions.

$= \dfrac{1}{2^2}$ • Use the Rule of Negative Exponents.

$= \dfrac{1}{4}$ • Simplify.

C. $(-49)^{\frac{3}{2}}$ • The base of the exponential expression is a negative expression, while the denominator of the exponent is a positive even number.

$(-49)^{\frac{3}{2}}$ is not a real number.

Problem 1 Simplify.

A. $64^{\frac{2}{3}}$ B. $16^{-\frac{3}{4}}$ C. $(-81)^{\frac{3}{4}}$

Solution See page A38.   A. 16   B. $\dfrac{1}{8}$ C. Not a real number

Example 2 Simplify.

A. $b^{\frac{1}{2}} \cdot b^{\frac{2}{3}} \cdot b^{-\frac{1}{4}}$ B. $(x^6 y^4)^{\frac{3}{2}}$ C. $\left(\dfrac{8a^3 b^{-4}}{64a^{-9} b^2}\right)^{\frac{2}{3}}$

Solution A. $b^{\frac{1}{2}} \cdot b^{\frac{2}{3}} \cdot b^{-\frac{1}{4}} = b^{\frac{1}{2}+\frac{2}{3}-\frac{1}{4}}$

• Use the Rule for Multiplying Exponential Expressions. (Do this step mentally.)

$$= b^{\frac{6}{12}+\frac{8}{12}-\frac{3}{12}}$$

$$= b^{\frac{11}{12}}$$

B. $(x^6 y^4)^{\frac{3}{2}} = x^{6\left(\frac{3}{2}\right)} y^{4\left(\frac{3}{2}\right)}$

• Use the Rule for Simplifying Powers of Products.

$$= x^9 y^6$$

C. $\left(\dfrac{8a^3 b^{-4}}{64a^{-9} b^2}\right)^{\frac{2}{3}} = \left(\dfrac{2^3 a^{12} b^{-6}}{2^6}\right)^{\frac{2}{3}}$

• Rewrite 8 as 2^3 and 64 as 2^6. Use the Rule for Dividing Exponential Expressions.

$$= \dfrac{2^2 a^8 b^{-4}}{2^4}$$

• Use the Rule for Simplifying Powers of Quotients.

$$= \dfrac{a^8}{2^2 b^4}$$

• Use the Rule of Negative Exponents.

$$= \dfrac{a^8}{4b^4}$$

Problem 2 Simplify.

A. $\dfrac{x^{\frac{1}{2}} y^{-\frac{5}{4}}}{x^{-\frac{4}{3}} y^{\frac{1}{3}}}$ B. $\left(x^{\frac{3}{4}} y^{\frac{1}{2}} z^{-\frac{2}{3}}\right)^{-\frac{4}{3}}$ C. $\left(\dfrac{16a^{-2} b^{\frac{4}{3}}}{9a^4 b^{-\frac{2}{3}}}\right)^{-\frac{1}{2}}$

Solution See page A39.

A. $\dfrac{x^{\frac{11}{6}}}{y^{\frac{19}{12}}}$ B. $\dfrac{z^{\frac{8}{9}}}{xy^{\frac{2}{3}}}$ C. $\dfrac{3a^3}{4b}$

2 Write exponential expressions as radical expressions and radical expressions as exponential expressions

Recall that $a^{\frac{1}{n}}$ **is the nth root of a.** The expression $\sqrt[n]{a}$ is another symbol for the nth root of a.

> **Definition of $\sqrt[n]{a}$**
>
> The expression $a^{\frac{1}{n}} = \sqrt[n]{a}$.

In the expression $\sqrt[n]{a}$, the symbol $\sqrt[n]{}$ is called a **radical**, n is the **index** of the radical, and a is the **radicand**. When $n = 2$, the radical expression represents a square root, and the index 2 is usually not written.

Any exponential expression with a rational exponent can be written as a radical expression.

If $a^{\frac{1}{n}}$ is a real number, then $a^{\frac{m}{n}} = a^{m \cdot \frac{1}{n}} = (a^m)^{\frac{1}{n}} = \sqrt[n]{a^m}$.

The expression $a^{\frac{m}{n}}$ can also be written $a^{\frac{m}{n}} = a^{\frac{1}{n} \cdot m} = (\sqrt[n]{a})^m$.

The exponential expression at the right has been written as a radical expression.

$$y^{\frac{2}{3}} = (y^2)^{\frac{1}{3}} = \sqrt[3]{y^2}$$

The radical expressions at the right have been written as exponential expressions.

$$\sqrt[5]{x^6} = (x^6)^{\frac{1}{5}} = x^{\frac{6}{5}}$$
$$\sqrt{17} = (17)^{\frac{1}{2}} = 17^{\frac{1}{2}}$$

Example 3 Rewrite the exponential expression as a radical expression.

A. $(5x)^{\frac{2}{5}}$ B. $-2x^{\frac{2}{3}}$

Solution A. $(5x)^{\frac{2}{5}} = \sqrt[5]{(5x)^2}$ • The denominator of the rational exponent is the index of the radical. The numerator is the power of the radicand.

$$= \sqrt[5]{25x^2}$$

B. $-2x^{\frac{2}{3}} = -2(x^2)^{\frac{1}{3}}$ • The -2 is not raised to the power.

$$= -2\sqrt[3]{x^2}$$

Problem 3 Rewrite the exponential expression as a radical expression.

A. $(2x^3)^{\frac{3}{4}}$ B. $-5a^{\frac{5}{6}}$

Solution See page A39.

A. $\sqrt[4]{8x^9}$ B. $-5\sqrt[6]{a^5}$

Example 4   Rewrite the radical expression as an exponential expression.

A. $\sqrt[5]{x^4}$ B. $\sqrt[3]{a^3 + b^3}$

Solution   A. $\sqrt[5]{x^4} = (x^4)^{\frac{1}{5}} = x^{\frac{4}{5}}$ • The index of the radical is the denominator of the rational exponent. The power of the radicand is the numerator of the rational exponent.

B. $\sqrt[3]{a^3 + b^3} = (a^3 + b^3)^{\frac{1}{3}}$ • Note that $(a^3 + b^3)^{\frac{1}{3}} \neq a + b$.

Problem 4   Rewrite the radical expression as an exponential expression.

A. $\sqrt[3]{3ab}$ B. $\sqrt[4]{x^4 + y^4}$

Solution   See page A39.

A. $(3ab)^{\frac{1}{3}}$ B. $(x^4 + y^4)^{\frac{1}{4}}$

3   Simplify expressions of the form $\sqrt[n]{a^n}$

Every positive number has two square roots, one a positive and one a negative number. For example, since $(5)^2 = 25$ and $(-5)^2 = 25$, there are two square roots of 25, 5 and -5.

The symbol $\sqrt{}$ is used to indicate the positive or **principle square root.** To indicate the negative square root of a number, a negative sign is placed in front of the radical.

$\sqrt{25} = 5$

$-\sqrt{25} = -5$

The square root of zero is zero.

$\sqrt{0} = 0$

The square root of a negative number is not a real number since the square of a real number must be positive.

$\sqrt{-25}$ is not a real number.

The square root of a squared positive number is a positive number.

$\sqrt{5^2} = \sqrt{25} = 5$

The square root of a squared negative number is a positive number.

$\sqrt{(-5)^2} = \sqrt{25} = 5$

For any real number a, $\sqrt{a^2} = |a|$ and $-\sqrt{a^2} = -|a|$.

Every number has only one cube root.

The cube root of a positive number is positive.

$\sqrt[3]{8} = 2$ since $2^3 = 8$.

The cube root of a negative number is negative.

$\sqrt[3]{-8} = -2$ since $(-2)^3 = -8$.

For any real number a, $\sqrt[3]{a^3} = a$.

The following properties hold true for finding the *n*th root of a real number.

If *n* is an even integer, then $\sqrt[n]{a^n} = |a|$ and $-\sqrt[n]{a^n} = -|a|$.

$\sqrt[6]{y^6} = |y|$

$-\sqrt[12]{x^{12}} = -|x|$

If *n* is an odd integer, then $\sqrt[n]{a^n} = a$.

$\sqrt[5]{b^5} = b$

Since it has been stated that all variables in this chapter represent positive numbers unless otherwise stated, it is not necessary to use the absolute value signs.

To simplify $\sqrt{121}$, write the prime factorization of the radicand in exponential form. The radicand is a perfect square since its exponent is an even number.

$\sqrt{121} = \sqrt{11^2}$

Write the radical expression as an exponential expression.

$= (11^2)^{\frac{1}{2}}$

Use the Rule for Simplifying Powers of Exponential Expressions.

$= 11$

The radicand of the radical expression $\sqrt[3]{x^6y^9}$ is a perfect cube since the exponents on the variables are divisible by 3. Write the radical expression as an exponential expression.

$\sqrt[3]{x^6y^9} = (x^6y^9)^{\frac{1}{3}}$

Use the Rule for Simplifying Powers of Products.

$= x^2y^3$

Example 5 Simplify.

A. $\sqrt[5]{x^{15}}$ B. $\sqrt{49x^2y^{12}}$ C. $-\sqrt[4]{16a^4b^8}$

Solution A. $\sqrt[5]{x^{15}} = (x^{15})^{\frac{1}{5}}$

• The radicand is a perfect fifth power since 15 is divisible by 5. Write the radical expression as an exponential expression. (Do this step mentally.)

$= x^3$

• Use the Rule for Simplifying Powers of Exponential Expressions.

B. $\sqrt{49x^2y^{12}} = \sqrt{7^2x^2y^{12}}$

$= 7xy^6$

• Write the prime factorization of 49.

C. $-\sqrt[4]{16a^4b^8} = -\sqrt[4]{2^4a^4b^8}$

$= -2ab^2$

Problem 5 Simplify.

A. $-\sqrt[4]{x^{12}}$ B. $\sqrt{121x^{10}y^4}$ C. $\sqrt[3]{-125a^6b^9}$

Solution See page A39.

A. $-x^3$ B. $11x^5y^2$ C. $-5a^2b^3$

EXERCISES 5.2

1 Simplify.

1. $8^{\frac{1}{3}}$ 2

2. $16^{\frac{1}{2}}$ 4

3. $9^{\frac{3}{2}}$ 27

4. $25^{\frac{3}{2}}$ 125

5. $27^{-\frac{2}{3}}$ $\dfrac{1}{9}$

6. $64^{-\frac{1}{3}}$ $\dfrac{1}{4}$

7. $32^{\frac{2}{5}}$ 4

8. $16^{\frac{3}{4}}$ 8

9. $(-25)^{\frac{5}{2}}$ Not a real number

10. $(-36)^{\frac{1}{4}}$ Not a real number

11. $\left(\dfrac{25}{49}\right)^{-\frac{3}{2}}$ $\dfrac{343}{125}$

12. $\left(\dfrac{8}{27}\right)^{-\frac{2}{3}}$ $\dfrac{9}{4}$

13. $x^{\frac{1}{2}}x^{\frac{1}{2}}$ x

14. $a^{\frac{1}{3}}a^{\frac{5}{3}}$ a^2

15. $y^{-\frac{1}{4}}y^{\frac{3}{4}}$ $y^{\frac{1}{2}}$

16. $x^{\frac{2}{5}} \cdot x^{-\frac{4}{5}}$ $\dfrac{1}{x^{\frac{2}{5}}}$

17. $x^{-\frac{2}{3}} \cdot x^{\frac{3}{4}}$ $x^{\frac{1}{12}}$

18. $x \cdot x^{-\frac{1}{2}}$ $x^{\frac{1}{2}}$

19. $a^{\frac{1}{3}} \cdot a^{\frac{3}{4}} \cdot a^{-\frac{1}{2}}$ $a^{\frac{7}{12}}$

20. $y^{-\frac{1}{6}} \cdot y^{\frac{2}{3}} \cdot y^{\frac{1}{2}}$ y

21. $\dfrac{a^{\frac{1}{2}}}{a^{\frac{3}{2}}}$ $\dfrac{1}{a}$

22. $\dfrac{b^{\frac{1}{3}}}{b^{\frac{4}{3}}}$ $\dfrac{1}{b}$

23. $\dfrac{y^{-\frac{3}{4}}}{y^{\frac{1}{4}}}$ $\dfrac{1}{y}$

24. $\dfrac{x^{-\frac{3}{5}}}{x^{\frac{1}{5}}}$ $\dfrac{1}{x^{\frac{4}{5}}}$

25. $\dfrac{y^{\frac{2}{3}}}{y^{-\frac{5}{6}}}$ $y^{\frac{3}{2}}$

26. $\dfrac{b^{\frac{3}{4}}}{b^{-\frac{3}{2}}}$ $b^{\frac{9}{4}}$

27. $(x^2)^{-\frac{1}{2}}$ $\dfrac{1}{x}$

28. $(a^8)^{-\frac{3}{4}}$ $a^{\frac{1}{6}}$

29. $\left(x^{-\frac{2}{3}}\right)^6$ $\dfrac{1}{x^4}$

30. $\left(y^{-\frac{5}{6}}\right)^{12}$ $\dfrac{1}{y^{10}}$

31. $\left(a^{-\frac{1}{2}}\right)^{-2}$ a

32. $\left(b^{-\frac{2}{3}}\right)^{-6}$ b^4

33. $\left(x^{-\frac{3}{8}}\right)^{-\frac{4}{5}}$ $x^{\frac{3}{10}}$

34. $\left(y^{-\frac{3}{2}}\right)^{-\frac{2}{9}}$ $y^{\frac{1}{3}}$

35. $\left(a^{\frac{1}{2}} \cdot a\right)^2$ a^3

36. $\left(b^{\frac{2}{3}} \cdot b^{\frac{1}{6}}\right)^6$ b^5

37. $\left(x^{-\frac{1}{2}} \cdot x^{\frac{3}{4}}\right)^{-2}$ $\dfrac{1}{x^{\frac{1}{2}}}$

38. $\left(a^{\frac{1}{2}} \cdot a^{-2}\right)^3$ $\dfrac{1}{a^{\frac{9}{2}}}$

39. $\left(y^{-\frac{1}{2}} \cdot y^{\frac{3}{2}}\right)^{\frac{2}{3}}$ $y^{\frac{2}{3}}$

40. $\left(b^{-\frac{2}{3}} \cdot b^{\frac{1}{4}}\right)^{-\frac{4}{3}}$ $b^{\frac{5}{9}}$

41. $(x^8y^2)^{\frac{1}{2}}$ x^4y

42. $(a^3b^9)^{\frac{2}{3}}$ a^2b^6

43. $(x^4y^2z^6)^{\frac{3}{2}}$ $x^6y^3z^9$

44. $(a^8b^4c^4)^{\frac{3}{4}}$ $a^6b^3c^3$

45. $(x^{-3}y^6)^{-\frac{1}{3}}$ $\dfrac{x}{y^2}$

46. $(a^2b^{-6})^{-\frac{1}{2}}$ $\dfrac{b^3}{a}$

47. $\left(x^{-2}y^{\frac{1}{3}}\right)^{-\frac{3}{4}}$ $\dfrac{x^{\frac{3}{2}}}{y^{\frac{1}{4}}}$

48. $\left(a^{-\frac{2}{3}}b^{\frac{2}{3}}\right)^{\frac{3}{2}}$ $\dfrac{b}{a}$

49. $\left(\dfrac{x^{\frac{1}{2}}}{y^{-2}}\right)^4$ $x^2 y^8$

50. $\left(\dfrac{b^{-\frac{3}{4}}}{a^{-\frac{1}{2}}}\right)^8$ $\dfrac{a^4}{b^6}$

51. $\dfrac{x^{\frac{1}{4}} \cdot x^{-\frac{1}{2}}}{x^{\frac{2}{3}}}$ $\dfrac{1}{x^{\frac{11}{12}}}$

52. $\dfrac{b^{\frac{1}{2}} \cdot b^{-\frac{3}{4}}}{b^{\frac{1}{4}}}$ $\dfrac{1}{b^{\frac{1}{2}}}$

53. $\left(\dfrac{y^{\frac{2}{3}} \cdot y^{-\frac{5}{6}}}{y^{\frac{1}{9}}}\right)^9$ $\dfrac{1}{y^{\frac{5}{2}}}$

54. $\left(\dfrac{a^{\frac{1}{3}} \cdot a^{-\frac{2}{3}}}{a^{\frac{1}{2}}}\right)^4$ $\dfrac{1}{a^{\frac{10}{3}}}$

55. $\left(\dfrac{b^2 \cdot b^{-\frac{3}{4}}}{b^{-\frac{1}{2}}}\right)^{-\frac{1}{2}}$ $\dfrac{1}{b^{\frac{7}{8}}}$

56. $\dfrac{\left(x^{-\frac{5}{6}} \cdot x^3\right)^{-\frac{2}{3}}}{x^{\frac{4}{3}}}$ $\dfrac{1}{x^{\frac{25}{9}}}$

57. $\left(a^{\frac{2}{3}}b^2\right)^6 \left(a^3 b^3\right)^{\frac{1}{3}}$ $a^5 b^{13}$

58. $\left(x^3 y^{-\frac{1}{2}}\right)^{-2} \left(x^{-3} y^2\right)^{\frac{1}{6}}$ $\dfrac{y^{\frac{4}{3}}}{x^{\frac{13}{2}}}$

59. $(16 m^{-2} n^4)^{-\frac{1}{2}} \left(mn^{\frac{1}{2}}\right)$ $\dfrac{m^2}{4 n^{\frac{3}{2}}}$

60. $(27 m^3 n^{-6})^{\frac{1}{3}} \left(m^{-\frac{1}{3}} n^{\frac{5}{6}}\right)^6$ $\dfrac{3 n^3}{m}$

61. $\left(\dfrac{x^{\frac{1}{2}} y^{-\frac{3}{4}}}{y^{\frac{2}{3}}}\right)^{-6}$ $\dfrac{y^{\frac{17}{2}}}{x^3}$

62. $\left(\dfrac{x^{\frac{1}{2}} y^{-\frac{5}{4}}}{y^{-\frac{3}{4}}}\right)^{-4}$ $\dfrac{y^2}{x^2}$

63. $\left(\dfrac{2^{-6} b^{-3}}{a^{-\frac{1}{2}}}\right)^{-\frac{2}{3}}$ $\dfrac{16 b^2}{a^{\frac{1}{3}}}$

64. $\left(\dfrac{49 c^{\frac{5}{3}}}{a^{-\frac{1}{4}} b^{\frac{5}{6}}}\right)^{-\frac{3}{2}}$ $\dfrac{b^{\frac{5}{4}}}{343 a^{\frac{3}{8}} c^{\frac{5}{2}}}$

65. $\dfrac{(x^{-2} y^4)^{\frac{1}{2}}}{\left(x^{\frac{1}{2}}\right)^4}$ $\dfrac{y^2}{x^3}$

66. $\dfrac{(x^{-3})^{\frac{1}{3}}}{(x^9 y^6)^{\frac{1}{6}}}$ $\dfrac{1}{x^{\frac{5}{2}} y}$

67. $a^{-\frac{1}{4}}\left(a^{\frac{5}{4}} - a^{\frac{9}{4}}\right)$ $a - a^2$

68. $x^{\frac{4}{3}}\left(x^{\frac{2}{3}} + x^{-\frac{1}{3}}\right)$ $x^2 + x$

69. $y^{\frac{2}{3}}\left(y^{\frac{1}{3}} + y^{-\frac{2}{3}}\right)$ $y + 1$

70. $b^{-\frac{2}{5}}\left(b^{-\frac{3}{5}} - b^{\frac{7}{5}}\right)$ $\dfrac{1}{b} - b$

71. $a^{\frac{1}{6}}\left(a^{\frac{5}{6}} - a^{-\frac{7}{6}}\right)$ $a - \dfrac{1}{a}$

72. $x^n \cdot x^{3n}$ x^{4n}

73. $a^{2n} \cdot a^{-5n}$ $\dfrac{1}{a^{3n}}$

74. $x^n \cdot x^{\frac{n}{2}}$ $x^{\frac{3n}{2}}$

75. $a^{\frac{n}{2}} \cdot a^{-\frac{n}{3}}$ $a^{\frac{n}{6}}$

76. $\dfrac{y^{\frac{n}{2}}}{y^{-n}}$ $y^{\frac{3n}{2}}$

77. $\dfrac{b^{\frac{m}{3}}}{b^m}$ $\dfrac{1}{b^{\frac{2m}{3}}}$

78. $(x^{2n})^n$ x^{2n^2}

79. $(x^{5n})^{2n}$ x^{10n^2}

80. $\left(x^{\frac{n}{4}} y^{\frac{n}{8}}\right)^8$ $x^{2n} y^n$

81. $\left(x^{\frac{n}{2}} y^{\frac{n}{3}}\right)^6$ $x^{3n} y^{2n}$

2   Rewrite the exponential expression as a radical expression.

82. $3^{\frac{1}{4}}$ $\sqrt[4]{3}$

83. $5^{\frac{1}{2}}$ $\sqrt{5}$

84. $a^{\frac{3}{2}}$ $\sqrt{a^3}$

85. $b^{\frac{4}{3}}$ $\sqrt[3]{b^4}$

86. $(2t)^{\frac{5}{2}}$ $\sqrt{32t^5}$

87. $(3x)^{\frac{2}{3}}$ $\sqrt[3]{9x^2}$

88. $-2x^{\frac{2}{3}}$ $-2\sqrt[3]{x^2}$

89. $-3a^{\frac{2}{5}}$ $-3\sqrt[5]{a^2}$

90. $(a^2b)^{\frac{2}{3}}$ $\sqrt[3]{a^4b^2}$

91. $(x^2y^3)^{\frac{3}{4}}$ $\sqrt[4]{x^6y^9}$

92. $(a^2b^4)^{\frac{3}{5}}$ $\sqrt[5]{a^6b^{12}}$

93. $(a^3b^7)^{\frac{3}{2}}$ $\sqrt{a^9b^{21}}$

94. $(4x + 3)^{\frac{3}{4}}$ $\sqrt[4]{(4x + 3)^3}$

95. $(3x - 2)^{\frac{1}{3}}$ $\sqrt[3]{3x - 2}$

96. $x^{-\frac{2}{3}}$ $\dfrac{1}{\sqrt[3]{x^2}}$

97. $b^{-\frac{3}{4}}$ $\dfrac{1}{\sqrt[4]{b^3}}$

Rewrite the radical expression as an exponential expression.

98. $\sqrt{14}$ $14^{\frac{1}{2}}$

99. $\sqrt{7}$ $7^{\frac{1}{2}}$

100. $\sqrt[3]{x}$ $x^{\frac{1}{3}}$

101. $\sqrt[4]{y}$ $y^{\frac{1}{4}}$

102. $\sqrt[3]{x^4}$ $x^{\frac{4}{3}}$

103. $\sqrt[4]{a^3}$ $a^{\frac{3}{4}}$

104. $\sqrt[5]{b^3}$ $b^{\frac{3}{5}}$

105. $\sqrt[4]{b^5}$ $b^{\frac{5}{4}}$

106. $\sqrt[3]{2x^2}$ $(2x^2)^{\frac{1}{3}}$

107. $\sqrt[5]{4y^7}$ $(4y^7)^{\frac{1}{5}}$

108. $-\sqrt{3x^5}$ $-(3x^5)^{\frac{1}{2}}$

109. $-\sqrt[4]{4x^5}$ $-(4x^5)^{\frac{1}{4}}$

110. $3x\sqrt[3]{y^2}$ $3xy^{\frac{2}{3}}$

111. $2y\sqrt{x^3}$ $2yx^{\frac{3}{2}}$

112. $\sqrt{a^2 + 2}$ $(a^2 + 2)^{\frac{1}{2}}$

113. $\sqrt{3 + y^2}$ $(3 + y^2)^{\frac{1}{2}}$

3   Simplify.

114. $\sqrt{x^{16}}$ x^8

115. $\sqrt{y^{14}}$ y^7

116. $-\sqrt{x^8}$ $-x^4$

117. $-\sqrt{a^6}$ $-a^3$

118. $\sqrt{x^2y^{10}}$ xy^5

119. $\sqrt{a^{14}b^6}$ a^7b^3

120. $\sqrt{25x^6}$ $5x^3$

121. $\sqrt{121y^{12}}$ $11y^6$

122. $\sqrt[3]{x^3y^9}$ xy^3

123. $\sqrt[3]{a^6b^{12}}$ a^2b^4

124. $-\sqrt[3]{x^{15}y^3}$ $-x^5y$

125. $-\sqrt[3]{a^9b^9}$ $-a^3b^3$

126. $\sqrt[3]{27a^9}$ $3a^3$

127. $\sqrt[3]{125b^{15}}$ $5b^5$

128. $\sqrt[3]{-8x^3}$ $-2x$

129. $\sqrt[3]{-a^6b^9}$ $-a^2b^3$

130. $\sqrt{16a^4b^{12}}$ $4a^2b^6$

131. $\sqrt{25x^8y^2}$ $5x^4y$

132. $\sqrt{-16x^4y^2}$
Not a real number

133. $\sqrt{-9a^6b^8}$
Not a real number

134. $\sqrt[3]{27x^9}$ $3x^3$

135. $\sqrt[3]{8a^{21}b^6}$ $2a^7b^2$

136. $\sqrt[3]{-64x^9y^{12}}$ $-4x^3y^4$

137. $\sqrt[3]{-27a^3b^{15}}$ $-3ab^5$

138. $\sqrt[4]{x^{16}}$ x^4

139. $\sqrt[4]{y^{12}}$ y^3

140. $\sqrt[4]{16x^{12}}$ $2x^3$

141. $\sqrt[4]{81a^{20}}$ $3a^5$

142. $-\sqrt[4]{x^8y^{12}}$ $-x^2y^3$

143. $-\sqrt[4]{a^{16}b^4}$ $-a^4b$

144. $\sqrt[5]{x^{20}y^{10}}$ x^4y^2

145. $\sqrt[5]{a^5b^{25}}$ ab^5

146. $\sqrt[4]{81x^4y^{20}}$ $3xy^5$

147. $\sqrt[4]{16a^8b^{20}}$ $2a^2b^5$

148. $\sqrt[5]{32a^5b^{10}}$ $2ab^2$

149. $\sqrt[5]{-32x^{15}y^{20}}$ $-2x^3y^4$

150. $\sqrt[5]{243x^{10}y^{40}}$ $3x^2y^8$

151. $\sqrt{\dfrac{16x^2}{y^{14}}}$ $\dfrac{4x}{y^7}$

152. $\sqrt{\dfrac{49a^4}{b^{24}}}$ $\dfrac{7a^2}{b^{12}}$

153. $\sqrt[3]{\dfrac{27b^3}{a^9}}$ $\dfrac{3b}{a^3}$

154. $\sqrt[3]{\dfrac{64x^{15}}{y^6}}$ $\dfrac{4x^5}{y^2}$

155. $\sqrt{(2x+3)^2}$ $2x+3$

156. $\sqrt{(4x+1)^2}$ $4x+1$

157. $\sqrt{x^2+2x+1}$ $x+1$

158. $\sqrt{x^2+4x+4}$ $x+2$

SUPPLEMENTAL EXERCISES 5.2

Which of the numbers is the largest, **a**, **b**, or **c**?

159. a. $16^{\frac{1}{2}}$ **b.** $16^{\frac{1}{4}}$ **c.** $16^{\frac{3}{2}}$
c

160. a. $27^{\frac{1}{3}}$ **b.** $27^{-\frac{4}{3}}$ **c.** $27^{\frac{2}{3}}$
c

161. a. $(-125)^{\frac{2}{3}}$ **b.** $(-125)^{-\frac{1}{3}}$ **c.** $(-125)^{\frac{1}{3}}$
a

162. a. $(-32)^{-\frac{1}{5}}$ **b.** $(-32)^{\frac{2}{5}}$ **c.** $(-32)^{\frac{1}{5}}$
b

163. a. $4^{\frac{1}{2}} \cdot 4^{\frac{3}{2}}$ **b.** $3^{\frac{4}{5}} \cdot 3^{\frac{6}{5}}$ **c.** $7^{\frac{1}{4}} \cdot 7^{\frac{3}{4}}$
a

164. a. $\dfrac{81^{\frac{3}{4}}}{81^{\frac{1}{4}}}$ **b.** $\dfrac{64^{\frac{2}{3}}}{64^{\frac{1}{3}}}$ **c.** $\dfrac{36^{\frac{5}{2}}}{36^{\frac{3}{2}}}$
c

Simplify.

165. $\sqrt[3]{\sqrt{x^6}}$ x

166. $\sqrt{\sqrt[3]{y^6}}$ y

167. $\sqrt[4]{\sqrt{a^8}}$ a

168. $\sqrt[5]{\sqrt[3]{b^{15}}}$ b

169. $\sqrt{\sqrt{16x^{12}}}$ $2x^3$

170. $\sqrt{\sqrt{81y^8}}$ $3y^2$

171. $\sqrt{\sqrt[n]{a^{4n}}}$ a^2

172. $\sqrt[n]{\sqrt{b^{6n}}}$ b^3

173. $\sqrt{\sqrt[3]{x^{12}y^{24}}}$ x^2y^4

174. $\sqrt[5]{\sqrt{a^{10}b^{20}}}$ ab^2

175. $\sqrt[3]{\sqrt{64x^{36}y^{30}}}$ $2x^6y^5$

176. $\sqrt[4]{\sqrt{256a^{16}b^{32}}}$ $2a^2b^4$

For what value of p is the given equation true?

177. $x^p x^{\frac{1}{4}} = x$ $\dfrac{3}{4}$

178. $y^p y^{\frac{2}{5}} = y$ $\dfrac{3}{5}$

179. $\sqrt[p]{x^8} = x^2$ 4

180. $\sqrt[p]{y^{15}} = y^5$ 3

181. $\dfrac{x^p}{x^{\frac{3}{5}}} = x^{\frac{1}{5}}$ $\dfrac{4}{5}$

182. $\dfrac{y^p}{y^{\frac{3}{4}}} = y^{\frac{1}{2}}$ $\dfrac{5}{4}$

183. $x^p x^{-\frac{1}{2}} = x^{\frac{1}{4}}$ $\dfrac{3}{4}$

184. $y^p y^{-\frac{5}{6}} = y^{\frac{1}{3}}$ $\dfrac{7}{6}$

185. $x^p x^{-\frac{2}{3}} = x^{-1}$ $-\dfrac{1}{3}$

186. $y^{-\frac{3}{4}} y^p = y^{-2}$ $-\dfrac{5}{4}$

187. $\dfrac{x^{\frac{1}{2}}}{x^p} = x^{\frac{5}{6}}$ $-\dfrac{1}{3}$

188. $\dfrac{y^{\frac{1}{3}}}{y^p} = y^{\frac{5}{9}}$ $-\dfrac{2}{9}$

SECTION 5.3

Operations on Radical Expressions

1 ### Simplify radical expressions

If a number is not a perfect power, its root can only be approximated, for example, $\sqrt{5}$ and $\sqrt[3]{3}$. These numbers are **irrational numbers.** Their decimal representations never terminate or repeat.

$$\sqrt{5} = 2.2360679\ldots \qquad \sqrt[3]{3} = 1.4422495\ldots$$

The approximate square roots and cube roots of the positive integers up to 100 can be found in the Appendix on page A4. The roots have been rounded to the nearest thousandth.

A radical expression is in simplest form when the radicand contains no factor that is a perfect power. The Product Property of Radicals is used to simplify radical expressions whose radicands are not perfect powers.

> **The Product Property of Radicals**
>
> If a and b are positive real numbers, then $\sqrt[n]{ab} = \sqrt[n]{a} \cdot \sqrt[n]{b}$ and $\sqrt[n]{a} \cdot \sqrt[n]{b} = \sqrt[n]{ab}$.

To simplify $\sqrt{48}$, write the prime factorization of the radicand in exponential form.

$$\sqrt{48} = \sqrt{2^4 \cdot 3}$$

Use the Product Property of Radicals to write the expression as a product.

$$= \sqrt{2^4} \sqrt{3}$$

Simplify.

$$= 2^2 \sqrt{3}$$
$$= 4\sqrt{3}$$

To simplify $\sqrt[3]{x^7}$, write the radicand as the product of a perfect cube and a factor that does not contain a perfect cube.

$$\sqrt[3]{x^7} = \sqrt[3]{x^6 \cdot x}$$

Use the Product Property of Radicals to write the expression as a product.

$$= \sqrt[3]{x^6} \sqrt[3]{x}$$

Simplify.

$$= x^2 \sqrt[3]{x}$$

Example 1   Simplify: $\sqrt[4]{32x^7}$

Solution $\sqrt[4]{32x^7} = \sqrt[4]{2^5 x^7}$

• Write the prime factorization of the coefficient of the radicand in exponential form.

$$= \sqrt[4]{2^4 x^4 (2x^3)}$$

• Write the radicand as the product of a perfect fourth power and factors that do not contain a perfect fourth power.

$$= \sqrt[4]{2^4 x^4} \sqrt[4]{2x^3}$$

• Use the Product Property of Radicals to write the expression as a product.

$$= 2x \sqrt[4]{2x^3}$$

• Simplify.

Problem 1   Simplify: $\sqrt[5]{x^7}$

Solution   See page A40.
$x \sqrt[5]{x^2}$

Example 2   Simplify $\sqrt{175}$. Then find the decimal approximation. Use the table on page A4.

Solution $\sqrt{175} = \sqrt{5^2 \cdot 7}$

• Write the prime factorization of the radicand in exponential form.

$$= \sqrt{5^2} \sqrt{7}$$

• Use the Product Property of Radicals to write the expression as a product.

$$= 5\sqrt{7}$$

• Simplify.

$$\approx 5(2.646)$$

• Replace the radical expression by the decimal approximation found on page A4.

$$\approx 13.230$$

• Simplify.

Problem 2   Simplify $\sqrt{216}$. Then find the decimal approximation. Use the table on page A4.

Solution   See page A40.
14.694

2 Add and subtract radical expressions

The Distributive Property is used to simplify the sum or difference of radical expressions that have the same radicand and the same index.

$$3\sqrt{5} + 8\sqrt{5} = (3 + 8)\sqrt{5} = 11\sqrt{5}$$

$$2\sqrt[3]{3x} - 9\sqrt[3]{3x} = (2 - 9)\sqrt[3]{3x} = -7\sqrt[3]{3x}$$

Radical expressions that are in simplest form and have unlike radicands or different indices cannot be simplified by the Distributive Property. The expressions shown below cannot be simplified by the Distributive Property.

$$3\sqrt[4]{2} - 6\sqrt[4]{3} \qquad 2\sqrt[4]{4x} + 3\sqrt[3]{4x}$$

Simplify: $3\sqrt{32x^2} - 2x\sqrt{2} + \sqrt{128x^2}$

Simplify each term.

$$3\sqrt{32x^2} - 2x\sqrt{2} + \sqrt{128x^2} =$$
$$3\sqrt{2^5x^2} - 2x\sqrt{2} + \sqrt{2^7x^2} =$$
$$3\sqrt{2^4x^2}\,\sqrt{2} - 2x\sqrt{2} + \sqrt{2^6x^2}\,\sqrt{2} =$$
$$3 \cdot 2^2x\sqrt{2} - 2x\sqrt{2} + 2^3x\sqrt{2} =$$
$$12x\sqrt{2} - 2x\sqrt{2} + 8x\sqrt{2} =$$

Simplify by using the Distributive Property.

$$18x\sqrt{2}$$

Example 3 Simplify.

A. $5b\sqrt[4]{32a^7b^5} - 2a\sqrt[4]{162a^3b^9}$ B. $5\sqrt[5]{2x^7y^{11}} - y\sqrt[5]{64x^7y^6}$

Solution A. $5b\sqrt[4]{32a^7b^5} - 2a\sqrt[4]{162a^3b^9} =$
$\quad 5b\sqrt[4]{2^5a^7b^5} - 2a\sqrt[4]{3^4 \cdot 2a^3b^9} =$
$\quad 5b\sqrt[4]{2^4a^4b^4}\,\sqrt[4]{2a^3b} - 2a\sqrt[4]{3^4b^8}\,\sqrt[4]{2a^3b} =$
$\quad 5b \cdot 2ab\sqrt[4]{2a^3b} - 2a \cdot 3b^2\sqrt[4]{2a^3b} =$
$\quad 10ab^2\sqrt[4]{2a^3b} - 6ab^2\sqrt[4]{2a^3b} =$
$\quad 4ab^2\sqrt[4]{2a^3b}$

B. $5\sqrt[5]{2x^7y^{11}} - y\sqrt[5]{64x^7y^6} =$
$\quad 5\sqrt[5]{2x^7y^{11}} - y\sqrt[5]{2^6x^7y^6} =$
$\quad 5\sqrt[5]{x^5y^{10}}\,\sqrt[5]{2x^2y} - y\sqrt[5]{2^5x^5y^5}\,\sqrt[5]{2x^2y} =$
$\quad 5 \cdot xy^2\sqrt[5]{2x^2y} - y \cdot 2xy\sqrt[5]{2x^2y} =$
$\quad 5xy^2\sqrt[5]{2x^2y} - 2xy^2\sqrt[5]{2x^2y} =$
$\quad 3xy^2\sqrt[5]{2x^2y}$

Problem 3   Simplify.

A. $3xy\sqrt[3]{81x^5y} - \sqrt[3]{192x^8y^4}$ B. $4a\sqrt[3]{81a^7b^9} + a^2b\sqrt[3]{192a^4b^6}$

Solution   See page A40.

A. $5x^2y\sqrt[3]{3x^2y}$ B. $16a^3b^3\sqrt[3]{3a}$

3 Multiply radical expressions

The Product Property of Radicals is used to multiply radical expressions with the same index.

$$\sqrt{3x} \cdot \sqrt{5y} = \sqrt{3x \cdot 5y} = \sqrt{15xy}$$

Simplify: $(\sqrt{x})^2$

Multiply the radicands.
Simplify.

$$(\sqrt{x})^2 = \sqrt{x}\,\sqrt{x}$$
$$= \sqrt{x \cdot x}$$
$$= \sqrt{x^2}$$
$$= x$$

Note: For $a > 0$, $(\sqrt{a})^2 = \sqrt{a^2} = a$.

To simplify $\sqrt[3]{2a^5b}\,\sqrt[3]{16a^2b^2}$, use the Product Property of Radicals to multiply the radicands. Then simplify.

$$\sqrt[3]{2a^5b}\,\sqrt[3]{16a^2b^2} = \sqrt[3]{32a^7b^3}$$
$$= \sqrt[3]{2^5a^7b^3}$$
$$= \sqrt[3]{2^3a^6b^3}\,\sqrt[3]{2^2a}$$
$$= 2a^2b\sqrt[3]{4a}$$

To simplify $\sqrt{2x}(\sqrt{8x} - \sqrt{3})$, use the Distributive Property to remove parentheses. Then simplify.

$$\sqrt{2x}(\sqrt{8x} - \sqrt{3}) = \sqrt{16x^2} - \sqrt{6x}$$
$$= \sqrt{2^4x^2} - \sqrt{6x}$$
$$= 2^2x - \sqrt{6x}$$
$$= 4x - \sqrt{6x}$$

Example 4   Simplify: $\sqrt{3x}(\sqrt{27x^2} - \sqrt{3x})$

Solution $\sqrt{3x}(\sqrt{27x^2} - \sqrt{3x}) = \sqrt{81x^3} - \sqrt{9x^2}$
$$= \sqrt{3^4x^3} - \sqrt{3^2x^2}$$
$$= \sqrt{3^4x^2}\sqrt{x} - \sqrt{3^2x^2}$$
$$= 3^2x\sqrt{x} - 3x$$
$$= 9x\sqrt{x} - 3x$$

Problem 4   Simplify: $\sqrt{5b}(\sqrt{3b} - \sqrt{10})$

Solution   See page A40.

$b\sqrt{15} - 5\sqrt{2b}$

To simplify $(\sqrt[3]{x} - 1)(\sqrt[3]{x} + 7)$, use the FOIL method to remove parentheses. Then simplify.

$$(\sqrt[3]{x} - 1)(\sqrt[3]{x} + 7) = \sqrt[3]{x^2} + 7\sqrt[3]{x} - \sqrt[3]{x} - 7$$
$$= \sqrt[3]{x^2} + 6\sqrt[3]{x} - 7$$

The expressions $a + b$ and $a - b$, which are the sum and difference of two terms, are called **conjugates** of each other. The product of conjugates of the form $(a + b)(a - b)$ is $a^2 - b^2$.

$$(\sqrt{x} - 3)(\sqrt{x} + 3) = (\sqrt{x})^2 - 3^2 = x - 9$$

Example 5 Simplify: $(2\sqrt[3]{x} - 3)(3\sqrt[3]{x} - 4)$

Solution $(2\sqrt[3]{x} - 3)(3\sqrt[3]{x} - 4) = 6\sqrt[3]{x^2} - 8\sqrt[3]{x} - 9\sqrt[3]{x} + 12$ • Use the FOIL method.
$$= 6\sqrt[3]{x^2} - 17\sqrt[3]{x} + 12$$

Problem 5 Simplify: $(2\sqrt[3]{2x} - 3)(\sqrt[3]{2x} - 5)$

Solution See page A40.
$$2\sqrt[3]{4x^2} - 13\sqrt[3]{2x} + 15$$

4 Divide radical expressions

The Quotient Property of Radicals is used to divide radical expressions with the same index.

The Quotient Property of Radicals

If a and b are positive real numbers, then $\sqrt[n]{\dfrac{a}{b}} = \dfrac{\sqrt[n]{a}}{\sqrt[n]{b}}$ and $\dfrac{\sqrt[n]{a}}{\sqrt[n]{b}} = \sqrt[n]{\dfrac{a}{b}}$.

To simplify $\sqrt[3]{\dfrac{81x^5}{y^6}}$, use the Quotient Property of Radicals.
Then simplify each radical expression.

$$\sqrt[3]{\frac{81x^5}{y^6}} = \frac{\sqrt[3]{81x^5}}{\sqrt[3]{y^6}}$$
$$= \frac{\sqrt[3]{3^4x^5}}{\sqrt[3]{y^6}}$$
$$= \frac{\sqrt[3]{3^3x^3}\,\sqrt[3]{3x^2}}{\sqrt[3]{y^6}}$$
$$= \frac{3x\sqrt[3]{3x^2}}{y^2}$$

To simplify $\dfrac{\sqrt{5a^4b^7c^2}}{\sqrt{ab^3c}}$, use the Quotient Property of Radicals. Then simplify the radicand.

$$\frac{\sqrt{5a^4b^7c^2}}{\sqrt{ab^3c}} = \sqrt{\frac{5a^4b^7c^2}{ab^3c}}$$

$$= \sqrt{5a^3b^4c}$$

$$= \sqrt{a^2b^4}\,\sqrt{5ac}$$

$$= ab^2\sqrt{5ac}$$

A radical expression is in simplest form when no radical remains in the denominator of the radical expression. The procedure used to remove a radical from the denominator is called **rationalizing the denominator.**

To simplify $\dfrac{2}{\sqrt{x}}$, multiply the expression by 1 in the form $\dfrac{\sqrt{x}}{\sqrt{x}}$. Then simplify.

$$\frac{2}{\sqrt{x}} = \frac{2}{\sqrt{x}} \cdot \frac{\sqrt{x}}{\sqrt{x}}$$

$$= \frac{2\sqrt{x}}{(\sqrt{x})^2}$$

$$= \frac{2\sqrt{x}}{x}$$

Example 6   Simplify.

A. $\dfrac{5}{\sqrt{5x}}$ B. $\dfrac{3x}{\sqrt[3]{4x}}$ C. $\dfrac{\sqrt{a}-\sqrt{a^3}}{\sqrt{a}}$

Solution   A. $\dfrac{5}{\sqrt{5x}} = \dfrac{5}{\sqrt{5x}} \cdot \dfrac{\sqrt{5x}}{\sqrt{5x}} =$ • Multiply the expression by $\dfrac{\sqrt{5x}}{\sqrt{5x}}$.

$$\frac{5\sqrt{5x}}{\sqrt{5^2x^2}} = \frac{5\sqrt{5x}}{5x} = \frac{\sqrt{5x}}{x}$$

B. $\dfrac{3x}{\sqrt[3]{4x}} = \dfrac{3x}{\sqrt[3]{2^2x}} \cdot \dfrac{\sqrt[3]{2x^2}}{\sqrt[3]{2x^2}} =$ • Multiply the expression by $\dfrac{\sqrt[3]{2x^2}}{\sqrt[3]{2x^2}}$.
$\sqrt[3]{4x} = \sqrt[3]{2^2x}$. $\sqrt[3]{2^2x} \cdot \sqrt[3]{2x^2} = \sqrt[3]{2^3x^3}$, a perfect cube.

$$\frac{3x\sqrt[3]{2x^2}}{\sqrt[3]{2^3x^3}} = \frac{3x\sqrt[3]{2x^2}}{2x} = \frac{3\sqrt[3]{2x^2}}{2}$$

C. $\dfrac{\sqrt{a}-\sqrt{a^3}}{\sqrt{a}} = \dfrac{\sqrt{a}-\sqrt{a^2}\,\sqrt{a}}{\sqrt{a}} =$ • Simplify the numerator.

$$\frac{\sqrt{a}-a\sqrt{a}}{\sqrt{a}} = \frac{\sqrt{a}(1-a)}{\sqrt{a}} = 1-a$$

Problem 6 Simplify.

A. $\dfrac{y}{\sqrt{3y}}$ B. $\dfrac{3}{\sqrt[3]{3x^2}}$ C. $\dfrac{\sqrt{x^5}+\sqrt{x}}{\sqrt{x}}$

Solution See page A40.

A. $\dfrac{\sqrt{3y}}{3}$ B. $\dfrac{\sqrt[3]{9x}}{x}$ C. x^2+1

To simplify a fraction that has a binomial radical expression in the denominator, multiply the numerator and denominator by the conjugate of the denominator.

$$\frac{\sqrt{x}-\sqrt{y}}{\sqrt{x}+\sqrt{y}}=\frac{\sqrt{x}-\sqrt{y}}{\sqrt{x}+\sqrt{y}}\cdot\frac{\sqrt{x}-\sqrt{y}}{\sqrt{x}-\sqrt{y}}$$

$$=\frac{(\sqrt{x})^2-\sqrt{xy}-\sqrt{xy}+(\sqrt{y})^2}{(\sqrt{x})^2-(\sqrt{y})^2}=\frac{x-2\sqrt{xy}+y}{x-y}$$

Example 7 Simplify: $\dfrac{3+\sqrt{y}}{3-\sqrt{y}}$

$$\frac{3+\sqrt{y}}{3-\sqrt{y}}=\frac{3+\sqrt{y}}{3-\sqrt{y}}\cdot\frac{3+\sqrt{y}}{3+\sqrt{y}}=\frac{3^2+3\sqrt{y}+3\sqrt{y}+(\sqrt{y})^2}{3^2-(\sqrt{y})^2}=\frac{9+6\sqrt{y}+y}{9-y}$$

Problem 7 Simplify: $\dfrac{\sqrt{2}+\sqrt{x}}{\sqrt{2}-\sqrt{x}}$

Solution See page A40.

$$\frac{2+2\sqrt{2x}+x}{2-x}$$

EXERCISES 5.3

1 Simplify.

1. $\sqrt{x^4y^3z^5}$ $x^2yz^2\sqrt{yz}$

2. $\sqrt{x^3y^6z^9}$ $xy^3z^4\sqrt{xz}$

3. $\sqrt{8a^3b^8}$ $2ab^4\sqrt{2a}$

4. $\sqrt{24a^9b^6}$ $2a^4b^3\sqrt{6a}$

5. $\sqrt{45x^2y^3z^5}$ $3xyz^2\sqrt{5yz}$

6. $\sqrt{60xy^7z^{12}}$ $2y^3z^6\sqrt{15xy}$

7. $\sqrt{-9x^3}$ Not a real number

8. $\sqrt{-x^2y^5}$ Not a real number

9. $\sqrt[3]{a^{16}b^8}$ $a^5b^2\sqrt[3]{ab^2}$

10. $\sqrt[3]{a^5b^8}$ $ab^2\sqrt[3]{a^2b^2}$

11. $\sqrt[3]{-125x^2y^4}$ $-5y\sqrt[3]{x^2y}$

12. $\sqrt[3]{-216x^5y^9}$ $-6xy^3\sqrt[3]{x^2}$

13. $\sqrt[3]{a^4b^5c^6}$ $abc^2\sqrt[3]{ab^2}$

14. $\sqrt[3]{a^8b^{11}c^{15}}$ $a^2b^3c^5\sqrt[3]{a^2b^2}$

15. $\sqrt[4]{16x^9y^5}$ $2x^2y\sqrt[4]{xy}$

16. $\sqrt[4]{64x^8y^{10}}$ $2x^2y^2\sqrt[4]{4y^2}$

Simplify. Then find the decimal approximation. Use the table on page A4.

17. $\sqrt{600}$ 24.49

18. $\sqrt{432}$ 20.784

19. $\sqrt{539}$ 23.219

20. $\sqrt{320}$ 17.888

21. $\sqrt{845}$ 29.068

22. $\sqrt{468}$ 21.636

23. $\sqrt{600,000}$ 774.6

24. $\sqrt{120,000}$ 346.4

2   Simplify.

25. $2\sqrt{x} - 8\sqrt{x}$ $-6\sqrt{x}$

26. $3\sqrt{y} + 12\sqrt{y}$ $15\sqrt{y}$

27. $\sqrt{8} - \sqrt{32}$ $-2\sqrt{2}$

28. $\sqrt{27} - \sqrt{75}$ $-2\sqrt{3}$

29. $\sqrt{128x} - \sqrt{98x}$ $\sqrt{2x}$

30. $\sqrt{48x} + \sqrt{147x}$ $11\sqrt{3x}$

31. $\sqrt{27a} - \sqrt{8a}$ $3\sqrt{3a} - 2\sqrt{2a}$

32. $\sqrt{18b} + \sqrt{75b}$ $3\sqrt{2b} + 5\sqrt{3b}$

33. $2\sqrt{2x^3} + 4x\sqrt{8x}$ $10x\sqrt{2x}$

34. $5y\sqrt{8y} + 2\sqrt{50y^3}$ $20y\sqrt{2y}$

35. $x\sqrt{75xy} - \sqrt{27x^3y}$ $2x\sqrt{3xy}$

36. $3\sqrt{8x^2y^3} - 2x\sqrt{32y^3}$ $-2xy\sqrt{2y}$

37. $2\sqrt{32x^2y^3} - xy\sqrt{98y}$ $xy\sqrt{2y}$

38. $6y\sqrt{x^3y} - 2\sqrt{x^3y^3}$ $4xy\sqrt{xy}$

39. $7b\sqrt{a^5b^3} - 2ab\sqrt{a^3b^3}$ $5a^2b^2\sqrt{ab}$

40. $2a\sqrt{27ab^5} + 3b\sqrt{3a^3b}$ $6ab^2\sqrt{3ab} + 3ab\sqrt{3ab}$

41. $\sqrt[3]{128} + \sqrt[3]{250}$ $9\sqrt[3]{2}$

42. $\sqrt[3]{16} - \sqrt[3]{54}$ $-\sqrt[3]{2}$

43. $2\sqrt[3]{3a^4} - 3a\sqrt[3]{81a}$ $-7a\sqrt[3]{3a}$

44. $2b\sqrt[3]{16b^2} + \sqrt[3]{128b^5}$ $8b\sqrt[3]{2b^2}$

45. $3\sqrt[3]{x^5y^7} - 8xy\sqrt[3]{x^2y^4}$ $-5xy^2\sqrt[3]{x^2y}$

46. $3\sqrt[4]{32a^5} - a\sqrt[4]{162a}$ $3a\sqrt[4]{2a}$

47. $2a\sqrt[4]{16ab^5} + 3b\sqrt[4]{256a^5b}$ $16ab\sqrt[4]{ab}$

48. $2\sqrt{50} - 3\sqrt{125} + \sqrt{98}$ $17\sqrt{2} - 15\sqrt{5}$

49. $3\sqrt{108} - 2\sqrt{18} - 3\sqrt{48}$
$6\sqrt{3} - 6\sqrt{2}$

50. $\sqrt{9b^3} - \sqrt{25b^3} + \sqrt{49b^3}$
$5b\sqrt{b}$

51. $\sqrt{4x^7y^5} + 9x^2\sqrt{x^3y^5} - 5xy\sqrt{x^5y^3}$
$6x^3y^2\sqrt{xy}$

52. $2x\sqrt{8xy^2} - 3y\sqrt{32x^3} + \sqrt{8x^3y^2}$
$-6xy\sqrt{2x}$

53. $5a\sqrt{3a^3b} + 2a^2\sqrt{27ab} - 4\sqrt{75a^5b}$
$-9a^2\sqrt{3ab}$

54. $\sqrt[3]{54xy^3} - 5\sqrt[3]{2xy^3} + y\sqrt[3]{128x}$
$2y\sqrt[3]{2x}$

55. $2\sqrt[3]{24x^3y^4} + 4x\sqrt[3]{81y^4} - 3y\sqrt[3]{24x^3y}$
$10xy\sqrt[3]{3y}$

56. $2a\sqrt[4]{32b^5} - 3b\sqrt[4]{162a^4b} + \sqrt[4]{2a^4b^5}$
$-4ab\sqrt[4]{2b}$

57. $6y\sqrt[4]{48x^5} - 2x\sqrt[4]{243xy^4} - 4\sqrt[4]{3x^5y^4}$ $2xy\sqrt[4]{3x}$

3   Simplify.

58. $\sqrt{8}\,\sqrt{32}$ 16

59. $\sqrt{14}\,\sqrt{35}$ $7\sqrt{10}$

60. $\sqrt[3]{4}\,\sqrt[3]{8}$ $2\sqrt[3]{4}$

61. $\sqrt[3]{6}\,\sqrt[3]{36}$ 6

62. $\sqrt{x^2y^5}\,\sqrt{xy}$ $xy^3\sqrt{x}$

63. $\sqrt{a^3b}\,\sqrt{ab^4}$ $a^2b^2\sqrt{b}$

64. $\sqrt{2x^2y}\,\sqrt{32xy}$ $8xy\sqrt{x}$

65. $\sqrt{5x^3y}\,\sqrt{10x^3y^4}$ $5x^3y^2\sqrt{2y}$

66. $\sqrt[3]{x^2y}\,\sqrt[3]{16x^4y^2}$ $2x^2y\sqrt[3]{2}$

67. $\sqrt[3]{4a^2b^3}\,\sqrt[3]{8ab^5}$ $2ab^2\sqrt[3]{4b^2}$

68. $\sqrt[4]{12ab^3}\,\sqrt[4]{4a^5b^2}$ $2ab\sqrt[4]{3a^2b}$

69. $\sqrt[4]{36a^2b^4}\,\sqrt[4]{12a^5b^3}$ $2ab\sqrt[4]{27a^3b^3}$

70. $\sqrt{3}\,(\sqrt{27} - \sqrt{3})$ 6

71. $\sqrt{10}\,(\sqrt{10} - \sqrt{5})$ $10 - 5\sqrt{2}$

72. $\sqrt{x}\,(\sqrt{x} - \sqrt{2})$ $x - \sqrt{2x}$

73. $\sqrt{y}\,(\sqrt{y} - \sqrt{5})$ $y - \sqrt{5y}$

74. $\sqrt{2x}\,(\sqrt{8x} - \sqrt{32})$ $4x - 8\sqrt{x}$

75. $\sqrt{3a}\,(\sqrt{27a^2} - \sqrt{a})$ $9a\sqrt{a} - a\sqrt{3}$

76. $(\sqrt{x} - 3)^2$ $x - 6\sqrt{x} + 9$

77. $(\sqrt{2x} + 4)^2$ $2x + 8\sqrt{2x} + 16$

78. $(4\sqrt{5} + 2)^2$ $84 + 16\sqrt{5}$

79. $2\sqrt{3x^2} \cdot 3\sqrt{12xy^3} \cdot \sqrt{6x^3y}$ $36x^3y^2\sqrt{6}$

80. $2\sqrt{14xy} \cdot 4\sqrt{7x^2y} \cdot 3\sqrt{8xy^2}$ $672x^2y^2$

81. $\sqrt[3]{8ab}\,\sqrt[3]{4a^2b^3}\,\sqrt[3]{9ab^4}$ $2ab^2\sqrt[3]{36ab^2}$

82. $\sqrt[3]{2a^2b}\,\sqrt[3]{4a^3b^2}\,\sqrt[3]{8a^5b^6}$ $4a^3b^3\sqrt[3]{a}$

83. $(\sqrt{2} - 3)(\sqrt{2} + 4)$ $-10 + \sqrt{2}$

84. $(\sqrt{5} - 5)(2\sqrt{5} + 2)$ $-8\sqrt{5}$

85. $(\sqrt{y} - 2)(\sqrt{y} + 2)$ $y - 4$

86. $(\sqrt{x} - y)(\sqrt{x} + y)$ $x - y^2$

87. $(\sqrt{2x} - 3\sqrt{y})(\sqrt{2x} + 3\sqrt{y})$ $2x - 9y$

88. $(2\sqrt{3x} - \sqrt{y})(2\sqrt{3x} + \sqrt{y})$ $12x - y$

89. $(\sqrt{a} - 2)(\sqrt{a} - 3)$ $a - 5\sqrt{a} + 6$

90. $(\sqrt{x} + 4)(\sqrt{x} - 7)$ $x - 3\sqrt{x} - 28$

91. $\sqrt[3]{a} + 2)(\sqrt[3]{a} + 3)$ $\sqrt[3]{a^2} + 5\sqrt[3]{a} + 6$

92. $(\sqrt[3]{x} - 4)(\sqrt[3]{x} + 5)$ $\sqrt[3]{x^2} + \sqrt[3]{x} - 20$

93. $(2\sqrt{x} - \sqrt{y})(3\sqrt{x} + \sqrt{y})$ $6x - \sqrt{xy} - y$

94. $(\sqrt{3x} - 2\sqrt{y})(3\sqrt{3x} - \sqrt{y})$ $9x - 7\sqrt{3xy} + 2y$

4   Simplify.

95. $\dfrac{\sqrt{32x^2}}{\sqrt{2x}}$ $4\sqrt{x}$

96. $\dfrac{\sqrt{60y^4}}{\sqrt{12y}}$ $y\sqrt{5y}$

97. $\dfrac{\sqrt{42a^3b^5}}{\sqrt{14a^2b}}$ $b^2\sqrt{3a}$

98. $\dfrac{\sqrt{65ab^4}}{\sqrt{5ab}}$ $b\sqrt{13b}$

99. $\dfrac{1}{\sqrt{5}}$ $\dfrac{\sqrt{5}}{5}$

100. $\dfrac{1}{\sqrt{2}}$ $\dfrac{\sqrt{2}}{2}$

101. $\dfrac{1}{\sqrt{2x}}$ $\dfrac{\sqrt{2x}}{2x}$

102. $\dfrac{2}{\sqrt{3y}}$ $\dfrac{2\sqrt{3y}}{3y}$

103. $\dfrac{5}{\sqrt{5x}}$ $\dfrac{\sqrt{5x}}{x}$

104. $\dfrac{9}{\sqrt{3a}}$ $\dfrac{3\sqrt{3a}}{a}$

105. $\sqrt{\dfrac{x}{5}}$ $\dfrac{\sqrt{5x}}{5}$

106. $\sqrt{\dfrac{y}{2}}$ $\dfrac{\sqrt{2y}}{2}$

107. $\dfrac{3}{\sqrt[3]{2}}$ $\dfrac{3\sqrt[3]{4}}{2}$

108. $\dfrac{5}{\sqrt[3]{9}}$ $\dfrac{5\sqrt[3]{3}}{3}$

109. $\dfrac{3}{\sqrt[3]{4x^2}}$ $\dfrac{3\sqrt[3]{2x}}{2x}$

110. $\dfrac{5}{\sqrt[3]{3y}}$ $\dfrac{5\sqrt[3]{9y^2}}{3y}$

111. $\dfrac{\sqrt{40x^3y^2}}{\sqrt{80x^2y^3}}$ $\dfrac{\sqrt{2xy}}{2y}$

112. $\dfrac{\sqrt{15a^2b^5}}{\sqrt{30a^5b^3}}$ $\dfrac{b\sqrt{2a}}{2a^2}$

113. $\dfrac{\sqrt{24a^2b}}{\sqrt{18ab^4}}$ $\dfrac{2\sqrt{3ab}}{3b^2}$

114. $\dfrac{\sqrt{12x^3y}}{\sqrt{20x^4y}}$ $\dfrac{\sqrt{15x}}{5x}$

115. $\dfrac{2}{\sqrt{5}+2}$ $2\sqrt{5}-4$

116. $\dfrac{5}{2-\sqrt{7}}$ $-\dfrac{10+5\sqrt{7}}{3}$

117. $\dfrac{3}{\sqrt{y}-2}$ $\dfrac{3\sqrt{y}+6}{y-4}$

118. $\dfrac{-7}{\sqrt{x}-3}$ $-\dfrac{7\sqrt{x}+21}{x-9}$

119. $\dfrac{\sqrt{2}-\sqrt{3}}{\sqrt{2}+\sqrt{3}}$ $-5+2\sqrt{6}$

120. $\dfrac{\sqrt{3}+\sqrt{4}}{\sqrt{2}+\sqrt{3}}$ $-\sqrt{6}+3-2\sqrt{2}+2\sqrt{3}$

121. $\dfrac{2\sqrt{27}-\sqrt{3}}{\sqrt{3}}$ 5

122. $\dfrac{3\sqrt{45}+8\sqrt{20}}{\sqrt{5}}$ 25

123. $\dfrac{\sqrt{x}-\sqrt{x^3}}{\sqrt{x}}$ $1-x$

124. $\dfrac{\sqrt{a^3}+\sqrt{a^5}}{\sqrt{a}}$ $a+a^2$

125. $\dfrac{3}{\sqrt[4]{8x^3}}$ $\dfrac{3\sqrt[4]{2x}}{2x}$

126. $\dfrac{-3}{\sqrt[4]{27y^2}}$ $-\dfrac{\sqrt[4]{3y^2}}{y}$

127. $\dfrac{4}{\sqrt[5]{16a^2}}$ $\dfrac{2\sqrt[5]{2a^3}}{a}$

128. $\dfrac{a}{\sqrt[5]{81a^4}}$ $\dfrac{\sqrt[5]{3a}}{3}$

129. $\dfrac{\sqrt{2a} - \sqrt{2b}}{\sqrt{ab}}$ $\dfrac{a\sqrt{2b} - b\sqrt{2a}}{ab}$

130. $\dfrac{\sqrt{5x} + \sqrt{5y}}{\sqrt{xy}}$ $\dfrac{x\sqrt{5y} + y\sqrt{5x}}{xy}$

131. $\dfrac{\sqrt{a} + a\sqrt{b}}{\sqrt{a} - a\sqrt{b}}$ $\dfrac{1 + 2\sqrt{ab} + ab}{1 - ab}$

132. $\dfrac{\sqrt{3} - 3\sqrt{y}}{\sqrt{3} + 3\sqrt{y}}$ $\dfrac{1 - 2\sqrt{3y} + 3y}{1 - 3y}$

133. $\dfrac{3\sqrt{xy} + 2\sqrt{xy}}{\sqrt{x} - \sqrt{y}}$ $\dfrac{5x\sqrt{y} + 5y\sqrt{x}}{x - y}$

SUPPLEMENTAL EXERCISES 5.3

Simplify.

134. $(\sqrt{8} - \sqrt{2})^3$ $2\sqrt{2}$

135. $(\sqrt{27} - \sqrt{3})^3$ $24\sqrt{3}$

136. $(\sqrt{2} - 2)^3$ $14\sqrt{2} - 20$

137. $(\sqrt{3} - 3)^3$ $30\sqrt{3} - 54$

138. $(\sqrt{2} - 3)^3$ $29\sqrt{2} - 45$

139. $(\sqrt{5} + 2)^3$ $38 + 17\sqrt{5}$

140. $\dfrac{3}{\sqrt{y + 1} + 1}$ $\dfrac{3\sqrt{y + 1} - 3}{y}$

141. $\dfrac{2}{\sqrt{x + 4} + 2}$ $\dfrac{2\sqrt{x + 4} - 4}{x}$

142. $\dfrac{\sqrt{a + 4} + 2}{\sqrt{a + 4} - 2}$ $\dfrac{a + 4\sqrt{a + 4} + 8}{a}$

143. $\dfrac{\sqrt{b + 9} - 3}{\sqrt{b + 9} + 3}$ $\dfrac{b - 6\sqrt{b + 9} + 18}{b}$

144. $\dfrac{3}{\sqrt{x + 3} - \sqrt{x}}$ $\sqrt{x + 3} + \sqrt{x}$

145. $\dfrac{4}{\sqrt{y + 4} - \sqrt{y}}$ $\sqrt{y + 4} + \sqrt{y}$

Rewrite as an expression with a single radical.

146. $\dfrac{\sqrt[3]{(x + y)^2}}{\sqrt{x + y}}$ $\sqrt[6]{(x + y)}$

147. $\dfrac{\sqrt[4]{(a + b)^3}}{\sqrt{a + b}}$ $\sqrt[4]{a + b}$

148. $\sqrt[4]{2y} \, \sqrt{x + 3}$ $\sqrt[4]{2y(x + 3)^2}$

149. $\sqrt[4]{2x} \, \sqrt{y - 2}$ $\sqrt[4]{2x(y - 2)^2}$

150. $\sqrt{a} \, \sqrt[3]{a + 3}$ $\sqrt[6]{a^3(a + 3)^2}$

151. $\sqrt{b} \, \sqrt[3]{b - 1}$ $\sqrt[6]{b^3(b - 1)^2}$

Factor over the set of real numbers.
EXAMPLE: $x^2 - 8$ is the difference of two squares and $\sqrt{8} = (2\sqrt{2})^2$.
$$x^2 - 8 = (x + 2\sqrt{2})(x - 2\sqrt{2}).$$

152. $y^2 - 27$ $(y + 3\sqrt{3})(y - 3\sqrt{3})$

153. $x^2 - 18$ $(x + 3\sqrt{2})(x - 3\sqrt{2})$

154. $a^2 - 12$ $(a + 2\sqrt{3})(a - 2\sqrt{3})$

155. $b^2 - 24$ $(b + 2\sqrt{6})(b - 2\sqrt{6})$

156. $y^2 + 4\sqrt{2}y + 8$ $(y + 2\sqrt{2})^2$

157. $x^2 + 6\sqrt{3}x + 27$ $(x + 3\sqrt{3})^2$

SECTION 5.4

Complex Numbers

1 Simplify complex numbers

The radical expression $\sqrt{-4}$ is not a real number since there is no real number whose square is -4. However, the solution of an algebraic equation is sometimes the square root of a negative number.

For example, the equation $x^2 + 1 = 0$ does not have a real number solution since there is no real number whose square is a negative number.

$$x^2 + 1 = 0$$
$$x^2 = -1$$

Around the seventeenth century, a new number, called an **imaginary number,** was defined so that a negative number would have a square root. The letter i was chosen to represent the number whose square is -1.

Definition of i

The number i, called the **imaginary unit,** has the property that

$$i^2 = -1.$$

Because $i^2 = -1$, i is the square root of -1.

$$i = \sqrt{-1}$$

An imaginary number is defined in terms of i.

If a is a positive real number, then the principal square root of negative a is the imaginary number $i\sqrt{a}$.

$$\sqrt{-a} = i\sqrt{a}$$

An imaginary number can be written as the product of a real number and i.

$$\sqrt{-4} = i\sqrt{4} = 2i$$
$$\sqrt{-13} = i\sqrt{13}$$
$$\sqrt{-7} = i\sqrt{7}$$

It is customary to write i in front of the radical to avoid confusing $\sqrt{a}i$ with $\sqrt{ai}$.

To simplify $\sqrt{-12}$, write $\sqrt{-12}$ as the product of a real number and i.
Simplify the radical factor.

$$\sqrt{-12} = i\sqrt{12}$$
$$= i\sqrt{2^2 \cdot 3}$$
$$= 2i\sqrt{3}$$

Example 1   Simplify: $\sqrt{-80}$

Solution $\sqrt{-80} = i\sqrt{80} = i\sqrt{2^4 \cdot 5} = 4i\sqrt{5}$

Problem 1   Simplify: $\sqrt{-45}$

Solution   See page A41.
$3i\sqrt{5}$

The real numbers and the imaginary numbers make up the complex numbers.

> ### Definition of a Complex Number
>
> A **complex number** is a number of the form $a + bi$, where a and b are real numbers, and $i = \sqrt{-1}$. The number a is the real part of $a + bi$, and b is the imaginary part.

Examples of complex numbers are shown below.

Real Part	Imaginary Part
a +	bi
3 +	$2i$
8 −	$10i$
$2x$ +	$3yi$

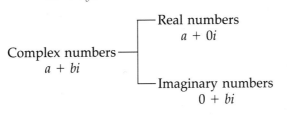

Complex numbers
$a + bi$

— Real numbers
$a + 0i$

— Imaginary numbers
$0 + bi$

A **real number** is a complex number in which $b = 0$.

An **imaginary number** is a complex number in which $a = 0$.

To simplify $\sqrt{20} - \sqrt{-50}$, write the complex number in the form $a + bi$.

$$\sqrt{20} - \sqrt{-50} = \sqrt{20} - i\sqrt{50}$$

Use the Product Property of Radicals to simplify each radical.

$$= \sqrt{2^2 \cdot 5} - i\sqrt{5^2 \cdot 2}$$
$$= 2\sqrt{5} - 5i\sqrt{2}$$

Example 2   Simplify: $\sqrt{25} + \sqrt{-40}$

Solution $\sqrt{25} + \sqrt{-40} = \sqrt{25} + i\sqrt{40} = \sqrt{5^2} + i\sqrt{2^2 \cdot 2 \cdot 5} = 5 + 2i\sqrt{10}$

Problem 2   Simplify: $\sqrt{98} - \sqrt{-60}$

Solution   See page A41.
$7\sqrt{2} - 2i\sqrt{15}$

2 Add and subtract complex numbers

To add two complex numbers, add the real parts and add the imaginary parts.

$$(a + bi) + (c + di) = (a + c) + (b + d)i$$

To subtract two complex numbers, subtract the real parts and subtract the imaginary parts.

$$(a + bi) - (c + di) = (a - c) + (b - d)i$$

For example, $(3 - 7i) - (4 - 2i) = (3 - 4) + [-7 - (-2)]i = -1 - 5i$.

Example 3   Simplify: $(3 + 2i) + (6 - 5i)$

Solution $(3 + 2i) + (6 - 5i) = (3 + 6) + (2 - 5)i$ • Add the real parts and add the imaginary parts. (Do this step mentally.)

$$= 9 - 3i$$

Problem 3   Simplify: $(-4 + 2i) - (6 - 8i)$

Solution   See page A41.
$-10 + 10i$

To simplify $(3 + \sqrt{-12}) + (7 - \sqrt{-27})$, write each complex number in the form $a + bi$.

$(3 + \sqrt{-12}) + (7 - \sqrt{-27}) =$
$(3 + i\sqrt{12}) + (7 - i\sqrt{27}) =$

Use the Product Property of Radicals to simplify each radical.

$(3 + i\sqrt{2^2 \cdot 3}) + (7 - i\sqrt{3^2 \cdot 3}) =$
$(3 + 2i\sqrt{3}) + (7 - 3i\sqrt{3}) =$

Add the complex numbers.

$10 - i\sqrt{3}$

Example 4   Simplify: $(9 - \sqrt{-8}) - (5 + \sqrt{-32})$

Solution $(9 - \sqrt{-8}) - (5 + \sqrt{-32}) =$
$(9 - i\sqrt{8}) - (5 + i\sqrt{32}) =$ • Write each complex number in the form $a + bi$.

$(9 - i\sqrt{2^2 \cdot 2}) - (5 + i\sqrt{2^4 \cdot 2}) =$ • Simplify each radical.
$(9 - 2i\sqrt{2}) - (5 + 4i\sqrt{2}) = 4 - 6i\sqrt{2}$

Problem 4   Simplify: $(16 - \sqrt{-45}) - (3 + \sqrt{-20})$

Solution   See page A41.
$13 - 5i\sqrt{5}$

3 Multiply complex numbers

When multiplying complex numbers, the term i^2 is frequently a part of the product. Recall that $i^2 = -1$.

To simplify $2i \cdot 3i$, multiply the imaginary numbers.

$$2i \cdot 3i = 6i^2$$

Replace i^2 by -1.

$$= 6(-1)$$

Simplify.

$$= -6$$

To simplify $\sqrt{-6} \cdot \sqrt{-24}$, write each radical as the product of a real number and i.

$$\sqrt{-6} \cdot \sqrt{-24} = i\sqrt{6} \cdot i\sqrt{24}$$

Multiply the imaginary numbers.

$$= i^2\sqrt{144}$$

Replace i^2 by -1.

$$= -\sqrt{144}$$

Simplify.

$$= -12$$

Note from this example that it would have been incorrect to multiply the radicands of the two radical expressions. To illustrate,

$$\sqrt{-6} \cdot \sqrt{-24} = \sqrt{(-6)(-24)} = \sqrt{144} = 12, \; not \; -12.$$

When simplifying square roots of negative numbers, first rewrite the radical expressions using i.

To simplify $4i(3 - 2i)$, use the Distributive Property to remove parentheses.

$$4i(3 - 2i) = 12i - 8i^2$$

Replace i^2 by -1.

$$= 12i - 8(-1)$$

Write the answer in the form $a + bi$.

$$= 8 + 12i$$

Example 5 Simplify: $\sqrt{-8}(\sqrt{6} - \sqrt{-2})$

Solution $\sqrt{-8}(\sqrt{6} - \sqrt{-2}) = i\sqrt{8}(\sqrt{6} - i\sqrt{2})$

- Write each complex number in the form $a + bi$.

$$= i\sqrt{48} - i^2\sqrt{16}$$

- Use the Distributive Property.

$$= i\sqrt{2^4 \cdot 3} - (-1)\sqrt{2^4}$$

- Simplify each radical. Replace i^2 by -1.

$$= 4i\sqrt{3} + 4$$
$$= 4 + 4i\sqrt{3}$$

- Write the answer in the form $a + bi$.

Problem 5 Simplify: $\sqrt{-3}(\sqrt{27} - \sqrt{-6})$

Solution See page A41.

$3\sqrt{2} + 9i$

The product of two complex numbers can be found by using the FOIL method. For example,

$$(2 + 4i)(3 - 5i) = 6 - 10i + 12i - 20i^2$$
$$= 6 + 2i - 20i^2$$
$$= 6 + 2i - 20(-1)$$
$$= 26 + 2i$$

The conjugate of $a + bi$ is $a - bi$.

The product of conjugates of the form $(a + bi)(a - bi)$ is $a^2 + b^2$.

$$(a + bi)(a - bi) = a^2 - b^2i^2 = a^2 - b^2(-1) = a^2 + b^2$$

For example, $(2 + 3i)(2 - 3i) = 2^2 + 3^2 = 4 + 9 = 13$.

Note that the product of a complex number and its conjugate is a real number.

Example 6   Simplify.

A. $(3 - 4i)(2 + 5i)$ B. $\left(\dfrac{9}{10} + \dfrac{3}{10}i\right)\left(1 - \dfrac{1}{3}i\right)$ C. $(4 + 5i)(4 - 5i)$

Solution   A. $(3 - 4i)(2 + 5i) =$
$6 + 15i - 8i - 20i^2 =$ • Use the FOIL method.
$6 + 7i - 20i^2 =$ • Combine like terms.
$6 + 7i - 20(-1) =$ • Replace i^2 by -1.
$26 + 7i$ • Write the answer in the form $a + bi$.

B. $\left(\dfrac{9}{10} + \dfrac{3}{10}i\right)\left(1 - \dfrac{1}{3}i\right) =$

$\dfrac{9}{10} - \dfrac{3}{10}i + \dfrac{3}{10}i - \dfrac{1}{10}i^2 =$ • Use the FOIL method.

$\dfrac{9}{10} - \dfrac{1}{10}i^2 = \dfrac{9}{10} - \dfrac{1}{10}(-1) =$ • Combine like terms. Replace i^2 by -1.

$\dfrac{9}{10} + \dfrac{1}{10} = 1$ • Simplify.

C. $(4 + 5i)(4 - 5i) = 4^2 + 5^2$ • The product of conjugates of the
$\qquad\qquad\qquad = 16 + 25 = 41$ form $(a + bi)(a - bi)$ is $a^2 + b^2$.

Problem 6   Simplify.

A. $(4 - 3i)(2 - i)$ B. $(3 - i)\left(\dfrac{3}{10} + \dfrac{1}{10}i\right)$ C. $(3 + 6i)(3 - 6i)$

Solution   See page A41.
A. $5 - 10i$ B. 1   C. 45

4   Divide complex numbers

A rational expression containing one or more complex numbers is in simplest form when no imaginary number remains in the denominator.

To simplify $\dfrac{2 - 3i}{2i}$, multiply the expression

by 1 in the form $\dfrac{i}{i}$.

Replace i^2 by -1.

Simplify.

Write the answer in the form $a + bi$.

$$\begin{aligned}
\frac{2 - 3i}{2i} &= \frac{2 - 3i}{2i} \cdot \frac{i}{i} \\
&= \frac{2i - 3i^2}{2i^2} \\
&= \frac{2i - 3(-1)}{2(-1)} \\
&= \frac{3 + 2i}{-2} \\
&= -\frac{3}{2} - i
\end{aligned}$$

Example 7   Simplify: $\dfrac{5 + 4i}{3i}$

Solution $\dfrac{5 + 4i}{3i} = \dfrac{5 + 4i}{3i} \cdot \dfrac{i}{i} = \dfrac{5i + 4i^2}{3i^2} = \dfrac{5i + 4(-1)}{3(-1)} = \dfrac{-4 + 5i}{-3} = \dfrac{4}{3} - \dfrac{5}{3}i$

Problem 7   Simplify: $\dfrac{2 - 3i}{4i}$

Solution   See page A41. $-\dfrac{3}{4} - \dfrac{1}{2}i$

To simplify a fraction that has a complex number in the denominator, multiply the numerator and denominator by the conjugate of the complex number.

$$\begin{aligned}
\frac{3 + 2i}{1 + i} &= \frac{(3 + 2i)}{(1 + i)} \cdot \frac{(1 - i)}{(1 - i)} = \frac{3 - 3i + 2i - 2i^2}{1^2 + 1^2} \\
&= \frac{3 - i - 2(-1)}{2} = \frac{5 - i}{2} = \frac{5}{2} - \frac{1}{2}i
\end{aligned}$$

Example 8   Simplify: $\dfrac{5 - 3i}{4 + 2i}$

Solution $\dfrac{5 - 3i}{4 + 2i} = \dfrac{(5 - 3i)}{(4 + 2i)} \cdot \dfrac{(4 - 2i)}{(4 - 2i)} = \dfrac{20 - 10i - 12i + 6i^2}{4^2 + 2^2} = \dfrac{20 - 22i + 6(-1)}{20}$

$= \dfrac{14 - 22i}{20} = \dfrac{7 - 11i}{10} = \dfrac{7}{10} - \dfrac{11}{10}i$

Problem 8 Simplify: $\dfrac{2 + 5i}{3 - 2i}$

Solution See page A42. $-\dfrac{4}{13} + \dfrac{19}{13}i$

EXERCISES 5.4

1 Simplify.

1. $\sqrt{-4}$
$2i$

2. $\sqrt{-64}$
$8i$

3. $\sqrt{-98}$
$7i\sqrt{2}$

4. $\sqrt{-72}$
$6i\sqrt{2}$

5. $\sqrt{-27}$
$3i\sqrt{3}$

6. $\sqrt{-75}$
$5i\sqrt{3}$

7. $\sqrt{16} + \sqrt{-4}$
$4 + 2i$

8. $\sqrt{25} + \sqrt{-9}$
$5 + 3i$

9. $\sqrt{12} - \sqrt{-18}$
$2\sqrt{3} - 3i\sqrt{2}$

10. $\sqrt{60} - \sqrt{-48}$
$2\sqrt{15} - 4i\sqrt{3}$

11. $\sqrt{160} - \sqrt{-147}$
$4\sqrt{10} - 7i\sqrt{3}$

12. $\sqrt{96} - \sqrt{-125}$
$4\sqrt{6} - 5i\sqrt{5}$

13. $\sqrt{-4a^2}$
$2ai$

14. $\sqrt{-16b^6}$
$4b^3i$

15. $\sqrt{-49x^{12}}$
$7x^6i$

16. $\sqrt{-32x^3y^2}$
$4xyi\sqrt{2x}$

17. $\sqrt{-144a^3b^5}$
$12ab^2i\sqrt{ab}$

18. $\sqrt{-81a^{10}b^9}$
$9a^5b^4i\sqrt{b}$

19. $\sqrt{4a} + \sqrt{-12a^2}$
$2\sqrt{a} + 2ai\sqrt{3}$

20. $\sqrt{25b} - \sqrt{-48b^2}$
$5\sqrt{b} - 4bi\sqrt{3}$

21. $\sqrt{18b^5} - \sqrt{-27b^3}$
$3b^2\sqrt{2b} - 3bi\sqrt{3b}$

22. $\sqrt{a^5b^2} - \sqrt{-a^5b^2}$
$a^2b\sqrt{a} - a^2bi\sqrt{a}$

23. $\sqrt{-50x^3y^3} + x\sqrt{25x^4y^3}$
$5x^3y\sqrt{y} + 5xyi\sqrt{2xy}$

24. $\sqrt{-121xy} + \sqrt{60x^2y^2}$
$2xy\sqrt{15} + 11i\sqrt{xy}$

25. $\sqrt{-49a^5b^2} - ab\sqrt{-25a^3}$
$2a^2bi\sqrt{a}$

26. $\sqrt{-16x^2y} - x\sqrt{-49y}$
$-3xi\sqrt{y}$

27. $\sqrt{12a^3} + \sqrt{-27b^3}$
$2a\sqrt{3a} + 3bi\sqrt{3b}$

2 Simplify.

28. $(2 + 4i) + (6 - 5i)$
$8 - i$

29. $(6 - 9i) + (4 + 2i)$
$10 - 7i$

30. $(-2 - 4i) - (6 - 8i)$
$-8 + 4i$

31. $(3 - 5i) + (8 - 2i)$
$11 - 7i$

32. $(8 - \sqrt{-4}) - (2 + \sqrt{-16})$
$6 - 6i$

33. $(5 - \sqrt{-25}) - (11 - \sqrt{-36})$
$-6 + i$

34. $(12 - \sqrt{-50}) + (7 - \sqrt{-8})$
$19 - 7i\sqrt{2}$

35. $(5 - \sqrt{-12}) - (9 + \sqrt{-108})$
$-4 - 8i\sqrt{3}$

36. $(\sqrt{8} + \sqrt{-18}) + (\sqrt{32} - \sqrt{-72})$
$6\sqrt{2} - 3i\sqrt{2}$

37. $(\sqrt{40} - \sqrt{-98}) - (\sqrt{90} + \sqrt{-32})$
$-\sqrt{10} - 11i\sqrt{2}$

38. $(5 - 3i) + 2i$
$5 - i$

39. $(6 - 8i) + 4i$
$6 - 4i$

40. $(7 + 2i) + (-7 - 2i)$
0

41. $(8 - 3i) + (-8 + 3i)$
0

42. $(9 + 4i) + 6$
$15 + 4i$

43. $(4 + 6i) + 7$
$11 + 6i$

3   Simplify.

44. $(7i)(-9i)$ 63

45. $(-6i)(-4i)$ -24

46. $\sqrt{-2}\,\sqrt{-8}$ -4

47. $\sqrt{-5}\,\sqrt{-45}$ -15

48. $\sqrt{-3}\,\sqrt{-6}$ $-3\sqrt{2}$

49. $\sqrt{-5}\,\sqrt{-10}$ $-5\sqrt{2}$

50. $2i(6 + 2i)$ $-4 + 12i$

51. $-3i(4 - 5i)$ $-15 - 12i$

52. $\sqrt{-2}\,(\sqrt{8} + \sqrt{-2})$ $-2 + 4i$

53. $\sqrt{-3}\,(\sqrt{12} - \sqrt{-6})$ $3\sqrt{2} + 6i$

54. $(5 - 2i)(3 + i)$ $17 - i$

55. $(2 - 4i)(2 - i)$ $-10i$

56. $(6 + 5i)(3 + 2i)$ $8 + 27i$

57. $(4 - 7i)(2 + 3i)$ $29 - 2i$

58. $(1 - i)\left(\dfrac{1}{2} + \dfrac{1}{2}i\right)$ 1

59. $\left(\dfrac{4}{5} - \dfrac{2}{5}i\right)\left(1 + \dfrac{1}{2}i\right)$ 1

60. $\left(\dfrac{6}{5} + \dfrac{3}{5}i\right)\left(\dfrac{2}{3} - \dfrac{1}{3}i\right)$ 1

61. $(2 - i)\left(\dfrac{2}{5} + \dfrac{1}{5}i\right)$ 1

62. $(4 - 3i)(4 + 3i)$ 25

63. $(8 - 5i)(8 + 5i)$ 89

64. $(3 - i)(3 + i)$ 10

65. $(7 - i)(7 + i)$ 50

66. $(6 - \sqrt{-2})^2$ $34 - 12i\sqrt{2}$

67. $(9 - \sqrt{-1})^2$ $80 - 18i$

4   Simplify.

68. $\dfrac{3}{i}$ $-3i$

69. $\dfrac{4}{5i}$ $-\dfrac{4}{5}i$

70. $\dfrac{2 - 3i}{-4i}$ $\dfrac{3}{4} + \dfrac{1}{2}i$

71. $\dfrac{16 + 5i}{-3i}$ $-\dfrac{5}{3} + \dfrac{16}{3}i$

72. $\dfrac{4}{5 + i}$ $\dfrac{10}{13} - \dfrac{2}{13}i$

73. $\dfrac{6}{5 + 2i}$ $\dfrac{30}{29} - \dfrac{12}{29}i$

74. $\dfrac{2}{2 - i}$ $\dfrac{4}{5} + \dfrac{2}{5}i$

75. $\dfrac{5}{4 - i}$ $\dfrac{20}{17} + \dfrac{5}{17}i$

76. $\dfrac{1 - 3i}{3 + i}$ $-i$

77. $\dfrac{2 + 12i}{5 + i}$ $\dfrac{11}{13} + \dfrac{29}{13}i$

78. $\dfrac{\sqrt{-10}}{\sqrt{8} - \sqrt{-2}}$ $-\dfrac{\sqrt{5}}{5} + \dfrac{2\sqrt{5}}{5}i$

79. $\dfrac{\sqrt{-2}}{\sqrt{12} - \sqrt{-8}}$ $-\dfrac{1}{5} + \dfrac{\sqrt{6}}{10}i$

80. $\dfrac{2 - 3i}{3 + i}$ $\dfrac{3}{10} - \dfrac{11}{10}i$

81. $\dfrac{3 + 5i}{1 - i}$ $-1 + 4i$

82. $\dfrac{5 + 3i}{3 - i}$ $\dfrac{6}{5} + \dfrac{7}{5}i$

SUPPLEMENTAL EXERCISES 5.4

Note the pattern when successive powers of i are simplified.

$i^1 = i$ $i^5 = i \cdot i^4 = i(1) = i$
$i^2 = -1$ $i^6 = i^2 \cdot i^4 = -1$
$i^3 = i^2 \cdot i = -i$ $i^7 = i^3 \cdot i^4 = -i$
$i^4 = i^2 \cdot i^2 = (-1)(-1) = 1$ $i^8 = i^4 \cdot i^4 = 1$

83. When the exponent on i is a multiple of 4, the power equals ___1___ .

Use the pattern above to simplify the power of i.

84. i^6 -1

85. i^9 i

86. i^{57} i

87. i^{65} i

88. i^{220} 1

89. i^{460} 1

90. i^0 1

91. i^{-2} -1

92. i^{-6} -1

93. i^{-34} -1

94. i^{-58} -1

95. i^{-180} 1

The property that the product of conjugates of the form $(a + bi)(a - bi) = a^2 + b^2$ can be used to factor the sum of two perfect squares over the set of complex numbers. For example, $x^2 + y^2 = (x + yi)(x - yi)$. Factor over the set of complex numbers.

96. $y^2 + 1$
$(y + i)(y - i)$

97. $a^2 + 4$
$(a + 2i)(a - 2i)$

98. $x^2 + 25$
$(x + 5i)(x - 5i)$

99. $4b^2 + 9$
$(2b + 3i)(2b - 3i)$

100. $16x^2 + y^2$
$(4x + yi)(4x - yi)$

101. $36a^2 + b^2$
$(6a + bi)(6a - bi)$

102. $49x^2 + 16$
$(7x + 4i)(7x - 4i)$

103. $9a^2 + 64$
$(3a + 8i)(3a - 8i)$

Solve.

104. **a.** Is $3i$ a solution of $2x^2 + 18 = 0$?
yes

 b. Is $-3i$ a solution of $2x^2 + 18 = 0$?
yes

105. **a.** Is $7i$ a solution of $x^2 + 49 = 0$?
yes

 b. Is $-7i$ a solution of $x^2 + 49 = 0$?
yes

106. **a.** Is $3 + i$ a solution of $x^2 - 6x + 10 = 0$?
yes

 b. Is $3 - i$ a solution of $x^2 - 6x + 10 = 0$?
yes

107. **a.** Is $1 + 3i$ a solution of $x^2 - 2x - 10 = 0$?
no

 b. Is $1 - 3i$ a solution of $x^2 - 2x - 10 = 0$?
no

108. **a.** Is $2 + 3i$ a solution of $x^2 - 4x + 13 = 0$?
yes

 b. Is $2 - 3i$ a solution of $x^2 - 4x + 13 = 0$?
yes

109. **a.** Is $-3 + 2i$ a solution of $x^2 + 6x + 13 = 0$?
yes

 b. Is $-3 - 2i$ a solution of $x^2 + 6x + 13 = 0$?
yes

SECTION 5.5

Equations Containing Radical Expressions

1 Solve equations containing one or more radical expressions

An equation that contains a variable expression in a radicand is a **radical equation.**

$$\sqrt{x + 2} = \sqrt{3x - 4}$$
$$\sqrt[3]{x - 4} = 2$$
$\left.\begin{array}{l}\end{array}\right\}$ Radical Equations

The following property of equality is used to solve radical equations:

> ## The Property of Raising Both Sides of an Equation to a Power
> If a and b are real numbers and $a = b$, then $a^n = b^n$.

Solve: $\sqrt{x - 2} - 6 = 0$

Rewrite the equation with the radical on one side of the equation and the constant on the other side.

$$\sqrt{x - 2} - 6 = 0$$
$$\sqrt{x - 2} = 6$$

Square both sides of the equation.

$$(\sqrt{x - 2})^2 = 6^2$$

Solve the resulting equation.

$$x - 2 = 36$$
$$x = 38$$

Check the solution.
When raising both sides of an equation to an even power, the resulting equation may have a solution that is not a solution of the original equation. Therefore, it is necessary to check the solution of a radical equation. Example 1C illustrates this.

Check:
$$\begin{array}{c|c} \sqrt{x - 2} - 6 = 0 & \\ \hline \sqrt{38 - 2} - 6 & 0 \\ \sqrt{36} - 6 & 0 \\ 6 - 6 & 0 \\ 0 = 0 & \end{array}$$

38 checks as a solution.
The solution is 38.

Example 1 Solve.

A. $\sqrt{3x - 2} - 8 = -3$ B. $\sqrt[3]{3x - 1} = -4$ C. $x + 2\sqrt{x - 1} = 9$

Solution A. $\sqrt{3x - 2} - 8 = -3$
$$\sqrt{3x - 2} = 5$$

• Rewrite the equation so that the radical is alone on one side of the equation.

$$(\sqrt{3x - 2})^2 = 5^2$$
$$3x - 2 = 25$$
$$3x = 27$$
$$x = 9$$

• Square both sides of the equation.
• Solve the resulting equation.

Check:
$$\begin{array}{c|c} \sqrt{3x - 2} - 8 = -3 & \\ \hline \sqrt{3 \cdot 9 - 2} - 8 & -3 \\ \sqrt{27 - 2} - 8 & -3 \\ \sqrt{25} - 8 & -3 \\ 5 - 8 & -3 \\ -3 = -3 & \end{array}$$

• Check the solution.

The solution is 9.

B. $\quad \sqrt[3]{3x - 1} = -4$

$\quad (\sqrt[3]{3x - 1})^3 = (-4)^3$ • Cube both sides of the equation.

$\quad\quad\quad 3x - 1 = -64$ • Solve the resulting equation.

$\quad\quad\quad\quad\quad 3x = -63$

$\quad\quad\quad\quad\quad\; x = -21$

Check: $\sqrt[3]{3x - 1} = -4$ • Check the solution.

$$
\begin{array}{c|c}
\sqrt[3]{3(-21) - 1} & -4 \\
\hline
\sqrt[3]{-63 - 1} & -4 \\
\sqrt[3]{-64} & -4 \\
\end{array}
$$

$\quad\quad\quad\quad\quad -4 = -4$

The solution is -21.

C. $\; x + 2\sqrt{x - 1} = 9$

$\quad\quad 2\sqrt{x - 1} = 9 - x$ • Rewrite the equation with the radical on one side of the equation.

$\quad (2\sqrt{x - 1})^2 = (9 - x)^2$ • Square both sides of the equation.

$\quad\quad 4(x - 1) = 81 - 18x + x^2$

$\quad\quad 4x - 4 = 81 - 18x + x^2$

$\quad\quad\quad\quad 0 = x^2 - 22x + 85$ • Write the quadratic equation in standard form.

$\quad\quad\quad\quad 0 = (x - 5)(x - 17)$ • Factor.

$x - 5 = 0 \quad\quad x - 17 = 0$ • Use of Principle of Zero Products.

$\quad\; x = 5 \quad\quad\quad\; x = 17$

Check:

$$
\begin{array}{c|c}
x + 2\sqrt{x - 1} = 9 & \\
\hline
5 + 2\sqrt{5 - 1} & 9 \\
5 + 2\sqrt{4} & 9 \\
5 + 2 \cdot 2 & 9 \\
5 + 4 & 9 \\
9 = 9 &
\end{array}
\qquad
\begin{array}{c|c}
x + 2\sqrt{x - 1} = 9 & \\
\hline
17 + 2\sqrt{17 - 1} & 9 \\
17 + 2\sqrt{16} & 9 \\
17 + 2 \cdot 4 & 9 \\
17 + 8 & 9 \\
25 \neq 9 &
\end{array}
$$

17 does not check as a solution.
The solution is 5.

Problem 1   Solve.

     A. $\sqrt{4x + 5} - 12 = -5$ B. $\sqrt[4]{x - 8} = 3$ C. $x + 3\sqrt{x + 2} = 8$

Solution   See page A42.

     A. 11   B. 89   C. 2

Example 2   Solve.

A. $\sqrt[3]{x + 8} = \sqrt[3]{5x - 20}$ B. $\sqrt{x + 7} = \sqrt{x} + 1$

Solution   A. $\sqrt[3]{x + 8} = \sqrt[3]{5x - 20}$ • A radical appears on each side of the equation.

$(\sqrt[3]{x + 8})^3 = (\sqrt[3]{5x - 20})^3$ • Cube both sides of the equation.
$x + 8 = 5x - 20$ • Solve the resulting equation.
$-4x + 8 = -20$
$-4x = -28$
$x = 7$

Check: $\dfrac{\sqrt[3]{x + 8} = \sqrt[3]{5x - 20}}{\begin{array}{c|c} \sqrt[3]{7 + 8} & \sqrt[3]{5 \cdot 7 - 20} \\ \sqrt[3]{15} & \sqrt[3]{35 - 20} \\ \sqrt[3]{15} = \sqrt[3]{15} \end{array}}$ • Check the solution.

The solution is 7.

B. $\sqrt{x + 7} = \sqrt{x} + 1$ • A radical appears on each side of the equation.

$(\sqrt{x + 7})^2 = (\sqrt{x} + 1)^2$ • Square both sides of the equation.
$x + 7 = x + 2\sqrt{x} + 1$ • Simplify the resulting equation.
$6 = 2\sqrt{x}$
$3 = \sqrt{x}$ • The equation contains a radical.
$3^2 = (\sqrt{x})^2$ • Square both sides of the equation.
$9 = x$

Check: $\dfrac{\sqrt{x + 7} = \sqrt{x} + 1}{\begin{array}{c|c} \sqrt{9 + 7} & \sqrt{9} + 1 \\ \sqrt{16} & 3 + 1 \\ 4 = 4 \end{array}}$ • Check the solution.

The solution is 9.

Problem 2   Solve.

A. $\sqrt{3x - 5} = \sqrt{x + 7}$ B. $\sqrt{x + 5} = 5 - \sqrt{x}$

Solution   See page A43.
A. 6   B. 4

2 Application problems

A right triangle contains one 90° angle. The side opposite the 90° angle is called the **hypotenuse**. The other two sides are called **legs**.

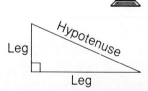

Pythagoras, a Greek mathematician, discovered that the square of the hypotenuse of a right triangle is equal to the sum of the squares of the two legs. This is called the **Pythagorean Theorem.**

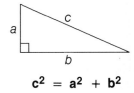

$$c^2 = a^2 + b^2$$

Example 3   A ladder 20 ft long is leaning against a building. How high on the building will the ladder reach when the bottom of the ladder is 8 ft from the building? Round to the nearest tenth.

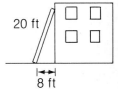

Strategy   To find the distance, use the Pythagorean Theorem. The hypotenuse is the length of the ladder. One leg is the distance from the bottom of the ladder to the base of the building. The distance along the building from the ground to the top of the ladder is the unknown leg.

Solution
$$c^2 = a^2 + b^2$$
$$20^2 = 8^2 + b^2$$
$$400 = 64 + b^2$$
$$336 = b^2$$
$$(336)^{\frac{1}{2}} = (b^2)^{\frac{1}{2}}$$
$$\sqrt{336} = b$$
$$18.3 \approx b$$

The distance is 18.3 ft.

Problem 3   Find the diagonal of a rectangle that is 6 cm in length and 3 cm in width. Round to the nearest tenth.

Solution   See page A43.
6.7 cm

Example 4   An object is dropped from a high building. Find the distance the object has fallen when the speed reaches 96 ft/s. Use the equation $v = \sqrt{64d}$, where v is the speed of the object, and d is the distance.

Strategy   To find the distance the object has fallen, replace v in the equation with the given value and solve for d.

Solution
$$v = \sqrt{64d}$$
$$96 = \sqrt{64d}$$
$$(96)^2 = (\sqrt{64d})^2$$
$$9216 = 64d$$
$$144 = d$$

The object has fallen 144 ft.

Problem 4 How far would a submarine periscope have to be above the water to locate a ship 5.5 mi away? The equation for the distance in miles that the lookout can see is $d = 1.4\sqrt{h}$, where h is the height in feet above the surface of the water. Round to the nearest thousandth.

Solution See page A43.

15.434 ft

EXERCISES 5.5

1 Solve.

1. $\sqrt{x} = 5$ 25

2. $\sqrt{y} = 2$ 4

3. $\sqrt[3]{a} = 3$ 27

4. $\sqrt[3]{y} = 5$ 125

5. $\sqrt{3x} = 12$ 48

6. $\sqrt{5x} = 10$ 20

7. $\sqrt[3]{4x} = -2$ -2

8. $\sqrt[3]{6x} = -3$ $-\dfrac{9}{2}$

9. $\sqrt{2x} = -4$ no solution

10. $\sqrt{5x} = -5$ no solution

11. $\sqrt{3x - 2} = 5$ 9

12. $\sqrt{5x - 4} = 9$ 17

13. $\sqrt{3 - 2x} = 7$ -23

14. $\sqrt{9 - 4x} = 4$ $-\dfrac{7}{4}$

15. $7 = \sqrt{1 - 3x}$ -16

16. $6 = \sqrt{8 - 7x}$ -4

17. $\sqrt[3]{4x - 1} = 2$ $\dfrac{9}{4}$

18. $\sqrt[3]{5x + 2} = 3$ 5

19. $\sqrt[3]{1 - 2x} = -3$ 14

20. $\sqrt[3]{3 - 2x} = -2$ $\dfrac{11}{2}$

21. $\sqrt[3]{9x + 1} = 4$ 7

22. $\sqrt{3x + 9} - 12 = 0$ 45

23. $\sqrt{4x - 3} - 5 = 0$ 7

24. $\sqrt{4x - 2} = \sqrt{3x + 9}$ 11

25. $\sqrt{2x + 4} = \sqrt{5x - 9}$ $\dfrac{13}{3}$

26. $\sqrt[3]{x - 2} = 3$ 29

27. $\sqrt[3]{2x - 6} = 4$ 35

28. $\sqrt[3]{3x - 9} = \sqrt[3]{2x + 12}$ 21

29. $\sqrt[3]{x - 12} = \sqrt[3]{5x + 16}$ -7

30. $\sqrt[4]{4x + 1} = 2$ $\dfrac{15}{4}$

31. $\sqrt[4]{2x - 9} = 3$ 45

32. $\sqrt{2x - 3} - 2 = 1$ 6

33. $\sqrt{3x - 5} - 5 = 3$ 23

34. $\sqrt[3]{2x - 3} + 5 = 2$ -12

35. $\sqrt[3]{x - 4} + 7 = 5$ -4

36. $\sqrt{5x - 16} + 1 = 4$ 35

37. $\sqrt{3x - 5} - 2 = 3$ 10

38. $\sqrt{2x - 1} - 8 = -5$ 5

39. $\sqrt{7x + 2} - 10 = -7$ 1

40. $\sqrt[3]{4x - 3} - 2 = 3$ 32

41. $\sqrt[3]{1 - 3x} + 5 = 3$ 3

42. $1 - \sqrt{4x + 3} = -5$ $\dfrac{33}{4}$

43. $7 - \sqrt{3x + 1} = -1$ 21

44. $\sqrt{x + 1} = 2 - \sqrt{x}$ $\dfrac{9}{16}$

45. $\sqrt{2x + 4} = 3 - \sqrt{2x}$ $\dfrac{25}{72}$

46. $\sqrt{x^2 + 3x - 2} - x = 1$ 3

47. $\sqrt{x^2 - 4x - 1} + 3 = x$ 5

48. $\sqrt{x^2 - 3x - 1} = 3$ 5, −2

49. $\sqrt{x^2 - 2x + 1} = 3$ 4, −2

50. $\sqrt[3]{x^2 + 2} - 3 = 0$ 5, −5

51. $\sqrt[3]{x^2 + 4} - 2 = 0$ 2, −2

52. $\sqrt[4]{x^2 + 2x + 8} - 2 = 0$ −4, 2

53. $\sqrt[4]{x^2 + x - 1} - 1 = 0$ −2, 1

54. $\sqrt[4]{x + 1} - x = 1$ 15, −1

55. $\sqrt[3]{x - 2} + 2 = x$ 2, 11

56. $x + 3\sqrt{x - 2} = 12$ 6

57. $x + 2\sqrt{x + 1} = 7$ 3

2   Solve.

58. Find the width of a rectangle that has a diagonal of 10 ft and a length of 8 ft.
6 ft

59. A 16-ft ladder is leaning against a building. How high on the building will the ladder reach when the bottom of the ladder is 5 ft from the building? Round to the nearest tenth.
15.2 ft

60. Find the length of a rectangle that has a diagonal of 13 m and a width of 5 m.
12 m

61. A 26-ft ladder is leaning against a building. How far is the bottom of the ladder from the wall when the ladder reaches a height of 24 ft on the building?
10 ft

62. How far would a submarine periscope have to be above the water to locate a ship 4.2 mi away? The equation for the distance in miles that the lookout can see is $d = 1.4\sqrt{h}$, where h is the height in feet above the surface of the water.
9 ft

63. How far would a submarine periscope have to be above the water to locate a ship 3.6 mi away? The equation for the distance in miles that the lookout can see is $d = 1.4\sqrt{h}$, where h is the height in feet above the surface of the water. Round to the nearest hundredth.
6.61 ft

64. An object is dropped from a bridge. Find the distance the object has fallen when the speed reaches 100 ft/s. Use the equation $v = \sqrt{64d}$, where v is the speed of the object, and d is the distance.
152.25 ft

65. An object is dropped from a high building. Find the distance the object has fallen when the speed reaches 400 ft/s. Use the equation $v = \sqrt{64d}$, where v is the speed of the object, and d is the distance.
2500 ft

66. Find the distance required for a car to reach a velocity of 40 m/s when the acceleration is 10 m/s². Use the equation $v = \sqrt{2as}$, where v is the velocity, a is the acceleration, and s is the distance.
80 m

67. Find the distance required for a car to reach a velocity of 60 ft/s when the acceleration is 15 ft/s². Use the equation $v = \sqrt{2as}$, where v is the velocity, a is the acceleration, and s is the distance.
120 ft

Use your calculator for the following exercises.

68. Find the length of a pendulum that makes one swing in 3 s. The equation for the time of one swing of a pendulum is given by $T = 2\pi\sqrt{\dfrac{L}{32}}$, where T is the time in seconds, and L is the length in feet. Use 3.14 for π. Round to the nearest hundredth.
7.30 ft

69. Find the length of a pendulum that makes one swing in 2.4 s. The equation for the time of one swing of a pendulum is given by $T = 2\pi\sqrt{\dfrac{L}{32}}$, where T is the time in seconds, and L is the length in feet. Use 3.14 for π. Round to the nearest hundredth.
4.67 ft

SUPPLEMENTAL EXERCISES 5.5

Solve.

70. $x^{\frac{3}{4}} = 8$ 16

71. $x^{\frac{2}{3}} = 9$ 27

72. $x^{\frac{5}{4}} = 32$ 16

73. $x^{\frac{3}{5}} = 27$ 243

74. $\sqrt{2x + 5} - \sqrt{3x - 2} = 1$ 2

75. $\sqrt{4x + 1} - \sqrt{2x + 4} = 1$ 6

76. $\sqrt{5x - 1} - \sqrt{3x - 2} = 1$ 2, 1

77. $\sqrt{5x + 4} - \sqrt{3x + 1} = 1$ 0, 1

78. $\sqrt{3x - 2} = \sqrt{2x - 3} + \sqrt{x - 1}$ 2

79. $\sqrt{2x + 3} + \sqrt{x + 2} = \sqrt{x + 5}$ -1

Solve the formula for the variable given.

80. $v = \sqrt{64d}; \ d$ $d = \dfrac{v^2}{64}$

81. $v = \sqrt{2as}; \ s$ $s = \dfrac{v^2}{2a}$

82. $a^2 + b^2 = c^2; \ a$ $a = \sqrt{c^2 - b^2}$

83. $A = \pi r^2; \ r$ $r = \dfrac{\sqrt{A\pi}}{\pi}$

84. $V = \pi r^2 h; \ r$ $r = \dfrac{\sqrt{V\pi h}}{\pi h}$

85. $V = \dfrac{4}{3}\pi r^3; \ r$ $r = \dfrac{\sqrt[3]{6V\pi^2}}{2\pi}$

Solve.

86. The length of a side of a square is 3 cm. Find the length of the diagonal of the square. Round to the nearest tenth.
4.2 cm

87. The length of a side of a square is 5 cm. Find the length of the diagonal of the square. Round to the nearest tenth.
7.1 cm

88. The perimeter of an isosceles right triangle is 2x. Find the area of the triangle in terms of x.
$3x^2 - 2\sqrt{2}x^2$

89. The area of an isoceles right triangle is $2x^2$. Find the perimeter of the triangle in terms of x.
$4x + 2x\sqrt{2}$

90. Two cyclists left an intersection at the same time. The first cyclist headed due south. The second cyclist headed due east. When the first cyclist had traveled 10 mi farther than the second cyclist, the cyclists were 50 mi apart. How far had each cyclist traveled?
first cyclist: 40 mi; second cyclist: 30 mi

91. Find three odd integers, a, b, and c, such that $a^2 + b^2 = c^2$.
Impossible; at least one of the integers must be even.

CALCULATORS AND COMPUTERS

 ## The $\boxed{y^x}$ Key on a Calculator

The $\boxed{y^x}$ key on a calculator is used to find powers of a number. For example, to find 4^7, enter the following key strokes:

$$4 \boxed{y^x} 7 \boxed{=}$$

The number 16384 should be in the display of your calculator.

Fractional powers can also be calculated by using the $\boxed{y^x}$ key. The memory key on your calculator is useful for this calculation. For example, to find $17^{\frac{2}{7}}$, first find the decimal equivalent for $\frac{2}{7}$. Store this result in the calculator's memory. Now the $\boxed{y^x}$ key is used to find the power. Here are the key strokes.

$$2 \boxed{\div} 7 \boxed{=} \boxed{M+} 17 \boxed{y^x} \boxed{MR} \boxed{=}$$

The number 2.2467608 should be in the display.

The symbol M+ is used here to mean store in memory, and the symbol MR is used to mean recall from memory. These symbols may be different on your calculator.

The calculation of an installment loan payment uses the $\boxed{y^x}$ key. Here is the formula and a sample calculation.

$$PMT = PRIN \cdot \frac{i}{1 - (1 + i)^{-n}}$$

In this formula, *PMT* is the monthly payment, *PRIN* is the amount borrowed, and *i* is the monthly interest rate as a decimal. The monthly interest rate is the annual rate divided by 12. The number of months of the loan is given by *n*.

A car is purchased and a loan of $8000 is secured at an annual interest rate of 8.5% for 5 years. Find the monthly payment.

First calculate $(1 + i)^{-n}$.

$(n = 5 \cdot 12 = 60; i = \dfrac{0.085}{12} = 0.0070833; 1 + i = 1.0070833)$

Enter: 1.0070833 $\boxed{y^x}$ 60 $\boxed{+/-}$ $\boxed{=}$ $\boxed{M+}$

The number 0.6547513 should be displayed. This number is also stored in memory because the $\boxed{M+}$ key was pressed.

Now the final calculation.

Enter: 8000 $\boxed{\times}$ 0.0070833 $\boxed{\div}$ $\boxed{(}$ 1 $\boxed{-}$ $\boxed{MR}$ $\boxed{)}$ $\boxed{=}$

The number 164.1321 should be in the display.

The monthly payment is $164.13.

CHAPTER SUMMARY

Key Words

A number written in **scientific notation** is a number written in the form $a \times 10^n$, where a is a number between 1 and 10.

The **nth root of a** is $a^{\frac{1}{n}}$. The expression $\sqrt[n]{a}$ is another symbol for the nth root of a. In the expression $\sqrt[n]{a}$, the symbol $\sqrt{}$ is called a **radical**, n is the **index** of the radical, and a is the **radicand**.

If $a^{\frac{1}{n}}$ is a real number, then $a^{\frac{m}{n}} = \sqrt[n]{a^m} = (\sqrt[n]{a})^m$.

The symbol $\sqrt{}$ is used to indicate the positive or **principal square root** of a number.

The expressions $a + b$ and $a - b$ are called **conjugates** of each other. The product of conjugates of the form $(a + b)(a - b) = a^2 - b^2$.

The procedure used to remove a radical from the denominator of a radical expression is called **rationalizing the denominator.**

A **complex number** is a number of the form $a + bi$, where a and b are real numbers and $i = \sqrt{-1}$. For the complex number $a + bi$, a is the **real part** of the complex number, and b is the **imaginary part** of the complex number.

A **radical equation** is an equation that contains a variable expression in a radicand.

Essential Rules

The Zero Property of Exponents

For any number a, $a \neq 0$, $a^0 = 1$.

The Rule of Negative Exponents

If n is a positive integer and $a \neq 0$, then
$$a^{-n} = \frac{1}{a^n} \text{ and } a^n = \frac{1}{a^{-n}}.$$

The Product Property of Radicals

If a and b are positive real numbers, then
$$\sqrt[n]{a}\,\sqrt[n]{b} = \sqrt[n]{ab} \text{ and } \sqrt[n]{ab} = \sqrt[n]{a}\,\sqrt[n]{b}.$$

The Quotient Property of Radicals

If a and b are positive real numbers, then
$$\frac{\sqrt[n]{a}}{\sqrt[n]{b}} = \sqrt[n]{\frac{a}{b}} \text{ and } \sqrt[n]{\frac{a}{b}} = \frac{\sqrt[n]{a}}{\sqrt[n]{b}}.$$

Addition of Complex Numbers

If $a + bi$ and $c + di$ are complex numbers, then
$$(a + bi) + (c + di) = (a + c) + (b + d)i.$$

Subtraction of Complex Numbers

If $a + bi$ and $c + di$ are complex numbers, then
$$(a + bi) - (c + di) = (a - c) + (b - d)i.$$

The Property of Raising Both Sides of an Equation to a Power

If a and b are real numbers and $a = b$, then
$$a^n = b^n.$$

The Pythagorean Theorem

The square of the hypotenuse of a right triangle is equal to the sum of the squares of the two legs.
$$c^2 = a^2 + b^2$$

CHAPTER REVIEW

1. Simplify: $\dfrac{(3x^{\frac{1}{2}}y^{\frac{3}{4}})^4}{(x^{-6}y^3)^{\frac{1}{3}}}$ $81x^4y^2$

2. Simplify: $\dfrac{2^3a^{-3}b^4}{2^{-1}a^2b^3}$ $\dfrac{16b}{a^5}$

3. Simplify: $\sqrt[3]{81x^6y^8}$ $3x^2y^2\sqrt[3]{3y^2}$

4. Write $-2\sqrt[3]{x^4}$ as an exponential expression. $-2x^{\frac{4}{3}}$

5. Simplify: $\left(\dfrac{3x^{-1}y}{x^{-2}y^2}\right)^{-3}\left(\dfrac{6x^{-2}y^{-3}}{x^{-3}y}\right)^2$ $\dfrac{4}{3xy^5}$

6. Simplify: $\sqrt{144x^4y^6}$
   $12x^2y^3$

7. Simplify: $\sqrt{2x}\,(\sqrt{16x}-\sqrt{x})$
   $3x\sqrt{2}$

8. Simplify: $\sqrt{25}-\sqrt{-4}$
   $5-2i$

9. Solve: $\sqrt[3]{2x+9}+8=5$
   -18

10. Simplify: $(1.5\times 10^5)(5\times 10^{-8})$
    0.0075

11. Simplify: $(3-8i)-(2-7i)$
    $1-i$

12. Simplify: $(\sqrt{5}-3)(\sqrt{5}+7)$
    $-16+4\sqrt{5}$

13. Write $2a^{\frac{2}{3}}$ as a radical expression.
    $2\sqrt[3]{a^2}$

14. Simplify: $\dfrac{6}{3-2i}$ $\dfrac{18}{13}+\dfrac{12}{13}i$

15. Simplify: $\dfrac{\sqrt{27x^3y^4}}{\sqrt{81xy^2}}$ $\dfrac{xy\sqrt{3}}{3}$

16. Solve: $\sqrt{5x-6}=7$
    11

17. Simplify: $y\sqrt{16x^2y}-2x\sqrt{9y^3}$
    $-2xy\sqrt{y}$

18. Simplify: $(9+8i)(2+3i)$
    $-6+43i$

19. Simplify: $\dfrac{a^{\frac{1}{2}}a^{-2}}{a^{-\frac{1}{3}}}$ $\dfrac{1}{a^{\frac{7}{6}}}$

20. Simplify: $\dfrac{\sqrt{2}}{\sqrt{y}-2}$ $\dfrac{\sqrt{2y}+2\sqrt{2}}{y-4}$

21. Simplify: $(2^{-1}x^2y^{-6})(2^{-1}y^{-4})^{-2}$
    $2x^2y^2$

22. Simplify: $\sqrt[3]{8a^3b^{12}}$
    $2ab^4$

23. Simplify: $\sqrt{40x^3}-x\sqrt{90x}$
    $-x\sqrt{10x}$

24. Simplify: $(\sqrt{3}-2)(\sqrt{3}-5)$
    $13-7\sqrt{3}$

25. Simplify: $-2i(7-4i)$
    $-8-14i$

26. Write $2\sqrt[4]{x^3}$ as an exponential expression.
    $2x^{\frac{3}{4}}$

27. Solve: $\sqrt[3]{3x-4}+5=1$
    -20

28. Simplify: $\sqrt{18}-\sqrt{-25}$
    $3\sqrt{2}-5i$

29. Simplify: $(7.5\times 10^{-12})(3\times 10^{20})$
    $2{,}250{,}000{,}000$

30. Simplify: $a^{-\frac{1}{2}}(a^{\frac{1}{2}}-a^{\frac{3}{2}})$
    $1-a$

31. Simplify: $\dfrac{\sqrt{x}}{\sqrt{x} - \sqrt{y}}$ $\dfrac{x + \sqrt{xy}}{x - y}$

32. Simplify: $\dfrac{2i}{3 - i}$ $-\dfrac{1}{5} + \dfrac{3}{5}i$

33. Simplify: $\sqrt{54x^3y^5}$
     $3xy^2\sqrt{6xy}$

34. Simplify: $(3 - \sqrt{-4}) + (4 + \sqrt{-9})$
     $7 + i$

35. Simplify: $\sqrt[3]{-4ab^4}\ \sqrt[3]{2a^3b^4}$
     $-2ab^2\sqrt[3]{ab^2}$

36. Solve: $\sqrt{3x + 10} = 1$
     -3

37. Write $(3x - 2)^{\frac{1}{3}}$ as a radical expression.
     $\sqrt[3]{3x - 2}$

38. Simplify: $\dfrac{\sqrt[3]{8x^4y^5}}{\sqrt[3]{16xy^6}}$ $\dfrac{x\sqrt[3]{4y^2}}{2y}$

39. A space vehicle travels 2.4×10^5 mi from the earth to the moon at an average velocity of 2×10^4 mph. How long does it take the space vehicle to reach the moon? Write the answer in scientific notation.
     1.2×10 h

40. An object is dropped from a high building. Find the distance the object has fallen when the speed reaches 192 ft/s. Use the equation $v = \sqrt{64d}$, where v is the speed of the object, and d is the distance.
     576 ft

CUMULATIVE REVIEW

1. Identify the property that justifies the statement.
    $(a + 2)b = ab + 2b$
    The Distributive Property

2. Simplify: $2x - 3[x - 2(x - 4) + 2x]$
    $-x - 24$

3. Find $A \cap B$ given $A = \{2, 4, 6\}$ and $B = \{1, 3, 5\}$.
    $A \cap B = \varnothing$

4. Graph the solution set of $\{x \mid x > -2\} \cup \{x \mid x < -4\}$.

5. Solve: $5 - \dfrac{2}{3}x = 4$ $\dfrac{3}{2}$

6. Solve: $2[4 - 2(3 - 2x)] = 4(1 - x)$ $\dfrac{2}{3}$

7. Solve: $3x - 4 \leq 8x + 1$
$\{x \mid x \geq -1\}$

8. Solve: $5 < 2x - 3 < 7$
$\{x \mid 4 < x < 5\}$

9. Solve: $2 + |4 - 3x| = 5$
$\dfrac{1}{3}$ and $\dfrac{7}{3}$

10. Solve: $|7 - 3x| > 1$
$\left\{x \mid x < 2 \text{ or } x > \dfrac{8}{3}\right\}$

11. Factor: $64a^2 - b^2$
$(8a + b)(8a - b)$

12. Factor: $x^5 + 2x^3 - 3x$
$x(x^2 + 3)(x + 1)(x - 1)$

13. Solve: $3x^2 + 13x - 10 = 0$
$\dfrac{2}{3}$ and -5

14. Graph the solution set of $x^2 - 2x - 8 < 0$.

15. Simplify: $\dfrac{4a^2 + 8a}{a^3 + a^2 - 2a}$ $\dfrac{4}{a - 1}$

16. Simplify: $\dfrac{1 - \dfrac{4}{y^2}}{\dfrac{1}{y} + \dfrac{2}{y^2}}$ $y - 2$

17. Solve $P = \dfrac{R - C}{n}$ for R.
$R = Pn + C$

18. Solve: $\dfrac{x - 5}{x + 2} \geq 0$
$\{x \mid x < -2 \text{ or } x \geq 5\}$

19. Simplify: $(3^{-1}x^3y^{-5})(3^{-1}y^{-2})^{-2}$ $\dfrac{3x^3}{y}$

20. Simplify: $\left(\dfrac{x^{-\frac{1}{2}}y^{\frac{3}{4}}}{y^{-\frac{5}{4}}}\right)^4$ $\dfrac{y^8}{x^2}$

21. Simplify: $\sqrt{20x^3} - x\sqrt{45x}$
$-x\sqrt{5x}$

22. Simplify: $(\sqrt{5} - 3)(\sqrt{5} - 2)$
$11 - 5\sqrt{5}$

23. Simplify: $\dfrac{\sqrt[3]{4x^5y^4}}{\sqrt[3]{8x^2y^5}}$ $\dfrac{x\sqrt[3]{4y^2}}{2y}$

24. Simplify: $(2 - \sqrt{-9})(5 + \sqrt{-16})$
$22 - 7i$

25. Simplify: $\dfrac{3i}{2 - i}$ $-\dfrac{3}{5} + \dfrac{6}{5}i$

26. Solve: $\sqrt[3]{2x - 5} + 3 = 6$
16

27. A collection of thirty stamps consists of 13¢ stamps and 18¢ stamps. The total value of the stamps is $4.85. Find the number of 18¢ stamps.
19 stamps

28. An investment of $2500 is made at an annual simple interest rate of 7.2%. How much additional money must be invested at an annual simple interest rate of 8.4% so that the total interest earned is $516?
$4000

29. The width of a rectangle is 6 ft less than the length. The area of the rectangle is 72 ft². Find the length and width of the rectangle.
length: 12 ft; width: 6 ft

30. A sales executive traveled 25 mi by car and then an additional 625 mi by plane. The rate by plane was five times faster than the rate by car. The total time of the trip was 3 h. Find the rate of the plane.
250 mph

31. How long does it take light to travel to the earth from the moon when the moon is 232,500 mi from the earth? Light travels 1.86×10^5 mi/s.
1.25 s

32. How far would a submarine periscope have to be above the water to locate a ship 7 mi away? The equation for the distance in miles that the lookout can see is $d = 1.4\sqrt{h}$, where h is the height in feet above the surface of the water.
25 ft

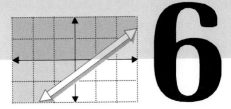

6

Linear Equations in Two Variables

OBJECTIVES

- Points on a rectangular coordinate system
- Determine a solution of a linear equation in two variables
- Graph a linear equation in two variables
- Graph the solution set of an inequality in two variables
- Find the slope of a line given two points
- Find the x- and y-intercepts of a straight line
- Graph a line given a point and the slope
- Find the equation of a line given a point and the slope or given two points
- Find parallel and perpendicular lines
- Obtain data from a graph

*B*rachistochrone Problem

Consider the diagram at the right. What curve should be
drawn so that a ball allowed to roll along the curve will travel
from *A* to *B* in the shortest time?

At first thought, one might conjecture that a straight line should connect the
two points since that shape is the shortest *distance* between the two points.
Actually, however, the answer is half of one arch of an inverted cycloid.

A cycloid is shown below as the graph in bold. One way to draw this curve
is to think of a wheel rolling along a straight line without slipping. Then a
point on the rim of the wheel traces a cycloid.

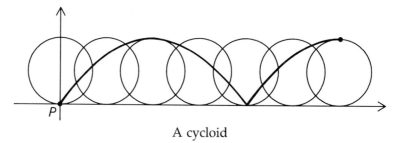

A cycloid

There are many applications of the idea of finding the shortest time between
two points. As the above problem illustrates, the path of shortest time is not
necessarily the path of shortest distance. Problems involving paths of shortest time are called *brachistochrone* problems.

The Rectangular Coordinate System

1 Points on a rectangular coordinate system

A **rectangular coordinate system** is formed by two number lines, one horizontal and one vertical, that intersect at the zero point of each line. The point of intersection is called the **origin**. The two lines are called the **coordinate axes,** or simply **axes**.

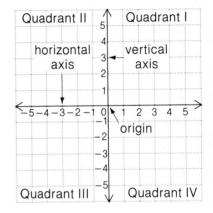

The axes determine a plane and divide the plane into four regions, called **quadrants**. The quadrants are numbered counterclockwise from I to IV.

Each point in the plane can be identified by a pair of numbers called an **ordered pair.** The first number of the pair measures a horizontal distance and is called the **abscissa**. The second number of the pair measures a vertical distance and is called the **ordinate**. The **coordinates** of a point are the numbers in the ordered pair associated with the point.

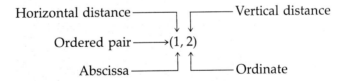

The **graph of an ordered pair** is a point in the plane. The graphs of the points $(-2, 3)$ and $(3, -2)$ are shown at the right. Notice that they are different points. The order in which the numbers in an ordered pair appear *is* important.

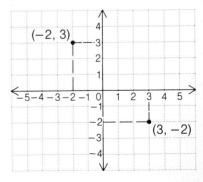

Example 1   A. Graph the ordered pairs $(2, -1)$ and $(-3, -4)$. Draw a line between the two points.

B. Draw a line through all points with an abscissa of 2.

Solution   A. 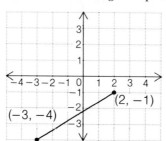   B.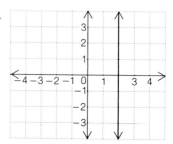

Problem 1   A. Graph the ordered pairs $(5, -2)$ and $(-2, 3)$. Draw a line between the two points.

B. Draw a line through all points with an ordinate of -1.

Solution   See page A45.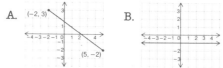

2   Determine a solution of a linear equation in two variables

An equation in the form $y = mx + b$, where m and b are constants, is a **linear equation in two variables.** Examples of linear equations in two variables are shown below.

$$y = 5x - 7 \qquad (m = 5, \quad b = -7)$$

$$y = \frac{2}{3}x - 8 \qquad \left(m = \frac{2}{3}, \quad b = -8\right)$$

$$y = -\frac{1}{2}x + 2 \qquad \left(m = -\frac{1}{2}, b = 2\right)$$

A **solution of an equation in two variables** is an ordered pair of numbers (x, y) that makes the equation a true statement.

Is $(2, 3)$ a solution of $y = \frac{1}{2}x + 2$?

Replace x with 2, the abscissa.

Replace y with 3, the ordinate.

$$
\begin{array}{c|c}
 & y = \frac{1}{2}x + 2 \\
\hline
3 & \frac{1}{2}(2) + 2 \\
 & 1 + 2 \\
\end{array}
$$

$$3 = 3$$

Compare the results. If the results are equal, the given ordered pair is a solution. If the results are not equal, the given ordered pair is not a solution.

Yes, (2, 3) is a solution of the equation $y = \frac{1}{2}x + 2$.

Besides the ordered pair (2, 3), there are many other ordered pair solutions of the equation $y = \frac{1}{2}x + 2$. For example, the method used above can be used to show that $(-4, 0)$, $(-2, 1)$, and $(4, 4)$ are also solutions.

In general, a linear equation in two variables has an infinite number of solutions. By choosing any value for x and substituting that value into the linear equation, a corresponding value of y can be found.

Example 2   Find the ordered pair solution of $y = \frac{2}{3}x + 3$ corresponding to $x = -5$.

Solution $y = \frac{2}{3}x + 3$

$y = \frac{2}{3}(-5) + 3 = -\frac{10}{3} + 3 = -\frac{1}{3}$ • Substitute -5 for x. Solve for y.

The ordered pair solution is $\left(-5, -\frac{1}{3}\right)$.

Problem 2   Find the ordered pair solution of $y = -2x + 5$ corresponding to $x = \frac{1}{3}$.

Solution   See page A45. $\left(\frac{1}{3}, \frac{13}{3}\right)$

3 Graph a linear equation in two variables

The **graph of an equation in two variables** is a drawing of the ordered pair solutions of the equation. For a linear equation in two variables, the graph is a straight line.

To graph a linear equation, find ordered pair solutions of the equation. Do this by choosing any value of x and finding the corresponding value of y. Repeat this procedure, choosing different values for x, until you have found the number of solutions desired. Since the graph of a linear equation in two variables is a straight line, and a straight line is determined by two points, it is necessary to find only two solutions. However, it is recommended that at least three solutions be used to ensure accuracy.

To graph $y = -2x + 1$, choose any values of x, and find the corresponding values of y. It is convenient to record these solutions in a table.

x	$y = -2x \quad + 1$	y
0	$-2(0) \quad + 1$	1
2	$-2(2) \quad + 1$	-3
-2	$-2(-2) + 1$	5

The horizontal axis is the x-axis. The vertical axis is the y-axis. Graph the ordered pair solutions $(0, 1)$, $(2, -3)$, and $(-2, 5)$. Draw a line through the ordered pair solutions.

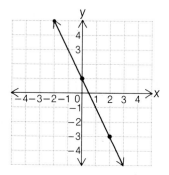

Remember that a graph is a drawing of the ordered pair solutions of the equation. Therefore, every point on the graph is a solution of the equation, and every solution of the equation is a point on the graph.

Example 3   Graph $y = -\dfrac{3}{2}x - 3$.

Solution

x	y
0	-3
-2	0
-4	3

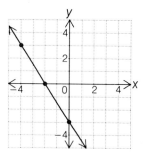

• Find at least three solutions. When the coefficient of x is a fraction, choose values of x that will simplify the evaluations. Display the ordered pairs in a table.

• Graph the ordered pairs on a rectangular coordinate system, and draw a straight line through the points.

Problem 3   Graph $y = -\dfrac{3}{4}x$.

Solution   See page A45.

An equation of the form $Ax + By = C$, where A, B, and C are constants, is also a linear equation. Examples of equations of the form $Ax + By = C$ are shown below.

$$3x - 2y = -5 \qquad (A = 3, \quad B = -2, C = -5)$$

$$-\frac{1}{2}x + 2y = 4 \qquad \left(A = -\frac{1}{2}, B = 2, \quad C = 4\right)$$

An equation of the form $Ax + By = C$ can be written in the form $y = mx + b$.

To write the equation $3x - 2y = -5$ in the form $y = mx + b$, add the additive inverse of $3x$ to both sides of the equation.

Multiply each side of the equation by the reciprocal of the coefficient -2.

$$3x - 2y = -5$$

$$-2y = -3x - 5$$

$$y = \frac{3}{2}x + \frac{5}{2}$$

To graph an equation of the form $Ax + By = C$, first solve the equation for y. Then follow the same procedure used for graphing an equation of the form $y = mx + b$.

Example 4 Graph $3x + 2y = 6$.

Solution $3x + 2y = 6$
$\qquad\qquad 2y = -3x + 6$ • Solve the equation for y.
$\qquad\qquad\ \ y = -\frac{3}{2}x + 3$

x	y
0	3
2	0
4	-3

• Find at least three solutions. Graph the ordered pairs on a rectangular coordinate system. Draw a straight line through the points.

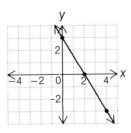

Problem 4 Graph $-3x + 2y = 4$.

Solution   See page A46.

The graph of an equation in which one of the variables is missing is either a horizontal or a vertical line.

The equation $y = -2$ could be written:

$$0 \cdot x + y = -2$$

No matter what value of x is chosen, y is always -2. Some solutions of the equation are $(3, -2)$, $(0, -2)$, and $(-2, -2)$. The graph is shown at the right.

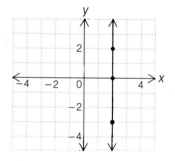

The **graph of $y = b$** is a horizontal line passing through point $(0, b)$.

The equation $x = 2$ could be written:

$$x + 0 \cdot y = 2$$

No matter what value of y is chosen, x is always 2. Some solutions of the equation are $(2, 2)$, $(2, 0)$, and $(2, -3)$. The graph is shown at the right.

The **graph of $x = a$** is a vertical line passing through point $(a, 0)$.

Example 5 Graph $x = -4$.

Solution

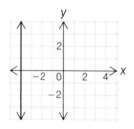

• The graph of an equation of the form $x = a$ is a vertical line passing through point $(a, 0)$.

Problem 5 Graph $y = 3$.

Solution See page A46.

4   Graph the solution set of an inequality in two variables

The graph of the linear equation $y = x - 1$ separates the plane into three sets:

the set of points on the line,
the set of points above the line,
the set of points below the line.

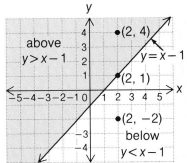

The point $(2, 1)$ is a solution of
$y = x - 1$.
The point $(2, 4)$ is a solution of
$y > x - 1$.
The point $(2, -2)$ is a solution
of $y < x - 1$.

The solution set of $y = x - 1$ is all points on the line. The solution set of the linear inequality $y > x - 1$ is all points above the line. The solution set of the linear inequality $y < x - 1$ is all points below the line.

The solution set of an inequality in two variables is a **half plane.**

The following illustrates the procedure for graphing a linear inequality:

Graph the solution set of $3x - 4y < 12$.
Solve the inequality for y.

$$3x - 4y < 12$$
$$-4y < -3x + 12$$
$$y > \frac{3}{4}x - 3$$
$$y = \frac{3}{4}x - 3$$

Change the inequality to an equality, and graph the line. If the inequality is $\leq$ **or** $\geq$, the line is in the solution set and is shown by a **solid line.** If the inequality is $<$ **or** $>$, the line is not part of the solution set and is shown by a **dotted line.**

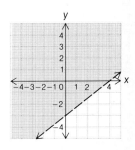

If the inequality is $>$ **or** $\geq$, shade the **upper half plane.** If the inequality is $<$ **or** $\leq$, shade the **lower half plane.**

As a check, the point $(0, 0)$ can be used to determine if the correct region of the plane has been shaded. If $(0, 0)$ is a solution of the inequality, then $(0, 0)$ should be in the shaded region. If $(0, 0)$ is not a solution of the inequality, then $(0, 0)$ should not be in the shaded region. In the above example, $(0, 0)$ is in the shaded region, and $(0, 0)$ is a solution of the inequality.

If the line passes through point $(0, 0)$, another point must be used as a check, for example, $(1, 0)$.

Example 6   Graph the solution set.

A. $x + 2y \leq 4$ B. $x \geq -1$

Solution   A. $x + 2y \leq 4$

$2y \leq -x + 4$

$y \leq -\dfrac{1}{2}x + 2$

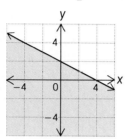

• Solve the inequality for y.

• Graph $y = -\dfrac{1}{2}x + 2$ as a solid line. Shade the lower half plane.

B. $x \geq -1$

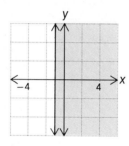

• Graph $x = -1$ as a solid line. The point $(0, 0)$ satisfies the inequality.

$x \geq -1$

$0 \geq -1$

Shade the half plane to the right of the line.

Problem 6   Graph the solution set.

A. $x + 3y > 6$ B. $y < 2$

Solution   See page A46.

A.   B.

EXERCISES 6.1

1

1. Graph the ordered pairs $(3, 2)$ and $(-1, 4)$. Draw a line between the two points.

2. Graph the ordered pairs $(-1, -3)$ and $(3, -2)$. Draw a line between the two points.

3. Graph the ordered pairs $(-3, -3)$ and $(2, -2)$. Draw a line between the two points.

4. Graph the ordered pairs $(-3, 2)$ and $(4, 2)$. Draw a line between the two points.

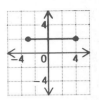

5. Find the coordinates of each of the points.

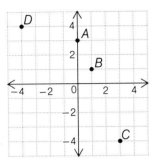

$A(0,3)$
$B(1,1)$
$C(3,-4)$
$D(-4,4)$

6. Find the coordinates of each of the points.

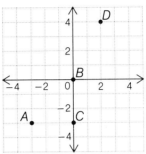

$A(-3,-3)$
$B(0,0)$
$C(0,-3)$
$D(2,4)$

7. Find the coordinates of each of the points.

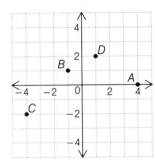

$A(4,0)$
$B(-1,1)$
$C(-4,-2)$
$D(1,2)$

8. Find the coordinates of each of the points.

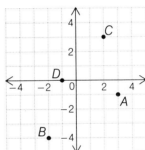

$A(3,-1)$
$B(-2,-4)$
$C(2,3)$
$D(-1,0)$

9. Draw a line through all points with an abscissa of 2.

10. Draw a line through all points with an abscissa of -3.

11. Draw a line through all points with an ordinate of -3.

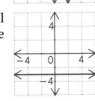

12. Draw a line through all points with an ordinate of 4.

2

13. Find the ordered pair solution of $y = \frac{2}{3}x - 4$ corresponding to $x = -3$.

$(-3, -6)$

14. Find the ordered pair solution of $y = \frac{1}{2}x - 5$ corresponding to $x = 4$.

$(4, -3)$

15. Find the ordered pair solution of $y = -\frac{2}{3}x + 4$ corresponding to $x = -6$.
$(-6, 8)$

16. Find the ordered pair solution of $y = -\frac{3}{4}x + 5$ corresponding to $x = -4$.
$(-4, 8)$

17. Find the ordered pair solution of $y = \frac{3}{2}x + 3$ corresponding to $x = -4$.
$(-4, -3)$

18. Find the ordered pair solution of $y = \frac{4}{3}x - 5$ corresponding to $x = 3$.
$(3, -1)$

19. Find the ordered pair solution of $y = -2x - 3$ corresponding to $x = -2$.
$(-2, 1)$

20. Find the ordered pair solution of $y = 3x + 5$ corresponding to $x = -3$.
$(-3, -4)$

21. Find the ordered pair solution of $y = -\frac{4}{3}x - 3$ corresponding to $x = -2$.
$\left(-2, -\frac{1}{3}\right)$

22. Find the ordered pair solution of $y = \frac{3}{2}x - 4$ corresponding to $x = 3$.
$\left(3, \frac{1}{2}\right)$

3 Graph.

23. $y = 3x - 4$

24. $y = -2x + 3$

25. $y = -\frac{2}{3}x$

26. $y = \frac{3}{2}x$

27. $y = \frac{2}{3}x - 4$

28. $y = \frac{3}{4}x + 2$

29. $y = -\frac{1}{3}x + 2$

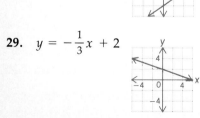

30. $y = -\frac{3}{2}x - 3$

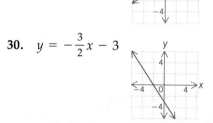

31. $2x - y = 3$

32. $2x + y = -3$

33. $2x + 5y = 10$

34. $x - 4y = 8$

35. $y = -2$

36. $y = \dfrac{1}{3}x$

37. $2x - 3y = 12$

38. $3x - y = -2$

4   Graph the solution set.

39. $3x - 2y \geq 6$

40. $4x - 3y \leq 12$

41. $x + 2y < 4$

42. $x + 3y < 6$

43. $2x - 5y \leq 10$

44. $2x + 3y \geq 6$

45. $3x - 5y > 15$

46. $4x - 5y > 10$

47. $y - 4 < 0$

48. $x + 2 \geq 0$

49. $6x + 5y < 15$

50. $3x - 5y < 10$

SUPPLEMENTAL EXERCISES 6.1

In each of the following exercises, three vertices of a rectangle are given. Find the coordinates of the fourth vertex.

51. $(3, 5)$, $(3, 0)$, $(0, 0)$
$(0,5)$

52. $(-2, 4)$, $(3, 4)$, $(3, -1)$
$(-2,-1)$

53. $(-3, -2)$, $(4, -2)$, $(4, 5)$
$(-3,5)$

54. $(6, -1)$, $(6, 3)$, $(-4, -1)$
$(-4,3)$

55. $(-7, -2)$, $(-1, -2)$, $(-7, 8)$
$(-1,8)$

56. $(5, -4)$, $(-2, -4)$, $(-2, -2)$
$(5,-2)$

For what value of k does the given point lie on the graph of the equation?

57. $3x - 2ky = 2$; $(2, 1)$
2

58. $kx + 3y = 4$; $(-1, 2)$
2

59. $x - ky = -1$; $(3, 1)$
4

60. $3x + 4ky = 2$; $(-2, 1)$
2

61. $kx + 6y = -2$; $(2, -1)$
2

62. $2kx + 3y = -3$; $(1, -5)$
6

Is the given point a solution of inequality **a**, inequality **b**, or both **a** and **b**?
a. $3x - 4y \leq 2$ **b.** $x - 2y \geq 1$

63. $(0, 2)$
a

64. $(4, 1)$
b

65. $(-4, -3)$
a and b

66. $\left(0, -\dfrac{1}{2}\right)$
a and b

67. $(4, -1)$
b

68. $(4, 4)$
a

Solve.

69. **a.** Show that the equation $y + 2 = 3(x + 5)$ is a linear equation by writing it in the form $y = mx + b$.

$y = 3x + 13$

b. Find the ordered pair solution corresponding to $x = -5$.

$(-5, -2)$

70. **a.** Show that the equation $y - 1 = \frac{1}{2}(x - 4)$ is a linear equation by writing it in the form $y = mx + b$.

$y = \frac{1}{2}x - 1$

b. Find the ordered pair solution corresponding to $x = 4$.

$(4, 1)$

71. For the linear equation $y = 3x - 2$, what is the increase in y that results when x is increased by 1?

3

72. For the linear equation $y = -x + 3$, what is the decrease in y that results when x is increased by 1?

1

73. A triangle has vertices whose coordinates are $(-4, -3)$, $(-4, 1)$, and $(1, -3)$. Find the number of square units in the area of the triangle.

10 square units

74. A triangle has vertices whose coordinates are $(-2, -3)$, $(-2, 4)$, and $(4, -3)$. Find the number of square units in the area of the triangle.

21 square units

75. A triangle has vertices whose coordinates are $(3, 2)$, $(6, 2)$, and $(3, 7)$. Find the number of units in the perimeter of the triangle.

$(8 + \sqrt{34})$ units

76. A triangle has vertices whose coordinates are $(-3, -2)$, $(-3, 3)$, and $(9, -2)$. Find the number of units in the perimeter of the triangle.

30 units

SECTION 6.2

Slopes and Intercepts of Straight Lines

1 Find the slope of a line given two points

The graphs of $y = 3x + 2$ and $y = \frac{2}{3}x + 2$
are shown at the right. Each graph crosses
the y-axis at the point $(0, 2)$, but the graphs
have different slants. The **slope** of a line is a
measure of the slant of a line. The symbol
for slope is m.

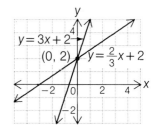

The slope of a line containing two points is
the ratio of the change in the y values of the
two points to the change in the x values. The
line containing the points $(-1, -3)$ and
$(5, 2)$ is graphed at the right.

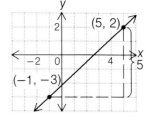

The change in the y values is the difference
between the two ordinates.

Change in $y = 2 - (-3) = 5$

The change in the x values is the difference
between the two abscissas.

Change in $x = 5 - (-1) = 6$

$$\text{Slope} = m = \frac{\text{Change in } y}{\text{Change in } x} = \frac{5}{6}$$

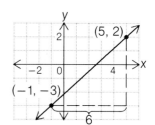

Slope Formula

The slope of a line containing two
points, P_1 and P_2, whose coordinates
are (x_1, y_1) and (x_2, y_2), is given by

$$\textbf{Slope} = m = \frac{y_2 - y_1}{x_2 - x_1}, \qquad x_1 \neq x_2.$$

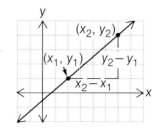

To find the slope of the line containing the points $(-2, 0)$ and $(4, 5)$, let $P_1 = (-2, 0)$ and $P_2 = (4, 5)$. It does not matter which point is named P_1 or P_2; the slope will be the same.

$$m = \frac{y_2 - y_1}{x_2 - x_1} = \frac{5 - 0}{4 - (-2)} = \frac{5}{6}$$

The slope is a positive number.

A line that slants upward to the right always has a **positive slope.**

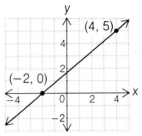

positive slope

To find the slope of the line containing the points $(-3, 4)$ and $(4, 2)$, let $P_1 = (-3, 4)$ and $P_2 = (4, 2)$.

$$m = \frac{y_2 - y_1}{x_2 - x_1} = \frac{2 - 4}{4 - (-3)} = \frac{-2}{7} = -\frac{2}{7}$$

The slope is a negative number.

A line that slants downward to the right always has a **negative slope.**

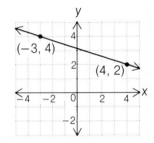

negative slope

To find the slope of the line containing the points $(-2, 2)$ and $(4, 2)$, let $P_1 = (-2, 2)$ and $P_2 = (4, 2)$.

$$m = \frac{y_2 - y_1}{x_2 - x_1} = \frac{2 - 2}{4 - (-2)} = \frac{0}{6} = 0$$

When $y_1 = y_2$, the graph is a horizontal line.

A horizontal line has **zero slope.**

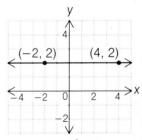

zero slope

To find the slope of the line containing the points $(1, -2)$ and $(1, 3)$, let $P_1 = (1, -2)$ and $P_2 = (1, 3)$.

$$m = \frac{y_2 - y_1}{x_2 - x_1} = \frac{3 - (-2)}{1 - 1} = \frac{5}{0}$$ Not a real number

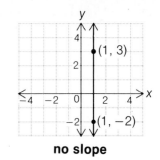

no slope

When $x_1 = x_2$, the denominator of $\frac{y_2 - y_1}{x_2 - x_1}$ is 0 and the graph is a vertical line. Because division by 0 is undefined, the line has no slope.

A vertical line has **no slope.**

Example 1   Find the slope of the line containing the points $(2, -5)$ and $(-4, 2)$.

Solution $m = \dfrac{y_2 - y_1}{x_2 - x_1} = \dfrac{2 - (-5)}{-4 - 2} = \dfrac{7}{-6}$ • Let $P_1 = (2, -5)$ and $P_2 = (-4, 2)$.

The slope is $-\dfrac{7}{6}$.

Problem 1   Find the slope of the line containing the points $(4, -3)$ and $(2, 7)$.

Solution   See page A47.
-5

2   Find the *x*- and *y*-intercepts of a straight line

The graph of the equation $x - 2y = 4$ is shown at the right. The graph crosses the *x*-axis at the point $(4, 0)$. This point is called the ***x*-intercept**. The graph also crosses the *y*-axis at the point $(0, -2)$. This point is called the ***y*-intercept**.

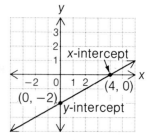

Some linear equations can be graphed by finding the *x*- and *y*-intercepts and then drawing a line through the two points. To find the *x*-intercept, let $y = 0$. (Any point on the *x*-axis has *y*-coordinate 0). To find the *y*-intercept, let $x = 0$. (Any point on the *y*-axis has *x*-coordinate 0.)

Example 2   Graph $4x - y = 4$ by using the *x*- and *y*-intercepts.

Solution   *x*-intercept: $4x - y = 4$ *y*-intercept: $4x - y = 4$

$$4x - 0 = 4 \qquad\qquad 4(0) - y = 4$$
$$4x = 4 \qquad\qquad -y = 4$$
$$x = 1 \qquad\qquad y = -4$$

- To find the *x*-intercept, let $y = 0$.
  To find the *y*-intercept, let $x = 0$.

The *x*-intercept is $(1, 0)$.   The *y*-intercept is $(0, -4)$.

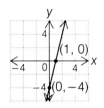

- Graph the points $(1, 0)$ and $(0, -4)$. Draw a line through the two points.

Problem 2   Graph $3x - y = 2$ by using the *x*- and *y*-intercepts.

Solution   See page A47.

$$\left(\frac{2}{3}, 0\right), (0, -2)$$

To find the *y*-intercept of $y = \frac{2}{3}x + 7$,

let $x = 0$.

The *y*-intercept is $(0, 7)$.

$$y = \frac{2}{3}x + 7$$
$$y = \frac{2}{3}(0) + 7$$
$$y = 7$$

For any equation of the form $y = mx + b$, the *y*-intercept is $(0, b)$.

Example 3   Graph $y = \frac{2}{3}x - 2$ by using the *x*- and *y*-intercepts.

Solution   *x*-intercept: $y = \frac{2}{3}x - 2$ *y*-intercept: $(0, b)$

$$0 = \frac{2}{3}x - 2 \qquad b = -2$$
$$-\frac{2}{3}x = -2$$
$$x = 3$$

- To find the *x*-intercept, let $y = 0$.
  For any equation of the form $y = mx + b$, the *y*-intercept is $(0, b)$.

The *x*-intercept is $(3, 0)$.   The *y*-intercept is $(0, -2)$.

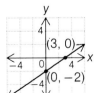

- Graph the points $(3, 0)$ and $(0, -2)$. Draw a line through the two points.

Problem 3   Graph $y = \frac{1}{4}x + 1$ by using the x- and y-intercepts.

Solution   See page A48.
(−4, 0), (0, 1)

3 Graph a line given a point and the slope

The graph of the equation $y = -\frac{3}{4}x + 4$ is shown at the right. The points $(-4, 7)$ and $(4, 1)$ are on the graph. The slope of the line is

$$m = \frac{7 - 1}{-4 - 4} = \frac{6}{-8} = -\frac{3}{4}$$

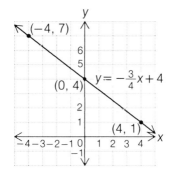

Note that the slope of the line has the same value as the coefficient of x. The y-intercept is $(0, 4)$, and the constant is 4.

Slope-Intercept Form of a Straight Line

For any equation of the form $y = mx + b$, the slope of the line is m, the coefficient of x. The y-intercept is $(0, b)$. The equation

$$y = mx + b$$

is called the **slope-intercept form of a straight line.**

When the equation of a straight line is in the form $y = mx + b$, the graph can be drawn using the slope and y-intercept. First locate the y-intercept. Use the slope to find a second point on the line. Then draw a line through the two points.

When the equation of a straight line is in the form $Ax + By = C$, first solve the equation for y. Then follow the same procedure used for an equation in the form $y = mx + b$.

To graph $x + 2y = 4$ by using the slope and y-intercept, solve the equation for y.

$$x + 2y = 4$$
$$2y = -x + 4$$
$$y = -\frac{1}{2}x + 2$$

The y-intercept is $(0, b) = (0, 2)$.

$$m = -\frac{1}{2} = \frac{-1}{2} = \frac{\text{Change in } y}{\text{Change in } x}$$

Beginning at the y-intercept $(0, 2)$, move right 2 units (change in x) and then down 1 unit (change in y).

$(2, 1)$ is a second point on the graph.

Draw a line through the points $(0, 2)$ and $(2, 1)$.

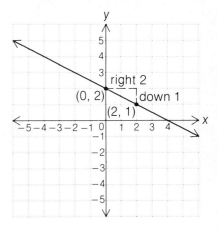

Example 4   Graph $y = -\frac{3}{2}x + 4$ by using the slope and y-intercept.

Solution y-intercept $= (0, 4)$ • Locate the y-intercept.

$$m = -\frac{3}{2} = \frac{-3}{2}$$ $\bullet\ m = \dfrac{\text{Change in } y}{\text{Change in } x}$

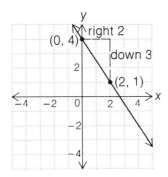

• Beginning at the y-intercept, $(0, 4)$, move right 2 units and then down 3 units.
$(2, 1)$ is a second point on the graph.
Draw a line through the points $(0, 4)$ and $(2, 1)$.

Problem 4   Graph $2x + 3y = 6$ by using the slope and y-intercept.

Solution   See page A48.

The graph of a line can be drawn when a point on the line and the slope of the line are given.

To graph the line that passes through point (2, 1) and has slope $\frac{2}{3}$, locate the point (2, 1) on the graph.

$$m = \frac{2}{3} = \frac{\text{Change in } y}{\text{Change in } x}$$

Beginning at the point (2, 1), move right 3 units and then up 2 units.

(5, 3) is a second point on the line.

Draw a line through the points (2, 1) and (5, 3).

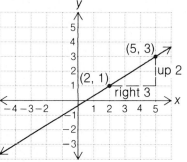

Example 5   Graph the line that passes through point $(-2, 3)$ and has slope $-\frac{4}{3}$.

Solution $(x_1, y_1) = (-2, 3)$

$$m = -\frac{4}{3} = \frac{-4}{3}$$

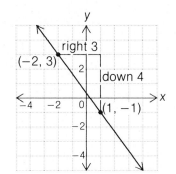

Problem 5   Graph the line that passes through point $(-3, -2)$ and has slope 3.

Solution   See page A48.

EXERCISES 6.2

1   Find the slope of the line containing the points.

1. $P_1(1, 3)$, $P_2(3, 1)$ -1

2. $P_1(2, 3)$, $P_2(5, 1)$ $-\dfrac{2}{3}$

3. $P_1(-1, 4)$, $P_2(2, 5)$ $\dfrac{1}{3}$

4. $P_1(3, -2)$, $P_2(1, 4)$ -3

5. $P_1(-1, 3)$, $P_2(-4, 5)$ $-\dfrac{2}{3}$

6. $P_1(-1, -2)$, $P_2(-3, 2)$ -2

7. $P_1(0, 3)$, $P_2(4, 0)$ $\quad -\dfrac{3}{4}$

8. $P_1(-2, 0)$, $P_2(0, 3)$ $\quad \dfrac{3}{2}$

9. $P_1(2, 4)$, $P_2(2, -2)$ $\quad$ no slope

10. $P_1(4, 1)$, $P_2(4, -3)$ $\quad$ no slope

11. $P_1(2, 5)$, $P_2(-3, -2)$ $\quad \dfrac{7}{5}$

12. $P_1(4, 1)$, $P_2(-1, -2)$ $\quad \dfrac{3}{5}$

13. $P_1(2, 3)$, $P_2(-1, 3)$ $\quad 0$

14. $P_1(3, 4)$, $P_2(0, 4)$ $\quad 0$

15. $P_1(0, 4)$, $P_2(-2, 5)$ $\quad -\dfrac{1}{2}$

16. $P_1(3, 0)$, $P_2(-1, -4)$ $\quad 1$

17. $P_1(-3, 4)$, $P_2(-2, 1)$ $\quad -3$

18. $P_1(4, -2)$, $P_2(2, -4)$ $\quad 1$

19. $P_1(-2, 3)$, $P_2(-2, 5)$ $\quad$ no slope

20. $P_1(-3, -1)$, $P_2(-3, 4)$ $\quad$ no slope

21. $P_1(-2, -5)$, $P_2(-4, -1)$ $\quad -2$

22. $P_1(-3, -2)$, $P_2(0, -5)$ $\quad -1$

23. $P_1(3, -1)$, $P_2(-2, -1)$ $\quad 0$

24. $P_1(0, -3)$, $P_2(-2, -3)$ $\quad 0$

2 Find the *x*- and *y*-intercepts and graph.

25. $x - 2y = -4$
x-intercept: $(-4, 0)$
y-intercept: $(0, 2)$

26. $3x + y = 3$
x-intercept: $(1, 0)$
y-intercept: $(0, 3)$

27. $4x - 2y = 5$
x-intercept: $\left(\dfrac{5}{4}, 0\right)$
y-intercept: $\left(0, -\dfrac{5}{2}\right)$

28. $2x - y = 4$
x-intercept: $(2, 0)$
y-intercept: $(0, -4)$

29. $3x + 2y = 5$
x-intercept: $\left(\dfrac{5}{3}, 0\right)$
y-intercept: $\left(0, \dfrac{5}{2}\right)$

30. $4x - 3y = 8$
x-intercept: $(2, 0)$
y-intercept: $\left(0, -\dfrac{8}{3}\right)$

31. $2x - 3y = 4$
x-intercept: $(2, 0)$
y-intercept: $\left(0, -\dfrac{4}{3}\right)$

32. $3x - 5y = 9$
x-intercept: $(3, 0)$
y-intercept: $\left(0, -\dfrac{9}{5}\right)$

33. $2x - 3y = 9$
x-intercept: $\left(\dfrac{9}{2}, 0\right)$
y-intercept: $(0, -3)$

34. $3x - 4y = 4$
x-intercept: $\left(\dfrac{4}{3}, 0\right)$
y-intercept: $(0, -1)$

35. $2x + y = 3$

x-intercept: $\left(\frac{3}{2}, 0\right)$

y-intercept: $(0, 3)$

36. $3x + y = -5$

x-intercept: $\left(-\frac{5}{3}, 0\right)$

y-intercept: $(0, -5)$

37. $3x + 2y = 4$

x-intercept: $\left(\frac{4}{3}, 0\right)$

y-intercept: $(0, 2)$

38. $3x + 4y = -12$

x-intercept: $(-4, 0)$

y-intercept: $(0, -3)$

39. $2x - 3y = -6$

x-intercept: $(-3, 0)$

y-intercept: $(0, 2)$

40. $4x - 3y = 6$

x-intercept: $\left(\frac{3}{2}, 0\right)$

y-intercept: $(0, -2)$

3 Graph by using the slope and the y-intercept.

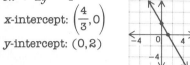

41. $y = \frac{1}{2}x + 2$

42. $y = \frac{2}{3}x - 3$

43. $y = -\frac{2}{3}x + 4$

44. $y = -\frac{1}{2}x + 2$

45. $y = -\frac{3}{2}x$

46. $y = \frac{3}{4}x$

47. $y = \frac{2}{3}x - 1$

48. $2x - 3y = 6$

49. $3x - y = 2$

50. $4x + y = 2$

51. $3x + 2y = 8$

52. $4x - 5y = 5$

53. $3x - 2y = 6$

54. $x - 3y = 3$

55. $x + 2y = 4$

56. Graph the line that passes through point $(2, 3)$ and has slope $\frac{1}{2}$.

57. Graph the line that passes through point $(-4, 1)$ and has slope $\frac{2}{3}$.

58. Graph the line that passes through point $(1, 4)$ and has slope $-\frac{2}{3}$.

59. Graph the line that passes through point $(0, -2)$ and has slope $-\frac{1}{3}$.

60. Graph the line that passes through point $(-3, 0)$ and has slope -3.

61. Graph the line that passes through point $(2, 0)$ and has slope -1.

SUPPLEMENTAL EXERCISES 6.2

From the graph of the equation, write the equation of the line in the form $y = mx + b$.

62.

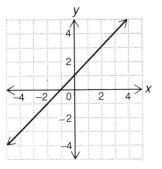

$y = x + 1$

63.

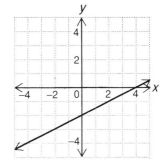

$y = \frac{1}{2}x - 2$

64.

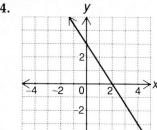

$$y = -\frac{3}{2}x + 3$$

65.

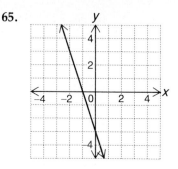

$$y = -3x - 3$$

Solve.

66. **a.** Show that the equation $\dfrac{x}{3} + \dfrac{y}{4} = 1$ is a linear equation by writing it in the form $y = mx + b$.

b. Find the x- and y-intercepts.

a. $y = -\dfrac{4}{3}x + 4$ b. $(3,0), (0,4)$

67. **a.** Show that the equation $\dfrac{x}{2} - \dfrac{y}{5} = 1$ is a linear equation by writing it in the form $y = mx + b$.

b. Find the x- and y-intercepts.

a. $y = \dfrac{5}{2}x - 5$ b. $(2,0), (0,-5)$

68. **a.** Show that the equation $\dfrac{y}{6} - \dfrac{x}{4} = 1$ is a linear equation by writing it in the form $y = mx + b$.

b. Find the x- and y-intercepts.

a. $y = \dfrac{3}{2}x + 6$ b. $(-4,0) (0,6)$

69. **a.** Show that the equation $x - \dfrac{y}{8} = 1$ is a linear equation by writing it in the form $y = mx + b$.

b. Find the x- and y-intercepts.

a. $y = 8x - 8$ b. $(1,0), (0,-8)$

70. What effect does increasing the coefficient of x have on the graph of $y = mx + b$?

The slope of the line rotates counterclockwise.

71. What effect does decreasing the coefficient of x have on the graph of $y = mx + b$?   The slope of the line rotates clockwise.

72. What effect does increasing the constant term have on the graph of
$y = mx + b$?
The graph of the line moves up.

73. What effect does decreasing the constant term have on the graph of
$y = mx + b$?
The graph of the line moves down.

SECTION 6.3

Finding Equations of Lines

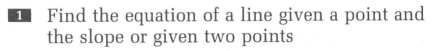 **1** Find the equation of a line given a point and
the slope or given two points

When the slope of a line and a point on the line are known, the equation of
the line can be determined.

To find the equation of the line that contains the point $(0, 3)$

and has slope $\frac{1}{2}$, use the point-slope form of the equation $\qquad y = mx + b$
since the given point is the y-intercept.

Replace m with $\frac{1}{2}$, the given slope.

Replace b with 3, the y-intercept. $\qquad\qquad\qquad\qquad\qquad y = \frac{1}{2}x + 3$

The equation of the line is $y = \frac{1}{2}x + 3$.

Example 1 Find the equation of the line that contains the point $(0, -2)$ and has
slope $-\frac{2}{3}$.

Solution $m = -\frac{2}{3} \qquad\qquad b = -2$

$y = mx + b$ $\qquad\qquad$ • Replace m with $-\frac{2}{3}$, the given slope.

$\qquad\qquad\qquad\qquad\qquad$ Replace b with -2, the y-intercept.
$y = -\frac{2}{3}x - 2$

The equation of the line is $y = -\frac{2}{3}x - 2$.

Problem 1 Find the equation of the line that contains the point $(0, 3)$ and has slope $-\frac{5}{4}$.

Solution See page A49. $y = -\frac{5}{4}x + 3$

One method for finding the equation of a line, given the slope and *any* point on the line, involves use of the point-slope formula. The point-slope formula is derived from the formula for slope.

Let (x_1, y_1) be the given point on the line and (x, y) be any other point on the line. Then the slope of the line is given by $\dfrac{y - y_1}{x - x_1} = m$.

Formula for slope

$$\dfrac{y - y_1}{x - x_1} = m$$

Multiply both sides of the equation by $(x - x_1)$.

$$\dfrac{y - y_1}{x - x_1}(x - x_1) = m(x - x_1)$$

Simplify.

$$y - y_1 = m(x - x_1)$$

Point-Slope Formula

The equation of the line with slope m and containing the point (x_1, y_1) can be found by the point-slope formula: $y - y_1 = m(x - x_1)$.

Example 2 Find the equation of the line that contains the point $(-2, 4)$ and has slope 2.

Solution
$$\begin{aligned} y - y_1 &= m(x - x_1) \\ y - 4 &= 2[x - (-2)] \\ y - 4 &= 2(x + 2) \\ y - 4 &= 2x + 4 \\ y &= 2x + 8 \end{aligned}$$

• Use the point-slope formula.
• Substitute the slope, 2, and the coordinates of the given point, $(-2, 4)$, into the point-slope formula.

The equation of the line is $y = 2x + 8$.

Problem 2 Find the equation of the line that contains the point $(4, -3)$ and has slope -3.

Solution See page A49.
$$y = -3x + 9$$

The point-slope formula and the formula for slope are used to find the equation of a line when two points are known.

Example 3   Find the equation of the line containing the given points.

A. $P_1(-2, 5)$, $P_2(-4, -1)$ B. $P_1(2, -3)$, $P_2(2, 5)$

Solution   A. $m = \dfrac{y_2 - y_1}{x_1 - x_2}$

$= \dfrac{-1 - 5}{-4 - (-2)} = \dfrac{-6}{-2} = 3$

$y - y_1 = m(x - x_1)$
$y - 5 = 3[x - (-2)]$
$y - 5 = 3(x + 2)$
$y - 5 = 3x + 6$
$y = 3x + 11$

• Find the slope. Let $(x_1, y_1) = (-2, 5)$ and $(x_2, y_2) = (-4, -1)$.

• Substitute the slope and the coordinates of either one of the known points into the point-slope formula.

The equation of the line is $y = 3x + 11$.

B. $m = \dfrac{y_2 - y_1}{x_2 - x_1} = \dfrac{5 - (-3)}{2 - 2} = \dfrac{8}{0}$

• The line has no slope. It is a vertical line passing through points $(2, -3)$ and $(2, 5)$. All points on the line have an abscissa of 2.

The equation of the line is $x = 2$.

Problem 3   Find the equation of the line containing the given points.

A. $P_1(4, -2)$, $P_2(-1, -7)$ B. $P_1(2, 3)$, $P_2(-5, 3)$

Solution   See pages A49 and A50.

A. $y = x - 6$ B. $y = 3$

---

2 # Find parallel and perpendicular lines

Two lines that have the same slope do not intersect and are called **parallel lines.** Two vertical lines are parallel lines. Two horizontal lines are parallel lines.

The slope of each of the lines at the right is $\dfrac{2}{3}$. The lines are parallel.

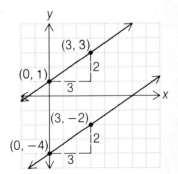

> ### Slopes of Parallel Lines
> For two nonvertical parallel lines, $m_1 = m_2$.

Is the line that contains the points $(-2, 1)$ and $(-5, -1)$ parallel to the line that contains the points $(1, 0)$ and $(4, 2)$?

Find the slope of each line.

$$m_1 = \frac{-1 - 1}{-5 - (-2)} = \frac{-2}{-3} = \frac{2}{3}$$

$$m_2 = \frac{2 - 0}{4 - 1} = \frac{2}{3}$$

$$m_1 = m_2 = \frac{2}{3}$$

The lines are parallel.

Are the lines $3x - 4y = 8$ and $6x - 8y = 2$ parallel?

Write each equation in slope-intercept form.

$$3x - 4y = 8 \qquad\qquad 6x - 8y = 2$$
$$-4y = -3x + 8 \qquad -8y = -6x + 2$$
$$y = \frac{3}{4}x - 2 \qquad\qquad y = \frac{3}{4}x - \frac{1}{4}$$

Find the slope of each line. $\qquad m_1 = \frac{3}{4} \qquad\qquad\qquad m_2 = \frac{3}{4}$

$$m_1 = m_2 = \frac{3}{4}$$

The lines are parallel.

Find the equation of the line containing the point $(-1, 4)$ and parallel to the line $2x - 3y = 2$.

Write the given equation in slope-intercept form to determine its slope.

$$2x - 3y = 2$$
$$-3y = -2x + 2$$
$$y = \frac{2}{3}x - \frac{2}{3} \qquad m = \frac{2}{3}$$

Parallel lines have the same slope.

$$y - y_1 = m(x - x_1)$$

Substitute the slope of the given line and the coordinates of the given point in the point-slope formula.

$$y - 4 = \frac{2}{3}[x - (-1)]$$

$$y - 4 = \frac{2}{3}x + \frac{2}{3}$$

$$y = \frac{2}{3}x + \frac{14}{3}$$

Example 4   Find the equation of the line containing the point $(3, -1)$ and parallel to the line $y = \frac{3}{2}x - 2$.

Solution $y - y_1 = m(x - x_1)$ • The slope of the given line is $\frac{3}{2}$. Substitute the slope of

$y - (-1) = \frac{3}{2}(x - 3)$ the given line and the coordinates of the given point in the point-slope formula.

$y + 1 = \frac{3}{2}x - \frac{9}{2}$

$y = \frac{3}{2}x - \frac{11}{2}$

The equation of the line is $y = \frac{3}{2}x - \frac{11}{2}$.

Problem 4   Are the lines $5x + 2y = 2$ and $5x + 2y = -6$ parallel?

Solution   See page A50.

Yes

Two lines that intersect at right angles are **perpendicular lines.**

Any horizontal line is perpendicular to any vertical line. For example, $x = 3$ is perpendicular to $y = -2$.

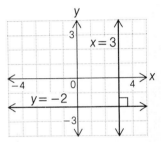

Two nonvertical lines are perpendicular if the product of their slopes is -1.

> ### Slopes of Perpendicular Lines
> For two nonvertical perpendicular lines, one with slope m_1 and one with slope m_2, $m_1 \cdot m_2 = -1$.

The line $y = \frac{1}{3}x - 2$ is perpendicular to

$y = -3x + 3$ since $\frac{1}{3}(-3) = -1$.

-3 is the **negative reciprocal** of $\frac{1}{3}$.

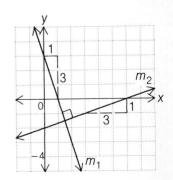

The line that contains the points $(4, 2)$ and $(-2, 5)$ is perpendicular to the line that contains the points $(-4, 3)$ and $(-3, 5)$ because the product of the slopes of the two lines is -1.

$$m_1 = \frac{5 - 2}{-2 - 4} = \frac{3}{-6} = -\frac{1}{2}$$

$$m_2 = \frac{5 - 3}{-3 - (-4)} = \frac{2}{1} = 2$$

$$m_1 \cdot m_2 = -\frac{1}{2}(2) = -1$$

The lines $3x + 4y = -8$ and $4x - 3y = -9$ are perpendicular because the product of the slopes of the two lines is -1.

$$\begin{array}{ll} 3x + 4y = -8 & 4x - 3y = -9 \\ 4y = -3x - 8 & -3y = -4x - 9 \\ y = -\frac{3}{4}x - 2 & y = \frac{4}{3}x + 3 \end{array}$$

$$m_1 = -\frac{3}{4} \qquad\qquad m_2 = \frac{4}{3}$$

$$m_1 \cdot m_2 = -\frac{3}{4}\left(\frac{4}{3}\right) = -1$$

Find the equation of the line containing the point $(3, -4)$ and perpendicular to the line $2x - y = -3$.

Solve the given equation for y to determine the slope of the given line.

$$\begin{array}{l} 2x - y = -3 \\ -y = -2x - 3 \\ y = 2x + 3 \qquad m_1 = 2 \end{array}$$

Substitute the value of m_1 into the equation $m_1 \cdot m_2 = -1$ and solve for m_2, the slope of the perpendicular line.

$$m_1 \cdot m_2 = -1$$
$$2m_2 = -1$$
$$m_2 = -\frac{1}{2}$$

Substitute the slope, m_2, and the coordinates of the given point in the point-slope formula.

$$y - y_1 = m(x - x_1)$$
$$y - (-4) = -\frac{1}{2}(x - 3)$$
$$y + 4 = -\frac{1}{2}x + \frac{3}{2}$$
$$y = -\frac{1}{2}x - \frac{5}{2}$$

Example 5   Are the lines $4x - y = -2$ and $x + 4y = -12$ perpendicular?

Solution

$$\begin{array}{ll} 4x - y = -2 & x + 4y = -12 \\ -y = -4x - 2 & 4y = -x - 12 \\ y = 4x + 2 & y = -\frac{1}{4}x - 3 \end{array}$$

$$m_1 = 4 \qquad\qquad m_2 = -\frac{1}{4}$$

$$m_1 \cdot m_2 = 4\left(-\frac{1}{4}\right) = -1$$

The lines are perpendicular.

• Solve each equation for y.

• Find the slope of each line.

• Find the product of the slopes.

Problem 5   Find the equation of the line containing the point $(-2, 2)$ and perpendicular to the line $x - 4y = 3$.

Solution   See page A50.
$y = -4x - 6$

EXERCISES 6.3

1   Find the equation of the line that contains the given point and has the given slope.

1. Point $(0, 5)$, $m = 2$ $y = 2x + 5$

2. Point $(0, 3)$, $m = 1$ $y = x + 3$

3. Point $(2, 3)$, $m = \dfrac{1}{2}$ $y = \dfrac{1}{2}x + 2$

4. Point $(5, 1)$, $m = \dfrac{2}{3}$ $y = \dfrac{2}{3}x - \dfrac{7}{3}$

5. Point $(-1, 4)$, $m = \dfrac{5}{4}$ $y = \dfrac{5}{4}x + \dfrac{21}{4}$

6. Point $(-2, 1)$, $m = \dfrac{3}{2}$ $y = \dfrac{3}{2}x + 4$

7. Point $(3, 0)$, $m = -\dfrac{5}{3}$ $y = -\dfrac{5}{3}x + 5$

8. Point $(-2, 0)$, $m = \dfrac{3}{2}$ $y = \dfrac{3}{2}x + 3$

9. Point $(2, 3)$, $m = -3$ $y = -3x + 9$

10. Point $(1, 5)$, $m = -\dfrac{4}{5}$ $y = -\dfrac{4}{5}x + \dfrac{29}{5}$

11. Point $(-1, 7)$, $m = -3$ $y = -3x + 4$

12. Point $(-2, 4)$, $m = -4$ $y = -4x - 4$

13. Point $(-1, -3)$, $m = \dfrac{2}{3}$ $y = \dfrac{2}{3}x - \dfrac{7}{3}$

14. Point $(-2, -4)$, $m = \dfrac{1}{4}$ $y = \dfrac{1}{4}x - \dfrac{7}{2}$

15. Point $(0, 0)$, $m = \dfrac{1}{2}$ $y = \dfrac{1}{2}x$

16. Point $(0, 0)$, $m = \dfrac{3}{4}$ $y = \dfrac{3}{4}x$

17. Point $(2, -3)$, $m = 3$ $y = 3x - 9$

18. Point $(4, -5)$, $m = 2$ $y = 2x - 13$

19. Point $(3, 5)$, $m = -\dfrac{2}{3}$ $y = -\dfrac{2}{3}x + 7$

20. Point $(5, 1)$, $m = -\dfrac{4}{5}$ $y = -\dfrac{4}{5}x + 5$

21. Point $(-2, 0)$, $m = 0$ $y = 0$

22. Point $(4, 0)$, $m = 0$ $y = 0$

23. Point $(-2, -3)$, $m = \dfrac{3}{2}$ $y = \dfrac{3}{2}x$

24. Point $(-3, -1)$, $m = \dfrac{4}{3}$ $y = \dfrac{4}{3}x + 3$

25. Point $(0, 2)$, no slope $x = 0$

26. Point $(0, 5)$, no slope $x = 0$

27. Point $(0, -3)$, $m = 3$ $y = 3x - 3$

28. Point $(0, -2)$, $m = 0$ $y = -2$

29. Point $(-5, 0)$, $m = -\dfrac{5}{2}$ $y = -\dfrac{5}{2}x - \dfrac{25}{2}$

30. Point $(-2, 0)$, $m = \dfrac{2}{5}$ $y = \dfrac{2}{5}x + \dfrac{4}{5}$

31. Point $(0, 5)$, $m = \dfrac{4}{3}$ $y = \dfrac{4}{3}x + 5$

32. Point $(0, -5)$, $m = \dfrac{6}{5}$ $y = \dfrac{6}{5}x - 5$

33. Point $(4, -1)$, $m = -\dfrac{2}{5}$ $y = -\dfrac{2}{5}x + \dfrac{3}{5}$

34. Point $(-3, 5)$, $m = -\dfrac{1}{4}$ $y = -\dfrac{1}{4}x + \dfrac{17}{4}$

35. Point $(3, -4)$, no slope $x = 3$

36. Point $(-2, 5)$, no slope $x = -2$

37. Point $(-2, -5)$, $m = -\dfrac{5}{4}$ $y = -\dfrac{5}{4}x - \dfrac{15}{2}$

38. Point $(-3, -2)$, $m = -\dfrac{2}{3}$ $y = -\dfrac{2}{3}x - 4$

39. Point $(-2, -3)$, $m = 0$ $y = -3$

40. Point $(-3, -2)$, $m = 0$ $y = -2$

Find the equation of the line containing the given points.

41. $P_1(0, 2)$, $P_2(3, 5)$ $y = x + 2$

42. $P_1(0, 4)$, $P_2(1, 5)$ $y = x + 4$

43. $P_1(0, -3)$, $P_2(-4, 5)$ $y = -2x - 3$

44. $P_1(0, -2)$, $P_2(-3, 4)$ $y = -2x - 2$

45. $P_1(2, 3)$, $P_2(5, 5)$ $y = \dfrac{2}{3}x + \dfrac{5}{3}$

46. $P_1(4, 1)$, $P_2(6, 3)$ $y = x - 3$

47. $P_1(-1, 3)$, $P_2(2, 4)$ $y = \dfrac{1}{3}x + \dfrac{10}{3}$

48. $P_1(-1, 1)$, $P_2(4, 4)$ $y = \dfrac{3}{5}x + \dfrac{8}{5}$

49. $P_1(-1, -2)$, $P_2(3, 4)$ $y = \dfrac{3}{2}x - \dfrac{1}{2}$

50. $P_1(-3, -1)$, $P_2(2, 4)$ $y = x + 2$

51. $P_1(0, 3)$, $P_2(2, 0)$ $y = -\dfrac{3}{2}x + 3$

52. $P_1(0, 4)$, $P_2(2, 0)$ $y = -2x + 4$

53. $P_1(-3, -1)$, $P_2(2, -1)$ $y = -1$

54. $P_1(-3, -5)$, $P_2(4, -5)$ $y = -5$

55. $P_1(-2, -3)$, $P_2(-1, -2)$ $y = x - 1$

56. $P_1(-4, -1)$, $P_2(-5, -2)$ $y = x + 3$

57. $P_1(-2, 3)$, $P_2(-1, 1)$ $y = -2x - 1$

58. $P_1(-3, 5)$, $P_2(-2, 3)$ $y = -2x - 1$

59. $P_1(-2, 5)$, $P_2(-2, 4)$ $x = -2$

60. $P_1(3, 6)$, $P_2(3, -2)$ $x = 3$

61. $P_1(3, 2)$, $P_2(-1, 5)$ $y = -\dfrac{3}{4}x + \dfrac{17}{4}$

62. $P_1(4, 1)$, $P_2(-2, 4)$ $y = -\dfrac{1}{2}x + 3$

63. $P_1(3, -1)$, $P_2(2, -4)$ $y = 3x - 10$

64. $P_1(4, 1)$, $P_2(3, -2)$ $y = 3x - 11$

65. $P_1(-2, 3)$, $P_2(2, -1)$ $y = -x + 1$

66. $P_1(3, 1)$, $P_2(-3, -2)$ $y = \dfrac{1}{2}x - \dfrac{1}{2}$

67. $P_1(-2, -3)$, $P_2(5, 0)$ $y = \dfrac{3}{7}x - \dfrac{15}{7}$

68. $P_1(7, 2)$, $P_2(4, 4)$ $y = -\dfrac{2}{3}x + \dfrac{20}{3}$

69. $P_1(2, 0)$, $P_2(0, -1)$ $y = \dfrac{1}{2}x - 1$

70. $P_1(0, 4)$, $P_2(-2, 0)$ $y = 2x + 4$

71. $P_1(3, -4)$, $P_2(-2, -4)$ $y = -4$

72. $P_1(-3, 3)$, $P_2(-2, 3)$ $y = 3$

73. $P_1(0, 0)$, $P_2(4, 3)$ $y = \dfrac{3}{4}x$

74. $P_1(2, -5)$, $P_2(0, 0)$ $y = -\dfrac{5}{2}x$

75. $P_1(2, -1)$, $P_2(-1, 3)$ $y = -\dfrac{4}{3}x + \dfrac{5}{3}$

76. $P_1(3, -5)$, $P_2(-2, 1)$ $y = -\dfrac{6}{5}x - \dfrac{7}{5}$

77. $P_1(-2, 5)$, $P_2(-2, -5)$ $x = -2$

78. $P_1(3, 2)$, $P_2(3, -4)$ $x = 3$

79. $P_1(2, 1)$, $P_2(-2, -3)$ $y = x - 1$

80. $P_1(-3, -2)$, $P_2(1, -4)$ $y = -\dfrac{1}{2}x - \dfrac{7}{2}$

2

81. Is the line $x = -2$ perpendicular to the line $y = 3$?
yes

82. Is the line $y = \frac{1}{2}$ perpendicular to the line $y = -4$?

    no

83. Is the line $x = -3$ parallel to the line $y = \frac{1}{3}$?

    no

84. Is the line $x = 4$ parallel to the line $x = -4$?
    yes

85. Is the line $y = \frac{2}{3}x - 4$ parallel to the line $y = -\frac{3}{2}x - 4$?

    no

86. Is the line $y = -2x + \frac{2}{3}$ parallel to the line $y = -2x + 3$?

    yes

87. Is the line $y = \frac{4}{3}x - 2$ perpendicular to the line $y = -\frac{3}{4}x + 2$?

    yes

88. Is the line $y = \frac{1}{2}x + \frac{3}{2}$ perpendicular to the line $y = -\frac{1}{2}x + \frac{3}{2}$?

    no

89. Are the lines $2x + 3y = 2$ and $2x + 3y = -4$ parallel?
    yes

90. Are the lines $2x - 4y = 3$ and $2x + 4y = -3$ parallel?
    no

91. Are the lines $x - 4y = 2$ and $4x + y = 8$ perpendicular?
    yes

92. Are the lines $4x - 3y = 2$ and $4x + 3y = -7$ perpendicular?
    no

93. Is the line that contains the points $(3, 2)$ and $(1, 6)$ parallel to the line that contains the points $(-1, 3)$ and $(-1, -1)$?
    no

94. Is the line that contains the points $(4, -3)$ and $(2, 5)$ parallel to the line that contains the points $(-2, -3)$ and $(-4, 1)$?
    no

95. Is the line that contains the points $(-3, 2)$ and $(4, -1)$ perpendicular to the line that contains the points $(1, 3)$ and $(-2, -4)$?
    yes

96. Is the line that contains the points $(-1, 2)$ and $(3, 4)$ perpendicular to the line that contains the points $(-1, 3)$ and $(-4, 1)$?
    no

97. Is the line that contains the points $(-5, 0)$ and $(0, 2)$ parallel to the line that contains the points $(5, 1)$ and $(0, -1)$?

yes

98. Is the line that contains the points $(3, 5)$ and $(-3, 3)$ perpendicular to the line that contains the points $(2, -5)$ and $(-4, 4)$?

no

99. Find the equation of the line containing the point $(-2, -4)$ and parallel to the line $2x - 3y = 2$.

$$y = \frac{2}{3}x - \frac{8}{3}$$

100. Find the equation of the line containing the point $(3, 2)$ and parallel to the line $3x + y = -3$.

$y = -3x + 11$

101. Find the equation of the line containing the point $(4, 1)$ and perpendicular to the line $y = -3x + 4$.

$$y = \frac{1}{3}x - \frac{1}{3}$$

102. Find the equation of the line containing the point $(2, -5)$ and perpendicular to the line $y = \frac{5}{2}x - 4$.

$$y = -\frac{2}{5}x - \frac{21}{5}$$

103. Find the equation of the line containing the point $(-1, -3)$ and perpendicular to the line $3x - 5y = 2$.

$$y = -\frac{5}{3}x - \frac{14}{3}$$

104. Find the equation of the line containing the point $(-1, 3)$ and perpendicular to the line $2x + 4y = -1$.

$y = 2x + 5$

105. Find the equation of the line containing the point $(-3, 1)$ and parallel to the line $y = \frac{2}{3}x - 1$.

$y = \frac{2}{3}x + 3$

106. Find the equation of the line containing the point $(4, -3)$ and parallel to the line $y = -\frac{4}{3}x + 2$.

$y = -\frac{4}{3}x + \frac{7}{3}$

107. Find the equation of the line containing the point $(-5, 4)$ and parallel to the line $y = -\frac{5}{3}x - 7$.

$y = -\frac{5}{3}x - \frac{13}{3}$

108. Find the equation of the line containing the point $(3, -4)$ and parallel to the line $y = \frac{3}{2}x$.

$y = \frac{3}{2}x - \frac{17}{2}$

109. Find the equation of the line containing the point $(-4, -2)$ and parallel to the line $5x - 2y = -4$. $\quad y = \dfrac{5}{2}x + 8$

110. Find the equation of the line containing the point $(-2, 0)$ and parallel to the line $3x - 4y = 6$. $\quad y = \dfrac{3}{4}x + \dfrac{3}{2}$

111. Find the equation of the line containing the point $(4, -3)$ and perpendicular to the line $y = 5x$. $\quad y = -\dfrac{1}{5}x - \dfrac{11}{5}$

112. Find the equation of the line containing the point $(-2, 5)$ and perpendicular to the line $y = \dfrac{2}{3}x - 5$. $\quad y = -\dfrac{3}{2}x + 2$

113. Find the equation of the line containing the point $(-3, -3)$ and perpendicular to the line $3x - 2y = 3$. $\quad y = -\dfrac{2}{3}x - 5$

114. Find the equation of the line containing the point $(-1, 6)$ and perpendicular to the line $4x + y = -3$. $\quad y = \dfrac{1}{4}x + \dfrac{25}{4}$

SUPPLEMENTAL EXERCISES 6.3

Is there a linear equation that contains all the given ordered pairs? If there is, find the equation.

115. $(2, 2)$, $(5, -1)$, $(3, 1)$
$y = -x + 4$

116. $(2, 1)$, $(-1, -5)$, $(4, 5)$
$y = 2x - 3$

117. $(4, 3)$, $(-6, -2)$, $(8, 5)$ $\quad y = \dfrac{1}{2}x + 1$

118. $(2, -4)$, $(-1, 5)$, $(3, 9)$ $\quad$ no

The given ordered pairs are solutions to the same linear equation. Find n.

119. $(0, 2)$, $(4, 6)$, $(-2, n)$
0

120. $(3, 3)$, $(-1, 4)$, $(-5, n)$
5

121. $(5, -2)$, $(4, 3)$, $(n, -7)$ $\quad$ 6

122. $(1, -2)$, $(4, 3)$, $(n, 7)$ $\quad \dfrac{32}{5}$

Find the value of k such that the line containing P_1 and P_2 is parallel to the line containing P_3 and P_4.

123. $P_1(3, 4)$, $P_2(-2, -1)$, $P_3(4, 1)$, $P_4(0, k)$ $\quad$ -3

124. $P_1(-3, 5)$, $P_2(6, -1)$, $P_3(-4, 1)$, $P_4(2, k)$ $\quad$ -3

125. $P_1(6, 2)$, $P_2(-3, -1)$, $P_3(1, 5)$, $P_4(k, 4)$ $\quad$ -2

126. $P_1(-4, 5)$, $P_2(2, 3)$, $P_3(5, -1)$, $P_4(k, 2)$ $\quad$ -4

Find the value of k such that the line containing P_1 and P_2 is perpendicular to the line containing P_3 and P_4.

127. $P_1(2, 5)$, $P_2(6, 4)$, $P_3(-2, 1)$, $P_4(-3, k)$ -3

128. $P_1(-3, 1)$, $P_2(3, -2)$, $P_3(-1, 5)$, $P_4(0, k)$ 7

129. $P_1(-1, 4)$, $P_2(2, 5)$, $P_3(6, 1)$, $P_4(k, 4)$ 5

130. $P_1(4, 2)$, $P_2(7, 6)$, $P_3(2, 5)$, $P_4(k, 8)$ -2

Solve.

131. Find the equations of the lines that form the sides of the triangle with vertices $(-3, 6)$, $(2, 0)$, and $(-2, -1)$.
$y = -\dfrac{6}{5}x + \dfrac{12}{5}; y = \dfrac{1}{4}x - \dfrac{1}{2}; y = -7x - 15$

132. Find the equations of the lines that form the sides of the triangle with vertices $(4, 2)$, $(-1, -3)$, and $(-5, 3)$.
$y = x - 2; y = -\dfrac{3}{2}x - \dfrac{9}{2}; y = -\dfrac{1}{9}x + \dfrac{22}{9}$

133. Show that the triangle with vertices $(-3, -2)$, $(1, 4)$, and $(3, -6)$ is a right triangle. (*Hint:* Show that two sides of the triangle are perpendicular).
Complete solution available in Solutions Manual.

134. Show that the triangle with vertices $(2, 5)$, $(6, 3)$, and $(-1, -1)$ is a right triangle. (*Hint:* Show that two sides of the triangle are perpendicular.)
Complete solution available in Solutions Manual.

135. Show that the points $(1, 6)$, $(3, 2)$, $(-1, -6)$ and $(-3, -2)$ are the vertices of a parallelogram. (*Hint:* Opposite sides of a parallelogram are parallel.)
Complete solution available in Solutions Manual.

136. Show that the points $(-2, 5)$, $(8, 0)$, $(-8, 2)$ and $(2, -3)$ are the vertices of a parallelogram. (*Hint:* Opposite sides of a parallelogram are parallel.)
Compete solution available in Solutions Manual.

SECTION 6.4

Applications of Linear Equations

1 Obtain data from a graph

The rectangular coordinate system is used in business, science, and mathematics to show a relationship between two variables. One variable is represented along the horizontal axis, and the other variable is represented along the vertical axis. A linear relationship between the variables is represented on the coordinate system as a straight line.

A company purchases a computer system for $10,000. The graph at the right shows the depreciation of the computer system over a five-year period.

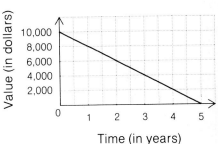

Information can be obtained from the graph. For example, the depreciated value of the computer after two years is $6000; the depreciated value of the computer after four years is $2000.

From the graph, an equation of the line that represents the depreciation can be written.

Use any two points shown on the graph to find the slope of the line. The points $(0, 10,000)$, $(1, 8000)$, $(2, 6000)$, $(3,4000)$, $(4,2000)$, and $(5,0)$ are all points on the graph. Points $(0, 10,000)$ and $(5,0)$ are used here.

$(x_1, y_1) = (0, 10,000)$
$(x_2, y_2) = (5,0)$
$$m = \frac{y_2 - y_1}{x_2 - x_1} = \frac{0 - 10,000}{5 - 0} = \frac{-10,000}{5} = -2000$$

Locate the y-intercept of the line on the graph.

The y-intercept is $(0, 10,000)$.

Use the slope-intercept form of an equation to write the equation of the line.

$y = mx + b$
$y = -2000x + 10,000$

The equation of the line that represents the depreciation is $y = -2000x + 10,000$.

In the equation, the slope represents the annual depreciation of the computer system. The *y*-intercept represents the computer's value at the time of purchase.

Once the equation of the line has been found, the equation can be used to determine the depreciated value of the computer system after any given number of years.

To find the depreciated value of the computer after two and one-half years, substitute 2.5 for *x* in the equation and solve for *y*.

$$y = -2000x + 10{,}000$$
$$y = -2000(2.5) + 10{,}000$$
$$y = -5000 + 10{,}000$$
$$y = 5000$$

The depreciated value of the computer system after two and one-half years is $5000.

Example 1   The graph below shows the relationship between the total cost of manufacturing toasters and the number of toasters manufactured. Write the equation of the line that represents the total cost of manufacturing the toasters. Use the equation to find the total cost of manufacturing 300 toasters.

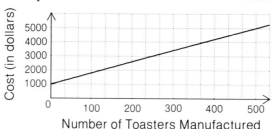

Strategy   To write the equation:
▶ Locate two points on the graph to find the slope of the line.
▶ Locate the *y*-intercept of the line on the graph.
▶ Use the slope-intercept form of an equation to write the equation of the line.

To find the cost of manufacturing 300 toasters, substitute 300 for *x* in the equation and solve for *y*.

Solution $m = \dfrac{y_2 - y_1}{x_2 - x_1} = \dfrac{5000 - 1000}{500 - 0} = \dfrac{4000}{500} = 8$ • Let $(x_1, y_1) = (0, 1000)$ and $(x_2, y_2) = (500, 5000)$. Find m.

The *y*-intercept is (0, 1000).

$$y = mx + b$$
$$y = 8x + 1000$$

• In the equation, the slope represents the unit cost, or the cost to manufacture one toaster. The *y*-intercept represents the fixed costs of operating the plant.

The equation of the line is $y = 8x + 1000$.

$y = 8x + 1000$
$y = 8(300) + 1000 = 3400$

The cost of manufacturing 300 toasters is $3400.

Problem 1   The relationship between Fahrenheit and Celsius temperature is shown on the graph below. Write the equation for the Fahrenheit temperature in terms of the Celsius temperature. Use the equation to find the Fahrenheit temperature when the Celsius temperature is 40°.

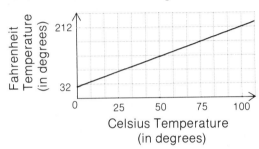

Solution   See pages A51 and A52.

$$y = \frac{9}{5}x + 32;\ 104°F$$

EXERCISES 6.4

1   Solve.

1. The graph on the right shows the relationship between the distance traveled by plane and the time of travel. Write an equation for the distance traveled in terms of the time of travel. Use the equation to find the distance traveled in two and one-half hours.
$y = 300x;\ 750\ mi$

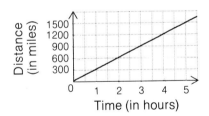

2. The graph on the right represents the relationship between the distance traveled by a cyclist and the time of travel. Write the equation for the distance traveled in terms of the time of travel. Use the equation to find the distance traveled in three hours.
$y = 8x;\ 24\ mi$

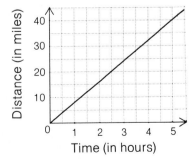

3. The relationship between the cost of a truck and the depreciation allowed for income tax purposes is shown in the graph on the right. Write the equation for the line that represents the depreciated value of the truck. Use the equation to find the value of the truck after four and one-half years.
    $y = -1200x + 12,000$; $6600

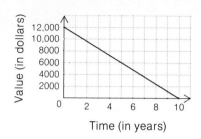

4. The graph on the right shows the relationship between a company's cost for new machinery and the depreciation allowed for income tax purposes. Write the equation for the line that represents the depreciated value of the machinery. Use the equation to find the depreciated value of the machinery after three and one-half years.
    $y = -120x + 600$; $180

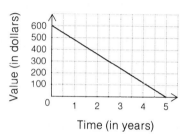

5. The graph on the right shows the relationship between the cost of manufacturing bicycles and the number of bicycles manufactured. Write the equation that represents the cost of manufacturing the bicycles. Use the equation to find the cost of manufacturing 200 bicycles.

    $y = 25x + 2000$; $7000

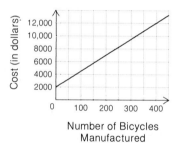

6. The relationship between the cost of manufacturing calculators and the number of calculators manufactured is shown in the graph on the right. Write the equation that represents the cost of manufacturing the calculators. Use the equation to find the cost of manufacturing 150 calculators.
    $y = 10x + 1500$; $3000

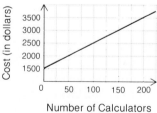

Economists frequently express the relationship between the price of a product and consumer demand for that product in a graph, called a demand curve.

7. Write the equation of the demand curve shown on the right. Use the equation to find the demand for the product when the price is $3.50 per unit.
    $y = -x + 5$; 1.5 units

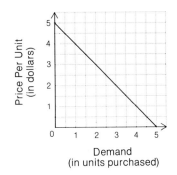

Demand
(in units purchased)

8. Write the equation of the demand curve shown on the right. Use the equation to find the demand for the product when the price is $5 per unit.
    $y = -2x + 10$; 2.5 units

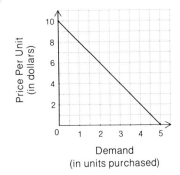

Demand
(in units purchased)

SUPPLEMENTAL EXERCISES 6.4

In 1988, the Social Security tax rate was 7.51% of gross earnings. The Social Security tax is deducted from the first $43,800 of gross earnings only.

9. Draw a graph to represent the amount of Social Security tax deducted from an employee's annual pay. Label the horizontal axis "Annual Pay." Include amounts from $0 to $40,000 on this axis.

Annual Pay

10. Write an equation for the amount of Social Security tax deducted from an employee's annual pay.
    $y = 0.0751x$

11. For the equation written for Exercise 10, what does the slope represent? What does the *y*-intercept represent?

    The slope represents the Social Security tax per $1.
    The *y*-intercept represents the amount of tax paid when $0 is earned.

12. If the horizontal axis were extended beyond $43,800, what would the graph of the line look like beyond this point? Write an equation for this portion of the graph.

    If the horizontal axis were extended beyond $43,800, the graph of the line would be a horizontal line.
    $y = 3289.38x$

CALCULATORS AND COMPUTERS

Linear Equations In Two Variables

The program EQUATION OF A STRAIGHT LINE on the Student Disk can be used to practice finding the equation of a line. Two types of problems are presented. One type asks you to find the equation of a line given the slope and a point on the line. The second type of problem is to find the equation of a line given two points on the line.

In each case, solve the problem using paper and pencil. When you are ready, press the RETURN key, and the answer will be displayed. The equation of the line is always given in the form $Ax + By = C$.

You may practice as long as you like. When you wish to quit, press the letter *Q* on the keyboard.

CHAPTER SUMMARY

Key Words

A **rectangular coordinate system** is formed by two number lines, one horizontal and one vertical, that intersect at the zero point of each line. The number lines that make up the coordinate system are called the **coordinate axes,** or simply **axes.** The **origin** is the point of intersection of the two coordinate axes.

A rectangular coordinate system divides the plane into four regions called **quadrants**.

An **ordered pair** (a, b) is used to locate a point in the plane. The first number in an ordered pair is called the **abscissa**. The second numer is called the **ordinate**.

An equation of the form $y = mx + b$ or $Ax + By = C$ is a **linear equation in two variables**. The **graph** of a linear equation in two variables is a straight line.

The solution set of an inequality in two variables is a **half plane.**

The **slope** of a line is a measure of the slant or tilt of the line. The symbol for slope is m. A line that slants upward to the right has a **positive slope.** A line that slants downward to the right has a **negative slope.** A horizontal line has **zero slope.** A vertical line has **no slope.**

The point at which a graph crosses the x-axis is called the **x-intercept.** The point at which a graph crosses the y-axis is called the **y-intercept.**

Two lines that have the same slope do not intersect and are called **parallel lines.**

Two lines that intersect at right angles are called **perpendicular lines.**

Essential Rules

Slope of a straight line $\text{Slope} = m = \dfrac{y_2 - y_1}{x_2 - x_1}$

Slope-intercept form of a straight line $y = mx + b$

Point-slope formula $y - y_1 = m(x - x_1)$

For two nonvertical parallel lines $m_1 = m_2$

For two nonvertical perpendicular lines $m_1 \cdot m_2 = -1$

CHAPTER REVIEW

1. Graph $y = \dfrac{2}{3}x - 4$.

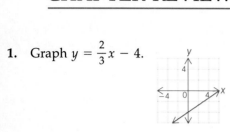

2. Find the equation of the line that contains the point $(-5, 2)$ and has slope $\dfrac{2}{5}$.

$$y = \dfrac{2}{5}x + 4$$

3. Find the equation of the line containing the point $(-3, 4)$ and parallel to the line $2x + 3y = 9$.

$$y = -\dfrac{2}{3}x + 2$$

4. Graph the ordered pairs $(3, -4)$ and $(4, -1)$. Draw a line between the two points.

5. Find the slope of the line containing the points $(-2, 3)$ and $(4, 2)$.

$$-\dfrac{1}{6}$$

6. Find the equation of the line containing the points $(3, -4)$ and $(-2, 3)$.

$$y = -\dfrac{7}{5}x + \dfrac{1}{5}$$

7. Find the ordered pair solution of $y = 2x + 6$ corresponding to $x = -3$.

$(-3, 0)$

8. Graph $2x - 3y = 6$ by using the x- and y-intercepts.

9. Graph the solution set of $5x - 2y \le 4$.

10. Find the equation of the line containing the point $(-2, -3)$ and perpendicular to the line $y = -\dfrac{1}{2}x - 3$.

$$y = 2x + 1$$

11. Find the equation of the line that contains the point $(0, 2)$ and has slope $-\dfrac{3}{4}$.

$$y = -\dfrac{3}{4}x + 2$$

12. Graph $2x + 3y = -3$.

13. Graph the line that passes through point $(-2, 3)$ and has slope $-\dfrac{3}{2}$.

14. Find the slope of the line containing the points $(2, 5)$ and $(-2, 5)$.

0

15. Find the equation of the line containing the points $(2, -3)$ and $(2, -5)$.
$x = 2$

16. Graph the solution set of $2x - 3y > 9$.

17. Find the x-intercept of the line $2x + 3y = 6$.
$(3, 0)$

18. Graph $y = -\frac{4}{3}x + 3$.

19. Find the equation of the line containing the point $(0, 0)$ and parallel to the line $y = -\frac{3}{2}x - 7$. $y = -\frac{3}{2}x$

20. Find the ordered pair solution of $y = -\frac{3}{4}x + 2$ corresponding to $x = 3$.
$\left(3, -\frac{1}{4}\right)$

21. Find the slope of the line containing the points $(-2, 4)$ and $(-2, -3)$.
no slope

22. Find the equation of the line containing the point $(0, 0)$ and perpendicular to the line $2x - 3y = -2$. $y = -\frac{3}{2}x$

23. Graph the solution set of $y > 3$.

24. Graph $3x + 2y = 1$.

25. Graph the line that passes through point $(-2, 2)$ and has slope 1.

26. Find the equation of the line that contains the point $(2, -4)$ and has slope $\frac{5}{2}$.
$y = \frac{5}{2}x - 9$

27. The graph at the right shows the relationship between the cost of a rental house and the depreciation allowed for income tax purposes. Write the equation for the depreciated value of the house.
$y = -\frac{10,000}{3}x + 50,000$

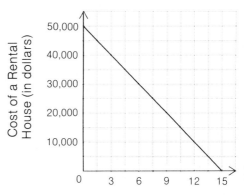

CUMULATIVE REVIEW

1. Simplify: $-4^2 \cdot (-3)^3$
   432

2. Simplify: $15 - 3[3 - (-2)]^2 \div 5$
   0

3. Evaluate $\dfrac{-a^2 - b^2}{2}$ when $a = 2$ and $b = -3$.
   $-\dfrac{13}{2}$

4. Solve: $3 - \dfrac{x}{2} = \dfrac{3}{4}$
   $\dfrac{9}{2}$

5. Solve: $2[\,y - 2(3 - y) + 4] = 4 - 3y$
   $\dfrac{8}{9}$

6. Solve: $4 < 3x + 1 < 10$
   $\{x \,|\, 1 < x < 3\}$

7. Solve: $8 - |2x - 1| = 4$
   $\dfrac{5}{2}$ and $-\dfrac{3}{2}$

8. Solve: $|4 - 5x| < 6$
   $\left\{x \,\middle|\, -\dfrac{2}{5} < x < 2\right\}$

9. Simplify: $\left(\dfrac{3ab^2}{2a}\right)^3 \left(\dfrac{b^2}{-3ab}\right)^2$
   $\dfrac{3b^8}{8a^2}$

10. Simplify: $3a^2 - a[2 - a(4 - a + a^2)]$
    $a^4 - a^3 + 7a^2 - 2a$

11. Factor: $8x^2y + 16x^2y^2 + 24xy^2$
    $8xy(x + 2xy + 3y)$

12. Factor: $6x^2 - 9bx + 4ax - 6ab$
    $(3x + 2a)(2x - 3b)$

13. Solve: $(y - 2)^2 = 9$
    5 and -1

14. Solve: $(x + 3)(x - 1)(x - 2) \geq 0$
    $\{x \,|\, -3 \leq x \leq 1 \text{ or } x \geq 2\}$

15. Simplify: $\dfrac{2x + 3}{x + 3} + \dfrac{x - 2}{5 - x} - \dfrac{15 - 3x}{x^2 - 2x - 15}$
    $\dfrac{x - 8}{x - 5}$

16. Simplify: $5b\sqrt[3]{16a^4b} - 2a\sqrt[3]{54ab^4}$
    $4ab\sqrt[3]{2ab}$

17. Simplify: $(3 - 4i)(6 - i)$
    $14 - 27i$

18. Solve: $\sqrt[3]{x + 3} = \sqrt[3]{4x - 9}$
    4

19. Find the ordered pair solution of $y = -\dfrac{5}{2}x + 4$ corresponding to $x = 2$.
    $(2, -1)$

20. Find the slope of the line containing the points $(-3, 4)$ and $(-4, 2)$.
    2

21. Find the equation of the line that contains the point $(-5, 2)$ and has slope -2.
    $y = -2x - 8$

22. Find the equation of the line containing the points $(1, -2)$ and $(-2, 4)$.
    $y = -2x$

23. Find the equation of the line containing the point $(-3, 4)$ and parallel to the line $2x - 3y = 6$.
    $y = \dfrac{2}{3}x + 6$

24. Find the equation of the line containing the point $(-1, 2)$ and perpendicular to the line $x - 4y = 6$.
    $y = -4x - 2$

25. Graph $2x + 3y = 6$.

26. Graph $2x - y \geq 4$.

27. Graph the line that passes through point $(-1, 3)$ and has slope $\dfrac{2}{3}$.

28. How many ounces of pure acid must be added to 500 oz of a 25% acid solution to make a 40% acid solution?

125 oz

29. A stamp collection consists of 42 stamps. The collection contains only 3¢ stamps and 5¢ stamps. The total value of the stamps is $1.58. Find the number of 3¢ stamps.

26 stamps

30. Two planes are 1800 mi apart and traveling toward each other. One plane is traveling twice as fast as the other plane. The planes meet in 3 h. Find the speed of each plane.

slower plane: 200 mph;
faster plane: 400 mph

31. If $6000 is invested at a simple annual interest rate of 7.8%, how much additional money must be invested at an annual simple interest rate of 11.2% so that the total interest earned is 8.8% of the total investment?

$2500

32. The graph below shows the relationship between the hours worked by a plumber and the plumber's wages. Write the equation that represents the plumber's wages. $y = 18x$

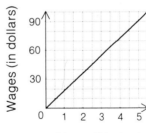

Hours Worked

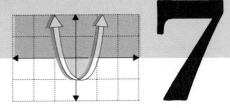

7

Quadratic Equations

OBJECTIVES

- Solve quadratic equations by factoring
- Write a quadratic equation given its solutions
- Solve quadratic equations by taking square roots
- Solve quadratic equations by completing the square
- Solve quadratic equations by using the quadratic formula
- Equations that are quadratic in form
- Radical equations
- Fractional equations
- Graph equations of the form $y = ax^2 + bx + c$
- Find the x-intercepts of a parabola
- Application problems

Complex Numbers

Negative numbers were not universally accepted in the mathematical community until well into the fourteenth century. It is no wonder then that *imaginary numbers* took an even longer time to gain acceptance.

Beginning in the mid-sixteenth century, mathematicians were beginning to integrate imaginary numbers into their writings. One notation for 3i was R (0   m   3). Literally this was interpreted as $\sqrt{0 - 3}$.

By the mid-eighteenth century, the symbol *i* was introduced. Still later it was shown that complex numbers could be thought of as points in the plane. The complex number 3 + 4*i* was associated with the point (3, 4).

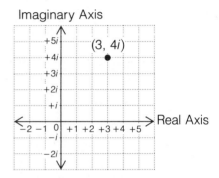

By the end of the nineteenth century, complex numbers were fully integrated into mathematics. This was due in large part to some eminent mathematicians who used complex numbers to prove theorems that had previously eluded proof.

Solving Quadratic Equations by Factoring or by Taking Square Roots

1 Solve quadratic equations by factoring

A **quadratic equation** is an equation of the form $ax^2 + bx + c = 0$, where a, b, and c are constants and $a \neq 0$.

$$3x^2 - x + 2 = 0, \quad a = 3, \quad b = -1, c = 2$$

$$-x^2 + 4 = 0, \quad a = -1, b = 0, \quad c = 4$$

$$6x^2 - 5x = 0, \quad a = 6, \quad b = -5, c = 0$$

A quadratic equation is in **standard form** when the polynomial is in descending order and equal to zero.

Since the degree of the polynomial $ax^2 + bx + c$ is 2, a quadratic equation is also called a **second-degree equation.**

Recall that quadratic equations sometimes can be solved by using the Principle of Zero Products.

> ### The Principle of Zero Products
>
> If the product of two factors is zero, then at least one of the factors must be zero.
>
> If $ab = 0$, then $a = 0$ or $b = 0$.

Solve by factoring: $x^2 - 6x = -9$

	$x^2 - 6x = -9$
Write the equation in standard form.	$x^2 - 6x + 9 = 0$
Use the Principle of Zero Products.	$(x - 3)(x - 3) = 0$
Solve each equation.	$x - 3 = 0 \qquad x - 3 = 0$
	$x = 3 \qquad\qquad x = 3$
Write the solutions.	The solution is 3.
	3 checks as a solution.

When a quadratic equation has two solutions that are the same number, the solution is called a **double root** of the equation. The solution 3 is a double root of the equation $x^2 - 6x = -9$.

Example 1   Solve for x by factoring: $x^2 - 4ax - 5a^2 = 0$

Solution $x^2 - 4ax - 5a^2 = 0$ • This is a literal equation. Solve for x in terms of a.

$(x + a)(x - 5a) = 0$

$x + a = 0 \qquad\qquad x - 5a = 0$
$\quad x = -a \qquad\qquad\quad x = 5a$

The solutions are $-a$ and $5a$.

Problem 1   Solve for x by factoring: $x^2 - 3ax - 4a^2 = 0$

Solution   See page A54.
$-a$ and $4a$

2 ■ Write a quadratic equation given its solutions

As shown below, the solutions of the equation $(x - r_1)(x - r_2) = 0$ are r_1 and r_2.

$(x - r_1)(x - r_2) = 0$ Check:

$x - r_1 = 0 \qquad x - r_2 = 0 \qquad \dfrac{(x - r_1)(x - r_2) = 0}{(r_1 - r_1)(r_1 - r_2)} \qquad \dfrac{(x - r_1)(x - r_2) = 0}{(r_2 - r_1)(r_2 - r_2)}$
$\quad x = r_1 \qquad\qquad x = r_2 \qquad\qquad 0 \cdot (r_1 - r_2) \qquad\qquad (r_2 - r_1) \cdot 0$
$\qquad\qquad\qquad\qquad\qquad\qquad\qquad\qquad\qquad 0 = 0 \qquad\qquad\qquad\qquad 0 = 0$

Using the equation $(x - r_1)(x - r_2) = 0$ and the fact that r_1 and r_2 are solutions of this equation, it is possible to write a quadratic equation given its solutions.

Write a quadratic equation that has solutions 4 and -5.

$\qquad\qquad\qquad\qquad\qquad\qquad\qquad\qquad\qquad (x - r_1)(x - r_2) = 0$
Replace r_1 by 4 and r_2 by -5. $\qquad\qquad\qquad (x - 4)[x - (-5)] = 0$

Simplify. $\qquad\qquad\qquad\qquad\qquad\qquad\qquad\qquad (x - 4)(x + 5) = 0$

Multiply. $\qquad\qquad\qquad\qquad\qquad\qquad\qquad\qquad\quad x^2 + x - 20 = 0$

Example 2 Write a quadratic equation that has integer coefficients and has solutions $\frac{2}{3}$ and $\frac{1}{2}$.

Solution $(x - r_1)(x - r_2) = 0$

$\left(x - \frac{2}{3}\right)\left(x - \frac{1}{2}\right) = 0$ • Replace r_1 by $\frac{2}{3}$ and r_2 by $\frac{1}{2}$.

$x^2 - \frac{7}{6}x + \frac{1}{3} = 0$ • Multiply.

$6\left(x^2 - \frac{7}{6}x + \frac{1}{3}\right) = 6 \cdot 0$ • Multiply each side of the equation by the LCM of the denominators.

$6x^2 - 7x + 2 = 0$

Problem 2 Write a quadratic equation that has integer coefficients and has solutions $-\frac{2}{3}$ and $\frac{1}{6}$.

Solution See page A55.
$18x^2 + 9x - 2 = 0$

3 Solve quadratic equations by taking square roots

The solution of the quadratic equation $x^2 = 16$ is shown at the right.

$$x^2 = 16$$
$$x^2 - 16 = 0$$
$$(x + 4)(x - 4) = 0$$
$$x + 4 = 0 \qquad x - 4 = 0$$
$$x = -4 \qquad x = 4$$

Note that the solution is the positive or the negative square root of 16, 4 or -4.

The solution can also be found by taking the square root of each side of the equation and writing the positive and the negative square roots of the number. The notation $x = \pm 4$ means $x = 4$ or $x = -4$.

$$x^2 = 16$$
$$\sqrt{x^2} = \sqrt{16}$$
$$x = \pm\sqrt{16} = \pm 4$$
The solutions are 4 and -4.

Solve by taking square roots: $3x^2 = 54$

Solve for x^2.

$$3x^2 = 54$$
$$x^2 = 18$$

Take the square root of each side of the equation.

$$\sqrt{x^2} = \sqrt{18}$$

Simplify.

$$x = \pm\sqrt{18} = \pm 3\sqrt{2}$$

Write the solutions.

The solutions are $3\sqrt{2}$ and $-3\sqrt{2}$.
$3\sqrt{2}$ and $-3\sqrt{2}$ check as solutions.

Solving a quadratic equation by taking the square root of each side of the equation can lead to solutions that are complex numbers.

Solve by taking square roots: $2x^2 + 18 = 0$

Solve for x^2.

$$\begin{aligned} 2x^2 + 18 &= 0 \\ 2x^2 &= -18 \\ x^2 &= -9 \\ \sqrt{x^2} &= \sqrt{-9} \end{aligned}$$

Take the square root of each side of the equation.

Simplify.

$$x = \pm\sqrt{-9} = \pm 3i$$

Write the solutions.

The solutions are $3i$ and $-3i$.
$3i$ and $-3i$ check as solutions.

An equation containing the square of a binomial can be solved by taking square roots.

Example 3   Solve by taking square roots: $3(x - 2)^2 + 12 = 0$

Solution
$$\begin{aligned} 3(x - 2)^2 + 12 &= 0 \\ 3(x - 2)^2 &= -12 \\ (x - 2)^2 &= -4 \\ \sqrt{(x - 2)^2} &= \sqrt{-4} \\ x - 2 &= \pm\sqrt{-4} = \pm 2i \end{aligned}$$

• Solve for $(x - 2)^2$.

• Take the square root of each side of the equation. Then simplify.

$$\begin{array}{ll} x - 2 = 2i & x - 2 = -2i \\ x = 2 + 2i & x = 2 - 2i \end{array}$$

• Solve for x.

The solutions are $2 + 2i$ and $2 - 2i$.

Problem 3   Solve by taking square roots: $2(x + 1)^2 + 24 = 0$

Solution   See page A55.
$-1 + 2i\sqrt{3}$ and $-1 - 2i\sqrt{3}$

EXERCISES 7.1

1   Solve by factoring.

1. $x^2 - 4x = 0$
   0 and 4

2. $y^2 + 6y = 0$
   -6 and 0

3. $t^2 - 25 = 0$
   -5 and 5

4. $p^2 - 81 = 0$
   9 and -9

5. $s^2 - s - 6 = 0$
   -2 and 3

6. $v^2 + 4v - 5 = 0$
   -5 and 1

7. $y^2 - 6y + 9 = 0$
   3

8. $x^2 + 10x + 25 = 0$
   -5

9. $9z^2 - 18z = 0$
   0 and 2

10. $4y^2 + 20y = 0$
    0 and -5

11. $r^2 - 3r = 10$
    -2 and 5

12. $p^2 + 5p = 6$
    1 and -6

13. $v^2 + 10 = 7v$

    2 and 5

14. $t^2 - 16 = 15t$

    -1 and 16

15. $2x^2 - 9x - 18 = 0$

    6 and $-\dfrac{3}{2}$

16. $3y^2 - 4y - 4 = 0$

    $-\dfrac{2}{3}$ and 2

17. $4z^2 - 9z + 2 = 0$

    $\dfrac{1}{4}$ and 2

18. $2s^2 - 9s + 9 = 0$

    $\dfrac{3}{2}$ and 3

19. $3w^2 + 11w = 4$

    -4 and $\dfrac{1}{3}$

20. $2r^2 + r = 6$

    $\dfrac{3}{2}$ and -2

21. $6x^2 = 23x + 18$

    $-\dfrac{2}{3}$ and $\dfrac{9}{2}$

22. $6x^2 = 7x - 2$

    $\dfrac{1}{2}$ and $\dfrac{2}{3}$

23. $4 - 15u - 4u^2 = 0$

    -4 and $\dfrac{1}{4}$

24. $3 - 2y - 8y^2 = 0$

    $-\dfrac{3}{4}$ and $\dfrac{1}{2}$

25. $x + 18 = x(x - 6)$

    -2 and 9

26. $t + 24 = t(t + 6)$

    -8 and 3

27. $4s(s + 3) = s - 6$

    -2 and $-\dfrac{3}{4}$

28. $3v(v - 2) = 11v + 6$

    $-\dfrac{1}{3}$ and 6

29. $u^2 - 2u + 4 = (2u - 3)(u + 2)$

    2 and -5

30. $(3v - 2)(2v + 1) = 3v^2 - 11v - 10$

    $-\dfrac{4}{3}$ and -2

31. $(3x - 4)(x + 4) = x^2 - 3x - 28$

    $-\dfrac{3}{2}$ and -4

Solve for x by factoring.

32. $x^2 + 14ax + 48a^2 = 0$
    $-6a$ and $-8a$

33. $x^2 - 9bx + 14b^2 = 0$
    $2b$ and $7b$

34. $x^2 + 9xy - 36y^2 = 0$
    $3y$ and $-12y$

35. $x^2 - 6cx - 7c^2 = 0$

    $-c$ and $7c$

36. $x^2 - ax - 20a^2 = 0$

    $-4a$ and $5a$

37. $2x^2 + 3bx + b^2 = 0$

    $-\dfrac{b}{2}$ and $-b$

38. $3x^2 - 4cx + c^2 = 0$

    $\dfrac{c}{3}$ and c

39. $3x^2 - 14ax + 8a^2 = 0$

    $\dfrac{2a}{3}$ and $4a$

40. $3x^2 - 11xy + 6y^2 = 0$

    $\dfrac{2y}{3}$ and $3y$

41. $3x^2 - 8ax - 3a^2 = 0$

    $-\dfrac{a}{3}$ and $3a$

42. $3x^2 - 4bx - 4b^2 = 0$

    $-\dfrac{2b}{3}$ and $2b$

43. $4x^2 + 8xy + 3y^2 = 0$

    $-\dfrac{3y}{2}$ and $-\dfrac{y}{2}$

44. $6x^2 - 11cx + 3c^2 = 0$
$\dfrac{3c}{2}$ and $\dfrac{c}{3}$

45. $6x^2 + 11ax + 4a^2 = 0$
$-\dfrac{a}{2}$ and $-\dfrac{4a}{3}$

46. $12x^2 - 5xy - 2y^2 = 0$
$-\dfrac{y}{4}$ and $\dfrac{2y}{3}$

2   Write a quadratic equation that has integer coefficients and has as solutions the given pair of numbers.

47. 2 and 5
$x^2 - 7x + 10 = 0$

48. 3 and 1
$x^2 - 4x + 3 = 0$

49. -2 and -4
$x^2 + 6x + 8 = 0$

50. -1 and -3
$x^2 + 4x + 3 = 0$

51. 6 and -1
$x^2 - 5x - 6 = 0$

52. -2 and 5
$x^2 - 3x - 10 = 0$

53. 3 and -3
$x^2 - 9 = 0$

54. 5 and -5
$x^2 - 25 = 0$

55. 4 and 4
$x^2 - 8x + 16 = 0$

56. 2 and 2
$x^2 - 4x + 4 = 0$

57. 0 and 5
$x^2 - 5x = 0$

58. 0 and -2
$x^2 + 2x = 0$

59. 0 and 3
$x^2 - 3x = 0$

60. 0 and -1
$x^2 + x = 0$

61. 3 and $\dfrac{1}{2}$
$2x^2 - 7x + 3 = 0$

62. 2 and $\dfrac{2}{3}$
$3x^2 - 8x + 4 = 0$

63. $-\dfrac{3}{4}$ and 2
$4x^2 - 5x - 6 = 0$

64. $-\dfrac{1}{2}$ and 5
$2x^2 - 9x - 5 = 0$

65. $-\dfrac{5}{3}$ and -2
$3x^2 + 11x + 10 = 0$

66. $-\dfrac{3}{2}$ and -1
$2x^2 + 5x + 3 = 0$

67. $-\dfrac{2}{3}$ and $\dfrac{2}{3}$
$9x^2 - 4 = 0$

68. $-\dfrac{1}{2}$ and $\dfrac{1}{2}$
$4x^2 - 1 = 0$

69. $\dfrac{1}{2}$ and $\dfrac{1}{3}$
$6x^2 - 5x + 1 = 0$

70. $\dfrac{3}{4}$ and $\dfrac{2}{3}$
$12x^2 - 17x + 6 = 0$

71. $\dfrac{6}{5}$ and $-\dfrac{1}{2}$
$10x^2 - 7x - 6 = 0$

72. $\dfrac{3}{4}$ and $-\dfrac{3}{2}$
$8x^2 + 6x - 9 = 0$

73. $-\dfrac{1}{4}$ and $-\dfrac{1}{2}$
$8x^2 + 6x + 1 = 0$

74. $-\dfrac{5}{6}$ and $-\dfrac{2}{3}$
$18x^2 + 27x + 10 = 0$

75. $\dfrac{3}{5}$ and $-\dfrac{1}{10}$
$50x^2 - 25x - 3 = 0$

76. $\dfrac{7}{2}$ and $-\dfrac{1}{4}$
$8x^2 - 26x - 7 = 0$

3   Solve by taking square roots.

77. $y^2 = 49$
7 and -7

78. $x^2 = 64$
8 and -8

79. $z^2 = -4$
$2i$ and $-2i$

80. $v^2 = -16$
$4i$ and $-4i$

81. $s^2 - 4 = 0$
2 and -2

82. $r^2 - 36 = 0$
6 and -6

83. $4x^2 - 81 = 0$
$\dfrac{9}{2}$ and $-\dfrac{9}{2}$

84. $9x^2 - 16 = 0$
$\dfrac{4}{3}$ and $-\dfrac{4}{3}$

85. $y^2 + 49 = 0$
$7i$ and $-7i$

86. $z^2 + 16 = 0$
$4i$ and $-4i$

87. $v^2 - 48 = 0$
$4\sqrt{3}$ and $-4\sqrt{3}$

88. $s^2 - 32 = 0$
$4\sqrt{2}$ and $-4\sqrt{2}$

89. $r^2 - 75 = 0$
$5\sqrt{3}$ and $-5\sqrt{3}$

90. $u^2 - 54 = 0$
$3\sqrt{6}$ and $-3\sqrt{6}$

91. $z^2 + 18 = 0$
$3i\sqrt{2}$ and $-3i\sqrt{2}$

92. $t^2 + 27 = 0$
$3i\sqrt{3}$ and $-3i\sqrt{3}$

93. $(x - 1)^2 = 36$
7 and -5

94. $(x + 2)^2 = 25$
-7 and 3

95. $3(y + 3)^2 = 27$
0 and -6

96. $4(s - 2)^2 = 36$
5 and -1

97. $5(z + 2)^2 = 125$
-7 and 3

98. $(x - 2)^2 = -4$
$2 + 2i$ and $2 - 2i$

99. $(x + 5)^2 = -25$
$-5 + 5i$ and $-5 - 5i$

100. $(x - 8)^2 = -64$
$8 + 8i$ and $8 - 8i$

101. $3(x - 4)^2 = -12$
$4 + 2i$ and $4 - 2i$

102. $5(x + 2)^2 = -125$
$-2 + 5i$ and $-2 - 5i$

103. $3(x - 9)^2 = -27$
$9 + 3i$ and $9 - 3i$

104. $2(y - 3)^2 = 18$

0 and 6

105. $\left(v - \dfrac{1}{2}\right)^2 = \dfrac{1}{4}$

0 and 1

106. $\left(r + \dfrac{2}{3}\right)^2 = \dfrac{1}{9}$

-1 and $-\dfrac{1}{3}$

107. $\left(x - \dfrac{2}{5}\right)^2 = \dfrac{9}{25}$

$-\dfrac{1}{5}$ and 1

108. $\left(y + \dfrac{1}{3}\right)^2 = \dfrac{4}{9}$

-1 and $\dfrac{1}{3}$

109. $\left(a + \dfrac{3}{4}\right)^2 = \dfrac{9}{16}$

$-\dfrac{3}{2}$ and 0

110. $4\left(x - \dfrac{1}{2}\right)^2 = 1$

0 and 1

111. $3\left(x - \dfrac{5}{3}\right)^2 = \dfrac{4}{3}$

$\dfrac{7}{3}$ and 1

112. $2\left(x + \dfrac{3}{5}\right)^2 = \dfrac{8}{25}$

-1 and $-\dfrac{1}{5}$

113. $(x + 5)^2 - 6 = 0$
$-5 + \sqrt{6}$ and $-5 - \sqrt{6}$

114. $(t - 1)^2 - 15 = 0$
$1 + \sqrt{15}$ and $1 - \sqrt{15}$

115. $(s - 2)^2 - 24 = 0$
$2 + 2\sqrt{6}$ and $2 - 2\sqrt{6}$

116. $(y + 3)^2 - 18 = 0$
$-3 + 3\sqrt{2}$ and $-3 - 3\sqrt{2}$

117. $(z + 1)^2 + 12 = 0$
$-1 + 2i\sqrt{3}$ and $-1 - 2i\sqrt{3}$

118. $(r - 2)^2 + 28 = 0$
$2 + 2i\sqrt{7}$ and $2 - 2i\sqrt{7}$

119. $(v - 3)^2 + 45 = 0$
$3 + 3i\sqrt{5}$ and $3 - 3i\sqrt{5}$

120. $(x + 5)^2 + 32 = 0$
$-5 + 4i\sqrt{2}$ and $-5 - 4i\sqrt{2}$

121. $\left(u + \dfrac{2}{3}\right)^2 - 18 = 0$

$\dfrac{-2 + 9\sqrt{2}}{3}$ and $\dfrac{-2 - 9\sqrt{2}}{3}$

122. $\left(z - \dfrac{1}{2}\right)^2 - 20 = 0$

$\dfrac{1 + 4\sqrt{5}}{2}$ and $\dfrac{1 - 4\sqrt{5}}{2}$

123. $\left(t - \dfrac{3}{4}\right)^2 - 27 = 0$

$\dfrac{3 + 12\sqrt{3}}{4}$ and $\dfrac{3 - 12\sqrt{3}}{4}$

124. $\left(y + \dfrac{2}{5}\right)^2 - 72 = 0$

$\dfrac{-2 + 30\sqrt{2}}{5}$ and $\dfrac{-2 - 30\sqrt{2}}{5}$

125. $\left(x + \dfrac{1}{2}\right)^2 + 40 = 0$

$-\dfrac{1}{2} + 2i\sqrt{10}$ and $-\dfrac{1}{2} - 2i\sqrt{10}$

126. $\left(r - \dfrac{3}{2}\right)^2 + 48 = 0$

$\dfrac{3}{2} + 4i\sqrt{3}$ and $\dfrac{3}{2} - 4i\sqrt{3}$

127. $\left(x - \dfrac{2}{3}\right)^2 + \dfrac{25}{9} = 0$

$\dfrac{2}{3} + \dfrac{5}{3}i$ and $\dfrac{2}{3} - \dfrac{5}{3}i$

128. $\left(y + \dfrac{5}{8}\right)^2 + \dfrac{25}{64} = 0$

$-\dfrac{5}{8} + \dfrac{5}{8}i$ and $-\dfrac{5}{8} - \dfrac{5}{8}i$

SUPPLEMENTAL EXERCISES 7.1

Write a quadratic equation that has as solutions the given pair of numbers.

129. $\sqrt{2}$ and $-\sqrt{2}$
$x^2 - 2 = 0$

130. $\sqrt{5}$ and $-\sqrt{5}$
$x^2 - 5 = 0$

131. i and $-i$
$x^2 + 1 = 0$

132. $2i$ and $-2i$
$x^2 + 4 = 0$

133. $2\sqrt{2}$ and $-2\sqrt{2}$
$x^2 - 8 = 0$

134. $3\sqrt{2}$ and $-3\sqrt{2}$
$x^2 - 18 = 0$

135. $2\sqrt{3}$ and $-2\sqrt{3}$
$x^2 - 12 = 0$

136. $i\sqrt{2}$ and $-i\sqrt{2}$
$x^2 + 2 = 0$

137. $2i\sqrt{3}$ and $-2i\sqrt{3}$
$x^2 + 12 = 0$

Solve for x.

138. $4a^2x^2 = 36b^2$

$-\dfrac{3b}{a}$ and $\dfrac{3b}{a}$

139. $2a^2x^2 = 32b^2$

$-\dfrac{4b}{a}$ and $\dfrac{4b}{a}$

140. $3y^2x^2 = 27z^2$

$-\dfrac{3z}{y}$ and $\dfrac{3z}{y}$

141. $5y^2x^2 = 125z^2$

$-\dfrac{5z}{y}$ and $\dfrac{5z}{y}$

142. $(x + a)^2 - 4 = 0$

$-a + 2$ and $-a - 2$

143. $(x - b)^2 - 1 = 0$

$b + 1$ and $b - 1$

144. $2(x - y)^2 - 8 = 0$

$y + 2$ and $y - 2$

145. $3(x + y)^2 - 27 = 0$

$-y + 3$ and $-y - 3$

146. $c^2(x + b)^2 = 9$

$-b + \dfrac{3}{c}$ and $-b - \dfrac{3}{c}$

147. $b^2(x - a)^2 = 16$

$a + \dfrac{4}{b}$ and $a - \dfrac{4}{b}$

148. $y^2(x - z)^2 - 12 = 0$

$z + \dfrac{2\sqrt{3}}{y}$ and $z - \dfrac{2\sqrt{3}}{y}$

149. $a^2(x - b)^2 - 24 = 0$

$b + \dfrac{2\sqrt{6}}{a}$ and $b - \dfrac{2\sqrt{6}}{a}$

Solve.

150. Show that the solutions of the equation $ax^2 + bx = 0$ are 0 and $-\dfrac{b}{a}$.
Complete solution available in Solutions Manual.

151. Show that the solutions of the equation $ax^2 + c = 0$, $a > 0$, $c > 0$, are $\dfrac{\sqrt{ca}}{a}i$ and $-\dfrac{\sqrt{ca}}{a}i$.
Complete solution available in Solutions Manual.

SECTION **7.2**

Solving Quadratic Equations by Completing the Square and by Using the Quadratic Formula

1 Solve quadratic equations by completing the square

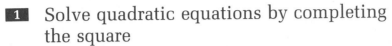

Recall that a perfect square trinomial is the square of a binomial.

Perfect Square Trinomial		Square of a Binomial
$x^2 + 8x + 16$	$=$	$(x + 4)^2$
$x^2 - 10x + 25$	$=$	$(x - 5)^2$
$x^2 + 2ax + a^2$	$=$	$(x + a)^2$

For each perfect square trinomial, the square of $\frac{1}{2}$ the coefficient of x equals the constant term.

$$\left(\frac{1}{2} \text{ coefficient of } x\right)^2 = \text{Constant term}$$

$$x^2 + 8x + 16, \quad \left(\frac{1}{2} \cdot 8\right)^2 = 16$$

$$x^2 - 10x + 25, \quad \left[\frac{1}{2}(-10)\right]^2 = 25$$

$$x^2 + 2ax + a^2, \quad \left(\frac{1}{2} \cdot 2a\right)^2 = a^2$$

To complete the square on $x^2 + bx$, add $\left(\frac{1}{2}b\right)^2$ to $x^2 + bx$.

Complete the square on $x^2 - 12x$. Write the resulting perfect square trinomial as the square of a binomial.

Find the constant term.
$$\left[\frac{1}{2}(-12)\right]^2 = (-6)^2 = 36$$

Complete the square on $x^2 - 12x$ by adding the constant term.
$$x^2 - 12x + 36$$

Write the resulting perfect square trinomial as the square of a binomial.
$$x^2 - 12x + 36 = (x - 6)^2$$

Complete the square on $z^2 + 3z$. Write the resulting perfect square trinomial as the square of a binomial.

Find the constant term.	$\left(\dfrac{1}{2} \cdot 3\right)^2 = \left(\dfrac{3}{2}\right)^2 = \dfrac{9}{4}$
Complete the square on $z^2 + 3z$ by adding the constant term.	$z^2 + 3z + \dfrac{9}{4}$
Write the resulting perfect square trinomial as the square of a binomial.	$z^2 + 3z + \dfrac{9}{4} = \left(z + \dfrac{3}{2}\right)^2$

While not all quadratic equations can be solved by factoring, any quadratic equation can be solved by completing the square. Add to each side of the equation the term that completes the square. Rewrite the equation in the form $(x + a)^2 = b$. Then take the square root of each side of the equation.

Solve by completing the square: $x^2 - 4x - 14 = 0$

$$x^2 - 4x - 14 = 0$$

Add the opposite of the constant term to each side of the equation.	$x^2 - 4x = 14$
Add the constant term that completes the square on $x^2 - 4x$ to each side of the equation. $\left[\dfrac{1}{2}(-4)\right]^2 = (-2)^2 = 4$	$x^2 - 4x + 4 = 14 + 4$
Factor the perfect square trinomial.	$(x - 2)^2 = 18$
Take the square root of each side of the equation.	$\sqrt{(x - 2)^2} = \sqrt{18}$
Simplify.	$x - 2 = \pm\sqrt{18} = \pm 3\sqrt{2}$
Solve for x.	$\begin{aligned} x - 2 &= 3\sqrt{2} & x - 2 &= -3\sqrt{2} \\ x &= 2 + 3\sqrt{2} & x &= 2 - 3\sqrt{2} \end{aligned}$
Write the solutions.	The solutions are $2 + 3\sqrt{2}$ and $2 - 3\sqrt{2}$.

Check:
$$x^2 - 4x - 14 = 0$$
$$\overline{(2 + 3\sqrt{2})^2 - 4(2 + 3\sqrt{2}) - 14}$$
$$4 + 12\sqrt{2} + 18 - 8 - 12\sqrt{2} - 14 \;\big|$$
$$0 = 0$$

$$x^2 - 4x - 14 = 0$$
$$\overline{(2 - 3\sqrt{2})^2 - 4(2 - 3\sqrt{2}) - 14}$$
$$4 - 12\sqrt{2} + 18 - 8 + 12\sqrt{2} - 14 \;\big|$$
$$0 = 0$$

When a, the coefficient of the x^2, is not 1, multiply each side of the equation by the reciprocal of a before completing the square.

Solve by completing the square: $2x^2 - x = 2$

$$2x^2 - x = 2$$

Multiply each side of the equation by the reciprocal of the coefficient of x^2.

$$\frac{1}{2}(2x^2 - x) = \frac{1}{2} \cdot 2$$

The coefficient of the x^2 term is now 1.

$$x^2 - \frac{1}{2}x = 1$$

Add the term that completes the square on $x^2 - \frac{1}{2}x$ to each side of the equation.

$$x^2 - \frac{1}{2}x + \frac{1}{16} = 1 + \frac{1}{16}$$

Factor the perfect square trinomial.

$$\left(x - \frac{1}{4}\right)^2 = \frac{17}{16}$$

Take the square root of each side of the equation.

$$\sqrt{\left(x - \frac{1}{4}\right)^2} = \sqrt{\frac{17}{16}}$$

Simplify.

$$x - \frac{1}{4} = \pm\frac{\sqrt{17}}{4}$$

Solve for x.

$$x - \frac{1}{4} = \frac{\sqrt{17}}{4} \qquad x - \frac{1}{4} = -\frac{\sqrt{17}}{4}$$
$$x = \frac{1}{4} + \frac{\sqrt{17}}{4} \qquad x = \frac{1}{4} - \frac{\sqrt{17}}{4}$$

Write the solutions.

The solutions are $\dfrac{1 + \sqrt{17}}{4}$ and $\dfrac{1 - \sqrt{17}}{4}$.

$\dfrac{1 + \sqrt{17}}{4}$ and $\dfrac{1 - \sqrt{17}}{4}$ check as solutions.

Example 1 Solve by completing the square.

A. $4x^2 - 8x + 1 = 0$ B. $x^2 + 4x + 5 = 0$

Solution A. $4x^2 - 8x + 1 = 0$

$$4x^2 - 8x = -1$$

• Add -1 to each side of the equation.

$$\frac{1}{4}(4x^2 - 8x) = \frac{1}{4}(-1)$$

• The coefficient of the x^2 term must be 1. Multiply each side of the equation by the reciprocal of the coefficient of x^2.

$$x^2 - 2x = -\frac{1}{4}$$

$$x^2 - 2x + 1 = -\frac{1}{4} + 1$$

• Complete the square.

$$(x - 1)^2 = \frac{3}{4}$$

• Factor the perfect square trinomial.

$$\sqrt{(x - 1)^2} = \sqrt{\frac{3}{4}}$$

• Take the square root of each side of the equation.

$$x - 1 = \pm\frac{\sqrt{3}}{2}$$

• Simplify.

$$x - 1 = \frac{\sqrt{3}}{2} \qquad x - 1 = -\frac{\sqrt{3}}{2}$$

• Solve for x.

$$x = 1 + \frac{\sqrt{3}}{2} \qquad x = 1 - \frac{\sqrt{3}}{2}$$

$$x = \frac{2 + \sqrt{3}}{2} \qquad x = \frac{2 - \sqrt{3}}{2}$$

The solutions are $\dfrac{2 + \sqrt{3}}{2}$ and $\dfrac{2 - \sqrt{3}}{2}$.

B. $x^2 + 4x + 5 = 0$

$$x^2 + 4x = -5$$

• Add -5 to each side of the equation.

$$x^2 + 4x + 4 = -5 + 4$$

• Complete the square.

$$(x + 2)^2 = -1$$

• Factor the perfect square trinomial.

$$\sqrt{(x + 2)^2} = \sqrt{-1}$$

• Take the square root of each side of the equation.

$$x + 2 = \pm i$$

• Simplify.

$$x + 2 = i \qquad\qquad x + 2 = -i$$

• Solve for x.

$$x = -2 + i \qquad\qquad x = -2 - i$$

The solutions are $-2 + i$ and $-2 - i$.

Problem 1 Solve by completing the square.

A. $4x^2 - 4x - 1 = 0$ B. $2x^2 + x - 5 = 0$

Solution See page A56.

A. $\dfrac{1 + \sqrt{2}}{2}$ and $\dfrac{1 - \sqrt{2}}{2}$ B. $\dfrac{-1 + \sqrt{41}}{4}$ and $\dfrac{-1 - \sqrt{41}}{4}$

2 ## Solve quadratic equations by using the quadratic formula

A general formula known as the **quadratic formula** can be derived by applying the method of completing the square to the standard form of a quadratic equation. This formula can be used to solve any quadratic equation.

The solution of the equation $ax^2 + bx + c = 0$ by completing the square is shown below.

$$ax^2 + bx + c = 0$$

Add the opposite of the constant term to each side of the equation.

$$ax^2 + bx + c + (-c) = 0 + (-c)$$

$$ax^2 + bx = -c$$

Multiply each side of the equation by the reciprocal of a, the coefficient of x^2.

$$\frac{1}{a}(ax^2 + bx) = \frac{1}{a}(-c)$$

$$x^2 + \frac{b}{a}x = -\frac{c}{a}$$

Complete the square by adding $\left(\dfrac{1}{2} \cdot \dfrac{b}{a}\right)^2$ to each side of the equation.

$$x^2 + \frac{b}{a}x + \left(\frac{1}{2} \cdot \frac{b}{a}\right)^2 = \left(\frac{1}{2} \cdot \frac{b}{a}\right)^2 - \frac{c}{a}$$

$$x^2 + \frac{b}{a}x + \frac{b^2}{4a^2} = \frac{b^2}{4a^2} - \frac{c}{a}$$

Simplify the right side of the equation.

$$x^2 + \frac{b}{a}x + \frac{b^2}{4a^2} = \frac{b^2}{4a^2} - \left(\frac{c}{a} \cdot \frac{4a}{4a}\right)$$

$$x^2 + \frac{b}{a}x + \frac{b^2}{4a^2} = \frac{b^2}{4a^2} - \frac{4ac}{4a^2}$$

$$x^2 + \frac{b}{a}x + \frac{b^2}{4a^2} = \frac{b^2 - 4ac}{4a^2}$$

Factor the perfect square trinomial on the left side of the equation.

$$\left(x + \frac{b}{2a}\right)^2 = \frac{b^2 - 4ac}{4a^2}$$

Take the square root of each side of the equation.

$$\sqrt{\left(x + \frac{b}{2a}\right)^2} = \sqrt{\frac{b^2 - 4ac}{4a^2}}$$

$$\left(x + \frac{b}{2a}\right) = \pm\frac{\sqrt{b^2 - 4ac}}{2a}$$

Solve for x.

$$x + \frac{b}{2a} = \frac{\sqrt{b^2 - 4ac}}{2a} \qquad x + \frac{b}{2a} = -\frac{\sqrt{b^2 - 4ac}}{2a}$$

$$x = -\frac{b}{2a} + \frac{\sqrt{b^2 - 4ac}}{2a} \qquad x = -\frac{b}{2a} - \frac{\sqrt{b^2 - 4ac}}{2a}$$

$$= \frac{-b + \sqrt{b^2 - 4ac}}{2a} \qquad = \frac{-b - \sqrt{b^2 - 4ac}}{2a}$$

The Quadratic Formula

The solutions of $ax^2 + bx + c = 0$, $a \neq 0$, are
$$\frac{-b + \sqrt{b^2 - 4ac}}{2a} \quad \text{and} \quad \frac{-b - \sqrt{b^2 - 4ac}}{2a}.$$
The quadratic formula is frequently written in the form

$$x = \frac{-b \pm \sqrt{b^2 - 4ac}}{2a}.$$

Solve by using the quadratic formula: $4x^2 = 8x - 13$

Write the equation in standard form.

$$4x^2 = 8x - 13$$
$$4x^2 - 8x + 13 = 0$$
$$a = 4, \ b = -8, \ c = 13$$

Replace a, b, and c in the quadratic formula by their values.

$$x = \frac{-b \pm \sqrt{b^2 - 4ac}}{2a}$$

$$= \frac{-(-8) \pm \sqrt{(-8)^2 - 4 \cdot 4 \cdot 13}}{2 \cdot 4}$$

Simplify.

$$= \frac{8 \pm \sqrt{64 - 208}}{8} = \frac{8 \pm \sqrt{-144}}{8}$$

$$= \frac{8 \pm 12i}{8} = \frac{2 \pm 3i}{2}$$

Write the solutions.

The solutions are $1 + \frac{3}{2}i$ and $1 - \frac{3}{2}i$.

Check:

$$
\begin{array}{c|c}
\multicolumn{2}{c}{4x^2 = 8x - 13} \\
\hline
4\left(1 + \frac{3}{2}i\right)^2 & 8\left(1 + \frac{3}{2}i\right) - 13 \\
4\left(1 + 3i - \frac{9}{4}\right) & 8 + 12i - 13 \\
4\left(-\frac{5}{4} + 3i\right) & -5 + 12i \\
\multicolumn{2}{c}{-5 + 12i = -5 + 12i}
\end{array}
\qquad
\begin{array}{c|c}
\multicolumn{2}{c}{4x^2 = 8x - 13} \\
\hline
4\left(1 - \frac{3}{2}i\right)^2 & 8\left(1 - \frac{3}{2}i\right) - 13 \\
4\left(1 - 3i - \frac{9}{4}\right) & 8 - 12i - 13 \\
4\left(-\frac{5}{4} - 3i\right) & -5 - 12i \\
\multicolumn{2}{c}{-5 - 12i = -5 - 12i}
\end{array}
$$

Example 2 Solve by using the quadratic formula.

A. $4x^2 + 12x + 9 = 0$ B. $2x^2 - x + 5 = 0$

Solution A. $4x^2 + 12x + 9 = 0$

$$x = \frac{-b \pm \sqrt{b^2 - 4ac}}{2a}$$

$$= \frac{-12 \pm \sqrt{12^2 - 4 \cdot 4 \cdot 9}}{2 \cdot 4}$$

$$= \frac{-12 \pm \sqrt{0}}{8} = \frac{-12}{8} = -\frac{3}{2}$$

The solution is $-\dfrac{3}{2}$.

- $a = 4$, $b = 12$, $c = 9$
- Replace a, b, and c in the quadratic formula by their values. Then simplify.
- The equation has a double root.

B. $2x^2 - x + 5 = 0$

$$x = \frac{-b \pm \sqrt{b^2 - 4ac}}{2a}$$

$$= \frac{-(-1) \pm \sqrt{(-1)^2 - 4(2)(5)}}{2 \cdot 2}$$

$$= \frac{1 \pm \sqrt{1 - 40}}{4} = \frac{1 \pm \sqrt{-39}}{4}$$

$$= \frac{1 \pm i\sqrt{39}}{4}$$

The solutions are $\dfrac{1}{4} + \dfrac{\sqrt{39}}{4}i$ and $\dfrac{1}{4} - \dfrac{\sqrt{39}}{4}i$.

- $a = 2$, $b = -1$, $c = 5$
- Replace a, b, and c in the quadratic formula by their values. Then simplify.

Problem 2 Solve by using the quadratic formula.

A. $x^2 + 6x - 9 = 0$ B. $4x^2 = 4x - 1$

Solution See page A57.

A. $-3 + 3\sqrt{2}$ and $-3 - 3\sqrt{2}$ B. $\dfrac{1}{2}$

In Example 2A, the solution of the equation is a double root, and in Example 2B, the solutions are complex numbers.

In the quadratic formula, the quantity $b^2 - 4ac$ is called the **discriminant**. When a, b, and c are real numbers, the discriminant determines whether a quadratic equation will have a double root, two real number solutions that are not equal, or two complex number solutions.

> **The Effect of the Discriminant on the Solutions of a Quadratic Equation**
> 1. If $b^2 - 4ac = 0$, the equation has one real number solution, a double root.
> 2. If $b^2 - 4ac > 0$, the equation has two real number solutions that are not equal.
> 3. If $b^2 - 4ac < 0$, the equation has two complex number solutions.

The equation $x^2 - 4x - 5 = 0$ has two real number solutions because the discriminant is greater than zero.

$a = 1, b = -4, c = -5$
$b^2 - 4ac$
$(-4)^2 - 4(1)(-5) = 16 + 20 = 36$
$36 > 0$

Example 3   Use the discriminant to determine whether $4x^2 - 2x + 5 = 0$ has one real number solution, two real number solutions, or two complex number solutions.

Solution $b^2 - 4ac$

$(-2)^2 - 4(4)(5) = 4 - 80 = -76$
$-76 < 0$

• $a = 4, b = -2, c = 5$

• The discriminant is less than 0.

The equation has two complex number solutions.

Problem 3   Use the discriminant to determine whether $3x^2 - x - 1 = 0$ has one real number solution, two real number solutions, or two complex number solutions.

Solution   See page A57.
two real number solutions

EXERCISES 7.2

1   Solve by completing the square.

1. $x^2 - 4x - 5 = 0$
   5 and -1

2. $y^2 + 6y + 5 = 0$
   -5 and -1

3. $v^2 + 8v - 9 = 0$
   -9 and 1

4. $w^2 - 2w - 24 = 0$
   6 and -4

5. $z^2 - 6z + 9 = 0$
   3

6. $u^2 + 10u + 25 = 0$
   -5

7. $r^2 + 4r - 7 = 0$
   $-2 + \sqrt{11}$ and $-2 - \sqrt{11}$

8. $s^2 + 6s - 1 = 0$
   $-3 + \sqrt{10}$ and $-3 - \sqrt{10}$

9. $x^2 - 6x + 7 = 0$
   $3 + \sqrt{2}$ and $3 - \sqrt{2}$

10. $y^2 + 8y + 13 = 0$
    $-4 + \sqrt{3}$ and $-4 - \sqrt{3}$

11. $z^2 - 2z + 2 = 0$
    $1 + i$ and $1 - i$

12. $t^2 - 4t + 8 = 0$
    $2 + 2i$ and $2 - 2i$

13. $s^2 - 5s - 24 = 0$
8 and -3

14. $v^2 + 7v - 44 = 0$
-11 and 4

15. $x^2 + 5x - 36 = 0$
4 and -9

16. $y^2 - 9y + 20 = 0$
5 and 4

17. $p^2 - 3p + 1 = 0$
$\dfrac{3 + \sqrt{5}}{2}$ and $\dfrac{3 - \sqrt{5}}{2}$

18. $r^2 - 5r - 2 = 0$
$\dfrac{5 + \sqrt{33}}{2}$ and $\dfrac{5 - \sqrt{33}}{2}$

19. $t^2 - t - 1 = 0$
$\dfrac{1 + \sqrt{5}}{2}$ and $\dfrac{1 - \sqrt{5}}{2}$

20. $u^2 - u - 7 = 0$
$\dfrac{1 + \sqrt{29}}{2}$ and $\dfrac{1 - \sqrt{29}}{2}$

21. $y^2 - 6y = 4$
$3 + \sqrt{13}$ and $3 - \sqrt{13}$

22. $w^2 + 4w = 2$
$-2 + \sqrt{6}$ and $-2 - \sqrt{6}$

23. $x^2 = 8x - 15$
3 and 5

24. $z^2 = 4z - 3$
3 and 1

25. $v^2 = 4v - 13$
$2 + 3i$ and $2 - 3i$

26. $x^2 = 2x - 17$
$1 + 4i$ and $1 - 4i$

27. $p^2 + 6p = -13$
$-3 + 2i$ and $-3 - 2i$

28. $x^2 + 4x = -20$
$-2 + 4i$ and $-2 - 4i$

29. $y^2 - 2y = 17$
$1 + 3\sqrt{2}$ and $1 - 3\sqrt{2}$

30. $x^2 + 10x = 7$
$-5 + 4\sqrt{2}$ and $-5 - 4\sqrt{2}$

31. $z^2 = z + 4$
$\dfrac{1 + \sqrt{17}}{2}$ and $\dfrac{1 - \sqrt{17}}{2}$

32. $r^2 = 3r - 1$
$\dfrac{3 + \sqrt{5}}{2}$ and $\dfrac{3 - \sqrt{5}}{2}$

33. $x^2 + 13 = 2x$
$1 + 2i\sqrt{3}$ and $1 - 2i\sqrt{3}$

34. $x^2 + 27 = 6x$
$3 + 3i\sqrt{2}$ and $3 - 3i\sqrt{2}$

35. $2y^2 + 3y + 1 = 0$
$-\dfrac{1}{2}$ and -1

36. $2t^2 + 5t - 3 = 0$
$\dfrac{1}{2}$ and -3

37. $4r^2 - 8r = -3$
$\dfrac{1}{2}$ and $\dfrac{3}{2}$

38. $4u^2 - 20u = -9$
$\dfrac{1}{2}$ and $\dfrac{9}{2}$

39. $6y^2 - 5y = 4$
$-\dfrac{1}{2}$ and $\dfrac{4}{3}$

40. $6v^2 - 7v = 3$
$-\dfrac{1}{3}$ and $\dfrac{3}{2}$

41. $4x^2 - 4x + 5 = 0$
$\dfrac{1}{2} + i$ and $\dfrac{1}{2} - i$

42. $4t^2 - 4t + 17 = 0$
$\dfrac{1}{2} + 2i$ and $\dfrac{1}{2} - 2i$

43. $9x^2 - 6x + 2 = 0$
$\dfrac{1}{3} + \dfrac{1}{3}i$ and $\dfrac{1}{3} - \dfrac{1}{3}i$

44. $9y^2 - 12y + 13 = 0$
$\dfrac{2}{3} + i$ and $\dfrac{2}{3} - i$

45. $2s^2 = 4s + 5$
$\dfrac{2 + \sqrt{14}}{2}$ and $\dfrac{2 - \sqrt{14}}{2}$

46. $3u^2 = 6u + 1$
$\dfrac{3 + 2\sqrt{3}}{3}$ and $\dfrac{3 - 2\sqrt{3}}{3}$

47. $2r^2 = 3 - r$
$-\dfrac{3}{2}$ and 1

48. $2x^2 = 12 - 5x$
$\dfrac{3}{2}$ and -4

49. $y - 2 = (y - 3)(y + 2)$
$1 + \sqrt{5}$ and $1 - \sqrt{5}$

50. $8s - 11 = (s - 4)(s - 2)$
$7 + \sqrt{30}$ and $7 - \sqrt{30}$

51. $6t - 2 = (2t - 3)(t - 1)$
$\dfrac{1}{2}$ and 5

52. $2z + 9 = (2z + 3)(z + 2)$
$\dfrac{1}{2}$ and -3

53. $(x - 4)(x + 1) = x - 3$
$2 + \sqrt{5}$ and $2 - \sqrt{5}$

54. $(y - 3)^2 = 2y + 10$
$4 + \sqrt{17}$ and $4 - \sqrt{17}$

▦ Solve by completing the square. Approximate the solutions to the
    nearest thousandth. Use the Table of Square Roots on page A4.

55. $z^2 + 2z = 4$
         −3.236 and 1.236

56. $t^2 - 4t = 7$
         5.317 and −1.317

57. $2x^2 = 4x - 1$
         1.707 and 0.293

58. $3y^2 = 5y - 1$
         1.434 and 0.232

59. $4z^2 + 2z - 1 = 0$
         0.309 and −0.809

60. $4w^2 - 8w = 3$
         2.323 and −0.323

2   Solve by using the quadratic formula.

61. $x^2 - 3x - 10 = 0$
         5 and −2

62. $z^2 - 4z - 8 = 0$
         $2 + 2\sqrt{3}$ and $2 - 2\sqrt{3}$

63. $y^2 + 5y - 36 = 0$
         4 and −9

64. $z^2 - 3z - 40 = 0$
         8 and −5

65. $w^2 = 8w + 72$
         $4 + 2\sqrt{22}$ and $4 - 2\sqrt{22}$

66. $t^2 = 2t + 35$
         7 and −5

67. $v^2 = 24 - 5v$

         3 and −8

68. $x^2 = 18 - 7x$

         2 and −9

69. $2y^2 + 5y - 3 = 0$
         $\dfrac{1}{2}$ and −3

70. $4p^2 - 7p + 3 = 0$
         $\dfrac{3}{4}$ and 1

71. $8s^2 = 10s + 3$
         $-\dfrac{1}{4}$ and $\dfrac{3}{2}$

72. $12t^2 = 5t + 2$
         $\dfrac{2}{3}$ and $-\dfrac{1}{4}$

73. $v^2 - 2v - 7 = 0$
         $1 + 2\sqrt{2}$ and $1 - 2\sqrt{2}$

74. $t^2 - 2t - 11 = 0$
         $1 + 2\sqrt{3}$ and $1 - 2\sqrt{3}$

75. $y^2 - 8y - 20 = 0$
         10 and −2

76. $x^2 = 14x - 24$

         2 and 12

77. $v^2 = 12v - 24$

         $6 + 2\sqrt{3}$ and $6 - 2\sqrt{3}$

78. $2z^2 - 2z - 1 = 0$
         $\dfrac{1 + \sqrt{3}}{2}$ and $\dfrac{1 - \sqrt{3}}{2}$

79. $4x^2 - 4x - 7 = 0$
         $\dfrac{1 + 2\sqrt{2}}{2}$ and $\dfrac{1 - 2\sqrt{2}}{2}$

80. $2p^2 - 8p + 5 = 0$
         $\dfrac{4 + \sqrt{6}}{2}$ and $\dfrac{4 - \sqrt{6}}{2}$

81. $2s^2 - 3s + 1 = 0$
         $\dfrac{1}{2}$ and 1

82. $4w^2 - 4w - 1 = 0$
         $\dfrac{1 + \sqrt{2}}{2}$ and $\dfrac{1 - \sqrt{2}}{2}$

83. $3x^2 + 10x + 6 = 0$
         $\dfrac{-5 + \sqrt{7}}{3}$ and $\dfrac{-5 - \sqrt{7}}{3}$

84. $3v^2 = 6v - 2$
         $\dfrac{3 + \sqrt{3}}{3}$ and $\dfrac{3 - \sqrt{3}}{3}$

85. $6w^2 = 19w - 10$
         $\dfrac{2}{3}$ and $\dfrac{5}{2}$

86. $z^2 + 2z + 2 = 0$

         $-1 + i$ and $-1 - i$

87. $p^2 - 4p + 5 = 0$

         $2 + i$ and $2 - i$

88. $y^2 - 2y + 5 = 0$
         $1 + 2i$ and $1 - 2i$

89. $x^2 + 6x + 13 = 0$
         $-3 + 2i$ and $-3 - 2i$

90. $s^2 - 4s + 13 = 0$
         $2 + 3i$ and $2 - 3i$

91. $t^2 - 6t + 10 = 0$

         $3 + i$ and $3 - i$

92. $2w^2 - 2w + 5 = 0$
         $\dfrac{1}{2} + \dfrac{3i}{2}$ and $\dfrac{1}{2} - \dfrac{3i}{2}$

93. $4v^2 + 8v + 3 = 0$

         $-\dfrac{3}{2}$ and $-\dfrac{1}{2}$

94. $2x^2 + 6x + 5 = 0$
         $-\dfrac{3}{2} + \dfrac{1}{2}i$ and $-\dfrac{3}{2} - \dfrac{1}{2}i$

95. $2y^2 + 2y + 13 = 0$
         $-\dfrac{1}{2} + \dfrac{5}{2}i$ and $-\dfrac{1}{2} - \dfrac{5}{2}i$

96. $4t^2 - 6t + 9 = 0$
         $\dfrac{3}{4} + \dfrac{3\sqrt{3}}{4}i$ and $\dfrac{3}{4} - \dfrac{3\sqrt{3}}{4}i$

97. $3v^2 + 6v + 1 = 0$
$\dfrac{-3 + \sqrt{6}}{3}$ and $\dfrac{-3 - \sqrt{6}}{3}$

98. $2r^2 = 4r - 11$
$1 + \dfrac{3\sqrt{2}}{2}i$ and $1 - \dfrac{3\sqrt{2}}{2}i$

99. $3y^2 = 6y - 5$
$1 + \dfrac{\sqrt{6}}{3}i$ and $1 - \dfrac{\sqrt{6}}{3}i$

100. $2x(x - 2) = x + 12$ $-\dfrac{3}{2}$ and 4

101. $10y(y + 4) = 15y - 15$ $-\dfrac{3}{2}$ and -1

102. $(3s - 2)(s + 1) = 2$ $-\dfrac{4}{3}$ and 1

103. $(2t + 1)(t - 3) = 9$ $-\dfrac{3}{2}$ and 4

Use the discriminant to determine whether the quadratic equation has one real number solution, two real number solutions, or two complex number solutions.

104. $2z^2 - z + 5 = 0$
two complex

105. $3y^2 + y + 1 = 0$
two complex

106. $9x^2 - 12x + 4 = 0$
one real

107. $4x^2 + 20x + 25 = 0$
one real

108. $2v^2 - 3v - 1 = 0$
two real

109. $3w^2 + 3w - 2 = 0$
two real

110. $2p^2 + 5p + 1 = 0$
two real

111. $2t^2 + 9t + 3 = 0$
two real

112. $5z^2 + 2 = 0$
two complex

Solve by using the quadratic formula. Approximate the solutions to the nearest thousandth. Use the Table of Square Roots on page A4.

113. $x^2 + 6x - 6 = 0$
0.873 and −6.873

114. $p^2 - 8p + 3 = 0$
7.606 and 0.394

115. $r^2 - 2r - 4 = 0$
3.236 and −1.236

116. $w^2 + 4w - 1 = 0$
0.236 and −4.236

117. $3t^2 = 7t + 1$
2.468 and −0.135

118. $2y^2 = y + 5$
1.851 and −1.351

SUPPLEMENTAL EXERCISES 7.2

Solve.

119. $\sqrt{2}y^2 + 3y - 2\sqrt{2} = 0$
$\dfrac{\sqrt{2}}{2}$ and $-2\sqrt{2}$

120. $\sqrt{3}z^2 + 10z - 3\sqrt{3} = 0$
$\dfrac{-5\sqrt{3} + \sqrt{102}}{3}$ and $\dfrac{-5\sqrt{3} - \sqrt{102}}{3}$

121. $\sqrt{2}x^2 + 5x - 3\sqrt{2} = 0$
$-3\sqrt{2}$ and $\dfrac{\sqrt{2}}{2}$

122. $\sqrt{3}w^2 + w - 2\sqrt{3} = 0$
$-\sqrt{3}$ and $\dfrac{2\sqrt{3}}{3}$

123. $t^2 - t\sqrt{3} + 1 = 0$
$\dfrac{\sqrt{3}}{2} + \dfrac{1}{2}i$ and $\dfrac{\sqrt{3}}{2} - \dfrac{1}{2}i$

124. $y^2 + y\sqrt{7} + 2 = 0$
$-\dfrac{\sqrt{7}}{2} + i$ and $-\dfrac{\sqrt{7}}{2} - i$

Solve for x.

125. $x^2 - ax - 2a^2 = 0$
$2a$ and $-a$

126. $x^2 - ax - 6a^2 = 0$
3a and −2a

127. $x^2 + 3ax - 4a^2 = 0$
a and $-4a$

128. $x^2 + 3ax - 10a^2 = 0$
$2a$ and $-5a$

129. $x^2 - 6ax + 8a^2 = 0$
$2a$ and $4a$

130. $x^2 + 9ax + 18a^2 = 0$
$-3a$ and $-6a$

131. $2x^2 + 3ax - 2a^2 = 0$ $\dfrac{a}{2}$ and $-2a$

132. $2x^2 - 7ax + 3a^2 = 0$ $\dfrac{a}{2}$ and $3a$

For what values of p does the quadratic equation have two real number solutions that are not equal? Write the answer in set builder notation.

133. $x^2 - 6x + p = 0$
$\{p \mid p < 9, p \in \text{real numbers}\}$

134. $x^2 + 10x + p = 0$
$\{p \mid p < 25, p \in \text{real numbers}\}$

135. $x^2 - 3x + p = 0$
$\left\{p \mid p < \dfrac{9}{4}, p \in \text{real numbers}\right\}$

136. $x^2 + 7x + p = 0$
$\left\{p \mid p < \dfrac{49}{4}, p \in \text{real numbers}\right\}$

For what values of p does the quadratic equation have two complex number solutions? Write the answer in set builder notation.

137. $x^2 - 2x + p = 0$
$\{p \mid p > 1, p \in \text{real numbers}\}$

138. $x^2 + 4x + p = 0$
$\{p \mid p > 4, p \in \text{real numbers}\}$

139. $x^2 - x + p = 0$
$\left\{p \mid p > \dfrac{1}{4}, p \in \text{real numbers}\right\}$

140. $x^2 + 5x + p = 0$
$\left\{p \mid p > \dfrac{25}{4}, p \in \text{real numbers}\right\}$

Solve.

141. Show that the equation $x^2 + bx - 1 = 0$ always has real number solutions regardless of the value of b.
$b^2 + 4 > 0$ for any real number b.

142. Show that the equation $2x^2 + bx - 2 = 0$ always has real number solutions regardless of the value of b.
$b^2 + 16 > 0$ for any real number b.

SECTION 7.3

Equations That Are Reducible to Quadratic Equations

1 Equations that are quadratic in form

Certain equations that are not quadratic can be expressed in quadratic form by making suitable substitutions. An equation is **quadratic in form** if it can be written as $au^2 + bu + c = 0$.

The equation at the right is quadratic in form.

$$x^4 - 4x^2 - 5 = 0$$
$$(x^2)^2 - 4(x^2) - 5 = 0$$
$$u^2 - 4u - 5 = 0$$

By letting $x^2 = u$, the equation can be written:
This equation is a quadratic equation.

The equation at the right is quadratic in form.

$$y - y^{\frac{1}{2}} - 6 = 0$$
$$(y^{\frac{1}{2}})^2 - (y^{\frac{1}{2}}) - 6 = 0$$

By letting $y^{\frac{1}{2}} = u$, the equation can be written:
This equation is a quadratic equation.

$$u^2 - u - 6 = 0$$

The key to recognizing equations that are quadratic in form: when the equation is written in standard form, the exponent on one variable term is $\frac{1}{2}$ the exponent on the other variable term.

The equation $z + 7z^{\frac{1}{2}} - 18 = 0$ is quadratic in form.

$$z + 7z^{\frac{1}{2}} - 18 = 0$$
$$(z^{\frac{1}{2}})^2 + 7(z^{\frac{1}{2}}) - 18 = 0$$

To solve this equation, let $z^{\frac{1}{2}} = u$.

$$u^2 + 7u - 18 = 0$$

Solve for u by factoring.

$$(u - 2)(u + 9) = 0$$

$u - 2 = 0$	$u + 9 = 0$
$u = 2$	$u = -9$

Replace u by $z^{\frac{1}{2}}$.

Solve for z by squaring both sides of the equation.

$z^{\frac{1}{2}} = 2$	$z^{\frac{1}{2}} = -9$
$(z^{\frac{1}{2}})^2 = 2^2$	$(z^{\frac{1}{2}})^2 = (-9)^2$
$z = 4$	$z = 81$

Check the solution. When squaring both sides of an equation, the resulting equation may have a solution that is not a solution of the original equation.

Check:

$$z + 7z^{\frac{1}{2}} - 18 = 0 \qquad\qquad z + 7z^{\frac{1}{2}} - 18 = 0$$

$$
\begin{array}{c|c}
4 + 7(4)^{\frac{1}{2}} - 18 & 0 \\
4 + 7 \cdot 2 - 18 & \\
4 + 14 - 18 & \\
& 0 = 0
\end{array}
\qquad
\begin{array}{c|c}
81 + 7(81)^{\frac{1}{2}} - 18 & 0 \\
81 + 7 \cdot 9 - 18 & \\
81 + 63 - 18 & \\
& 126 \neq 0
\end{array}
$$

4 checks as a solution, but 81 does not check as a solution.

The solution is 4.

Example 1   Solve.

A. $x^4 + x^2 - 12 = 0$

B. $x^{\frac{2}{3}} - 2x^{\frac{1}{3}} - 3 = 0$

Solution

$x^4 + x^2 - 12 = 0$

$(x^2)^2 + (x^2) - 12 = 0$

$u^2 + u - 12 = 0$

$(u - 3)(u + 4) = 0$

• The equation is quadratic in form.

• Let $x^2 = u$.

• Solve for u by factoring.

$u - 3 = 0 \qquad\qquad u + 4 = 0$

$u = 3 \qquad\qquad\quad u = -4$

$x^2 = 3 \qquad\qquad\quad x^2 = -4$

$\sqrt{x^2} = \sqrt{3} \qquad\qquad \sqrt{x^2} = \sqrt{-4}$

$x = \pm\sqrt{3} \qquad\qquad x = \pm 2i$

• Replace u by x^2.

• Solve for x by taking square roots.

The solutions are $\sqrt{3}$, $-\sqrt{3}$, $2i$, and $-2i$.

B. $x^{\frac{2}{3}} - 2x^{\frac{1}{3}} - 3 = 0$

$(x^{\frac{1}{3}})^2 - 2(x^{\frac{1}{3}}) - 3 = 0$

$u^2 - 2u - 3 = 0$

$(u - 3)(u + 1) = 0$

• The equation is quadratic in form.

• Let $x^{\frac{1}{3}} = u$.

• Solve for u by factoring.

$u - 3 = 0 \qquad\qquad u + 1 = 0$

$u = 3 \qquad\qquad\quad u = -1$

$x^{\frac{1}{3}} = 3 \qquad\qquad\quad x^{\frac{1}{3}} = -1$

$(x^{\frac{1}{3}})^3 = 3^3 \qquad\qquad (x^{\frac{1}{3}})^3 = (-1)^3$

$x = 27 \qquad\qquad\quad x = -1$

• Replace u by $x^{\frac{1}{3}}$.

• Solve for x by cubing both sides of the equation.

The solutions are 27 and -1.

Problem 1   Solve.

A. $x - 5x^{\frac{1}{2}} + 6 = 0$

B. $4x^4 + 35x^2 - 9 = 0$

Solution   See page A58.   A. 4 and 9   B. $\frac{1}{2}$, $-\frac{1}{2}$, $3i$ and $-3i$

2 Radical equations

Certain equations containing a radical can be solved by first solving the equation for the radical expression and then squaring each side of the equation.

Remember that when squaring both sides of an equation, the resulting equation may have a solution that is not a solution of the original equation. Therefore, the solutions to a radical equation must be checked.

Solve: $\sqrt{x + 2} + 4 = x$

	$\sqrt{x + 2} + 4 = x$
Solve for the radical expression.	$\sqrt{x + 2} = x - 4$
Square each side of the equation.	$(\sqrt{x + 2})^2 = (x - 4)^2$
Simplify.	$x + 2 = x^2 - 8x + 16$
Write the equation in standard form.	$0 = x^2 - 9x + 14$
Solve for x by factoring.	$0 = (x - 7)(x - 2)$

$$x - 7 = 0 \qquad\qquad x - 2 = 0$$
$$x = 7 \qquad\qquad\qquad x = 2$$

Check the solution.

Check:

$$\sqrt{x + 2} + 4 = x \qquad\qquad \sqrt{x + 2} + 4 = x$$

$$\begin{array}{c|c} \sqrt{7 + 2} + 4 & 7 \\ \sqrt{9} + 4 & \\ 3 + 4 & \\ \hline & 7 = 7 \end{array} \qquad \begin{array}{c|c} \sqrt{2 + 2} + 4 & 2 \\ \sqrt{4} + 4 & \\ 2 + 4 & \\ \hline & 6 \neq 2 \end{array}$$

7 checks as a solution, but 2 does not check as a solution.

The solution is 7.

Example 2 Solve: $\sqrt{7y - 3} + 3 = 2y$

Solution

$$\sqrt{7y - 3} + 3 = 2y$$
$$\sqrt{7y - 3} = 2y - 3 \qquad \text{• Solve for the radical expression.}$$
$$(\sqrt{7y - 3})^2 = (2y - 3)^2 \qquad \text{• Square both sides of the equation.}$$
$$7y - 3 = 4y^2 - 12y + 9$$
$$0 = 4y^2 - 19y + 12 \qquad \text{• Write the equation in standard form.}$$
$$0 = (4y - 3)(y - 4) \qquad \text{• Solve by y by factoring.}$$

$$4y - 3 = 0 \qquad\qquad y - 4 = 0$$
$$4y = 3 \qquad\qquad\qquad y = 4 \qquad \text{• 4 checks as a solution.}$$
$$y = \frac{3}{4} \qquad\qquad\qquad\qquad \frac{3}{4} \text{ does not check as a solution.}$$

The solution is 4.

Problem 2   Solve: $\sqrt{2x + 1} + x = 7$

Solution   See page A58.

4

If an equation contains more than one radical, the procedure of solving for the radical expression and squaring each side of the equation may have to be repeated.

Example 3   Solve: $\sqrt{2y + 1} - \sqrt{y} = 1$

Solution $\sqrt{2y + 1} - \sqrt{y} = 1$

$\sqrt{2y + 1} - \sqrt{y} + 1$ • Solve for one of the radical expressions.

$(\sqrt{2y + 1})^2 = (\sqrt{y} + 1)^2$ • Square both sides of the equation.

$2y + 1 = y + 2\sqrt{y} + 1$

$y = 2\sqrt{y}$ • Solve for the radical expression.

$y^2 = (2\sqrt{y})^2$ • Square both sides of the equation.

$y^2 = 4y$

$y^2 - 4y = 0$

$y(y - 4) = 0$

$y = 0 \qquad y - 4 = 0$

$ y = 4$ • 0 and 4 check as solutions.

The solutions are 0 and 4.

Problem 3   Solve: $\sqrt{2x - 1} + \sqrt{x} = 2$

Solution   See page A59.

1

3   Fractional equations

After each side of a fractional equation has been multiplied by the LCM of the denominators, the resulting equation is sometimes a quadratic equation. The solutions to the resulting equation must be checked because multiplying by a variable expression may produce an equation that has a solution that is not a solution of the original equation.

Solve: $\dfrac{1}{r} + \dfrac{1}{r+1} = \dfrac{3}{2}$

$$\dfrac{1}{r} + \dfrac{1}{r+1} = \dfrac{3}{2}$$

Multiply each side of the equation by the LCM of the denominators.

$$2r(r+1)\left(\dfrac{1}{r} + \dfrac{1}{r+1}\right) = 2r(r+1) \cdot \dfrac{3}{2}$$
$$2(r+1) + 2r = r(r+1) \cdot 3$$
$$2r + 2 + 2r = 3r(r+1)$$
$$4r + 2 = 3r^2 + 3r$$
$$0 = 3r^2 - r - 2$$

Write the equation in standard form.

Solve for r by factoring.

$$0 = (3r+2)(r-1)$$

$$
\begin{array}{ll}
3r + 2 = 0 & r - 1 = 0 \\
3r = -2 & r = 1 \\
r = -\dfrac{2}{3} &
\end{array}
$$

$-\dfrac{2}{3}$ and 1 check as solutions.

Write the solutions.

The solutions are $-\dfrac{2}{3}$ and 1.

Example 4   Solve.

A. $\dfrac{9}{x-3} = 2x + 1$ B. $\dfrac{18}{2a-1} + 3a = 17$

Solution   A.

$$\dfrac{9}{x-3} = 2x + 1$$ • The LCM is $x - 3$.

$$(x-3)\dfrac{9}{x-3} = (x-3)(2x+1)$$
$$9 = 2x^2 - 5x - 3$$
$$0 = 2x^2 - 5x - 12$$ • Write the equation in standard form.
$$0 = (2x+3)(x-4)$$ • Solve for x by factoring.

$$
\begin{array}{ll}
2x + 3 = 0 & x - 4 = 0 \\
2x = -3 & x = 4 \\
x = -\dfrac{3}{2} &
\end{array}
$$

The solutions are $-\dfrac{3}{2}$ and 4.

B.
$$\frac{18}{2a - 1} + 3a = 17$$

$$(2a - 1)\left(\frac{18}{2a - 1} + 3a\right) = (2a - 1)17$$

$$(2a - 1)\frac{18}{2a - 1} + (2a - 1)(3a) = (2a - 1)17$$

$$18 + 6a^2 - 3a = 34a - 17$$
$$6a^2 - 37a + 35 = 0$$
$$(6a - 7)(a - 5) = 0$$

$$6a - 7 = 0 \qquad a - 5 = 0$$
$$6a = 7 \qquad a = 5$$
$$a = \frac{7}{6}$$

The solutions are $\frac{7}{6}$ and 5.

Problem 4 Solve.

A. $3y + \dfrac{25}{3y - 2} = -8$ B. $\dfrac{5}{x + 2} = 2x - 5$

Solution See page A59.

A. -1 B. $-\dfrac{5}{2}$ and 3

EXERCISES 7.3

1   Solve.

1. $x^4 - 13x^2 + 36 = 0$
   3, -3, 2, -2

2. $y^4 - 5y^2 + 4 = 0$
   2, -2, 1, -1

3. $z^4 - 6z^2 + 8 = 0$
   2, -2, $\sqrt{2}$, $-\sqrt{2}$

4. $t^4 - 12t^2 + 27 = 0$
   3, -3, $\sqrt{3}$, $-\sqrt{3}$

5. $p - 3p^{\frac{1}{2}} + 2 = 0$
   1 and 4

6. $v - 7v^{\frac{1}{2}} + 12 = 0$
   9 and 16

7. $x - x^{\frac{1}{2}} - 12 = 0$
   16

8. $w - 2w^{\frac{1}{2}} - 15 = 0$
   25

9. $z^4 + 3z^2 - 4 = 0$
   $2i$, $-2i$, 1, -1

10. $y^4 + 5y^2 - 36 = 0$
    $3i$, $-3i$, 2, -2

11. $x^4 + 12x^2 - 64 = 0$
    $4i$, $-4i$, 2, -2

12. $x^4 - 81 = 0$
    $3i$, $-3i$, 3, -3

13. $p + 2p^{\frac{1}{2}} - 24 = 0$
    16

14. $v + 3v^{\frac{1}{2}} - 4 = 0$
    1

15. $y^{\frac{2}{3}} - 9y^{\frac{1}{3}} + 8 = 0$
    512 and 1

16. $z^{\frac{2}{3}} - z^{\frac{1}{3}} - 6 = 0$
27 and −8

17. $x^6 - 9x^3 + 8 = 0$
2 and 1

18. $y^6 + 9y^3 + 8 = 0$
−2 and −1

19. $z^8 - 17z^4 + 16 = 0$
1, −1, 2, −2

20. $v^4 - 15v^2 - 16 = 0$
4, −4, i, −i

21. $p^{\frac{2}{3}} + 2p^{\frac{1}{3}} - 8 = 0$
−64 and 8

22. $w^{\frac{2}{3}} + 3w^{\frac{1}{3}} - 10 = 0$
−125 and 8

23. $2x - 3x^{\frac{1}{2}} + 1 = 0$
$\frac{1}{4}$ and 1

24. $3y - 5y^{\frac{1}{2}} - 2 = 0$
4

2 Solve.

25. $\sqrt{x + 1} + x = 5$
3

26. $\sqrt{x - 4} + x = 6$
5

27. $x = \sqrt{x} + 6$
9

28. $\sqrt{2y - 1} = y - 2$
5

29. $\sqrt{3w + 3} = w + 1$
2 and −1

30. $\sqrt{2s + 1} = s - 1$
4

31. $\sqrt{4y + 1} - y = 1$
0 and 2

32. $\sqrt{3s + 4} + 2s = 12$
4

33. $\sqrt{10x + 5} - 2x = 1$
$-\frac{1}{2}$ and 2

34. $\sqrt{t + 8} = 2t + 1$
1

35. $\sqrt{p + 11} = 1 - p$
−2

36. $x - 7 = \sqrt{x - 5}$
9

37. $\sqrt{x - 1} - \sqrt{x} = -1$
1

38. $\sqrt{y} + 1 = \sqrt{y + 5}$
4

39. $\sqrt{2x - 1} = 1 - \sqrt{x - 1}$
1

40. $\sqrt{x + 6} + \sqrt{x + 2} = 2$
−2

41. $\sqrt{t + 3} + \sqrt{2t + 7} = 1$
−3

42. $\sqrt{5 - 2x} = \sqrt{2 - x} + 1$
−2 and 2

3 Solve.

43. $x = \dfrac{10}{x - 9}$
10 and −1

44. $z = \dfrac{5}{z - 4}$
5 and −1

45. $\dfrac{t}{t + 1} = \dfrac{-2}{t - 1}$
$-\frac{1}{2} + \frac{\sqrt{7}}{2}i$ and $-\frac{1}{2} - \frac{\sqrt{7}}{2}i$

46. $\dfrac{2v}{v - 1} = \dfrac{5}{v + 2}$
$\frac{1}{4} + \frac{\sqrt{39}}{4}i$ and $\frac{1}{4} - \frac{\sqrt{39}}{4}i$

47. $\dfrac{y - 1}{y + 2} + y = 1$
1 and −3

48. $\dfrac{2p - 1}{p - 2} + p = 8$
3 and 5

49. $\dfrac{3r + 2}{r + 2} - 2r = 1$
0 and −1

50. $\dfrac{2v + 3}{v + 4} + 3v = 4$
$-\frac{13}{3}$ and 1

51. $\dfrac{2}{2x + 1} + \dfrac{1}{x} = 3$
$\frac{1}{2}$ and $-\frac{1}{3}$

52. $\dfrac{3}{s} - \dfrac{2}{2s - 1} = 1$

$\dfrac{3}{2}$ and 1

53. $\dfrac{16}{z - 2} + \dfrac{16}{z + 2} = 6$

6 and $-\dfrac{2}{3}$

54. $\dfrac{2}{y + 1} + \dfrac{1}{y - 1} = 1$

0 and 3

55. $\dfrac{t}{t - 2} + \dfrac{2}{t - 1} = 4$

$\dfrac{4}{3}$ and 3

56. $\dfrac{4t + 1}{t + 4} + \dfrac{3t - 1}{t + 1} = 2$

1 and $-\dfrac{11}{5}$

57. $\dfrac{5}{2p - 1} + \dfrac{4}{p + 1} = 2$

$-\dfrac{1}{4}$ and 3

58. $\dfrac{3w}{2w + 3} + \dfrac{2}{w + 2} = 1$

0 and -3

59. $\dfrac{2v}{v + 2} + \dfrac{3}{v + 4} = 1$

$\dfrac{-5 + \sqrt{33}}{2}$ and $\dfrac{-5 - \sqrt{33}}{2}$

60. $\dfrac{x + 3}{x + 1} - \dfrac{x - 2}{x + 3} = 5$

$\dfrac{-13 + \sqrt{89}}{10}$ and $\dfrac{-13 - \sqrt{89}}{10}$

SUPPLEMENTAL EXERCISES 7.3

Solve.

61. $\dfrac{3x^2}{2} = 4x - 2$

$\dfrac{2}{3}$ and 2

62. $\dfrac{2x^2}{3} = 9 - x$

$-\dfrac{9}{2}$ and 3

63. $\dfrac{x^2}{4} + \dfrac{x}{2} = 6$

4 and -6

64. $\dfrac{x^2}{9} + \dfrac{2x}{3} = 3$

-9 and 3

65. $3\left(\dfrac{x + 1}{2}\right)^2 = 54$

$-1 + 6\sqrt{2}$ and $-1 - 6\sqrt{2}$

66. $2\left(\dfrac{x - 2}{3}\right)^2 = 24$

$2 + 6\sqrt{3}$ and $2 - 6\sqrt{3}$

67. $p + 2p^{\frac{1}{2}} = 8$

4

68. $p + 3p^{\frac{1}{2}} = 9$

$\dfrac{27 + 9\sqrt{5}}{2}$ and $\dfrac{27 - 9\sqrt{5}}{2}$

69. $\sqrt{x + 3} = x - 3$

6

70. $\sqrt{x + 2} = 4 - x$

2

71. $\sqrt{2x + 3} + x = 6$

3

72. $\sqrt{3x + 4} + x = 8$

4

73. $\dfrac{x + 2}{3} + \dfrac{2}{x - 2} = 3$

5 and 4

74. $\dfrac{x - 1}{2} + \dfrac{4}{x + 1} = 2$

3 and 1

75. $\dfrac{x^4}{4} + 1 = \dfrac{5x^2}{4}$

2, −2, 1, −1

76. $\dfrac{x^4}{4} + 2 = \dfrac{9x^2}{4}$

$2\sqrt{2}$, $-2\sqrt{2}$, 1, −1

77. $\dfrac{x^4}{3} - \dfrac{8x^2}{3} = 3$

3, −3, i, $-i$

78. $\dfrac{x^4}{6} + \dfrac{x^2}{6} = 2$

$2i$, $-2i$, $\sqrt{3}$, $-\sqrt{3}$

79. $\dfrac{x^2}{4} + \dfrac{x}{2} + \dfrac{1}{8} = 0$

$\dfrac{-2 + \sqrt{2}}{2}$ and $\dfrac{-2 - \sqrt{2}}{2}$

80. $\dfrac{x^2}{2} + \dfrac{x}{3} + \dfrac{1}{6} = 0$

$-\dfrac{1}{3} + \dfrac{\sqrt{2}}{3}i$ and $-\dfrac{1}{3} - \dfrac{\sqrt{2}}{3}i$

81. $p + 4p^{\frac{1}{2}} = 2$

$10 + 4\sqrt{6}$ and $10 - 4\sqrt{6}$

82. $p + 6p^{\frac{1}{2}} = 3$

$21 + 12\sqrt{3}$ and $21 - 12\sqrt{3}$

83. $\dfrac{x^2}{3} - \dfrac{x}{6} = 1$

$-\dfrac{3}{2}$ and 2

84. $\dfrac{x^2}{2} - \dfrac{x}{4} = 1$

$\dfrac{1 + \sqrt{33}}{4}$ and $\dfrac{1 - \sqrt{33}}{4}$

85. $\dfrac{x^4}{8} + \dfrac{x^2}{4} = 3$

$i\sqrt{6}$, $-i\sqrt{6}$, 2, −2

86. $\dfrac{x^4}{8} - \dfrac{x^2}{2} = 4$

$2\sqrt{2}$, $-2\sqrt{2}$, $2i$, $-2i$

87. $\sqrt{x + 3} = x + 2$

$\dfrac{-3 + \sqrt{5}}{2}$ and $\dfrac{-3 - \sqrt{5}}{2}$

88. $\sqrt{x + 5} = x - 4$

$\dfrac{9 + \sqrt{37}}{2}$ and $\dfrac{9 - \sqrt{37}}{2}$

89. $\sqrt{2x + 3} = x - 1$

$2 + \sqrt{6}$ and $2 - \sqrt{6}$

90. $\sqrt{3x - 1} = x - 2$

$\dfrac{7 + \sqrt{29}}{2}$ and $\dfrac{7 - \sqrt{29}}{2}$

91. $\sqrt{x^4 - 2} = x$

$\sqrt{2}$ and i

92. $\sqrt{x^4 + 4} = 2x$

$\sqrt{2}$

93. $\dfrac{x + 2}{4} - \dfrac{4}{x - 2} = 3$

$6 + 4\sqrt{2}$ and $6 - 4\sqrt{2}$

94. $\dfrac{x + 3}{2} - \dfrac{2}{x - 3} = 5$

$5 + 2\sqrt{2}$ and $5 - 2\sqrt{2}$

95. $\dfrac{2}{x + 2} + \dfrac{2}{x - 2} = 2$

$1 + \sqrt{5}$ and $1 - \sqrt{5}$

96. $\dfrac{1}{x + 3} + \dfrac{1}{x + 4} = 1$

$\dfrac{-5 + \sqrt{5}}{2}$ and $\dfrac{-5 - \sqrt{5}}{2}$

Solve for x.

97. $a\left(\dfrac{x - 4}{b}\right)^2 = a^3$

$4 + ab$ and $4 - ab$

98. $y\left(\dfrac{x - 6}{z}\right)^2 = y^3$

$6 + yz$ and $6 - yz$

99. $a\left(\dfrac{x - 2}{b}\right)^2 = a^3$

$2 + ab$ and $2 - ab$

SECTION 7.4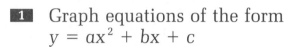

Graphing Quadratic Equations in Two Variables

1 Graph equations of the form
$y = ax^2 + bx + c$

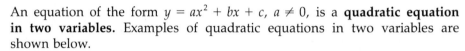

An equation of the form $y = ax^2 + bx + c$, $a \neq 0$, is a **quadratic equation in two variables.** Examples of quadratic equations in two variables are shown below.

$$y = 2x^2 - x - 3$$

$$y = -x^2 + 4$$

$$y = 3x^2 - 2x$$

The graph of a quadratic equation in two variables is a **parabola.** The graph is "cup" shaped and opens either up or down. The coefficient of x^2 determines whether the parabola opens up or down. When a is **positive**, the parabola **opens up.** When a is **negative**, the parabola **opens down.** The graphs of two parabolas are shown below.

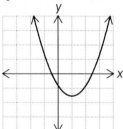

Parabola that opens up
$y = ax^2 + bx + c$, $a > 0$

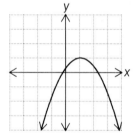

Parabola that opens down
$y = ax^2 + bx + c$, $a < 0$

Since the graph of a parabola is cup shaped, drawing its graph requires finding enough ordered pair solutions of the equation so that its cup shape can be determined. Remember that when a is positive, the parabola will open up, and when a is negative, the parabola will open down.

For example, the graph of $y = x^2 - x - 2$ will open up because a is positive ($a = 1$).

To graph $y = x^2 - x - 2$, find enough ordered pair solutions to determine the cup shape. These ordered pairs can be recorded in a table.

x	y
0	-2
1	-2
-1	0
2	0
-2	4
3	4

Graph the ordered pair solutions on a rectangular coordinate system.

Draw a parabola through the points.

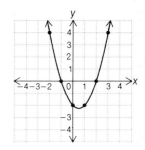

Example 1   Graph.

A. $y = -2x^2 + x + 4$

B. $y = -2x^2 - 4x$

Solution   A.

x	y
0	4
1	3
−1	1
2	−2
−2	−6

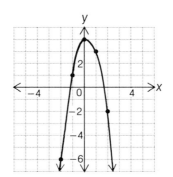

• Since a is negative ($a = -2$), the parabola will open down.

B.

x	y
0	0
1	−6
−1	2
−2	0
−3	−6

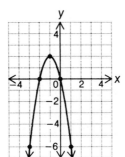

Problem 1   Graph.

A. $y = \dfrac{1}{2}x^2 - 2x - 1$

B. $y = -x^2 + 3x - 4$

Solution   See page A60.

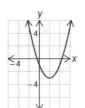

A.

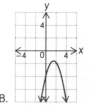

B.

2 Find the x-intercepts of a parabola

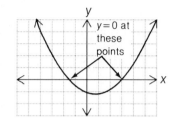

The points at which a graph crosses or touches a coordinate axis are called the **intercepts** of the graph.

When the graph of a parabola crosses the x-axis, the y-coordinate is zero.

To find the x-intercepts for the parabola, $y = ax^2 + bx + c$, let $y = 0$. Then solve for x.

Example 2   Find the x-intercepts of the parabola.

     A. $y = x^2 - 2x - 1$ B. $y = 4x^2 - 4x + 1$

Solution   A. $y = x^2 - 2x - 1$

       $0 = x^2 - 2x - 1$ • Let $y = 0$.

$$x = \frac{-b \pm \sqrt{b^2 - 4ac}}{2a}$$

• The equation is difficult to factor. Use the quadratic formula to solve for x.

$$= \frac{-(-2) \pm \sqrt{(-2)^2 - 4(1)(-1)}}{2 \cdot 1}$$

$$= \frac{2 \pm \sqrt{4 + 4}}{2} = \frac{2 \pm \sqrt{8}}{2}$$

$$= \frac{2 \pm 2\sqrt{2}}{2} = 1 \pm \sqrt{2}$$

The x-intercepts are $(1 + \sqrt{2}, 0)$ and $(1 - \sqrt{2}, 0)$.

     B. $y = 4x^2 - 4x + 1$

       $0 = 4x^2 - 4x + 1$ • Let $y = 0$.

       $0 = (2x - 1)(2x - 1)$ • Solve for x by factoring.

     $2x - 1 = 0$ $2x - 1 = 0$

         $2x = 1$ $2x = 1$

            $x = \frac{1}{2}$ $x = \frac{1}{2}$ • The equation has a double root.

The x-intercept is $\left(\frac{1}{2}, 0\right)$.

Problem 2   Find the x-intercepts of the parabola.

     A. $y = 2x^2 - 5x + 2$ B. $y = x^2 + 4x + 5$

Solution   See page A60.

                    A. $\left(\frac{1}{2}, 0\right)$ and $(2, 0)$ B. There are no x-intercepts.

In Example 2B, the parabola only touches the x-axis. In this case, the parabola is said to be **tangent** to the x-axis at $x = \dfrac{1}{2}$.

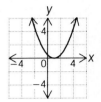

A parabola can have one x-intercept, as in Example 2B; two x-intercepts, as in the graph at the top of the previous page; or no x-intercepts, as in the graph at the right.

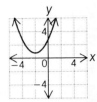

The parabola $y = ax^2 + bx + c$ will have one x-intercept if the equation $0 = ax^2 + bx + c$ has one real number solution, two x-intercepts if the equation has two real number solutions, and no x-intercepts if the equation has no real number solutions.

Since the discriminant determines whether there are one, two, or no real number solutions of the equation $0 = ax^2 + bx + c$, it can also be used to determine whether there are one, two, or no x-intercepts of the parabola $y = ax^2 + bx + c$.

The Effect of the Discriminant on the Number of x-Intercepts of a Parabola

1. If $b^2 - 4ac = 0$, the parabola has one x-intercept.
2. If $b^2 - 4ac > 0$, the parabola has two x-intercepts.
3. If $b^2 - 4ac < 0$, the parabola has no x-intercepts.

The parabola $y = 2x^2 - x + 2$ has no x-intercepts since the discriminant is less than zero.

$a = 2, b = -1, c = 2$
$b^2 - 4ac$
$(-1)^2 - 4(2)(2) = 1 - 16 = -15$
$-15 < 0$

Example 3   Use the discriminant to determine the number of x-intercepts of the parabola $y = x^2 - 6x + 9$.

Solution $b^2 - 4ac$ • $a = 1, b = -6, c = 9$
$(-6)^2 - 4(1)(9) = 36 - 36 = 0$ • The discriminant is equal to zero.

The parabola has one x-intercept.

Problem 3   Use the discriminant to determine the number of x-intercepts of the parabola $y = x^2 - x - 6$.

Solution   See page A61.

two x-intercepts

EXERCISES 7.4

1   Graph.

1. $y = x^2$

2. $y = -x^2$

3. $y = x^2 - 2$

4. $y = x^2 + 2$

5. $y = -x^2 + 3$

6. $y = -x^2 - 1$

7. $y = \dfrac{1}{2}x^2$

8. $y = 2x^2$

9. $y = 2x^2 - 1$

10. $y = -\dfrac{1}{2}x^2 + 2$

11. $y = x^2 - 2x$

12. $y = x^2 + 2x$

13. $y = -2x^2 + 4x$

14. $y = \frac{1}{2}x^2 - x$

15. $y = x^2 - x - 2$

16. $y = x^2 - 3x + 2$

17. $y = 2x^2 - x - 5$

18. $y = 2x^2 - x - 3$

19. $y = -2x^2 - 3x + 2$

20. $y = 2x^2 - 7x + 3$

21. $y = x^2 - 4x + 4$

22. $y = -x^2 + 6x - 9$

23. $y = x^2 + 4x + 5$

24. $y = -x^2 - 2x - 1$

2   Find the *x*-intercepts of the parabola.

25. $y = x^2 - 4$
(2,0) and (−2,0)

26. $y = x^2 - 9$
(3,0) and (−3,0)

27. $y = 2x^2 - 4x$
(0,0) and (2,0)

28. $y = 3x^2 + 6x$
(0,0) and (−2,0)

29. $y = x^2 - x - 2$
(2,0) and (−1,0)

30. $y = x^2 - 2x - 8$
(4, 0) and (−2,0)

31. $y = 2x^2 - 5x - 3$
(3,0) and $\left(-\frac{1}{2},0\right)$

32. $y = 4x^2 + 11x + 6$
$\left(-\frac{3}{4},0\right)$ and (−2,0)

33. $y = 3x^2 - 19x - 14$
$\left(-\frac{2}{3},0\right)$ and (7,0)

34. $y = 6x^2 + 7x + 2$
$\left(-\frac{1}{2},0\right)$ and $\left(-\frac{2}{3},0\right)$

35. $y = 3x^2 - 19x + 20$
(5,0) and $\left(\frac{4}{3},0\right)$

36. $y = 3x^2 + 19x + 28$
(−4,0) and $\left(-\frac{7}{3},0\right)$

37. $y = 9x^2 - 12x + 4$
$\left(\dfrac{2}{3}, 0\right)$

38. $y = x^2 - 2$
$(\sqrt{2}, 0)$ and $(-\sqrt{2}, 0)$

39. $y = 9x^2 - 2$
$\left(\dfrac{\sqrt{2}}{3}, 0\right)$ and $\left(-\dfrac{\sqrt{2}}{3}, 0\right)$

40. $y = 2x^2 - x - 1$
$\left(-\dfrac{1}{2}, 0\right)$ and $(1, 0)$

41. $y = 2x^2 - 5x - 3$
$\left(-\dfrac{1}{2}, 0\right)$ and $(3, 0)$

42. $y = x^2 + 2x - 1$
$(-1 + \sqrt{2}, 0)$ and $(-1 - \sqrt{2}, 0)$

43. $y = x^2 + 4x - 3$
$(-2 + \sqrt{7}, 0)$ and $(-2 - \sqrt{7}, 0)$

44. $y = x^2 + 6x + 10$
no x-intercepts

45. $y = -x^2 - 4x - 5$
no x-intercepts

46. $y = x^2 - 2x - 2$
$(1 + \sqrt{3}, 0)$ and $(1 - \sqrt{3}, 0)$

47. $y = -x^2 - 2x + 1$
$(-1 + \sqrt{2}, 0)$ and $(-1 - \sqrt{2}, 0)$

48. $y = -x^2 + 4x + 1$
$(2 + \sqrt{5}, 0)$ and $(2 - \sqrt{5}, 0)$

49. $y = x^2 + 4x + 2$
$(-2 + \sqrt{2}, 0)$ and $(-2 - \sqrt{2}, 0)$

50. $y = x^2 + 2x - 4$
$(-1 + \sqrt{5}, 0)$ and $(-1 - \sqrt{5}, 0)$

51. $y = x^2 + 8x + 14$
$(-4 + \sqrt{2}, 0)$ and $(-4 - \sqrt{2}, 0)$

52. $y = x^2 - 6x + 7$
$(3 + \sqrt{2}, 0)$ and $(3 - \sqrt{2}, 0)$

53. $y = x^2 - 2x - 4$
$(1 + \sqrt{5}, 0)$ and $(1 - \sqrt{5}, 0)$

54. $y = x^2 - 2x + 2$
no x-intercepts

Use the discriminant to determine the number of x-intercepts of the parabola.

55. $y = 2x^2 + x + 1$
no x-intercepts

56. $y = 2x^2 + 2x - 1$
two

57. $y = -x^2 - x + 3$
two

58. $y = -2x^2 + x + 1$
two

59. $y = x^2 - 8x + 16$
one

60. $y = x^2 - 10x + 25$
one

61. $y = -3x^2 - x - 2$
no x-intercepts

62. $y = -2x^2 + x - 1$
no x-intercepts

63. $y = 4x^2 - x - 2$
two

64. $y = 2x^2 + x + 4$
no x-intercepts

65. $y = -2x^2 - x - 5$
no x-intercepts

66. $y = -3x^2 + 4x - 5$
no x-intercepts

67. $y = x^2 + 8x + 16$
one

68. $y = x^2 - 12x + 36$
one

69. $y = x^2 + x - 3$
two

70. $y = x^2 + 7x - 2$
two

71. $y = x^2 + 5x - 4$
two

72. $y = x^2 - 5x + 8$
no x-intercepts

73. $y = 3x^2 - 2x + 4$
no x-intercepts

74. $y = 4x^2 - 9x + 1$
two

75. $y = 5x^2 + 2x + 3$
no x-intercepts

76. $y = 4x^2 - x + 1$
no x-intercepts

77. $y = 2x^2 + 3x + 4$
no x-intercepts

78. $y = -5x^2 + x - 1$
no x-intercepts

79. $y = -10x^2 + x + 1$
two

80. $y = -5x^2 - 7x - 3$
no x-intercepts

81. $y = -7x^2 + 3x + 7$
two

SUPPLEMENTAL EXERCISES 7.4

Graph.

82. $y = (x - 2)^2$

83. $y = (x + 3)^2$

84. $y = 2(x - 1)^2$

85. $y = -3(x + 1)^2$

86. $y = (x - 3)^2 + 1$

87. $y = (x + 2)^2 - 4$

88. $y = -3(x - 2)^2 + 2$

89. $y = 2(x - 1)^2 + 3$

90. $y = \frac{1}{2}(x - 4)^2 - 3$

91. $y = \frac{1}{2}(x + 2)^2 - 3$

Find the value of k such that the parabola contains the given point.

92. $y = x^2 - 3x + k;\ (2, 5)$
7

93. $y = x^2 + 2x + k;\ (-3, 1)$
−2

94. $y = 2x^2 + 3x + k;\ (-4, 8)$
−12

95. $y = 3x^2 - 4x + k;\ (2, -6)$
−10

96. $y = x^2 + kx - 4;\ (-1, 3)$
−6

97. $y = x^2 + kx + 2;\ (3, -1)$
−4

98. $y = 2x^2 + kx - 3;\ (4, -3)$
−8

99. $y = 3x^2 + kx - 6;\ (-2, 4)$
1

Solve.

100. The point (x_1, y_1) lies in Quadrant I and is a solution of the equation
$y = 3x^2 - 2x - 1$. Given $y_1 = 5$, find x_1.
$$\frac{1 + \sqrt{19}}{3}$$

101. The point (x_1, y_1) lies in Quadrant II and is a solution of the equation
$y = 2x^2 + 5x - 3$. Given $y_1 = 9$, find x_1.
-4

102. The point (x_1, y_1) lies in Quadrant II and is a solution of the equation
$y = 2x^2 + 11x + 4$. Given $y_1 = 10$, find x_1.
-6

103. The point (x_1, y_1) lies in Quadrant II and is a solution of the equation
$y = 3x^2 + 7x + 2$. Given $y_1 = 8$, find x_1.
-3

104. What effect does increasing the coefficient of x^2 have on the graph of
$y = ax^2 + bx + c$?
The graph becomes thinner.

105. What effect does decreasing the coefficient of x^2 have on the graph of
$y = ax^2 + bx + c$?
The graph becomes wider.

106. What effect does increasing the constant term have on the graph of
$y = ax^2 + bx + c$?
The graph is higher on the rectangular coordinate system.

107. What effect does decreasing the constant term have on the graph of
$y = ax^2 + bx + c$?
The graph is lower on the rectangular coordinate system.

SECTION 7.5

Applications of Quadratic Equations

1 Application problems

The application problems in this section are similar to those problems that were solved earlier in the text. Each of the strategies for the problems in this section will result in a quadratic equation.

A small pipe takes 16 min longer to empty a tank than does a larger pipe. Working together, the pipes can empty the tank in 6 min. How long would it take each pipe working alone to empty the tank?

STRATEGY *for solving an application problem*

▶ **Determine the type of problem. Is it a uniform motion problem, a geometry problem, an integer problem, or a work problem?**

The problem is a work problem.

▶ **Choose a variable to represent the unknown quantity. Write numerical or variable expressions for all the remaining quantities. These results can be recorded in a table.**

The unknown time of the larger pipe: t
The unknown time of the smaller pipe: $t + 16$

	Rate of work	·	Time worked	=	Part of task completed
Larger pipe	$\dfrac{1}{t}$	·	6	=	$\dfrac{6}{t}$
Smaller pipe	$\dfrac{1}{t + 16}$	·	6	=	$\dfrac{6}{t + 16}$

▶ **Determine how the quantities are related. If necessary, review the strategies presented in Chapter 2.**

The sum of the parts of the task completed must equal 1.

$$\frac{6}{t} + \frac{6}{t + 16} = 1$$

$$t(t + 16)\left(\frac{6}{t} + \frac{6}{t + 16}\right) = t(t + 16) \cdot 1$$

$$(t + 16)6 + 6t = t^2 + 16t$$

$$6t + 96 + 6t = t^2 + 16t$$

$$0 = t^2 + 4t - 96$$

$$0 = (t + 12)(t - 8)$$

$$t + 12 = 0 \qquad\qquad t - 8 = 0$$
$$t = -12 \qquad\qquad\quad t = 8$$

The solution $t = -12$ is not possible since time cannot be a negative number.

The time for the smaller pipe is $t + 16$. Replace t by 8 and evaluate.

$$t + 16 = 8 + 16 = 24$$

The larger pipe requires 8 min to empty the tank.
The smaller pipe requires 24 min to empty the tank.

Example 1 In 8 h, two campers rowed 15 mi down a river and then rowed back to their campsite. The rate of the river's current was 1 mph. Find the rate at which the campers row in calm water.

Strategy ▶ This is a uniform motion problem.
▶ Unknown rowing rate of the campers: r

	Distance	Rate	Time
Down river	15	$r + 1$	$\dfrac{15}{r + 1}$
Up river	15	$r - 1$	$\dfrac{15}{r - 1}$

▶ The total time of the trip was 8 h.

Solution

$$\frac{15}{r + 1} + \frac{15}{r - 1} = 8$$

$$(r + 1)(r - 1)\left(\frac{15}{r + 1} + \frac{15}{r - 1}\right) = (r + 1)(r - 1)8$$

$$(r - 1)15 + (r + 1)15 = (r^2 - 1)8$$

$$15r - 15 + 15r + 15 = 8r^2 - 8$$

$$30r = 8r^2 - 8$$

$$0 = 8r^2 - 30r - 8$$

$$0 = 2(4r^2 - 15r - 4)$$

$$0 = 2(4r + 1)(r - 4)$$

$$4r + 1 = 0 \qquad\qquad r - 4 = 0$$
$$4r = -1 \qquad\qquad r = 4$$
$$r = -\frac{1}{4}$$

• The solution $r = -\dfrac{1}{4}$ is not possible because the rate cannot be a negative number.

The rowing rate is 4 mph.

Problem 1 The length of a rectangle is 3 m more than the width. The area is 54 m². Find the length of the rectangle.

Solution See page A62.
9 m

EXERCISES 7.5

1   Solve.

1. The length of a rectangle is one foot more than twice the width. The area of the rectangle is 10 ft^2. Find the length and width of the rectangle.
    width: 2 ft; length: 5 ft

2. The height of a triangle is twice the length of the base. The area of the triangle is 25 cm^2. Find the length of the base and the height of the triangle.
    height: 10 cm; base: 5 cm

3. The height of a triangle is two meters less than the length of the base. The area of the triangle is 12 m^2. Find the length of the base and the height of the triangle.
    height: 4 m; base: 6 m

4. The length of a rectangle is one meter less than three times the width. The area of the rectangle is 44 m^2. Find the length and width of the rectangle.
    length: 11 m; width: 4 m

5. The difference between the squares of two consecutive integers is nineteen. Find the two integers.
    9 and 10

6. The difference between the squares of two consecutive even integers is twenty-eight. Find the two integers.
    6 and 8 or −6 and −8

7. The sum of the squares of two consecutive even integers is twenty. Find the two integers.
    2 and 4 or −2 and −4

8. The sum of the squares of three consecutive odd integers is eleven. Find the three integers.
    −3, −1, and 1 or −1, 1, and 3

9. The sum of five times an integer and the product of three and the square of the integer is two. Find the integer.
    −2

10. Seven times an integer plus twice the square of the integer is four. Find the integer.
    −4

11. An old pump requires 3 h longer to empty a pool than does a new pump. With both pumps working, the pool can be emptied in 2 h. Find the time required for the new pump working alone to empty the pool.
    3 h

12. A chemistry experiment requires that a vacuum be created in a chamber. A small vacuum pump requires 15 s longer than does a second larger pump to evacuate the chamber. Working together, the chamber can be evacuated in 4 s. Find the time required for the larger vacuum pump working alone to evacuate the chamber.
5 s

13. A large pipe can fill a tank in 6 min less time than it takes a smaller pipe to fill the same tank. Working together, both pipes can fill the tank in 4 min. How long would it take each pipe working alone to fill the tank?
larger pipe: 6 min; smaller pipe: 12 min

14. A small air conditioner requires 16 min longer to cool a room 3° than does a larger air conditioner. Working together, the two air conditioners can cool the room 3° in 6 min. How long would it take each air conditioner working alone to cool the room 3°?
larger air conditioner: 8 min; smaller air conditioner: 24 min

15. A small heating unit takes 8 h longer to melt a piece of iron than does a larger unit. Working together, the heating units can melt the iron in 3 h. How long would it take each heating unit working alone to melt the iron?
smaller unit: 12 h; larger unit: 4 h

16. An old mechanical mail sorter takes 21 min longer to sort a batch of mail than does a second, newer model. With both sorters working, a batch of mail can be sorted in 10 min. How long would it take each sorter working alone to sort the batch of mail?
new sorter: 14 min; old sorter: 35 min

17. The rate of a small plane in calm air is 100 mph. Flying with a wind, the plane can fly 240 mi in one hour less time than is required to make the return trip of 240 mi. Find the rate of the wind.
20 mph

18. A cabin cruiser made a trip of 100 mi in 8 h. The boat traveled the first 40 mi at a constant rate before increasing its speed by 5 mph. Another 60 mi was traveled at the increased speed. Find the rate of the cabin cruiser for the first 40 mi.
10 mph

19. A car travels 120 mi. A second car, traveling 10 mph faster than the first car, makes the same trip in 1 h less time. Find the speed of each car.
1st car: 30 mph; 2nd car: 40 mph

20. A cyclist traveled 60 mi at a constant rate before reducing the speed by 2 mph. Another 40 mi was traveled at the reduced speed. The total time for the 100-mi trip was 9 h. Find the rate during the first 60 mi.
     12 mph

SUPPLEMENTAL EXERCISES 7.5

Solve.

21. The sum of a number and twice its reciprocal is $\frac{33}{4}$. Find the number.
     $\frac{1}{4}$ or 8

22. The difference between a number and twice its reciprocal is $\frac{49}{5}$. Find the number.
     $-\frac{1}{5}$ or 10

23. The numerator of a fraction is 2 less than the denominator. The sum of the fraction and three times its reciprocal is $\frac{13}{2}$. Find the fraction.
     $\frac{2}{4}$

24. The numerator of a fraction is 3 less than the denominator. The sum of the fraction and four times its reciprocal is $\frac{17}{2}$. Find the fraction.
     $\frac{3}{6}$

25. Find two consecutive integers whose cubes differ by 127.
     6 and 7 or −7 and −6

26. Find two consecutive even integers whose cubes differ by 488.
     8 and 10 or −10 and −8

27. An open box is formed from a rectangular piece of cardboard whose length is 8 cm more than its width by cutting squares 2 cm in length from each corner and then folding up the sides. Find the dimensions of the box if its volume is 256 cm^3.
     2 cm × 8 cm × 16 cm

28. An open box is formed from a rectangular piece of cardboard whose length is 6 cm more than its width by cutting squares 4 cm in length from each corner and then folding up the sides. Find the dimensions of the box if its volume is 288 cm^3.
     4 cm × 6 cm × 12 cm

29. The volumes of two spheres differ by 684π cm^3. The radius of the larger sphere is 3 cm more than the radius of the smaller sphere. Find the radius of the larger sphere. (*Hint:* The formula for the volume of a sphere is $V = \dfrac{4}{3}\pi r^3$.)
    9 cm

30. The volumes of two spheres differ by 372π cm^3. The radius of the larger sphere is 3 cm more than the radius of the smaller sphere. Find the radius of the larger sphere. (*Hint:* The formula for the volume of a sphere is $V = \dfrac{4}{3}\pi r^3$.)
    7 cm

31. The lengths of the three sides of a right triangle are x cm, $(x + 1)$ cm, and $(x + 2)$ cm. Find the lengths of the sides of the triangle.
    3 cm, 4 cm, 5 cm

32. The lengths of the three sides of a right triangle are x cm, $(2x + 2)$ cm, and $(2x + 3)$ cm. Find the lengths of the sides of the triangle.
    5 cm, 12 cm, 13 cm

33. What number is equal to one less than its square?
    $\dfrac{1 + \sqrt{5}}{2}$ and $\dfrac{1 - \sqrt{5}}{2}$

34. Show that no real number can be equal to one more than its square.
    Complete solution available in Solutions Manual.

CALCULATORS AND COMPUTERS

Solving Quadratic Equations

The solutions of all quadratic equations can be found by using the quadratic formula. Although this formula is available, sometimes the coefficients of the equation make the use of the formula difficult because the computations are very tedious.

Consider trying to solve the equation

$$2.984x^2 + 9834.1x - 509.0023 = 0$$

by using the quadratic formula. The computations would take a long time even with a calculator.

The program QUADRATIC EQUATIONS on the Student Disk will solve any quadratic equation with real number coefficients, including those whose solutions are complex numbers. You can use the program to test your ability to solve equations by comparing your answer to that of the program.

CHAPTER SUMMARY

Key Words

A **quadratic equation** is an equation of the form $ax^2 + bx + c = 0$, where a, b, and c are constants and $a \neq 0$. A quadratic equation is also called a **second-degree equation.**

A quadratic equation is in **standard form** when the polynomial is in descending order and equal to zero.

When a quadratic equation has two solutions that are the same number, the solution is called a **double root** of the equation.

Adding to a binomial the constant term that makes it a perfect square trinomial is called **completing the square.**

For an equation of the form $ax^2 + bx + c = 0$, the quantity $b^2 - 4ac$ is called the **discriminant**.

An equation of the form $y = ax^2 + bx + c$, where $a \neq 0$, is a **quadratic equation in two variables**. The graph of a quadratic equation in two variables is a **parabola**.

The points at which a graph crosses a coordinate axis are called the **intercepts** of the graph.

Essential Rules

The Principle of Zero Products	If $ab = 0$, then $a = 0$ or $b = 0$.
The Quadratic Formula	$x = \dfrac{-b \pm \sqrt{b^2 - 4ac}}{2a}$

CHAPTER REVIEW

1. Solve: $6x^2 - 5x - 6 = 0$ $-\dfrac{2}{3}$ and $\dfrac{3}{2}$

2. Solve: $3(x - 2)^2 - 24 = 0$
   $2 + 2\sqrt{2}$ and $2 - 2\sqrt{2}$

3. Solve: $3x^2 - 6x = 2$ $\dfrac{3 + \sqrt{15}}{3}$ and $\dfrac{3 - \sqrt{15}}{3}$

4. Solve: $x^2 + 4x + 12 = 0$
   $-2 + 2i\sqrt{2}$ and $-2 - 2i\sqrt{2}$

5. Use the discriminant to determine whether $3x^2 - 4x = 1$ has one real number solution, two real number solutions, or two complex number solutions.
   two real

6. Write a quadratic equation that has integer coefficients and has solutions $\dfrac{1}{2}$ and -4.
   $2x^2 + 7x - 4 = 0$

7. Solve: $\sqrt{2x + 1} + 5 = 2x$ 4

8. Graph $y = x^2 - 5x + 5$.

9. Find the x-intercepts of the parabola $y = x^2 + 2x - 4$.
   $(-1 + \sqrt{5}, 0)$ and $(-1 - \sqrt{5}, 0)$

10. Solve: $x^4 - 4x^2 + 3 = 0$
    $1, -1, \sqrt{3}, -\sqrt{3}$

11. Solve: $3x^2 + 10x = 8$
    $\dfrac{2}{3}$ and -4

12. Solve: $x^2 - 6x - 2 = 0$
    $3 + \sqrt{11}$ and $3 - \sqrt{11}$

13. Solve: $\dfrac{2x}{x - 3} + \dfrac{5}{x - 1} = 1$
    -9 and 2

14. Solve: $2x^2 - 2x = 1$
    $\dfrac{1 + \sqrt{3}}{2}$ and $\dfrac{1 - \sqrt{3}}{2}$

15. Use the discriminant to determine the number of x-intercepts of the parabola $y = 2x^2 + 7x - 12$.
    two

16. Write a quadratic equation that has integer coefficients and has solutions $\dfrac{1}{3}$ and -3.
    $3x^2 + 8x - 3 = 0$

17. Solve: $2x^2 + 9x = 5$ $\dfrac{1}{2}$ and -5

18. Solve: $2(x + 1)^2 - 36 = 0$
    $-1 + 3\sqrt{2}$ and $-1 - 3\sqrt{2}$

19. Solve: $x^2 + 6x + 10 = 0$
    $-3 + i$ and $-3 - i$

20. Graph $y = -x^2 - 2x + 3$.

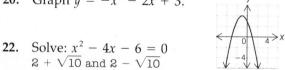

21. Solve: $x^4 - 6x^2 + 8 = 0$
    $2, -2, \sqrt{2}, -\sqrt{2}$

22. Solve: $x^2 - 4x - 6 = 0$
    $2 + \sqrt{10}$ and $2 - \sqrt{10}$

23. Use the discriminant to determine whether $4x^2 - 3x = 5$ has one real number solution, two real number solutions, or two complex number solutions.
    two real

24. Find the x-intercepts of the parabola $y = -2x^2 + x + 1$.
    $(1, 0)$ and $\left(-\dfrac{1}{2}, 0\right)$

25. Solve: $\sqrt{3x + 1} - 1 = x$
    0 and 1

26. Solve: $3x^2 + 7x = 6$ $\dfrac{2}{3}$ and -3

27. Solve: $2x^2 + 2x - 3 = 0$
    $\dfrac{-1 + \sqrt{7}}{2}$ and $\dfrac{-1 - \sqrt{7}}{2}$

28. Solve: $\dfrac{x}{x + 2} - \dfrac{4x}{x + 3} = 1$ $-\dfrac{3}{2}$ and -1

29. Use the discriminant to determine the number of x-intercepts of the parabola $y = -2x^2 - x - 2$.
    no x-intercepts

30. Solve: $3x^2 = 2x + 2$
    $\dfrac{1 + \sqrt{7}}{3}$ and $\dfrac{1 - \sqrt{7}}{3}$

31. The length of a rectangle is two feet more than twice the width. The area of the rectangle is 40 ft². Find the length and width of the rectangle.
    length: 10 ft; width: 4 ft

32. The height of a triangle is twice the length of the base. The area of the triangle is 36 m². Find the length of the base of the triangle.
    6 m

33. The rate of a boat in calm water is 10 mph. Traveling with the current, the boat travels 28 mi in 4 h less time than is required to travel 36 mi against the current. Find the rate of the current.
    4 mph

34. It took a motorboat two hours more to travel 16 mi against the current than it did to travel 16 mi with the current. The rate of the current was 2 mph. Find the rate of the motorboat in calm water.
    6 mph

CUMULATIVE REVIEW

1. Evaluate $2a^2 - b^2 \div c^2$ when $a = 3$, $b = -4$, and $c = -2$.
   14

2. Graph the solution set of $\{x \mid x > -3\} \cap \{x \mid x \le -1\}$.

3. Solve: $\dfrac{2x - 3}{4} - \dfrac{x + 4}{6} = \dfrac{3x - 2}{8}$
   -28

4. Solve: $2 - 4(1 - x) < 3(2x - 1) + 5$
   $\{x \mid x > -2\}$

5. Solve: $|3 - 4x| = 7$ -1 and $\dfrac{5}{2}$

6. Find the slope of the line containing the points $(3, -4)$ and $(-1, 2)$. $-\dfrac{3}{2}$

7. Find the equation of the line containing the point $(1, 2)$ and parallel to the line $x - y = 1$.
    $y = x + 1$

8. Solve: $x^2 + 5x + 6 < 0$
    $\{x \mid -3 < x < -2\}$

9. Factor: $-3x^3y + 6x^2y^2 - 9xy^3$
    $-3xy(x^2 - 2xy + 3y^2)$

10. Factor: $6x^2 - 7x - 20$
    $(2x - 5)(3x + 4)$

11. Factor: $a^nx + a^ny - 2x - 2y$
    $(a^n - 2)(x + y)$

12. Solve: $\dfrac{2x - 3}{x + 2} \le 1$ $\{x \mid -2 < x \le 5\}$

13. Simplify: $(3x^3 - 13x^2 + 10) \div (3x - 4)$
    $x^2 - 3x - 4 - \dfrac{6}{3x - 4}$

14. Simplify: $\dfrac{x^2 + 2x + 1}{8x^2 + 8x} \cdot \dfrac{4x^3 - 4x^2}{x^2 - 1}$
    $\dfrac{x}{2}$

15. Solve: $\dfrac{x}{2x + 3} - \dfrac{3}{4x^2 - 9} = \dfrac{x}{2x - 3}$ $-\dfrac{1}{2}$

16. Solve: $S = \dfrac{n}{2}(a + b)$ for b. $b = \dfrac{2S - an}{n}$

17. Simplify: $-2i(7 - 4i)$
    $-8 - 14i$

18. Simplify: $a^{-\frac{1}{2}}(a^{\frac{1}{2}} - a^{\frac{3}{2}})$
    $1 - a$

19. Simplify: $\dfrac{\sqrt[3]{8x^4y^5}}{\sqrt[3]{16xy^6}}$
    $\dfrac{x\sqrt[3]{4y^2}}{2y}$

20. Graph $3x + 2y > 8$.

21. Solve: $3x^2 + 8x = 16$ $\dfrac{4}{3}$ and -4

22. Solve: $x^2 + 4x + 9 = 0$
    $-2 + i\sqrt{5}$ and $-2 - i\sqrt{5}$

23. Use the discriminant to determine whether $3x^2 - 4x = 2$ has one real number solution, two real number solutions, or two complex number solutions.
    two real

24. Solve: $x^4 + 4x^2 - 32 = 0$
    $2, -2, -2i\sqrt{2}, 2i\sqrt{2}$

25. Solve: $\sqrt{5x + 1} - 1 = x$
    0 and 3

26. Solve: $\dfrac{x}{x + 3} + \dfrac{3x}{x + 8} = 1$
    2 and -4

27. Graph $y = -x^2 + 3x + 1$.

28. Find the x-intercepts of the parabola $y = -3x^2 + x + 2$.
    $(1, 0)$ and $\left(-\dfrac{2}{3}, 0\right)$

29. How many pounds of a 20% aluminum alloy must be mixed with 1000 lb of a 75% aluminum alloy to make a 40% aluminum alloy?
    1750 lb

30. The length of a rectangle is $(4x - 3)$ ft. The width is $(2x + 1)$ ft. Find the area of the rectangle in terms of the variable x.
    $(8x^2 - 2x - 3)$ ft^2

31. The inlet pipe can fill a water tank in 40 min. The outlet pipe can empty the tank in 25 min. How long would it take to empty a full tank when both pipes are open?
    $66\frac{2}{3}$ min

32. A driver travels 300 mi. A second driver, traveling 10 mph faster than the first driver, makes the same trip in one hour less time. At what speed did the second driver travel?
    60 mph

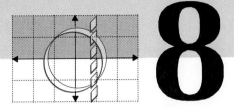

8

Functions and Relations

OBJECTIVES

- Evaluate functions
- Find the domain and range of a function
- Graph functions
- Determine whether or not a relation is a function
- Determine from a graph whether a function is one-to-one
- Composite functions
- Inverse functions
- Variation problems

History of Equal Signs

A portion of a page of the first book that used an equals sign, =, is shown below. This book was written in 1557 by Robert Recorde and was titled *The Whetstone of Witte.*

> Howbeit, for eafic alteratiõ of *equations.* I will propounde a fewe exãples, bicaufe the extraction of their rootes, maie the more aptly bee wroughte. And to auoide the tedioufe repetition of thefe woordes : is equalle to : I will fette as I doe often in woorke bfe, a paire of paralleles, or Gemowe lines of one lengthe, thus: ——————, bicaufe noe. 2. thynges, can be moare equalle.   And now marke thefe nombers.

1. $14.ze.--|--.15.9=====71.9.$

2. $20.ze.-------18.9====.102.9.$

3. $26.3--|--10ze.====9.3-----10ze--|--213.9.$

4. $19.ze--|--192.9.====103--|--1089----19ze$

5. $18.ze--|--24.9.====8.3.--|--2.ze.$

6. $343-------12ze====40ze--|--4809---9.3$

Notice in the illustration the words "bicause noe 2 thyngs can be moare equalle." Recorde decided that two things could not be more equal than two parallel lines of the same length. Therefore, it made sense to use this symbol to show equality.

This page also illustrates the use of the plus sign, +, and the minus sign, −. These symbols had been widely used only for about 100 years when this book was written.

8.1

Introduction to Functions and Relations

1 Evaluate functions

There are many situations in science, business, and mathematics where a correspondence exists between two quantities. The correspondence can be described by a formula, in a table, or in a graph, and then recorded as ordered pairs.

The formula $d = 16t^2$ describes a correspondence between the distance (d) a rock will fall and the time (t) of its fall. For each value of t, the formula assigns only one value for the distance. The ordered pair $(2, 64)$ indicates that in 2 seconds a rock will fall 64 feet. Some of the other ordered pairs determined by the correspondence are shown at the right.

Time in seconds ↓	Distance in feet ↓
(1, 16)	
(3, 144)	
(4, 256)	
(5, 400)	

The table at the right describes a grading scale that defines a correspondence between a percent score and a letter grade. For any percent score, the table assigns only one letter grade. The ordered pair $(86, B)$ indicates that a score of 86% receives a letter grade of B.

Score	Grade
90–100	A
80–89	B
70–79	C
60–69	D
0–59	F

The graph at the right defines a correspondence between the depth of a scuba diver and the pressure on the diver. For each depth, the graph assigns only one pressure. The ordered pair $(10, 20)$ indicates that when at a depth of 10 feet, the pressure is 20 pounds per square inch.

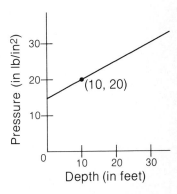

In each of these examples, a correspondence or a rule determines a set of ordered pairs. The set of ordered pairs determined by the correspondence is called a function.

> ### Definition of a Function
>
> A **function** is a set of ordered pairs in which no two ordered pairs that have the same first component have different second components.

For example, the ordered pair $(2, 64)$ was determined by the formula $d = 16t^2$. The first component, 2, can only have a second component of 64; no other second component can be paired with this first component. There is exactly one value of d that can be paired with a value of t.

The ordered pairs of a function can have different first components paired with the same second component. For example, the ordered pairs determined by the grading scale shown on the previous page include $(80, B)$, $(83, B)$, and $(87, B)$. The fact that 80, or any number between 80 and 89, cannot be paired with any letter other than B is the condition that makes this correspondence a function.

Not every correspondence between two sets is a function. Consider the correspondence that assigns to a positive real number a square root of that number. This is not a function because each positive real number can be paired with the positive or the negative square root of that number. For example, 9 can be paired with 3 or with -3. Thus the set of ordered pairs would contain $(9, 3)$ and $(9, -3)$. But the definition of function requires that no two ordered pairs with the same first component have different second components.

A **relation** assigns to each member of a first set one or more members of a second set.

The correspondence that pairs a positive real number with a square root of that number is a relation. In general, a relation is <u>any</u> set of ordered pairs. A function is a special kind of relation.

Although a function can always be described in terms of ordered pairs, frequently functions are described by an equation. The letter f is commonly used to represent a function, but any letter can be used.

The "square" function assigns to each real number its square. The square function is described by the equation

$$f(x) = x^2. \qquad \text{Read } f(x) \text{ as "}f\text{ of }x\text{" or "the value of } f \text{ at } x\text{."}$$

$f(x)$ is the symbol for the number that is paired with x. In terms of ordered pairs, this is written $(x, f(x))$.

It is important to remember that $f(x)$ does not mean f times x. The letter f stands for the function, and $f(x)$ is the number that is paired with x.

To **evaluate a function** means to find the number that is paired with a given number. To evaluate the square function $f(x) = x^2$ at 4 means to find the number that is paired with 4.

The notation $f(4)$ is used to indicate the number that is paired with 4. To evaluate the function $f(x) = x^2$ at 4, replace x by 4 and simplify.

$$f(x) = x^2$$
$$f(4) = 4^2$$
$$f(4) = 16$$

The **value** of the function at 4 is 16.

Example 1   Evaluate the function $f(x) = 2x + 1$ at $x = 3$.

Solution $f(x) = 2x + 1$ • $f(3)$ is the number that is paired with 3. Replace x by 3 and
$f(3) = 2(3) + 1$ simplify.
$f(3) = 7$

Problem 1   Evaluate the function $s(t) = 2t^2 + 3t - 4$ at $t = -3$.

Solution   See page A64.
$s(-3) = 5$

Example 2   Evaluate the function $f(x) = x^2$ at $x = a + h$.

Solution $f(x) = x^2$
$f(a + h) = (a + h)^2$ • Replace x by its value.
$f(a + h) = a^2 + 2ah + h^2$ • Simplify.

Problem 2   Evaluate the function $f(x) = x^2$ at $x = a + 1$.

Solution   See page A64.
$f(a + 1) = a^2 + 2a + 1$

The basic operations of addition and subtraction can be performed on functions.

Example 3   For $f(x) = x^2$ and $g(x) = 3x$, evaluate $f(x) + g(x)$ at $x = 5$.

Solution $f(x) + g(x) = x^2 + 3x$
$f(5) + g(5) = 5^2 + 3(5)$ • Replace x in each function by its value.
$f(5) + g(5) = 25 + 15$ • Simplify.
$f(5) + g(5) = 40$

Problem 3   For $f(x) = 2x^2$ and $g(x) = 4x - 1$, evaluate $f(x) - g(x)$ at $x = -1$.

Solution   See page A64.
$f(-1) - g(-1) = 7$

2   Find the domain and range of a function

The basic concept of a function is a set of ordered pairs. The ordered pairs may be formed by using a correspondence or rule that pairs a member of a first set with only one member of a second set. The first set is the **domain** of the function. The second set is the **range** of the function.

In the grading scale shown at the right, the correspondence pairs a percent score with one of the letters A, B, C, D, or F.

The domain of the grading scale is the percent scores 0 to 100.

Score	Grade
90–100	A
80–89	B
70–79	C
60–69	D
0–59	F

The range of the grading scale is the set of letters A, B, C, D, and F.

Example 4   Find the domain and range of the function $\{(1, 0), (2, 3), (3, 8), (4, 15)\}$.

Solution   The domain is $\{1, 2, 3, 4\}$.   • The domain of the function is the set of the first components in the ordered pairs.

The range is $\{0, 3, 8, 15\}$.   • The range of the function is the set of the second components in the ordered pairs.

Problem 4   Find the domain and range of the function $\{(0, 1), (1, 3), (2, 5), (3, 7), (4, 9)\}$.

Solution   See page A64.
domain: $\{0, 1, 2, 3, 4\}$; range: $\{1, 3, 5, 7, 9\}$

When a function is described by an equation, the range of a function can be found by evaluating the function at each point of the domain.

Example 5   Find the range of the function $f(x) = 2x - 3$ if the domain is $\{0, 1, 2, 3\}$.

Solution $f(x) = 2x - 3$
$f(0) = 2(0) - 3 = -3$ • Replace x by each member of the domain. The range
$f(1) = 2(1) - 3 = -1$ includes the values of $f(0)$, $f(1)$, $f(2)$, and $f(3)$.
$f(2) = 2(2) - 3 = 1$
$f(3) = 2(3) - 3 = 3$

The range is $\{-3, -1, 1, 3\}$.

Problem 5 Find the range of the function $f(x) = x^2 - 2x + 1$ if the domain is $\{-2, -1, 0, 1, 2\}$.

Solution See page A64.
$\{0, 1, 4, 9\}$

Given an element a in the range of a function, it is possible to find an element in the domain that corresponds to a.

The number 3 is in the range of the function $f(x) = x^2 - 1$. Find an element in the domain that corresponds to 3 and write an ordered pair that belongs to the function.

Because 3 is in the range, $f(x) = 3$.
Replace $f(x)$ by $x^2 - 1$.
Solve for x.

$$f(x) = 3$$
$$x^2 - 1 = 3$$
$$x^2 = 4$$
$$x = \pm 2$$

There are two values in the domain that can be paired with the range element 3. The two values are 2 and -2. Two ordered pairs that belong to the function are $(2, 3)$ and $(-2, 3)$. *Remember:* A function can have different first elements paired with the same second element. A function cannot have the same first element paired with different second elements.

Example 6 The number -2 is in the range of the function $f(x) = 3x + 1$. Find an element in the domain that corresponds to -2, and write an ordered pair that belongs to the function.

Solution
$$f(x) = -2$$
$$3x + 1 = -2$$
$$3x = -3$$
$$x = -1$$
$$(-1, -2)$$

• Because -2 is in the range, $f(x) = -2$.
• Replace $f(x)$ by $3x + 1$.
• Solve for x.
• The element in the domain that corresponds to -2 is -1.
• The ordered pair $(-1, -2)$ belongs to the function.

Problem 6 The number -4 is in the range of the function $f(x) = 2x - 1$. Find an element in the domain that corresponds to -4, and write an ordered pair that belongs to the function.

Solution See page A64.
$$-\frac{3}{2}; \left(-\frac{3}{2}, -4\right)$$

When an equation is used to define a function and the domain is not stated, the domain is the set of real numbers for which the equation produces real numbers. For example:

For all real numbers x, $f(x) = x^2 + 5$ is a real number. The domain of f is the set of real numbers.

The domain of the function $g(x) = \dfrac{1}{x - 4}$ is all real numbers except 4. When $x = 4$, $g(x)$ is undefined.

The domain of the function $h(x) = \sqrt{x + 3}$ is all real numbers greater than or equal to -3. When x is less than -3, $h(x)$ is not a real number. For example, $h(-5) = \sqrt{-2}$, which is not a real number.

> ### Domain of a Function
>
> Unless otherwise stated, the domain of a function is all real numbers except:
> (a) those values for which the denominator of the function is zero; or
> (b) those values for which the value of the function is not a real number.

Example 7   What values are excluded from the domain of the function $f(x) = \dfrac{5}{x^2 - 6x}$?

Solution $x^2 - 6x = 0$ • Find all values of x for which the denominator equals zero.

$x(x - 6) = 0$

$x = 0 \qquad x - 6 = 0$
$\qquad\qquad\quad\ x = 6$

0 and 6 are excluded from the domain of the function.

Problem 7   What values are excluded from the domain of the function $g(x) = \sqrt{2x - 5}$?

Solution   See page A65. $\left\{ x \mid x < \dfrac{5}{2} \right\}$

EXERCISES 8.1

1   For the function $f(x) = 3x^2$, find:

1. $f(2)$
    12

2. $f(1)$
    3

3. $f(-1)$
    3

4. $f(-2)$
    12

5. $f(a)$
    $3a^2$

6. $f(w)$
    $3w^2$

For the function $g(x) = x^2 - x + 1$, find:

7. $g(0)$
    1

8. $g(1)$
    1

9. $g(-2)$
    7

10. $g(-1)$
3

11. $g(t)$
$t^2 - t + 1$

12. $g(s)$
$s^2 - s + 1$

For the function $f(x) = 4x - 3$, find:

13. $f(-2)$
-11

14. $f(3)$
9

15. $f(2 + h)$
$4h + 5$

16. $f(3 + h)$
$4h + 9$

17. $f(1 + h) - f(1)$
$4h$

18. $f(-1 + h) - f(-1)$
$4h$

For the function $g(x) = x^2 - 1$, find:

19. $g(2)$
3

20. $g(-3)$
8

21. $g(1 + h)$
$h^2 + 2h$

22. $g(2 + h)$
$h^2 + 4h + 3$

23. $g(3 + h) - g(3)$
$h^2 + 6h$

24. $g(-1 + h) - g(-1)$
$h^2 - 2h$

For $f(x) = 2x^2 - 3$ and $g(x) = -2x + 3$, find:

25. $f(2) - g(2)$
6

26. $f(3) - g(3)$
18

27. $f(0) + g(0)$
0

28. $f(1) + g(1)$
0

29. $f(1 + h) - f(1)$
$2h^2 + 4h$

30. $f(2 + h) - f(2)$
$2h^2 + 8h$

31. $\dfrac{g(1 + h) - g(1)}{h}$
-2

32. $\dfrac{g(-2 + h) - g(-2)}{h}$
-2

33. $\dfrac{f(3 + h) - f(3)}{h}$
$2h + 12$

34. $\dfrac{f(-1 + h) - f(-1)}{h}$
$2h - 4$

35. $\dfrac{g(a + h) - g(a)}{h}$
-2

36. $\dfrac{f(a + h) - f(a)}{h}$
$4a + 2h$

2   Find the domain and range of the function.

37. $\{(1, 1), (2, 4), (3, 7), (4, 10), (5, 13)\}$ D: $\{1, 2, 3, 4, 5\}$; R: $\{1, 4, 7, 10, 13\}$

38. $\{(2, 6), (4, 18), (6, 38), (8, 66), (10, 102)\}$ D: $\{2, 4, 6, 8, 10\}$; R: $\{6, 18, 38, 66, 102\}$

39. $\{(0, 1), (2, 2), (4, 3), (6, 4)\}$ D: $\{0, 2, 4, 6\}$; R: $\{1, 2, 3, 4\}$

40. $\{(0, 1), (1, 2), (4, 3), (9, 4)\}$ D: $\{0, 1, 4, 9\}$; R: $\{1, 2, 3, 4\}$

41. $\{(1, 0), (3, 0), (5, 0), (7, 0), (9, 0)\}$ D $= \{1, 3, 5, 7, 9\}$; R $= \{0\}$

42. $\{(-2, -4), (2, 4), (-1, 1), (1, 1), (-3, 9), (3, 9)\}$ D: $\{-3, -2, -1, 1, 2, 3\}$; R: $\{-4, 1, 4, 9\}$

43. $\{(0, 0), (1, 1), (-1, 1), (2, 2), (-2, 2)\}$ D: $\{-2, -1, 0, 1, 2\}$; R: $\{0, 1, 2\}$

44. $\{(0, -5), (5, 0), (10, 5), (15, 10)\}$ D: $\{0, 5, 10, 15\}$; R: $\{-5, 0, 5, 10\}$

Find the range of the function.

45. $f(x) = 4x - 3$; domain = $\{0, 1, 2, 3, 4\}$ $\{-3, 1, 5, 9, 13\}$

46. $g(x) = x^2 + 2x - 1$; domain = $\{-2, -1, 0, 1, 2\}$ $\{-2, -1, 2, 7\}$

47. $h(x) = \dfrac{x}{2} + 3$; domain = $\{-4, -2, 0, 2, 4\}$ $\{1, 2, 3, 4, 5\}$

48. $F(x) = \sqrt{x + 1}$; domain = $\{-1, 0, 3, 8, 15\}$ $\{0, 1, 2, 3, 4\}$

49. $G(x) = \dfrac{2}{x + 3}$; domain = $\{-2, -1, 0, 1, 2\}$ $\left\{\dfrac{2}{5}, \dfrac{1}{2}, \dfrac{2}{3}, 1, 2\right\}$

50. $f(x) = |5x - 2|$; domain = $\{-10, -5, 0, 5, 10\}$ $\{2, 23, 27, 48, 52\}$

51. $g(a) = \dfrac{a^2 + 1}{3a - 1}$; domain = $\{-1, 0, 1, 2\}$ $\left\{-1, -\dfrac{1}{2}, 1\right\}$

52. $h(a) = (a^2 + 3a)^2$; domain = $\{-2, -1, 0, 1, 2\}$ $\{0, 4, 16, 100\}$

In the exercises below, a function and a number in the range of that function are given. Find an element in the domain that corresponds to the number, and write an ordered pair that belongs to the function.

53. $f(x) = x + 5$; -3
  -8; $(-8, -3)$

54. $g(x) = x - 4$; 6
  10; $(10, 6)$

55. $h(a) = 3a + 2$; -1

  -1; $(-1, -1)$

56. $f(a) = 2a - 5$; 0
  $\dfrac{5}{2}$; $\left(\dfrac{5}{2}, 0\right)$

57. $g(x) = \dfrac{2}{3}x + \dfrac{1}{3}$; 1
  1; $(1, 1)$

58. $h(x) = \dfrac{3}{2}x - 1$; -4
  -2; $(-2, -4)$

59. $h(x) = x^2 + 3$; 7
  ± 2; $(2, 7)$ or $(-2, 7)$

60. $g(x) = x^2 - 3$; -2
  ± 1; $(1, -2)$ or $(-1, -2)$

61. $f(x) = \dfrac{x + 1}{5}; 7$

34; (34,7)

62. $g(a) = \dfrac{a - 3}{4}; 5$

23; (23,5)

What values of x are excluded from the domain of the function?

63. $f(x) = x^2 + 3x - 4$

64. $g(x) = 2x^2 - x + 5$

65. $h(x) = \dfrac{2x}{3x^2 - x}$

$0, \dfrac{1}{3}$

66. $F(x) = \dfrac{6}{5x^2 - 2x}$

$0, \dfrac{2}{5}$

67. $G(x) = \dfrac{x + 1}{x^2 + x - 6}$

$-3, 2$

68. $f(x) = \dfrac{x - 2}{x^2 - 3x - 4}$

$4, -1$

69. $g(x) = \sqrt{6x - 2}$

$\left\{ x \mid x < \dfrac{1}{3} \right\}$

70. $h(x) = \sqrt{3x - 9}$

$\{ x \mid x < 3 \}$

71. $F(x) = |3x - 7|$

72. $G(x) = \dfrac{x}{|x|}$

0

SUPPLEMENTAL EXERCISES 8.1

Solve.

73. Let $f(x)$ be the digit in the xth decimal place of the repeating digit $0.\overline{387}$. For example, $f(1) = 3$ because 3 is the digit in the first decimal place. Find $f(14)$.

$f(14) = 8$

74. Let $f(x)$ be the digit in the xth decimal place of the repeating digit $0.\overline{018}$. For example, $f(1) = 0$ because 0 is the digit in the first decimal place. Find $f(21)$.

$f(21) = 8$

75. If $f(x) = \sqrt{x} - 2$ and $f(a) = 4$, find a.

$a = 18$

76. If $f(x) = \sqrt{x + 5}$, and $f(a) = 3$, find a.

$a = 4$

77. If $f(a) = a^2 + a$ and $g(b) = b^2$, find $f(3) - g(3)$.

$f(3) - g(3) = 3$

78. If $f(a) = 2a^2 + 2$ and $g(b) = 3b - 1$, find $f(2) - g(2)$.

$f(2) - g(2) = 5$

79. Given $f(x) = [x]$, where $[x]$ means the greatest integer less than or equal to x, find $f(3\pi)$.
    $f(3\pi) = 9$

80. Given $f(x) = [x]$, where $[x]$ means the greatest integer less than or equal to x, find $f(2\sqrt{3})$.
    $f(2\sqrt{3}) = 3$

81. Given $f(x) = (x + 1)(x - 1)$, for what values of x is $f(x)$ negative? Write your answer in set builder notation.
    $\{x \mid -1 < x < 1\}$

82. Given $f(x) = (x + 2)(x - 2)$, for what values of x is $f(x)$ negative? Write your answer in set builder notation.
    $\{x \mid -2 < x < 2\}$

83. Given $f(x) = -|x + 3|$, for what value of x is $f(x)$ maximum?
    $x = -3$

84. Given $f(x) = -|2x - 2|$, for what value of x is $f(x)$ maximum?
    $x = 1$

85. $f(a, b) =$ the sum of a and b
    $g(a, b) =$ the product of a and b
    Find $f(2, 5) + g(2, 5)$.
    $f(2,5) + g(2,5) = 17$

86. $f(a, b) =$ the greatest common divisor of a and b
    $g(a, b) =$ the least common multiple of a and b
    Find $f(14, 35) + g(14, 35)$.
    $f(14,35) + g(14,35) = 77$

87. Given $f(n) = n(n - 1) \cdots 3 \cdot 2 \cdot 1$, find $f(5) - f(4)$.
    $f(5) - f(4) = 96$

88. Given $f(n) = n(n - 1) \cdots 3 \cdot 2 \cdot 1$, find $f(6) - f(2)$.
    $f(6) - f(2) = 718$

89. The sale price of an item is a function, s, of the original price, p, where $s(p) = 0.80p$. If an item's original price is $200, what is the sale price of the item?
    $160

90. The markup on an item is a function, m, of its cost, c, where $m(c) = 0.25c$. If the cost of an item is $150, what is the markup on the item?
    $37.50

SECTION 8.2

Graphs of Functions

1 Graph functions

A function is a set of ordered pairs in which no two ordered pairs that have the same first component have different second components. The ordered pairs of a function can be written as $(x, f(x))$. However, often the value of the function, $f(x)$, is labeled y, and the ordered pairs are written (x, y), where $y = f(x)$. The **graph of a function** is a graph of the ordered pairs (x, y) of the function.

Since the graph of the equation $y = mx + b$ is a straight line, a function of the form $f(x) = mx + b$ is a **linear function.**

To graph the function $f(x) = 2x + 1$, think of the function as the equation $y = mx + b$.

This is the equation of a straight line. The slope is 2, and the y-intercept is 1.

Graph the straight line.

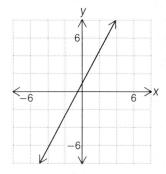

In the equation $y = 2x + 1$, the value of the variable y *depends* on the value of x. Thus y is called the **dependent variable,** and x is called the **independent variable.** In the function $f(x) = 2x + 1$, $f(x)$ is a symbol for the dependent variable.

In graphing functions, it is important to remember that $f(x)$ is the y-coordinate of an ordered pair.

The graph of the linear function $f(x) = x + 2$ is shown at the right.

When $x = -6$, $y = f(-6) = -4$.

When $x = 5$, $y = f(5) = 7$.

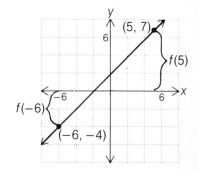

By associating the horizontal axis with the domain and the vertical axis with the range of a function, the graph of a function can be used to help determine its domain and range. For example, because the graph of the function $f(x) = x + 2$ extends infinitely in both directions, the domain is the real numbers, and the range is the real numbers. In general, for any linear function of the form $f(x) = mx + b$, $m \neq 0$, the domain is the real numbers, and the range is the real numbers.

To graph the quadratic function $f(x) = x^2 - 1$, think of the function as the equation $y = x^2 - 1$.

This is a quadratic equation in two variables; the graph is a parabola.

The coefficient of x^2 is positive; the graph will open up.

Find enough ordered pairs to draw the graph.

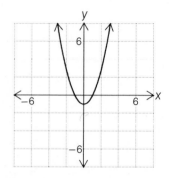

Because the function $f(x) = x^2 - 1$ is a real number for all values of x, the domain of the function is the real numbers. But note from the graph of this function that no portion of the parabola is below a y value of -1. For any value of x, $y \geq -1$. The range of this function is $\{y \mid y \geq -1\}$.

To graph the function $f(x) = -x^2 - 2x + 2$, think of the function as the equation $y = -x^2 - 2x + 2$.

This is a quadratic equation in two variables; the graph is a parabola.

The coefficient of x^2 is negative; the graph will open down.

Find enough ordered pairs to draw the graph.

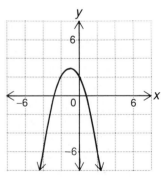

The domain of the function $f(x) = -x^2 - 2x + 2$ is all real numbers. To find the range of this function, complete the square on $-x^2 - 2x + 2$.

$$
\begin{aligned}
-x^2 - 2x + 2 &= -(x^2 + 2x) + 2 \\
&= -(x^2 + 2x + 1) + 1 + 2 \\
&= -(x + 1)^2 + 3
\end{aligned}
$$

Because $-(x + 1)^2 \leq 0$ for all x, $-(x + 1)^2 + 3 \leq 3$ for all values of x. The range of the function is $\{y \mid y \leq 3\}$.

To graph the function $f(x) = |x - 2|$, think of the function as the equation $y = |x - 2|$.

Find enough ordered pairs to draw the graph.

x	y
-1	3
0	2
1	1
2	0
3	1
4	2

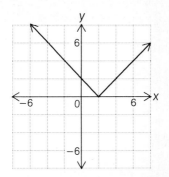

This function is an example of a linear **absolute value function.** The graphs of these functions are V shaped.

The domain of the function $f(x) = |x - 2|$ is all real numbers. The range is $\{y \mid y \geq 0\}$.

To graph the function $f(x) = x^3 + 1$, think of the function as the equation $y = x^3 + 1$.

Find enough ordered pairs to draw the graph.

x	y
-2	-7
-1	0
0	1
1	2
2	9

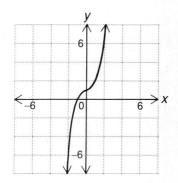

This function is an example of a **cubic function;** the degree of the polynomial $x^3 + 1$ is 3. The domain of the function $f(x) = x^3 + 1$ is all real numbers. The range of the function is all real numbers.

To graph the function $f(x) = 2$, think of the function as the equation $y = 2$.

The graph of an equation of the form $y = b$ is a horizontal line passing through the point $(0, b)$.

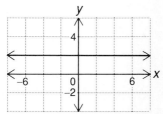

This function is an example of a **constant function;** for any value in the domain, the corresponding value of $f(x)$ is the constant b. The domain of the function $f(x) = 2$ is all real numbers. The range is $\{2\}$.

Example 1   Graph. State the domain and range of the function.

A. $f(x) = |x| + 2$ B. $f(x) = -x^3 - 2x^2 + 2$

Solution   A.

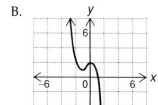

• This is an absolute value function. The graph is V shaped.

The domain is all real numbers.
The range is $\{y \mid y \geq 2\}$.

B.

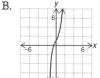

• This is a cubic function.
Some ordered pairs are $(-2, 2)$, $(-1, 1)$, $(0, 2)$, and $(1, -1)$.

The range is all real numbers. The domain is all real numbers.

Problem 1   Graph. State the domain and range of the function.

A. $f(x) = |x + 2|$ B. $f(x) = x^3 + 2x + 1$

Solution   See page A65.

A.

B.

D: all real numbers   D: all real numbers
R: $\{y \mid y \geq 0\}$ R: all real numbers

2 # Determine whether or not a relation is a function

As stated previously, not every correspondence between two sets is a function. A function assigns to each member of the domain one and <u>only one</u> member of the range. A relation assigns to each member of the domain <u>one or more</u> members of the range.

Is the relation that pairs a positive real number with its fourth root a function?

This is not a function because each positive real number can be paired with the positive or the negative fourth root of that number. For example, since 16 can be paired with 2 or −2, the ordered pairs would contain $(16, 2)$ and $(16, -2)$. By the definition of a function, no two ordered pairs with the same first component can have different second components.

Is the relation that pairs a number with its absolute value a function?

Since any real number has only one absolute value, the correspondence is a function. Some of the ordered pairs determined by the function are $(-3, 3)$, $(0, 0)$ and $(4, 4)$. This function can be described by the equation $f(x) = |x|$.

Example 2 Is the set of ordered pairs a function?

A. $\{(0, 0), (2, 0), (4, 0), (6, 0)\}$ B. $\{(0, 0), (0, 2), (0, 4), (0, 6)\}$

Solution A. Each member of the domain is paired with only one member of the range.
The relation is a function.

B. The number 0 in the domain is paired with different members of the range.
The relation is not a function.

Problem 2 Is the set of ordered pairs a function?

A. $\{(0, 0), (2, 2), (4, 0), (6, 2)\}$ B. $\{(1, 2), (2, 2), (3, 4), (4, 4)\}$

Solution See page A65.
A. yes   B. yes

The ordered pairs in Problem 2A are graphed at the right. The domain is associated with the horizontal axis, and the range is associated with the vertical axis.

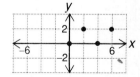

A function is a set of ordered pairs in which no two ordered pairs that have the same first component have different second components. Graphically, this means that a graph is the graph of a function if any vertical line intersects the graph at no more than one point. This is called the **vertical line test** for a function.

Problem 2A is a function, and a vertical line cannot intersect the graph of this function at more than one point.

The ordered pairs in Example 2B are graphed at the right. Because the ordered pairs have the same first component and different second components, Example 2B is not a function, and a vertical line can intersect the graph at more than one point. This is the graph of a relation.

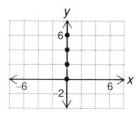

Example 3 Use the vertical line test to determine if the graph shown at the right is the graph of a function.

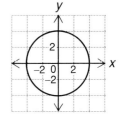

Solution

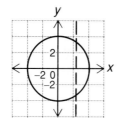

• A vertical line can intersect the graph at more than one point. Therefore, the relation includes ordered pairs with the same first component and different second components; for example, $(0, 4)$ and $(0, -4)$.

The graph is not the graph of a function.

Problem 3 Use the vertical line test to determine if the graph shown at the right is a function.

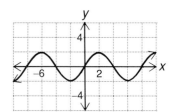

Solution See page A65.
yes

3 Determine from a graph whether a function is one-to-one

Recall that a function is a set of ordered pairs in which no two ordered pairs that have the same first component have different second components. This means that given any x there is only one y that can be paired with that x. A **one-to-one function** satisfies the additional condition that given any y, there is only one x that can be paired with the given y. One-to-one functions are commonly expressed by writing 1–1.

The function $f(x) = x^2$ is not a 1–1 function since, given $y = 4$, there are two possible values of x, 2 and -2, which can be paired with the given y value.

Just as the vertical line test can be used to determine if a graph represents a function, a **horizontal line test** can be used to determine if the graph of a function represents a 1–1 function. The graph of a function represents a 1–1 function if any horizontal line intersects the graph at no more than one point.

The graph of a quadratic function is shown at the right. Note that a horizontal line would intersect the graph at more than one point. Therefore, this function is not a 1–1 function. In general, the function $f(x) = ax^2 + bx + c$, $a \neq 0$, is not a 1–1 function.

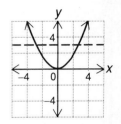

Since any vertical line will intersect the graph at the right at no more than one point, the graph is the graph of a function. Since any horizontal line will intersect the graph at no more than one point, the graph is the graph of a 1–1 function.

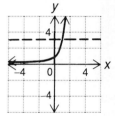

Example 4   Determine if the graph represents the graph of a 1–1 function.

A.

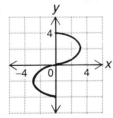

B.

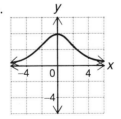

Solution   A.

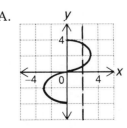

• A vertical line would intersect the graph at more than one point.
The graph does not represent a function.

It is not the graph of a 1–1 function.

B.

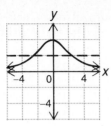

• A horizontal line would intersect the curve at more than one point.

It is not the graph of a 1–1 function.

Problem 4   Determine if the graph represents the graph of a 1–1 function.

A.

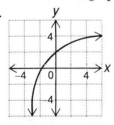

B.

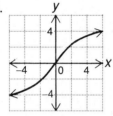

Solution   See page A66.

A. yes   B. yes

EXERCISES 8.2

1   Graph. State the domain and range of the function.

1. $f(x) = 2x - 1$
D: all reals; R: all reals

2. $f(x) = 4$
D: all reals; R: $y = 4$

3. $f(x) = 2x^2 - 1$
D: all reals; R: $y \geq -1$

4. $f(x) = 2x^2 - 3$
D: all reals; R: $y \geq -3$

5. $f(x) = |x - 1|$
D: all reals; R: $y \geq 0$

6. $f(x) = |x - 3|$
D: all reals; R: $y \geq 0$

7. $f(x) = x^3 - 1$
D: all reals; R: all reals

8. $f(x) = -x^3 + 1$
D: all reals; R: all reals

9. $f(x) = -3x - 1$

D: all reals; R: all reals

10. $f(x) = -\frac{1}{2}x + 1$

D: all reals; R: all reals

11. $f(x) = \frac{1}{2}x^2$

D: all reals; R: $y \geq 0$

12. $f(x) = \frac{1}{3}x^2$

D: all reals; R: $y \geq 0$

13. $f(x) = |x| + 1$
D: all reals; R: $y \geq 1$

14. $f(x) = |x| - 1$
D: all reals; R: $y \geq -1$

15. $f(x) = x^3 + 2x^2$
D: all reals; R: all reals

16. $f(x) = x^3 - 3x^2$
D: all reals; R: all reals

17. $f(x) = \frac{2}{3}x + 4$

D: all reals; R: all reals

18. $f(x) = \frac{3}{4}x + 1$

D: all reals; R: all reals

19. $f(x) = 2x^2 - 4x - 3$
D: all reals; R: $y \geq -5$

20. $f(x) = -1$
D: all reals; R: $y = -1$

21. $f(x) = 2|x| - 1$
D: all reals; R: $y \geq -1$

22. $f(x) = 2|x| + 2$
D: all reals; R: $y \geq 2$

23. $f(x) = 2x^3 + 3x$
D: all reals; R: all reals

24. $f(x) = -2x^3 + 4x$
D: all reals; R: all reals

25. $f(x) = -\dfrac{1}{2}x - 2$
D: all reals; R: all reals

26. $f(x) = -3$
D: all reals; R: $y = -3$

27. $f(x) = x^2 + 2x - 4$
D: all reals; R: $y \geq -5$

28. $f(x) = -x^2 + 2x - 1$
D: all reals; R: $y \leq 0$

29. $f(x) = -|x|$
D: all reals; R: $y \leq 0$

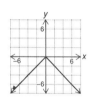

30. $f(x) = -|x + 2|$
D: all reals; R: $y \leq 0$

31. $f(x) = x^3 - x^2 + x - 1$
D: all reals; R: all reals

32. $f(x) = x^3 + x^2 - x + 1$
D: all reals; R: all reals

2   Is the set of ordered pairs a function?

33. $\{(0, 0), (1, 0), (2, 0), (3, 0)\}$
yes

34. $\{(1, 1), (1, 2), (1, 3), (1, 4)\}$
no

35. $\{(-1, 1), (0, 1), (1, 1), (2, 1)\}$
yes

36. $\{(2, 2), (4, 4), (6, 6), (8, 8)\}$
yes

37. $\{(-3, 3), (-2, 2), (-1, 1), (-2, -2)\}$
no

38. $\{(-5, -5), (5, 5), (-5, 5), (5, -5)\}$
no

39. $\{(1, 4), (2, 3), (3, 2), (4, 1)\}$
yes

40. $\{(-4, -1), (-3, -2), (-2, -3), (-1, -4)\}$
yes

Use the vertical line test to determine if the graph is the graph of a function.

41.
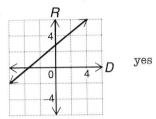
yes

42.
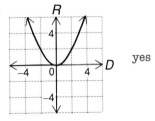
yes

43.
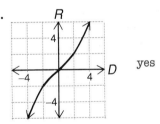
no

44.
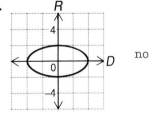
yes

45.
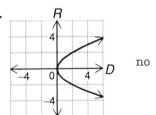
yes

46.
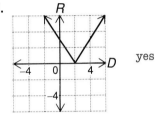
no

3   Determine if the graph represents the graph of a 1–1 function.

47.
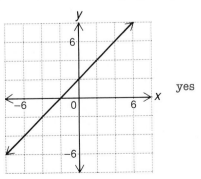
yes

48.
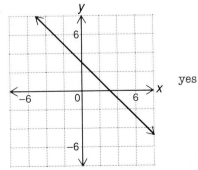
yes

49.

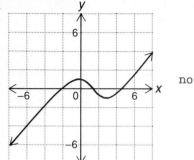

no

50.

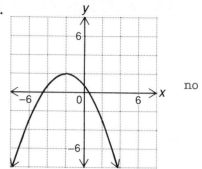

no

51.

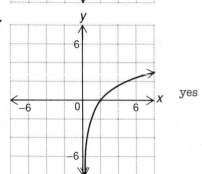

yes

52.

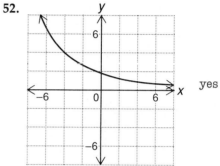

yes

53.

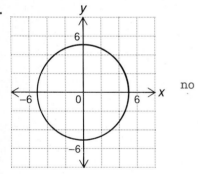

no

54.

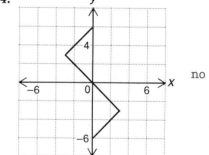

no

SUPPLEMENTAL EXERCISES 8.2

Name (**a**) the dependent variable and (**b**) the independent variable in each equation.

55. $C = \pi d$
    a. C b. d

56. $C = 2\pi r$
    a. C b. r

57. $A = \pi r^2$
    a. A b. r

58. $A = \frac{4}{3}\pi r^3$
    a. A b. r

59. $d = 16t^2$
    a. d b. t

60. $A = s^2$
    a. A b. s

61. $F = \dfrac{9}{5}C + 32$

    a. F   b. C

62. $C = \dfrac{5}{9}F - 32$

    a. C   b. F

63. $y = 3x - 4$

    a. y   b. x

64. $y = x^2 + 3x + 1$

    a. y   b. x

65. $y = 3x + 1$

    a. y   b. x

66. $y = |5x|$

    a. y   b. x

Which of the following relations are not functions?

67. **a.** $f(x) = x$ **b.** $f(x) = \left|\dfrac{x}{2}\right|$ **c.** $\{(3, 1), (1, 3), (3, 0), (0, 3)\}$

    c

68. **a.** $f(x) = -x$ **b.** $f(x) = \dfrac{2}{\sqrt{x}}$ **c.** $\{(1, 4), (4, 1), (1, -4), (-4, 1)\}$

    c

Solve.

69. A linear function includes the ordered pairs $(2, 4)$ and $(4, 10)$. Find $f(-1)$.

    $f(-1) = -5$

70. A linear function includes the ordered pairs $(2, 3)$ and $(4, 1)$. Find $f(7)$.

    $f(7) = -2$

71. For every increase of one unit in the value of x, the value of the linear function $f(x)$ increases by 4 units. Given $f(3) = 8$, find $f(0)$.

    $f(0) = -4$

72. For every increase of one unit in the value of x, the value of the linear function $f(x)$ increases by 6 units. Given $f(4) = 21$, find $f(0)$.

    $f(0) = -3$

73. f is a linear function with $f(2) = 0$ and $f(-1) = 0$. Find $f(1)$.

    $f(1) = 0$

74. f is a linear function with $f(1) = 2$ and $f(-1) = 8$. Find $f(2)$.

    $f(2) = -1$

75. f is a constant function with $f(3) = 5$. Find $f(-1)$.

    $f(-1) = 5$

76. f is a constant function with $f(1) = 4$. Find $f(6)$.

    $f(6) = 4$

77. **a.** For what real value of x does $x^3 = 0$?
    b. What is the x-intercept of the graph of $f(x) = x^3$?

      a. $x = 0$ b. $(0,0)$

78. **a.** For what real value of x does $x^3 - 1 = 0$?
    b. What is the x-intercept of the graph of $f(x) = x^3 - 1$?

      a. $x = 1$ b. $(1,0)$

79. **a.** For what real value of x does $x^3 + 1 = 0$?
     b. What is the x-intercept of the graph of $f(x) = x^3 + 1$?
        a. $x = -1$ b. $(-1, 0)$

80. **a.** For what real value of x does $x^3 - 8 = 0$?
     b. What is the x-intercept of the graph of $f(x) = x^3 - 8$?
        a. $x = 2$ b. $(2, 0)$

81. Find a solution of the equation $x^3 + 2x^2 - x - 2 = 0$. (*Hint:* Graph the function $f(x) = x^3 + 2x^2 - x - 2$, and locate the x-intercept.)
     $(1, 0)$, $(-1, 0)$, and $(-2, 0)$

82. Find a solution of the equation $x^3 - 3x - 2 - 0$. (*Hint:* Graph the function $f(x) = x^3 - 3x - 2$, and locate the x-intercept.)
     $(-1, 0)$ and $(2, 0)$

SECTION 8.3

Composite Functions and Inverse Functions

1 Composite functions

Recall from Section 8.1 that to evaluate a function, replace the variable by the given value and then simplify. For example, to evaluate the function $f(x) = x^2 - 3x$ at 5, replace x by 5 and then simplify.

$$f(x) = x^2 - 3x$$
$$f(5) = 5^2 - 3(5) = 25 - 15 = 10$$

The value of the function at 5 is 10.

A function can be evaluated at the value of another function. For example, consider the functions $f(x) = 2x + 7$ and $g(x) = x - 3$. The expression $f(g(x))$ means to evaluate the function f at $g(x)$. Replace x by $g(x)$ and then simplify.

$$f(x) = 2x + 7$$
$$f(g(x)) = 2(g(x)) + 7 = 2(x - 3) + 7 = 2x + 1$$

The function $f(g(x))$ is called the **composition of the functions f and g.** The function that results from the composition of two functions is called the **composite function.**

The composition of the functions g and f is written $g(f(x))$.

$$g(x) = x - 3$$
$$g(f(x)) = f(x) - 3 = 2x + 7 - 3 = 2x + 4$$

It is important to note from the example above that $f(g(x)) \neq g(f(x))$.

Consider the functions

$$f(x) = x^2 - 2x \quad \text{and} \quad g(x) = 3x - 2.$$

To evaluate $f(g(2))$, first evaluate $g(2)$.

$$g(x) = 3x - 2$$
$$g(2) = 3(2) - 2 = 4$$

$g(2) = 4$. Evaluate $f(x)$ at 4.

$$f(x) = x^2 - 2x$$
$$f(4) = 4^2 - 2(4) = 16 - 8 = 8$$

Thus $f(g(2)) = 8$.

Example 1   Given $f(x) = x^2 - 1$ and $g(x) = 3x + 4$, evaluate the composite functions.
A. $f(g(0))$ B. $g(f(x))$

Solution   A. $g(x) = 3x + 4$ • To evaluate $f(g(0))$, first evaluate $g(0)$.
$\ g(0) = 3(0) + 4 = 4$

$\ f(x) = x^2 - 1$ • Substitute the value of $g(0)$ for x in $f(x)$.
$\ f(4) = 4^2 - 1 = 15$ $\quad g(0) = 4$

$\ f(g(0)) = 15$

B. $g(f(x)) = g(x^2 - 1)$ • $f(x) = x^2 - 1$
$ = 3(x^2 - 1) + 4$ • Substitute $x^2 - 1$ for x in the function $g(x)$.
$\qquad\qquad\qquad\qquad\qquad\qquad g(x) = 3x + 4$

$ = 3x^2 - 3 + 4$
$ = 3x^2 + 1$

Problem 1   Evaluate the composite functions given $g(x) = 3x - 2$ and $h(x) = x^2 + 1$.
A. $g(h(0))$ B. $h(g(x))$

Solution   See page A67.
A. $g(h(0)) = 1$ B. $h(g(x)) = 9x^2 - 12x + 5$

2 Inverse functions

The **inverse of a function** is a function in which the components of each ordered pair are reversed.

For example, some of the ordered pairs of the function defined by $f(x) = 2x$ are $(0, 0)$, $(-1, -2)$, $(3, 6)$, and $\left(\frac{1}{2}, 1\right)$.

The inverse of this function would contain the ordered pairs $(0, 0)$, $(-2, -1)$, $(6, 3)$, and $\left(1, \frac{1}{2}\right)$.

Now consider the function defined by $g(x) = x^2$. Some of the ordered pairs of this function are $(0, 0)$, $(-1, 1)$, $(1, 1)$, $(-3, 9)$, and $(3, 9)$. Reversing the ordered pairs gives $(0, 0)$, $(1, -1)$, $(1, 1)$, $(9, -3)$, and $(9, 3)$. These ordered pairs do not satisfy the definition of a function because there are ordered pairs with the same first element and different second elements. This example illustrates that not all functions have an inverse function.

The graphs of the two functions $f(x) = 2x$ and $g(x) = x^2$ are shown below.

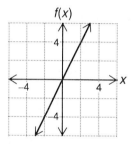
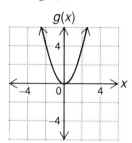

By the horizontal line test, the function f is a 1–1 function while the function g is not a 1–1 function.

Condition for an Inverse Function

A function has an inverse function if and only if it is a 1–1 function.

The symbol f^{-1} is used to denote the inverse of a function. $f^{-1}(x)$ is read "f inverse of x." Note that this is not the reciprocal of $f(x)$ but is the notation used for the inverse of a 1–1 function.

By the definition of an inverse function, the domain of f^{-1} is the range of f, and the range of f^{-1} is the domain of f.

To find the inverse of a linear function, interchange x and y. Then solve for y.

To find the inverse of the function $f(x) = 3x + 6$, think of the function as the equation $y = 3x + 6$.

$f(x) = 3x + 6$
$y = 3x + 6$

Interchange x and y.

$x = 3y + 6$

Solve for y.

$3y = x - 6$

$$y = \frac{1}{3}x - 2$$

Replace y with $f^{-1}(x)$.

$$f^{-1}(x) = \frac{1}{3}x - 2$$

The inverse of the linear function $f(x) = 3x + 6$ is the linear function $f^{-1}(x) = \frac{1}{3}x - 2$.

The inverse of a linear function with $m \neq 0$ will always be another linear function.

Example 2   Find the inverse of the function $f(x) = 2x - 4$.

Solution $f(x) = 2x - 4$
$y = 2x - 4$ • Think of the function as the equation $y = 2x - 4$.
$x = 2y - 4$ • Interchange x and y.
$2y = x + 4$ • Solve for y.

$$y = \frac{1}{2}x + 2$$

$$f^{-1}(x) = \frac{1}{2}x + 2$$

Problem 2   Find the inverse of the function $f(x) = 4x + 2$.

Solution   See page A67. $f^{-1}(x) = \frac{1}{4}x - \frac{1}{2}$

The graph of the linear function $f(x) = 2x - 4$ and its inverse, $f^{-1}(x) = \frac{1}{2}x + 2$, found in Example 2, is shown at the right.

The inverse function is the mirror image of f with respect to the line $y = x$.

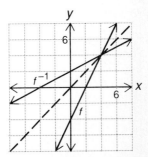

If two functions are inverses, then their graphs are mirror images of each other with respect to the line $y = x$.

The composite functions $f(f^{-1}(x))$ and $f^{-1}(f(x))$ have the following property:

$$f(f^{-1}(x)) = f^{-1}(f(x)) = x$$

For the functions in Example 2,

$$f(f^{-1}(x)) = f\left(\frac{1}{2}x + 2\right) \qquad f^{-1}(f(x)) = f^{-1}(2x - 4)$$

$$= 2\left(\frac{1}{2}x + 2\right) - 4 \qquad\qquad = \frac{1}{2}(2x - 4) + 2$$

$$= x + 4 - 4 \qquad\qquad\qquad = x - 2 + 2$$

$$= x \qquad\qquad\qquad\qquad\quad = x$$

This concept of inverse functions is similar to the additive inverse and multiplicative inverse used in arithmetic operations. For example, adding the number a and its additive inverse $(-a)$ to an expression results in the original expression.

$$x = x + a + (-a) = x$$

Inverse functions operate in a similar manner; one undoes the other.

For the functions $f(x) = 2x - 4$ and $f^{-1}(x) = \frac{1}{2}x + 2$,

$$f^{-1}(f(2)) = f^{-1}(0) \qquad f(f^{-1}(2)) = f(3)$$

$$= 2 \qquad\qquad\qquad\qquad = 2$$

Example 3   Are the functions $f(x) = -2x + 3$ and $g(x) = -\frac{1}{2}x + \frac{3}{2}$ inverses of each other?

Solution $f(g(x)) = f\left(-\frac{1}{2}x + \frac{3}{2}\right)$ • Use the property that for inverses
$$f(f^{-1}(x)) = f^{-1}(f(x)) = x.$$

$$= -2\left(-\frac{1}{2}x + \frac{3}{2}\right) + 3$$

$$= x - 3 + 3$$

$$= x \qquad\qquad\qquad\qquad • f(g(x)) = x$$

$$g(f(x)) = g(-2x + 3)$$

$$= -\frac{1}{2}(-2x + 3) + \frac{3}{2}$$

$$= x - \frac{3}{2} + \frac{3}{2}$$

$$= x \qquad\qquad\qquad\qquad • g(f(x)) = x$$

The functions are inverses of each other.

Problem 3   Are the functions $h(x) = 4x + 2$ and $g(x) = \dfrac{1}{4}x - \dfrac{1}{2}$ inverses of each other?

Solution   See page A67.
yes

As stated previously, the function $f(x) = x^2$ does not have an inverse. Two of the ordered pair solutions of this function are $(3, 9)$ and $(-3, 9)$.

The graph of the function $f(x) = x^2$ is shown at the right. This graph does not pass the horizontal line test for the graph of a 1–1 function. The mirror image of the graph with respect to the line $y = x$ is also shown. This graph does not pass the vertical line test for the graph of a function.

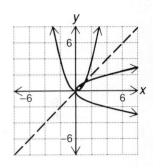

A quadratic function with domain the real numbers does not have an inverse function.

EXERCISES 8.3

1   Given $f(x) = 2x - 3$ and $g(x) = 4x - 1$, evaluate the composite function.

1. $f(g(0))$
-5

2. $g(f(0))$
-13

3. $f(g(2))$
11

4. $g(f(-2))$
-29

5. $f(g(x))$
$8x - 5$

6. $g(f(x))$
$8x - 13$

Given $h(x) = 2x + 4$ and $f(x) = \dfrac{1}{2}x + 2$, evaluate the composite function.

7. $h(f(0))$
8

8. $f(h(0))$
4

9. $h(f(2))$
10

10. $f(h(-1))$
3

11. $h(f(x))$
$x + 8$

12. $f(h(x))$
$x + 4$

Given $g(x) = x^2 + 3$ and $h(x) = x - 2$, evaluate the composite function.

13. $g(h(0))$
7

14. $h(g(0))$
1

15. $g(h(4))$
7

16. $h(g(-2))$
5

17. $g(h(x))$
$x^2 - 4x + 7$

18. $h(g(x))$
$x^2 + 1$

Given $f(x) = x^2 + x + 1$ and $h(x) = 3x + 2$, evaluate the composite function.

19. $f(h(0))$

7

20. $h(f(0))$

5

21. $f(h(-1))$

1

22. $h(f(-2))$

11

23. $f(h(x))$

$9x^2 + 15x + 7$

24. $h(f(x))$

$3x^2 + 3x + 5$

2 Find the inverse of the function. If the function does not have an inverse, write "no inverse."

25. $\{(1, 0), (2, 3), (3, 8), (4, 15)\}$

$\{(0,1),(3,2),(8,3),(15,4)\}$

26. $\{(1, 0), (2, 1), (-1, 0), (-2, 1)\}$

no inverse

27. $\{(3, 5), (-3, -5), (2, 5), (-2, -5)\}$

no inverse

28. $\{(-5, -5), (-3, -1), (-1, 3), (1, 7)\}$

$\{(-5,-5),(-1,-3),(3,-1),(7,1)\}$

29. $f(x) = 4x - 8$

$f^{-1}(x) = \dfrac{1}{4}x + 2$

30. $f(x) = 3x + 6$

$f^{-1}(x) = \dfrac{1}{3}x - 2$

31. $f(x) = x^2 - 1$

no inverse

32. $f(x) = 2x + 4$

$f^{-1}(x) = \dfrac{1}{2}x - 2$

33. $f(x) = x - 5$

$f^{-1}(x) = x + 5$

34. $f(x) = \dfrac{1}{2}x - 1$

$f^{-1}(x) = 2x + 2$

35. $f(x) = \dfrac{1}{3}x + 2$

$f^{-1}(x) = 3x - 6$

36. $f(x) = -2x + 2$

$f^{-1}(x) = -\dfrac{1}{2}x + 1$

37. $f(x) = -3x - 9$

$f^{-1}(x) = -\dfrac{1}{3}x - 3$

38. $f(x) = 2x^2 + 2$

no inverse

39. $f(x) = \dfrac{2}{3}x + 4$

$f^{-1}(x) = \dfrac{3}{2}x - 6$

40. $f(x) = \dfrac{3}{4}x - 4$

$f^{-1}(x) = \dfrac{4}{3}x + \dfrac{16}{3}$

41. $f(x) = -\dfrac{1}{3}x + 1$

$f^{-1}(x) = -3x + 3$

42. $f(x) = -\dfrac{1}{2}x + 2$

$f^{-1}(x) = -2x + 4$

43. $f(x) = 2x - 5$

$f^{-1}(x) = \dfrac{1}{2}x + \dfrac{5}{2}$

44. $f(x) = 3x + 4$

$f^{-1}(x) = \dfrac{1}{3}x - \dfrac{4}{3}$

45. $f(x) = x^2 + 3$

no inverse

46. $f(x) = 5x - 2$

$f^{-1}(x) = \dfrac{1}{5}x + \dfrac{2}{5}$

47. $f(x) = 4x - 2$

$f^{-1}(x) = \dfrac{1}{4}x + \dfrac{1}{2}$

48. $f(x) = 6x - 3$

$f^{-1}(x) = \dfrac{1}{6}x + \dfrac{1}{2}$

49. $f(x) = -8x + 4$

$f^{-1}(x) = -\dfrac{1}{8}x + \dfrac{1}{2}$

50. $f(x) = -6x + 2$

$f^{-1}(x) = -\dfrac{1}{6}x + \dfrac{1}{3}$

51. $f(x) = 8x + 6$

$f^{-1}(x) = \dfrac{1}{8}x - \dfrac{3}{4}$

52. $f(x) = \dfrac{1}{2}x^2 - 4$

no inverse

Are the functions inverses of each other?

53. $f(x) = 4x; g(x) = \dfrac{x}{4}$

yes

54. $g(x) = x + 5; h(x) = x - 5$

yes

55. $f(x) = 3x; h(x) = \dfrac{1}{3x}$

no

56. $h(x) = x + 2; g(x) = 2 - x$

no

57. $g(x) = 3x + 2; f(x) = \dfrac{1}{3}x - \dfrac{2}{3}$

yes

58. $h(x) = 4x - 1; f(x) = \dfrac{1}{4}x + \dfrac{1}{4}$

yes

59. $f(x) = \dfrac{1}{2}x - \dfrac{3}{2}; g(x) = 2x + 3$

yes

60. $g(x) = -\dfrac{1}{2}x - \dfrac{1}{2}; h(x) = -2x + 1$

no

Complete.

61. The domain of the inverse function f^{-1} is the _____ of f.

range

62. The range of the inverse function f^{-1} is the _____ of f.

domain

63. For any function f and its inverse f^{-1}, $f(f^{-1}(3)) = $ _____ .

3

64. For any function f and its inverse f^{-1}, $f^{-1}(f(-4)) = $ _____ .

-4

SUPPLEMENTAL EXERCISES 8.3

If f is a 1–1 function and $f(0) = 5$, $f(1) = 7$, and $f(2) = 9$, find:

65. $f^{-1}(5)$

0

66. $f^{-1}(7)$

1

67. $f^{-1}(9)$

2

If f is a 1–1 function and $f(1) = 2$, $f(2) = 5$, and $f(3) = 8$, find:

68. $f^{-1}(2)$

1

69. $f^{-1}(5)$

2

70. $f^{-1}(8)$

3

Given $f(x) = -x + 4$, find:

71. $f^{-1}(2)$

2

72. $f^{-1}(4)$

0

73. $f^{-1}(6)$

-2

Given $f(x) = 3x - 5$, find:

74. $f^{-1}(0)$

$\dfrac{5}{3}$

75. $f^{-1}(2)$

$\dfrac{7}{3}$

76. $f^{-1}(4)$

3

The graphs of the functions f and g are shown at the right. Use the graphs to determine the values of the composite functions.

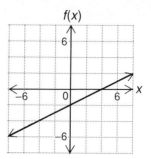

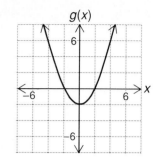

77. $f(g(0))$
−3

78. $f(g(2))$
−2

79. $f(g(4))$
1

80. $f(g(−2))$
−2

81. $f(g(−4))$
1

82. $g(f(0))$
0

83. $g(f(4))$
−2

84. $g(f(−4))$
6

Graph the function. Then use the property that the inverse function is the mirror image of f with respect to the line $y = x$ to graph the inverse function on the same coordinate axes. (*Hint:* Interchange the two coordinates in the ordered pair solutions of f to find ordered pair solutions of f^{-1}.)

85. $f(x) = 3x - 4$

86. $f(x) = 3x + 2$

87. $f(x) = \dfrac{1}{4}x + 2$

88. $f(x) = \dfrac{1}{3}x - 1$

89. $f(x) = x^3 + 1$

90. $f(x) = x^3 - 1$

91. $f(x) = 2x^3 - 2$

92. $f(x) = 2x^3 + 3$

SECTION **8.4**

Variation

1 Variation problems

Direct variation is a special function that can be expressed as the equation $y = kx$, where k is a constant. The equation $y = kx$ is read "y varies directly as x" or "y is proportional to x." The constant k is called the **constant of variation** or the **constant of proportionality.**

The circumference (C) of a circle varies directly as the diameter (d). The direct variation equation is written $C = \pi d$. The constant of variation is π.

A nurse makes $10 per hour. The nurse's total wage (w) is directly proportional to the number of hours (h) worked. The equation of variation is $w = 10h$. The constant of proportionality is 10.

A direct variation equation can be written in the form $y = kx^n$, where n is a positive number. For example, the equation $y = kx^2$ is read "y varies directly as the square of x."

The area (A) of a circle varies directly as the square of the radius (r) of the circle. The direct variation equation is $A = \pi r^2$.

Given that V varies directly as r and that $V = 20$ when $r = 4$, the constant of a variation can be found by writing the basic direct variation equation, replacing V and r by the given values and solving for the constant of variation.

$$V = kr$$
$$20 = k \cdot 4$$
$$5 = k$$

The direct variation equation can then be written by substituting the value of k into the basic direct variation equation.

$$V = 5r$$

Example 1 The amount (A) of medication prescribed for a person is directly related to the person's weight (W). For a 50-kg person, 2 ml of medication are prescribed. How many milliliters of medication are required for a person who weighs 75 kg?

Strategy To find the required amount of medication:
▶ Write the basic direct variation equation, replace the variables by the given values, and solve for k.
▶ Write the direct variation equation, replacing k by its value. Substitute 75 for W, and solve for A.

Solution $A = kW$

$2 = k \cdot 50$

$\dfrac{1}{25} = k$

$A = \dfrac{1}{25} W$

$= \dfrac{1}{25} \cdot 75 = 3$

The required amount of medication is 3 ml.

Problem 1   The distance (s) a body falls from rest varies directly as the square of the time (t) of the fall. An object falls 64 ft in 2 s. How far will it fall in 5 s?

Solution   See page A68.

400 ft

Joint variation is a variation in which a variable varies directly as the product of two or more other variables. A joint variation can be expressed as the equation $z = kxy$, where k is a constant. The equation $z = kxy$ is read "z varies jointly as x and y."

The area (A) of a triangle varies jointly as the base (b) and the height (h). The joint variation equation is written $A = \dfrac{1}{2} bh$. The constant of variation is $\dfrac{1}{2}$.

Inverse variation is a function that can be expressed as the equation $y = \dfrac{k}{x}$, where k is a constant. The equation $y = \dfrac{k}{x}$ is read "y varies inversely as x" or "y is inversely proportional to x."

In general, an inverse variation equation can be written $y = \dfrac{k}{x^n}$, where n is a positive number. For example, the equation $y = \dfrac{k}{x^2}$ is read "y varies inversely as the square of x."

Given that P varies inversely as the square of x and that $P = 5$ when $x = 2$, the variation constant can be found by writing the basic inverse variation equation, replacing P and x by the given values, and solving for the constant of variation.

$$P = \dfrac{k}{x^2}$$

$$5 = \dfrac{k}{2^2}$$

$$5 = \dfrac{k}{4}$$

$$20 = k$$

The inverse variation equation can then be found by substituting the value of k into the basic inverse variation equation.

$$P = \dfrac{20}{x^2}$$

Example 2   A company that produces personal computers has determined that the number of computers it can sell (s) is inversely proportional to the price (P) of the computer. Two thousand computers can be sold when the price is $2500. How many computers can be sold if the price of a computer is $2000?

Strategy   To find the number of computers:
- ▶ Write the basic inverse variation equation, replace the variables by the given values, and solve for k.
- ▶ Write the inverse variation equation, replacing k by its value. Substitute 2000 for P, and solve for s.

Solution
$$s = \frac{k}{P}$$

$$2000 = \frac{k}{2500}$$

$$5{,}000{,}000 = k$$

$$s = \frac{5{,}000{,}000}{P}$$

$$= \frac{5{,}000{,}000}{2000} = 2500$$

At a price of $2000, 2500 computers can be sold.

Problem 2   The resistance (R) to the flow of electric current in a wire of fixed length is inversely proportional to the square of the diameter (d) of a wire. If a wire of diameter 0.01 cm has a resistance of 0.5 ohms, what is the resistance on a wire that is 0.02 cm in diameter?

Solution   See page A68.
0.125 ohms

A **combined variation** is a variation in which two or more types of variation occur at the same time. For example, in physics, the volume (V) of a gas varies directly as the temperature (T) and inversely as the pressure (P). This combined variation is written $V = \dfrac{kT}{P}$.

Example 3   The pressure (P) of a gas varies directly as the temperature (T) and inversely as the volume (V). When $T = 50°$ and $V = 275$ in.^{3}, $P = 20$ lb/in.^{2}. Find the pressure of a gas when $T = 60°$ and $V = 250$ in.^{3}.

Strategy   To find the pressure:
- ▶ Write the basic combined variation equation, replace the variables by the given values, and solve for k.
- ▶ Write the combined variation equation, replacing k by its value. Substitute 60 for T and 250 for V, and solve for P.

Solution $P = \dfrac{kT}{V}$

$20 = \dfrac{k(50)}{275}$

$110 = k$

$P = \dfrac{110T}{V}$

$= \dfrac{110(60)}{250} = 26.4$

The pressure is 26.4 lb/in.^{2}.

Problem 3   The strength (s) of a rectangular beam varies directly as its width (w) and inversely as the square of its depth (d). If the strength of a beam 2 in. wide and 12 in. deep is 1200 lb, find the strength of a beam 4 in. wide and 8 in. deep.

Solution   See page A69.
5400 lb

EXERCISES 8.4

1   Solve.

1. The distance (d) a spring will stretch varies directly as the force (f) applied to the spring. If a force of 6 lb is required to stretch a spring 3 in., what force is required to stretch the spring 4 in.?
    8 lb

2. The pressure (p) on a diver in the water varies directly as the depth (d). If the pressure is 4.5 lb/in.^{2} when the depth is 10 ft, what is the pressure when the depth is 15 ft?
    6.75 lb/in.^{2}

3. The number of bushels of wheat (b) produced by a farm is directly proportional to the number of acres (A) planted in wheat. If a 20-acre farm yields 450 bushels of wheat, what is the yield of a farm that has 30 acres of wheat?
    675 bushels

4. The profit (P) realized by a company varies directly as the number of products it sells (s). If a company makes a profit of $4000 on the sale of 250 products, what is the profit when the company sells 5000 products?
    $80,000

5. The stopping distance (*s*) of a car varies directly as the square of its speed (*v*). If a car traveling 50 mph requires 170 ft to stop, find the stopping distance for a car traveling 60 mph.
244.8 ft

6. The distance (*s*) a ball will roll down an inclined plane is directly proportional to the square of the time (*t*). If the ball rolls 6 ft in one second, how far will it roll in 3 s?
54 ft

7. The period (*p*) of a pendulum, or the time it takes for a pendulum to make one complete swing, varies directly as the square root of the length (*L*) of the pendulum. If the period of a pendulum is 1.5 s when the length is 2 ft, find the period when the length is 5 ft. Round to the nearest hundredth.
2.37 s

8. The distance (*d*) a person can see to the horizon from a point above the surface of the earth varies directly as the square root of the height (*H*). If, for a height of 500 ft, the horizon is 19 mi away, how far is the horizon from a point that is 800 ft high? Round to the nearest hundredth.
24.04 mi

9. The speed (*v*) of a gear varies inversely as the number of teeth (*t*). If a gear that has 45 teeth makes 24 revolutions per minute, how many revolutions will a gear that has 36 teeth make?
30 revolutions/min

10. For a constant temperature, the pressure (*P*) of a gas varies inversely as the volume (*V*). If the pressure is 30 lb/in.^{2} when the volume is 500 ft^3, find the pressure when the volume is 200 ft^3.
75 lb/in.^{2}

11. The number of items (*n*) that can be purchased for a given amount of money is inversely proportional to the cost (*C*) of an item. If 60 items can be purchased when the cost per item is $.25, how many items can be purchased when the cost per item is $.20?
75 items

12. The length (*L*) of a rectangle of fixed area varies inversely as the width (*w*). If the length of a rectangle is 8 ft when the width is 5 ft, find the length of the rectangle when the width is 4 ft.
10 ft

13. The intensity (*l*) of a light source is inversely proportional to the square of the distance (*d*) from the source. If the intensity is 12 lumens at a distance of 10 ft, what is the intensity when the distance is 5 ft?
48 lumens

14. The repulsive force (f) between two north poles of a magnet is in-
    versely proportional to the square of the distance (d) between them. If
    the repulsive force is 20 lb when the distance is 4 in., find the repulsive
    force when the distance is 2 in.
    80 lb

15. The current (l) in a wire varies directly as the voltage (v) and inversely
    as the resistance (r). If the current is 22 amps when the voltage is
    110 volts and the resistance is 5 ohms, find the current when the
    voltage is 195 volts and the resistance is 15 ohms.
    13 amps

16. The pressure (p) in a liquid varies jointly as the depth (d) and the den-
    sity (D) of the liquid. If the pressure is 150 lb/in.^{2} when the depth is
    100 in. and the density is 1.2, find the pressure when the density re-
    mains the same and the depth is 75 in.
    112.5 lb/in.^{2}

17. The power (P) in an electric circuit varies jointly as the current (l) and
    the square of the resistance (R). If the power is 100 watts when the cur-
    rent is 4 amps and the resistance is 5 ohms, find the power when the
    current is 2 amps and the resistance is 10 ohms.
    200 watts

18. The wind force (w) on a vertical surface varies jointly as the area (A) of
    the surface and the square of the wind velocity (v). When the wind is
    blowing at 30 mph, the force on a 10-ft^2 area is 45 lb. Find the force on
    this area when the wind is blowing at 60 mph.
    180 lb

19. The frequency of vibration (f) of a string varies directly as the square
    root of tension (T) and inversely as the length (L) of the string. If the
    frequency is 40 vibrations per second when the tension is 25 lb and
    the length of the string is 3 ft, find the frequency when the tension is
    36 lb and the string is 4 ft.
    36 vibrations/s

20. The resistance (R) of a wire varies directly as the length (L) of the wire
    and inversely as the square of the diameter (d). If the resistance is
    9 ohms in 50 ft of wire that has a diameter of 0.05 in., find the resis-
    tance in 50 ft of a similar wire that has a diameter of 0.02 in.
    56.25 ohms

SUPPLEMENTAL EXERCISES 8.4

Solve.

21. **a.** Graph $y = kx$ when $k = 2$.
   b. What kind of function does the graph represent?
   linear function

22. **a.** Graph $y = kx$ when $k = \dfrac{1}{2}$.
   b. What kind of function does the graph represent?
   linear function

23. **a.** Graph $y = kx^2$ when $k = 2$.
   b. What kind of function does the graph represent?
   a quadratic function

24. **a.** Graph $y = kx^2$ when $k = \dfrac{1}{2}$.
   b. What kind of function does the graph represent?
   a quadratic function

25. **a.** Graph $y = \dfrac{k}{x}$ when $k = 2$ and $x > 0$.
   b. Is this the graph of a function?
   yes

26. **a.** Graph $y = \dfrac{k}{x^2}$ when $k = 2$ and $x > 0$.
   b. Is this the graph of a function?
   yes

27. In the inverse variation equation $y = \dfrac{k}{x}$, what is the effect on x if y doubles?
   x is halved

28. In the direct variation equation $y = kx$, what is the effect on y when x doubles?
y doubles

Complete using the word *directly* or *inversely*.

29. If a varies directly as b and inversely as c, then c varies _____ as b and _____ as a.
directly, inversely

30. If a varies _____ as b and c, then abc is constant.
inversely

31. If the length of a rectangle is held constant, the area of the rectangle varies _____ as the width.
directly

32. If the area of a rectangle is held constant, the length of the rectangle varies _____ as the width.
inversely

CALCULATORS AND COMPUTERS

 Graphing Functions

Once the program GRAPH A FUNCTION on the Apple version of the Student Disk is loaded, you will be asked to enter the function to be graphed. The cursor will be blinking following the notation "F(X) =".

The operational signs used to enter a function are $+$, $-$, $*$ (times), and $/$ (divided by). The symbol $\wedge$ is used to enter an exponent. Use "ABS" for an absolute value function. The computer program uses the Order of Operations Agreement.

Some examples of functions and the corresponding keystroking are shown below.

Function	Keystroking
$f(x) = x^3 + 2x - 1$	x^3 + 2*x − 1
$f(x) = 3x^2 + 2x - 4$	3*x^2 + 2*x − 4
$f(x) = \lvert x - 1 \rvert$	ABS(x − 1)
$f(x) = \lvert x \rvert + 2$	ABS(x) + 2
$f(x) = 5x - 3$	5*x − 3
$f(x) = \dfrac{1}{2}x - \dfrac{3}{4}$	(1/2)*x − (3/4)
$f(x) = -4$	−4

After entering the function you wish to see graphed, the following prompts will appear one by one on the screen:

MINIMUM X =   MAXIMUM X =

MINIMUM Y =   MAXIMUM Y =

You might begin by entering −5 for the minimum x value, 5 for the maximum x value, −5 for the minimum y value, and 5 for the maximum y value. As you get more familiar with the program, you might want to experiment with other values. After answering each of the four prompts, press RETURN.

On the IBM version of the Student Disk, you will be asked to enter the type of function to be graphed and then to enter the coefficients of the variable terms and the constant.

For both the Apple and IBM versions of the program after a function has been graphed, the following prompt will appear at the bottom of the computer screen:

(E)RASE   (D)RAW   (M)ENU

Type "E" to enter a new function to be graphed, type "D" to draw the graph of another function on the grid currently on the screen, type "M" to return to the main menu.

CHAPTER SUMMARY

Key Words

A **relation** is a set of ordered pairs and assigns to each member of a first set one or more members of a second set.

A **function** is a relation in which each element of the first set is assigned to one and only one member of the second set. No two ordered pairs that have the same first component have different second components.

The **domain** of a function is the set of first components of the ordered pairs of the function. The **range** of a function is the set of second components of the ordered pairs of the function.

A function of the form $f(x) = mx + b$ is a **linear function.**

A **one-to-one function** is a function that satisfies the additional condition that given any y, there is only one x that can be paired with the given y.

The **vertical line test** is used to determine whether or not a graph is the graph of a function. The **horizontal line test** is used to determine whether or not the graph of a function is the graph of a 1–1 function.

A **composite function** is a function that involves more than one function.

The **inverse of a one-to-one function** is a function in which the components of each ordered pair are reversed.

Direct variation is a special function that can be expressed as the equation $y = kx^n$, where k is a constant called the **constant of variation** or the **constant of proportionality**.

Joint variation is a variation in which a variable varies directly as the product of two or more variables. A joint variation can be expressed as the equation $z = kxy$, where k is a constant.

Inverse variation is a function that can be expressed as the equation $y = \dfrac{k}{x^n}$, where k is a constant.

Combined variation is a variation in which two or more types of variation occur at the same time.

Essential Rules

For the function $f(x)$ and its inverse $f^{-1}(x)$, $f(f^{-1}(x)) = f^{-1}(f(x)) = x$

CHAPTER REVIEW

1. For the function $f(x) = x^2 - 1$, find $f(4)$.
$f(4) = 15$

2. Graph the function $f(x) = -x^2 + 1$.

3. Is the set of ordered pairs a function?
$\{(1, 3), (3, 3), (2, 4), (4, 4)\}$
yes

4. Evaluate the function $g(x) = -x^2 + 3x - 2$ at $x = -2$.
$g(-2) = -12$

5. Graph the function $f(x) = \dfrac{1}{2}x - 2$.

6. What values of x are excluded from the domain of the function $f(x) = \dfrac{3}{x^2 - 8x}$?
$0, 8$

7. Find the range of the function $f(x) = 4x - 3$ if the domain is $\{-2, -1, 0, 1, 2\}$.
$\{-11, -7, -3, 1, 5\}$

8. Given $f(x) = 2x^2 - 3$ and $g(x) = 4x + 1$, evaluate $f(g(0))$.
$f(g(0)) = -1$

9. Find the inverse of the function

   $f(x) = \dfrac{1}{2}x - 1.$ $f^{-1}(x) = 2x + 2$

10. Graph the function
    $f(x) = -x^2 + 2x.$
    State the domain and
    range of the function.
    D: all reals; R: $y \le 1$

    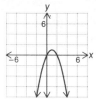

11. Given $g(x) = -3x + 2$ and $h(x) = x - 4$,
    evaluate $g(h(x))$. $g(h(x)) = -3x + 14$

12. Are the functions $f(x) = 2x - 8$ and

    $g(x) = \dfrac{1}{2}x + 4$ inverses of each other?

    yes

13. Graph the function
    $f(x) = |x + 1|.$
    State the domain and
    range of the function.
    D: all reals; R: $y \ge 0$

    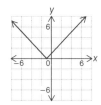

14. Find the domain of the function
    $\{(3, -1), (5, -3), (7, -5), (9, -7)\}.$
    $\{3, 5, 7, 9\}$

15. Evaluate the function $h(x) = 2x^3 + 1$ at
    $x = -1.$
    $h(-1) = -1$

16. Graph the function
    $f(x) = x^3 + 2.$

17. Are the functions $f(x) = 3x + 6$ and

    $g(x) = \dfrac{1}{3}x + 2$ inverses of each

    other?

    no

18. For the function $g(x) = -2x + 3$, find
    $g(1 + h).$
    $g(1 + h) = -2h + 1$

19. What values of x are excluded from the
    domain of the function $f(x) = \sqrt{x - 7}$?
    $\{x \mid x < 7\}$

20. Find the inverse of the function
    $f(x) = 3x + 6.$

    $f^{-1}(x) = \dfrac{1}{3}x - 2$

21. The number -2 is in the range of the
    function $f(x) = 3x + 2$. Find an element in
    the domain that corresponds to -2.
    $-\dfrac{4}{3}$

22. Evaluate the function $f(x) = \dfrac{x}{x^2 + 1}$ at
    $x = -2.$

    $f(-2) = -\dfrac{2}{5}$

23. For $g(x) = 2x^2 + 1$ and $h(x) = 3x + 4$, find
    $g(0) + h(0).$
    $g(0) + h(0) = 5$

24. Graph the function
    $f(x) = |x| - 2.$
    State the domain and
    range of the function.
    D: all reals; R: $y \ge -2$

25. Evaluate $f(x) = 3x^3 - 1$ at $x = -1$.
$f(-1) = -4$

26. Find the range of the function
$g(a) = \dfrac{a^2 - 1}{2a - 1}$ if the domain is $\{0, 1, 2\}$.
$\{0, 1\}$

27. Determine if the graph represents the graph of a function.
no

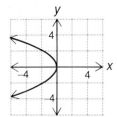

28. Determine if the graph represents the graph of a 1–1 function.
yes

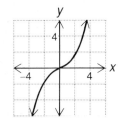

29. The stopping distance (s) of a car varies directly as the square of the speed (v) of the car. For a car traveling at 50 mph, the stopping distance is 170 ft. Find the stopping distance of a car that is traveling at 30 mph.
61.2 ft

30. The current (I) in an electric circuit varies inversely as the resistance (R). If the current in the circuit is 4 amps when the resistance is 50 ohms, find the current in the circuit when the resistance is 100 ohms.
2 amps

CUMULATIVE REVIEW

1. Evaluate $-3a + \left| \dfrac{3b - ab}{3b - c} \right|$ when $a = 2$, $b = 2$, and $c = -2$.
$-\dfrac{23}{4}$

2. Graph the solution set of $\{x \mid x < -3\} \cap \{x \mid x > -4\}$.

3. Solve: $\dfrac{3x - 1}{6} - \dfrac{5 - x}{4} = \dfrac{5}{6}$
3

4. Solve: $4x - 2 < -10$ or $3x - 1 > 8$
$\{x \mid x < -2 \text{ or } x > 3\}$

5. Solve: $|8 - 2x| \geq 0$
all real numbers

6. Simplify: $\left(\dfrac{3a^3 b}{2a} \right)^2 \left(\dfrac{a^2}{-3b^2} \right)^3 - \dfrac{a^{10}}{12b^4}$

7. Simplify: $(x - 4)(2x^2 + 4x - 1)$
$2x^3 - 4x^2 - 17x + 4$

8. Factor: $a^4 - 2a^2 - 8$
$(a + 2)(a - 2)(a^2 + 2)$

9. Factor: $x^3 y + x^2 y^2 - 6xy^3$
$xy(x + 3y)(x - 2y)$

10. Solve: $(b + 2)(b - 5) = 2b + 14$
$-3, 8$

11. Solve: $x^2 - 2x > 15$
$\{x \mid x < -3 \text{ or } x > 5\}$

12. Simplify: $\dfrac{x^2 + 4x - 5}{2x^2 - 3x + 1} - \dfrac{x}{2x - 1}$
$\dfrac{5}{2x - 1}$

13. Solve: $\dfrac{5}{x^2 + 7x + 12} = \dfrac{9}{x + 4} - \dfrac{2}{x + 3}$

-2

14. Simplify: $\dfrac{4 - 6i}{2i}$

$-3 - 2i$

15. Solve: $\sqrt[3]{5x - 2} = 2$

2

16. Graph the solution set of $3x - 4y \geq 8$.

17. Find the equation of the line containing the points $(-3, 4)$ and $(2, -6)$.

$y = -2x - 2$

18. Find the equation of the line containing the point $(-3, 1)$ and perpendicular to the line $2x - 3y = 6$.

$y = -\dfrac{3}{2}x - \dfrac{7}{2}$

19. Solve: $3x^2 = 3x - 1$

$\dfrac{1}{2} + \dfrac{\sqrt{3}}{6}i$ and $\dfrac{1}{2} - \dfrac{\sqrt{3}}{6}i$

20. Solve: $\sqrt{8x + 1} = 2x - 1$

3

21. Evaluate the function $f(x) = 2x^2 - 3$ at $x = -2$.

$f(-2) = 5$

22. Find the range of the function $f(x) = |3x - 4|$ if the domain is $\{0, 1, 2, 3\}$.

$\{1, 2, 4, 5\}$

23. Is the set of ordered pairs a function? $\{(-3, 0), (-2, 0), (-1, 1), (0, 1)\}$

yes

24. Graph the function

$f(x) = \dfrac{1}{4}x^2$.

25. Given $g(x) = 3x - 5$ and $h(x) = \dfrac{1}{2}x + 4$, find $g(h(2))$.

$g(h(2)) = 10$

26. Find the inverse of the function $f(x) = -3x + 9$.

$f^{-1}(x) = -\dfrac{1}{3}x + 3$

27. Find the cost per pound of a tea mixture made from 30 lb of tea costing $4.50 per pound and 45 lb of tea costing $3.60 per pound.

$3.96

28. How many pounds of an 80% copper alloy must be mixed with 50 lb of a 20% copper alloy to make an alloy that is 40% copper?

25 lb

29. Six ounces of an insecticide are mixed with 16 gal of water to make a spray for spraying an orange grove. How much additional insecticide is required to be mixed with 28 gal of water?

4.5 oz

30. A large pipe can fill a tank in 8 min less time than it takes a smaller pipe to fill the same tank. Working together, both pipes can fill the tank in 3 min. How long would it take the larger pipe working alone to fill the tank?
4 min

31. The distance (d) a spring stretches varies directly as the force (f) used to stretch the spring. If a force of 50 lb can stretch the spring 30 in., how far will a force of 40 lb stretch the spring?
24 in.

32. The frequency of vibration (f) in an open pipe organ varies inversely as the length (L) of the pipe. If the air in a pipe 2 m long vibrates 60 times per minute, find the frequency in a pipe that is 1.5 m long.
80 vibrations/min

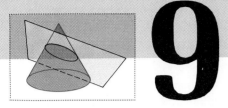

9

Conic Sections

OBJECTIVES

- Graph parabolas
- Find the minimum or maximum of a quadratic function
- Application problems
- Find the distance between two points in the plane
- Find the equation of a circle and then graph the circle
- Write the equation of a circle in standard form and then graph the circle
- Graph an ellipse with center at the origin
- Graph a hyperbola with center at the origin
- Graph the solution set of a quadratic inequality in two variables

Conic Sections

The graphs of three curves — the ellipse, the parabola, and the hyperbola — are discussed in this chapter. These curves were studied by the Greeks and were known prior to 400 B.C. The names of these curves were first used by Apollonius around 250 B.C. in *Conic Sections*, the most authoritative Greek discussion of these curves. Apollonius borrowed the names from a school founded by Pythagoras.

The diagram at the right shows the path of a planet around the sun. The curve traced out by the planet is an ellipse. The **aphelion** is the position of the planet when it is farthest from the sun. The **perihelion** is the position when the planet is nearest to the sun.

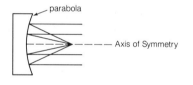

A telescope, like the one at the Palomar Observatory, has a cross section that is in the shape of a parabola. A parabolic mirror has the unusual property that all light rays parallel to the axis of symmetry that hit the mirror are reflected to the same point. This point is called the **focus of the parabola.**

Some comets, unlike Halley's Comet, travel with such speed that they are not captured by the sun's gravitational field. The path of the comet as it comes around the sun is in the shape of a hyperbola.

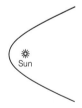

SECTION 9.1

The Parabola

1 ■ Graph parabolas

A parabola is one of a number of curves called conic sections. The graph of a **conic section** can be represented by the intersection of a plane and a cone.

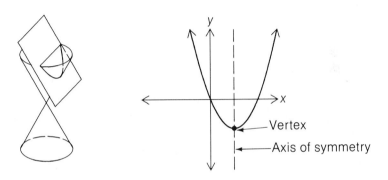

Every parabola has an axis of symmetry and a vertex that is on the axis of symmetry. To understand the axis of symmetry, think of folding the paper along that axis. The two halves of the curve will match up.

The graph of the equation $y = ax^2 + bx + c$, $a \neq 0$, is a parabola with the axis of symmetry parallel to the y-axis. The parabola opens up when $a > 0$ and opens down when $a < 0$. When the parabola open up, the vertex is the lowest point on the parabola. When the parabola opens down, the vertex is the highest point on the parabola.

The coordinates of the vertex of a parabola can be found by completing the square.

To find the vertex of the parabola $y = x^2 - 4x + 5$, group the variable terms.

$$y = x^2 - 4x + 5$$
$$y = (x^2 - 4x) + 5$$

Complete the square on $x^2 - 4x$. Note that 4 is added and subtracted. Since $4 - 4 = 0$, the equation is not changed.

$$y = (x^2 - 4x + 4) - 4 + 5$$

Factor the trinomial and combine like terms.

$$y = (x - 2)^2 + 1$$

439

Since the coefficient of x^2 is positive, the parabola opens up. The vertex is the lowest point on the parabola, or that point which has the least y-coordinate.

Since $(x - 2)^2 \geq 0$ for any x, the least y-coordinate occurs when $(x - 2)^2 = 0$. $(x - 2)^2 = 0$ when $x = 2$.

To find the y-coordinate of the vertex, replace x by 2, and solve for y.

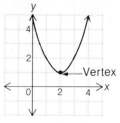

$$y = (x - 2)^2 + 1$$
$$= (2 - 2)^2 + 1$$
$$= 1$$

The vertex is $(2, 1)$.

By following the procedure of this example and completing the square on the equation $y = ax^2 + bx + c$, the **x-coordinate of the vertex** is $-\dfrac{b}{2a}$. The y-coordinate of the vertex can then be determined by substituting this value of x into $y = ax^2 + bx + c$ and solving for y.

Since the axis of symmetry is parallel to the y-axis and passes through the vertex, the equation of the **axis of symmetry** is $x = -\dfrac{b}{2a}$. For example, if the x-coordinate of the vertex of a parabola is 4, then the axis of symmetry is the line $x = 4$.

Example 1 Find the axis of symmetry and the vertex of the parabola $y = -3x^2 + 6x + 1$. Then sketch its graph.

Solution $-\dfrac{b}{2a} = -\dfrac{6}{2(-3)} = 1$

• Find the x-coordinate of the vertex. $a = -3$, $b = 6$.

The axis of symmetry is the line $x = 1$.

• The axis of symmetry is the line $x = -\dfrac{b}{2a}$.

$$y = -3x^2 + 6x + 1$$
$$= -3(1)^2 + 6(1) + 1$$
$$= 4$$

• Find the y-coordinate of the vertex by replacing x by 1 and solving for y.

The vertex is $(1, 4)$.

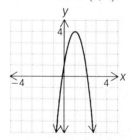

• Since a is negative, the parabola opens down. Find a few ordered pairs and use symmetry to sketch the graph.

Problem 1   Find the axis of symmetry and the vertex of the parabola $y = x^2 - 2$. Then sketch its graph.

Solution   See page A71.
Axis of symmetry: $x = 0$
Vertex: $(0, -2)$

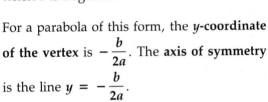

The graph of an equation of the form $x = ay^2 + by + c$, $a \neq 0$, is also a parabola. In this case, the parabola opens to the right when a is positive and opens to the left when a is negative.

For a parabola of this form, the **y-coordinate of the vertex** is $-\dfrac{b}{2a}$. The **axis of symmetry** is the line $y = -\dfrac{b}{2a}$.

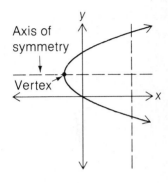

Using the vertical line test, the graph of a parabola of this form is not the graph of a function. The graph of $x = ay^2 + by + c$ is a relation.

Example 2   Find the axis of symmetry and the vertex of the parabola $x = 2y^2 - 8y + 5$. Then sketch its graph.

Solution

$-\dfrac{b}{2a} = -\dfrac{-8}{2(2)} = 2$

• Find the y-coordinate of the vertex.
  $a = 2$, $b = -8$.

The axis of symmetry is the line $y = 2$.

• The axis of symmetry is the line $y = -\dfrac{b}{2a}$.

$x = 2y^2 - 8y + 5$
$ = 2(2)^2 - 8(2) + 5$
$ = -3$

• Find the x-coordinate of the vertex by replacing y by 2 and solving for x.

The vertex is $(-3, 2)$.

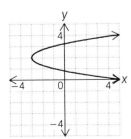

• Since a is positive, the parabola opens to the right.
  Find a few ordered pairs, and use symmetry to sketch the graph.

Problem 2   Find the axis of symmetry and the vertex of the parabola $x = -2y^2 - 4y - 3$. Then sketch its graph.

Solution   See page A72.
Axis of symmetry: $y = -1$
Vertex: $(-1, -1)$

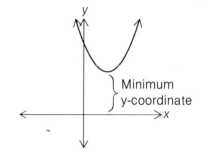

2   # Find the minimum or maximum of a quadratic function

The graph of $f(x) = x^2 - 2x + 3$ is shown at the right. Since a is positive, the parabola opens up. The vertex of the parabola is the lowest point on the parabola. It is the point that has the minimum y-coordinate. Therefore, the value of the function at this point is a **minimum**.

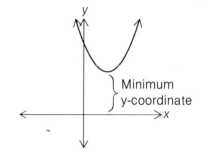

The graph of $f(x) = -x^2 + 2x + 1$ is shown at the right. Since a is negative, the parabola opens down. The vertex of the parabola is the highest point on the parabola. It is the point that has the maximum y-coordinate. Therefore, the value of the function at this point is a **maximum**.

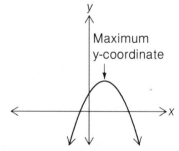

To find the minimum or maximum value of a quadratic function, first find the x-coordinate of the vertex. Then evaluate the function at that value.

Example 3   Find the maximum value of the function $f(x) = -2x^2 + 4x + 3$.

Solution $x = -\dfrac{b}{2a} = -\dfrac{4}{2(-2)} = 1$ • Find the x-coordinate of the vertex.
$a = -2$, $b = 4$

$f(x) = -2x^2 + 4x + 3$ • Evaluate the function at $x = 1$.
$f(1) = -2(1)^2 + 4(1) + 3$
$\quad\,\, = 5$

The maximum value of the function is 5.

Problem 3 Find the minimum value of the function $f(x) = 2x^2 - 3x + 1$.

Solution See page A72. $-\dfrac{1}{8}$

3 Application problems

Example 4 A mining company has determined that the cost in dollars (c) per ton of mining a mineral is given by the function $c(x) = 0.2x^2 - 2x + 12$, where x is the number of tons of the mineral that is mined. Find the number of tons of the mineral that should be mined to minimize the cost. What is the minimum cost?

Strategy ▶ To find the number of tons that will minimize the cost, find the x-coordinate of the vertex.
▶ To find the minimum cost, evaluate the function at the x-coordinate of the vertex.

Solution $x = -\dfrac{b}{2a} = -\dfrac{-2}{2(0.2)} = 5$

To minimize the cost, 5 tons should be mined.

$c(x) = 0.2x^2 - 2x + 12$
$c(5) = 0.2(5)^2 - 2(5) + 12 = 5 - 10 + 12 = 7$

The minimum cost per ton is $7.

Problem 4 The height in feet (s) of a ball thrown straight up is given by the function $s(t) = -16t^2 + 64t$, where t is the time in seconds. Find the time it takes the ball to reach its maximum height. What is the maximum height?

Solution See page A72.
The ball reaches its maximum height in 2 s. The maximum height is 64 ft.

Example 5 Find two numbers whose difference is 10 and whose product is a minimum.

Strategy Let x represent one number. Since the difference between the two numbers is 10, $x + 10$ represents the other number. Then their product is represented by $x^2 + 10x$.
▶ To find the first of the two numbers, find the x-coordinate of the vertex of the function $f(x) = x^2 + 10x$.
▶ To find the other number, replace x in $x + 10$ by the x-coordinate of the vertex and evaluate.

Solution $x = -\dfrac{b}{2a} = -\dfrac{10}{2(1)} = -5$

$x + 10 = -5 + 10 = 5$

The numbers are -5 and 5.

> **Problem 5**   A mason is forming a rectangular floor for a storage shed. The perimeter of the rectangle is 44 ft. What dimensions would give the floor a maximum area?
>
> **Solution**   See page A72.
> length: 11 ft; width: 11 ft

EXERCISES 9.1

1 Find the vertex and axis of symmetry of the parabola. Then sketch its graph.

1. $y = x^2 - 2x - 4$
Vertex: $(1, -5)$
Axis of
symmetry: $x = 1$

2. $y = x^2 + 4x - 4$
Vertex: $(-2, -8)$
Axis of
symmetry: $x = -2$

3. $y = -x^2 + 2x - 3$
Vertex: $(1, -2)$
Axis of
symmetry: $x = 1$

4. $y = -x^2 + 4x - 5$
Vertex: $(2, -1)$
Axis of
symmetry: $x = 2$

5. $x = y^2 + 6y + 5$

Vertex: $(-4, -3)$
Axis of
symmetry: $y = -3$

6. $x = y^2 - y - 6$
Vertex: $\left(-\dfrac{25}{4}, \dfrac{1}{2}\right)$
Axis of
symmetry: $y = \dfrac{1}{2}$

7. $y = 2x^2 - 4x + 1$
Vertex: $(1, -1)$
Axis of
symmetry: $x = 1$

8. $y = 2x^2 + 4x - 5$
Vertex: $(-1, -7)$
Axis of
symmetry: $x = -1$

9. $y = x^2 - 5x + 4$

Vertex: $\left(\dfrac{5}{2}, -\dfrac{9}{4}\right)$
Axis of
symmetry: $x = \dfrac{5}{2}$

10. $y = x^2 + 5x + 6$

Vertex: $\left(-\dfrac{5}{2}, -\dfrac{1}{4}\right)$
Axis of
symmetry: $x = -\dfrac{5}{2}$

11. $x = y^2 - 2y - 5$

Vertex: $(-6, 1)$
Axis of
symmetry: $y = 1$

12. $x = y^2 - 3y - 4$

Vertex: $\left(-\dfrac{25}{4}, \dfrac{3}{2}\right)$
Axis of
symmetry: $y = \dfrac{3}{2}$

13. $y = -3x^2 - 9x$

Vertex: $\left(-\dfrac{3}{2}, \dfrac{27}{4}\right)$

Axis of symmetry: $x = -\dfrac{3}{2}$

14. $y = -2x^2 + 6x$

Vertex: $\left(\dfrac{3}{2}, \dfrac{9}{2}\right)$

Axis of symmetry: $x = \dfrac{3}{2}$

15. $x = -\dfrac{1}{2}y^2 + 4$

Vertex: $(4, 0)$

Axis of symmetry: $y = 0$

16. $x = -\dfrac{1}{4}y^2 - 1$

Vertex: $(-1, 0)$

Axis of symmetry: $y = 0$

17. $x = \dfrac{1}{2}y^2 - y + 1$

Vertex: $\left(\dfrac{1}{2}, 1\right)$

Axis of symmetry: $y = 1$

18. $x = -\dfrac{1}{2}y^2 + 2y - 3$

Vertex: $(-1, 2)$

Axis of symmetry: $y = 2$

19. $y = \dfrac{1}{2}x^2 + 2x - 6$

Vertex: $(-2, -8)$

Axis of symmetry: $x = -2$

20. $y = -\dfrac{1}{2}x^2 + x - 3$

Vertex: $\left(1, -\dfrac{5}{2}\right)$

Axis of symmetry: $x = 1$

2 Find the minimum or maximum value of the quadratic function.

21. $f(x) = x^2 - 2x + 3$

minimum: 2

22. $f(x) = x^2 + 3x - 4$

minimum: $-\dfrac{25}{4}$

23. $f(x) = -2x^2 + 4x - 3$

maximum: -1

24. $f(x) = -x^2 - x + 2$

maximum: $\dfrac{9}{4}$

25. $f(x) = 3x^2 + 3x - 2$

minimum: $-\dfrac{11}{4}$

26. $f(x) = x^2 - 5x + 3$

minimum: $-\dfrac{13}{4}$

27. $f(x) = -3x^2 + 4x - 2$

maximum: $-\dfrac{2}{3}$

28. $f(x) = -2x^2 - 5x + 1$

maximum: $\dfrac{33}{8}$

3 Solve.

29. The height in feet (s) of a rock thrown upward at an initial speed of 64 ft/s from a cliff 50 ft high is given by the function $s(t) = -16t^2 + 64t + 50$, where t is the time in seconds. Find the maximum height above the ground that the rock will attain.
114 ft

30. The height in feet (s) of a ball thrown upward at an initial speed of 80 ft/s from a platform 50 ft high is given by the function $s(t) = -16t^2 + 80t + 50$, where t is the time in seconds. Find the maximum height above the ground that the ball will attain.
150 ft

31. A pool is treated with a chemical to reduce the amount of algae. The amount of algae in the pool t days after the treatment can be approximated by the function $A(t) = 40t^2 - 400t + 500$. How many days after treatment will the pool have the least amount of algae?
5 days

32. The suspension cable that supports a small footbridge hangs in the shape of a parabola. The height in feet of the cable above the bridge is given by the function $h(x) = 0.25x^2 - 0.8x + 25$, where x is the distance from one end of the bridge. What is the minimum height of the cable above the bridge?
24.36 ft

33. A manufacturer of microwave ovens believes that the revenue the company receives is related to the price (p) of an oven by the function $R(p) = 125p - \frac{1}{4}p^2$. What price will give the maximum revenue?
$250

34. A manufacturer of camera lenses estimated that the average monthly cost of producing lenses can be given by the function $C(x) = 0.1x^2 - 20x + 2000$, where x is the number of lenses produced each month. Find the number of lenses to produce in order to minimize the average cost.
100 lenses

35. Find two numbers whose sum is 20 and whose product is a maximum.
10 and 10

36. Find two numbers whose sum is 50 and whose product is a maximum.
25 and 25

37. Find two numbers whose difference is 24 and whose product is a minimum.
12 and −12

38. Find two numbers whose difference is 14 and whose product is a minimum.
7 and −7

SUPPLEMENTAL EXERCISES 9.1

Use the vertex and the direction in which the parabola opens to determine the domain and range of the relation.

39. $y = x^2 - 4x - 2$
D: the real numbers; R: $y \geq -6$

40. $y = x^2 - 6x + 1$
D: the real numbers; R: $y \geq -8$

41. $y = -x^2 + 2x - 3$
D: the real numbers; R: $y \leq -2$

42. $y = -x^2 - 2x + 4$
D: the real numbers: R: $y \leq 5$

43. $x = y^2 + 6y - 5$
D: $x \geq -14$; R: the real numbers

44. $x = y^2 + 4y - 3$
D: $x \geq -7$; R: the real numbers

45. $x = -y^2 - 2y + 6$
D: $x \leq 7$; R: the real numbers

46. $x = -y^2 - 6y + 2$
D: $x \leq 11$; R: the real numbers

47. $y = -2x^2 + 4x - 1$
D: the real numbers; R: $y \leq 1$

48. $y = -\frac{1}{2}x^2 - x + 5$
D: the real numbers; R: $y \leq \frac{11}{2}$

49. $x = -\frac{1}{2}y^2 + 2y - 4$
D: $x \leq -2$; R: the real numbers

50. $x = -2y^2 + 8y - 1$
D: $x \leq 7$; R: the real numbers

An equation of the form $y = ax^2 + bx + c$ can be written in the form $y = a(x - h)^2 + k$, where (h, k) are the coordinates of the vertex of the parabola. Use the process of completing the square to rewrite the equation in the form $y = a(x - h)^2 + k$. Find the vertex. (*Hint:* Review the example at the bottom of page 439.)

51. $y = x^2 - 4x + 5$
$y = (x - 2)^2 + 1$: vertex: $(2, 1)$

52. $y = x^2 - 2x - 2$
$y = (x - 1)^2 - 3$; vertex: $(1, -3)$

53. $y = x^2 - 6x + 3$
$y = (x - 3)^2 - 6$; vertex: $(3, -6)$

54. $y = x^2 + 4x - 1$
$y = (x + 2)^2 - 5$; vertex: $(-2, -5)$

55. $y = x^2 + x + 2$
$y = \left(x + \frac{1}{2}\right)^2 + \frac{7}{4}$; vertex: $\left(-\frac{1}{2}, \frac{7}{4}\right)$

56. $y = x^2 - x - 3$
$y = \left(x - \frac{1}{2}\right)^2 - \frac{13}{4}$; vertex: $\left(\frac{1}{2}, -\frac{13}{4}\right)$

SECTION 9.2
The Circle

1 Find the distance between two points in the plane

The distance between two points on a horizontal or vertical number line is the absolute value of the difference between the coordinates of the two points.

Find the distance between the points -2 and 4 on the vertical number line shown at the right.

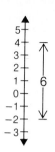

The distance is the absolute value of the difference between the coordinates.

$$\text{distance} = |4 - (-2)| = |6| = 6$$

Absolute value is used so that the coordinates can be subtracted in either order.

$$\text{distance} = |-2 - 4| = |-6| = 6$$

Now consider the points and the right triangle in the coordinate plane shown at the right. The vertical distance between the two points (x_1, y_1) and (x_2, y_2) is $|y_1 - y_2|$.

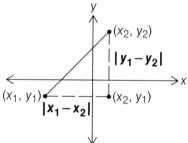

The horizontal distance between these two points is $|x_1 - x_2|$.

The Pythagorean Theorem is used to find the distance between two points (x_1, y_1) and (x_2, y_2).

$$d^2 = |x_1 - x_2|^2 + |y_1 - y_2|^2$$

Since the square of a number is always nonnegative, the absolute value signs are not necessary.

$$d = \sqrt{(x_1 - x_2)^2 + (y_1 - y_2)^2}$$

The Distance Formula

If (x_1, y_1) and (x_2, y_2) are two points in the plane, then the distance between the two points is given by $d = \sqrt{(x_1 - x_2)^2 + (y_1 - y_2)^2}$.

Example 1   Find the distance between the points $(2, -1)$ and $(-4, 3)$.

Solution $(x_1, y_1) = (2, -1)$ $(x_2, y_2) = (-4, 3)$

$$d = \sqrt{(x_1 - x_2)^2 + (y_1 - y_2)^2} = \sqrt{[2 - (-4)]^2 + (-1 - 3)^2}$$
$$= \sqrt{6^2 + (-4)^2} = \sqrt{36 + 16} = \sqrt{52} = 2\sqrt{13}$$

Problem 1   Find the distance between the points $(3, -2)$ and $(-1, -5)$.

Solution   See page A73.

5

2 Find the equation of a circle and then
graph the circle

A **circle** is a conic section that is formed by the intersection of a cone by a
plane parallel to the base of the cone.

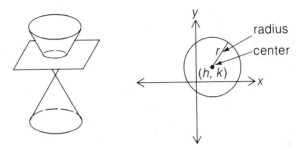

A **circle** can be defined as the set of all points (x, y) in the plane that are a
fixed distance from a given point (h, k) called the center. The fixed distance is
the **radius** of the circle.

Using the distance formula, the equation of a circle can be determined.

Let (h, k) represent the center of the circle, r the
radius, and (x, y) any point on the circle.
Then $r = \sqrt{(x - h)^2 + (y - k)^2}$.

Squaring both sides of the equation gives the
equation of a circle.

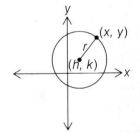

The Standard Form of an Equation of a Circle

Let r be the radius of a circle and (h, k) the center of the circle. Then the
equation of the circle is given by

$$(x - h)^2 + (y - k)^2 = r^2$$

To find the equation of the circle with radius 4 and center $(-1, 2)$, use the standard
form of the equation of a circle.

$$(x - h)^2 + (y - k)^2 = r^2$$

Replace r by 4, h by -1, and k by 2.

$$[(x - (-1)]^2 + (y - 2)^2 = 4^2$$
$$(x + 1)^2 + (y - 2)^2 = 16$$

To sketch the graph of this circle, draw a circle with center $(-1, 2)$ and radius 4.

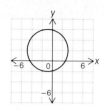

Recall that the graph of a circle is not the graph of a function. The graph of a circle is the graph of a relation.

A circle passes through point $(3, -4)$ and has as its center the point $(2, 1)$. To find the equation of this circle, use the distance formula to find the radius of the circle since a point on the circle and the center is known.

$$\sqrt{(x_1 - x_2)^2 + (y_1 - y_2)^2} = r$$
$$\sqrt{(3 - 2)^2 + (-4 - 1)^2} = r$$
$$\sqrt{1^2 + (-5)^2} = r$$
$$\sqrt{1 + 25} = r$$
$$\sqrt{26} = r$$

The radius is $\sqrt{26}$. The equation of the circle is $(x - 2)^2 + (y - 1)^2 = 26$.

Example 2   Sketch a graph of the circle $(x + 2)^2 + (y - 1)^2 = 4$.

Solution   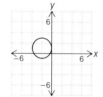   • Draw a circle with center $(-2, 1)$ and radius 2.

Problem 2   Sketch a graph of the circle $(x - 2)^2 + (y + 3)^2 = 9$.

Solution   See page A74.

Example 3   Find the equation of the circle with radius 5 and center $(-1, 3)$. Then sketch its graph.

Solution
$$(x - h)^2 + (y - k)^2 = r^2$$
$$[x - (-1)]^2 + (y - 3)^2 = 5^2$$
$$(x + 1)^2 + (y - 3)^2 = 25$$

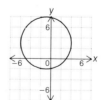

Problem 3   Find the equation of the circle with radius 4 and center $(2, -3)$. Then sketch its graph.

Solution   See page A74.
$$(x - 2)^2 + (y + 3)^2 = 16$$

3   Write the equation of a circle in standard form and then graph the circle

The equation of a circle can also be expressed as $x^2 + y^2 + ax + by + c = 0$. To rewrite this equation in standard form, it is necessary to complete the square on the x and y terms.

Write the equation of the circle $x^2 + y^2 + 4x + 2y + 1 = 0$ in standard form.

$$x^2 + y^2 + 4x + 2y + 1 = 0$$

Subtract the constant term from each side of the equation.

$$x^2 + y^2 + 4x + 2y = -1$$

Rewrite the equation by grouping terms involving x and terms involving y.

$$(x^2 + 4x) + (y^2 + 2y) = -1$$

Complete the square on $x^2 + 4x$ and $y^2 + 2y$.

$$(x^2 + 4x + 4) + (y^2 + 2y + 1) = -1 + 4 + 1$$
$$(x^2 + 4x + 4) + (y^2 + 2y + 1) = 4$$

Factor each trinomial.

$$(x + 2)^2 + (y + 1)^2 = 4$$

Example 4   Write the equation of the circle $x^2 + y^2 + 3x - 2y = 1$ in standard form. Then sketch its graph.

Solution

$$x^2 + y^2 + 3x - 2y = 1$$

$$(x^2 + 3x) + (y^2 - 2y) = 1$$

- Group terms involving x and terms involving y.

$$\left(x^2 + 3x + \frac{9}{4}\right) + (y^2 - 2y + 1) = 1 + \frac{9}{4} + 1$$

- Complete the square on $x^2 + 3x$ and $y^2 - 2y$.

$$\left(x + \frac{3}{2}\right)^2 + (y - 1)^2 = \frac{17}{4}$$

- Factor each trinomial.

- Draw a circle with center $\left(-\frac{3}{2}, 1\right)$ and radius $\sqrt{\frac{17}{4}} = \frac{\sqrt{17}}{2}$.

> **Problem 4** Write the equation of the circle $x^2 + y^2 - 4x + 8y + 15 = 0$ in standard form. Then sketch its graph.
>
> **Solution** See page A74.
> $(x - 2)^2 + (y + 4)^2 = 5$

EXERCISES 9.2

1 Find the distance between the two points.

1. $P_1 = (3, 2)$; $P_2 = (4, 5)$ $\sqrt{10}$

2. $P_1 = (4, 5)$; $P_2 = (1, 1)$ 5

3. $P_1 = (-4, 2)$; $P_2 = (1, -1)$ $\sqrt{34}$

4. $P_1 = (3, -2)$; $P_2 = (-1, 4)$ $2\sqrt{13}$

5. $P_1 = (-1, -2)$; $P_2 = (1, 5)$ $\sqrt{53}$

6. $P_1 = (-2, -3)$; $P_2 = (3, 1)$ $\sqrt{41}$

7. $P_1 = (-1, -1)$; $P_2 = (-4, -5)$ 5

8. $P_1 = (-4, 1)$; $P_2 = (3, -2)$ $\sqrt{58}$

2 Sketch a graph of the circle.

9. $(x - 2)^2 + (y + 2)^2 = 9$

10. $(x + 2)^2 + (y - 3)^2 = 16$

11. $(x + 3)^2 + (y - 1)^2 = 25$

12. $(x - 2)^2 + (y + 3)^2 = 4$

13. Find the equation of the circle with radius 2 and center $(2, -1)$. Then sketch its graph.
$(x - 2)^2 + (y + 1)^2 = 4$

14. Find the equation of the circle with radius 3 and center $(-1, -2)$. Then sketch its graph.
$(x + 1)^2 + (y + 2)^2 = 9$

15. Find the equation of the circle that passes through point $(1, 2)$ and whose center is $(-1, 1)$. Then sketch its graph.
$(x + 1)^2 + (y - 1)^2 = 5$

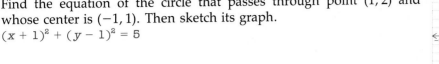

16. Find the equation of the circle that passes through point $(-1, 3)$ and whose center is $(-2, 1)$. Then sketch its graph.
$(x + 2)^2 + (y - 1)^2 = 5$

3 Write the equation of the circle in standard form. Then sketch its graph.

17. $x^2 + y^2 - 2x + 4y - 20 = 0$
$(x - 1)^2 + (y + 2)^2 = 25$

18. $x^2 + y^2 - 4x + 8y + 4 = 0$
$(x - 2)^2 + (y + 4)^2 = 16$

19. $x^2 + y^2 + 6x + 8y + 9 = 0$
$(x + 3)^2 + (y + 4)^2 = 16$

20. $x^2 + y^2 - 6x + 10y + 25 = 0$
$(x - 3)^2 + (y + 5)^2 = 9$

SUPPLEMENTAL EXERCISES 9.2

The vertices of a triangle are the given points. Use the Pythagorean Theorem to determine whether the triangle is a right triangle.

21. $(0, 2)$, $(3, 5)$, $(6, 2)$
yes

22. $(-3, 4)$, $(-1, 4)$, $(7, -2)$
no

23. $(-1, 0)$, $(7, -8)$, $(0, -7)$
no

24. $(-3, 6)$, $(6, -6)$, $(-6, 0)$
yes

Write the equation of the circle in standard form.

25. The circle has center $(3, 0)$ and passes through the origin.
$(x - 3)^2 + y^2 = 9$

26. The circle has center $(-2, 0)$ and passes through the origin.
$(x + 2)^2 + y^2 = 4$

27. The diameter of the circle has endpoints $(-1, 3)$ and $(5, 5)$.
$(x - 2)^2 + (y - 4)^2 = 10$

28. The diameter of the circle has endpoints $(-2, 4)$ and $(2, -2)$.
$x^2 + (y - 1)^2 = 13$

29. The circle has radius 1, is tangent to both the x- and y-axes, and lies in Quadrant II.
$(x + 1)^2 + (y - 1)^2 = 1$

30. The circle has radius 1, is tangent to both the x- and y-axes, and lies in Quadrant IV.
$(x - 1)^2 + (y + 1)^2 = 1$

SECTION 9.3

The Ellipse and the Hyperbola

1 Graph an ellipse with center at the origin

The orbits of the planets around the sun are oval shaped. This oval shape can be described as an **ellipse**, which is another of the conic sections.

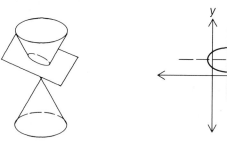

There are two **axes of symmetry** for an ellipse. The intersection of the two axes is the **center** of the ellipse.

An ellipse with center at the origin is shown at the right. Note that there are two x-intercepts and two y-intercepts.

Using the vertical line test, the graph of an ellipse is not the graph of a function. The graph of an ellipse is the graph of a relation.

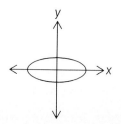

> **The Standard Form of the Equation of an Ellipse with Center at the Origin**
>
> The equation of an ellipse with center at the origin is $\dfrac{x^2}{a^2} + \dfrac{y^2}{b^2} = 1$.
>
> The x-intercepts are $(a, 0)$ and $(-a, 0)$.
> The y-intercepts are $(0, b)$ and $(0, -b)$.

By finding the x- and y-intercepts for an ellipse and using the fact that the ellipse is oval shaped, a graph of an ellipse can be sketched.

Example 1 Sketch a graph of the ellipse.

A. $\dfrac{x^2}{9} + \dfrac{y^2}{4} = 1$ B. $\dfrac{x^2}{16} + \dfrac{y^2}{12} = 1$

Solution A. $\dfrac{x^2}{9} + \dfrac{y^2}{4} = 1$

x-intercepts: $(3, 0)$ and $(-3, 0)$

y-intercepts: $(0, 2)$ and $(0, -2)$

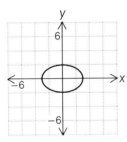

- $a^2 = 9,\ b^2 = 4$
- The x-intercepts are $(a, 0)$ and $(-a, 0)$.
- The y-intercepts are $(0, b)$ and $(0, -b)$.
- Use the intercepts and symmetry to sketch the graph of the ellipse.

B. $\dfrac{x^2}{16} + \dfrac{y^2}{12} = 1$

x-intercepts: $(4, 0)$ and $(-4, 0)$

y-intercepts: $(0, 2\sqrt{3})$ and $(0, -2\sqrt{3})$

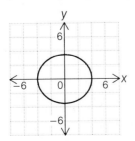

- $a^2 = 16,\ b^2 = 12$
- The x-intercepts are $(a, 0)$ and $(-a, 0)$.
- The y-intercepts are $(0, b)$ and $(0, -b)$.
- Use the intercepts and symmetry to sketch the graph of the ellipse. $2\sqrt{3} \approx 3.5$

Problem 1   Sketch a graph of the ellipse.

A. $\dfrac{x^2}{4} + \dfrac{y^2}{25} = 1$ B. $\dfrac{x^2}{18} + \dfrac{y^2}{9} = 1$

Solution   See page A75.

A.

B.

The *x*-intercepts of the graph of the ellipse $\dfrac{x^2}{16} + \dfrac{y^2}{16} = 1$ are $(4, 0)$ and $(-4, 0)$. The *y*-intercepts are $(0, 4)$ and $(0, -4)$.

This is the graph of a circle. A circle is a special case of an ellipse and occurs when $a^2 = b^2$ in the equation $\dfrac{x^2}{a^2} + \dfrac{y^2}{b^2} = 1$.

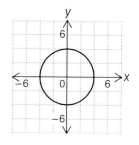

2 Graph a hyperbola with center at the origin

A **hyperbola** is a conic section that is formed by the intersection of a cone by a plane perpendicular to the base of the cone.

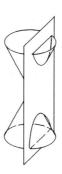
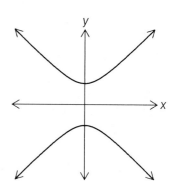

The hyperbola has two **vertices** and an **axis of symmetry** that passes through the vertices. The **center** of a hyperbola is the point halfway between the two vertices.

The graphs at the top of the next page show two possible graphs of a hyperbola with center at the origin. In the first graph, the axis of symmetry is the *x*-axis, and the vertices are *x*-intercepts. In the second graph, the axis of symmetry is the *y*-axis, and the vertices are *y*-intercepts.

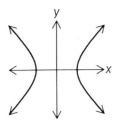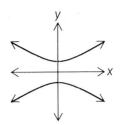

In either case, the graph of a hyperbola is not the graph of a function. The graph of a hyperbola is the graph of a relation.

The Standard Form of the Equation of a Hyperbola with Center at the Origin

The equation of a hyperbola for which the axis of symmetry is the x-axis is $\dfrac{x^2}{a^2} - \dfrac{y^2}{b^2} = 1$. The vertices are $(a, 0)$ and $(-a, 0)$.

The equation of a hyperbola for which the axis of symmetry is the y-axis is $\dfrac{y^2}{a^2} - \dfrac{x^2}{b^2} = 1$. The vertices are $(0, a)$ and $(0, -a)$.

To sketch a hyperbola, it is helpful to draw two lines that are "approached" by the hyperbola. These two lines are called **asymptotes**. As the hyperbola gets farther from the origin, the hyperbola "gets closer to" the asymptotes.

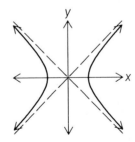

Since the asymptotes are straight lines, their equations are linear equations.

The equations of the asymptotes for the hyperbola $\dfrac{x^2}{a^2} - \dfrac{y^2}{b^2} = 1$ are $y = \dfrac{b}{a}x$ and $y = -\dfrac{b}{a}x$.

The equations of the asymptotes for the hyperbola $\dfrac{y^2}{a^2} - \dfrac{x^2}{b^2} = 1$ are $y = \dfrac{a}{b}x$ and $y = -\dfrac{a}{b}x$.

Example 2   Sketch a graph of the hyperbola.

A. $\dfrac{x^2}{16} - \dfrac{y^2}{4} = 1$ B. $\dfrac{y^2}{16} - \dfrac{x^2}{25} = 1$

Solution   A. $\dfrac{x^2}{16} - \dfrac{y^2}{4} = 1$

• $a^2 = 16,\ b^2 = 4$

Axis of symmetry: x-axis

Vertices: $(4, 0)$ and $(-4, 0)$

• The vertices are $(a, 0)$ and $(-a, 0)$.

Asymptotes: $y = \dfrac{1}{2}x$ and $y = -\dfrac{1}{2}x$

• The asymptotes are $y = \dfrac{b}{a}x$ and $y = -\dfrac{b}{a}x.$

• Sketch the asymptotes. Use symmetry and the fact that the hyperbola will approach the asymptotes to sketch its graph.

B. $\dfrac{y^2}{16} - \dfrac{x^2}{25} = 1$

• $a^2 = 16,\ b^2 = 25$

Axis of symmetry: y-axis

Vertices: $(0, 4)$ and $(0, -4)$

• The vertices are $(0, a)$ and $(0, -a)$.

Asymptotes: $y = \dfrac{4}{5}x$ and $y = -\dfrac{4}{5}x$

• The asymptotes are $y = \dfrac{a}{b}x$ and $y = -\dfrac{a}{b}x.$

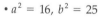

• Sketch the asymptotes. Use symmetry and the fact that the hyperbola will approach the asymptotes to sketch its graph.

Problem 2   Sketch a graph of the hyperbola.

A. $\dfrac{x^2}{9} - \dfrac{y^2}{25} = 1$ B. $\dfrac{y^2}{9} - \dfrac{x^2}{9} = 1$

Solution   See page A75.

A.   B.

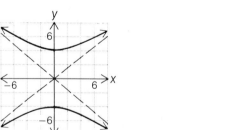

EXERCISES 9.3

1   Sketch a graph of the ellipse.

1. $\dfrac{x^2}{4} + \dfrac{y^2}{9} = 1$

2. $\dfrac{x^2}{25} + \dfrac{y^2}{16} = 1$

3. $\dfrac{x^2}{25} + \dfrac{y^2}{9} = 1$

4. $\dfrac{x^2}{16} + \dfrac{y^2}{9} = 1$

5. $\dfrac{x^2}{36} + \dfrac{y^2}{16} = 1$

6. $\dfrac{x^2}{49} + \dfrac{y^2}{64} = 1$

7. $\dfrac{x^2}{9} + \dfrac{y^2}{25} = 1$

8. $\dfrac{x^2}{8} + \dfrac{y^2}{25} = 1$

9. $\dfrac{x^2}{12} + \dfrac{y^2}{4} = 1$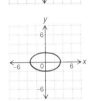

2   Sketch a graph of the hyperbola.

10. $\dfrac{x^2}{9} - \dfrac{y^2}{16} = 1$

11. $\dfrac{x^2}{25} - \dfrac{y^2}{4} = 1$

12. $\dfrac{y^2}{16} - \dfrac{x^2}{9} = 1$

13. $\dfrac{y^2}{4} - \dfrac{x^2}{9} = 1$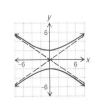

14. $\dfrac{x^2}{4} - \dfrac{y^2}{25} = 1$

15. $\dfrac{x^2}{9} - \dfrac{y^2}{49} = 1$

16. $\dfrac{y^2}{25} - \dfrac{x^2}{9} = 1$

17. $\dfrac{y^2}{4} - \dfrac{x^2}{16} = 1$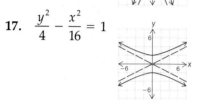

18. $\dfrac{x^2}{25} - \dfrac{y^2}{16} = 1$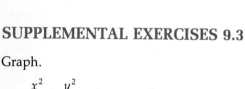

SUPPLEMENTAL EXERCISES 9.3

Graph.

19. $\dfrac{x^2}{4} + \dfrac{y^2}{16} = 1$

20. $\dfrac{x^2}{9} - \dfrac{y^2}{9} = 1$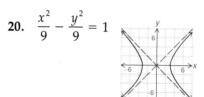

21. $\dfrac{y^2}{16} - \dfrac{x^2}{4} = 1$

22. $\dfrac{x^2}{16} + \dfrac{y^2}{36} = 1$

23. $\dfrac{x^2}{25} - \dfrac{y^2}{9} = 1$

24. $\dfrac{y^2}{9} - \dfrac{x^2}{36} = 1$

25. $\dfrac{x^2}{36} + \dfrac{y^2}{9} = 1$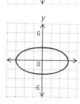

26. $\dfrac{x^2}{16} - \dfrac{y^2}{25} = 1$

SECTION 9.4 _____

Quadratic Inequalities

1 Graph the solution set of a quadratic inequality in two variables

The **graph of a quadratic inequality** in two variables is a region of the plane that is bounded by one of the conic sections (parabola, circle, ellipse, or hyperbola). When graphing an inequality of this type, use the point $(0, 0)$ to determine which portion of the plane to shade.

To graph the solution set of $x^2 + y^2 > 9$, change the inequality to an equality.

$$x^2 + y^2 > 9$$
$$x^2 + y^2 = 9$$

This is the equation of a circle with center $(0, 0)$ and radius 3.

Since the inequality is $>$, the graph is drawn as a dotted circle.

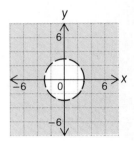

Substitute the point $(0, 0)$ in the inequality. Since $0^2 + 0^2 > 9$ is not true, the point $(0, 0)$ should not be in the shaded region.

Example 1 Graph the solution set.

A. $y \leq x^2 + 2x + 2$ B. $\dfrac{y^2}{9} - \dfrac{x^2}{4} \geq 1$

Solution A. $y \leq x^2 + 2x + 2$
$y = x^2 + 2x + 2$

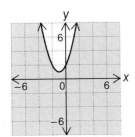

• Change the inequality to an equality.
This is the equation of a parabola that opens up. The vertex is $(-1, 1)$.
The axis of symmetry is the line $x = -1$.

• Since the inequality is $\leq$, the graph is drawn as a solid line.
Substitute the point $(0, 0)$ into the inequality.
Since $0 < 0^2 + 2(0) + 2$ is true, the point $(0, 0)$ should be in the shaded region.

B. $\dfrac{y^2}{9} - \dfrac{x^2}{4} \geq 1$

$\dfrac{y^2}{9} - \dfrac{x^2}{4} = 1$

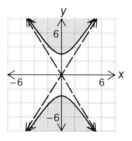

• Change the inequality to an equality.
This is the equation of a hyperbola.
The vertices are $(0, -3)$ and $(0, 3)$. The equations of
the asymptotes are $y = \dfrac{3}{2}x$ and $y = -\dfrac{3}{2}x$.

• Since the inequality is $\geq$, the graph is drawn as a
solid line.
Substitute the point $(0, 0)$ into the inequality.
Since $\dfrac{0^2}{9} - \dfrac{0^2}{4} \geq 1$ is not true, the point $(0, 0)$ should
not be in the shaded region.

Problem 1   Graph the solution set.

A. $\dfrac{x^2}{9} + \dfrac{y^2}{16} \leq 1$ B. $\dfrac{x^2}{9} - \dfrac{y^2}{4} \leq 1$

A.   B.

Solution   See page A76.

EXERCISES 9.4

1   Graph the solution set.

1. $y < x^2 - 4x + 3$

2. $y < x^2 - 2x - 3$

3. $(x - 1)^2 + (y + 2)^2 \leq 9$

4. $(x + 2)^2 + (y - 3)^2 > 4$

5. $\dfrac{x^2}{16} + \dfrac{y^2}{25} < 1$

6. $\dfrac{x^2}{9} + \dfrac{y^2}{4} \geq 1$

7. $\dfrac{x^2}{25} - \dfrac{y^2}{9} \le 1$

8. $\dfrac{y^2}{25} - \dfrac{x^2}{36} > 1$

SUPPLEMENTAL EXERCISES 9.4

Graph the solution set.

9. $x < y^2 - 6y + 1$

10. $x \ge 2y^2 - 8y + 7$

11. $x^2 + y^2 + 2x + 4y + 1 > 0$

12. $x^2 + y^2 + 6x - 7 \le 0$

13. $\dfrac{x^2}{4} + \dfrac{y^2}{8} < 1$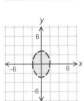

14. $\dfrac{x^2}{16} + \dfrac{y^2}{18} > 1$

CALCULATORS AND COMPUTERS

   Graph of an Ellipse or Hyperbola

The program GRAPH OF AN ELLIPSE OR HYPERBOLA on the Student Disk will graph an ellipse or hyperbola with the center at the origin. The program asks you to select from one of the following:

   1) $\dfrac{x^2}{a^2} + \dfrac{y^2}{b^2} = 1$ 2) $\dfrac{x^2}{a^2} - \dfrac{y^2}{b^2} = 1$ 3) $\dfrac{y^2}{a^2} - \dfrac{x^2}{b^2} = 1$

You provide the values for a and b, and the program will draw the graph. The values of a and b should be kept between 1 and 8. After the graph has been drawn, you may draw another graph without erasing the first graph, erase the graph and then draw another graph, or quit the program.

By choosing various values for a and b and viewing the resulting graphs, you can better understand the effect these numbers have on the graph.

CHAPTER SUMMARY

Key Words

The graph of a **conic section** can be represented by the intersection of a plane and a cone.

A **circle** is the set of all points (x, y) in the plane that are a fixed distance from a given point (h, k) called the center. The fixed distance is the **radius** of the circle.

The **asymptotes** of a hyperbola are two straight lines that are "approached" by the hyperbola. As the hyperbola gets farther from the origin, the hyperbola gets "closer to" the asymptotes.

The **graph of a quadratic inequality** in two variables is a region of the plane that is bounded by one of the conic sections.

Essential Rules

Equation of a Parabola
$y = ax^2 + bx + c$
When $a > 0$, the parabola opens up.
When $a < 0$, the parabola opens down.

The x-coordinate of the vertex is $-\dfrac{b}{2a}$.

The axis of symmetry is the line $x = -\dfrac{b}{2a}$.

$x = ay^2 + by + c$
When $a > 0$, the parabola opens to the right.
When $a < 0$, the parabola opens to the left.

The y-coordinate of the vertex is $-\dfrac{b}{2a}$.

The axis of symmetry is the line $y = -\dfrac{b}{2a}$.

Distance Formula	If (x_1, y_1) and (x_2, y_2) are two points in the plane, then the distance between the two points is given by $d = \sqrt{(x_1 - x_2)^2 + (y_1 - y_2)^2}$.
Equation of a Circle	$(x - h)^2 + (y - k)^2 = r^2$ The center is (h, k), and the radius is r.
Equation of an Ellipse	$\dfrac{x^2}{a^2} + \dfrac{y^2}{b^2} = 1$ The x-intercepts are $(a, 0)$ and $(-a, 0)$. The y-intercepts are $(0, b)$ and $(0, -b)$.
Equation of a Hyperbola	$\dfrac{x^2}{a^2} - \dfrac{y^2}{b^2} = 1$ The axis of symmetry is the x-axis. The vertices are $(a, 0)$ and $(-a, 0)$. The equations of the asymptotes are $y = \pm\dfrac{b}{a}x$. $\dfrac{y^2}{a^2} - \dfrac{x^2}{b^2} = 1$ The axis of symmetry is the y-axis. The vertices are $(0, a)$ and $(0, -a)$. The equations of the asymptotes are $y = \pm\dfrac{a}{b}x$.

CHAPTER REVIEW

1. Sketch a graph of
   $y = \dfrac{1}{2}x^2 + x - 4$.

2. Find the equation of the circle that passes through point $(1, -2)$ and whose center is $(4, 1)$.
   $(x - 4)^2 + (y - 1)^2 = 18$

3. Find the minimum value of the function
   $f(x) = 2x^2 - 6x + 1$.
   $-\dfrac{7}{2}$

4. Sketch a graph of
   $\dfrac{x^2}{4} + \dfrac{y^2}{36} = 1$.

5. Find the axis of symmetry of the parabola
$y = -x^2 + 6x - 5$.
$x = 3$

6. Graph the solution
set of
$(x + 3)^2 + (y - 1)^2 \geq 16$.

7. Sketch a graph of
$\dfrac{y^2}{25} - \dfrac{x^2}{16} = 1$.

8. Find the distance between the points
$(3, -4)$ and $(0, 3)$.
$\sqrt{58}$

9. Write the equation
$x^2 + y^2 - 4x + 2y - 4 = 0$
in standard form.
Then sketch its graph.
$(x - 2)^2 + (y + 1)^2 = 9$

10. Sketch a graph of
$x = y^2 + 2y - 5$.

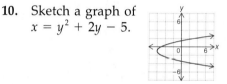

11. Find two numbers whose difference is 30
and whose product is a minimum.
15 and −15

12. Graph the solution set
of $\dfrac{x^2}{16} - \dfrac{y^2}{25} \geq 1$.

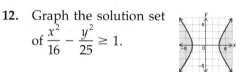

13. Find the equation of the circle with radius
3 and center $(-2, 4)$.
$(x + 2)^2 + (y - 4)^2 = 9$

14. Find the vertex of the parabola
$x = y^2 - 4y + 1$.
$(-3, 2)$

15. Sketch a graph of
$x = y^2 - y - 2$.

16. Find the maximum product of two numbers whose sum is 20.
100

17. Find the equation of the circle that passes
through point $(2, 5)$ and whose center is
$(-2, 1)$.
$(x + 2)^2 + (y - 1)^2 = 32$

18. Sketch a graph of
$\dfrac{y^2}{9} - \dfrac{x^2}{4} = 1$.

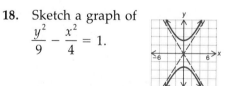

19. Sketch a graph of
$\dfrac{x^2}{16} + \dfrac{y^2}{4} = 1$.

20. Write the equation
$x^2 + y^2 - 4x + 2y + 1 = 0$
in standard form.
Then sketch its graph.
$(x - 2)^2 + (y + 1)^2 = 4$

21. Find the maximum value of the function
$f(x) = -3x^2 + 6x + 1$.
4

22. Sketch a graph of
$y = -x^2 - 2x + 3$.

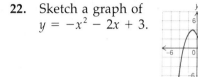

23. Graph the solution set
of $\dfrac{x^2}{9} - \dfrac{y^2}{16} \le 1$.

24. Find the distance between the points
$(-5, 1)$ and $(2, 0)$.
5√2

25. The perimeter of a rectangle is 48 cm. What dimensions would give the rectangle a maximum area?
12 cm × 12 cm

26. Find the vertex of the parabola
$y = -x^2 + 3x - 2$.
$\left(\dfrac{3}{2}, \dfrac{1}{4}\right)$

CUMULATIVE REVIEW

1. Graph the solution set of
$\{x \mid x < 4\} \cap \{x \mid x > 2\}$.

2. Solve: $\dfrac{5x - 2}{3} - \dfrac{1 - x}{5} = \dfrac{x + 4}{10}$
$\dfrac{38}{53}$

3. Solve: $4 + |3x + 2| < 6$
$\left\{x \mid -\dfrac{4}{3} < x < 0\right\}$

4. Find the equation of the line that contains the point $(2, -3)$ and has slope $-\dfrac{3}{2}$.
$y = -\dfrac{3}{2}x$

5. Find the equation of the line containing the point $(4, -2)$ and perpendicular to the line $y = -x + 5$.
$y = x - 6$

6. Simplify: $x^{2n}(x^{2n} + 2x^n - 3x)$
$x^{4n} + 2x^{3n} - 3x^{2n+1}$

7. Factor: $(x - 1)^3 - y^3$
$(x - y - 1)(x^2 - 2x + 1 + xy - y + y^2)$

8. Solve: $\dfrac{3x - 2}{x + 4} \le 1$
$\{x \mid -4 < x \le 3\}$

9. Simplify: $\dfrac{\dfrac{ax - bx}{ax + ay - bx - by}}{\dfrac{x}{x + y}}$

10. Simplify: $\dfrac{\dfrac{x - 4}{x - 5}}{3x - 2} - \dfrac{1 + x}{3x^2 + x - 2}$
$\dfrac{}{3x - 2}$

11. Solve: $\dfrac{6x}{2x-3} - \dfrac{1}{2x-3} = 7$

$\dfrac{5}{2}$

12. Graph the solution set of $5x + 2y > 10$.

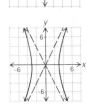

13. Simplify: $\left(\dfrac{12a^2 b^{-2}}{a^{-3}b^{-4}}\right)^{-1}\left(\dfrac{ab}{4^{-1}a^{-2}b^4}\right)^2$

$\dfrac{4a}{3b^8}$

14. Write $2\sqrt[4]{x^3}$ as an exponential expression.

$2x^{\frac{3}{4}}$

15. Simplify: $\sqrt{18} - \sqrt{-25}$

$3\sqrt{2} - 5i$

16. Solve: $2x^2 + 2x - 3 = 0$

$\dfrac{-1+\sqrt{7}}{2}$ and $\dfrac{-1-\sqrt{7}}{2}$

17. Solve: $a^{\frac{2}{3}}\ \ 2a^{\frac{1}{3}}\ \ 3 = 0$

27 and −1

18. Solve: $x - \sqrt{2x - 3} = 3$

6

19. Evaluate the function $f(x) = -x^2 + 3x - 2$ at $x = -3$.

−20

20. Find the inverse of the function $f(x) = 4x + 8$.

$f^{-1}(x) = \dfrac{1}{4}x - 2$

21. Find the maximum value of the function $f(x) = -2x^2 + 4x - 2$.

0

22. Find the maximum product of two numbers whose sum is 40.

400

23. Find the distance between the points $(2, 4)$ and $(-1, 0)$.

5

24. Find the equation of the circle that passes through $(3, 1)$ and whose center is $(-1, 2)$.

$(x + 1)^2 + (y - 2)^2 = 17$

25. Graph $x = y^2 - 2y + 3$.

26. Graph $\dfrac{x^2}{25} + \dfrac{y^2}{4} = 1$.

27. Graph $\dfrac{y^2}{4} - \dfrac{x^2}{25} = 1$.

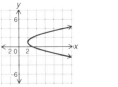

28. Graph $\dfrac{x^2}{9} - \dfrac{y^2}{36} = 1$.

29. Tickets for a school play sold for $4.00 for each adult and $1.50 for each child. The total receipt for the 192 tickets sold was $493. Find the number of adult tickets sold.

82 tickets

30. A motorcycle travels 180 mi in the same amount of time that a car travels 144 mi. The rate of the motorcycle is 12 mph faster than the rate of the car. Find the rate of the motorcycle.
60 mph

31. The rate of a river's current is 1.5 mph. A rowing crew can row 12 mi down this river and 12 mi back in 6 h. Find the rowing rate of the crew in calm water.
4.5 mph

32. The speed (v) of a gear varies inversely as the number of teeth (t). If a gear that has 36 teeth makes 30 revolutions per minute, how many revolutions will a gear that has 60 teeth make?
18 revolutions/min

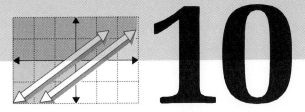

10

Systems of Equations and Inequalities

OBJECTIVES

- Solve systems of linear equations by graphing
- Solve systems of linear equations by the substitution method
- Solve systems of two linear equations in two variables by the addition method
- Solve systems of three linear equations in three variables by the addition method
- Evaluate determinants
- Solve systems of equations by using Cramer's Rule
- Solve systems of equations by using matrices
- Rate-of-wind and water current problems
- Application problems
- Solve nonlinear systems of equations
- Graph the solution set of a system of inequalities

*F*our-Color Theorem

The Four-Color Theorem asserts that any map on a plane or sphere needs at most four colors to color it so that no two countries sharing a common boundary will have the same color. This conjecture was first proposed around 1850 and has been the subject of extensive mathematical research.

The theorem was finally proved in 1976, more than 125 years after it was first proposed. One of the unique aspects of the proof of this theorem is that it was based on an intricate computer-assisted analysis. It was the first time a computer was used to prove a long-standing, previously unsolved problem.

The mathematicians who proved the result were affiliated with the University of Illinois. In tribute to their efforts, a postage-meter stamp was designed to commemorate the event.

FOUR COLORS

SUFFICE

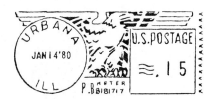

Solving Systems of Linear Equations by Graphing and by the Substitution Method

1 Solve systems of linear equations by graphing

A **system of equations** is two or more equations considered together. The system shown below is a system of two linear equations in two variables.

$$3x + 4y = 7$$
$$2x - 3y = 6$$

The graphs of the equations are straight lines.

A **solution of a system of equations in two variables** is an ordered pair that is a solution of each equation of the system.

Is $(3, -2)$ a solution of the system
$$2x - 3y = 12$$
$$5x + 2y = 11?$$

$2x - 3y = 12$	
$2(3) - 3(-2)$	12
$6 - (-6)$	
	$12 = 12$

$5x + 2y = 11$	
$5(3) + 2(-2)$	11
$15 + (-4)$	
	$11 = 11$

Yes, since $(3, -2)$ is a solution of each equation, it is a solution of the system of equations.

A solution of a system of linear equations can be found by graphing the lines of the system on the same coordinate axes. The point of intersection of the lines is the ordered pair that lies on both lines. It is the solution of the system of equations.

Solve by graphing: $x + 2y = 4$
$2x + y = -1$

Graph each line.

Find the point of intersection.

The ordered pair $(-2, 3)$ lies on each line.

The solution is $(-2, 3)$.

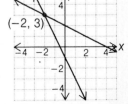

When the graphs of a system of equations intersect at only one point, the equations are called **independent equations**.

Solve by graphing: $2x + 3y = 6$
$4x + 6y = -12$

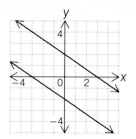

Graph each line.

The lines are parallel and therefore do not intersect.
The system of equations has no solution.

When a system of equations has no solution, it is called an **inconsistent system of equations.**

Solve by graphing: $x - 2y = 4$
$2x - 4y = 8$

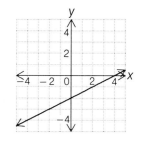

Graph each line.

The two equations represent the same line. Any ordered pair that is a solution of one equation is also a solution of the other equation.

The equations in this system are **dependent equations.**

Example 1   Solve by graphing.
  A. $2x - y = 3$ B. $2x + 3y = 6$
  $3x + y = 2$ $y = -\dfrac{2}{3}x + 1$

Solution   A.

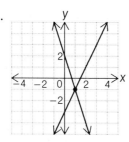

B.

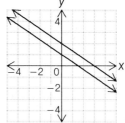

The solution is $(1, -1)$.   The system is inconsistent.

Problem 1   Solve by graphing.
  A. $x + y = 1$ B. $3x - 4y = 12$
  $2x + y = 0$ $y = \dfrac{3}{4}x - 3$

Solution   See page A79.
  A. $(-1, 2)$ B. dependent equations

2 Solve systems of linear equations by the substitution method

Graphical methods of solving systems of equations are often unsatisfactory since it can be difficult to read exact values of a point of intersection. For example, the point $\left(\dfrac{2}{3}, -\dfrac{4}{7}\right)$ would be difficult to read from a graph.

An algebraic method, called the **substitution method,** can be used to find an exact solution of a system of equations.

In the system of equations at the right, equation (2) states that $y = 3x - 4$.

$$\begin{aligned}(1) && 2x - 3y &= 5 \\ (2) && y &= 3x - 4\end{aligned}$$

Substitute $3x - 4$ for y in equation (1).

$$2x - 3(3x - 4) = 5$$

Solve for x.

$$\begin{aligned}2x - 9x + 12 &= 5 \\ -7x + 12 &= 5 \\ -7x &= -7 \\ x &= 1\end{aligned}$$

Substitute the value of x into equation (2), and solve for y.

$$\begin{aligned}y &= 3x - 4 \\ y &= 3(1) - 4 \\ y &= 3 - 4 \\ y &= -1\end{aligned}$$

The solution is $(1, -1)$.

Solve by substitution: $\begin{aligned}3x - y &= 5 \\ 2x + 5y &= 9\end{aligned}$

$$\begin{aligned}(1) && 3x - y &= 5 \\ (2) && 2x + 5y &= 9\end{aligned}$$

Solve equation (1) for y. Equation (1) is chosen because it is the easier equation to solve for one variable in terms of the other.

$$\begin{aligned}3x - y &= 5 \\ -y &= -3x + 5 \\ y &= 3x - 5\end{aligned}$$

Substitute $3x - 5$ for y in equation (2).

$$\begin{aligned}2x + 5y &= 9 \\ 2x + 5(3x - 5) &= 9 \\ 2x + 15x - 25 &= 9 \\ 17x - 25 &= 9 \\ 17x &= 34 \\ x &= 2\end{aligned}$$

Substitute the value of x into the equation $y = 3x - 5$ and solve for y.

$$\begin{aligned}y &= 3x - 5 \\ y &= 3(2) - 5 \\ y &= 6 - 5 \\ y &= 1\end{aligned}$$

The solution is $(2, 1)$.

Check:

$3x - y = 5$	$2x + 5y = 9$
$3(2) - 1 \mid 5$	$2(2) + 5(1) \mid 9$
$6 - 1 \mid$	$4 + 5 \mid$
$5 = 5$	$9 = 9$

Example 2 Solve by substitution.

A. $3x - 3y = 2$ B. $9x + 3y = 12$
     $y = x + 2$ $y = -3x + 4$

Solution A. $3x - 3y = 2$ (1)   • Equation (2) states that $y = x + 2$.
     $y = x + 2$ (2)

$3x - 3(x + 2) = 2$ • Substitute $x + 2$ for y in equation (1).
$3x - 3x - 6 = 2$
         $-6 = 2$ • This is not a true equation. The lines are parallel.

The system is inconsistent.

B. $9x + 3y = 12$ (1)   • Equation (2) states that $y = -3x + 4$.
    $y = -3x + 4$ (2)

$9x + 3(-3x + 4) = 12$ • Substitute $-3x + 4$ for y in equation (1).
$9x - 9x + 12 = 12$
         $12 = 12$ • This is a true equation. Any ordered pair that is a solution of one equation is a solution of the other equation.

The equations are dependent.

Problem 2 Solve by substitution.

A. $3x - y = 3$ B. $6x - 3y = 6$
      $6x + 3y = -4$ $2x - y = 2$

Solution See page A80.   A. $\left(\dfrac{1}{3}, -2\right)$ B. dependent equations

EXERCISES 10.1

1 Solve by graphing.

1. $x + y = 2$
   $x - y = 4$
   $(3, -1)$

2. $x + y = 1$
   $3x - y = -5$
   $(-1, 2)$

3. $x - y = -2$
   $x + 2y = 10$
   $(2, 4)$

4. $2x - y = 5$
   $3x + y = 5$
   $(2, -1)$

5. $3x - 2y = 6$
$y = 3$
$(4, 3)$

6. $x = 4$
$3x - 2y = 4$
$(4, 4)$

7. $x = 4$
$y = -1$
$(4, -1)$

8. $x + 2 = 0$
$y - 1 = 0$
$(-2, 1)$

9. $y = x - 5$
$2x + y = 4$
$(3, -2)$

10. $2x - 5y = 4$
$y = x + 1$
$(-3, -2)$

11. $y = \frac{1}{2}x - 2$
$x - 2y = 8$

no solution

12. $2x + 3y = 6$
$y = -\frac{2}{3}x + 1$

no solution

13. $2x - 5y = 10$
$y = \frac{2}{5}x - 2$

dependent equations

14. $3x - 2y = 6$
$y = \frac{3}{2}x - 3$

dependent equations

15. $3x - 4y = 12$
$5x + 4y = -12$
$(0, -3)$

16. $2x - 3y = 6$
$2x - 5y = 10$
$(0, -2)$

2   Solve by substitution.

17. $3x - 2y = 4$
$x = 2$
$(2, 1)$

18. $y = -2$
$2x + 3y = 4$
$(5, -2)$

19. $y = 2x - 1$
$x + 2y = 3$
$(1, 1)$

20. $y = -x + 1$
$2x - y = 5$
$(2, -1)$

21. $x = 3y + 1$
$x - 2y = 6$
$(16, 5)$

22. $x = 2y - 3$
$3x + y = 5$
$(1, 2)$

23. $4x - 3y = 5$
$y = 2x - 3$
$(2, 1)$

24. $3x + 5y = -1$
$y = 2x - 8$
$(3, -2)$

25. $5x - 2y = 9$
$y = 3x - 4$
$(-1, -7)$

26. $4x - 3y = 2$
$y = 3x + 1$
$(-1, -2)$

27. $x = 2y + 4$
$4x + 3y = -17$
$(-2, -3)$

28. $3x - 2y = -11$
$x = 2y - 9$
$(-1, 4)$

29. $5x + 4y = -1$
$y = 2 - 2x$
$(3, -4)$

30. $3x + 2y = 4$
$y = 1 - 2x$
$(-2, 5)$

31. $2x - 5y = -9$
$y = 9 - 2x$
$(3, 3)$

32. $5x + 2y = 15$
$x = 6 - y$
$(1, 5)$

33. $7x - 3y = 3$
$x = 2y + 2$
$(0, -1)$

34. $3x - 4y = 6$
$x = 3y + 2$
$(2, 0)$

35. $2x + 2y = 7$
$y = 4x + 1$
$\left(\dfrac{1}{2}, 3\right)$

36. $3x + 7y = -5$
$y = 6x - 5$
$\left(\dfrac{2}{3}, -1\right)$

37. $3x + y = 5$
$2x + 3y = 8$
$(1, 2)$

38. $4x + y = 9$
$3x - 4y = 2$
$(2, 1)$

39. $x + 3y = 5$
$2x + 3y = 4$
$(-1, 2)$

40. $x - 4y = 2$
$2x - 5y = 1$
$(-2, -1)$

41. $3x - y = 10$
$5x + 2y = 2$
$(2, -4)$

42. $2x - y = 5$
$3x - 2y = 9$
$(1, -3)$

43. $3x + 4y = 14$
$2x + y = 1$
$(-2, 5)$

44. $5x + 3y = 8$
$3x + y = 8$
$(4, -4)$

45. $3x + 5y = 0$
$x - 4y = 0$
$(0, 0)$

46. $2x - 7y = 0$
$3x + y = 0$
$(0, 0)$

47. $5x - 3y = -2$
$-x + 2y = -8$
$(-4, -6)$

48. $2x + 7y = 1$
$-x + 4y = 7$
$(-3, 1)$

49. $y = 3x + 2$
$y = 2x + 3$
$(1, 5)$

50. $y = 3x - 7$
$y = 2x - 5$

$(2, -1)$

51. $y = 3x + 1$
$y = 6x - 1$
$\left(\dfrac{2}{3}, 3\right)$

52. $y = 2x - 3$
$y = 4x - 4$
$\left(\dfrac{1}{2}, -2\right)$

53. $x = 2y + 1$
$x = 3y - 1$
$(5, 2)$

54. $x = 4y + 1$
$x = -2y - 5$
$(-3, -1)$

55. $y = 5x - 1$
$y = 5 - x$
$(1, 4)$

56. $y = 3 - 2x$
$y = 2 - 3x$
$(-1, 5)$

57. $-x + 2y = 13$
$5x + 3y = 13$
$(-1, 6)$

58. $3x - y = 8$
$4x - 7y = 5$
$(3, 1)$

SUPPLEMENTAL EXERCISES 10.1

For what values of k will the system of equations be inconsistent?

59. $2x - 2y = 5$
$kx - 2y = 3$ 2

60. $6x - 3y = 4$ $\dfrac{3}{2}$
$3x - ky = 1$

61. $x = 6y + 6$ $\dfrac{1}{2}$
$kx - 3y = 6$

62. $x = 2y + 2$
$kx - 8y = 2$ 4

Solve. (*Hint:* These equations are not linear equations. First rewrite the equations as linear equations by substituting x for $\dfrac{1}{a}$ and y for $\dfrac{1}{b}$.)

63. $\dfrac{2}{a} + \dfrac{3}{b} = 4$

$\dfrac{4}{a} + \dfrac{1}{b} = 3$

$(2, 1)$

64. $\dfrac{2}{a} + \dfrac{1}{b} = 1$

$\dfrac{8}{a} - \dfrac{2}{b} = 0$

$\left(6, \dfrac{3}{2}\right)$

65. $\dfrac{1}{a} + \dfrac{3}{b} = 2$

$\dfrac{4}{a} - \dfrac{1}{b} = 3$

$\left(\dfrac{13}{11}, \dfrac{13}{5}\right)$

66. $\dfrac{3}{a} + \dfrac{4}{b} = -1$

$\dfrac{1}{a} + \dfrac{6}{b} = 2$

$(-1, 2)$

67. $\dfrac{6}{a} + \dfrac{1}{b} = -1$

$\dfrac{3}{a} - \dfrac{2}{b} = -3$

$(-3, 1)$

68. $\dfrac{1}{a} - \dfrac{4}{b} = 1$

$\dfrac{5}{a} - \dfrac{2}{b} = -3$

$\left(-\dfrac{9}{7}, -\dfrac{9}{4}\right)$

Solve.

69. The sum of two numbers is 44. One number is 8 less than the other number. Find the numbers.
26 and 18

70. The sum of two numbers is 76. One number is 12 less than the other number. Find the numbers.
44 and 32

71. The sum of two numbers is 128. One number is 16 more than the other number. Find the numbers.
56 and 72

72. The sum of two numbers is 116. One number is 14 more than the other number. Find the numbers.
51 and 65

73. The sum of two numbers is 19. Five less than twice the first number equals the second number. Find the numbers.
8 and 11

74. The sum of two numbers is 22. Three times the first plus 2 equals the second number. Find the numbers.
5 and 17

SECTION $\mathbf{10.2}$ _____

Solving Systems of Linear Equations by the Addition Method

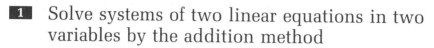 Solve systems of two linear equations in two variables by the addition method

The **addition method** is an alternative method for solving a system of equations. This method is based on the Addition Property of Equations. It is appropriate when it is not convenient to solve one equation for one variable in terms of another variable.

Note, for the system of equations at the right, the effect of adding equation (2) to equation (1). Since $-3y$ and $3y$ are additive inverses, adding the equations results in an equation with only one variable.

$$\begin{array}{ll} (1) & 5x - 3y = 14 \\ (2) & 2x + 3y = -7 \\ \hline & 7x + 0y = 7 \\ & 7x = 7 \end{array}$$

The solution of the resulting equation is the first component of the ordered pair solution of the system.

$$\begin{array}{l} 7x = 7 \\ x = 1 \end{array}$$

The second component is found by substituting the value of x into equation (1) or (2) and then solving for y. Equation (1) is used here.

$$(1) \quad 5x - 3y = 14$$
$$\begin{array}{l} 5(1) - 3y = 14 \\ 5 - 3y = 14 \\ -3y = 9 \\ y = -3 \end{array}$$

The solution is $(1, -3)$.

Sometimes adding the two equations does not eliminate one of the variables. In this case, use the Multiplication Property of Equations to rewrite one or both of the equations so that when the equations are added, one of the variables is eliminated. To do this, first choose which variable to eliminate. The coefficients of that variable must be additive inverses. Multiply each equation by a constant that will produce coefficients that are additive inverses.

Solve by the addition method: $3x + 4y = 2$ (1) $3x + 4y = 2$
$2x + 5y = -1$ (2) $2x + 5y = -1$

Eliminate x. Multiply equation (1) by 2 and equation (2) by -3. Note how the constants are selected.

$$2 \diagdown (3x + 4y) = 2(2)$$
$$-3 \diagup (2x + 5y) = -3(-1)$$

—The negative sign is used so that the coefficients will be additive inverses.

The coefficients of the x terms are additive inverses.

$$6x + 8y = 4$$
$$-6x - 15y = 3$$

Add the equations.
Solve for y.

$$-7y = 7$$
$$y = -1$$

Substitute the value of y into one of the equations, and solve for x. Equation (1) is used here.

$$(1) \quad 3x + 4y = 2$$
$$3x + 4(-1) = 2$$
$$3x - 4 = 2$$
$$3x = 6$$
$$x = 2$$

The solution is $(2, -1)$.

Solve by the addition method: $\dfrac{2}{3}x + \dfrac{1}{2}y = 4$

$\dfrac{1}{4}x - \dfrac{3}{8}y = -\dfrac{3}{4}$

$$(1) \quad \dfrac{2}{3}x + \dfrac{1}{2}y = 4$$
$$(2) \quad \dfrac{1}{4}x - \dfrac{3}{8}y = -\dfrac{3}{4}$$

Clear the fractions. Multiply each equation by the LCM of the denominators.

$$6\left(\dfrac{2}{3}x + \dfrac{1}{2}y\right) = 6(4)$$
$$8\left(\dfrac{1}{4}x - \dfrac{3}{8}y\right) = 8\left(-\dfrac{3}{4}\right)$$

$$4x + 3y = 24$$
$$2x - 3y = -6$$

Eliminate y. Add the equations.
Solve for x.

$$6x = 18$$
$$x = 3$$

Substitute the value of x into equation (1), and solve for y.

$$\dfrac{2}{3}x + \dfrac{1}{2}y = 4$$

$$\dfrac{2}{3}(3) + \dfrac{1}{2}y = 4$$

$$2 + \dfrac{1}{2}y = 4$$

$$\dfrac{1}{2}y = 2$$

$$y = 4$$

The solution is $(3, 4)$.

To solve the system of equations at the right, eliminate x. Multiply equation (1) by -2 and add to equation (2).

(1) $3x - 2y = 5$
(2) $6x - 4y = 1$

$$\begin{aligned} -6x + 4y &= -10 \\ 6x - 4y &= 1 \\ \hline 0 &= -9 \end{aligned}$$

This is not a true equation.

The system is inconsistent.

The graph of the equations of the system is shown at the right. Note that the lines are parallel and therefore do not intersect.

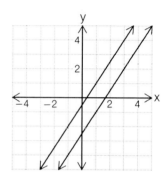

Example 1   Solve by the addition method.

A. $3x - 2y = 2x + 5$ B. $4x - 8y = 36$
   $2x + 3y = -4$ $3x - 6y = 27$

Solution   A. $3x - 2y = 2x + 5$ (1)
                     $2x + 3y = -4$ (2)

$x - 2y = 5$ $2x + 3y = -4$	• Write equation (1) in the form $Ax + By = C$.
$-2(x - 2y) = -2(5)$ $2x + 3y = -4$	• To eliminate x, multiply each side of equation (1) by -2.
$\begin{aligned} -2x + 4y &= -10 \\ 2x + 3y &= -4 \\ \hline 7y &= -14 \\ y &= -2 \end{aligned}$	• Add the equations.
$\begin{aligned} 2x + 3y &= -4 \\ 2x + 3(-2) &= -4 \\ 2x - 6 &= -4 \\ 2x &= 2 \\ x &= 1 \end{aligned}$	• Replace y in equation (2) by its value.

The solution is $(1, -2)$.

B. $4x - 8y = 36$ (1)
$3x - 6y = 27$ (2)

$3(4x - 8y) = 3(36)$
$-4(3x - 6y) = -4(27)$

• To eliminate x, multiply each side of equation (1) by 3 and each side of equation (2) by -4.

$12x - 24y = 108$
$-12x + 24y = -108$
$0 = 0$

• Add the equations. This is a true equation. Any ordered pair that is a solution of one equation is a solution of the other equation.

The equations are dependent.

Problem 1 Solve by the addition method.
A. $2x + 5y = 6$ B. $2x + y = 5$
$3x - 2y = 6x + 2$ $4x + 2y = 6$

Solution See page A81.
A. $(-2, 2)$ B. inconsistent

2 Solve systems of three linear equations in three variables by the addition method

An equation of the form $Ax + By + Cz = D$, where A, B, C, and D are constants, is a **linear equation in three variables.** Examples of linear equations in three variables are shown below.

$$2x + 4y - 3z = 7 \qquad\qquad x - 6y + z = -8$$

The graph of a linear equation in three variables is a plane.

Just as a solution of an equation in two variables is an ordered pair (x, y), a **solution of an equation in three variables** is an ordered triple (x, y, z). For example, $(2, 1, -3)$ and $(3, 1, -2)$ are two solutions of the equation

$$2x - y - 2z = 9 .$$

The ordered triples $(1, 3, 2)$ and $(2, -1, 3)$ are not solutions of this equation.

A **system of linear equations in three variables** is shown below.

$$x - 2y + z = 6$$
$$3x + y - 2z = 2$$
$$2x - 3y + 5z = 1$$

A **solution of a system of equations in three variables** is an ordered triple that is a solution of each equation of the system.

A system of linear equations in three variables can be solved by using the addition method. First, eliminate one variable from any two of the given equations. Then eliminate the same variable from any other two equations. The result will be a system of two equations in two variables. Solve this system by the addition method.

Solve:
$$x + 4y - z = 10$$
$$3x + 2y + z = 4$$
$$2x - 3y + 2z = -7$$

(1) $x + 4y - z = 10$
(2) $3x + 2y + z = 4$
(3) $2x - 3y + 2z = -7$

Eliminate z from equations (1) and (2) by adding the two equations.

$x + 4y - z = 10$
$3x + 2y + z = 4$
(4) $4x + 6y = 14$

Eliminate z from equations (1) and (3). Multiply equation (1) by 2 and add to equation (3).

$2x + 8y - 2z = 20$
$2x - 3y + 2z = -7$
(5) $4x + 5y = 13$

Solve the system of two equations in two variables.

(4) $4x + 6y = 14$
(5) $4x + 5y = 13$

Eliminate x. Multiply equation (5) by -1 and add to equation (4). Solve for y.

$4x + 6y = 14$
$-4x - 5y = -13$
$y = 1$

Substitute the value of y into equation (4) or (5), and solve for x. Equation (4) is used here.

$4x + 6y = 14$
$4x + 6(1) = 14$
$4x + 6 = 14$
$4x = 8$
$x = 2$

Substitute the value of y and the value of x into one of the equations in the original system. Equation (2) is used here.

$3x + 2y + z = 4$
$3(2) + 2(1) + z = 4$
$6 + 2 + z = 4$
$8 + z = 4$
$z = -4$

The solution is $(2, 1, -4)$.

Solve:
$$2x - 3y - z = 1$$
$$x + 4y + 3z = 2$$
$$4x - 6y - 2z = 5$$

(1) $2x - 3y - z = 1$
(2) $x + 4y + 3z = 2$
(3) $4x - 6y - 2z = 5$

Eliminate x from equations (1) and (2). Multiply equation (2) by -2 and add to equation (1).

$2x - 3y - z = 1$
$-2x - 8y - 6z = -4$
$-11y - 7z = -3$

Eliminate x from equations (1) and (3). Multiply equation (1) by -2 and add to equation (3).

$$-4x + 6y + 2z = -2$$
$$4x - 6y - 2z = 5$$
$$0 = 3$$

This is not a true equation. The system of equations is inconsistent and therefore has no solution.

Example 2   Solve: $3x - y + 2z = 1$
$2x + 3y + 3z = 4$
$x + y - 4z = -9$

Solution $3x - y + 2z = 1$ (1)
$2x + 3y + 3z = 4$ (2)
$x + y - 4z = -9$ (3)

$3x - y + 2z = 1$ • Eliminate y. Add equations (1) and (3).
$\underline{x + y - 4z = -9}$
$4x \qquad - 2z = -8$
$2x - z = -4$ • Multiply each side of the equation by $\frac{1}{2}$.

$9x - 3y + 6z = 3$ • Multiply equation (1) by 3 and add to equation (2).
$\underline{2x + 3y + 3z = 4}$
$11x \qquad + 9z = 7$

(4) $2x - z = -4$ • Solve the system of two equations.
(5) $11x + 9z = 7$

$18x - 9z = -36$ • Multiply equation (4) by 9 and add to equation (5).
$\underline{11x + 9z = 7}$
$29x \qquad = -29$
$x = -1$

$2x - z = -4$ • Replace x by -1 in equation (4).
$2(-1) - z = -4$
$-2 - z = -4$
$-z = -2$
$z = 2$

$x + y - 4z = -9$ • Replace x by -1 and z by 2 in equation (3).
$-1 + y - 4(2) = -9$
$-1 + y - 8 = -9$
$-9 + y = -9$
$y = 0$

The solution is $(-1, 0, 2)$.

> **Problem 2**   Solve: $x - y + z = 6$
> $2x + 3y - z = 1$
> $x + 2y + 2z = 5$
>
> **Solution**   See page A82.
> $(3, -1, 2)$

EXERCISES 10.2

1   Solve by the addition method.

1. $x - y = 5$
$x + y = 7$
$(6, 1)$

2. $x + y = 1$
$2x - y = 5$
$(2, -1)$

3. $3x + y = 4$
$x + y = 2$
$(1, 1)$

4. $x - 3y = 4$
$x + 5y = -4$
$(1, -1)$

5. $3x + y = 7$
$x + 2y = 4$
$(2, 1)$

6. $x - 2y = 7$
$3x - 2y = 9$
$(1, -3)$

7. $2x + 3y = -1$
$x + 5y = 3$
$(-2, 1)$

8. $x + 5y = 7$
$2x + 7y = 8$
$(-3, 2)$

9. $3x - y = 4$
$6x - 2y = 8$
dependent equations

10. $x - 2y = -3$
$-2x + 4y = 6$

dependent equations

11. $2x + 5y = 9$
$4x - 7y = -16$
$\left(-\dfrac{1}{2}, 2\right)$

12. $8x - 3y = 21$
$4x + 5y = -9$
$\left(\dfrac{3}{2}, -3\right)$

13. $4x - 6y = 5$
$2x - 3y = 7$
inconsistent

14. $3x + 6y = 7$
$2x + 4y = 5$
inconsistent

15. $3x - 5y = 7$
$x - 2y = 3$
$(-1, -2)$

16. $3x + 4y = 25$
$2x + y = 10$
$(3, 4)$

17. $x + 3y = 7$
$-2x + 3y = 22$
$(-5, 4)$

18. $2x - 3y = 14$
$5x - 6y = 32$
$(4, -2)$

19. $3x + 2y = 16$
$2x - 3y = -11$

$(2, 5)$

20. $2x - 5y = 13$
$5x + 3y = 17$

$(4, -1)$

21. $4x + 4y = 5$
$2x - 8y = -5$
$\left(\dfrac{1}{2}, \dfrac{3}{4}\right)$

22. $3x + 7y = 16$
$4x - 3y = 9$
$(3, 1)$

23. $5x + 4y = 0$
$3x + 7y = 0$
$(0, 0)$

24. $3x - 4y = 0$
$4x - 7y = 0$
$(0, 0)$

25. $5x + 2y = 1$
$2x + 3y = 7$

$(-1, 3)$

26. $3x + 5y = 16$
$5x - 7y = -4$

$(2, 2)$

27. $3x - 6y = 6$
$9x - 3y = 8$
$\left(\dfrac{2}{3}, -\dfrac{2}{3}\right)$

28. $4x - 8y = 5$
$8x + 2y = 1$
$\left(\dfrac{1}{4}, -\dfrac{1}{2}\right)$

29. $5x + 2y = 2x + 1$
$2x - 3y = 3x + 2$
$(1, -1)$

30. $3x + 3y = y + 1$
$x + 3y = 9 - x$
$(-3, 5)$

31. $\dfrac{2}{3}x - \dfrac{1}{2}y = 3$
$\dfrac{1}{3}x - \dfrac{1}{4}y = \dfrac{3}{2}$
dependent equations

32. $\dfrac{3}{4}x + \dfrac{1}{3}y = -\dfrac{1}{2}$
$\dfrac{1}{2}x - \dfrac{5}{6}y = -\dfrac{7}{2}$
$(-2, 3)$

33. $\dfrac{2}{5}x - \dfrac{1}{3}y = 1$
$\dfrac{3}{5}x + \dfrac{2}{3}y = 5$
$(5, 3)$

34. $\dfrac{5}{6}x + \dfrac{1}{3}y = \dfrac{4}{3}$
$\dfrac{2}{3}x - \dfrac{1}{2}y = \dfrac{11}{6}$
$(2, -1)$

35. $\dfrac{3}{4}x + \dfrac{2}{5}y = -\dfrac{3}{20}$
$\dfrac{3}{2}x - \dfrac{1}{4}y = \dfrac{3}{4}$
$\left(\dfrac{1}{3}, -1\right)$

36. $\dfrac{2}{5}x - \dfrac{1}{2}y = \dfrac{13}{2}$
$\dfrac{3}{4}x - \dfrac{1}{5}y = \dfrac{17}{2}$
$(10, -5)$

37. $4x - 5y = 3y + 4$
$2x + 3y = 2x + 1$
$\left(\dfrac{5}{3}, \dfrac{1}{3}\right)$

38. $5x - 2y = 8x - 1$
$2x + 7y = 4y + 9$
$(-3, 5)$

39. $2x + 5y = 5x + 1$
$3x - 2y = 3y + 3$
inconsistent

2 Solve by the addition method.

40. $x + 2y - z = 1$
$2x - y + z = 6$
$x + 3y - z = 2$
$(2, 1, 3)$

41. $x + 3y + z = 6$
$3x + y - z = -2$
$2x + 2y - z = 1$
$(-1, 2, 1)$

42. $2x - y + 2z = 7$
$x + y + z = 2$
$3x - y + z = 6$
$(1, -1, 2)$

43. $x - 2y + z = 6$
$x + 3y + z = 16$
$3x - y - z = 12$
$(6, 2, 4)$

44. $3x + y = 5$
$3y - z = 2$
$x + z = 5$
$(1, 2, 4)$

45. $2y + z = 7$
$2x - z = 3$
$x - y = 3$
$(4, 1, 5)$

46. $x - y + z = 1$
$2x + 3y - z = 3$
$-x + 2y - 4z = 4$
$(2, -1, -2)$

47. $2x + y - 3z = 7$
$x - 2y + 3z = 1$
$3x + 4y - 3z = 13$
$(3, 1, 0)$

48. $2x + 3z = 5$
$3y + 2z = 3$
$3x + 4y = -10$
$(-2, -1, 3)$

49. $3x + 4z = 5$
$2y + 3z = 2$
$2x - 5y = 8$
$(-1, -2, 2)$

50. $2x + 4y - 2z = 3$
$x + 3y + 4z = 1$
$x + 2y - z = 4$
inconsistent

51. $x - 3y + 2z = 1$
$x - 2y + 3z = 5$
$2x - 6y + 4z = 3$
inconsistent

52. $2x + y - z = 5$
$x + 3y + z = 14$
$3x - y + 2z = 1$
$(1, 4, 1)$

53. $3x - y - 2z = 11$
$2x + y - 2z = 11$
$x + 3y - z = 8$
$(2, 1, -3)$

54. $3x + y - 2z = 2$
$x + 2y + 3z = 13$
$2x - 2y + 5z = 6$
$(1, 3, 2)$

55. $4x + 5y + z = 6$
$2x - y + 2z = 11$
$x + 2y + 2z = 6$
$(2, -1, 3)$

56. $2x - y + z = 6$
$3x + 2y + z = 4$
$x - 2y + 3z = 12$
$(1, -1, 3)$

57. $3x + 2y - 3z = 8$
$2x + 3y + 2z = 10$
$x + y - z = 2$
$(6, -2, 2)$

58. $3x - 2y + 3z = -4$
$2x + y - 3z = 2$
$3x + 4y + 5z = 8$
$(0, 2, 0)$

59. $3x - 3y + 4z = 6$
$4x - 5y + 2z = 10$
$x - 2y + 3z = 4$
$(0, -2, 0)$

60. $3x - y + 2z = 2$
$4x + 2y - 7z = 0$
$2x + 3y - 5z = 7$
$(1, 5, 2)$

61. $2x + 2y + 3z = 13$
$-3x + 4y - z = 5$
$5x - 3y + z = 2$
$(2, 3, 1)$

62. $2x - 3y + 7z = 0$
$x + 4y - 4z = -2$
$3x + 2y + 5z = 1$
$(-2, 1, 1)$

63. $5x + 3y - z = 5$
$3x - 2y + 4z = 13$
$4x + 3y + 5z = 22$
$(1, 1, 3)$

SUPPLEMENTAL EXERCISES 10.2

Solve. (*Hint:* Multiply both sides of each equation in the system by a multiple of 10 so that the coefficients and constants are integers.)

64. $0.2x - 0.3y = 0.5$
$0.3x - 0.2y = 0.5$
$(1, -1)$

65. $0.4x - 0.9y = -0.1$
$0.3x + 0.2y = 0.8$
$(2, 1)$

66. $0.5x - 0.4y = 1.8$
$0.3x - 0.5y = 1.6$
$(2, -2)$

67. $0.1x + 0.6y = -1.7$
$0.4x + 0.5y = -1.1$
$(1, -3)$

68. $1.25x - 0.25y = -1.5$
$1.5x + 2.5y = 1$
$(-1, 1)$

69. $2.25x + 1.5y = 3$
$1.75x + 2.25y = 1.25$
$(2, -1)$

70. $1.5x + 2.5y + 1.5z = 8$
$0.5x - 2y - 1.5z = -1$
$2.5x - 1.5y + 2z = 2.5$
$(3, 2, -1)$

71. $1.6x - 0.9y + 0.3z = 2.9$
$1.6x + 0.5y - 0.1z = 3.3$
$0.8x - 0.7y + 0.1z = 1.5$
$(2, 0, -1)$

72. $0.6x + 0.5y - 0.2z = 0.6$
$0.3x + 0.4y + 0.2z = 0.3$
$0.5x + 0.6y - 0.4z = 1.1$
$(-1, 2, -1)$

73. $1.2x - 1.4y + 1.1z = 3.8$
$1.6x + 1.5y + 1.2z = 1.7$
$1.2x + 1.2y + 0.5z = 1.2$
$(2, -1, 0)$

Solve.

74. The point of intersection of the graphs of the equations $Ax + 5y = 7$ and $2x + By = 8$ is $(2, -1)$. Find A and B.
$A = 6, B = -4$

75. The point of intersection of the graphs of the equations $Ax + 3y = 6$ and $2x + By = -4$ is $(3, -2)$. Find A and B.
$A = 4, B = 5$

76. The point of intersection of the graphs of the equations $Ax + 2y - 3z = 13$, $3x + By + z = 11$, and $2x - 3y + Cz = 0$ is $(2, 3, -1)$. Find A, B, and C.
$A = 2, B = 2, C = -5$

77. The point of intersection of the graphs of the equations $Ax + 3y + 2z = 8$, $2x + By - 3z = -12$, and $3x - 2y + Cz = 1$ is $(3, -2, 4)$. Find A, B, and C.
$A = 2, B = 3, C = -3$

78. The sum of three numbers is 62. The sum of the first and second numbers is equal to 10 less than the third. The first number minus the second number is equal to 34 less than the third number. Find the three numbers.
12, 14, and 36

79. The sum of three numbers is 66. The sum of the first and second numbers is equal to 2 more than the third. The first number minus the second number is equal to 30 less than the third number. Find the three numbers.
16, 18, and 32

SECTION **10.3**

Solving Systems of Equations by Using Determinants and by Using Matrices

1 Evaluate determinants

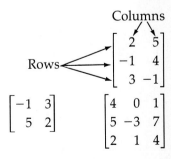

The rectangular array of numbers shown at the right is called a **matrix**. The matrix has three rows and two columns and is called a 3×2 (read "three by two") matrix.

Columns

$$\begin{bmatrix} 2 & 5 \\ -1 & 4 \\ 3 & -1 \end{bmatrix}$$

Rows

A **square matrix** is one that has the same number of rows as columns. An example of a 2×2 and a 3×3 matrix is shown at the right.

$$\begin{bmatrix} -1 & 3 \\ 5 & 2 \end{bmatrix}$$

$$\begin{bmatrix} 4 & 0 & 1 \\ 5 & -3 & 7 \\ 2 & 1 & 4 \end{bmatrix}$$

Associated with every square matrix is a number called its **determinant**.

The determinant of a 2×2 matrix $\begin{bmatrix} a_1 & a_2 \\ b_1 & b_2 \end{bmatrix}$ is written $\begin{vmatrix} a_1 & a_2 \\ b_1 & b_2 \end{vmatrix}$. The value of this determinant is given by the formula

$$\begin{vmatrix} a_1 & a_2 \\ b_1 & b_2 \end{vmatrix} = a_1 b_2 - a_2 b_1$$

Note that vertical bars are used to represent the determinant, and brackets are used to represent the matrix.

Find the value of the determinant.

$$\begin{vmatrix} 3 & 4 \\ -1 & 2 \end{vmatrix}$$

Use the formula.

$$\begin{vmatrix} 3 & 4 \\ -1 & 2 \end{vmatrix} = 3 \cdot 2 - 4(-1) = 6 - (-4) = 10$$

The value of the determinant is 10.

The evaluation of the determinant of a 3×3 matrix is accomplished by rewriting that determinant in terms of 2×2 determinants, called **minors**.

The **minor of an element** in a 3×3 determinant is the 2×2 determinant that is obtained by eliminating the row and column that contain the element.

To find the minor of -3 for the determinant $\begin{vmatrix} 2 & -3 & 4 \\ 0 & 4 & 8 \\ -1 & 3 & 6 \end{vmatrix}$,

eliminate the row and column that contain -3.

$$\begin{array}{|ccc|} \hline 2 & -3 & 4 \\ 0 & 4 & 8 \\ -1 & 3 & 6 \\ \hline \end{array}$$

The resulting determinant is the minor of -3.

The minor of -3 is $\begin{vmatrix} 0 & 8 \\ -1 & 6 \end{vmatrix}$.

To find the minor of 5 for the determinant $\begin{vmatrix} 0 & -3 & 4 \\ 2 & 1 & -4 \\ 3 & 2 & 5 \end{vmatrix}$,

eliminate the row and column that contain 5.

$$\begin{array}{ccc} 0 & -3 & 4 \\ 2 & 1 & -4 \\ 3 & 2 & 5 \end{array}$$

The resulting determinant is the minor of 5.

The minor of 5 is $\begin{vmatrix} 0 & -3 \\ 2 & 1 \end{vmatrix}$.

The value of a 3×3 determinant can be found by **expanding the minors.** The expansion by the minors of the elements of the first row of the determinant is shown below.

$$
\begin{array}{ccc}
 & \text{minor of } a_1 & \text{minor of } a_2 & \text{minor of } a_3
\end{array}
$$

$$
\begin{vmatrix} a_1 & a_2 & a_3 \\ b_1 & b_2 & b_3 \\ c_1 & c_2 & c_3 \end{vmatrix} = a_1 \cdot \begin{vmatrix} b_2 & b_3 \\ c_2 & c_3 \end{vmatrix} - a_2 \cdot \begin{vmatrix} b_1 & b_3 \\ c_1 & c_3 \end{vmatrix} + a_3 \cdot \begin{vmatrix} b_1 & b_2 \\ c_1 & c_2 \end{vmatrix}
$$

To find the value of the determinant $\begin{vmatrix} 2 & -1 & -3 \\ 1 & 2 & 0 \\ 3 & -1 & 2 \end{vmatrix}$,

expand by the minors of the first row.

$$
\begin{vmatrix} 2 & -1 & -3 \\ 1 & 2 & 0 \\ 3 & -1 & 2 \end{vmatrix} = 2\begin{vmatrix} 2 & 0 \\ -1 & 2 \end{vmatrix} - (-1)\begin{vmatrix} 1 & 0 \\ 3 & 2 \end{vmatrix} + (-3)\begin{vmatrix} 1 & 2 \\ 3 & -1 \end{vmatrix}
$$

$$
\begin{aligned}
&= 2(4 - 0) - (-1)(2 - 0) + (-3)(-1 - 6) \\
&= 2(4) - (-1)(2) + (-3)(-7) \\
&= 8 + 2 + 21 \\
&= 31
\end{aligned}
$$

The value of the determinant is 31.

Any row or column can be used in the expansion by minors. The signs of the terms in an expansion are given by the pattern shown at the right.

$$
\begin{vmatrix} + & - & + \\ - & + & - \\ + & - & + \end{vmatrix}
$$

For the example above, when the value of the determinant is found by expanding about the elements of the third column, the answer is the same.

$$
\begin{vmatrix} 2 & -1 & -3 \\ 1 & 2 & 0 \\ 3 & -1 & 2 \end{vmatrix} = -3\begin{vmatrix} 1 & 2 \\ 3 & -1 \end{vmatrix} - 0\begin{vmatrix} 2 & -1 \\ 3 & -1 \end{vmatrix} + 2\begin{vmatrix} 2 & -1 \\ 1 & 2 \end{vmatrix}
$$

$$
\begin{aligned}
&= -3(-1 - 6) - 0 + 2(4 + 1) \\
&= -3(-7) + 2(5) \\
&= 21 + 10 \\
&= 31
\end{aligned}
$$

To find the value of a 3×3 determinant, expand by the minors of the elements of any row or column. However, note that by choosing a row or column that contains a zero, the computation is simplified.

Example 1   Evaluate the determinant.

A. $\begin{vmatrix} 3 & -2 \\ 6 & -4 \end{vmatrix}$ B. $\begin{vmatrix} -2 & 3 & 1 \\ 4 & -2 & 0 \\ 1 & -2 & 3 \end{vmatrix}$

Solution   A. $\begin{vmatrix} 3 & -2 \\ 6 & -4 \end{vmatrix} = 3(-4) - 6(-2)$

$= -12 + 12 = 0$

The value of the determinant is 0.

B. $\begin{vmatrix} -2 & 3 & 1 \\ 4 & -2 & 0 \\ 1 & -2 & 3 \end{vmatrix} = -4\begin{vmatrix} 3 & 1 \\ -2 & 3 \end{vmatrix} + (-2)\begin{vmatrix} -2 & 1 \\ 1 & 3 \end{vmatrix} + 0$ • Expand by minors of the second row.

$= -4(9 + 2) - 2(-6 - 1)$
$= -4(11) - 2(-7)$
$= -44 + 14$
$= -30$

The value of the determinant is -30.

Problem 1   Evaluate the determinant.

A. $\begin{vmatrix} -1 & -4 \\ 3 & -5 \end{vmatrix}$ B. $\begin{vmatrix} 1 & 4 & -2 \\ 3 & 1 & 1 \\ 0 & -2 & 2 \end{vmatrix}$

Solution   See page A83.
A. 17   B. -8

2 Solve systems of equations by using Cramer's Rule

The connection between determinants and systems of equations can be understood by solving a general system of linear equations.

Solve: $a_1 x + b_1 y = c_1$
$\quad\quad\quad a_2 x + b_2 y = c_2$

Eliminate y. Multiply equation (1) by b_2 and equation (2) by $-b_1$.

Add the equations.

Assuming $a_1 b_2 - a_2 b_1 \neq 0$, solve for x.

(1) $a_1 x + b_1 y = c_1$
(2) $a_2 x + b_2 y = c_2$

$a_1 b_2 x + b_1 b_2 y = c_1 b_2$
$-a_2 b_1 x - b_1 b_2 y = -c_2 b_1$

$a_1 b_2 x - a_2 b_1 x = c_1 b_2 - c_2 b_1$
$(a_1 b_2 - a_2 b_1)x = c_1 b_2 - c_2 b_1$

$x = \dfrac{c_1 b_2 - c_2 b_1}{a_1 b_2 - a_2 b_1}$

The denominator $a_1b_2 - a_2b_1$ is the determinant of the coefficients of x and y. This is called the **coefficient determinant.**

$$a_1b_2 - a_2b_1 = \begin{vmatrix} a_1 & b_1 \\ a_2 & b_2 \end{vmatrix}$$

Coefficients of x ⟶
Coefficients of y ⟶

The numerator for x, $c_1b_2 - c_2b_1$, is the determinant obtained by replacing the first column in the coefficient determinant by the constants c_1 and c_2. This is called the **numerator determinant.**

$$c_1b_2 - c_2b_1 = \begin{vmatrix} c_1 & b_1 \\ c_2 & b_2 \end{vmatrix}$$

Constants of
the equations ⟶

Following a similar procedure and eliminating x, the y-component of the solution can also be expressed in determinant form. These results are summarized in Cramer's Rule.

Cramer's Rule

The solution of the system of equations $\begin{aligned} a_1x + b_1y &= c_1 \\ a_2x + b_2y &= c_2 \end{aligned}$ is given by

$x = \dfrac{D_x}{D}$ and $y = \dfrac{D_y}{D}$, where

$$D = \begin{vmatrix} a_1 & b_1 \\ a_2 & b_2 \end{vmatrix}, \quad D_x = \begin{vmatrix} c_1 & b_1 \\ c_2 & b_2 \end{vmatrix}, \quad D_y = \begin{vmatrix} a_1 & c_1 \\ a_2 & c_2 \end{vmatrix}, \text{ and } D \neq 0.$$

Example 2 Solve by using Cramer's Rule: $\begin{aligned} 2x - 3y &= 8 \\ 5x + 6y &= 11 \end{aligned}$

Solution $D = \begin{vmatrix} 2 & -3 \\ 5 & 6 \end{vmatrix} = 27$

• Find the value of the coefficient determinant.

$D_x = \begin{vmatrix} 8 & -3 \\ 11 & 6 \end{vmatrix} = 81 \qquad D_y = \begin{vmatrix} 2 & 8 \\ 5 & 11 \end{vmatrix} = -18$

• Find the value of each of the numerator determinants.

$x = \dfrac{D_x}{D} = \dfrac{81}{27} = 3 \qquad y = \dfrac{D_y}{D} = \dfrac{-18}{27} = \dfrac{-2}{3}$

• Use Cramer's Rule to write the solutions.

The solution is $\left(3, -\dfrac{2}{3}\right)$.

Problem 2 Solve by using Cramer's Rule: $\begin{aligned} 6x - 6y &= 5 \\ 2x - 10y &= -1 \end{aligned}$

Solution See page A83. $\left(\dfrac{7}{6}, \dfrac{1}{3}\right)$

For the system shown at the right, $D = 0$. Therefore, $\dfrac{D_x}{D}$ is undefined.

$$6x - 9y = 5$$
$$4x - 6y = 4$$

When $D = 0$, the system of equations is dependent if both D_x and D_y are zero. The system of equations is inconsistent if either D_x or D_y is not zero.

$$D = \begin{vmatrix} 6 & -9 \\ 4 & -6 \end{vmatrix} = 0$$

A procedure similar to that followed for two equations in two variables can be used to extend Cramer's Rule to three equations in three variables.

Cramer's Rule for Three Equations in Three Variables

$$a_1x + b_1y + c_1z = d_1$$

The solution of the system of equations $a_2x + b_2y + c_2z = d_2$ is given by

$$a_3x + b_3y + c_3z = d_3$$

$x = \dfrac{D_x}{D}, y = \dfrac{D_y}{D}$, and $z = \dfrac{D_z}{D}$, where

$$D = \begin{vmatrix} a_1 & b_1 & c_1 \\ a_2 & b_2 & c_2 \\ a_3 & b_3 & c_3 \end{vmatrix}, D_x = \begin{vmatrix} d_1 & b_1 & c_1 \\ d_2 & b_2 & c_2 \\ d_3 & b_3 & c_3 \end{vmatrix}, D_y = \begin{vmatrix} a_1 & d_1 & c_1 \\ a_2 & d_2 & c_2 \\ a_3 & d_3 & c_3 \end{vmatrix}, D_z = \begin{vmatrix} a_1 & b_1 & d_1 \\ a_2 & b_2 & d_2 \\ a_3 & b_3 & d_3 \end{vmatrix},$$

and $D \neq 0$.

Example 3   Solve by using Cramer's Rule: $3x - y + z = 5$
$$x + 2y - 2z = -3$$
$$2x + 3y + z = 4$$

Solution

$$D = \begin{vmatrix} 3 & -1 & 1 \\ 1 & 2 & -2 \\ 2 & 3 & 1 \end{vmatrix} = 28$$ • Find the value of the coefficient determinant.

$$D_x = \begin{vmatrix} 5 & -1 & 1 \\ -3 & 2 & -2 \\ 4 & 3 & 1 \end{vmatrix} = 28$$ • Find the value of each of the numerator determinants.

$$D_y = \begin{vmatrix} 3 & 5 & 1 \\ 1 & -3 & -2 \\ 2 & 4 & 1 \end{vmatrix} = 0$$

$$D_z = \begin{vmatrix} 3 & -1 & 5 \\ 1 & 2 & -3 \\ 2 & 3 & 4 \end{vmatrix} = 56$$

$$x = \frac{D_x}{D} = \frac{28}{28} = 1$$ • Use Cramer's Rule to write the solution.

$$y = \frac{D_y}{D} = \frac{0}{28} = 0$$

$$z = \frac{D_z}{D} = \frac{56}{28} = 2$$

The solution is $(1, 0, 2)$.

Problem 3   Solve by using Cramer's Rule: $2x - y + z = -1$
$$3x + 2y - z = 3$$
$$x + 3y + z = -2$$

Solution   See page A83. $\left(\frac{3}{7}, -\frac{1}{7}, -2\right)$

3   Solve systems of equations by using matrices

As previously stated in this section, a **matrix** is a rectangular array of numbers. Each number in the matrix is called an **element** of the matrix. The matrix shown below, with 3 rows and 4 columns, is a 3×4 matrix.

$$\begin{bmatrix} 1 & 4 & -3 & 6 \\ -2 & 5 & 2 & 0 \\ -1 & 3 & 7 & -4 \end{bmatrix}$$

The notation a_{ij} refers to the element of a matrix in the ith row and jth column. For the matrix shown above, $a_{23} = 2$ (2nd row, 3rd column), $a_{31} = -1$ (3rd row, 1st column), and $a_{13} = -3$ (1st row, 3rd column).

The elements $a_{11}, a_{22}, a_{33}, \ldots, a_{nn}$ form the **main diagonal** of a matrix. The elements 1, 5, and 7 form the main diagonal of the matrix shown above.

By considering only the coefficients and constants for the system of equations below, the corresponding 3×4 **augmented matrix** can be formed.

$$3x - 2y + z = 2 \qquad \begin{bmatrix} 3 & -2 & 1 & 2 \\ 1 & 0 & -3 & -2 \\ 2 & -1 & 4 & 5 \end{bmatrix}$$
$$x \qquad - 3z = -2$$
$$2x - y + 4z = 5$$

Note that when a term is missing from one of the equations of the system, the coefficient of that term is 0, and a 0 is entered in the matrix.

A system of equations can be written from an augmented matrix.

$$\begin{bmatrix} 2 & -1 & 4 & 1 \\ 1 & 1 & 0 & 3 \\ 3 & -2 & -1 & 5 \end{bmatrix} \qquad \begin{aligned} 2x - y + 4z &= 1 \\ x + y &= 3 \\ 3x - 2y - z &= 5 \end{aligned}$$

A system of equations can be solved by writing the system in matrix form and then performing operations on the matrix similar to those performed on the equations of the system. These operations are called **elementary row operations.**

Elementary Row Operations

1. Interchange two rows.
2. Multiply all the elements in a row by the same nonzero number.
3. Replace a row by the sum of that row and a multiple of any other row.

The goal is to use the elementary row operations to rewrite the matrix with 1's down the main diagonal and 0's to the left of the 1's in all rows except the first. This is called the **echelon form** of the matrix. Examples of echelon form are shown below.

$$\begin{bmatrix} 1 & 2 & 2 \\ 0 & 1 & -1 \end{bmatrix} \qquad \begin{bmatrix} 1 & -3 & -1 & -6 \\ 0 & 1 & -2 & 7 \\ 0 & 0 & 1 & -2 \end{bmatrix} \qquad \begin{bmatrix} 1 & \frac{1}{2} & -\frac{1}{3} & 2 \\ 0 & 1 & -\frac{2}{3} & -2 \\ 0 & 0 & 1 & 6 \end{bmatrix}$$

A system of equations can be solved by using elementary row operations to rewrite the augmented matrix of a system of equations in echelon form.

The system of equations at the right can be solved by first writing the system in matrix form.

$$\begin{aligned} 2x + 5y &= 8 \\ 3x + 4y &= 5 \end{aligned} \qquad \begin{bmatrix} 2 & 5 & 8 \\ 3 & 4 & 5 \end{bmatrix}$$

Element a_{11} must be a 1. Multiply row 1 by $\frac{1}{2}$.

$$\begin{bmatrix} 1 & \frac{5}{2} & 4 \\ 3 & 4 & 5 \end{bmatrix}$$

Element a_{21} must be a 0. Multiply row 1 by -3, and add it to row 2. Replace row 2 by the sum.

$$\begin{bmatrix} 1 & \frac{5}{2} & 4 \\ 0 & -\frac{7}{2} & -7 \end{bmatrix}$$

Element a_{22} must be a 1. Multiply row 2 by $-\frac{2}{7}$. The matrix is now in echelon form.

$$\begin{bmatrix} 1 & \frac{5}{2} & 4 \\ 0 & 1 & 2 \end{bmatrix}$$

Write the system of equations represented by the matrix. $x + \dfrac{5}{2}y = 4$

$y = 2$

Substitute the value of y into equation 1, and solve for x. $x + \dfrac{5}{2}(2) = 4$

$x + 5 = 4$

$x = -1$

The solution is $(-1, 2)$.

The order in which the elements in the 2×3 matrix were changed is important.

1. Change a_{11} to a 1.
2. Change a_{21} to a 0.
3. Change a_{22} to a 1.

$$\begin{bmatrix} a_{11} & a_{12} & a_{13} \\ a_{21} & a_{22} & a_{23} \end{bmatrix}$$

Example 4   Solve by using a matrix: $3x + 2y = 3$
$2x - 3y = 15$

Solution

$$\begin{bmatrix} 3 & 2 & 3 \\ 2 & -3 & 15 \end{bmatrix} \qquad \begin{bmatrix} 1 & \frac{2}{3} & 1 \\ 2 & -3 & 15 \end{bmatrix}$$

• Write the system in matrix form.

Multiply row 1 by $\dfrac{1}{3}$.

$$\begin{bmatrix} 1 & \frac{2}{3} & 1 \\ 0 & -\frac{13}{3} & 13 \end{bmatrix} \qquad \begin{bmatrix} 1 & \frac{2}{3} & 1 \\ 0 & 1 & -3 \end{bmatrix}$$

• Multiply row 1 by -2 and add to row 2. Multiply row 3 by $-\dfrac{3}{13}$.

(1) $x + \dfrac{2}{3}y = 1$ $x + \dfrac{2}{3}(-3) = 1$

(2) $y = -3$ $x - 2 = 1$

$x = 3$

• Write the system of equations represented by the matrix.
• Substitute the value of y into equation 1, and solve for x.

The solution is $(3, -3)$.

Problem 4   Solve by using a matrix: $3x - 5y = -12$
$4x - 3y = -5$

Solution   See page A84.
$(1, 3)$

The matrix method of solving systems of equations can be extended to larger systems. A system of three equations in three unknowns is written as a 3×4 augmented matrix.

The order in which the elements in a 3×4 matrix are changed is

1. Change a_{11} to a 1.
2. Change a_{21} and a_{31} to 0's.
3. Change a_{22} to a 1.
4. Change a_{32} to a 0.
5. Change a_{33} to a 1.

$$\begin{bmatrix} a_{11} & a_{12} & a_{13} & a_{14} \\ a_{21} & a_{22} & a_{23} & a_{24} \\ a_{31} & a_{32} & a_{33} & a_{34} \end{bmatrix}$$

To solve the system shown at the right, write the system in matrix form.

$$\begin{aligned} 2x + 3y + 3z &= -2 \\ x + 2y - 3z &= 9 \\ 3x - 2y - 4z &= 1 \end{aligned}$$

$$\begin{bmatrix} 2 & 3 & 3 & -2 \\ 1 & 2 & -3 & 9 \\ 3 & -2 & -4 & 1 \end{bmatrix}$$

Element a_{11} must be a 1. Interchange rows 1 and 2.

$$\begin{bmatrix} 1 & 2 & -3 & 9 \\ 2 & 3 & 3 & -2 \\ 3 & -2 & -4 & 1 \end{bmatrix}$$

Element a_{21} must be a 0. Multiply row 1 by -2 and add to row 2. Replace row 2 by the sum. Element a_{31} must be a 0. Multiply row 1 by -3 and add to row 3. Replace row 3 by the sum.

$$\begin{bmatrix} 1 & 2 & -3 & 9 \\ 0 & -1 & 9 & -20 \\ 0 & -8 & 5 & -26 \end{bmatrix}$$

Element a_{22} must be a 1. Multiply row 2 by -1.

$$\begin{bmatrix} 1 & 2 & -3 & 9 \\ 0 & 1 & -9 & 20 \\ 0 & -8 & 5 & -26 \end{bmatrix}$$

Element a_{32} must be a 0. Multiply row 2 by 8 and add to row 3. Replace row 3 by the sum.

$$\begin{bmatrix} 1 & 2 & -3 & 9 \\ 0 & 1 & -9 & 20 \\ 0 & 0 & -67 & 134 \end{bmatrix}$$

Element a_{33} must be a 1. Multiply row 3 by $-\dfrac{1}{67}$.

$$\begin{bmatrix} 1 & 2 & -3 & 9 \\ 0 & 1 & -9 & 20 \\ 0 & 0 & 1 & -2 \end{bmatrix}$$

Write the system represented by the matrix.

$$\begin{aligned} (1) \quad x + 2y - 3z &= 9 \\ (2) \qquad y - 9z &= 20 \\ (3) \qquad\qquad z &= -2 \end{aligned}$$

Substitute the value of z into equation 2, and solve for y.

$$y - 9z = 20$$
$$y - 9(-2) = 20$$
$$y + 18 = 20$$
$$y = 2$$

Substitute the values of y and z into equation 1, and solve for x.

$$x + 2y - 3z = 9$$
$$x + 2(2) - 3(-2) = 9$$
$$x + 4 + 6 = 9$$
$$x = -1$$

The solution is $(-1, 2, -2)$.

Solve by using matrices: $x - y + z = 2$
$x + 2y - z = 3$
$3x + 3y - z = 6$

Write the system in matrix form.

$$\begin{bmatrix} 1 & -1 & 1 & 2 \\ 1 & 2 & -1 & 3 \\ 3 & 3 & -1 & 6 \end{bmatrix}$$

Element a_{11} is a 1. Element a_{21} must be a 0. Multiply row 1 by -1 and add to row 2. Replace row 2 by the sum. Element a_{31} must be a 0. Multiply row 1 by -3 and add to row 3. Replace row 3 by the sum.

$$\begin{bmatrix} 1 & -1 & 1 & 2 \\ 0 & 3 & -2 & 1 \\ 0 & 6 & -4 & 0 \end{bmatrix}$$

Element a_{22} must be a 1. Multiply row 2 by $\frac{1}{3}$.

$$\begin{bmatrix} 1 & -1 & 1 & 2 \\ 0 & 1 & -\frac{2}{3} & \frac{1}{3} \\ 0 & 6 & -4 & 0 \end{bmatrix}$$

Element a_{32} must be a 0. Multiply row 2 by -6 and add to row 3. Replace row 3 by the sum.

$$\begin{bmatrix} 1 & -1 & 1 & 2 \\ 0 & 1 & -\frac{2}{3} & \frac{1}{3} \\ 0 & 0 & 0 & -2 \end{bmatrix}$$

The matrix cannot be written in echelon form. Write the system represented by the matrix. The equation $0 = -2$ is not a true equation. The system of equations is inconsistent.

$$x - y + z = 2$$
$$y - \frac{2}{3}z = \frac{1}{3}$$
$$0 = -2$$

The system of equations has no solution.

Example 5   Solve by using a matrix: $3x + 2y + 3z = 2$
$$2x - 3y + 4z = 5$$
$$x + 4y + 2z = 8$$

Solution
$$\begin{bmatrix} 3 & 2 & 3 & 2 \\ 2 & -3 & 4 & 5 \\ 1 & 4 & 2 & 8 \end{bmatrix}$$

• Write the system in matrix form.

$$\begin{bmatrix} 1 & 4 & 2 & 8 \\ 2 & -3 & 4 & 5 \\ 3 & 2 & 3 & 2 \end{bmatrix}$$

• Interchange rows 1 and 3.

$$\begin{bmatrix} 1 & 4 & 2 & 8 \\ 0 & -11 & 0 & -11 \\ 0 & -10 & -3 & -22 \end{bmatrix}$$

• Multiply row 1 by -2 and add to row 2.
Multiply row 1 by -3 and add to row 3.

$$\begin{bmatrix} 1 & 4 & 2 & 8 \\ 0 & 1 & 0 & 1 \\ 0 & -10 & -3 & -22 \end{bmatrix}$$

• Multiply row 2 by $-\dfrac{1}{11}$.

$$\begin{bmatrix} 1 & 4 & 2 & 8 \\ 0 & 1 & 0 & 1 \\ 0 & 0 & -3 & -12 \end{bmatrix}$$

• Multiply row 2 by 10 and add to row 3.

$$\begin{bmatrix} 1 & 4 & 2 & 8 \\ 0 & 1 & 0 & 1 \\ 0 & 0 & 1 & 4 \end{bmatrix}$$

• Multiply row 3 by $-\dfrac{1}{3}$.

$$x + 4y + 2z = 8$$
$$y = 1$$
$$z = 4$$

• Write the system of equations represented by the matrix.

$$x + 4y + 2z = 8$$
$$x + 4(1) + 2(4) = 8$$
$$x + 4 + 8 = 8$$
$$x = -4$$

• Substitute the values of y and z into equation 1, and solve for x.

The solution is $(-4, 1, 4)$.

Problem 5   Solve by using a matrix: $3x - 2y - 3z = 5$
$$x + 3y - 2z = -4$$
$$2x + 6y + 3z = 6$$

Solution   See page A84.
$(3, -1, 2)$

EXERCISES 10.3

1 Evaluate the determinant.

1. $\begin{vmatrix} 2 & -1 \\ 3 & 4 \end{vmatrix}$ 11

2. $\begin{vmatrix} 5 & 1 \\ -1 & 2 \end{vmatrix}$ 11

3. $\begin{vmatrix} 6 & -2 \\ -3 & 4 \end{vmatrix}$ 18

4. $\begin{vmatrix} -3 & 5 \\ 1 & 7 \end{vmatrix}$ -26

5. $\begin{vmatrix} 3 & 6 \\ 2 & 4 \end{vmatrix}$ 0

6. $\begin{vmatrix} 5 & -10 \\ 1 & -2 \end{vmatrix}$ 0

7. $\begin{vmatrix} 1 & -1 & 2 \\ 3 & 2 & 1 \\ 1 & 0 & 4 \end{vmatrix}$ 15

8. $\begin{vmatrix} 4 & 1 & 3 \\ 2 & -2 & 1 \\ 3 & 1 & 2 \end{vmatrix}$ 3

9. $\begin{vmatrix} 3 & -1 & 2 \\ 0 & 1 & 2 \\ 3 & 2 & -2 \end{vmatrix}$ -30

10. $\begin{vmatrix} 4 & 5 & -2 \\ 3 & -1 & 5 \\ 2 & 1 & 4 \end{vmatrix}$ -56

11. $\begin{vmatrix} 4 & 2 & 6 \\ -2 & 1 & 1 \\ 2 & 1 & 3 \end{vmatrix}$ 0

12. $\begin{vmatrix} 3 & 6 & -3 \\ 4 & -1 & 6 \\ -1 & -2 & 3 \end{vmatrix}$ -54

2 Solve by using Cramer's Rule.

13. $2x - 5y = 26$
$5x + 3y = 3$
$(3, -4)$

14. $3x + 7y = 15$
$2y + 5y = 11$
$(-2, 3)$

15. $x - 4y = 8$
$3x + 7y = 5$
$(4, -1)$

16. $5x + 2y = -5$
$3x + 4y = 11$

$(-3, 5)$

17. $2x + 3y = 4$
$6x - 12y = -5$

$\left(\dfrac{11}{14}, \dfrac{17}{21}\right)$

18. $5x + 4y = 3$
$15x - 8y = -21$

$\left(-\dfrac{3}{5}, \dfrac{3}{2}\right)$

19. $2x + 5y = 6$
$6x - 2y = 1$

$\left(\dfrac{1}{2}, 1\right)$

20. $7x + 3y = 4$
$5x - 4y = 9$

$(1, -1)$

21. $-2x + 3y = 7$
$4x - 6y = 9$

no solution

22. $9x + 6y = 7$
$3x + 2y = 4$

no solution

23. $2x - 5y = -2$
$3x - 7y = -3$

$(-1, 0)$

24. $8x + 7y = -3$
$2x + 2y = 5$

$\left(-\dfrac{41}{2}, 23\right)$

25. $2x - y + 3z = 9$
$x + 4y + 4z = 5$
$3x + 2y + 2z = 5$
$(1, -1, 2)$

26. $3x - 2y + z = 2$
$2x + 3y + 2z = -6$
$3x - y + z = 0$
$(-1, -2, 1)$

27. $3x - y + z = 11$
$x + 4y - 2z = -12$
$2x + 2y - z = -3$
$(2, -2, 3)$

28. $x + 2y + 3z = 8$
$2x - 3y + z = 5$
$3x - 4y + 2z = 9$
$(-3, -2, 5)$

29. $4x - 2y + 6z = 1$
$3x + 4y + 2z = 1$
$2x - y + 3z = 2$
no solution

30. $x - 3y + 2z = 1$
$2x + y - 2z = 3$
$3x - 9y + 6z = -3$
no solution

31. $5x - 4y + 2z = 4$
$3x - 5y + 3z = -4$
$3x + y - 5z = 12$
$\left(\dfrac{68}{25}, \dfrac{56}{25}, -\dfrac{8}{25}\right)$

32. $2x + 4y + z = 7$
$x + 3y - z = 1$
$3x + 2y - 2z = 5$
$\left(\dfrac{53}{19}, -\dfrac{1}{19}, \dfrac{31}{19}\right)$

3   Solve by using matrices.

33. $3x + y = 6$
$2x - y = -1$
$(1,3)$

34. $2x + y = 3$
$x - 4y = 6$
$(2,-1)$

35. $x - 3y = 8$
$3x - y = 0$
$(-1,-3)$

36. $2x + 3y = 16$
$x - 4y = -14$
$(2,4)$

37. $y = 4x - 10$
$2y = 5x - 11$
$(3,2)$

38. $2y = 4 - 3x$
$y = 1 - 2x$
$(-2,5)$

39. $2x - y = -4$
$y = 2x - 8$

no solution

40. $3x - 2y = -8$
$y = \dfrac{3}{2}x - 2$

no solution

41. $4x - 3y = -14$
$3x + 4y = 2$

$(-2,2)$

42. $5x + 2y = 3$
$3x + 4y = 13$

$(-1,4)$

43. $5x + 4y + 3z = -9$
$x - 2y + 2z = -6$
$x - y - z = 3$
$(0,0,-3)$

44. $x - y - z = 0$
$3x - y + 5z = -10$
$x + y - 4z = 12$
$(1,3,-2)$

45. $5x - 5y + 2z = 8$
$2x + 3y - z = 0$
$x + 2y - z = 0$
$(1,-1,-1)$

46. $2x + y - 5z = 3$
$3x + 2y + z = 15$
$5x - y - z = 5$
$(2,4,1)$

47. $2x + 3y + z = 5$
$3x + 3y + 3z = 10$
$4x + 6y + 2z = 5$
no solution

48. $x - 2y + 3z = 2$
$2x + y + 2z = 5$
$2x - 4y + 6z = -4$

no solution

49. $3x + 2y + 3z = 2$
$6x - 2y + z = 1$
$3x + 4y + 2z = 3$
$\left(\dfrac{1}{3}, \dfrac{1}{2}, 0\right)$

50. $2x + 3y - 3z = -1$
$2x + 3y + 3z = 3$
$4x - 4y + 3z = 4$
$\left(\dfrac{1}{2}, 0, \dfrac{2}{3}\right)$

51. $5x - 5y - 5z = 2$
$5x + 5y - 5z = 6$
$10x + 10y + 5z = 3$
$\left(\dfrac{1}{5}, \dfrac{2}{5}, -\dfrac{3}{5}\right)$

52. $3x - 2y + 2z = 5$
$6x + 3y - 4z = -1$
$3x - y + 2z = 4$
$\left(\dfrac{2}{3}, -1, \dfrac{1}{2}\right)$

53. $4x + 4y - 3z = 3$
$8x + 2y + 3z = 0$
$4x - 4y + 6z = -3$
$\left(\dfrac{1}{4}, 0, -\dfrac{2}{3}\right)$

SUPPLEMENTAL EXERCISES 10.3

Solve for x.

54. $\begin{vmatrix} 3 & 2 \\ 4 & x \end{vmatrix} = -11$
-1

55. $\begin{vmatrix} -1 & 4 \\ 2 & x \end{vmatrix} = -11$
3

56. $\begin{vmatrix} -2 & 3 \\ 5 & x \end{vmatrix} = -3$
-6

57. $\begin{vmatrix} 1 & 0 & 2 \\ 4 & 3 & -1 \\ 0 & 2 & x \end{vmatrix} = -24$
-14

58. $\begin{vmatrix} -2 & 1 & 3 \\ 0 & x & 4 \\ -1 & 2 & -3 \end{vmatrix} = -24$
-4

59. $\begin{vmatrix} 3 & -2 & 1 \\ -1 & 0 & x \\ 2 & 4 & 0 \end{vmatrix} = 44$
-3

Complete.

60. If all the elements in one row or one column of a 2×2 matrix are zeros, the value of the determinant of the matrix is _____.
0

61. If all the elements in one row or one column of a 3×3 matrix are zeros, the value of the determinant of the matrix is _____.
0

62. **a.** The value of the determinant $\begin{vmatrix} x & x & a \\ y & y & b \\ z & z & c \end{vmatrix}$ is _____.
0

b. If two columns of a matrix contain identical elements, the value of the determinant is _____.
0

SECTION 10.4

Application Problems in Two Variables

1 Rate-of-wind and water current problems

Motion problems that involve an object moving with or against a wind or current normally require two variables to solve.

A motorboat traveling with the current can go 24 mi in 2 h. Against the current, it takes 3 h to go the same distance. Find the rate of the motorboat in calm water and the rate of the current.

STRATEGY *for solving rate-of-wind or water current problems*

▶ Choose one variable to represent the rate of the object in calm conditions and a second variable to represent the rate of the wind or current. Using these variables, express the rate of the object with and against the wind or current. Use the equation rt = d to write expressions for the distance traveled by the object. The results can be recorded in a table.

Rate of the boat in calm water: x
Rate of the current: y

	Rate	·	Time	=	Distance
With current	$x + y$	·	2	=	$2(x + y)$
Against current	$x - y$	·	3	=	$3(x - y)$

▶ Determine how the expressions for distance are related.

The distance traveled with the current is 24 mi.
The distance traveled against the current is 24 mi.

$2(x + y) = 24$
$3(x - y) = 24$

Solve the system of equations.

$2(x + y) = 24$ $\frac{1}{2} \cdot 2(x + y) = \frac{1}{2} \cdot 24$ $x + y = 12$

$3(x - y) = 24$ $\frac{1}{3} \cdot 3(x - y) = \frac{1}{3} \cdot 24$ $x - y = 8$

$$2x = 20$$
$$x = 10$$

Replace x by 10 in the equation $x + y = 12$. Solve for y.

$x + y = 12$
$10 + y = 12$
$y = 2$

The rate of the boat in calm water is 10 mph.
The rate of the current is 2 mph.

Example 1   Flying with the wind, a plane flew 1000 mi in 5 h. Flying against the wind, the plane could fly only 500 mi in the same amount of time. Find the rate of the plane in calm air and the rate of the wind.

Strategy   ▶ Rate of the plane in still air: p
Rate of the wind: w

	Rate	Time	Distance
With wind	$p + w$	5	$5(p + w)$
Against wind	$p - w$	5	$5(p - w)$

▶ The distance traveled with the wind is 1000 mi.
The distance traveled against the wind is 500 mi.

Solution $5(p + w) = 1000$
$5(p - w) = 500$

$p + w = 200$ • Multiply each side of the equations by $\frac{1}{5}$.
$p - w = 100$
$2p = 300$ • Add the equations.
$p = 150$

$p + w = 200$ • Substitute the value of p into one of the equations.
$150 + w = 200$
$w = 50$

The rate of the plane in calm air is 150 mph.
The rate of the wind is 50 mph.

Problem 1   A rowing team rowing with the current traveled 18 mi in 2 h. Against the current, the team rowed 10 mi in 2 h. Find the rate of the rowing team in calm water and the rate of the current.

Solution   See page A85.
rowing team: 7 mph; current: 2 mph

2 Application problems

The application problems in this section are varieties of those problems solved earlier in the text. Each of the strategies for the problems in this section will result in a system of equations.

A store owner purchased twenty 60-watt light bulbs and 30 fluorescent lights for a total cost of $40. A second purchase, at the same prices, included thirty 60-watt bulbs and 10 fluorescent lights for a total cost of $25. Find the cost of a 60-watt bulb and a fluorescent light.

STRATEGY *for solving an application problem in two variables*

▶ Choose one variable to represent one of the unknown quantities and a second variable to represent the other unknown quantity. Write numerical or variable expressions for all the remaining quantities. These results can be recorded in two tables, one for each of the conditions.

Cost of a 60-watt bulb: b
Cost of a fluorescent light: f

First purchase

	Amount	·	Unit cost	=	Value
60-watt	20	·	b	=	$20b$
Fluorescent	30	·	f	=	$30f$

Second purchase

	Amount	·	Unit cost	=	Value
60-watt	30	·	b	=	$30b$
Fluorescent	10	·	f	=	$10f$

▶ Determine a system of equations. The strategies presented in Chapter 2 can be used to determine the relationships between the expressions in the tables. Each table wili give one equation of the system.

The total of the first purchase was $40.
The total of the second purchase was $25.

$$20b + 30f = 40$$
$$30b + 10f = 25$$

Solve the system of equations.

$$\begin{array}{lll} 20b + 30f = 40 & 3(20b + 30f) = 3 \cdot 40 & 60b + 90f = 120 \\ 30b + 10f = 25 & -2(30b + 10f) = -2 \cdot 25 & -60b - 20f = -50 \\ & & 70f = 70 \\ & & f = 1 \end{array}$$

Replace f by 1 in the equation $20b + 30f = 40$. Solve for b.

$$20b + 30f = 40$$
$$20b + 30(1) = 40$$
$$20b + 30 = 40$$
$$20b = 10$$
$$b = 0.5$$

The cost of a 60-watt bulb was \$.50.
The cost of a fluorescent light was \$1.00.

Example 2 The total value of the nickels and dimes in a coin bank is \$2.50. If the nickels were dimes and the dimes were nickels, the total value of the coins would be \$3.05. Find the number of dimes and the number of nickels in the bank.

Strategy ▶ Number of nickels in the bank: n
Number of dimes in the bank: d

Coins in the bank now:

Coin	Number	Value	Total Value
Nickels	n	5	$5n$
Dimes	d	10	$10d$

Coins in the bank if the nickels were dimes and the dimes were nickels:

Coin	Number	Value	Total Value
Nickels	d	5	$5d$
Dimes	n	10	$10n$

▶ The value of the nickels and dimes in the bank is \$2.50.
The value of the nickels and dimes in the bank would be \$3.05.

Solution
$$5n + 10d = 250 \quad (1)$$
$$10n + 5d = 305 \quad (2)$$

$$10n + 20d = 500$$ • Multiply equation (1) by 2 and equation (2) by -1.
$$-10n - 5d = -305$$ • Add the equations.
$$15d = 195$$
$$d = 13$$

$$5n + 10d = 250$$ • Substitute the value of d into one of the equations.
$$5n + 10(13) = 250$$
$$5n + 130 = 250$$
$$5n = 120$$
$$n = 24$$

There are 24 nickels and 13 dimes in the bank.

Problem 2   A citrus fruit grower purchased 25 orange trees and 20 grapefruit trees for $290. The next week, at the same prices, the grower bought 20 orange trees and 30 grapefruit trees for $330. Find the cost of an orange tree and the cost of a grapefruit tree.

Solution   See page A86.
orange tree: $6; grapefruit tree: $7

EXERCISES 10.4

1   Solve.

1. Flying with the wind, a small plane flew 300 mi in 2 h. Against the wind, the plane could fly only 270 mi in the same amount of time. Find the rate of the plane in calm air and the rate of the wind.
plane: 142.5 mph; wind: 7.5 mph

2. A jet plane flying with the wind went 2200 mi in 4 h. Against the wind, the plane could fly only 1820 mi in the same amount of time. Find the rate of the plane in calm air and the rate of the wind.
plane: 502.5 mph; wind: 47.5 mph

3. A cabin cruiser traveling with the current went 45 mi in 3 h. Against the current, it took 5 h to travel the same distance. Find the rate of the cabin cruiser in calm water and the rate of the current.
cabin cruiser: 12 mph; current: 3 mph

4. A motorboat traveling with the current went 36 mi in 2 h. Against the current, it took 3 h to travel the same distance. Find the rate of the boat in calm water and the rate of the current.
motorboat: 15 mph; current: 3 mph

5. Flying with the wind, a pilot flew 600 mi between two cities in 4 h. The return trip against the wind took 5 h. Find the rate of the plane in calm air and the rate of the wind.
plane: 135 mph; wind: 15 mph

6. A turbo-prop plane flying with the wind flew 800 mi in 4 h. Flying against the wind, the plane required 5 h to travel the same distance. Find the rate of the wind and the rate of the plane in calm air.
plane: 180 mph; wind: 20 mph

7. A motorboat traveling with the current went 72 km in 3 h. Against the current, the boat could go only 48 km in the same amount of time. Find the rate of the boat in calm water and the rate of the current.
motorboat: 20 km/h; current: 4 km/h

8. A rowing team rowing with the current traveled 20 km in 2 h. Rowing against the current, the team rowed 12 km in the same amount of time. Find the rate of the team in calm water and the rate of the current.
rowing team: 8 km/h; current: 2 km/h

9. A plane flying with a tailwind flew 600 mi in 5 h. Against the wind, the plane required 6 h to fly the same distance. Find the rate of the plane in calm air and the rate of the wind.
plane: 110 mph; wind: 10 mph

10. Flying with the wind, a plane flew 720 mi in 3 h. Against the wind, the plane required 4 h to fly the same distance. Find the rate of the plane in calm air and the rate of the wind.
plane: 210 mph; wind: 30 mph

11. A motorboat traveling with the current went 48 mi in 3 h. Against the current, it took 4.8 h to travel the same distance. Find the rate of the boat in calm water and the rate of the current.
motorboat: 13 mph; current: 3 mph

12. A plane traveling with the wind flew 3625 mi in 6.25 h. Against the wind, the plane required 7.25 h to fly the same distance. Find the rate of the plane in calm air and the rate of the wind.
plane: 540 mph; wind: 40 mph

13. A cabin cruiser traveling with the current went 45 mi in 2.5 h. Against the current, the boat could go only 30 mi in the same amount of time. Find the rate of the cabin cruiser and the rate of the current.
cabin cruiser: 15 mph; current: 3 mph

14. Flying with the wind, a plane flew 450 mi in 3 h. Against the wind, the plane could fly only 270 mi in the same amount of time. Find the rate of the plane in calm air and the rate of the wind.
plane: 120 mph; wind: 30 mph

2 Solve.

15. A carpenter purchased 60 ft of redwood and 80 ft of pine for a total cost of $27. A second purchase, at the same prices, included 100 ft of redwood and 60 ft of pine for a total cost of $34. Find the cost per foot of redwood and pine.
pine: $.15/ft; redwood: $.25/ft

16. A merchant mixed 10 lb of a cinnamon tea with 5 lb of spice tea. The 15-lb mixture cost $40. A second mixture included 12 lb of the cinnamon tea and 8 lb of the spice tea. The 20-lb mixture cost $54. Find the cost per pound of the cinnamon tea and the spice tea.
cinnamon: $2.50/lb; spice: $3/lb

17. During one month, a homeowner used 500 units of electricity and 100 units of gas for a total cost of $88. The next month, 400 units of electricity and 150 units of gas were used for a total cost of $76. Find the cost per unit of gas.
    $.08

18. A contractor buys 16 yd of nylon carpet and 20 yd of wool carpet for $920. A second purchase, at the same prices, includes 18 yd of nylon carpet and 25 yd of wool carpet for $1100. Find the cost per yard of the wool carpet.
    $26/yd

19. The total value of the quarters and dimes in a coin bank is $5.75. If the quarters were dimes and the dimes were quarters, the total value of the coins would be $6.50. Find the number of quarters in the bank.
    15 quarters

20. A coin bank contains only nickels and dimes. The total value of the coins in the bank is $2.50. If the nickels were dimes and the dimes were nickels, the total value of the coins would be $3.50. Find the number of nickels in the bank.
    30 nickels

21. A company manufacturers both color and black-and-white television sets. The cost of materials for a black-and-white TV is $20, while the cost of materials for a color TV is $80. The cost of labor to manufacture a black-and-white TV is $30, while the cost of labor to manufacture a color TV is $50. During a week when the company has budgeted $4200 for materials and $2800 for labor, how many color TV's does the company plan to manufacture?
    50 color TV's

22. A company manufactures both 10-speed and standard model bicycles. The cost of materials for a 10-speed bicycle is $35, while the cost of materials for a standard bicycle is $25. The cost of labor to manufacture a 10-speed bicycle is $40, while the cost of labor to manufacture a standard bicycle is $20. During a week when the company has budgeted $1250 for materials and $1300 for labor, how many 10-speed bicycles does the company plan to manufacture?
    25 10-speed bicycles

Use your calculator for the following exercises.

23. A pharmacist has two vitamin-supplement powders. The first powder is 20% vitamin B_1 and 10% vitamin B_2. The second is 15% vitamin B_1 and 20% vitamin B_2. How many milligrams of each of the two powders should the pharmacist use to make a mixture that contains 130 mg of vitamin B_1 and 80 mg of vitamin B_2?
    1st powder: 560 mg; 2nd powder: 120 mg

24. A chemist has two alloys, one of which is 10% gold and 15% lead, the other of which is 30% gold and 40% lead. How many grams of each of the two alloys should be used to make an alloy that contains 60 g of gold and 88 g of lead?
1st alloy: 480 g; 2nd alloy: 40 g

SUPPLEMENTAL EXERCISES 10.4

Solve.

25. Two angles are complementary. The larger angle is 9° more than eight times the measure of the smaller angle. Find the measure of the two angles. (Complementary angles are two angles whose sum is 90°.)
9° and 81°

26. Two angles are supplementary. The larger angle is 40° more than three times the measure of the smaller angle. Find the measure of the two angles. (Supplementary angles are two angles whose sum is 180°.)
35° and 145°

27. The sum of the digits of a two-digit number equals $\frac{1}{7}$ of the number. If the digits of the number are reversed, the new number is equal to 36 less than the original number. Find the original number.
84

28. The sum of the digits of a two-digit number equals $\frac{1}{5}$ of the number. If the digits of the number are reversed, the new number is equal to 9 more than the original number. Find the original number.
45

29. The sum of the ages of a gold coin and a silver coin is 75 years. The age of the gold coin ten years from now is five years less than the age of the silver coin ten years ago. Find the present ages of the two coins.
gold coin: 25 years; silver coin: 50 years

30. The difference between the ages of an oil painting and a watercolor is 35 years. The age of the oil painting five years from now is twice the age of the watercolor five years ago. Find the present ages of each.
oil painting: 85 years; watercolor: 50 years

31. The total value of the nickels, dimes, and quarters in a coin bank is $3.50. If the nickels were dimes and the dimes were nickels, the total value of the coins would be $4.25. If the quarters were dimes and the dimes were quarters, the total value of the coins would be $4.25. Find the number of nickels, dimes, and quarters in the bank.
25 nickels, 10 dimes, and 5 quarters

32. The total value of the nickels, dimes, and quarters in a coin bank is $4.00. If the nickels were dimes and the dimes were nickels, the total value of the coins would be $3.75. If the quarters were dimes and the dimes were quarters, the total value of the coins would be $6.25. Find the number of nickels, dimes, and quarters in the bank.
15 nickels, 20 dimes, and 5 quarters

SECTION 10.5 ────────────────────

Solving Systems of Quadratic Equations and Systems of Inequalities

1 Solve nonlinear systems of equations

Nonlinear systems of equations can be solved by using either a substitution method or an addition method.

The system of equations at the right contains a linear equation and a quadratic equation. When a system contains both a linear and a quadratic equation, the substitution method is used.

$$(1) \quad 2x - y = 4$$
$$(2) \quad y^2 = 4x$$

Solve equation (1) for y.

$$2x - y = 4$$
$$-y = -2x + 4$$
$$y = 2x - 4$$

Substitute $2x - 4$ for y into equation (2).

$$y^2 = 4x$$
$$(2x - 4)^2 = 4x$$

Write the equation in standard form.

$$4x^2 - 16x + 16 = 4x$$
$$4x^2 - 20x + 16 = 0$$

Solve for x by factoring.

$$4(x^2 - 5x + 4) = 0$$
$$4(x - 4)(x - 1) = 0$$

$$x - 4 = 0 \qquad\qquad x - 1 = 0$$
$$x = 4 \qquad\qquad\qquad x = 1$$

Substitute the values of x into the equation $y = 2x - 4$, and solve for y.

$$y = 2x - 4 \qquad\qquad y = 2x - 4$$
$$y = 2(4) - 4 \qquad\quad y = 2(1) - 4$$
$$y = 4 \qquad\qquad\qquad y = -2$$

The solutions are $(4, 4)$ and $(1, -2)$.

The graph of the system that was solved on the previous page is shown at the right. Note that the line intersects the parabola at two points. These points correspond to the solutions.

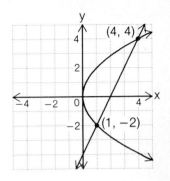

The system of equations at the right contains a linear equation. The substitution method is used to solve the system.

$$(1) \qquad x^2 + y^2 = 4$$
$$(2) \qquad \qquad y = x + 3$$

Substitute the expression for y into equation (1).

$$x^2 + y^2 = 4$$
$$x^2 + (x + 3)^2 = 4$$

Write the equation in standard form.

$$x^2 + x^2 + 6x + 9 = 4$$
$$2x^2 + 6x + 5 = 0$$

Since the discriminant of the quadratic equation is less than zero, the equation has two complex number solutions. Therefore, the system of equations has no real number solutions.

$$b^2 - 4ac = 6^2 - 4(2)(5)$$
$$= 36 - 40$$
$$= -4$$

The graphs of the equations of this system are shown at the right. Note that the two graphs do not intersect.

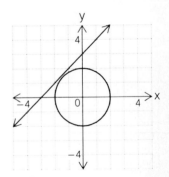

The addition method is used to solve the system of equations shown at the right.

$$(1) \quad 4x^2 + y^2 = 16$$
$$(2) \quad x^2 + y^2 = 4$$

Multiply equation (2) by -1 and add to equation (1).

$$4x^2 + y^2 = 16$$
$$-x^2 - y^2 = -4$$
$$3x^2 = 12$$

Solve for x.

$$x^2 = 4$$
$$x = \pm 2$$

Substitute the values of x into equation (2), and solve for y.

$$x^2 + y^2 = 4$$
$$2^2 + y^2 = 4$$
$$y^2 = 0$$
$$y = 0$$

$$x^2 + y^2 = 4$$
$$(-2)^2 + y^2 = 4$$
$$y^2 = 0$$
$$y = 0$$

The solutions are $(2, 0)$ and $(-2, 0)$.

The graphs of the equations in this system are shown at the right. Note that the graphs intersect at two points.

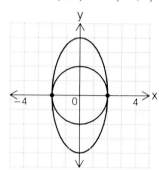

Example 1   Solve.

A. $y = 2x^2 - 3x - 1$
   $y = x^2 - 2x + 5$

B. $3x^2 - 2y^2 = 26$
   $x^2 - y^2 = 5$

Solution   A. $y = 2x^2 - 3x - 1$
   $y = x^2 - 2x + 5$

$$2x^2 - 3x - 1 = x^2 - 2x + 5$$
$$x^2 - x - 6 = 0$$
$$(x - 3)(x + 2) = 0$$

• Use the substitution method.

$$x - 3 = 0 \qquad x + 2 = 0$$
$$x = 3 \qquad\quad x = -2$$

$$\begin{aligned} y &= 2x^2 - 3x - 1 \\ y &= 2(3)^2 - 3(3) - 1 \\ y &= 18 - 9 - 1 \\ y &= 8 \end{aligned}$$

$$\begin{aligned} y &= 2x^2 - 3x - 1 \\ y &= 2(-2)^2 - 3(-2) - 1 \\ y &= 8 + 6 - 1 \\ y &= 13 \end{aligned}$$

• Substitute the values of x into equation (1).

The solutions are $(3, 8)$ and $(-2, 13)$.

B. $3x^2 - 2y^2 = 26$ (1)
   $x^2 - y^2 = 5$ (2)

$$3x^2 - 2y^2 = 26$$
$$-2x^2 + 2y^2 = -10$$
$$x^2 = 16$$
$$x = \pm\sqrt{16} = \pm 4$$

• Use the addition method. Multiply equation (2) by -2.

$$x^2 - y^2 = 5$$
$$4^2 - y^2 = 5$$
$$16 - y^2 = 5$$
$$-y^2 = -11$$
$$y^2 = 11$$
$$y = \pm\sqrt{11}$$

$$x^2 - y^2 = 5$$
$$(-4)^2 - y^2 = 5$$
$$16 - y^2 = 5$$
$$-y^2 = -11$$
$$y^2 = 11$$
$$y = \pm\sqrt{11}$$

• Substitute the values of x into equation (2).

The solutions are $(4, \sqrt{11})$, $(4, -\sqrt{11})$, $(-4, \sqrt{11})$, and $(-4, -\sqrt{11})$.

Problem 1   Solve.

A. $y = 2x^2 + x - 3$
   $y = 2x^2 - 2x + 9$

B. $x^2 - y^2 = 10$
   $x^2 + y^2 = 8$

Solution   See page A87.

A. $(4, 33)$

B. No real number solution

2   Graph the solution set of a system of inequalities

The **solution set of a system of inequalities** is the intersection of the solution sets of each inequality. To graph the solution set of a system of inequalities, first graph the solution set for each inequality. The solution set of the system of inequalities is the region of the plane represented by the intersection of the two shaded regions.

To graph the solution set of $2x - y \leq 3$
                              $3x + 2y \geq 8$,
graph the solution set of each inequality.

The solution set is the region of the plane represented by the intersection of the solution sets of each inequality.

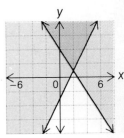

Example 2   Graph the solution set.

A. $\dfrac{x^2}{9} + \dfrac{y^2}{4} \geq 1$

   $\dfrac{x^2}{4} - \dfrac{y^2}{9} > 1$

B. $y > x^2$
   $y < x + 2$

Solution   A.

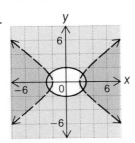

B.

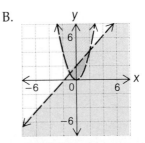

Problem 2   Graph the solution set.

A. $x^2 + y^2 < 16$ B. $y \geq x - 1$
     $y^2 < x$ $y < -2x$

Solution   See page A87.

A.

B.

EXERCISES 10.5

1   Solve.

1. $y = x^2 - x - 1$
$y = 2x + 9$
$(-2, 5)$ and $(5, 19)$

2. $y = x^2 - 3x + 1$
$y = x + 6$
$(5, 11)$ and $(-1, 5)$

3. $y^2 = -x + 3$
$x - y = 1$
$(-1, -2)$ and $(2, 1)$

4. $y^2 = 4x$
$x - y = -1$
$(1, 2)$

5. $y^2 = 2x$
$x + 2y = -2$
$(2, -2)$

6. $y^2 = 2x$
$x - y = 4$
$(2, -2)$ and $(8, 4)$

7. $x^2 + 2y^2 = 12$
$2x - y = 2$
$(2, 2)$ and $\left(-\dfrac{2}{9}, -\dfrac{22}{9}\right)$

8. $x^2 + 4y^2 = 37$
$x - y = -4$
$\left(-\dfrac{27}{5}, -\dfrac{7}{5}\right)$ and $(-1, 3)$

9. $x^2 + y^2 = 13$
$x + y = 5$
$(3, 2)$ and $(2, 3)$

10. $x^2 + y^2 = 16$
$x - 2y = -4$
$(-4, 0)$ and $\left(\dfrac{12}{5}, \dfrac{16}{5}\right)$

11. $4x^2 + y^2 = 12$
$y = 4x^2$
$\left(\dfrac{\sqrt{3}}{2}, 3\right)$ and $\left(-\dfrac{\sqrt{3}}{2}, 3\right)$

12. $2x^2 + y^2 = 6$
$y = 2x^2$
$(1, 2)$ and $(-1, 2)$

13. $y = x^2 - 2x - 3$
$y = x - 6$
no real number solution

14. $y = x^2 + 4x + 5$
$y = -x - 3$
no real number solution

15. $3x^2 - y^2 = -1$
$x^2 + 4y^2 = 17$
$(1, 2), (1, -2), (-1, 2),$ and $(-1, -2)$

16. $x^2 + y^2 = 10$
$x^2 + 9y^2 = 18$
$(3, 1), (-3, 1), (3, -1),$ and $(-3, -1)$

17. $2x^2 + 3y^2 = 30$
  $x^2 + y^2 = 13$
  $(3,2), (3,-2), (-3,2),$ and $(-3,-2)$

18. $x^2 + y^2 = 61$
  $x^2 - y^2 = 11$
  $(6,5), (6,-5), (-6,5),$ and $(-6,-5)$

19. $y = 2x^2 - x + 1$
  $y = x^2 - x + 5$

  $(2,7)$ and $(-2,11)$

20. $y = -x^2 + x - 1$
  $y = x^2 + 2x - 2$

  $\left(\dfrac{1}{2}, -\dfrac{3}{4}\right)$ and $(-1,-3)$

21. $2x^2 + 3y^2 = 24$
  $x^2 - y^2 = 7$
  $(3,\sqrt{2}), (3,-\sqrt{2}), (-3,\sqrt{2})$ and $(-3,-\sqrt{2})$

22. $2x^2 + 3y^2 = 21$
  $x^2 + 2y^2 = 12$
  $(\sqrt{6},\sqrt{3}), (\sqrt{6},-\sqrt{3}), (-\sqrt{6},\sqrt{3}),$
  and $(-\sqrt{6},-\sqrt{3})$

23. $x^2 + y^2 = 36$
  $4x^2 + 9y^2 = 36$
  no real number solution

24. $2x^2 + 3y^2 = 12$
   $x^2 - y^2 = 25$
  no real number solution

25. $11x^2 - 2y^2 = 4$
   $3x^2 + y^2 = 15$
  $(\sqrt{2},3), (\sqrt{2},-3), (-\sqrt{2},3),$ and $(-\sqrt{2},-3)$

26. $x^2 + 4y^2 = 25$
   $x^2 - y^2 = 5$
  $(3,2), (3,-2), (-3,2),$ and $(-3,-2)$

27. $2x^2 - y^2 = 7$
   $2x - y = 5$

  $(2,-1)$ and $(8,11)$

28. $3x^2 + 4y^2 = 7$
   $x - 2y = -3$

  $(-1,1)$ and $\left(-\dfrac{1}{2}, \dfrac{5}{4}\right)$

29. $y = 3x^2 + x - 4$
  $y = 3x^2 - 8x + 5$
  $(1,0)$

30. $y = 2x^2 + 3x + 1$
  $y = 2x^2 + 9x + 7$
  $(-1,0)$

---

2   Graph the solution set.

31. $2x + y \geq 1$
   $3x + 2y < 6$

32. $3x - 4y < 12$
   $x + 2y < 6$

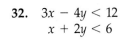

33. $x - 2y \leq 6$
   $2x + 3y \leq 6$

34. $x - 3y > 6$
   $2x + 3y > 9$

35. $2x + 3y \le 15$
$3x - y \le 6$
$y \ge 0$

36. $x + y \le 6$
$x - y \le 2$
$x \ge 0$

37. $x^2 + y^2 < 16$
$y > x + 1$

38. $y > x^2 - 4$
$y < x - 2$

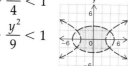

39. $x^2 + y^2 < 25$

$\dfrac{x^2}{9} + \dfrac{y^2}{36} < 1$

40. $\dfrac{x^2}{9} - \dfrac{y^2}{4} < 1$

$\dfrac{x^2}{25} + \dfrac{y^2}{9} < 1$

41. $x^2 + y^2 > 4$
$x^2 + y^2 < 25$

42. $\dfrac{x^2}{25} + \dfrac{y^2}{16} \le 1$

$\dfrac{x^2}{4} + \dfrac{y^2}{4} \ge 1$

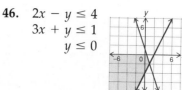

SUPPLEMENTAL EXERCISES 10.5

Solve.

43. Is it possible for two ellipses with centers at the origin to intersect in exactly three points?
no

44. Is it possible for two circles with centers at the origin to intersect in exactly two points?
no

Graph the solution set.

45. $x - 3y \le 6$
$5x - 2y \ge 4$
$y \ge 0$

46. $2x - y \le 4$
$3x + y \le 1$
$y \le 0$

47. $y > x^2 - 3$
$y < x + 3$
$x \geq 0$

48. $x^2 + y^2 \leq 25$
$y > x + 1$
$x > 0$,

49. $x^2 + y^2 < 3$
$x > y^2 - 1$
$y \leq 0$

50. $\dfrac{x^2}{4} - \dfrac{y^2}{25} \leq 1$
$\dfrac{x^2}{25} + \dfrac{y^2}{4} \leq 1$
$y \geq 0$

51. $\dfrac{x^2}{16} + \dfrac{y^2}{4} \leq 1$
$x^2 + y^2 \leq 4$
$x > 0$
$y < 0$

52. $\dfrac{x^2}{4} + \dfrac{y^2}{25} \leq 1$
$x > y^2 - 4$
$x < 0$
$y > 0$

CALCULATORS AND COMPUTERS

Cramer's Rule

In this chapter, determinants were used as one method of solving a system of linear equations. By evaluating some determinants and then using Cramer's Rule, the solution to a system of equations could be found.

The difficulty with using Cramer's Rule when there are more than three variables in the system of equations is the number of computations necessary to evaluate a single determinant. There are 2 terms in the evaluation of a 2×2 determinant, 6 terms in a 3×3, 24 terms in a 4×4, 120 in a 5×5, 720 in a 6×6, and 5040 in a 7×7.

Thus to solve a linear system of seven equations with seven unknowns would require evaluating eight 7×7 determinants, which requires evaluating more than 41,000 terms. And that does not count all the multiplications within each term. Happily, however, computer programs can be written to do the evaluations of determinants. One such program, CRAMER'S RULE, can be found on the Student Disk.

This program gives the solution to equations in terms of the variables x_1, x_2, x_3, and so on instead of the usual x, y, z. This is more convenient since it is impossible to run out of variables. Variables written in this way are called subscripted variables.

You are encouraged to use this program on some of the exercises in your text or make up some systems and try the program. The coefficients you use can be any real numbers, but remember to convert any fraction to a decimal before you enter that number.

CHAPTER SUMMARY

Key Words

Equations considered together are called a **system of equations.**

A **solution of a system of equations in two variables** is an ordered pair that is a solution of each equation of the system.

When the graphs of a system of equations intersect at only one point, the equations are called **independent equations.** When a system of equations has no solution, it is called an **inconsistent system of equations.** When the graphs of a system coincide, the equations are called **dependent equations.**

An equation of the form $Ax + By + Cz = D$, where A, B, C, and D are constants, is called a **linear equation in three variables.** A **solution of a system of equations in three variables** is an ordered triple that is a solution of each equation of the system.

A **matrix** is a rectangular array of numbers. Each number in the matrix is an **element** of the matrix. A **square matrix** has the same number of rows as columns.

A **determinant** is a number associated with a square matrix. The **minor of an element** in a 3 × 3 determinant is the 2 × 2 determinant that is obtained by eliminating the row and column that contain that element.

The **solution set of a system of inequalities** is the intersection of the solution sets of each inequality.

Essential Rules

Cramer's Rule is a method of solving a system of equations by using determinants.

Elementary row operations include
1. Interchange two rows.
2. Multiply all the elements in a row by the same nonzero number.
3. Replace a row by the sum of that row and a multiple of any other row.

CHAPTER REVIEW

1. Solve by substitution:
$$4x + 3y = 11$$
$$3x - y = 5$$
$(2,1)$

2. Solve by the addition method:
$$2x - y - z = 4$$
$$x + 2y + z = 3$$
$$x - 3y - 2z = 1$$
inconsistent equations

3. Solve by using Cramer's Rule:
$$2x - 2y + z = -5$$
$$4x + 3y + 6z = 8$$
$$x + y + 2z = 3$$
$(-1,2,1)$

4. Solve: $3x^2 - y^2 = 2$
$$x^2 + 2y^2 = 3$$
$(1,1), (1,-1), (-1,1)$ and $(-1,-1)$

5. Solve by graphing: $2x - 3y = -6$
$$x + 3y = -12$$
$(-6,-2)$

6. Solve by the addition method:
$$3x - 4y = 8$$
$$5x + 3y = -6$$
$(0,-2)$

7. Evaluate the determinant: $\begin{vmatrix} 5 & -2 \\ 3 & 4 \end{vmatrix}$

26

8. Solve by using a matrix: $2x + 3y = 0$
$$3x - 2y = 13$$
$(3,-2)$

9. Graph the solution set:
$$4x - 3y > 6$$
$$x + 2y < 4$$

10. Solve by substitution: $2x - 5y = 23$
$$y = 2x - 3$$
$(-1,-5)$

11. Solve by the addition method:
$$x - 2y + z = 7$$
$$3x - z = -1$$
$$3y + z = 1$$
$(1,-1,4)$

12. Solve by using Cramer's Rule:
$$3x - 2y = 2$$
$$-2x + 3y = 1$$
$\left(\dfrac{8}{5}, \dfrac{7}{5}\right)$

13. Solve: $2x + y = 9$

$4y^2 = x$ $\left(\dfrac{81}{16}, -\dfrac{9}{8}\right)$ and $(4,1)$

14. Solve by graphing: $x - 2y = 2$

$3x + 2y = 6$

$(2,0)$

15. Solve by using a matrix:

$2x - 2y - 6z = 1$

$4x + 2y + 3z = 1$ $\left(\dfrac{1}{2}, -1, \dfrac{1}{3}\right)$

$2x - 3y - 3z = 3$

16. Evaluate the determinant:

$$\begin{vmatrix} 3 & -2 & 5 \\ 4 & 6 & 3 \\ 1 & 2 & 1 \end{vmatrix}$$

12

17. Graph the solution set:

$2x - 5y \leq 10$

$x + 2y > 2$

18. Solve by the addition method:

$5x + 2y = 5$

$6x + 3y = 4$ $\left(\dfrac{7}{3}, -\dfrac{10}{3}\right)$

19. Solve by using Cramer's Rule:

$4x - 3y = 17$

$3x - 2y = 12$

$(2,-3)$

20. Solve by using a matrix:

$3x + 2y - z = -1$

$x + 2y + 3z = -1$

$3x + 4y + 6z = 0$

$(2,-3,1)$

21. A cabin cruiser traveling with the current went 48 mi in 3 h. The return trip against the current required 6 h. Find the rate of the boat in calm water and the rate of the current.

cabin cruiser: 12 mph; current: 4 mph

22. Flying with the wind, a small plane required 2 h to fly 150 mi. Against the wind, it took 3 h to fly the same distance. Find the rate of the wind.

12.5 mph

23. A sheetmetal shop owner ordered 50 lb of tin and 25 lb of a zinc alloy for a total cost of $350. A second purchase, at the same prices, included 30 lb of tin and 60 lb of the zinc alloy. The total cost was $570. Find the cost per pound of the tin and of the zinc alloy.

tin alloy: $3/lb; zinc alloy: $8/lb

24. A restaurant manager buys 100 lb of hamburger and 50 lb of steak for a total cost of $270. A second purchase, at the same prices, includes 150 lb of hamburger and 100 lb of steak. The total cost is $480. Find the price of one pound of steak.

$3.00

CUMULATIVE REVIEW

1. Solve: $\frac{3}{2}x - \frac{3}{8} + \frac{1}{4}x = \frac{7}{12}x - \frac{5}{6}$

$-\frac{11}{28}$

2. Solve: $3x + 2 \le 5$ and $x + 5 \ge 1$
$\{x \mid -4 \le x \le 1\}$

3. Find the equation of the line containing the points $(2, -1)$ and $(3, 4)$.
$y = 5x - 11$

4. Factor: $6x^2 - 19x + 10$
$(2x - 5)(3x - 2)$

5. Simplify: $\dfrac{2x}{x^2 - 5x + 6} - \dfrac{3}{x^2 - 2x - 3}$

$\dfrac{2x^2 - x + 6}{(x - 2)(x - 3)(x + 1)}$

6. Solve: $\dfrac{3}{x^2 - 5x + 6} - \dfrac{x}{x - 3} = \dfrac{2}{x - 2}$

-3

7. Simplify: $a^{-1} + a^{-1}b$
$\dfrac{1 + b}{a}$

8. Simplify: $\sqrt[3]{-4ab^4}\ \sqrt[3]{2a^3b^4}$
$-2ab^2\sqrt[3]{ab^2}$

9. Solve: $\sqrt{3x + 10} = 1$
-3

10. Solve: $2x^2 + 9x = 5$
$\dfrac{1}{2}$ and -5

11. Solve: $3x^2 = 2x + 2$
$\dfrac{1 + \sqrt{7}}{3}$ and $\dfrac{1 - \sqrt{7}}{3}$

12. Solve: $x^{-2} - 5x^{-1} + 6 = 0$
$\dfrac{1}{2}$ and $\dfrac{1}{3}$

13. Use the discriminant to determine the number of x-intercepts of the parabola $y = -2x^2 - x - 2$.
no x-intercepts

14. Graph $f(x) = |x| - 2$.

15. Find the inverse of the function
$f(x) = \dfrac{2}{3}x - 1.$
$f^{-1}(x) = \dfrac{3}{2}x + \dfrac{3}{2}$

16. Find the distance between the points $(-4, 0)$ and $(2, 2)$.
$2\sqrt{10}$

17. Write the equation
$x^2 + y^2 - 4x + 2y - 4 = 0$
in standard form.
Then sketch its graph.
$(x - 2)^2 + (y + 1)^2 = 9$

18. Solve by graphing: $3x - y = 4$
$3x - 2y = 2$
$(2, 2)$

19. Solve by substitution: $3x - 2y = 7$
$$y = 2x - 1$$

$(-5, -11)$

20. Solve by the addition method:
$$3x + 2z = 1$$
$$2y - z = 1$$
$$x + 2y = 1$$

$(1, 0, -1)$

21. Evaluate the determinant: $\begin{vmatrix} 2 & -5 & 1 \\ 3 & 1 & 2 \\ 6 & -1 & 4 \end{vmatrix}$

3

22. Solve by using Cramer's Rule:
$$3x - 3y = 2$$
$$6x - 4y = 5$$

$\left(\dfrac{7}{6}, \dfrac{1}{2} \right)$

23. Solve by using a matrix:
$$3x + 3y + 2z = 1$$
$$x + 2y + z = 1$$
$$2x - 2y + z = 0$$

$(-1, 0, 2)$

24. Solve: $x^2 + y^2 = 5$
$$y^2 = 4x$$

$(1, 2)$ and $(1, -2)$

25. A coin purse contains 40 coins in nickels, dimes, and quarters. There are three times as many dimes as quarters. The total value of the coins is $4.10. How many nickels are in the coin purse?

16 nickels

26. The distance (d) a spring stretches varies directly as the force (f) used to stretch the spring. If a force of 40 lb can stretch the spring 24 in., how far will a force of 60 lb stretch the spring?

36 in.

27. The height of a triangle is twice the length of the base. The area of the triangle is 36 m^2. Find the length of the base.

6 m

28. How many milliliters of pure water must be added to 100 ml of a 4% salt solution to make a 2.5% solution?

60 ml

29. An average score of 75–84 in a chemistry class receives a C grade. A student has grades of 70, 79, 76, and 73 on four chemistry tests. Find the range of scores on the fifth test that will give the student a C grade for the course.

77 or better

30. A rowboat traveling with the current went 22.5 mi in 3 h. Against the current, it took 5 h to travel the same distance. Find the rate of the rowboat in calm water and the rate of the current.

rowboat: 6 mph; current: 1.5 mph

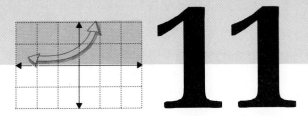

11

Exponential and Logarithmic Functions

OBJECTIVES

- Evaluate exponential functions
- Graph exponential functions
- Write equivalent exponential and logarithmic expressions
- Graph the logarithmic function $f(x) = \log_b x$
- The Properties of Logarithms
- Find common logarithms
- Find common antilogarithms
- Use interpolation to find a common logarithm or antilogarithm
- Evaluate numerical expressions by using common logarithms
- Solve exponential equations
- Solve logarithmic equations
- Find the logarithm of a number other than base 10
- Application problems

Napier Rods

The labor that is involved in calculating the products of large numbers has led many people to devise ways to short-cut the procedure. One such way was first described in the early 1600s by John Napier and is based on Napier Rods.

A Napier Rod consists of placing a number and the first 9 multiples of that number on a rectangular piece of paper (Napier's Rod). This is done for the first 9 positive integers. It is necessary to have more than one rod for each number. The rod for 7 is shown at the right.

	7	
row 1	0 / 7	$1 \times 7 - 7$
row 2	1 / 4	$2 \times 7 = 14$
row 3	2 / 1	$3 \times 7 = 21$
row 4	2 / 8	$4 \times 7 = 28$
row 5	3 / 5	$5 \times 7 = 35$
row 6	4 / 2	$6 \times 7 = 42$
row 7	4 / 9	$7 \times 7 = 49$
row 8	5 / 6	$8 \times 7 = 56$
row 9	6 / 3	$9 \times 7 = 63$

To illustrate how these rods were used to multiply, an example will be used.

Multiply: 2893
          × 246

Place the rods for 2, 8, 9, and 3 next to one another. The products for 2, 4, and 6 are found by using the numbers along the 2nd, 4th, and 6th rows.

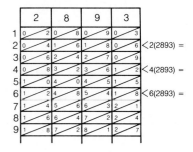

Each number is found by adding the digits diagonally downward on the diagonal and carrying to the next diagonal when necessary. The products for 2 and 6 are shown at the right.

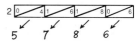

Notice that the sum of the 3rd diagonal is 13, thus carrying to the next diagonal is necessary.

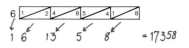

The final product is then found by addition.

Before the invention of the electronic calculator, logarithms were used to ease the drudgery of lengthy calculations. John Napier is also credited with the invention of logarithms.

$$
\begin{array}{r}
2893 \\
\times \quad 246 \\
\hline
5786 \\
11572 \\
+ \ 17358 \\
\hline
711678
\end{array}
$$

The Exponential and Logarithmic Functions

1 Evaluate exponential functions

A function of the form $f(x) = b^x$, where b is a positive real number not equal to 1, is an **exponential function.** The number b is the **base** of the exponential function.

The value of $f(x) = 2^x$ at $x = 3$ is $f(3) = 2^3 = 8$.

The value of $f(x) = 3^x$ at $x = -2$ is $f(-2) = 3^{-2} = \dfrac{1}{3^2} = \dfrac{1}{9}$.

If the base of an exponential function were allowed to be a negative number, the value of the function could be a complex number. For example,

the value of $f(x) = (-4)^x$ at $x = \dfrac{1}{2}$ is $f\!\left(\dfrac{1}{2}\right) = (-4)^{\frac{1}{2}} = \sqrt{-4} = 2i$.

For this reason, the base of an exponential function is required to be a positive real number.

Using a calculator, the value of the function $f(x) = 4^x$ at $x = \sqrt{2}$ can be found to the desired degree of accuracy by using approximations for $\sqrt{2}$.

$$
\begin{aligned}
f(\sqrt{2}) &\approx f(1.41) &&= 4^{1.41} &&\approx 7.06 \\
f(\sqrt{2}) &\approx f(1.414) &&= 4^{1.414} &&\approx 7.101 \\
f(\sqrt{2}) &\approx f(1.4142) &&= 4^{1.4142} &&\approx 7.1029 \\
f(\sqrt{2}) &\approx f(1.41421) &&= 4^{1.41421} &&\approx 7.10296
\end{aligned}
$$

Example 1 Evaluate the function $f(x) = \left(\dfrac{1}{2}\right)^x$ at $x = 2$ and $x = -3$.

Solution $f(x) = \left(\dfrac{1}{2}\right)^x$

$f(2) = \left(\dfrac{1}{2}\right)^2 = \dfrac{1}{4}$ $\qquad\qquad\qquad$ $f(-3) = \left(\dfrac{1}{2}\right)^{-3} = 2^3 = 8$

Problem 1 Evaluate the function $f(x) = \left(\dfrac{2}{3}\right)^x$ at $x = 3$ and $x = -2$.

Solution See page A90.

$f(3) = \dfrac{8}{27};\ f(-2) = \dfrac{9}{4}$

Example 2 Evaluate the function $f(x) = 2^{3x-1}$ at $x = 1$ and $x = -1$.

 Solution $f(x) = 2^{3x-1}$

$f(1) = 2^{3(1)-1} = 2^2 = 4$ $\qquad$ $f(-1) = 2^{3(-1)-1} = 2^{-4} = \dfrac{1}{2^4} = \dfrac{1}{16}$

Problem 2 Evaluate the function $f(x) = 2^{2x+1}$ at $x = 0$ and $x = -2$.

 Solution See page A90.

$$f(0) = 2;\ f(-2) = \frac{1}{8}$$

The base of an exponential function can be any positive real number not equal to 1. Therefore, the base of an exponential function can be a positive irrational number. For example, the base of the exponential function $f(x) = \pi^x$ is π.

In order to evaluate a function whose base is an irrational number, it is necessary to approximate the base to a certain degree of accuracy. The result will be an approximation.

Using the approximation 3.14 for π, the value of $f(x) = \pi^x$ at $x = 2$ is $f(2) \approx 3.14^2 = 9.8596 \approx 9.86$, to the nearest hundredth.

Example 3 Using the approximation 2.24 for $\sqrt{5}$, evaluate the function $f(x) = (\sqrt{5})^{2x}$ at $x = 4$. Round to the nearest hundredth.

 Solution $f(x) = (\sqrt{5})^{2x}$
$f(4) \approx 2.24^{2(4)} = 2.24^8 = 633.84657 \approx 633.85$

Problem 3 Using the approximation 1.414 for $\sqrt{2}$, evaluate the function $f(x) = (\sqrt{2})^x$ at $x = 4$. Round to the nearest thousandth.

 Solution See page A90.
3.998

A base that is frequently used in applications is an irrational number designated by e. The number e is an irrational number and is approximately equal to 2.718281828.

To evaluate the exponential function $f(x) = e^x$, replace e by a decimal approximation of e, and evaluate the function.

Using the approximation 2.718 for e, the value of $f(x) = e^x$ at $x = 3$ is $f(3) \approx 2.718^3 = 20.07929 \approx 20.079$, to the nearest thousandth.

Example 4 Using the approximation 2.718 for e, evaluate the function $f(x) = 2e^x$ at $x = 1$.

 Solution $f(x) = 2e^x$
$f(1) \approx 2(2.718)^1 = 2(2.718) = 5.436$

Problem 4   Using the approximation 2.718 for e, evaluate the function $f(x) = e^{-x}$ at $x = -1$.

Solution   See page A90.
2.718

2   Graph exponential functions

Some of the properties of an exponential function can be seen by considering its graph.

To graph the function $f(x) = 2^x$, think of the function as the equation $y = 2^x$.

Choose values of x, and find the corresponding values of y. The results can be recorded in a table.

Graph the ordered pairs on a rectangular coordinate system.

Connect the points with a smooth curve.

x	$f(x) = y$	
-2	2^{-2}	$= \dfrac{1}{4}$
-1	2^{-1}	$= \dfrac{1}{2}$
0	2^0	$= 1$
1	2^1	$= 2$
2	2^2	$= 4$
3	2^3	$= 8$

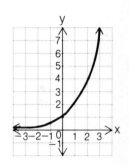

Note that a vertical line would intersect the graph at only one point. Therefore, by the vertical line test, $f(x) = 2^x$ is a function. Also notice that a horizontal line would intersect the graph at only one point. Therefore, $f(x) = 2^x$ is a one-to-one function.

To graph $f(x) = \left(\dfrac{1}{2}\right)^x$, think of the function as the equation $y = \left(\dfrac{1}{2}\right)^x$.

Choose values of x, and find the corresponding values of y.

Graph the ordered pairs on a rectangular coordinate system.

Connect the points with a smooth curve.

x	$f(x) = y$	
-3	$\left(\dfrac{1}{2}\right)^{-3}$	$= 8$
-2	$\left(\dfrac{1}{2}\right)^{-2}$	$= 4$
-1	$\left(\dfrac{1}{2}\right)^{-1}$	$= 2$
0	$\left(\dfrac{1}{2}\right)^{0}$	$= 1$
1	$\left(\dfrac{1}{2}\right)^{1}$	$= \dfrac{1}{2}$
2	$\left(\dfrac{1}{2}\right)^{2}$	$= \dfrac{1}{4}$

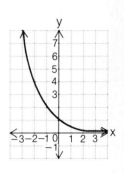

Applying the vertical and horizontal line tests, $f(x) = \left(\dfrac{1}{2}\right)^x$ is also a one-to-one function.

The graph of $f(x) = 2^{-x}$ is shown at the right.

Note that since $2^{-x} = (2^{-1})^x = \left(\frac{1}{2}\right)^x$, the

graphs of $f(x) = 2^{-x}$ and $f(x) = \left(\frac{1}{2}\right)^x$ are

the same.

x	y
-3	8
-2	4
-1	2
0	1
1	$\frac{1}{2}$
2	$\frac{1}{4}$

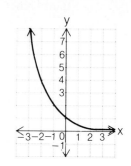

Example 5   Graph.

A. $f(x) = 3^{\frac{1}{2}x-1}$ B. $f(x) = 2^x - 1$

Solution   A.

x	y
-2	$\frac{1}{9}$
0	$\frac{1}{3}$
2	1
4	3

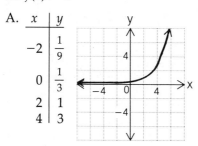

B.

x	y
-2	$-\frac{3}{4}$
-1	$-\frac{1}{2}$
0	0
1	1
2	3
3	7

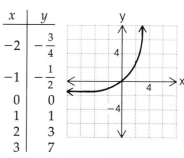

Problem 5   Graph.

A. $f(x) = 2^{-\frac{1}{2}x}$ B. $f(x) = 2^x + 1$

Solution   See page A90.

A.

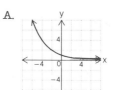

B.

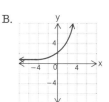

3   Write equivalent exponential and logarithmic expressions

Because $y = b^x$ is a 1–1 function, $y = b^x$ has an inverse function. Recall that the inverse of a function is formed by interchanging x and y. Therefore, the inverse of the exponential function $y = 2^x$ is $x = 2^y$.

For the function $y = 2^x$, the inverse function is $x = 2^y$, and y is called the **logarithm** of x to the base 2, written $y = \log_2 x$, where the abbreviation log is used for logarithm.

The graph of the function $x = 2^y$ or $y = \log_2 x$ is shown at the right. Using this graph, it is possible to estimate the logarithm of a number.

When $x = 4$, $y = 2$. This can be written $\log_2 4 = 2$. Read $\log_2 4$ as "the logarithm base 2 of 4" or "the log base 2 of 4."

When $x = 6$, $y \approx 2.6$. Therefore, $\log_2 6 \approx 2.6$.

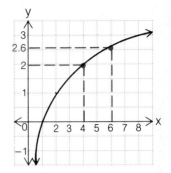

Definition of a Logarithm

For $b > 0$, $b \neq 1$, $y = \log_b x$ is equivalent to $x = b^y$.

The table below shows equivalent statements written in both exponential and logarithmic form.

Exponential Form	Logarithmic Form
$2^4 = 16$	$\log_2 16 = 4$
$\left(\dfrac{2}{3}\right)^2 = \dfrac{4}{9}$	$\log_{\frac{2}{3}}\left(\dfrac{4}{9}\right) = 2$
$10^{-1} = 0.1$	$\log_{10} 0.1 = -1$

Example 6   Write $3^4 = 81$ in logarithmic form.

Solution $3^4 = 81$ is equivalent to $\log_3 81 = 4$.

Problem 6   Write $7^3 = 343$ in logarithmic form.

Solution   See page A90.
$\log_7 343 = 3$

Example 7   Write $\log_5 125 = 3$ in exponential form.

Solution $\log_5 125 = 3$ is equivalent to $5^3 = 125$.

Problem 7   Write $\log_{\frac{1}{2}}\left(\dfrac{1}{8}\right) = 3$ in exponential form.

Solution   See page A90.
$$\left(\dfrac{1}{2}\right)^3 = \dfrac{1}{8}$$

Logarithms to the base 10 are called **common logarithms.** Usually the base, 10, is omitted when writing the common logarithm of a number. Therefore, $\log_{10} x$ is written $\log x$.

Example 8   Write $\log 100 = 2$ in exponential form.

Solution $\log 100 = 2$ is equivalent to $10^2 = 100$.

Problem 8   Write $\log 0.1 = -1$ in exponential form.

Solution   See page A90.
$10^{-1} = 0.1$

When e is used as a base of the logarithm, the logarithm is referred to as the **natural logarithm** and is abbreviated $\ln x$.

Because $e^1 = e$, $\ln e = 1$.

Example 9   Write $\ln 0.368 = -1$ in exponential form.

Solution $\ln 0.368 = -1$ is equivalent to $e^{-1} = 0.368$.

Problem 9   Write $\ln 7.389 = 2$ in exponential form.

Solution   See page A90.
$e^2 = 7.389$

Evaluate: $\log_2 8$

Write an equation. $\log_2 8 = x$

Write the equation in its equivalent exponential form. $8 = 2^x$

Write 8 in exponential form using 2 as the base. $2^3 = 2^x$

Solve for x using the fact that **if $b^x = b^y$, then $x = y$.** $3 = x$

                                                                                                                           $\log_2 8 = 3$

Example 10   Evaluate: $\log_3\left(\dfrac{1}{9}\right)$

Solution $\log_3\left(\dfrac{1}{9}\right) = x$ • Write an equation.

$\dfrac{1}{9} = 3^x$ • Write the equation in its equivalent exponential form.

$3^{-2} = 3^x$ • Write $\dfrac{1}{9}$ in exponential form using 3 as the base.

$-2 = x$ • Solve for x using the fact that if $b^x = b^y$, then $x = y$.

$\log_3\left(\dfrac{1}{9}\right) = -2$

Problem 10   Evaluate: $\log_4 64$

Solution   See page A90.
$\log_4 64 = 3$

To solve $\log_4 x = -2$ for x, write the equation in its equivalent exponential form.

$$\log_4 x = -2$$
$$4^{-2} = x$$

Solve for x.

$$\frac{1}{16} = x$$

The solution is $\frac{1}{16}$.

Example 11   Solve for x: $\log_5 x = 2$

Solution $\log_5 x = 2$
$\quad\ 5^2 = x$ • Write the equation in its equivalent exponential form.
$\quad\ 25 = x$

The solution is 25.

Problem 11   Solve for x: $\log_2 x = -4$

Solution   See page A91. $\dfrac{1}{16}$

4 Graph the logarithmic function $f(x) = \log_b x$

The graph of a logarithmic function can be found by using the relationship between the exponential and logarithmic functions.

To graph $f(x) = \log_2 x$, think of the function as the equation $y = \log_2 x$.

$$f(x) = \log_2 x$$
$$y = \log_2 x$$

Write the equivalent exponential equation.

$$x = 2^y$$

Since the equation is solved for x in terms of y, it is easier to choose values of y and find the corresponding values of x. The results can be recorded in a table.

Graph the ordered pairs on a rectangular coordinate system.

Connect the points with a smooth curve.

x	y
$\frac{1}{4}$	-2
$\frac{1}{2}$	-1
1	0
2	1
4	2

Applying the vertical and horizontal line tests, $f(x) = \log_2 x$ is a one-to-one function.

Recall that the graph of the inverse of a function f is the mirror image of f with respect to the line $y = x$. The graph of $f(x) = 2^x$ is shown on page 529. Since $g(x) = \log_2 x$ is the inverse of the function $f(x) = 2^x$, the graphs of these functions are mirror images of each other with respect to the line $y = x$.

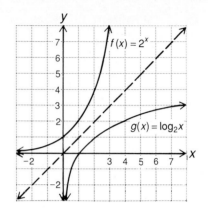

To graph $f(x) = \log_2 x + 1$, think of the function as the equation $y = \log_2 x + 1$.

$$f(x) = \log_2 x + 1$$
$$y = \log_2 x + 1$$

Solve the equation for $\log_2 x$.

$$y - 1 = \log_2 x$$

Write the equivalent exponential equation.

$$2^{y-1} = x$$

Choose values of y, and find the corresponding values of x.

Graph the ordered pairs on a rectangular coordinate system.

Connect the points with a smooth curve.

x	y
$\frac{1}{4}$	-1
$\frac{1}{2}$	0
1	1
2	2
4	3

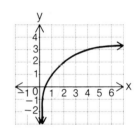

Example 12   Graph.

A. $f(x) = \log_3 x$ B. $f(x) = 2 \log_3 x$

Solution   A. $f(x) = \log_3 x$

$$y = \log_3 x$$
$$x = 3^y$$

• Substitute y for $f(x)$.
• Write the equivalent exponential equation.

x	y
$\frac{1}{9}$	-2
$\frac{1}{3}$	-1
1	0
3	1

• Choose values of y, and find the corresponding values of x. Graph the ordered pairs on a rectangular coordinate system. Connect the points with a smooth curve.

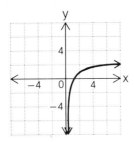

B. $f(x) = 2 \log_3 x$

$\quad\quad y = 2 \log_3 x$ • Substitute y for $f(x)$.

$\quad\quad \dfrac{y}{2} = \log_3 x$ • Solve the equation for $\log_3 x$.

$\quad\quad x = 3^{\frac{y}{2}}$ • Write the equivalent exponential equation.

x	y
$\dfrac{1}{9}$	-4
$\dfrac{1}{3}$	-2
1	0
3	2

• Choose values of y, and find the corresponding values of x. Graph the ordered pairs on a rectangular coordinate system. Connect the points with a smooth curve.

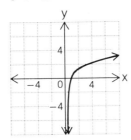

Problem 12   Graph.

A. $f(x) = \log_2(x - 1)$

B. $f(x) = \log_3 2x$

A.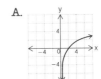

B.

Solution   See page A91.

EXERCISES 11.1

1   Given the function $f(x) = 3^x$, find:

1. $f(2)$ 9

2. $f(3)$ 27

3. $f(-2)$ $\dfrac{1}{9}$

4. $f(-1)$ $\dfrac{1}{3}$ **5.** $f(0)$ 1   **6.** $f(1)$ 3

Given the function $f(x) = 2^{x+1}$, find:

7. $f(3)$ 16   **8.** $f(1)$ 4   **9.** $f(-3)$ $\dfrac{1}{4}$

10. $f(-4)$ $\dfrac{1}{8}$ **11.** $f(-1)$ 1   **12.** $f(0)$ 2

Given the function $f(x) = \left(\dfrac{1}{2}\right)^{2x}$, find:

13. $f(0)$ 1   **14.** $f(1)$ $\dfrac{1}{4}$ **15.** $f(-2)$ 16

16. $f(-1)$ 4   **17.** $f\left(\dfrac{3}{2}\right)$ $\dfrac{1}{8}$ **18.** $f\left(-\dfrac{1}{2}\right)$ 2

Given the function $f(x) = \left(\dfrac{1}{3}\right)^{x-1}$, find:

19. $f(2)$ $\dfrac{1}{3}$ **20.** $f(3)$ $\dfrac{1}{9}$ **21.** $f(-1)$ 9

22. $f(-2)$ 27   **23.** $f(0)$ 3   **24.** $f(1)$ 1

Given the function $f(x) = 2^{x^2}$, find:

25. $f(1)$ 2   **26.** $f(0)$ 1   **27.** $f(2)$ 16

28. $f(3)$ 512   **29.** $f(-1)$ 2   **30.** $f(-2)$ 16

Use the approximation 1.4142 for $\sqrt{2}$. Round to the nearest ten-thousandth.

31. Given the function $f(x) = (\sqrt{2})^{2x}$, find $f(1)$.   2

32. Given the function $f(x) = 3(\sqrt{2})^{x}$, find $f(2)$.   5.9999

33. Given the function $f(x) = (\sqrt{2})^{-x}$, find $f(-1)$.   1.4142

34. Given the function $f(x) = (\sqrt{2})^{-x}$, find $f(-2)$.   2.0000

Use the approximation 2.718 for e. Round to the nearest thousandth.

35. Given the function $f(x) = e^{2x}$, find $f\left(\dfrac{1}{2}\right)$.   2.718

36. Given the function $f(x) = e^{3x}$, find $f(1)$.   20.079

37. Given the function $f(x) = 3e^{x-1}$, find $f(2)$.   8.154

38. Given the function $f(x) = e^{-2x}$, find $f(-1)$.   7.388

2   Graph.

39. $f(x) = 3^x$

40. $f(x) = 3^{-x}$

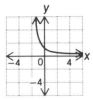

41. $f(x) = 2^{x+1}$

42. $f(x) = 2^{x-1}$

43. $f(x) = \left(\dfrac{1}{3}\right)^x$

44. $f(x) = \left(\dfrac{2}{3}\right)^x$

45. $f(x) = 2^{-x} + 1$

46. $f(x) = 2^x - 3$

47. $f(x) = \left(\dfrac{1}{2}\right)^{2x}$

48. $f(x) = 2^{\frac{1}{2}x}$

3   Write the exponential expression in logarithmic form.

49. $2^5 = 32$
$\log_2 32 = 5$

50. $3^4 = 81$
$\log_3 81 = 4$

51. $5^2 = 25$
$\log_5 25 = 2$

52. $10^3 = 1000$
$\log 1000 = 3$

53. $4^{-2} = \dfrac{1}{16}$
$\log_4 \dfrac{1}{16} = -2$

54. $3^{-3} = \dfrac{1}{27}$
$\log_3 \dfrac{1}{27} = -3$

55. $\left(\dfrac{1}{2}\right)^2 = \dfrac{1}{4}$
$\log_{\frac{1}{2}} \dfrac{1}{4} = 2$

56. $\left(\dfrac{1}{3}\right)^4 = \dfrac{1}{81}$
$\log_{\frac{1}{3}} \dfrac{1}{81} = 4$

57. $3^0 = 1$
$\log_3 1 = 0$

58. $10^0 = 1$
$\log 1 = 0$

59. $a^x = w$
$\log_a w = x$

60. $b^y = c$
$\log_b c = y$

Write the logarithmic expression in exponential form.

61. $\log_3 9 = 2$
$3^2 = 9$

62. $\log_2 32 = 5$
$2^5 = 32$

63. $\log_4 4 = 1$
$4^1 = 4$

64. $\log_7 7 = 1$
$7^1 = 7$

65. $\log 1 = 0$

$10^0 = 1$

66. $\log_8 1 = 0$

$8^0 = 1$

67. $\log 0.01 = -2$

$10^{-2} = 0.01$

68. $\log_5 \frac{1}{5} = -1$

$5^{-1} = \frac{1}{5}$

69. $\log_{\frac{1}{3}}\left(\frac{1}{9}\right) = 2$

$\left(\frac{1}{3}\right)^2 = \frac{1}{9}$

70. $\log_{\frac{1}{4}}\left(\frac{1}{16}\right) = 2$

$\left(\frac{1}{4}\right)^2 = \frac{1}{16}$

71. $\log_b u = v$

$b^v = u$

72. $\log_c x = y$

$c^y = x$

73. $\log 1000 = 3$
$10^3 = 1000$

74. $\ln 20.086 = 3$
$e^3 = 20.086$

75. $\ln 1 = 0$
$e^0 = 1$

76. $\ln 0.135 = -2$
$e^{-2} = 0.135$

Evaluate.

77. $\log_4 16$
2

78. $\log_3 27$
3

79. $\log_2 32$
5

80. $\log 1000$
3

81. $\log 100$
2

82. $\log_5 125$
3

83. $\log_6 216$
3

84. $\log_7 1$
0

Solve for x.

85. $\log_3 x = 2$
9

86. $\log_5 x = 1$
5

87. $\log_4 x = 3$
64

88. $\log_2 x = 6$
64

89. $\log_7 x = -1$
$\frac{1}{7}$

90. $\log_8 x = -2$
$\frac{1}{64}$

91. $\log_6 x = 3$

216

92. $\log_4 x = 0$

1

4 Graph.

93. $f(x) = \log_4 x$

94. $f(x) = \log_2(x + 1)$

95. $f(x) = \log_3(2x - 1)$

96. $f(x) = \log_2\left(\frac{1}{2}x\right)$

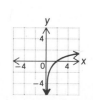

97. $f(x) = 3 \log_2 x$

98. $f(x) = \dfrac{1}{2} \log_2 x$

99. $f(x) = -\log_2 x$

100. $f(x) = -\log_3 x$

SUPPLEMENTAL EXERCISES 11.1

Use your calculator to evaluate the function. Round to the nearest hundredth.

101. $f(x) = 2^x$ at $x = \sqrt{2}$ 2.67

102. $f(x) = 2^x$ at $x = \sqrt{3}$ 3.32

103. $f(x) = 3^x$ at $x = \sqrt{2}$ 4.73

104. $f(x) = 3^x$ at $x = \sqrt{3}$ 6.70

105. $f(x) = \left(\dfrac{1}{2}\right)^x$ at $x = \sqrt{2}$ 0.38

106. $f(x) = \left(\dfrac{1}{2}\right)^x$ at $x = \sqrt{5}$ 0.21

107. $f(x) = 4^x$ at $x = \pi$ 77.88

108. $f(x) = 4^x$ at $x = -\pi$ 0.01

Find the range of the function.

109. $f(x) = 2^x$ $y > 0$

110. $f(x) = \left(\dfrac{1}{2}\right)^x$ $y > 0$

111. $f(x) = 3^{x-1}$ $y > 0$

112. $f(x) = 2^x - 1$ $y > -1$

Find the domain of the function.

113. $f(x) = \log_2 x$ $x > 0$

114. $f(x) = \log_2 x + 1$ $x > 0$

115. $f(x) = 2 \log_3 x$ $x > 0$

116. $f(x) = \log_2(x + 1)$ $x > -1$

Graph both functions on the same rectangular coordinate system.

117. $f(x) = 3^x; g(x) = \log_3 x$

118. $f(x) = 4^x; g(x) = \log_4 x$

119. $f(x)10^x$; $g(x) = \log_{10} x$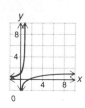

120. $f(x) = \left(\dfrac{1}{2}\right)^x$; $g(x) = \log_{\frac{1}{2}} x$

Graph.

121. $f(x) = e^x$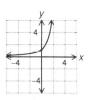

122. $f(x) = \ln x$

SECTION 11.2 —————————————————

The Properties of Logarithms

1 ### The Properties of Logarithms

Since a logarithm is a special kind of exponent, the Properties of Logarithms are similar to the Properties of Exponents.

The table at the right shows some powers of 2 and the equivalent logarithmic form.

The table can be used to show that $\log_2 4 + \log_2 8$ equals $\log_2 32$.

$2^0 = 1$	$\log_2 1 = 0$
$2^1 = 2$	$\log_2 2 = 1$
$2^2 = 4$	$\log_2 4 = 2$
$2^3 = 8$	$\log_2 8 = 3$
$2^4 = 16$	$\log_2 16 = 4$
$2^5 = 32$	$\log_2 32 = 5$

$$\log_2 4 + \log_2 8 = 2 + 3 = 5$$
$$\log_2 32 = 5$$
$$\log_2 4 + \log_2 8 = \log_2 32$$

Note that $\log_2 32 = \log_2 (4 \times 8) = \log_2 4 + \log_2 8$.

The property of logarithms that states that the logarithm of the product of two numbers equals the sum of the logarithms of the two numbers is similar to the property of exponents that states that to multiply two exponential expressions with the same base, add the exponents.

> ### The Logarithm Property of the Product of Two Numbers
> For any positive real numbers, x, y, and b, $b \neq 1$,
> $$\log_b xy = \log_b x + \log_b y.$$

Proof:
Let $\log_b x = m$ and $\log_b y = n$.

Write each equation in its equivalent exponential form. $\qquad x = b^m \qquad\qquad y = b^n$

Use substitution and the Properties of Exponents.
$$xy = b^m b^n$$
$$xy = b^{m+n}$$

Write the equation in its equivalent logarithmic form. $\qquad \log_b xy = m + n$

Substitute $\log_b x$ for m and $\log_b y$ for n. $\qquad \log_b xy = \log_b x + \log_b y$

The Logarithm Property of Products is used to rewrite logarithmic expressions.

The $\log_b 6z$ is written in **expanded form** as $\log_b 6 + \log_b z$.

$$\log_b 6z = \log_b 6 + \log_b z$$

The $\log_b 12 + \log_b r$ is written as a single logarithm as $\log_b 12r$.

$$\log_b 12 + \log_b r = \log_b 12r$$

The Logarithm Property of Products can be extended to include the logarithm of the product of more than two factors. For example,

$$\log_b xyz = \log_b (xy)z = \log_b xy + \log_b z = \log_b x + \log_b y + \log_b z$$

To write $\log_b 5st$ in expanded form, use the Logarithm Property of Products. $\qquad \log_b 5st = \log_b 5 + \log_b s + \log_b t$

A second property of logarithms involves the logarithm of the quotient of two numbers. This property of logarithms is also based on the fact that a logarithm is an exponent and that to divide two exponential expressions with the same base, the exponents are subtracted.

> ### The Logarithm Property of the Quotient of Two Numbers
>
> For any positive real numbers, x, y, and b, $b \neq 1$, $\log_b \dfrac{x}{y} = \log_b x - \log_b y$.

Proof:
Let $\log_b x = m$ and $\log_b y = n$.

Write each equation in its equivalent exponential form.

$$x = b^m \qquad\qquad y = b^n$$

Use substitution and the Properties of Exponents.

$$\frac{x}{y} = \frac{b^m}{b^n}$$

$$\frac{x}{y} = b^{m-n}$$

Write the equation in its equivalent logarithmic form.

$$\log_b \frac{x}{y} = m - n$$

Substitute $\log_b x$ for m and $\log_b y$ for n.

$$\log_b \frac{x}{y} = \log_b x - \log_b y$$

The Logarithm Property of Quotients is used to rewrite logarithmic expressions.

The $\log_b \dfrac{p}{8}$ is written in expanded form as $\log_b p - \log_b 8$.

$$\log_b \frac{p}{8} = \log_b p - \log_b 8$$

The $\log_b y - \log_b v$ is written as a single logarithm as $\log_b \dfrac{y}{v}$.

$$\log_b y - \log_b v = \log_b \frac{y}{v}$$

A third property of logarithms, which is especially useful in the computation of the power of a number, is based on the fact that a logarithm is an exponent, and the power of an exponential expression is found by multiplying the exponents.

The table of the powers of 2 shown on page 540 can be used to show that $\log_2 2^3$ equals $3\log_2 2$.

$$\log_2 2^3 = \log_2 8 = 3$$
$$3 \log_2 2 = 3 \cdot 1 = 3$$
$$\log_2 2^3 = 3 \log_2 2$$

The Logarithmic Property of the Power of a Number

For any positive real numbers x and b, $b \neq 1$, and for any real number, r, $\log_b x^r = r \log_b x$.

Proof:

Let $\log_b x = m$.

Write the equation in its equivalent exponential form. $x = b^m$

Raise both sides to the r power. $x^r = (b^m)^r$
$x^r = b^{mr}$

Write the equation in its equivalent logarithmic form. $\log_b x^r = mr$

Substitute $\log_b x$ for m. $\log_b x^r = r \log_b x$

The Logarithm Property of Powers is used to rewrite logarithmic expressions.

The $\log_b x^3$ is written in terms of $\log_b x$ as $3 \log_b x$.

$$\log_b x^3 = 3 \log_b x$$

$\dfrac{2}{3} \log_4 z$ is written with a coefficient of 1 as $\log_4 z^{\frac{2}{3}}$.

$$\frac{2}{3} \log_4 z = \log_4 z^{\frac{2}{3}}$$

The Properties of Logarithms can be used in combination to simplify expressions containing logarithms.

Example 1   Write the logarithm in expanded form.

A. $\log_b \dfrac{xy}{z}$ B. $\log_b \dfrac{x^2}{y^3}$ C. $\log_8 \sqrt{x^3 y}$

Solution   A. $\log_b \dfrac{xy}{z} = \log_b(xy) - \log_b z$ • Use the Logarithm Property of Quotients.

$= \log_b x + \log_b y - \log_b z$ • Use the Logarithm Property of Products.

B. $\log_b \dfrac{x^2}{y^3} = \log_b x^2 - \log_b y^3$ • Use the Logarithm Property of Quotients.

$= 2 \log_b x - 3 \log_b y$ • Use the Logarithm Property of Powers.

C. $\log_8 \sqrt{x^3 y} = \log_8 (x^3 y)^{\frac{1}{2}}$ · Write the radical expression as an exponential expression.

$= \dfrac{1}{2} \log_8 x^3 y$ · Use the Logarithm Property of Powers.

$= \dfrac{1}{2} (\log_8 x^3 + \log_8 y)$ · Use the Logarithm Property of Products.

$= \dfrac{1}{2} (3 \log_8 x + \log_8 y)$ · Use the Logarithm Property of Powers.

$= \dfrac{3}{2} \log_8 x + \dfrac{1}{2} \log_8 y$ · Use the Distributive Property.

Problem 1   Write the logarithm in expanded form.

A. $\log_b \dfrac{x^2}{y}$ B. $\log_b y^{\frac{1}{3}} z^3$ C. $\log_8 \sqrt[3]{xy^2}$

Solution   See page A92.

A. $2 \log_b x - \log_b y$ B. $\dfrac{1}{3} \log_b y + 3 \log_b z$ C. $\dfrac{1}{3} \log_8 x + \dfrac{2}{3} \log_8 y$

Example 2   Express as a single logarithm with a coefficient of 1.

A. $3 \log_5 x + \log_5 y - 2 \log_5 z$

B. $\dfrac{1}{2} (\log_3 x - 3 \log_3 y + \log_3 z)$

Solution   A. $3 \log_5 x + \log_5 y - 2 \log_5 z =$

$\log_5 x^3 + \log_5 y - \log_5 z^2 =$ · Use the Logarithm Property of Powers.

$\log_5 x^3 y - \log_5 z^2 =$ · Use the Logarithm Property of Products.

$\log_5 \dfrac{x^3 y}{z^2}$ · Use the Logarithm Property of Quotients.

B. $\dfrac{1}{2} (\log_3 x - 3 \log_3 y + \log_3 z) =$

$\dfrac{1}{2} (\log_3 x - \log_3 y^3 + \log_3 z) =$ · Use the Logarithm Property of Powers.

$\dfrac{1}{2} \left(\log_3 \dfrac{x}{y^3} + \log_3 z \right) =$ · Use the Logarithm Property of Quotients.

$\dfrac{1}{2} \left(\log_3 \dfrac{xz}{y^3} \right) =$ · Use the Logarithm Property of Products.

$\log_3 \left(\dfrac{xz}{y^3} \right)^{\frac{1}{2}} = \log_3 \sqrt{\dfrac{xz}{y^3}}$ · Use the Logarithm Property of Powers. Write the exponential expression as a radical expression.

Problem 2   Express as a single logarithm with a coefficient of 1.

A. $2 \log_b x - 3 \log_b y - \log_b z$ B. $\dfrac{1}{3}(\log_4 x - 2 \log_4 y + \log_4 z)$

Solution   See page A93.

A. $\log_b \dfrac{x^2}{y^3 z}$

B. $\log_4 \sqrt[3]{\dfrac{xz}{y^2}}$

Two other properties of logarithmic functions can be established directly from the equivalence of the expressions $\log_b x = y$ and $x = b^y$.

For any positive real number b, $b \neq 1$, and any real number n, $\mathbf{\log_b b^n = n}$.

For any positive real number b, $b \neq 1$, $\mathbf{\log_b 1 = 0}$.

Example 3   Find $8 \log_4 4$.

Solution $8 \log_4 4 = \log_4 4^8 = 8$ • Since $\log_b b^n = n$, $\log_4 4^8 = 8$.

Problem 3   Find $\log_9 1$.

Solution   See page A93.

0

EXERCISES 11.2

1   Write the logarithm in expanded form.

1. $\log_8(xz)$
    $\log_8 x + \log_8 z$

2. $\log_7(4y)$
    $\log_7 4 + \log_7 y$

3. $\log_3 x^5$
    $5 \log_3 x$

4. $\log_2 y^7$
    $7 \log_2 y$

5. $\log_b\left(\dfrac{r}{s}\right)$
    $\log_b r - \log_b s$

6. $\log_c\left(\dfrac{z}{4}\right)$
    $\log_c z - \log_c 4$

7. $\log_3(x^2 y^6)$
    $2 \log_3 x + 6 \log_3 y$

8. $\log_4(t^4 u^2)$
    $4 \log_4 t + 2 \log_4 u$

9. $\log_7\left(\dfrac{u^3}{v^4}\right)$
    $3 \log_7 u - 4 \log_7 v$

10. $\log\left(\dfrac{s^5}{t^2}\right)$
    $5 \log s - 2 \log t$

11. $\log_2(rs)^2$
    $2 \log_2 r + 2 \log_2 s$

12. $\log_3(x^2 y)^3$
    $6 \log_3 x + 3 \log_3 y$

13. $\log_9 x^2 yz$
    $2 \log_9 x + \log_9 y + \log_9 z$

14. $\log_6 xy^2 z^3$
    $\log_6 x + 2 \log_6 y + 3 \log_6 z$

15. $\log_5\left(\dfrac{xy^2}{z^4}\right)$
    $\log_5 x + 2 \log_5 y - 4 \log_5 z$

16. $\log_b\left(\dfrac{r^2 s}{t^3}\right)$
    $2 \log_b r + \log_b s - 3 \log_b t$

17. $\log_8\left(\dfrac{x^2}{yz^2}\right)$

$2\log_8 x - \log_8 y - 2\log_8 z$

18. $\log_9\left(\dfrac{x}{y^2z^3}\right)$

$\log_9 x - 2\log_9 y - 3\log_9 3$

19. $\log_7\sqrt{xy}$

$\dfrac{1}{2}\log_7 x + \dfrac{1}{2}\log_7 y$

20. $\log_8\sqrt[3]{xz}$

$\dfrac{1}{3}\log_8 x + \dfrac{1}{3}\log_8 z$

21. $\log_2\sqrt{\dfrac{x}{y}}$

$\dfrac{1}{2}\log_2 x - \dfrac{1}{2}\log_2 y$

22. $\log_3\sqrt[3]{\dfrac{r}{s}}$

$\dfrac{1}{3}\log_3 r - \dfrac{1}{3}\log_3 s$

23. $\log_4\sqrt{x^3 y}$

$\dfrac{3}{2}\log_4 x + \dfrac{1}{2}\log_4 y$

24. $\log_3\sqrt{x^5 y^3}$

$\dfrac{5}{2}\log_3 x + \dfrac{3}{2}\log_3 y$

25. $\log_7\sqrt{\dfrac{x^3}{y}}$

$\dfrac{3}{2}\log_7 x - \dfrac{1}{2}\log_7 y$

26. $\log_b\sqrt[3]{\dfrac{r^2}{t}}$

$\dfrac{2}{3}\log_b r - \dfrac{1}{3}\log_b t$

27. $\log_b x\sqrt{\dfrac{y}{z}}$

$\log_b x + \dfrac{1}{2}\log_b y - \dfrac{1}{2}\log_b z$

28. $\log_4 y\sqrt[3]{\dfrac{r}{s}}$

$\log_4 y + \dfrac{1}{3}\log_4 r - \dfrac{1}{3}\log_4 s$

29. $\log_3\dfrac{t}{\sqrt{x}}$

$\log_3 t - \dfrac{1}{2}\log_3 x$

30. $\log_4\left(\dfrac{\sqrt{uv}}{x}\right)$

$\dfrac{1}{2}\log_4 u + \dfrac{1}{2}\log_4 v - \log_4 x$

Express as a single logarithm with a coefficient of 1.

31. $\log_3 x^3 - \log_3 y$

$\log_3\dfrac{x^3}{y}$

32. $\log_7 t + \log_7 v^2$

$\log_7 tv^2$

33. $\log_8 x^4 + \log_8 y^2$

$\log_8 x^4 y^2$

34. $\log_2 r^2 + \log_2 s^3$

$\log_2 r^2 s^3$

35. $3\log_7 x$

$\log_7 x^3$

36. $4\log_8 y$

$\log_8 y^4$

37. $3\log_5 x + 4\log_5 y$

$\log_5 x^3 y^4$

38. $2\log_6 x + 5\log_6 y$

$\log_6 x^2 y^5$

39. $-2\log_4 x$

$\log_4\dfrac{1}{x^2}$

40. $-3\log_2 y$

$\log_2\dfrac{1}{y^3}$

41. $2 \log_3 x - \log_3 y + 2 \log_3 z$
$\log_3 \dfrac{x^2 z^2}{y}$

42. $4 \log_5 r - 3 \log_5 s + \log_5 t$
$\log_5 \dfrac{r^4 t}{s^3}$

43. $\log_b x - (2 \log_b y + \log_b z)$
$\log_b \dfrac{x}{y^2 z}$

44. $2 \log_2 x - (3 \log_2 y + \log_2 z)$
$\log_2 \dfrac{x^2}{y^3 z}$

45. $2(\log_4 x + \log_4 y)$
$\log_4 x^2 y^2$

46. $3(\log_5 r + \log_5 t)$
$\log_5 r^3 t^3$

47. $\dfrac{1}{2}(\log_6 x - \log_6 y)$
$\log_6 \sqrt{\dfrac{x}{y}}$

48. $\dfrac{1}{3}(\log_8 x - \log_8 y)$
$\log_8 \sqrt[3]{\dfrac{x}{y}}$

49. $2(\log_4 s - 2 \log_4 t + \log_4 r)$
$\log_4 \dfrac{s^2 r^2}{t^4}$

50. $3(\log_9 x + 2 \log_9 y - 2 \log_9 z)$
$\log_9 \dfrac{x^3 y^6}{z^6}$

51. $\log_5 x - 2(\log_5 y + \log_5 z)$
$\log_5 \dfrac{x}{y^2 z^2}$

52. $\log_4 t - 3(\log_4 u + \log_4 v)$
$\log_4 \dfrac{t}{u^3 v^3}$

53. $3 \log_2 t - 2(\log_2 r - \log_2 v)$
$\log_2 \dfrac{t^3 v^2}{r^2}$

54. $2 \log_b x - 3(\log_b y - \log_b z)$
$\log_b \dfrac{x^2 z^3}{y^3}$

55. $\dfrac{1}{2}(3 \log_4 x - 2 \log_4 y + \log_4 z)$
$\log_4 \sqrt{\dfrac{x^3 z}{y^2}}$

56. $\dfrac{1}{3}(4 \log_5 t - 3 \log_5 u - 3 \log_5 v)$
$\log_5 \sqrt[3]{\dfrac{t^4}{u^3 v^3}}$

57. $\dfrac{1}{2} \log_b x - \dfrac{2}{3} \log_b y + \dfrac{1}{2} \log_b z$
$\log_b \dfrac{\sqrt{xz}}{\sqrt[3]{y^2}}$

58. $\dfrac{2}{3} \log_b x + \dfrac{1}{3} \log_b y - \dfrac{1}{2} \log_b z$
$\log_b \dfrac{\sqrt[3]{x^2 y}}{\sqrt{z}}$

59. $\log_b x - \dfrac{2}{3} \log_b y + 4 \log_b z$
$\log_b \dfrac{xz^4}{\sqrt[3]{y^2}}$

60. $2 \log_b x - 3 \log_b y - 4 \log_b z$
$\log_b \dfrac{x^2}{y^3 z^4}$

Simplify.

61. $9 \log_3 3$
9

62. $5 \log_6 6$
5

63. $\log_7 1$
0

64. $\log_4 1$
0

65. $8 \log_5 5$
8

66. $2 \log_7 7$
2

SUPPLEMENTAL EXERCISES 11.2

Use the Properties of Logarithms to solve for x.

67. $\log_8 x = 3 \log_8 2$ 8

68. $\log_5 x = 2 \log_5 3$ 9

69. $\log_4 x = \log_4 2 + \log_4 3$ 6

70. $\log_3 x = \log_3 4 + \log_3 7$ 28

71. $\log_7 x = \log_7 8 - \log_7 4$ 2

72. $\log x = \log 64 - \log 8$ 8

73. $\log_6 x = 3 \log_6 2 - \log_6 4$ 2

74. $\log_9 x = 5 \log_9 2 - \log_9 8$ 4

75. $\log x = \frac{1}{3} \log 27$ 3

76. $\log_2 x = \frac{3}{2} \log_2 4$ 8

Given that $\log 2 = 0.3010$ and $\log 4 = 0.6021$, evaluate the logarithm.
(Example: $\log 8 = \log (2 \cdot 4) = \log 2 + \log 4 = 0.3010 + 0.6021 = 0.9031$.)

77. $\log 16$ 1.2042

78. $\log 32$ 1.5050

79. $\log 64$ 1.8063

80. $\log 0.5$ $\left(Hint: 0.5 = \frac{2}{4} \right)$ −0.3011

81. $\log 20$ (*Hint:* $20 = 10 \cdot 2$)   1.3010

82. $\log 50$ $\left(Hint: 50 = 10 \cdot \frac{10}{2} \right)$ 1.6990

SECTION 11.3 _____

Computations with Logarithms

1 ## Find common logarithms

Recall that logarithms to the base 10 are called **common logarithms** and that the base, 10, is omitted when writing the common logarithm of a number. Throughout this chapter, $\log x$ is used for $\log_{10} x$.

Using the definition of logarithms, the examples below show that the logarithm of a power of 10 is the exponent on 10.

$$\log 100 = \log 10^2 = 2$$
$$\log 10 = \log 10^1 = 1$$
$$\log 1 = \log 10^0 = 0$$
$$\log 0.1 = \log 10^{-1} = -1$$
$$\log 0.01 = \log 10^{-2} = -2$$

To find the base 10 logarithm of a number other than a power of 10, it is necessary to use a calculator or a table of logarithms. Since most logarithms are irrational numbers, a calculator or table of logarithms gives approximate values of logarithms. Despite the fact that the logarithms are approximate values, it is customary to use the equal sign (=) rather than the approximately equal sign ($\approx$) when writing logarithms.

The base 10 logarithms of numbers from 1 to 9.99 can be found in the Table of Common Logarithms on pages A2–A3. The logarithms in this table have been rounded to the nearest ten-thousandth. A portion of this table is shown below.

x	0	1	2	3	4	5	6
2.0	.3010	.3032	.3054	.3075	.3096	.3118	.3139
2.1	.3222	.3243	.3263	.3284	.3304	.3324	.3345
2.2	.3424	.3444	.3464	.3483	.3502	.3522	.3541
2.3	.3617	.3636	.3655	.3674	.3692	.3711	.3729
2.4	.3802	.3820	.3838	.3856	.3874	.3892	.3909
2.5	.3979	.3997	.4014	.4031	.4048	.4065	.4082
2.6	.4150	.4166	.4183	.4200	.4216	.4232	.4249

To find log 2.35, locate 2.3 under x in the left-hand column of the table. Move right to the column headed by 5.

$$\log 2.35 = 0.3711$$

Remember that the common logarithm of a number, N, is the power to which 10 must be raised to equal N.

Note from the table above that log 2 = 0.3010. Therefore, $2 = 10^{0.3010}$.

From the example above, log 2.35 = 0.3711. Therefore, $2.35 = 10^{0.3711}$.

To find the common logarithm of a number not in the Table of Common Logarithms, first write the number in scientific notation. Then use the table and the Properties of Logarithms to find the logarithm of the number.

To find log 247, write 247 in scientific notation.

$$\log 247 = \log (2.47 \times 10^2)$$

Use the Logarithm Property of Products.

$$= \log 2.47 + \log 10^2$$

Locate log 2.47 in the table.

$$= 0.3927 + \log 10^2$$

By the definition of logarithms, the logarithm of a power of 10 is the exponent on 10.

$$= 0.3927 + 2$$

Add.

$$= 2.3927$$

To find log 2470, write 2470 in scientific notation.

$$\log 2470 = \log (2.47 \times 10^3)$$

Use the Logarithm Property of Products.

$$= \log 2.47 + \log 10^3$$

Use the table and the definition of logarithms.

$$= 0.3927 + 3$$

Add.

$$= 3.3927$$

Note that for log 247 and log 2470, the decimal part of the logarithm, 0.3927, is the same. The integer part is the exponent on 10 when the number is written in scientific notation.

The decimal part of the logarithm is called the **mantissa**. The integer part is called the **characteristic**.

To find log 0.00247, write 0.00247 in scientific notation.

$$\log 0.00247 = \log (2.47 \times 10^{-3})$$

Use the Logarithm Property of Products.

$$= \log 2.47 + \log 10^{-3}$$

Use the table and the definition of logarithms.

$$= 0.3927 + (-3)$$

In this last example, adding the characteristic to the mantissa would result in a negative logarithm. This form of a logarithm is inconvenient to use with logarithm tables since these tables contain only positive logarithms. Therefore, when the characteristic of a logarithm is negative, it is customary to leave the logarithm as written or to rewrite the logarithm so that the mantissa remains unchanged.

For example: $\log 0.00247 = 0.3927 + (-3)$

$$\log 0.00247 = 0.3927 + (7 - 10) = 7.3927 - 10$$

$$\log 0.00247 = 0.3927 + (12 - 15) = 12.3927 - 15$$

Of these forms, $\log 0.00247 = 7.3927 - 10$ is most common.

Example 1   Find the logarithm.

A. log 8360   B. log 0.0217

Solution   A. $\log 8360 = \log (8.36 \times 10^3)$

$$= \log 8.36 + \log 10^3$$

$$= 0.9222 + 3$$

$$= 3.9222$$

- Write the number in scientific notation.
- Use the Logarithm Property of Products.
- Use the table and the definition of logarithms.

B. log $0.0217 = \log (2.17 \times 10^{-2})$

          $= \log 2.17 + \log 10^{-2}$

          $= 0.3365 + (-2)$

          $= 0.3365 + (8 - 10)$

          $= 8.3365 - 10$

- Write the number in scientific notation.
- Use the Logarithm Property of Products.
- Use the table and the definition of logarithms.
- Rewrite -2 as $8 - 10$. (Do this step mentally.)

Problem 1   Find the logarithm.

    A. log 93,000   B. log 0.0006

Solution   See page A93.

    A. 4.9685   B. 6.7782 − 10

2   Find common antilogarithms

In the last objective, given a number, N, the logarithm of N was found. In this objective, the reverse process is examined. Given the *logarithm* of N, the number N will be found.

For example, to find N given log $N = 0.3636$, locate 0.3636 in the body of the Table of Common Logarithms.

Find the number that corresponds to this mantissa.

The number is 2.31.

log $2.31 = 0.3636$

x	0	1	2	3
2.0	.3010	.3032	.3054	.3075
2.1	.3222	.3243	.3263	.3284
2.2	.3424	.3444	.3464	.3483
2.3	.3617	.3636	.3655	.3674
2.4	.3802	.3820	.3838	.3856
2.5	.3979	.3997	.4014	.4031
2.6	.4150	.4166	.4183	.4200

In the equation log $2.31 = 0.3636$, the number 2.31 is called the **antilogarithm** of 0.3636. This is written antilog $0.3636 = 2.31$, where the abbreviation antilog is used for antilogarithm.

The common logarithm of a number, N, is the power to which 10 must be raised to equal N.

log $2 = 0.3010$ because $2 = 10^{0.3010}$.

The common antilogarithm of a number, N, is the Nth power of 10.

antilog $0.3010 = 2$ because $10^{0.3010} = 2$.

To solve log N = 1.1206, write 1.1206 as the sum of the mantissa and the characteristic.

$\log N = 1.1206$
$\log N = 0.1206 + 1$

Locate the mantissa in the Table of Common Logarithms. Find the number that corresponds to this mantissa. Since the characteristic is 1, the exponent on 10 is 1 when the number is written in scientific notation.

$N = 1.32 \times 10^1$

Simplify.

$N = 13.2$

Check: $\log 13.2 =$
$\log (1.32 \times 10^1) =$
$\log 1.32 + \log 10^1 =$
$0.1206 + 1 = 1.1206$

Example 2   Find the antilogarithm.
A. antilog 1.9745

B. antilog (8.7332 − 10)

Solution   A. antilog 1.9745 =
antilog (0.9745 + 1) =

• Write the number as the sum of the mantissa and the characteristic. (Do this step mentally.)

$9.43 \times 10^1 =$

• Use the table to find the number that corresponds to the mantissa. The characteristic is the exponent on 10.

94.3

• Simplify.

B. antilog (8.7332 − 10) =
antilog [0.7332 + (−2)] =

• Write the number as the sum of the mantissa and the characteristic. (Do this step mentally.)

$5.41 \times 10^{-2} =$

• Use the table to find the number that corresponds to the mantissa. The characteristic is the exponent on 10.

0.0541

• Simplify.

Problem 2   Find the antilogarithm.
A. antilog 2.3365

B. antilog (9.7846 − 10)

Solution   See page A93.
A. 217   B. 0.609

Given log $N = x$, then antilog $x = N$. Since log $N = x$ is equivalent to $N = 10^x$ and to antilog $x = N$, 10^x can be substituted for N in the equation antilog $x = N$. The resulting equation is

$$\textbf{antilog } x = \textbf{10}^x$$

This equation is especially useful when using a calculator to find antilogarithms. To find antilog x, enter x. Then press the $\boxed{10^x}$ key.

3   Use interpolation to find a common logarithm or antilogarithm

Since 3.257 is not included in the Table of Common Logarithms on pages A2–A3, it is not possible to find log 3.257 directly from the table. However, using the first number less than 3.257 that can be found in the table (3.250) and the first number greater than 3.257 (3.260), it is possible to obtain an approximation for log 3.257 by using linear interpolation. Linear interpolation is based on the fact that many functions can be approximated, over small intervals, by a straight line.

The graph of the equation $y = \log x$ is shown at the right. The units on the x- and y-axes have been distorted to illustrate linear interpolation.

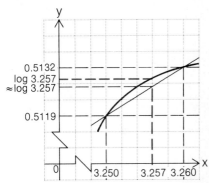

When $x = 3.250$, $y = 0.5119$. When $x = 3.260$, $y = 0.5132$. When $x = 3.257$, the value of y is greater than 0.5119 and less than 0.5132. The actual value of log 3.257 is on the curve. An approximation for this value is on the straight line.

A proportion can be written using the fact that between any two points on a straight line, the slope is the same.

$$\frac{y_2 - y_1}{x_2 - x_1} = \frac{y - y_1}{x - x_1}$$

Let $(x_1, y_1) = (3.250, 0.5119)$ and $(x_2, y_2) = (3.260, 0.5132)$. (x, y) is the point where $x = 3.257$.

$$\frac{0.5132 - 0.5119}{3.260 - 3.250} = \frac{y - 0.5119}{3.257 - 3.250}$$

$$\frac{0.0013}{0.01} = \frac{y - 0.5119}{0.007}$$

Solve the proportion for y.

$$0.00091 = y - 0.5119$$
$$0.51281 = y$$

log 3.257 $\approx$ 0.5128, rounded to the nearest ten-thousandth.

A more convenient method of using linear interpolation is illustrated in the following example:

Find log 32.57.

$$\log 32.57 = \log (3.257 \times 10^1)$$
$$= \log 3.257 + \log 10^1$$

Arrange the numbers 3.250, 3.257, and 3.260 and their corresponding mantissas in a table. Indicate the differences between the numbers and the mantissas, as shown at the right, using d to represent the unknown difference between the mantissa of 3.250 and the mantissa of 3.257.

	Number	Mantissa
	3.250	0.5119
0.007 / 0.01	3.257	0.5119 + d
	3.260	0.5132

d 0.0013

Write a proportion.

$$\frac{0.007}{0.01} = \frac{d}{0.0013}$$

Solve for d, rounding to the nearest ten-thousandth.

$$0.0009 = d$$

Add the value of d to the mantissa 0.5119.

$$\log 3.257 + \log 10^1 = (0.5119 + d) + 1$$
$$= (0.5119 + 0.0009) + 1$$
$$= 0.5128 + 1$$
$$= 1.5128$$

Linear interpolation can also be used to approximate an antilogarithm.

To find antilog $(7.3431 - 10)$, use the table to find the first mantissa less than 0.3431 and the first mantissa greater than 0.3431. Arrange the mantissas and their corresponding antilogs in a table. Indicate the differences between the mantissas and the antilogs using d to represent the unknown difference.

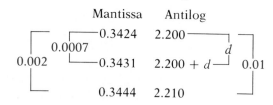

	Mantissa	Antilog
	0.3424	2.200
0.0007 / 0.002	0.3431	2.200 + d
	0.3444	2.210

d 0.01

Write a proportion.

$$\frac{0.0007}{0.002} = \frac{d}{0.01}$$

Solve for d, rounding to the nearest thousandth.

$$0.004 = d$$

Add the value of d to the antilog 2.20.

$$\begin{aligned}
\text{antilog } (7.3431 - 10) &= (2.20 + d) \times 10^{-3} \\
&= (2.20 + 0.004) \times 10^{-3} \\
&= 2.204 \times 10^{-3} \\
&= 0.002204
\end{aligned}$$

Example 3   Find log 0.02903.

Solution $\log 0.02903 = \log (2.903 \times 10^{-2}) = \log 2.903 + \log 10^{-2}$

Number	Mantissa
2.900	0.4624
2.903	0.4624 + d
2.910	0.4639

with 0.01, 0.003, d, 0.0015

$$\frac{0.003}{0.01} = \frac{d}{0.0015}$$
$$0.0005 = d$$

$\log 2.903 + \log 10^{-2} = (0.4624 + d) + (-2) = (0.4624 + 0.0005) + (-2) = 0.4629 + (-2) = 8.4629 - 10$

Problem 3   Find log 31,560.

Solution   See page A94.
4.4991

Example 4   Find antilog $(9.8465 - 10)$.

Solution

Mantissa	Antilog
0.8463	7.02
0.8465	7.02 + d
0.8470	7.03

with 0.0007, 0.0002, d, 0.01

antilog $(9.8465 - 10) = (7.02 + d) \times 10^{-1} = (7.02 + 0.003) \times 10^{-1} = 7.023 \times 10^{-1} = 0.7023$

Problem 4   Find antilog 1.4907.

Solution   See page A94.
30.95

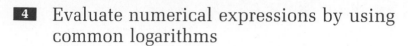

4 Evaluate numerical expressions by using common logarithms

Logarithms were developed in the seventeenth century as an aid in scientific calculations. Today, with the widespread use of calculators, the necessity of using logarithms for computations has been minimized. However, logarithms do play an important role in the analysis of economic and scientific problems.

To solve a numerical problem by using logarithms, use the fact that **if $x = y$, then $\log x = \log y$.**

Example 5 Evaluate $\dfrac{(2.46)^2(0.144)}{\sqrt{15.3}}$ using common logarithms.

Solution $N = \dfrac{(2.46)^2(0.144)}{\sqrt{15.3}}$ • Write an equation.

$$\log N = \log \left[\frac{(2.46)^2(0.144)}{\sqrt{15.3}} \right]$$ • Take the common logarithm of each side of the equation.

$$= \log (2.46)^2 + \log (0.144) - \log (15.3)^{\frac{1}{2}}$$ • Use the Properties of Logarithms.

$$= 2 \log (2.46) + \log (0.144) - \frac{1}{2} \log (15.3)$$

$$= 2(0.3909) + 9.1584 - 10 - \frac{1}{2}(1.1847)$$ • Use the Table of Common Logarithms.

$$= 0.7818 + 9.1584 - 10 - 0.5924$$ • Simplify.
$$= 9.3478 - 10$$

$$N = \text{antilog} (9.3478 - 10)$$ • Find the antilogarithm.
$$= 2.227 \times 10^{-1}$$
$$= 0.2227$$

Problem 5 Evaluate $\dfrac{125}{\sqrt{(1.1)(9.8)}}$ using common logarithms.

Solution See page A94.
38.07

EXERCISES 11.3

1   Find the logarithm.

1. 5.87	0.7686	**2.** 6.42	0.8075	**3.** 34.5	1.5378
4. 10.9	1.0374	**5.** 389	2.5899	**6.** 592	2.7723

7. 1030   3.0128
8. 9060   3.9571
9. 47,300   4.6749
10. 68,200   4.8338
11. 0.114   9.0569 − 10
12. 0.209   9.3201 − 10
13. 0.002   7.3010 − 10
14. 0.034   8.5315 − 10
15. 0.00175   7.2430 − 10
16. 0.0806   8.9063 − 10
17. 0.984   9.9930 − 10
18. 0.79   9.8976 − 10
19. 27,300   4.4362
20. 10,800   4.0334
21. 0.00571   7.7566 − 10
22. 0.00367   7.5647 − 10
23. 0.909   9.9586 − 10
24. 20.3   1.3075

2 Find the antilogarithm.

25. 0.7396   5.49
26. 0.6474   4.44
27. 1.5353   34.3
28. 1.4829   30.4
29. 3.9015   7970
30. 3.9557   9030
31. 2.8007   632
32. 2.0128   103
33. 9.6580 − 10   0.455
34. 9.7033 − 10   0.505
35. 8.3892 − 10   0.0245
36. 8.8344 − 10   0.0683
37. 7.8704 − 10   0.00742
38. 7.9791 − 10   0.00953
39. 2.6767   475
40. 0.1492 + (−1)   0.141
41. 0.6920 + (−1)   0.492

3 Find the logarithm.

42. 3.845   0.5849
43. 4.965   0.6960
44. 84.52   1.9270
45. 67.38   1.8286
46. 499.3   2.6984
47. 699.6   2.8449
48. 0.07071   8.8495 − 10
49. 0.01453   8.1623 − 10
50. 75,930   4.8804
51. 83,090   4.9196
52. 0.8407   9.9247 − 10
53. 0.9813   9.9918 − 10

Find the antilogarithm.

54. 0.9006   7.954
55. 0.8466   7.024
56. 2.7360   544.5
57. 2.5375   344.8
58. 9.7408 − 10   0.5505
59. 9.7320 − 10   0.5395
60. 2.6003   398.4
61. 2.7004   501.7
62. 8.3088 − 10   0.02036
63. 8.5372 − 10   0.03445
64. 1.6525   44.93
65. 1.6809   47.96

4 Evaluate the expression using common logarithms. Interpolate if necessary.

66. $(16)(19.8)$ 316.8
67. $(5.47)(81.7)$ 446.9
68. $(0.45)(0.611)$ 0.2749
69. $(0.98)(0.0153)$ 0.01499
70. $\dfrac{79.3}{26.9}$ 2.948
71. $\dfrac{40.9}{36.3}$ 1.127
72. $\dfrac{(57)(31.4)}{107}$ 16.73
73. $\dfrac{(68)(12.3)}{217}$ 3.854
74. $(8.7)^3$ 658.4
75. $(9.03)^4$ 6650
76. $\dfrac{\sqrt{19.8} \cdot 52}{7.09}$ 32.65
77. $\dfrac{\sqrt[3]{145} \cdot 7.62}{17.3}$ 2.315

78. $\sqrt{(48.5)(3.13)}$ 12.32   **79.** $\sqrt[3]{(70.7)(91.4)}$ 18.63   **80.** $\left(\dfrac{83}{257}\right)^4$ 0.01089

81. $\sqrt[3]{0.688}$ 0.8829   **82.** $\sqrt[3]{0.719}$ 0.8958   **83.** $\sqrt{\dfrac{76.3}{89.7}}$ 0.9224

SUPPLEMENTAL EXERCISES 11.3

Complete. Use the fact that log 3.14 = 0.4969.

84. $10^{0.4969} = $ _____   3.14

85. antilog 0.4969 = _____   3.14

86. antilog (log 3.14) = _____   3.14

87. log (antilog 0.4969) = _____   0.4969

88. $10^{\log 3.14} = $ _____   3.14

89. 10 antilog 0.4969 = _____   31.4

Solve. Use the Table of Common Logarithms on pages A2–A3.

90. log $N = 3.8561$ 7180

91. log $N = 4.9499$ 89,100

92. log $N = 1.6928$ 49.3

93. log $N = 2.4314$ 270.0

94. log $N = 7.7959 - 10$ 0.00625

95. log $N = 9.5105 - 10$ 0.3240

96. log $N = 8.2878 - 10$ 0.0194

97. log $N = 6.3927 - 10$ 0.0002470

98. $N = 10^{2.8463}$ 702

99. $N = 10^{4.9325}$ 85,600

100. $N = 10^{7.5855-10}$ 0.00385

101. $N = 10^{9.6972-10}$ 0.4980

SECTION 11.4
Exponential and Logarithmic Equations

1 Solve exponential equations

An **exponential equation** is one in which the variable occurs in the exponent. The examples at the right are exponential equations.

$$6^{2x+1} = 6^{3x-2}$$
$$4^x = 3$$
$$2^{x+1} = 7$$

An exponential equation in which each side of the equation can be expressed in terms of the same base can be solved by using the property that the exponential function is one-to-one.

The one-to-one property of the exponential function:

If $b^x = b^y$, then $x = y$.

Example 1   Solve and check: $9^{x+1} = 27^{x-1}$

Solution
$$9^{x+1} = 27^{x-1}$$
$$(3^2)^{x+1} = (3^3)^{x-1}$$ • Rewrite each side of the equation using the same base.
$$3^{2x+2} = 3^{3x-3}$$
$$2x + 2 = 3x - 3$$ • Use the one-to-one property of the exponential function to equate the exponents.
$$2 = x - 3$$ • Solve the resulting equation.
$$5 = x$$

The solution is 5.

Check: $9^{x+1} = 27^{x-1}$
$$\begin{array}{c|c} 9^{5+1} & 27^{5-1} \\ 9^6 & 27^4 \\ (3^2)^6 & (3^3)^4 \\ 3^{12} & = 3^{12} \end{array}$$

Problem 1   Solve and check: $10^{3x+5} = 10^{x-3}$

Solution   See page A95.
-4

When each side of an exponential equation cannot easily be expressed in terms of the same base, logarithms are used to solve the exponential equation. Recall that **if $x = y$, then $\log x = \log y$.**

Example 2   Solve for x. Round to the nearest ten-thousandth.

A. $4^x = 7$

B. $3^{2x} = 4$

Solution   A. $4^x = 7$
$$\log 4^x = \log 7$$ • Take the common logarithm of each side of the equation.
$$x \log 4 = \log 7$$ • Rewrite using the Properties of Logarithms.
$$x = \frac{\log 7}{\log 4}$$ • Solve for x.
$$= \frac{0.8451}{0.6021} = 1.4036$$ • Note that $\dfrac{\log 7}{\log 4} \neq \log 7 - \log 4$.

The solution is 1.4036.

B. $3^{2x} = 4$

$\log 3^{2x} = \log 4$ • Take the common logarithm of each side of the equation.

$2x \log 3 = \log 4$ • Rewrite using the Properties of Logarithms.

$2x = \dfrac{\log 4}{\log 3}$ • Solve for $2x$.

$2x = \dfrac{0.6021}{0.4771}$

$2x = 1.2620$

$x = 0.6310$

The solution is 0.6310.

Problem 2   Solve for x. Round to the nearest ten-thousandth.

A. $4^{3x} = 25$ B. $(1.06)^x = 1.5$

Solution   See page A95.

A. 0.7739   B. 6.9605

2   Solve logarithmic equations

A logarithmic equation can be solved by using the Properties of Logarithms.

To solve $\log_9 x + \log_9(x - 8) = 1$, use the Logarithm Property of Products to rewrite the left side of the equation.

$\log_9 x + \log_9(x - 8) = 1$

$\log_9 x(x - 8) = 1$

Write the equation in exponential form.

$9^1 = x(x - 8)$

Simplify and solve for x.

$9 = x^2 - 8x$

$0 = x^2 - 8x - 9$

$0 = (x - 9)(x + 1)$

$x - 9 = 0 \qquad\qquad x + 1 = 0$

$x = 9 \qquad\qquad x = -1$

When x is replaced by 9 in the original equation, 9 checks as a solution. Replacing x by -1 in the original equation results in the expression $\log_9(-1)$. Since the logarithm of a negative number is not a real number, -1 does not check as a solution. Therefore, the solution of the equation is 9.

Some logarithmic equations can be solved by using the fact that **if $\log_b u = \log_b v$, then $u = v$.** The use of this property is illustrated in Example 3B.

Example 3   Solve for x.

A. $\log_3(2x - 1) = 2$

B. $\log_2 x - \log_2(x - 1) = \log_2 2$

Solution   A. $\log_3(2x - 1) = 2$

$\qquad 3^2 = 2x - 1$ • Rewrite in exponential form.

$\qquad\ \ 9 = 2x - 1$ • Solve for x.

$\qquad 10 = 2x$

$\qquad\ \ 5 = x$

The solution is 5.

B. $\log_2 x - \log_2(x - 1) = \log_2 2$

$$\log_2\left(\frac{x}{x - 1}\right) = \log_2 2$$ • Use the Logarithm Property of Quotients.

$$\frac{x}{x - 1} = 2$$ • Use the fact that if $\log_b u = \log_b v$, then $u = v$.

$$(x - 1)\left(\frac{x}{x - 1}\right) = (x - 1)2$$ • Solve for x.

$$x = 2x - 2$$

$$-x = -2$$

$$x = 2$$

The solution is 2.

Problem 3   Solve for x.

A. $\log_4(x^2 - 3x) = 1$

B. $\log_3 x + \log_3(x + 3) = \log_3 4$

Solution   See page A95.

A. -1 and 4   B. 1

3   # Find the logarithm of a number other than base 10

Once a table of logarithms for a given base has been developed, it is possible to find the logarithm of a number to any positive base other than 1.

Since the base 10 logarithms are given on pages A2–A3, this table will be used to find $\log_b x$ for any base b, $b > 0$, $b \neq 1$.

For example, to find $\log_2 7$, write an equation.

$$\log_2 7 = x$$

Rewrite in exponential form.

$$2^x = 7$$

Solve the exponential equation using common logarithms.

$$\log 2^x = \log 7$$
$$x \log 2 = \log 7$$
$$x = \frac{\log 7}{\log 2} = \frac{0.8451}{0.3010} = 2.8076$$

$$\log_2 7 = 2.8076$$

Using a procedure similar to the one shown above, a formula can be written that makes it possible to use any given table of logarithms to find the logarithm of a number with a different base.

The Change of Base Formula

$$\log_a N = \frac{\log_b N}{\log_b a}$$

Example 4   A. Find $\log_3 12$.   B. Find $\ln 5$. Use 2.72 for e.

Solution   A. $\log_3 12 = \dfrac{\log_{10} 12}{\log_{10} 3}$

$\bullet$ Use the Change of Base Formula. $N = 12$, $a = 3$, $b = 10$.

$$= \frac{1.0792}{0.4771} = 2.2620$$

B. $\ln 5 = \dfrac{\log_{10} 5}{\log_{10} e}$

$\bullet$ Use the Change of Base Formula. $N = 5$, $a = e$, $b = 10$.

$$= \frac{\log_{10} 5}{\log_{10} 2.72}$$

$\bullet$ Replace e with 2.72.

$$= \frac{0.6990}{0.4346} = 1.6084$$

Problem 4   A. Find $\log_9 23$.   B. Find $\ln 3$. Use 2.72 for e.

Solution   See page A95.
A. 1.4271   B. 1.0978

The natural logarithm of a number was found in Example 4B and Problem 4B. To find the natural logarithm of a number, N, using a calculator, enter N. Then press the $\boxed{\ln}$ key.

EXERCISES 11.4

1 Solve for x. Round to the nearest ten-thousandth.

1. $5^{4x-1} = 5^{x+2}$ 1

2. $7^{4x-3} = 7^{2x+1}$ 2

3. $8^{x-4} = 8^{5x+8}$ -3

4. $10^{4x-5} = 10^{x+4}$ 3

5. $5^x = 6$ 1.1133

6. $7^x = 10$ 1.1833

7. $12^x = 6$ 0.7211

8. $10^x = 5$ 0.6990

9. $\left(\frac{1}{2}\right)^x = 3$ -1.5850

10. $\left(\frac{1}{3}\right)^x = 2$ -0.6309

11. $(1.5)^x = 2$ 1.7093

12. $(2.7)^x = 3$ 1.1059

13. $10^x = 21$ 1.3222

14. $10^x = 37$ 1.5682

15. $2^{-x} = 7$ -2.8076

16. $3^{-x} = 14$ -2.4022

17. $2^{x-1} = 6$ 3.5854

18. $4^{x+1} = 9$ 0.5848

19. $3^{2x-1} = 4$ 1.1310

20. $4^{-x+2} = 12$ 0.2076

21. $9^x = 3^{x+1}$ 1

22. $2^{x-1} = 4^x$ -1

23. $8^{x+2} = 16^x$ 6

24. $9^{3x} = 81^{x-4}$ -8

25. $5^{x^2} = 21$ 1.3753 and -1.3753

26. $3^{x^2} = 40$ 1.8325 and -1.8325

27. $2^{4x-2} = 20$ 1.5806

28. $4^{3x+8} = 12$ -2.0692

29. $3^{-x+2} = 18$ -0.6310

30. $5^{-x+1} = 15$ -0.6825

31. $4^{2x} = 100$ 1.6610

32. $3^{3x} = 1000$ 2.0960

33. $2.5^{-x} = 4$ -1.5132

34. $3.25^{x+1} = 4.2$ 0.2174

35. $0.25^x = 0.125$ 1.4999

36. $0.1^{5x} = 1 \times 10^{-2}$ $\dfrac{2}{5}$

2 Solve for x.

37. $\log_3(x + 1) = 2$ 8

38. $\log_5(x - 1) = 1$ 6

39. $\log_2(2x - 3) = 3$ $\dfrac{11}{2}$

40. $\log_4(3x + 1) = 2$ 5

41. $\log_2(x^2 + 2x) = 3$ 2 and -4

42. $\log_3(x^2 + 6x) = 3$ -9 and 3

43. $\log_5\left(\dfrac{2x}{x - 1}\right) = 1$ $\dfrac{5}{3}$

44. $\log_6\dfrac{3x}{x + 1} = 1$ -2

45. $\log_7 x = \log_7(1 - x)$ $\dfrac{1}{2}$

46. $\dfrac{3}{4}\log x = 3$ 10,000

47. $\dfrac{2}{3}\log x = 6$ 1,000,000,000

48. $\log(x - 2) - \log x = 3$ no solution

49. $\log_2(x - 3) + \log_2(x + 4) = 3$ 4

50. $\log x - 2 = \log(x - 4)$ $\dfrac{400}{99}$

51. $\log_3 x + \log_3(x - 1) = \log_3 6$ 3

52. $\log_4 x + \log_4(x - 2) = \log_4 15$ 5

53. $\log_2 8x - \log_2(x^2 - 1) = \log_2 3$ 3

54. $\log_5 3x - \log_5(x^2 - 1) = \log_5 2$ 2

55. $\log_9 x + \log_9(2x - 3) = \log_9 2$ 2

56. $\log_6 x + \log_6(3x - 5) = \log_6 2$ 2

57. $\log_8 6x = \log_8 2 + \log_8(x - 4)$
no solution

58. $\log_7 5x = \log_7 3 + \log_7(2x + 1)$
no solution

59. $\log_9 7x = \log_9 2 + \log_9(x^2 - 2)$ 4

60. $\log_3 x = \log_3 2 + \log_3(x^2 - 3)$ 2

61. $\log (x^2 + 3) - \log (x + 1) = \log 5$
$\dfrac{5 + \sqrt{33}}{2}$

62. $\log (x + 3) + \log (2x - 4) = \log 3$
$\dfrac{-1 + \sqrt{31}}{2}$

3   Find the logarithm. Round to the nearest ten thousandth.

63. $\log_3 7$ 1.7713

64. $\log_2 9$ 3.1701

65. $\log_5 12$ 1.5439

66. $\log_5 10$ 1.4306

67. $\log_8 4$ 0.6667

68. $\log_7 6$ 0.9208

69. $\log_3 15$ 2.4651

70. $\log_{12} 9$ 0.8842

71. $\log_8 6$ 0.8617

72. $\log_4 8$ 1.4999

73. $\log_5 30$ 2.1132

74. $\log_6 28$ 1.8597

75. $\log_3 8.6$ 1.9587

76. $\log_5 9.2$ 1.3788

77. $\log_9 4$ 0.6310

78. $\log_8 5$ 0.7740

79. $\log_3(0.5)$ −0.6309

80. $\log_5(0.6)$ −0.3173

81. $\log_7(1.7)$ 0.2726

82. $\log_6(2.3)$ 0.4648

83. $\log_7(1.4)$ 0.1729

84. $\log_4 121$ 3.4592

85. $\log_7 432$ 3.1186

86. $\log_2 0.27$ −1.8890

87. $\log_3 0.38$ −0.6012

88. $\log_5 300$ 3.5438

89. $\log_8 432$ 2.9183

90. $\log_6 0.0053$ −2.9243

91. $\log_8 0.0079$ −2.3280

92. $\log_6 72$ 2.3867

93. $\log_7 91$ 2.3181

94. $\log_3 121$ 4.3653

95. $\log_4 832$ 4.8499

96. $\log_2 0.3$ −1.7372

97. $\log_3 0.4$ −0.8340

98. $\log_2 17.3$ 4.1130

99. $\log_5 18.6$ 1.8162

100. $\log_3 1.79$ 0.5301

101. $\log_6 7.84$ 1.1492

102. $\log_9 43$ 1.7119

103. $\log_8 61$ 1.9769

104. $\log_5 44.2$ 2.3539

SUPPLEMENTAL EXERCISES 11.4

Solve for x. Round to the nearest ten-thousandth.

105. $8^{\frac{x}{2}} = 6$ 1.7234

106. $4^{\frac{x}{3}} = 2$ 1.5

107. $5^{\frac{3x}{2}} = 7$ 0.8060

108. $9^{\frac{2x}{3}} = 8$ 1.4197

109. $1.2^{\frac{x}{2}-1} = 1.4$ 5.6894

110. $5.6^{\frac{x}{3}+1} = 7.8$ 0.5769

Given that $\ln\ 2 = 0.6931$, $\ln\ 3 = 1.0986$, and $\ln\ 10 = 2.3026$, evaluate the natural logarithm.
(Example: $\ln\ 30 = \ln\ (3 \cdot 10) = \ln\ 3 + \ln\ 10 = 1.0986 + 2.3026 = 3.4012$.)

111. $\ln\ 6$
1.7917

112. $\ln\ 20$
2.9957

113. $\ln\ 5$ $\left(Hint:\ 5 = \dfrac{10}{2}\right)$
1.6095

114. $\ln\ 0.2$ $\left(Hint:\ 0.2 = \dfrac{2}{10}\right)$
-1.6095

115. $\ln\ 0.3$ $\left(Hint:\ 0.3 = \dfrac{3}{10}\right)$
-1.2040

116. $\ln\ 1.5$ $\left(Hint:\ 1.5 = \dfrac{3}{2}\right)$
0.4055

Solve the system of equations.

117. $2^x = 8^y$
$x + y = 4$
$(3, 1)$

118. $3^{2y} = 27^x$
$y - x = 2$
$(4, 6)$

119. $\log\ (x + y) = 3$
$x = y + 4$
$(502, 498)$

120. $\log\ (x + y) = 3$
$x - y = 20$
$(510, 490)$

121. $8^{3x} = 4^{2y}$
$x - y = 5$
$(-4, -9)$

122. $9^{3x} = 81^{3y}$
$x + y = 3$
$(2, 1)$

The following "proof" shows that $0.04 < 0.008$. Find the error.

123.
$$2 < 3$$
$$2 \log\ (0.2) < 3 \log\ (0.2)$$
$$\log\ (0.2)^2 < \log\ (0.2)^3$$
$$(0.2)^2 < (0.2)^3$$
$$0.04 < 0.008$$

The error is in step 2.
$\log\ 0.2$ is < 0. When both sides of an inequality are multiplied by a negative number, the inequality symbol must be reversed.

SECTION 11.5

Applications of Exponential and Logarithmic Functions: A Calculator Approach

1 Application problems

Solving problems that involve exponential or logarithmic equations can be quite tedious without a calculator. Each of the problems in this section was solved by using a calculator.

A biologist places one single-celled bacterium in a culture, and each hour that particular species of bacteria divides into two bacteria. After one hour, there will be two bacteria. After two hours, each of the two bacteria will divide, and there will be four bacteria. After three hours, each of the four bacteria will divide, and there will be eight bacteria.

The table at the right shows the number of bacteria in the culture after various intervals of time, t, in hours. Values in this table could also be found by using the exponential equation $N = 2^t$.

Time, t	Number of bacteria, N
0	1
1	2
2	4
3	8
4	16

The equation $N = 2^t$ is an example of an **exponential growth equation.** In general, any equation that can be written in the form $A = A_0 b^{kt}$, where A is the size at time t, A_0 is the initial size, $b > 1$, and k is a positive real number, is an exponential growth equation. These equations play an important role not only in population growth studies but also in physics, chemistry, psychology, and economics.

Recall from Chapter 2 that interest is the amount of money paid or received when borrowing or investing money. **Compound interest** is interest that is computed not only on the original principal but also on the interest already earned. The compound interest formula is an exponential equation.

The **compound interest formula** is $P = A(1 + i)^n$, where A is the original value of an investment, i is the interest rate per compounding period, n is the total number of compounding periods, and P is the value of the investment after n periods.

An investment broker deposits $1000 into an account that earns 12% annual interest compounded quarterly. What is the value of the investment after two years?

Find i, the interest rate per quarter. The quarterly rate is the annual rate divided by 4, the number of quarters in one year.

$$i = \frac{12\%}{4} = \frac{0.12}{4} = 0.03$$

Find n, the number of compounding periods. The investment is compounded quarterly, 4 times a year, for 2 years.

$$n = 4 \cdot 2 = 8$$

Use the compound interest formula.

$$P = A(1 + i)^n$$

Replace A, i, and n by their values.

$$P = 1000(1 + 0.03)^8$$

Solve for P.

$$P \approx 1267$$

The value of the investment after two years is $1267.

Exponential decay is another important example of an exponential equation. One of the most common illustrations is the decay of a radioactive substance.

An isotope of cobalt has a half-life of approximately 5 years. This means that one half of any given amount of a cobalt isotope will disintegrate in 5 years.

The table at the right indicates the amount of the initial 10 mg of a cobalt isotope that remains after various intervals of time, t, in years. Values in this table could also be found by using the exponential equation $A = 10\left(\dfrac{1}{2}\right)^{\frac{t}{5}}$.

Time, t	Amount, A
0	10
5	5
10	2.5
15	1.25
20	0.625

The equation $A = 10\left(\dfrac{1}{2}\right)^{\frac{t}{5}}$ is an example of an **exponential decay equation.** Comparing this equation to the equation on exponential growth, note that for exponential growth, the base of the exponential equation is greater than 1, while for exponential decay, the base is between 0 and 1.

A method by which an archeologist can measure the age of a bone is called **carbon dating.** Carbon dating is based on a radioactive isotope of carbon called carbon 14, which has a half-life of approximately 5570 years. The exponential decay equation is given by $A = A_0\left(\dfrac{1}{2}\right)^{\frac{t}{5570}}$, where A_0 is the original amount of carbon 14 present in the bone, t is the age of the bone, and A is the amount present after t years.

A bone that originally contained 100 mg of carbon 14 now has 70 mg of carbon 14. What is the approximate age of the bone?

Use the exponential decay equation.

$$A = A_0\left(\dfrac{1}{2}\right)^{\frac{t}{5570}}$$

Replace A_0 and A by their given values, and solve for t.

$$70 = 100\left(\dfrac{1}{2}\right)^{\frac{t}{5570}}$$

$$70 = 100(0.5)^{\frac{t}{5570}}$$

Divide each side of the equation by 100.

$$0.7 = (0.5)^{\frac{t}{5570}}$$

Take the common logarithm of each side of the equation. Then simplify.

$$\log 0.7 = \log (0.5)^{\frac{t}{5570}}$$

$$\log 0.7 = \dfrac{t}{5570} \log 0.5$$

$$\dfrac{5570 \log 0.7}{\log 0.5} = t$$

$$2866 = t$$

The age of the bone is approximately 2866 years.

A chemist measures the acidity or alkalinity of a solution by the concentration of hydrogen ions, H^+, in the solution by the formula $pH = -\log (H^+)$. A neutral solution such as distilled water has a pH of 7, acids have a pH of less than 7, and alkaline solutions (also called basic solutions) have a pH of greater than 7.

Find the pH of vinegar for which $H^+ = 1.26 \times 10^{-3}$. Round to the nearest tenth.

Use the pH equation.
$H^+ = 1.26 \times 10^{-3}$.

$$
\begin{aligned}
pH &= -\log (H^+) \\
&= -\log (1.26 \times 10^{-3}) \\
&= -(\log 1.26 + \log 10^{-3}) \\
&= -[0.1004 + (-3)] = 2.8996
\end{aligned}
$$

The pH of vinegar is 2.9.

The **Richter scale** measures the magnitude, M, of an earthquake in terms of the intensity, l, of its shock waves. This can be expressed as the logarithmic equation $M = \log \dfrac{l}{l_0}$, where l_0 is a constant.

How many times stronger is an earthquake that has magnitude 4 on the Richter scale than one that has magnitude 2 on the scale?

Let l_1 represent the intensity of the earthquake that has magnitude 4, and let l_2 represent the intensity of the earthquake that has magnitude 2. The ratio of l_1 to l_2, $\dfrac{l_1}{l_2}$, measures how much stronger l_1 is than l_2.

$$4 = \log \dfrac{l_1}{l_0}$$
$$2 = \log \dfrac{l_2}{l_0}$$

Use the Ricther equation to write a system of equations, one equation for magnitude 4 and one for magnitude 2. Then rewrite the system using the Properties of Logarithms.

$$4 = \log l_1 - \log l_0$$
$$2 = \log l_2 - \log l_0$$

Use the addition method to eliminate $\log l_0$.

$$2 = \log l_1 - \log l_2$$

Rewrite the equation using the Properties of Logarithms.

$$2 = \log \dfrac{l_1}{l_2}$$

Solve for the ratio using the relationship between logarithms and exponents.

$$\dfrac{l_1}{l_2} = 10^2 = 100$$

An earthquake that has magnitude 4 on the Richter scale is 100 times stronger than an earthquake that has magnitude 2.

The percent of light that will pass through a substance is given by the equation **log $P = -kd$**, where P is the percent of light passing through the substance, k is a constant that depends on the substance, and d is the thickness of the substance.

Find the percent of light that will pass through opaque glass for which $k = 0.4$, and d is 0.5 cm.

Replace k and d in the equation by their given values, and solve for P.

$$\log P = -kd$$
$$\log P = -(0.4)(0.5)$$
$$\log P = -0.2$$

Use the relationship between the logarithmic and exponential function.

$$P = 10^{-0.2}$$
$$P = 0.6310$$

Approximately 63.1% of the light will pass through the glass.

Example 1   An investment of $3000 is placed into an account that earns 12% annual interest compounded monthly. In approximately how many years will the investment be worth twice the original amount? Round to the nearest whole number.

Strategy   To find the time, solve the compound interest formula for n. Use

$$P = 6000, \ A = 3000, \text{ and } i = \frac{12\%}{12} = \frac{0.12}{12} = 0.01.$$

Solution
$$P = A(1 + i)^n$$
$$6000 = 3000(1 + 0.01)^n$$
$$6000 = 3000(1.01)^n$$
$$2 = (1.01)^n$$
$$\log 2 = \log (1.01)^n$$
$$\log 2 = n \log 1.01$$
$$\frac{\log 2}{\log 1.01} = n$$
$$70 = n \qquad \qquad \bullet \ n \text{ is the number of months.}$$

70 months $\div$ 12 months ≈ 5.8

In approximately 6 years the investment will be worth $6000.

Problem 1   Find the pH of sodium carbonate for which $H^+ = 2.51 \times 10^{-12}$. Round to the nearest tenth.

Solution   See page A96.
11.6

Example 2   The number of words per minute a student can type will increase with practice and can be approximated by the equation $N = 100[1 - (0.9)^t]$, where N is the number of words typed per minute after t days of instruction. Find the number of words a student will type per minute after 8 days of instruction.

Strategy   To find the number of words per minute, replace t by its given value in the equation, and solve for N.

Solution $N = 100[1 - (0.9)^t] = 100[1 - (0.9)^8] = 56.95$

After 8 days of instruction, a student will type approximately 57 words per minute.

Problem 2   An earthquake that measures 5 on the Richter scale can cause serious damage to buildings. The San Francisco earthquake of 1906 would have measured about 7.8 on this scale. How many times stronger was the San Francisco earthquake than one that can cause serious damage? Round to the nearest whole number.

Solution   See page A96.
631 times stronger

EXERCISES 11.5

1   Solve. Round to the nearest whole number.

Use the compound interest formula $P = A(1 + i)^n$, where A is the original value of an investment, i is the interest rate per compounding period, n is the total number of compounding periods, and P is the value of the investment after n periods.

1. An investor deposits $5000 into an account that earns 10% interest compounded semiannually. What is the value of the investment after 6 years?
$8979

2. A deposit of $1500 is made into an account that earns 10% annual interest compounded daily. What is the value of the investment after one year? (1 year = 365 days)
$1658

3. An insurance broker deposits $2500 into an account that earns 8% annual interest compounded semiannually. In approximately how many years will the investment be worth $5000?
9 years

4. An investor deposits $2500 into an account that earns 9% annual interest compounded semiannually. In approximately how many years will the investment be worth twice the original amount?
8 years

5. A company president has determined that it will be necessary to purchase a new computer in three years. The estimated cost of the computer is $15,000. How much money must be deposited in an account that earns 8% annual interest compounded quarterly so that the value of the account in three years will be $15,000?
    $11,827

6. The maintenance supervisor for a company estimates that in two years the company will need to purchase a new lathe at a cost of $10,000. How much money must be deposited into an account that earns 12% annual interest compounded monthly so that the value of the account in two years will be $10,000?
    $7876

Use the exponential decay equation $A = A_0\left(\dfrac{1}{2}\right)^{\frac{t}{k}}$, where A is the amount of a radioactive material present after a time t, k is the half-life, and A_0 is the original amount of radioactive material.

7. An isotope has a half-life of approximately 50 days. How many days are required for a 5-mg sample of this isotope to decay to 1 mg?
    116 days

8. An isotope of sodium has a half-life of approximately 3 years. How long will it take an original sample of 9 mg of this isotope to decay to 5 mg?
    3 years

9. A technician measured the amount of a radioactive substance as 8 mg. Ten hours later, another measurement showed that there were 6 mg remaining. What is the half-life of this substance?
    24 h

10. A scientist measures the amount of a radioactive material as 12 mg. Three hours later, a second measurement shows that there are 10 mg of the material remaining. What is the half-life of this radioactive material?
    11 h

Solve. Round to the nearest whole number or percent.

The percent of correct welds a student welder can make will increase with practice and can be approximated by the equation $P = 100[1 - (0.75)^t]$, where P is the percent of correct welds, and t is the number of weeks of practice.

11. Find the percent of correct welds a student will make after three weeks of practice.
    58%

12. How many weeks of practice are necessary before a student will make 90% of the welds correctly?   8 weeks

Use the pH equation pH = $-\log (H^+)$, where H^+ is the hydrogen ion concentration of a solution.

13. Find the pH of a hydrogen chloride solution for which the hydrogen ion concentration is 8.5×10^{-3}.
     2

14. Find the pH of a sodium hydroxide solution for which the hydrogen ion concentration is 3.4×10^{-9}.
     8

The percent of light that will pass through a material is given by the equation $\log P = -kd$, where P is the percent of light passing through the material, k is a constant that depends on the material, and d is the thickness of the material in centimeters.

15. The constant k for a piece of tinted glass is 0.4. How thick a piece of this glass is needed so that 40% of the light that is incident to the glass will pass through it?
     1 cm

16. The constant k for a piece of opaque glass that is 0.6 cm thick is 0.3. Find the percent of light that will pass through the glass.
     66%

Use the Richter equation $M = \log \dfrac{l}{l_0}$, where M is the magnitude of an earthquake, l is the intensity of its shock waves, and l_0 is a constant.

17. How many times stronger is an earthquake that has magnitude 7.5 on the Richter scale than one that has magnitude 3.5 on the scale?
     10,000 times stronger

18. How many times stronger is an earthquake that has magnitude 6.3 on the Richter scale than one that has magnitude 4 on the scale?
     200 times stronger

The number of decibels, D, of a sound can be given by the equation $D = 10(\log l + 16)$, where l is the power of the sound measured in watts.

19. Find the number of decibels from a jet engine for which the power of the sound is 0.0025 watts.
     134 decibels

20. Find the number of decibels of normal conversation. The power of the sound of normal conversation is approximately 3.2×10^{-10} watts.

65 decibels

SUPPLEMENTAL EXERCISES 11.5

Use the exponential growth equation $A = A_0 b^{kt}$, where A is the size at time t, A_0 is the initial size, and $b = 2$.

21. At 9 A.M., a culture of bacteria had a population of 1.5×10^6. At noon, the population was 3×10^6. If the population is growing exponentially, at what time will the population be 9×10^6? Round to the nearest hour.

5 P.M.

22. The population of a colony of bacteria was 2×10^4 at 10 A.M. At noon, the population had grown to 3×10^4. If the population is growing exponentially, at what time will the population be 8×10^4? Round to the nearest hour.

5 P.M.

Use the compound interest formula $P = A(1 + i)^n$.

23. If the average annual rate of inflation is 8%, in how many years will prices double? Round to the nearest whole number.

9 years

24. If the average annual rate of inflation is 5%, in how many years will prices double? Round to the nearest whole number.

14 years

25. An investment of $1000 earns $177.23 in interest in two years. If the interest is compounded annually, find the annual interest rate. Round to the nearest tenth of a percent. (*Hint:* The solution will require taking the common logarithm of each side of the equation and taking the common antilogarithm of each side of the equation.)

8.5%

26. An investment of $1000 earns $242.30 in interest in two years. If the interest is compounded annually, find the annual interest rate. Round to the nearest tenth of a percent. (*Hint:* The solution will require taking the common logarithm of each side of the equation and taking the common antilogarithm of each side of the equation.)

11.5%

CALCULATORS AND COMPUTERS

 The $\boxed{10^x}$ and $\boxed{\log}$ Keys on a Calculator

The $\boxed{\log}$ key on a calculator gives the logarithm base 10 of a number. For this discussion, log x will mean $\log_{10}x$. The $\boxed{10^x}$ key gives the antilogarithm of x.

Here are some examples along with the keystrokes.

Find log 18.   Enter: 18 $\boxed{\log}$
                     The answer is 1.2552725.

Find log 0.0235.   Enter: 0.0235 $\boxed{\log}$
                     The answer is −1.6289321.

Find antilog 1.23514.   Enter: 1.23514 $\boxed{10^x}$
                     The answer is 17.184623.

Find antilog (−2.351). Enter: 2.351 $\boxed{+/-}$ $\boxed{10^x}$ Note that in this last
                     The answer is 0.0044566.   example, a negative
                                                   mantissa can be
                                                   used.

For those calculators that do not have a $\boxed{10^x}$ key, the antilogarithm can be found by using the $\boxed{INV}$ or $\boxed{2nd}$ keys. The $\boxed{INV}$ key will be used here, but the keystrokes for the $\boxed{2nd}$ key would be the same.

Find antilog 2.9132.   Enter 2.9132 $\boxed{INV}$ $\boxed{\log}$
                     The answer is 818.84179.

Here is a further illustration of the use of the $\boxed{\log}$ key.

An investor wishes to double an investment of $4000 in 5 years. What interest rate, compounded quarterly, is necessary to reach this goal?

Use the compound interest equation $P = A(1 + i)^n$, where $P = 8000$, $A = 4000$, and $n = 5 \cdot 4 = 20$.

$$P = A(1 + i)^n$$
$$8000 = 4000(1 + i)^{20}$$
$$2 = (1 + i)^{20}$$
$$\log 2 = 20 \log (1 + i)$$
$$\frac{\log 2}{20} = \log (1 + i)$$

Use your calculator to find $\dfrac{\log 2}{20}$.

Enter: 2 $\boxed{\log}$ $\boxed{\div}$ 20 $\boxed{=}$

$$0.0150515 = \log(1 + i)$$
$$10^{0.0150515} = 1 + i$$

Recall that $\log_{10} a = b$ means $10^b = a$. Use your calculator to find $10^{0.0150515}$.
Enter: $0.0150515 \boxed{10^x}$.

$$1.0352649 = 1 + i$$
$$0.0352649 = i$$

The quarterly interest rate is 3.53%.
The annual interest rate is $4 \times 3.53\% = 14.12\%$.

CHAPTER SUMMARY

Key Words

A function of the form $f(x) = b^x$ is an **exponential function,** where b is a positive real number not equal to 1. The number b is the **base** of the exponential function.

For $b > 0$, $b \neq 1$, $y = \log_b x$ **is equivalent to** $x = b^y$.

$\log_b x$ is the logarithm of x to the base b.

Common logarithms are logarithms to the base 10. (Usually the base 10 is omitted when writing the common logarithm of a number.)

Natural logarithms are logarithms to the base e. The number e is an irrational number approximately equal to 2.718281828. The natural logarithm is abbreviated ln x.

The **mantissa** is the decimal part of a logarithm.

The **characteristic** is the integer part of a logarithm.

The **common antilogarithm** of a number, N, is the Nth power of 10, written antilog $N = 10^N$.

An **exponential equation** is one in which the variable occurs in the exponent.

Essential Rules

The Logarithm Property of the Product of Two Numbers

For any positive real numbers x, y, and b, $b \neq 1$, $\log_b xy = \log_b x + \log_b y$.

The Logarithm Property of the Quotient of Two Numbers	For any positive real numbers x, y, and b, $b \neq 1$, $\log_b \dfrac{x}{y} = \log_b x - \log_b y$.
The Logarithm Property of the Power of a Number	For any positive real numbers x and b, $b \neq 1$, and any real number r, $\log_b x^r = r \log_b x$.
Additional Properties of Logarithms	For any positive real number b, $b \neq 1$, and any real number n, $\log_b b^n = n$. For any positive real number b, $b \neq 1$, $\log_b 1 = 0$.
One-to-One Property of the Exponential Function	If $b^x = b^y$, then $x = y$.
To Solve Numerical Problems Using Logarithms	If $x = y$, then $\log_b x = \log_b y$.
The Change of Base Formula	$\log_a N = \dfrac{\log_b N}{\log_b a}$

CHAPTER REVIEW

1. Evaluate: $\log_4 16$

2

2. Write $\dfrac{1}{2}(\log_3 x - \log_3 y)$ as a single logarithm with a coefficient of 1.

$\log_3 \sqrt{\dfrac{x}{y}}$

3. Find antilog 2.8523.

711.7

4. Solve for x: $8^x = 2^{x-6}$

-3

5. Evaluate the function $f(x) = \left(\dfrac{2}{3}\right)^x$ at $x = 0$.

1

6. Solve for x: $\log_3 x = -2$

$\dfrac{1}{9}$

7. Find log 0.00367.

$7.5647 - 10$

8. Evaluate $\dfrac{2.74}{42.6}$ by using common logarithms.

0.06433

9. Solve for x: $\log x + \log (x - 4) = \log 12$

6

10. Graph $f(x) = 2^x - 3$.

11. Write $\log_6 \sqrt{xy^3}$ in expanded form.

$\frac{1}{2}\log_6 x + \frac{3}{2}\log_6 y$

12. Find log 17.48.

1.2425

13. Solve for x: $3^{7x+1} = 3^{4x-5}$

−2

14. Evaluate the function $f(x) = 3^{x+1}$ at $x = -2$.

$\frac{1}{3}$

15. Graph $f(x) = \log_2(2x)$.

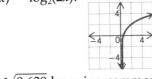

16. Find antilog $(8.6064 - 10)$.

0.0404

17. Evaluate $\sqrt{0.638}$ by using common logarithms.

0.7787

18. Find $\log_3 19$.

2.6804

19. Evaluate: $\log_2 16$

4

20. Find log 0.0491

$8.6911 - 10$

21. Find $\log_2 5$.

2.3223

22. Find antilog $(9.8463 - 10)$.

0.702

23. Solve for x: $4^x = 8^{x-1}$

3

24. Evaluate the function $f(x) = \left(\frac{1}{4}\right)^x$ at $x = -1$.

4

25. Write $\log_5 \sqrt{\dfrac{x}{y}}$ in expanded form.

$\frac{1}{2}\log_5 x - \frac{1}{2}\log_5 y$

26. Find log 86.52.

1.9371

27. Solve for x: $\log_5 x = 3$

125

28. Evaluate $\sqrt{0.297}$ by using common logarithms.

0.545

29. Graph $f(x) = \left(\frac{1}{2}\right)^x + 1$.

30. Solve for x: $\log x + \log (2x + 3) = \log 2$

$\frac{1}{2}$

31. Write $3 \log_b x - 5 \log_b y$ as a single logarithm with a coefficient of 1.

$\log_b \dfrac{x^3}{y^5}$

32. Evaluate the function $f(x) = 2^{-x-1}$ at $x = -3$.

4

33. Graph $f(x) = \log_2 x - 1$.

34. Find antilog 1.7048.

50.68

35. Evaluate $\dfrac{49.1}{68.3}$ by using common logarithms.

0.719

36. Solve for x: $4^{5x-2} = 4^{3x+2}$

2

37. Use the exponential decay equation $A = A_0\left(\dfrac{1}{2}\right)^{\frac{t}{k}}$, where A is the amount of a radioactive material after time t, k is the half-life, and A_0 is the original amount of radioactive material, to find the half-life of a material that decays from 10 mg to 9 mg in 5 h. Round to the nearest whole number.

33 h

38. The percent of light that will pass through an opaque material is given by the equation $\log P = -0.5d$, where P is the percent of light that passes through the material, and d is the thickness of the material in centimeters. How thick must this opaque material be so that only 50% of the light that is incident to the material will pass through it?

0.602 cm

CUMULATIVE REVIEW

1. Graph the solution set of $\{x \,|\, x < 0\} \cap \{x \,|\, x > -4\}$.

2. Solve: $4 - 2[x - 3(2 - 3x) - 4x] = 2x$

$\dfrac{8}{7}$

3. Solve: $|2x - 5| \le 3$

$\{x \,|\, 1 \le x \le 4\}$

4. Factor: $4x^{2n} + 7x^n + 3$

$(4x^n + 3)(x^n + 1)$

5. Solve: $x^2 + 4x - 5 \le 0$

$\{x \mid -5 \le x \le 1\}$

6. Simplify: $\dfrac{1 - \dfrac{5}{x} + \dfrac{6}{x^2}}{1 + \dfrac{1}{x} - \dfrac{6}{x^2}}$ $\dfrac{x-3}{x+3}$

7. Graph the solution set of $\dfrac{x+2}{x-1} \ge 0$.

8. Solve $S = 2wh + 2wL + 2Lh$ for L.

$L = \dfrac{S - 2wh}{2w + 2h}$

9. Simplify: $\dfrac{\sqrt{xy}}{\sqrt{x} - \sqrt{y}}$

$\dfrac{x\sqrt{y} + y\sqrt{x}}{x - y}$

10. Simplify: $y\sqrt{18x^5y^4} - x\sqrt{98x^3y^6}$

$-4x^2y^3\sqrt{2x}$

11. Simplify: $\dfrac{i}{2 - i}$

$-\dfrac{1}{5} + \dfrac{2}{5}i$

12. Find the equation of the line containing the point $(2, -2)$ and parallel to the line $2x - y = 5$.

$y = 2x - 6$

13. Write a quadratic equation that has integer coefficients and has as solutions $\dfrac{1}{3}$ and -3.

$3x^2 + 8x - 3 = 0$

14. Solve: $x^2 - 4x - 6 = 0$

$2 + \sqrt{10}$ and $2 - \sqrt{10}$

15. Graph $y = -x^2 - 2x + 3$.

16. Find the range of the function $f(x) = x^2 - 3x - 4$ if the domain is $\{-1, 0, 1, 2, 3\}$.

$\{-6, -4, 0\}$

17. Given $f(x) = x^2 + 2x + 1$ and $g(x) = 2x - 3$, find $f(g(0))$.

4

18. Graph $\dfrac{x^2}{4} - \dfrac{y^2}{16} = 1$.

19. Solve by the addition method:

$3x - y + z = 3$

$x + y + 4z = 7$

$3x - 2y + 3z = 8$

$(0, -1, 2)$

20. Solve: $y = x^2 + x - 3$

$y = 2x - 1$

$(2, 3)$ and $(-1, -3)$

21. Graph $f(x) = \left(\dfrac{1}{2}\right)^x - 1$.

22. Graph $f(x) = \log_2 x + 1$.

23. Evaluate the function $f(x) = 3^{-x+1}$ at $x = -4$.

243

24. Solve for x: $\log_4 x = 3$

64

25. Solve for x: $2^{3x+2} = 4^{x+5}$

8

26. Solve for x: $\log x + \log (3x + 2) = \log 5$

1

27. An alloy containing 25% tin is mixed with an alloy containing 50% tin. How much of each were used to make 2000 lb of an alloy containing 40% tin?

25% alloy: 800 lb; 50% alloy: 1200 lb

28. An account executive earns $500 per month plus 8% commission on the amount of sales. The executive's goal is to earn a minimum of $3000 per month. What amount of sales will enable the executive to earn $3000 or more per month?

$31,250 or more

29. A new printer can print checks three times faster than an older printer. The old printer can print the checks in 30 min. How long would it take to print the checks when both printers are operating?

7.5 min

30. For a constant temperature, the pressure (P) of a gas varies inversely as the volume (V). If the pressure is 50 lb/in.^{2} when the volume is 250 ft^3, find the pressure when the volume is 25 ft^3.

500 lb/in.^{2}

31. A contractor buys 45 yd of nylon carpet and 30 yd of wool carpet for $1170. A second purchase, at the same prices, includes 25 yd of nylon carpet and 80 yd of wool carpet for $1410. Find the cost per yard of the wool carpet.

$12

32. An investor deposits $10,000 into an account that earns 9% interest compounded monthly. What is the value of the investment after 5 years? Use the compound interest formula $P = A(1 + i)^n$, where A is the original value of an investment, i is the interest rate per compounding period, n is the total number of compounding periods, and P is the value of the investment after n periods.

$15,656.81

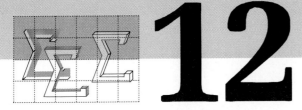

12

Sequences and Series

OBJECTIVES

- Write the terms of a sequence
- Find the sum of a series
- Find the nth term of an arithmetic sequence
- Find the sum of an arithmetic series
- Application problems
- Find the nth term of a geometric sequence
- Finite geometric series
- Infinite geometric series
- Application problems
- Expand $(a + b)^n$

*T*ower of Hanoi

The Tower of Hanoi is a puzzle that has the following form: Three pegs are placed in a board. A number of disks, graded in size, are stacked on one of the pegs with the largest disk on the bottom and the succeedingly smaller disks placed on top. The disks are moved according to the following rules:

1. Only one disk at a time may be moved.
2. A larger disk cannot be placed over a smaller disk.

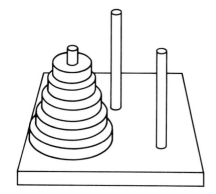

The object of the puzzle is to transfer all the disks from one peg to one of the other two pegs. If initially there is only one disk, then only one move would be required. With two disks initially, three moves are required, and with three disks, seven moves are required.

You can try this puzzle using playing cards. Select a number of playing cards, in sequence, from a deck. The number of cards corresponds to the number of disks, and the number on the card corresponds to the size of the disk. For example, an "ace" would correspond to the smallest disk, a "two" would correspond to the next largest disk, and so on for five cards. Now place three coins on a table to be used as the pegs. Pile the cards on one of the coins (in order), and try to move the pile to a second coin. Now a numerically larger card cannot be placed on a numerically smaller card. Try this for 5 cards. The minimum number of moves is 31.

The chart below shows the minimum number of moves required for an initial number of disks. The difference between the number of moves for each succeeding disk is also given.

1	2	3	4	5	6	7	8
1	3	7	15	31	63	127	255

$$2 \quad 4 \quad 8 \quad 16 \quad 32 \quad 64 \quad 128$$

For this last list of numbers, each succeeding number can be found by multiplying the preceding number by a constant (in this case 2). Such a list of numbers is called a *geometric sequence*. The formula for the minimum number of moves is given by $M = 2^n - 1$, where M is the number of moves, and n is the number of disks. This equation is an *exponential equation*.

Introduction to Sequences and Series

1 ## Write the terms of a sequence

An investor deposits $100 in an account that earns 10% interest compounded annually. The amount of interest earned each year can be determined by using the compound interest formula used in Chapter 11. The amount of interest earned in each of the first four years of the investment is shown below.

Year	1	2	3	4
Interest earned	$10	$11	$12.10	$13.31

The list of numbers 10, 11, 12.10, 13.31 is called a sequence. A **sequence** is an ordered list of numbers. The list 10, 11, 12.10, 13.31 is ordered because the position of a number in this list indicates the year in which that amount of interest was earned. Each of the numbers of a sequence is called a **term** of the sequence.

Examples of other sequences are shown at the right. These sequences are separated into two groups. A **finite sequence** contains a finite number of terms. An **infinite sequence** contains an infinite number of terms.

$$1, 1, 2, 3, 5, 8$$
$$1, 2, 3, 4, 5, 6, 7, 8$$
$$1, -1, 1, -1$$
Finite Sequences

$$1, 3, 5, 7, \ldots$$
$$1, \frac{1}{2}, \frac{1}{4}, \frac{1}{8}, \ldots$$
$$1, 1, 2, 3, 5, 8, \ldots$$
Infinite Sequences

For the sequence at the right, the first term is 2, the second term is 4, the third term is 6, and the fourth term is 8.

$$2, 4, 6, 8, \ldots$$

A general sequence is shown at the right. The first term is a_1, the second term is a_2, the third term is a_3, and the nth term, also called the **general term** of the sequence, is a_n. Note that each term of the sequence is paired with a natural number.

$$a_1, a_2, a_3, \ldots a_n, \ldots$$

Frequently, a sequence has a definite pattern that can be expressed by a formula.

Each term of the sequence shown at the right is paired with a natural number by the formula $a_n = 3n$. The first term, a_1, is 3. The second term, a_2, is 6. The third term, a_3, is 9. The nth term, a_n, is $3n$.

$$a_n = 3n$$

$$a_1, \quad a_2, \quad a_3, \ldots \quad a_n, \ldots$$
$$3(1), 3(2), 3(3), \ldots 3(n), \ldots$$
$$3, \quad 6, \quad 9, \ldots \quad 3n, \ldots$$

Example 1 Write the first three terms of the sequence whose nth term is given by the formula $a_n = 2n - 1$.

Solution $a_n = 2n - 1$
$a_1 = 2(1) - 1 = 1$ • Replace n by 1.
$a_2 = 2(2) - 1 = 3$ • Replace n by 2.
$a_3 = 2(3) - 1 = 5$ • Replace n by 3.

The first term is 1, the second term is 3, and the third term is 5.

Problem 1 Write the first four terms of the sequence whose nth term is given by the formula $a_n = n(n + 1)$.

Solution See page A99.
$2, 6, 12, 20$

Example 2 Find the eighth and tenth terms of the sequence whose nth term is given by the formula $a_n = \dfrac{n}{n + 1}$.

Solution $a_n = \dfrac{n}{n + 1}$

$a_8 = \dfrac{8}{8 + 1} = \dfrac{8}{9}$ • Replace n by 8.

$a_{10} = \dfrac{10}{10 + 1} = \dfrac{10}{11}$ • Replace n by 10.

The eighth term is $\dfrac{8}{9}$, and the tenth term is $\dfrac{10}{11}$.

Problem 2 Find the sixth and ninth terms of the sequence whose nth term is given by the formula $a_n = \dfrac{1}{n(n + 2)}$.

Solution See page A99.
$a_6 = \dfrac{1}{48}; \; a_9 = \dfrac{1}{99}$

2 Find the sum of a series

On page 583, the sequence 10, 11, 12.10, 13.31 was shown to represent the amount of interest earned in each of 4 years of an investment.

10, 11, 12.10, 13.31

The sum of the terms of this sequence represents the total interest earned by the investment over the four-year period.

$10 + 11 + 12.10 + 13.31 = 46.41$

The total interest earned over the four-year period is $46.41.

The indicated sum of the terms of a sequence is called a **series**. Given the sequence 10, 11, 12.10, 13.31, the series $10 + 11 + 12.10 + 13.31$ can be written.

S_n is used to indicate the sum of the first n terms of a sequence.

For the example above, the sums of the series S_1, S_2, S_3, and S_4 represent the total interest earned for 1, 2, 3, and 4 years, respectively.

$$S_1 = 10 \qquad\qquad\qquad\quad = 10$$
$$S_2 = 10 + 11 \qquad\qquad\quad = 21$$
$$S_3 = 10 + 11 + 12.10 \qquad = 33.10$$
$$S_4 = 10 + 11 + 12.10 + 13.31 = 46.41$$

For the general sequence $a_1, a_2, a_3, \ldots a_n$, the series S_1, S_2, S_3, and S_n are shown at the right.

$$S_1 = a_1$$
$$S_2 = a_1 + a_2$$
$$S_3 = a_1 + a_2 + a_3$$
$$S_n = a_1 + a_2 + a_3 + \cdots + a_n$$

It is convenient to represent a series in a compact form called **summation notation,** or **sigma notation.** The Greek letter sigma, Σ, is used to indicate a sum.

The first four terms of the sequence whose nth term is given by the formula $a_n = 2n$ are 2, 4, 6, 8. The corresponding series is shown at the right written in summation notation and is read "the summation from 1 to 4 of $2n$." The letter n is called the **index** of the summation.

$$\sum_{n=1}^{4} 2n$$

To write the terms of the series, replace n by the consecutive integers from 1 to 4.

$$\sum_{n=1}^{4} 2n = 2(1) + 2(2) + 2(3) + 2(4)$$

The series is $2 + 4 + 6 + 8$.

$$= 2 + 4 + 6 + 8$$

The sum of the series is 20.

$$= 20$$

Example 3   Find the sum of the series.

A. $\displaystyle\sum_{i=1}^{3} (2i - 1)$ B. $\displaystyle\sum_{n=3}^{6} \frac{1}{2} n$

Solution   A. $\displaystyle\sum_{i=1}^{3} (2i - 1) =$

$[2(1) - 1] + [2(2) - 1] + [2(3) - 1] =$ • Replace i by 1, 2, and 3.
$1 + 3 + 5 =$ • Write the series.
9 • Find the sum of the series.

B. $\displaystyle\sum_{n=3}^{6} \frac{1}{2} n =$

$\frac{1}{2}(3) + \frac{1}{2}(4) + \frac{1}{2}(5) + \frac{1}{2}(6) =$ • Replace n by 3, 4, 5, and 6.

$\frac{3}{2} + 2 + \frac{5}{2} + 3 =$ • Write the series.

9 • Find the sum of the series.

Problem 3   Find the sum of the series.

A. $\displaystyle\sum_{n=1}^{4} (7 - n)$ B. $\displaystyle\sum_{i=3}^{6} (i^2 - 2)$

Solution   See page A99.
A.   18   B.   78

Example 4   Write $\displaystyle\sum_{i=1}^{5} x^i$ in expanded form.

Solution $\displaystyle\sum_{i=1}^{5} x^i =$ • This is a variable series.

$x + x^2 + x^3 + x^4 + x^5$ • Replace i by 1, 2, 3, 4, and 5.

Problem 4   Write $\displaystyle\sum_{n=1}^{5} nx$ in expanded form.

Solution   See page A99.
$x + 2x + 3x + 4x + 5x$

EXERCISES 12.1

1   Write the first four terms of the sequence whose nth term is given by the formula.

1. $a_n = n + 1$
2, 3, 4, 5

2. $a_n = n - 1$
0, 1, 2, 3

3. $a_n = 2n + 1$
3, 5, 7, 9

4. $a_n = 3n - 1$
2, 5, 8, 11

5. $a_n = 2 - 2n$
0, -2, -4, -6

6. $a_n = 1 - 2n$
-1, -3, -5, -7

7. $a_n = 2^n$
2, 4, 8, 16

8. $a_n = 3^n$
3, 9, 27, 81

9. $a_n = n^2 + 1$
2, 5, 10, 17

10. $a_n = n^2 - 1$
0, 3, 8, 15

11. $a_n = \dfrac{n}{n^2 + 1}$
$\dfrac{1}{2}, \dfrac{2}{5}, \dfrac{3}{10}, \dfrac{4}{17}$

12. $a_n = \dfrac{n^2 - 1}{n}$
$0, \dfrac{3}{2}, \dfrac{8}{3}, \dfrac{15}{4}$

13. $a_n = n - \dfrac{1}{n}$
$0, \dfrac{3}{2}, \dfrac{8}{3}, \dfrac{15}{4}$

14. $a_n = n^2 - \dfrac{1}{n}$
$0, \dfrac{7}{2}, \dfrac{26}{3}, \dfrac{63}{4}$

15. $a_n = (-1)^{n+1} n$
1, -2, 3, -4

16. $a_n = \dfrac{(-1)^{n+1}}{n + 1}$
$\dfrac{1}{2}, -\dfrac{1}{3}, \dfrac{1}{4}, -\dfrac{1}{5}$

17. $a_n = \dfrac{(-1)^{n+1}}{n^2 + 1}$
$\dfrac{1}{2}, -\dfrac{1}{5}, \dfrac{1}{10}, -\dfrac{1}{17}$

18. $a_n = (-1)^n (n^2 + 2n + 1)$
-4, 9, -16, 25

19. $a_n = (-1)^n 2^n$
-2, 4, -8, 16

20. $a_n = \dfrac{1}{3} n^3 + 1$
$\dfrac{4}{3}, \dfrac{11}{3}, 10, \dfrac{67}{3}$

21. $a_n = 2\left(\dfrac{1}{3}\right)^{n+1}$
$\dfrac{2}{9}, \dfrac{2}{27}, \dfrac{2}{81}, \dfrac{2}{243}$

Find the indicated term of the sequence whose *n*th term is given by the formula.

22. $a_n = 3n + 4$; a_{12} 40

23. $a_n = 2n - 5$; a_{10} 15

24. $a_n = n(n - 1)$; a_{11} 110

25. $a_n = \dfrac{n}{n + 1}$; a_{12} $\dfrac{12}{13}$

26. $a_n = (-1)^{n-1} n^2$; a_{15} 225

27. $a_n = (-1)^{n-1}(n - 1)$; a_{25} 24

28. $a_n = \left(\dfrac{1}{2}\right)^n$; a_8 $\dfrac{1}{256}$

29. $a_n = \left(\dfrac{2}{3}\right)^n$; a_5 $\dfrac{32}{243}$

30. $a_n = (n + 2)(n + 3)$; a_{17} 380

31. $a_n = (n + 4)(n + 1)$; a_7 88

32. $a_n = \dfrac{(-1)^{2n-1}}{n^2}$; a_6 $-\dfrac{1}{36}$

33. $a_n = \dfrac{(-1)^{2n}}{n + 4}$; a_{16} $\dfrac{1}{20}$

34. $a_n = \dfrac{3}{2} n^2 - 2$; a_8 94

35. $a_n = \dfrac{1}{3} n + n^2$; a_6 38

2   Find the sum of the series.

36. $\displaystyle\sum_{n=1}^{5}(2n+3)$ 45

37. $\displaystyle\sum_{i=1}^{7}(i+2)$ 42

38. $\displaystyle\sum_{i=1}^{4}2i$ 20

39. $\displaystyle\sum_{n=1}^{7}n$ 28

40. $\displaystyle\sum_{i=1}^{6}i^2$ 91

41. $\displaystyle\sum_{i=1}^{5}(i^2+1)$ 60

42. $\displaystyle\sum_{n=1}^{6}(-1)^n$ 0

43. $\displaystyle\sum_{n=1}^{4}\frac{1}{2n}$ $\dfrac{25}{24}$

44. $\displaystyle\sum_{i=3}^{6}i^3$ 432

45. $\displaystyle\sum_{n=2}^{4}2^n$ 28

46. $\displaystyle\sum_{n=3}^{7}\frac{n}{n-1}$ $\dfrac{129}{20}$

47. $\displaystyle\sum_{i=3}^{6}\frac{i+1}{i}$ $\dfrac{99}{20}$

48. $\displaystyle\sum_{i=1}^{4}\frac{1}{2^i}$ $\dfrac{15}{16}$

49. $\displaystyle\sum_{i=1}^{5}\frac{1}{2i}$ $\dfrac{137}{120}$

50. $\displaystyle\sum_{n=1}^{4}(-1)^{n-1}n^2$ -10

51. $\displaystyle\sum_{i=1}^{4}(-1)^{i-1}(i+1)$ -2

52. $\displaystyle\sum_{n=3}^{5}\frac{(-1)^{n-1}}{n-2}$ $\dfrac{5}{6}$

53. $\displaystyle\sum_{n=4}^{7}\frac{(-1)^{n-1}}{n-3}$ $-\dfrac{7}{12}$

Write the series in expanded form.

54. $\displaystyle\sum_{n=1}^{5}2x^n$

$2x+2x^2+2x^3+2x^4+2x^5$

55. $\displaystyle\sum_{n=1}^{4}\frac{2n}{x}$

$\dfrac{2}{x}+\dfrac{4}{x}+\dfrac{6}{x}+\dfrac{8}{x}$

56. $\displaystyle\sum_{i=1}^{5}\frac{x^i}{i}$

$x+\dfrac{x^2}{2}+\dfrac{x^3}{3}+\dfrac{x^4}{4}+\dfrac{x^5}{5}$

57. $\displaystyle\sum_{i=1}^{4}\frac{x^i}{i+1}$

$\dfrac{x}{2}+\dfrac{x^2}{3}+\dfrac{x^3}{4}+\dfrac{x^4}{5}$

58. $\displaystyle\sum_{i=3}^{5}\frac{x^i}{2i}$

$\dfrac{x^3}{6}+\dfrac{x^4}{8}+\dfrac{x^5}{10}$

59. $\displaystyle\sum_{i=2}^{4}\frac{x^i}{2i-1}$

$\dfrac{x^2}{3}+\dfrac{x^3}{5}+\dfrac{x^4}{7}$

60. $\displaystyle\sum_{n=1}^{5}x^{2n}$

$x^2+x^4+x^6+x^8+x^{10}$

61. $\displaystyle\sum_{n=1}^{4}x^{2n-1}$

$x+x^3+x^5+x^7$

62. $\displaystyle\sum_{i=1}^{4}\frac{x^i}{i^2}$

$x+\dfrac{x^2}{4}+\dfrac{x^3}{9}+\dfrac{x^4}{16}$

63. $\displaystyle\sum_{n=1}^{4}(2n)x^n$

$2x+4x^2+6x^3+8x^4$

64. $\displaystyle\sum_{n=1}^{4} nx^{n-1}$

$1 + 2x + 3x^2 + 4x^3$

65. $\displaystyle\sum_{i=1}^{5} x^{-i}$

$\dfrac{1}{x} + \dfrac{1}{x^2} + \dfrac{1}{x^3} + \dfrac{1}{x^4} + \dfrac{1}{x^5}$

SUPPLEMENTAL EXERCISES 12.1

Write a formula for the nth term of the sequence.

66. The sequence of the natural numbers
$a_n = n$

67. The sequence of the odd natural numbers
$a_n = 2n - 1$

68. The sequence of the negative even integers
$a_n = -2n$

69. The sequence of the negative odd integers
$a_n = -2n + 1$

70. The sequence of the multiples of 7
$a_n = 7n$

71. The sequence of the positive integers that are divisible by 4
$a_n = 4n$

72. The sequence of the negative integers less than -5
$a_n = -(n + 5)$

73. The sequence of the positive odd integers greater than 9
$a_n = 2n + 9$

74. The sequence for which a_1 is 8 and each term is 6 more than the preceding term
$a_n = 6n + 2$

75. The sequence for which a_1 is 3 and each term is 10 less than the preceding term
$a_n = -10n + 13$

Find the sum of the series. Write your answer as a single logarithm with a coefficient of 1.

76. $\displaystyle\sum_{n=1}^{2} \log n$
$\log 2$

77. $\displaystyle\sum_{i=1}^{2} \log 2i$
$\log 8$

78. $\displaystyle\sum_{n=1}^{5} n \log x$
$15 \log x$

79. $\displaystyle\sum_{i=3}^{4} 2i \log x$
$14 \log x$

Solve.

80. The first 22 numbers in the sequence $4, 44, 444, 4444, \ldots$ are added together. What digit is in the thousands' place of the sum?
4

81. The first 31 numbers in the sequence $6, 66, 666, 6666, \ldots$ are added together. What digit is in the hundreds' place of the sum?
3

SECTION 12.2

Arithmetic Sequences and Series

1 Find the nth term of an arithmetic sequence

A company's expenses for training a new employee are quite high. To encourage employees to continue their employment with the company, a company that has a six-month training program offers a starting salary of $400 a month and then a $100-per-month pay increase each month during the training period.

The sequence below shows the employee's monthly salaries during the training period. Each term of the sequence is found by adding $100 to the previous term.

Month	1	2	3	4	5	6
Salary	400	500	600	700	800	900

The sequence $400, 500, 600, 700, 800, 900$ is called an arithmetic sequence. An **arithmetic sequence,** or **arithmetic progression,** is one in which the difference between any two consecutive terms is constant. The difference between consecutive terms is called the **common difference** of the sequence.

Each of the sequences shown at the right is an arithmetic sequence. To find the common difference of an arithmetic sequence, subtract the first term from the second term.

$2, 7, 12, 17, 22, \ldots$ Common difference: 5

$3, 1, -1, -3, -5, \ldots$ Common difference: -2

$1, \dfrac{3}{2}, 2, \dfrac{5}{2}, 3, \dfrac{7}{2}, \ldots$ Common difference: $\dfrac{1}{2}$

Consider an arithmetic sequence in which the first term is a_1 and the common difference is d. By adding the common difference to each successive term of the arithmetic sequence, a formula for the nth term can be found.

The first term is a_1.

$$a_1 = a_1$$

To find the second term, add the common difference d to the first term.

$$a_2 = a_1 + d$$

To find the third term, add the common difference d to the second term.

$$a_3 = a_2 + d = (a_1 + d) + d$$
$$a_3 = a_1 + 2d$$

To find the fourth term, add the common difference d to the third term.

$$a_4 = a_3 + d = (a_1 + 2d) + d$$
$$a_4 = a_1 + 3d$$

Note the relationship between the term number and the number that multiplies d. The multiplier of d is one less than the term number.

$$a_n = a_1 + (n - 1)d$$

The Formula for the nth Term of an Arithmetic Sequence

The nth term of an arithmetic sequence with a common difference of d is given by $a_n = a + (n - 1)d$.

Example 1   Find the 27th term of the arithmetic sequence $-4, -1, 2, 5, 8, \ldots$.

Solution $d = a_2 - a_1 = -1 - (-4) = 3$ • Find the common difference.

$a_n = a_1 + (n - 1)d$ • Use the Formula for the nth Term of an
$a_{27} = -4 + (27 - 1)3$ Arithmetic Sequence to find the 27th term.
$\quad = -4 + (26)3 = -4 + 78$ $n = 27, a_1 = -4, d = 3$
$\quad = 74$

Problem 1   Find the 15th term of the arithmetic sequence $9, 3, -3, -9, \ldots$.

Solution   See page A100.
-75

Example 2   Find the formula for the nth term of the arithmetic sequence $-5, -2, 1, 4, \ldots$.

Solution $d = a_2 - a_1 = -2 - (-5) = 3$ • Find the common difference.

$a_n = a_1 + (n - 1)d$ • Use the Formula for the nth Term of an
$a_n = -5 + (n - 1)3$ Arithmetic Sequence. $a_1 = -5, d = 3$
$a_n = -5 + 3n - 3$
$a_n = 3n - 8$

Problem 2   Find the formula for the nth term of the arithmetic sequence $-3, 1, 5, 9, \ldots$.

Solution   See page A100.
$a_n = 4n - 7$

Example 3   Find the number of terms in the finite arithmetic sequence $7, 10, 13, \ldots, 55$.

Solution $d = a_2 - a_1 = 10 - 7 = 3$ • Find the common difference.

$a_n = a_1 + (n - 1)d$ • Use the Formula for the nth Term of an Arith-
$55 = 7 + (n - 1)3$ metic Sequence. $a_n = 55, a_1 = 7, d = 3$
$55 = 7 + 3n - 3$ • Solve for n.
$55 = 3n + 4$
$51 = 3n$
$17 = n$

There are 17 terms in the sequence.

Problem 3   Find the number of terms in the finite arithmetic sequence $7, 9, 11, \ldots, 59$.

Solution   See page A100.

27 terms

2   Find the sum of an arithmetic series

The indicated sum of the terms of an arithmetic sequence is called an **arithmetic series.** The sum of an arithmetic series can be found by using a formula.

> **The Formula for the Sum of n Terms of an Arithmetic Series**
>
> Let a_1 be the first term of a finite arithmetic sequence, n the number of terms, and a_n the last term of the sequence. Then the sum of the series S_n is given by $S_n = \dfrac{n}{2}(a_1 + a_n)$.

Each term of the arithmetic sequence shown at the $2, 5, 8, \ldots, 17, 20$
right was found by adding 3 to the previous term.

Each term of the reverse arithmetic sequence can be $20, 17, 14, \ldots, 5, 2$
found by subtracting 3 from the previous term.

This idea is used in the following proof of the Formula for the Sum of n Terms of an Arithmetic Series:

Let S_n represent the sum of the sequence.

$$S_n = a_1 + (a_1 + d) + (a_1 + 2d) + \cdots + a_n$$

Write the terms of the sum of the sequence in reverse order. The sum will be the same.

$$S_n = a_n + (a_n - d) + (a_n - 2d) + \cdots + a_1$$

Add the two equations.

$$2S_n = (a_1 + a_n) + (a_1 + a_n) + (a_1 + a_n) + \cdots + (a_1 + a_n)$$

Simplify the right side of the equation by using the fact that there are n terms in the sequence.

$$2S_n = n(a_1 + a_n)$$

Solve for S_n.

$$S_n = \frac{n}{2}(a_1 + a_n)$$

Example 4 Find the sum of the first 10 terms of the arithmetic sequence $2, 4, 6, 8, \ldots$.

Solution $d = a_2 - a_1 = 4 - 2 = 2$ • Find the common difference.

$$a_n = a_1 - (n - 1)d$$
$$a_{10} = 2 + (10 - 1)2 = 2 + (9)2$$
$$= 2 + 18 = 20$$

• Use the Formula for the nth Term of an Arithmetic Sequence to find the 10th term.

$$S_n = \frac{n}{2}(a_1 + a_n)$$

• Use the Formula for the Sum of n Terms of an Arithmetic Series. $n = 10, a_1 = 2, a_n = 20$

$$S_{10} = \frac{10}{2}(2 + 20) = 5(22) = 110$$

Problem 4 Find the sum of the first 25 terms of the arithmetic sequence $-4, -2, 0, 2, 4, \ldots$.

Solution See page A100.

500

Example 5   Find the sum of the arithmetic series $\sum_{n=1}^{25} (3n + 1)$.

Solution $a_n = 3n + 1$
$a_1 = 3(1) + 1 = 4$ • Find the first term.
$a_{25} = 3(25) + 1 = 76$ • Find the 25th term.

$S_n = \dfrac{n}{2}(a_1 + a_n)$ • Use the Formula for the Sum of n Terms of an Arithmetic Series. $n = 25, a_1 = 4, a_n = 76$

$S_{25} = \dfrac{25}{2}(4 + 76) = \dfrac{25}{2}(80)$

$\quad\ = 1000$

Problem 5   Find the sum of the arithmetic series $\sum_{n=1}^{18} (3n - 2)$.

Solution   See page A101.
477

3   Application problems

Example 6   The distance a ball rolls down a ramp each second is given by an arithmetic sequence. The distance in feet traveled by the ball during the nth second is given by $2n - 1$. Find the distance the ball will travel during the 10th second.

Strategy   To find the distance:
▶ Find the first and second terms of the sequence.
▶ Find the common difference of the arithmetic sequence.
▶ Use the Formula for the nth Term of an Arithmetic Sequence to find the 10th term.

Solution $a_n = 2n - 1$
$a_1 = 2(1) - 1 = 1$
$a_2 = 2(2) - 1 = 3$

$d = a_2 - a_1 = 3 - 1 = 2$

$a_n = a_1 + (n - 1)d$
$a_{10} = 1 + (10 - 1)2 = 1 + (9)2 = 19$

The ball will travel 19 ft during the 10th second.

Problem 6   A contest offers 20 prizes. The first prize is \$10,000, and each successive prize is \$300 less than the preceding prize. What is the value of the 20th-place prize? What is the total amount of prize money that is being awarded?

Solution   See page A101.

20th-place prize: \$4300   total amount awarded: \$143,000

EXERCISES 12.2

1   Find the indicated term of the arithmetic sequence.

1. $1, 11, 21, \ldots; a_{15}$ 141

2. $3, 8, 13, \ldots; a_{20}$ 98

3. $-6, -2, 2, \ldots; a_{15}$ 50

4. $-7, -2, 3, \ldots; a_{14}$ 58

5. $3, 7, 11, \ldots; a_{18}$ 71

6. $-13, -6, 1, \ldots; a_{31}$ 197

7. $-\frac{3}{4}, 0, \frac{3}{4}, \ldots; a_{11}$ $8\frac{1}{4}$

8. $\frac{3}{8}, 1, \frac{13}{8}, \ldots; a_{17}$ $10\frac{3}{8}$

9. $2, \frac{5}{2}, 3, \ldots; a_{31}$ 17

10. $1, \frac{5}{4}, \frac{3}{2}, \ldots; a_{17}$ 5

11. $6, 5.75, 5.50, \ldots; a_{10}$ 3.75

12. $4, 3.7, 3.4, \ldots; a_{12}$ 0.7

Find the formula for the nth term of the arithmetic sequence.

13. $1, 2, 3, \ldots$
$a_n = n$

14. $1, 4, 7, \ldots$
$a_n = 3n - 2$

15. $6, 2, -2, \ldots$
$a_n = -4n + 10$

16. $3, 0, -3, \ldots$
$a_n = -3n + 6$

17. $2, \frac{7}{2}, 5, \ldots$
$a_n = \dfrac{3n + 1}{2}$

18. $7, 4.5, 2, \ldots$
$a_n = -2.5n + 9.5$

19. $-8, -13, -18, \ldots$
$a_n = -5n - 3$

20. $17, 30, 43, \ldots$
$a_n = 13n + 4$

21. $26, 16, 6, \ldots$
$a_n = -10n + 36$

Find the number of terms in the finite arithmetic sequence.

22. $-2, 1, 4, \ldots, 73$ 26

23. $7, 11, 15, \ldots, 171$ 42

24. $-\frac{1}{2}, \frac{3}{2}, \frac{7}{2}, \ldots, \frac{71}{2}$ 19

25. $\frac{1}{3}, \frac{5}{3}, 3, \ldots, \frac{61}{3}$ 16

26. $1, 5, 9, \ldots, 81$ 21

27. $3, 8, 13, \ldots, 98$ 20

28. $2, 0, -2, \ldots, -56$ 30

29. $1, -3, -7, \ldots, -75$ 20

30. $\frac{5}{2}, 3, \frac{7}{2}, \ldots, 13$ 22

31. $\frac{7}{3}, \frac{13}{3}, \frac{19}{3}, \ldots, \frac{79}{3}$ 13

32. $1, 0.75, 0.50, \ldots, -4$ 21

33. $3.5, 2, 0.5, \ldots, -25$ 20

2 Find the sum of the indicated number of terms of the arithmetic sequence.

34. $1, 3, 5, \ldots; n = 50$ 2500

35. $2, 4, 6, \ldots; n = 25$ 650

36. $20, 18, 16, \ldots; n = 40$ -760

37. $25, 20, 15, \ldots; n = 22$ -605

38. $\frac{1}{2}, 1, \frac{3}{2}, \ldots; n = 27$ 189

39. $2, \frac{11}{4}, \frac{7}{2}, \ldots; n = 10$ $\frac{215}{4}$

Find the sum of the arithmetic series.

40. $\sum\limits_{i=1}^{15} (3i - 1)$ 345

41. $\sum\limits_{i=1}^{15} (3i + 4)$ 420

42. $\sum\limits_{n=1}^{17} \left(\frac{1}{2}n + 1\right)$ $\frac{187}{2}$

43. $\sum\limits_{n=1}^{10} (1 - 4n)$ -210

44. $\sum\limits_{i=1}^{15} (4 - 2i)$ -180

45. $\sum\limits_{n=1}^{10} (5 - n)$ -5

3 Solve.

46. The distance that an object dropped from a cliff will fall is 16 ft the first second, 48 ft the next second, 80 ft the third second, and so on in an arithmetic sequence. What is the total distance the object will fall in 6 s? 576 ft

47. An exercise program calls for walking 10 min each day for a week. Each week thereafter, the amount of time spent walking increases by 5 min per day. In how many weeks will a person be walking 60 min each day? 11 weeks

48. A display of cans in a grocery store consists of 24 cans in the bottom row, 21 cans in the next row, and so on in an arithmetic sequence. The top row has 3 cans. Find the total number of cans in the display. 108 cans

49. A theater in the round has 68 seats in the first row, 80 seats in the second row, 92 seats in the third row, and so on in an arithmetic sequence. Find the total number of seats in the theater if there are 22 rows of seats. 4268 seats

50. The loge seating section in a concert hall consists of 27 rows of chairs. There are 73 seats in the first row, 79 seats in the second row, 85 seats in the third row, and so on in an arithmetic sequence. How many seats are in the loge seating section? 4077 seats

51. The salary schedule for an engineering assistant is $700 for the first month and a $35-per-month salary increase for the next eight months. Find the monthly salary during the eighth month. Find the total salary for the eight-month period. $945; $6580

SUPPLEMENTAL EXERCISES 12.2

Solve.

52. Find the sum of the first 50 positive integers.
1275

53. Find the sum of the first 100 natural numbers.
5050

54. How many terms of the arithmetic sequence $-6, -2, 2, \ldots$ must be added together for the sum of the series to be 90?
9

55. How many terms of the arithmetic sequence $-3, 2, 7, \ldots$ must be added together for the sum of the series to be 116?
8

56. Given $a_1 = -9$, $a_n = 21$, and $S_n = 36$, find d and n.
$d = 6; n = 6$

57. Given $a_1 = -5$, $a_n = 19$, and $S_n = 49$, find d and n.
$d = 4; n = 7$

58. The third term of an arithmetic sequence is 4, and the eighth term is 30. Find the first term.
$-\dfrac{32}{5}$

59. The fourth term of an arithmetic sequence is 9, and the ninth term is 29. Find the first term.
-3

SECTION 12.3

Geometric Sequences and Series

1 Find the nth term of a geometric sequence

An ore sample contains 20 mg of a radioactive material with a half-life of one week. The amount of the radioactive material that the sample contains at the beginning of each week can be determined by using the exponential decay equation used in Chapter 11.

The sequence below represents the amount in the sample at the beginning of each week. Each term of the sequence is found by multiplying the preceding term by $\frac{1}{2}$.

Week	1	2	3	4	5
Amount	20	10	5	2.5	1.25

The sequence $20, 10, 5, 2.5, 1.25$ is called a geometric sequence. A **geometric sequence,** or **geometric progression,** is one in which each successive term of the sequence is the same nonzero constant multiple of the preceding term. The common multiple is called the **common ratio** of the sequence.

Each of the sequences shown at the right is a geometric sequence. To find the common ratio of a geometric sequence, divide the second term of the sequence by the first term.

$3, 6, 12, 24, 48, \ldots$ Common ratio: 2

$4, -12, 36, -108, 324, \ldots$ Common ratio: -3

$6, 4, \dfrac{8}{3}, \dfrac{16}{9}, \dfrac{32}{27}, \ldots$ Common ratio: $\dfrac{2}{3}$

Consider a geometric sequence in which the first term is a_1 and the common ratio is r. By multiplying each successive term of the geometric sequence by the common ratio, a formula for the nth term can be found.

The first term is a_1.

$$a_1 = a_1$$

To find the second term, multiply the first term by the common ratio r.

$$a_2 = a_1 r$$

To find the third term, multiply the second term by the common ratio r.

$$a_3 = (a_1 r)r$$
$$a_3 = a_1 r^2$$

To find the fourth term, multiply the third term by the common ratio r.

$$a_4 = (a_1 r^2)r$$
$$a_4 = a_1 r^3$$

Note the relationship between the term number and the number that is the exponent on r. The exponent on r is one less than the term number.

$$a_n = a_1 r^{n-1}$$

The Formula for the nth Term of a Geometric Sequence

The nth term of a geometric sequence with first term a_1 and common ratio r is given by $a_n = a_1 r^{n-1}$.

Example 1 Find the 6th term of the geometric sequence $3, 6, 12, \ldots$.

Solution $r = \dfrac{a_2}{a_1} = \dfrac{6}{3} = 2$ • Find the common ratio.

$a_n = a_1 r^{n-1}$ • Use the Formula for the nth Term of a Geo-
$a_6 = 3(2)^{6-1} = 3(2)^5 = 3(32)$ metric Sequence. $n = 6, a_1 = 3, r = 2$
   $= 96$

Problem 1 Find the 5th term of the geometric sequence $5, 2, \dfrac{4}{5}, \ldots$.

Solution See page A102. $\dfrac{16}{125}$

Example 2 Find a_3 for the geometric sequence $8, a_2, a_3, 27, \ldots$.

Solution $a_n = a_1 r^{n-1}$ • Find the common ratio. $a_4 = 27, a_1 = 8, n = 4$

$a_4 = a_1 r^{4-1}$
$27 = 8r^{4-1}$

$\dfrac{27}{8} = r^3$

$\dfrac{3}{2} = r$

$a_3 = 8\left(\dfrac{3}{2}\right)^{3-1} = 8\left(\dfrac{3}{2}\right)^2 = 8\left(\dfrac{9}{4}\right)$ • Use the Formula for the nth Term of a Geo-
$\qquad = 18$ metric Sequence.

Problem 2 Find a_3 for the geometric sequence $3, a_2, a_3, -192, \ldots$.

Solution See page A102.
48

2 Finite geometric series

The indicated sum of the terms of a geometric sequence is called a **geometric series.** The sum of a geometric series can be found by a formula.

> **The Formula for the Sum of n Terms of a Finite Geometric Series**
>
> Let a_1 be the first term of a finite geometric sequence, n the number of terms, r the common ratio, and $r \neq 1$. The sum of the series S_n is given by $S_n = \dfrac{a_1(1 - r^n)}{1 - r}$.

Proof of the Formula for the Sum of n Terms of a Finite Geometric Series:

Let S_n represent the sum of n terms of the sequence.

$$S_n = a_1 + a_1 r + a_1 r^2 + \cdots + a_1 r^{n-2} + a_1 r^{n-1}$$

Multiply each side of the equation by r.

$$rS_n = a_1 r + a_1 r^2 + a_1 r^3 + \cdots + a_1 r^{n-1} + a_1 r^n$$

Subtract the two equations.

$$S_n - rS_n = a_1 - a_1 r^n$$

Assuming $r \neq 1$, solve for S_n.

$$(1 - r)S_n = a_1(1 - r^n)$$
$$S_n = \dfrac{a_1(1 - r^n)}{1 - r}$$

Example 3   Find the sum of the geometric sequence $2, 8, 32, 128, 512$.

Solution $r = \dfrac{a_2}{a_1} = \dfrac{8}{2} = 4$ • Find the common ratio.

$S_n = \dfrac{a_1(1 - r^n)}{1 - r}$ • Use the Formula for the Sum of n Terms of a Finite Geometric Series. $n = 5, a_1 = 2, r = 4$

$S_5 = \dfrac{2(1 - 4^5)}{1 - 4} = \dfrac{2(1 - 1024)}{-3}$

$= \dfrac{2(-1023)}{-3} = \dfrac{-2046}{-3} = 682$

Problem 3   Find the sum of the geometric sequence $1, -\dfrac{1}{3}, \dfrac{1}{9}, -\dfrac{1}{27}$.

Solution   See page A102.

$\dfrac{20}{27}$

Example 4   Find the sum of the geometric series $\displaystyle\sum_{n=1}^{10} (-20)(-2)^{n-1}$.

Solution $a_n = (-20)(-2)^{n-1}$
$a_1 = (-20)(-2)^{1-1} = (-20)(-2)^0$ • Find the first term.
$= (-20)(1) = -20$

$a_2 = (-20)(-2)^{2-1} = (-20)(-2)^1$ • Find the second term.
$= (-20)(-2) = 40$

$r = \dfrac{a_2}{a_1} = \dfrac{40}{-20} = -2$ • Find the common ratio.

$S_n = \dfrac{a_1(1 - r^n)}{1 - r}$ • Use the Formula for the Sum of n Terms of a Finite Geometric Series. $n = 10, a_1 = -20, r = -2$

$S_{10} = \dfrac{-20[1 - (-2)^{10}]}{1 - (-2)} = \dfrac{-20(1 - 1024)}{3}$

$= \dfrac{-20(-1023)}{3} = \dfrac{20{,}460}{3} = 6820$

Problem 4   Find the sum of the geometric series $\displaystyle\sum_{n=1}^{5} \left(\dfrac{1}{2}\right)^n$.

Solution   See page A102.

$\dfrac{31}{32}$

3   Infinite geometric series

When the absolute value of the common ratio of a geometric sequence is less than 1, $|r| < 1$, then as n becomes larger, r^n becomes closer to zero.

Examples of geometric sequences for which $|r| < 1$ are shown at the right. As the number of terms increases, the value of the last term listed is closer to zero.

$$1, \frac{1}{3}, \frac{1}{9}, \frac{1}{27}, \frac{1}{81}, \frac{1}{243}, \cdots$$

$$1, -\frac{1}{2}, \frac{1}{4}, -\frac{1}{8}, \frac{1}{16}, -\frac{1}{32}, \cdots$$

The indicated sum of the terms of an infinite geometric sequence is called an **infinite geometric series.**

An example of an infinite geometric series is shown at the right. The first term is 1. The common ratio is $\frac{1}{3}$.

$$1 + \frac{1}{3} + \frac{1}{9} + \frac{1}{27} + \frac{1}{81} + \frac{1}{243} + \cdots$$

The sum of the first 5, 7, 12, and 15 terms, along with the values of r^n, are shown at the right. As n increases, the sum of the terms is closer to 1.5, and the value of r^n is closer to zero.

n	S_n	r^n
5	1.4938272	0.0041152
7	1.4993141	0.0004572
12	1.4999972	0.0000019
15	1.4999999	0.0000001

Using the Formula for the Sum of n Terms of a Geometric Series and the fact that r^n approaches zero when $|r| < 1$ and n increases, a formula for an infinite geometric series can be found.

The sum of the first n terms of a geometric series is shown at the right. If $|r| < 1$, then r^n can be made very close to zero by using larger and larger values of n. Therefore, the sum of the first n terms is approximately $\frac{a_1}{1 - r}$.

Approximately zero

$$S_n = \frac{a_1(1 - r^n)}{1 - r}$$

$$S_n \approx \frac{a_1(1 - 0)}{1 - r} \approx \frac{a_1}{1 - r}$$

> **The Formula for the Sum of an Infinite Geometric Series**
>
> The sum of an infinite geometric series in which $|r| < 1$ and a_1 is the first term is given by $S = \dfrac{a_1}{1 - r}$.

When $|r| \geq 1$, the infinite geometric series does not have a sum. For example, the sum of the infinite geometric series $1 + 2 + 4 + 8 + \cdots$ increases without limit.

Example 5   Find the sum of the infinite geometric sequence $1, -\dfrac{1}{2}, \dfrac{1}{4}, -\dfrac{1}{8}, \ldots$.

Solution $S = \dfrac{a_1}{1 - r} = \dfrac{1}{1 - \left(-\dfrac{1}{2}\right)} = \dfrac{1}{\dfrac{3}{2}} = \dfrac{2}{3}$ • The common ratio is $-\dfrac{1}{2}$. $\left|-\dfrac{1}{2}\right| < 1$.
Use the Formula for the Sum of an Infinite Geometric Series.

Problem 5   Find the sum of the infinite geometric sequence $3, -2, \dfrac{4}{3}, -\dfrac{8}{9}, \ldots$.

Solution   See page A103. $\dfrac{9}{5}$

The sum of an infinite geometric series can be applied to nonterminating, repeating decimals.

The repeating decimal shown at the right has been rewritten as an infinite geometric series, with the first term $\dfrac{3}{10}$ and common ratio $\dfrac{1}{10}$.

$0.33\overline{3} = 0.3 + 0.03 + 0.003 + \cdots$
$= \dfrac{3}{10} + \dfrac{3}{100} + \dfrac{3}{1000} + \cdots$

Use the Formula for the Sum of an Infinite Geometric Series.

$S = \dfrac{a_1}{1 - r} = \dfrac{\dfrac{3}{10}}{1 - \dfrac{1}{10}} = \dfrac{\dfrac{3}{10}}{\dfrac{9}{10}} = \dfrac{3}{9} = \dfrac{1}{3}$

This method can be used to find an equivalent numerical fraction for any repeating decimal.

Example 6   Find an equivalent fraction for $0.122\overline{2}$.

Solution $0.122\overline{2}$
$0.1 + 0.02 + 0.002 + 0.0002 + \cdots =$
$\dfrac{1}{10} + \dfrac{2}{100} + \dfrac{2}{1000} + \dfrac{2}{10,000} + \cdots$

• Write the decimal as an infinite geometric series. The geometric series does not begin with the first term. The series begins with $\dfrac{2}{100}$. The common ratio is $\dfrac{1}{10}$.

$S = \dfrac{a_1}{1 - r} = \dfrac{\dfrac{2}{100}}{1 - \dfrac{1}{10}} = \dfrac{\dfrac{2}{100}}{\dfrac{9}{10}} = \dfrac{2}{90}$

• Use the Formula for the Sum of an Infinite Geometric Series.

$0.122\overline{2} = \dfrac{1}{10} + \dfrac{2}{90} = \dfrac{11}{90}$

• Add $\dfrac{1}{10}$ to the sum of the series.

An equivalent fraction is $\dfrac{11}{90}$.

Problem 6   Find an equivalent fraction for $0.36\overline{36}$.

Solution   See page A103.

$$\frac{4}{11}$$

4 Application problems

Example 7   On the first swing, the length of the arc through which a pendulum swings is 16 in. The length of each successive swing is $\frac{7}{8}$ of the preceding swing. Find the length of the arc on the fifth swing. Round to the nearest tenth.

Strategy   To find the length of the arc on the fifth swing, use the Formula for the nth Term of a Geometric Sequence. $n = 5, a_1 = 16, r = \frac{7}{8}$

Solution $a_n = a_1 r^{n-1}$

$$a_5 = 16\left(\frac{7}{8}\right)^{5-1} = 16\left(\frac{7}{8}\right)^4 = 16\left(\frac{2401}{4096}\right) = \frac{38,416}{4096} \approx 9.4$$

The length of the arc on the fifth swing is 9.4 in.

Problem 7   You start a chain letter and send it to three friends. Each of the three friends sends the letter to three other friends, and the sequence is repeated. Assuming no one breaks the chain, how many letters will have been mailed from the first through the sixth mailings?

Solution   See page A103.
1092 letters

EXERCISES 12.3

1   Find the indicated term of the geometric sequence.

1. $2, 8, 32, \ldots; a_9$ 131,072

2. $4, 3, \frac{9}{4}, \ldots; a_8$ $\frac{2187}{4096}$

3. $6, -4, \frac{8}{3}, \ldots; a_7$ $\frac{128}{243}$

4. $-5, 15, -45, \ldots; a_7$ -3645

5. $1, \sqrt{2}, 2, \ldots; a_9$ 16

6. $3, 3\sqrt{3}, 9, \ldots; a_8$ $81\sqrt{3}$

Find a_2 and a_3 for the geometric sequence.

7. $9, a_2, a_3, \dfrac{8}{3}, \ldots$ 6, 4

8. $8, a_2, a_3, \dfrac{27}{8}, \ldots$ $6, \dfrac{9}{2}$

9. $3, a_2, a_3, -\dfrac{8}{9}, \ldots$ $-2, \dfrac{4}{3}$

10. $6, a_2, a_3, -48, \ldots$ $-12, 24$

11. $-3, a_2, a_3, 192, \ldots$ $12, -48$

12. $5, a_2, a_3, 625, \ldots$ $25, 125$

2   Find the sum of the indicated number of terms of the geometric sequence.

13. $2, 6, 18, \ldots; n = 7$ 2186

14. $-4, 12, -36, \ldots; n = 7$ -2188

15. $12, 9, \dfrac{27}{4}, \ldots; n = 5$ $\dfrac{2343}{64}$

16. $3, 3\sqrt{2}, 6, \ldots; n = 12$ $189 + 189\sqrt{2}$

Find the sum of the geometric series.

17. $\displaystyle\sum_{i=1}^{5} (2)^i$ 62

18. $\displaystyle\sum_{n=1}^{6} \left(\dfrac{3}{2}\right)^n$ $\dfrac{1995}{64}$

19. $\displaystyle\sum_{i=1}^{5} \left(\dfrac{1}{3}\right)^i$ $\dfrac{121}{243}$

20. $\displaystyle\sum_{n=1}^{6} \left(\dfrac{2}{3}\right)^n$ $\dfrac{1330}{729}$

21. $\displaystyle\sum_{i=1}^{5} (4)^i$ 1364

22. $\displaystyle\sum_{n=1}^{8} (3)^n$ 9840

23. $\displaystyle\sum_{i=1}^{4} (7)^i$ 2800

24. $\displaystyle\sum_{n=1}^{5} (5)^n$ 3905

25. $\displaystyle\sum_{i=1}^{5} \left(\dfrac{3}{4}\right)^i$ $\dfrac{2343}{1024}$

26. $\displaystyle\sum_{n=1}^{3} \left(\dfrac{7}{4}\right)^n$ $\dfrac{651}{64}$

27. $\displaystyle\sum_{i=1}^{4} \left(\dfrac{5}{3}\right)^i$ $\dfrac{1360}{81}$

28. $\displaystyle\sum_{n=1}^{6} \left(\dfrac{1}{2}\right)^n$ $\dfrac{63}{64}$

3   Find the sum of the infinite geometric series.

29. $3 + 2 + \dfrac{4}{3} + \cdots$ 9

30. $2 - \dfrac{1}{4} + \dfrac{1}{32} + \cdots$ $\dfrac{16}{9}$

31. $6 - 4 + \dfrac{8}{3} + \cdots$ $\dfrac{18}{5}$

32. $\dfrac{1}{10} + \dfrac{1}{100} + \dfrac{1}{1000} + \cdots$ $\dfrac{1}{9}$

33. $\dfrac{7}{10} + \dfrac{7}{100} + \dfrac{7}{1000} + \cdots$ $\dfrac{7}{9}$

34. $\dfrac{5}{100} + \dfrac{5}{10,000} + \dfrac{5}{1,000,000} + \cdots$ $\dfrac{5}{99}$

Find an equivalent fraction for the repeating decimal.

35. $0.88\overline{8}$ $\dfrac{8}{9}$

36. $0.55\overline{5}$ $\dfrac{5}{9}$

37. $0.22\overline{2}$ $\dfrac{2}{9}$

38. $0.99\overline{9}$ 1

39. $0.45\overline{45}$ $\dfrac{5}{11}$

40. $0.18\overline{18}$ $\dfrac{2}{11}$

41. $0.166\overline{6}$ $\dfrac{1}{6}$

42. $0.833\overline{3}$ $\dfrac{5}{6}$

4 Solve.

Use your calculator for the following exercises.

43. A laboratory ore sample contains 400 mg of a radioactive material with a half-life of 1 h. Find the amount of radioactive material in the sample at the beginning of the fourth hour.
50 mg

44. On the first swing, the length of the arc through which a pendulum swings is 20 in. The length of each successive swing is $\frac{4}{5}$ of the preceding swing. What is the total distance the pendulum has traveled during the four swings? Round to the nearest tenth.
59.0 in.

45. To test the bounce of a tennis ball, the ball is dropped from a height of 10 ft. The ball bounces 75% of its previous height with each bounce. How high does the ball bounce on the sixth bounce? Round to the nearest tenth.
1.8 ft

46. The temperature of a hot water spa is 80°F. Each hour the temperature is 5% higher than during the previous hour. Find the temperature of the spa after 5 h. Round to the nearest tenth.
102.1°F

47. A real estate broker estimates that a piece of land will increase in value at a rate of 15% each year. If the original value of the land is $25,000, what will be its value in 10 years?
$101,138.92

48. Suppose an employee receives a wage of 1¢ for the first day of work, 2¢ the second day, 4¢ the third day, and so on in a geometric sequence. Find the total amount of money earned for working 30 days.
$10,737,418.23

49. Assume the average value of a home increases 5% per year. How much would a house costing $100,000 be worth in 30 years?
$411,613.50

50. A culture of bacteria doubles every 3 h. If there were 1000 bacteria at the beginning, how many bacteria will there be after 24 h?
128,000 bacteria

SUPPLEMENTAL EXERCISES 12.3

State whether the sequence is arithmetic (A), geometric (G), or neither (N),
and write the next term in the sequence.

51. $4, -2, 1, \ldots$ (G), $-\dfrac{1}{2}$

52. $-8, 0, 8, \ldots$ (A), 16

53. $5, 6.5, 8, \ldots$ (A), 9.5

54. $-7, 14, -28, \ldots$ (G), 56

55. $1, 4, 9, 16, \ldots$ (N), 25

56. $\sqrt{1}, \sqrt{2}, \sqrt{3}, \sqrt{4}$ (N), $\sqrt{5}$

57. $x^8, x^6, x^4, \ldots$ (G), x^2

58. $5a^2, 3a^2, a^2, \ldots$ (A), $-a^2$

59. $\log x, 2 \log x, 3 \log x, \ldots$ (A), $4 \log x$

60. $\log x, 3 \log x, 9 \log x, \ldots$ (G), $27 \log x$

Solve.

61. The third term of a geometric sequence is 3, and the sixth term is $\dfrac{1}{9}$. Find the first term.
27

62. The fourth term of a geometric sequence is -8, and the seventh term is -64. Find the first term.
-1

63. Given $a_n = 5$, $r = \dfrac{1}{2}$, and $S_n = 155$ for a geometric sequence, find a_1 and n.
$a_1 = 80$; $n = 5$

64. Given $a_n = 162$, $r = -3$, and $S_n = 122$ for a geometric sequence, find a_1 and n.
$a_1 = 2$; $n = 5$

SECTION 12.4 _____

Binomial Expansions

1 Expand $(a + b)^n$

By carefully observing the series for each expansion of the binomial $(a + b)^n$ shown below, it is possible to identify some interesting patterns.

$$(a + b)^1 = a + b$$

$$(a + b)^2 = a^2 + 2ab + b^2$$

$$(a + b)^3 = a^3 + 3a^2b + 3ab^2 + b^3$$

$$(a + b)^4 = a^4 + 4a^3b + 6a^2b^2 + 4ab^3 + b^4$$

$$(a + b)^5 = a^5 + 5a^4b + 10a^3b^2 + 10a^2b^3 + 5ab^4 + b^5$$

> **Patterns for the Variable Part of the Expansion of $(a + b)^n$**
> 1. The first term is a^n. The exponent on a decreases by 1 for each successive term.
> 2. The exponent on b increases by 1 for each successive term. The last term is b^n.
> 3. The degree of each term is n.

The variable parts of the terms of the expansion of $(a + b)^6$ are shown below.

$$a^6, a^5b, a^4b^2, a^3b^3, a^2b^4, ab^5, b^6$$

The first term is a^6. For each successive term, the exponent on a decreases by 1, and the exponent on b increases by 1. The last term is b^6.

The variable parts of the general expression of $(a + b)^n$ are

$$a^n, a^{n-1}b, a^{n-2}b^2, \ldots ab^{n-1}, b^n$$

A pattern for the coefficients of the terms of the expanded binomial can be found by writing the coefficients in a triangular array, which is known as **Pascal's Triangle.**

Each row begins and ends with the number 1. Any other number in a row is the sum of the two closest numbers above it. For example, $4 + 6 = 10$.

For $(a + b)^1$:   1   1
For $(a + b)^2$:   1   2   1
For $(a + b)^3$:   1   3   3   1
For $(a + b)^4$:   1   4   6   4   1
For $(a + b)^5$: 1   5   10   10   5   1

To write the sixth row of Pascal's Triangle, first write the numbers of the fifth row. The first and last numbers of the sixth row are 1. Each of the other numbers of the sixth row can be obtained by finding the sum of the two closest numbers above it in the fifth row.

1   5   10   10   5   1
1   6   15   20   15   6   1

These numbers will be the coefficients of the terms of the expansion of $(a + b)^6$.

Using the numbers of the sixth row of Pascal's Triangle for the coefficients and the pattern for the variable part of each term, the expanded form of $(a + b)^6$ can be written

$$(a + b)^6 = a^6 + 6a^5b + 15a^4b^2 + 20a^3b^3 + 15a^2b^4 + 6ab^5 + b^6$$

Although Pascal's Triangle can be used to find the coefficients for the expanded form of the power of any binomial, this method is inconvenient when the power of the binomial is large. An alternative method for determining these coefficients is based on the concept of a **factorial**.

Definition of Factorial

$n!$ (read "n factorial") is the product of the first n consecutive natural numbers.

$$n! = n(n - 1)(n - 2)\cdots 3 \cdot 2 \cdot 1$$

0! is defined to be 1: $0! = 1$

$$5! = 5 \cdot 4 \cdot 3 \cdot 2 \cdot 1 = 120$$

$$7! = 7 \cdot 6 \cdot 5 \cdot 4 \cdot 3 \cdot 2 \cdot 1 = 5040$$

To evaluate 6!, write the factorial as a product. Then simplify.

$$6! = 6 \cdot 5 \cdot 4 \cdot 3 \cdot 2 \cdot 1$$
$$= 720$$

Example 1   Evaluate: $\dfrac{7!}{4!\,3!}$

Solution $\dfrac{7!}{4!\,3!} = \dfrac{7 \cdot 6 \cdot 5 \cdot 4 \cdot 3 \cdot 2 \cdot 1}{(4 \cdot 3 \cdot 2 \cdot 1)(3 \cdot 2 \cdot 1)}$ • Write each factorial as a product.

$= 35$ • Simplify.

Problem 1   Evaluate: $\dfrac{12!}{7!\,5!}$

Solution   See page A104.
792

The coefficients in a binomial expansion can be given in terms of factorials.

Note that in the expansion of $(a + b)^5$, the coefficient of a^2b^3 can be given by $\dfrac{5!}{2!\,3!}$. The numerator is the factorial of the power of the binomial. The denominator is the product of the factorials of the exponents on a and b.

$(a + b)^5 =$
$a^5 + 5a^4b + 10a^3b^2 + 10a^2b^3 + 5ab^4 + b^5$

$\dfrac{5!}{2!\,3!} = \dfrac{5 \cdot 4 \cdot 3 \cdot 2 \cdot 1}{(2 \cdot 1)(3 \cdot 2 \cdot 1)} = 10$

In general, the coefficients of $(a + b)^n$ are given as the quotients of factorials. The **coefficient of** $a^{n-r}b^r$ **is** $\dfrac{n!}{(n-r)!\,r!}$. The symbol $\dbinom{n}{r}$ is used to express this quotient of factorials.

$$\binom{n}{r} = \frac{n!}{(n-r)!\,r!}$$

Example 2   Evaluate: $\dbinom{8}{5}$

Solution $\dbinom{8}{5} = \dfrac{8!}{(8-5)!\,5!} = \dfrac{8!}{3!\,5!}$ • Write the quotient of the factorials.

$$= \frac{8 \cdot 7 \cdot 6 \cdot 5 \cdot 4 \cdot 3 \cdot 2 \cdot 1}{(3 \cdot 2 \cdot 1)(5 \cdot 4 \cdot 3 \cdot 2 \cdot 1)} = 56 \quad \text{• Simplify.}$$

Problem 2   Evaluate: $\dbinom{7}{0}$

Solution   See page A104.
1

Using factorials and the pattern for the variable part of each term, a formula for any natural number power of a binomial can be written.

> **The Binomial Expansion Formula**
>
> $(a + b)^n =$
>
> $$\binom{n}{0}a^n + \binom{n}{1}a^{n-1}b + \binom{n}{2}a^{n-2}b^2 + \cdots + \binom{n}{r}a^{n-r}b^r + \cdots + \binom{n}{n}b^n$$

The Binomial Expansion Formula is used below to expand $(a + b)^7$.

$(a + b)^7 =$

$$\binom{7}{0}a^7 + \binom{7}{1}a^6b + \binom{7}{2}a^5b^2 + \binom{7}{3}a^4b^3 + \binom{7}{4}a^3b^4 + \binom{7}{5}a^2b^5 + \binom{7}{6}ab^6 + \binom{7}{7}b^7 =$$

$$a^7 + 7a^6b + 21a^5b^2 + 35a^4b^3 + 35a^3b^4 + 21a^2b^5 + 7ab^6 + b^7$$

Example 3   Write $(4x + 3y)^3$ in expanded form.

Solution $(4x + 3y)^3 = \binom{3}{0}(4x)^3 + \binom{3}{1}(4x)^2(3y) + \binom{3}{2}(4x)(3y)^2 + \binom{3}{3}(3y)^3$

$= 1(64x^3) + 3(16x^2)(3y) + 3(4x)(9y^2) + 1(27y^3)$

$= 64x^3 + 144x^2y + 108xy^2 + 27y^3$

Problem 3   Write $(3m - n)^4$ in expanded form.

Solution   See page A104.
$81m^4 - 108m^3n + 54m^2n^2 - 12mn^3 + n^4$

Example 4   Find the first 3 terms in the expansion of $(x + 3)^{15}$.

Solution $(x + 3)^{15}$

$= \binom{15}{0}x^{15} + \binom{15}{1}x^{14}(3) + \binom{15}{2}x^{13}(3)^2 + \cdots$

$= 1x^{15} + 15x^{14}(3) + 105x^{13}(9) + \cdots$

$= x^{15} + 45x^{14} + 945x^{13} + \cdots$

Problem 4   Find the first 3 terms in the expansion of $(y - 2)^{10}$.

Solution   See page A104.
$y^{10} - 20y^9 + 180y^8 + \cdots$

The Binomial Theorem can also be used to write any term of a binomial expansion.

Note that in the expansion of $(a + b)^5 =$
$(a + b)^5$, the exponent on b is $a^5 + 5a^4b + 10a^3b^2 + 10a^2b^3 + 5ab^4 + b^5$
one less than the term number.

The Formula for the rth Term in a Binomial Expansion

The rth term of $(a + b)^n$ is $\binom{n}{r-1}a^{n-r+1}b^{r-1}$.

Example 5   Find the 4th term in the expansion of $(x + 3)^7$.

Solution $\binom{n}{r-1}a^{n-r+1}b^{r-1}$ • Use the Formula for the rth Term in a Bino-
                                                          mial Expansion. $r = 4, n = 7$

$\binom{7}{4-1}x^{7-4+1}(3)^{4-1} = \binom{7}{3}x^4(3)^3$

$= 35x^4(27)$

$= 945x^4$

> **Problem 5** Find the 3rd term in the expansion of $(r - 2s)^7$.
>
> **Solution** See page A104.
> $84r^5s^2$

EXERCISES 12.4

1 Evaluate.

1. $3!$ 6

2. $4!$ 24

3. $8!$ 40,320

4. $9!$ 362,880

5. $0!$ 1

6. $1!$ 1

7. $\dfrac{5!}{2!\,3!}$ 10

8. $\dfrac{8!}{5!\,3!}$ 56

9. $\dfrac{6!}{6!\,0!}$ 1

10. $\dfrac{10!}{10!\,0!}$ 1

11. $\dfrac{9!}{6!\,3!}$ 84

12. $\dfrac{10!}{2!\,8!}$ 45

Evaluate.

13. $\dbinom{7}{2}$ 21

14. $\dbinom{8}{6}$ 28

15. $\dbinom{10}{2}$ 45

16. $\dbinom{9}{6}$ 84

17. $\dbinom{9}{0}$ 1

18. $\dbinom{10}{10}$ 1

19. $\dbinom{6}{3}$ 20

20. $\dbinom{7}{6}$ 7

21. $\dbinom{11}{1}$ 11

22. $\dbinom{13}{1}$ 13

23. $\dbinom{4}{2}$ 6

24. $\dbinom{8}{4}$ 70

Write in expanded form.

25. $(x + y)^4$
    $x^4 + 4x^3y + 6x^2y^2 + 4xy^3 + y^4$

26. $(r - s)^3$
    $r^3 - 3r^2s + 3rs^2 - s^3$

27. $(x - y)^5$
    $x^5 - 5x^4y + 10x^3y^2 - 10x^2y^3 + 5xy^4 - y^5$

28. $(y - 3)^4$
    $y^4 - 12y^3 + 54y^2 - 108y + 81$

29. $(2m + 1)^4$
    $16m^4 + 32m^3 + 24m^2 + 8m + 1$

30. $(2x + 3y)^3$
    $8x^3 + 36x^2y + 54xy^2 + 27y^3$

31. $(2r - 3)^5$
    $32r^5 - 240r^4 + 720r^3 - 1080r^2 + 810r - 243$

32. $(x + 3y)^4$
    $x^4 + 12x^3y + 54x^2y^2 + 108xy^3 + 81y^4$

Find the first three terms in the expansion.

33. $(a + b)^{10}$
    $a^{10} + 10a^9b + 45a^8b^2$

34. $(a + b)^9$
    $a^9 + 9a^8b + 36a^7b^2$

35. $(a - b)^{11}$
    $a^{11} - 11a^{10}b + 55a^9b^2$

36. $(a - b)^{12}$
    $a^{12} - 12a^{11}b + 66a^{10}b^2$

37. $(2x + y)^8$
$256x^8 + 1024x^7y + 1792x^6y^2$

38. $(x + 3y)^9$
$x^9 + 27x^8y + 324x^7y^2$

39. $(4x - 3y)^8$
$65,536x^8 - 393,216x^7y + 1,032,192x^6y^2$

40. $(2x - 5)^7$
$128x^7 - 2240x^6 + 16,800x^5$

41. $\left(x + \dfrac{1}{x}\right)^7$
$x^7 + 7x^5 + 21x^3$

42. $\left(x - \dfrac{1}{x}\right)^8$
$x^8 - 8x^6 + 28x^4$

43. $(x^2 + 3)^5$
$x^{10} + 15x^8 + 90x^6$

44. $(x^2 - 2)^6$
$x^{12} - 12x^{10} + 60x^8$

Find the indicated term in the expansion.

45. $(2x - 1)^7$; 4th term $\quad -560x^4$

46. $(x + 4)^5$; 3rd term $\quad 160x^3$

47. $(x^2 - y^2)^6$; 2nd term $\quad -6x^{10}y^2$

48. $(x^2 + y^2)^7$; 6th term $\quad 21x^4y^{10}$

49. $(y - 1)^9$; 5th term $\quad 126y^5$

50. $(x - 2)^8$; 8th term $\quad -1024x$

51. $\left(n + \dfrac{1}{n}\right)^5$; 2nd term $\quad 5n^3$

52. $\left(x + \dfrac{1}{2}\right)^6$; 3rd term $\quad \dfrac{15}{4}x^4$

53. $\left(\dfrac{x}{2} + 2\right)^5$; 1st term $\quad \dfrac{x^5}{32}$

54. $\left(y - \dfrac{2}{3}\right)^6$; 3rd term $\quad \dfrac{20}{3}y^4$

SUPPLEMENTAL EXERCISES 12.4

Solve.

55. Write the 7th row of Pascal's Triangle.
$1 \quad 7 \quad 21 \quad 35 \quad 35 \quad 21 \quad 7 \quad 1$

56. Write the 8th row of Pascal's Triangle.
$1 \quad 8 \quad 28 \quad 56 \quad 70 \quad 56 \quad 28 \quad 8 \quad 1$

57. Evaluate $\dfrac{n!}{(n - 2)!}$ for $n = 50$.
2450

58. Evaluate $\dfrac{n!}{(n - 3)!}$ for $n = 20$.
6840

59. Simplify $\dfrac{n!}{(n - 1)!}$.
n

60. Simplify $\dfrac{(n + 1)!}{(n - 1)!}$.
$n^2 + n$

61. Write the term that contains an x^3 in the expansion of $(x + a)^7$.
$35x^3a^4$

62. Write the term that contains a y^4 in the expansion of $(y - a)^8$.
$70y^4a^4$

Expand the binomial.

63. $(x^{\frac{1}{2}} + 2)^4$
$x^2 + 8x^{\frac{3}{2}} + 24x + 32x^{\frac{1}{2}} + 16$

64. $(x^{\frac{1}{3}} + 3)^3$
$x + 9x^{\frac{2}{3}} + 27x^{\frac{1}{3}} + 27$

65. $(x^{-1} + y^{-1})^3$
$\dfrac{1}{x^3} + \dfrac{3}{x^2 y} + \dfrac{3}{xy^2} + \dfrac{1}{y^3}$

66. $(x^{-1} - y^{-1})^3$
$\dfrac{1}{x^3} - \dfrac{3}{x^2 y} + \dfrac{3}{xy^2} - \dfrac{1}{y^3}$

67. $(1 + i)^6$
$-8i$

68. $(1 - i)^6$
$8i$

Use binomial expansion to evaluate the expression.

69. $(1.01)^4$ [*Hint:* $(1.01)^4 = (1 + 0.01)^4$]
1.04060401

70. $(1.02)^3$ [*Hint:* $(1.02)^3 = (1 + 0.02)^3$]
1.061208

CALCULATORS AND COMPUTERS

 Continued Fractions

An example of a **continued fraction** is $1 + \cfrac{1}{2 + \cfrac{9}{2 + \cfrac{25}{2 + \cfrac{49}{2 + \cdots}}}}$

The pattern continues with each numerator the square of an odd integer. Approximations to these fractions can be found by using a calculator.

1. To evaluate the continued fraction above, begin by dividing 49 by 2.

    49 $\boxed{\div}$ 2

2. Now add 2. Store the result in memory.

    $\boxed{+}$ 2 $\boxed{=}$ $\boxed{\text{M+}}$

3. Divide 25 by the result in memory. Then add 2 and store in memory. The MC clears the memory before the new result is stored.

    25 $\boxed{\div}$ $\boxed{\text{MR}}$ $\boxed{+}$ 2 $\boxed{\text{MC}}$ $\boxed{=}$ $\boxed{\text{M+}}$

4. Repeat step 3, but divide 9 by the result in memory.

5. Repeat step 3, but divide 1 by the result in memory and add 1 instead of 2.

The result is 1.197719. This value approximates $\dfrac{4}{\pi}$. In fact, as the fraction is continued, the approximation becomes closer and closer to $\dfrac{4}{\pi}$.

Here are two other continued fractions. Evaluate them by using your calculator.

$$1 + \cfrac{1}{2 + \cfrac{1}{2 + \cfrac{1}{2 + \cdots}}} \qquad\qquad 1 + \cfrac{1}{1 + \cfrac{1}{1 + \cfrac{1}{1 + \cdots}}}$$

The first continued fraction is approximately $\sqrt{2}$. As the fraction is continued, the approximation becomes closer to $\sqrt{2}$. The second continued fraction is approximately $1 + \sqrt{5}$. This number is called the golden ratio and has played an important role in art and architecture.

CHAPTER SUMMARY

Key Words

A **sequence** is an ordered list of numbers.

Each of the numbers of a sequence is called a **term** of the sequence.

A **finite sequence** contains a finite number of terms.

An **infinite sequence** contains an infinite number of terms.

The indicated sum of a sequence is a **series**.

An **arithmetic sequence,** or arithmetic progression, is one in which the difference between any two consecutive terms is constant. The difference between consecutive terms is called the **common difference** of the sequence.

A **geometric sequence,** or geometric progression, is one in which each successive term of the sequence is the same nonzero constant multiple of the preceding term. The common multiple is called the **common ratio** of the sequence.

n **factorial,** written *n*!, is the product of the first *n* positive integers.

Essential Rules

The Formula for the *n*th Term of an Arithmetic Sequence
The *n*th term of an arithmetic sequence with a common difference of d is given by $a_n = a_1 + (n - 1)d$.

The Formula for the Sum of n Terms of an Arithmetic Series
Let a_1 be the first term of a finite arithmetic sequence, n the number of terms, and a_n the last term of the sequence. Then the sum of the series S_n is given by $S_n = \dfrac{n}{2}(a_1 + a_n)$.

The Formula for the nth Term of a Geometric Sequence
The nth term of a geometric sequence with first term a_1 and common ratio r is given by $a_n = a_1 r^{n-1}$.

The Formula for the Sum of n Terms of a Finite Geometric Series
Let a_1 be the first term of a finite geometric sequence, n the number of terms, r the common ratio, and $r \neq 1$. Then the sum of the series, S_n is given by
$$S_n = \frac{a_1(1 - r^n)}{1 - r}.$$

The Formula for the Sum of an Infinite Geometric Series
The sum of an infinite geometric series in which $|r| < 1$ and a_1 is the first term is given by $S = \dfrac{a_1}{1 - r}$.

The Binomial Expansion Formula
$$(a + b)^n = \binom{n}{0}a^n + \binom{n}{1}a^{n-1}b + \binom{n}{2}a^{n-2}b^2 + \cdots + \binom{n}{n}b^n$$

The Formula for the rth Term in a Binomial Expansion
The rth term of $(a + b)^n$ is $\binom{n}{r - 1}a^{n-r+1}b^{r-1}$.

CHAPTER REVIEW

1. Write $\displaystyle\sum_{i=1}^{4} 3x^i$ in expanded form.

$3x + 3x^2 + 3x^3 + 3x^4$

2. Find the number of terms in the finite arithmetic sequence $-5, -8, -11, \ldots, -50$.

16

3. Find the 7th term of the geometric sequence $4, 4\sqrt{2}, 8, \ldots$.

32

4. Find the sum of the infinite geometric sequence $4, 3, \dfrac{9}{4}, \ldots$.

16

5. Evaluate: $\binom{9}{3}$

84

6. Write the 14th term of the sequence whose nth term is given by the formula $a_n = \dfrac{8}{n+2}$.

$\dfrac{1}{2}$

7. Find the 10th term of the arithmetic sequence $-10, -4, 2, \ldots$.

44

8. Find the sum of the first 18 terms of the arithmetic sequence $-25, -19, -13, \ldots$.

468

9. Find the sum of the first five terms of the geometric sequence $-6, 12, -24, \ldots$.

-66

10. Evaluate: $\dfrac{8!}{4!\,4!}$

70

11. Find the 7th term in the expansion of $(3x + y)^9$.

$2268x^3y^6$

12. Find the sum of the series $\sum\limits_{n=1}^{4} (3n + 1)$.

34

13. Write the 6th term of the sequence whose nth term is given by the formula
$$a_n = \dfrac{n+1}{n}.$$

$\dfrac{7}{6}$

14. Find the formula for the nth term of the arithmetic sequence $12, 9, 6, \ldots$.

$a_n = -3n + 15$

15. Find the 5th term of the geometric sequence $6, 2, \dfrac{2}{3}, \ldots$.

$\dfrac{2}{27}$

16. Find an equivalent fraction for $0.233\overline{3}$.

$\dfrac{7}{30}$

17. Find the 35th term of the arithmetic sequence $-13, -16, -19, \ldots$.

-115

18. Find the sum of the first six terms of the geometric sequence $1, \dfrac{3}{2}, \dfrac{9}{4}, \ldots$.

$\dfrac{665}{32}$

19. Find the sum of the first 21 terms of the arithmetic sequence $5, 12, 19, \ldots$.

1575

20. Find the 4th term in the expansion of $(x - 2y)^7$.

$-280x^4y^3$

21. Find the number of terms in the finite arithmetic sequence $1, 7, 13, \ldots, 121$.

21

22. Find the 8th term of the geometric sequence $\dfrac{3}{8}, \dfrac{3}{4}, \dfrac{3}{2}, \ldots$.

48

23. Find the sum of the series $\sum_{i=1}^{5} 2i$.

30

24. Find the sum of the first five terms of the geometric series $1, 4, 16, \ldots$.

341

25. Evaluate: $5!$

120

26. Find the 3rd term in the expansion of $(x - 4)^6$.

$240x^4$

27. Find the 30th term of the arithmetic sequence $-2, 3, 8, \ldots$.

143

28. Find the sum of the first 25 terms of the arithmetic sequence $25, 21, 17, \ldots$.

-575

29. An inventory of supplies for a fabric manufacturer indicated that 7500 yd of material were in stock on January 1. On February 1, and on the first of the month for each successive month, the manufacturer sent 550 yd of material to retail outlets. How much material was in stock after the shipment on October 1?

2550 yd

30. An ore sample contains 320 mg of a radioactive substance with a half-life of one day. Find the amount of radioactive material in the sample at the beginning of the fifth day.

20 mg

CUMULATIVE REVIEW

1. Factor: $2x^6 + 16$
$2(x^2 + 2)(x^4 - 2x^2 + 4)$

2. Simplify: $\dfrac{4x^2}{x^2 + x - 2} - \dfrac{3x - 2}{x + 2}$
$\dfrac{x^2 + 5x - 2}{(x + 2)(x - 1)}$

3. Solve: $2x - 1 > 3$ or $1 - 3x > 7$
$\{x \mid x < -2 \text{ or } x > 2\}$

4. Simplify: $\left(\dfrac{x^{-\frac{3}{4}}x^{\frac{3}{2}}}{x^{-\frac{5}{2}}}\right)^{-8}$
$\dfrac{1}{x^{26}}$

5. Simplify: $\sqrt{2y}\left(\sqrt{8xy} - \sqrt{y}\right)$
$4y\sqrt{x} - y\sqrt{2}$

6. Solve: $2x^2 - x + 7 = 0$
$\dfrac{1}{4} + \dfrac{\sqrt{55}}{4}i$ and $\dfrac{1}{4} - \dfrac{\sqrt{55}}{4}i$

7. Solve: $5 - \sqrt{x} = \sqrt{x + 5}$
4

8. Graph $3x - 2y = -4$.

9. Graph $2x - 3y < 9$.

10. Graph the function $f(x) = |x - 3|$.

11. Find the inverse of the function $f(x) = 2x - 4$.
$f^{-1}(x) = \dfrac{1}{2}x + 2$

12. Find the equation of the circle that passes through the point $(4, 2)$ and whose center is $(-1, -1)$.
$(x + 1)^2 + (y + 1)^2 = 34$

13. Solve: $\dfrac{x}{x + 3} + \dfrac{3x}{x + 8} = 1$
-4 and 2

14. Graph $\dfrac{x^2}{16} + \dfrac{y^2}{9} = 1$.

15. Solve by the addition method:
$3x - 3y = 2$
$6x - 4y = 5$
$\left(\dfrac{7}{6}, \dfrac{1}{2}\right)$

16. Solve by using a matrix: $4x - 3y = 0$
$5x - 3y = 3$
$(3, 4)$

17. Evaluate the determinant: $\begin{vmatrix} -3 & 1 \\ 4 & 2 \end{vmatrix}$
-10

18. Write $\log_6 xy^3z$ in expanded form.
$\log_6 x + 3\log_6 y + \log_6 z$

19. Solve for x: $\log_4 x = -3$
$\dfrac{1}{64}$

20. Solve for x: $4^{x+2} = 8^{x-1}$
7

21. Write the 5th term of the sequence whose nth term is given by the formula $a_n = n(n - 1)$.
20

22. Find the sum of the series $\displaystyle\sum_{n=1}^{7} (-1)^{n-1}(n + 2)$.
6

23. Find the 33rd term of the arithmetic sequence $-7, -10, -13, \ldots.$
-103

24. Find the sum of the infinite geometric sequence $3, -2, \dfrac{4}{3}, \ldots.$

$\dfrac{9}{5}$

25. Find an equivalent fraction for $0.4\overline{6}$.
$\dfrac{7}{15}$

26. Find the 6th term in the expansion of $(2x + y)^6$.
$12xy^5$

27. A plane can fly at a rate of 215 mph in calm air. Traveling with the wind, the plane flew 750 mi in the same amount of time that it flew 540 mi against the wind. Find the rate of the wind.
35 mph

28. Flying with the wind, a plane flew 360 mi in 3 h. Against the wind, the plane required 4 h to fly the same distance. Find the rate of the plane in calm air and the rate of the wind.
plane: 105 mph; wind: 15 mph

29. Find three consecutive even integers whose sum is between 60 and 72.
20, 22, and 24

30. Use the exponential decay equation $A = A_0\left(\dfrac{1}{2}\right)^{\frac{t}{k}}$, where A is the amount of radioactive material present after time t, k is the half-life, and A_0 is the original amount of radioactive material, to find the half-life of a material that decays from 25 mg to 20 mg in 10 days. Round to the nearest whole number.
31 days

31. Boxes were stacked in a warehouse so that there are 37 boxes in the bottom row, 32 boxes in the next row, and so on in an arithmetic sequence. There are 2 boxes on the top row. How many boxes are in the stack?
156 boxes

32. A chain letter is sent to five people. Each of the five people in turn mail the letter to another five people, and the process is repeated. What is the total number of people who have received the letter after four mailings?
625 people

FINAL EXAM

1. Simplify:
$12 - 8[3 - (-2)]^2 \div 5 - 3$
-31

2. Evaluate $\dfrac{a^2 - b^2}{a - b}$ when $a = 3$ and $b = -4$.
-1

3. Simplify:
$5 - 2[3x - 7(2 - x) - 5x]$
$33 - 10x$

4. Solve: $\dfrac{3}{4}x - 2 = 4$
8

5. Solve:
$\dfrac{2 - 4x}{3} - \dfrac{x - 6}{12} = \dfrac{5x - 2}{6}$
$\dfrac{2}{3}$

6. Solve:
$2 - 3x < 6$ and $2x + 1 > 4$
$\left\{ x \mid x > \dfrac{3}{2} \right\}$

7. Solve:
$8 - |5 - 3x| = 1$
$-\dfrac{2}{3}$ and 4

8. Solve: $|2x + 5| < 3$
$\{ x \mid -4 < x < -1 \}$

9. Simplify:
$2a[5 - a(2 - 3a) - 2a] + 3a^2$
$6a^3 - 5a^2 + 10a$

10. Factor: $8 - x^3 y^3$
$(2 - xy)(4 + 2xy + x^2 y^2)$

11. Factor: $x - y - x^3 + x^2 y$
$(x - y)(1 + x)(1 - x)$

12. Solve: $2x^2 + 9x - 5 > 0$
$\left\{ x \mid x < -5 \text{ or } x > \dfrac{1}{2} \right\}$

13. Simplify:
$(2x^3 - 7x^2 + 4) \div (2x - 3)$
$x^2 - 2x - 3 - \dfrac{5}{2x - 3}$

14. Simplify:
$\dfrac{x^2 - 3x}{2x^2 - 3x - 5} \div \dfrac{4x - 12}{4x^2 - 4}$
$\dfrac{x(x - 1)}{2x - 5}$

15. Simplify:
$\dfrac{x - 2}{x + 2} - \dfrac{x + 3}{x - 3}$
$-\dfrac{10x}{(x + 2)(x - 3)}$

16. Simplify: $\dfrac{\dfrac{3}{x} + \dfrac{1}{x + 4}}{\dfrac{1}{x} + \dfrac{3}{x + 4}}$
$\dfrac{x + 3}{x + 1}$

17. Solve:

$$\frac{5}{x - 2} - \frac{5}{x^2 - 4} = \frac{1}{x + 2}$$

$$-\frac{7}{4}$$

18. Solve: $\dfrac{(x + 2)(x - 3)}{x + 1} \geq 0$

$\{x \mid -2 \leq x < -1 \text{ or } x \geq 3\}$

19. Solve $a_n = a_1 + (n - 1)d$ for d.

$$d = \frac{a_n - a_1}{n - 1}$$

20. Simplify:

$$\left(\frac{4x^2y^{-1}}{3x^{-1}y}\right)^{-2}\left(\frac{2x^{-1}y^2}{9x^{-2}y^2}\right)^3$$

$$\frac{y^4}{162x^3}$$

21. Simplify: $\left(\dfrac{3x^{\frac{2}{3}}y^{\frac{1}{2}}}{6x^2y^{\frac{4}{3}}}\right)^6$

$$\frac{1}{64x^8y^5}$$

22. Simplify: $x\sqrt{18x^2y^3} - y\sqrt{50x^4y}$
$-2x^2y\sqrt{2y}$

23. Simplify: $\dfrac{\sqrt{16x^5y^4}}{\sqrt{32xy^7}}$

$$\frac{x^2\sqrt{2y}}{2y^2}$$

24. Simplify: $\dfrac{3}{2 + i}$

$$\frac{6}{5} - \frac{3}{5}i$$

25. Graph $2x - 3y = 9$ by using the x- and y-intercepts.

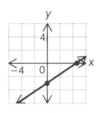

26. Graph the solution set of $3x + 2y > 6$.

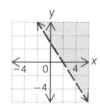

27. Find the equation of the line containing the points $(3, -2)$ and $(1, 4)$.
$y = -3x + 7$

28. Find the equation of the line containing the point $(-2, 1)$ and perpendicular to the line $3x - 2y = 6$.

$$y = -\frac{2}{3}x - \frac{1}{3}$$

29. Write a quadratic equation that has integer coefficients and has solutions $-\dfrac{1}{2}$ and 2.
$2x^2 - 3x - 2 = 0$

30. Solve: $2x^2 - 3x - 1 = 0$
$\dfrac{3 + \sqrt{17}}{4}$ and $\dfrac{3 - \sqrt{17}}{4}$

31. Solve: $x^{\frac{2}{3}} - x^{\frac{1}{3}} - 6 = 0$
27 and -8

32. Solve: $\dfrac{2}{x} - \dfrac{2}{2x + 3} = 1$

$\dfrac{3}{2}$ and -2

33. For $f(x) = 3x^2 - 1$ and $g(x) = -2x + 4$, find $f(2) - g(2)$.

11

34. Graph the function $f(x) = -x^2 + 4$.

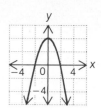

35. For $g(x) = x^2 - 2x + 1$ and $h(x) = 3x - 4$, find $g(h(0))$.

25

36. Find the inverse of the function

$$f(x) = \frac{2}{3}x - 4.$$

$$f^{-1}(x) = \frac{3}{2}x + 6$$

37. Graph $\dfrac{x^2}{16} + \dfrac{y^2}{4} = 1$

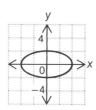

38. Graph the solution set of $\dfrac{y^2}{9} - \dfrac{x^2}{25} \geq 1$.

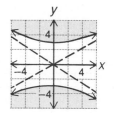

39. Solve by the addition method:
$$3x - 2y = 1$$
$$5x - 3y = 3$$
(3, 4)

40. Solve by using a matrix:
$$4x + 3y = 3$$
$$6x + 6y = 5$$
$\left(\dfrac{1}{2}, \dfrac{1}{3}\right)$

41. Evaluate the determinant:
$$\begin{vmatrix} 3 & 4 \\ -1 & 2 \end{vmatrix}$$

10

42. Solve: $x^2 - y^2 = 4$
$$x + y = 1$$
$\left(\dfrac{5}{2}, -\dfrac{3}{2}\right)$

43. Graph $f(x) = \log_2(x + 1)$.

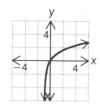

44. Write $2(\log_2 a - \log_2 b)$ as a single logarithm with a coefficient of 1.

$\log_2 \dfrac{a^2}{b^2}$

45. Solve for x:
$$\log_3 x - \log_3(x - 3) = \log_3 2$$

6

46. Write $\displaystyle\sum_{i=1}^{5} 2y^i$ in expanded form.

$2y + 2y^2 + 2y^3 + 2y^4 + 2y^5$

47. Find an equivalent fraction for $0.5\overline{1}$.
$\dfrac{23}{45}$

48. Find the third term in the expansion of $(x - 2y)^9$.
$144x^7y^2$

49. The sum of two integers is twenty-four. Seven times the smaller integer is two less than three times the larger integer. Find the integers.
7 and 17

50. Find the cost per pound of a tea mixture made from 50 lb of tea costing $7.56 per pound and 80 lb of tea costing $3.40 per pound.
$5

51. An investment of $9000 is deposited into two simple interest accounts. On one account, the annual simple interest is 8%. On the other, the annual simple interest rate is 6.4%. How much should be invested in each account so that the interest earned by each account is the same?
$4000 @ 8%, $5000 @ 6.4%

52. How many ounces of pure water must be added to 75 oz of a 6% salt solution to make a 2.5% salt solution?
105 oz

53. A pre-election survey showed that 7 out of every 12 voters would vote in an election. At this rate, how many people would be expected to vote in a city of 138,000?
80,500 people

54. An experienced electrician can wire a room twice as fast as an apprentice electrician. Working together, they can wire a room in 16 h. How long would it take the apprentice working alone to wire the room?
48 h

55. Find the distance required for a car to reach a velocity of 72 ft/s when the acceleration is 15 ft/s^2. Use the equation $v = \sqrt{2as}$, where v is the velocity, a is the acceleration, and s is the distance.
172.8 ft

56. An old pump requires 6 h longer to empty a pool than does a new pump. With both pumps working, the pool can be emptied in 4 h. Find the time required for the new pump working alone to empty the pool.
6 h

57. The length (L) of a rectangle of fixed area varies inversely as the width (W). If the length of a rectangle is 12 ft when the width is 7 ft, find the length of the rectangle when the width is 24 ft.
3.5 ft

58. During one month, a homeowner used 400 units of electricity and 150 units of gas for a total cost of $57. The next month, 280 units of electricity and 110 units of gas were used for a total cost of $40.60. Find the cost per unit of gas.
$.14

59. An average score of 82–90 in a physics class receives a B grade. A student has grades of 77, 91, and 81 on three physics tests. Find the range of scores on the fourth test that will give the student a B grade for the course.
79 or better

60. A real estate broker estimates that a piece of land will increase in value at a rate of 10% each year. If the original value of the land is $20,000, what will be its value in 10 years?
$51,874.85

Appendix

Table of Common Logarithms ────────────

Decimal approximations have been rounded to the nearest ten-thousandth.

x	0	1	2	3	4	5	6	7	8	9
1.0	.0000	.0043	.0086	.0128	.0170	.0212	.0253	.0294	.0334	.0374
1.1	.0414	.0453	.0492	.0531	.0569	.0607	.0645	.0682	.0719	.0755
1.2	.0792	.0828	.0864	.0899	.0934	.0969	.1004	.1038	.1072	.1106
1.3	.1139	.1173	.1206	.1239	.1271	.1303	.1335	.1367	.1399	.1430
1.4	.1461	.1492	.1523	.1553	.1584	.1614	.1644	.1673	.1703	.1732
1.5	.1761	.1790	.1818	.1847	.1875	.1903	.1931	.1959	.1987	.2014
1.6	.2041	.2068	.2095	.2122	.2148	.2175	.2201	.2227	.2253	.2279
1.7	.2304	.2330	.2355	.2380	.2405	.2430	.2455	.2480	.2501	.2529
1.8	.2553	.2577	.2601	.2625	.2648	.2672	.2695	.2718	.2742	.2765
1.9	.2788	.2810	.2833	.2856	.2878	.2900	.2923	.2945	.2967	.2989
2.0	.3010	.3032	.3054	.3075	.3096	.3118	.3139	.3160	.3181	.3201
2.1	.3222	.3243	.3263	.3284	.3304	.3324	.3345	.3365	.3385	.3404
2.2	.3424	.3444	.3464	.3483	.3502	.3522	.3541	.3560	.3579	.3598
2.3	.3617	.3636	.3655	.3674	.3692	.3711	.3729	.3747	.3766	.3784
2.4	.3802	.3820	.3838	.3856	.3874	.3892	.3909	.3927	.3945	.3962
2.5	.3979	.3997	.4014	.4031	.4048	.4065	.4082	.4099	.4116	.4133
2.6	.4150	.4166	.4183	.4200	.4216	.4232	.4249	.4265	.4281	.4298
2.7	.4314	.4330	.4346	.4362	.4378	.4393	.4409	.4425	.4440	.4456
2.8	.4472	.4487	.4502	.4518	.4533	.4548	.4564	.4579	.4594	.4609
2.9	.4624	.4639	.4654	.4669	.4683	.4698	.4713	.4728	.4742	.4757
3.0	.4771	.4786	.4800	.4814	.4829	.4843	.4857	.4871	.4886	.4900
3.1	.4914	.4928	.4942	.4955	.4969	.4983	.4997	.5011	.5024	.5038
3.2	.5051	.5065	.5079	.5092	.5105	.5119	.5132	.5145	.5159	.5172
3.3	.5185	.5198	.5211	.5224	.5237	.5250	.5263	.5276	.5289	.5302
3.4	.5315	.5328	.5340	.5353	.5366	.5378	.5391	.5403	.5416	.5428
3.5	.5441	.5453	.5465	.5478	.5490	.5502	.5514	.5527	.5539	.5551
3.6	.5563	.5575	.5587	.5599	.5611	.5623	.5635	.5647	.5658	.5670
3.7	.5682	.5694	.5705	.5717	.5729	.5740	.5752	.5763	.5775	.5786
3.8	.5798	.5809	.5821	.5832	.5843	.5855	.5866	.5877	.5888	.5899
3.9	.5911	.5922	.5933	.5944	.5955	.5966	.5977	.5988	.5999	.6010
4.0	.6021	.6031	.6042	.6053	.6064	.6075	.6085	.6096	.6107	.6117
4.1	.6128	.6138	.6149	.6160	.6170	.6180	.6191	.6201	.6212	.6222
4.2	.6232	.6243	.6253	.6263	.6274	.6284	.6294	.6304	.6314	.6325
4.3	.6335	.6345	.6355	.6365	.6375	.6385	.6395	.6405	.6415	.6425
4.4	.6435	.6444	.6454	.6464	.6474	.6484	.6493	.6503	.6513	.6522
4.5	.6532	.6542	.6551	.6561	.6571	.6580	.6590	.6599	.6609	.6618
4.6	.6628	.6637	.6646	.6656	.6665	.6675	.6684	.6693	.6702	.6712
4.7	.6721	.6730	.6739	.6749	.6758	.6767	.6776	.6785	.6794	.6803
4.8	.6812	.6821	.6830	.6839	.6848	.6857	.6866	.6875	.6884	.6893
4.9	.6902	.6911	.6920	.6928	.6937	.6946	.6955	.6964	.6972	.6981
5.0	.6990	.6998	.7007	.7016	.7024	.7033	.7042	.7050	.7059	.7067
5.1	.7076	.7084	.7093	.7101	.7110	.7118	.7126	.7135	.7143	.7152
5.2	.7160	.7168	.7177	.7185	.7193	.7202	.7210	.7218	.7226	.7235
5.3	.7243	.7251	.7259	.7267	.7275	.7284	.7292	.7300	.7308	.7316
5.4	.7324	.7332	.7340	.7348	.7356	.7364	.7372	.7380	.7388	.7396

Table of Common Logarithms ——————————

x	0	1	2	3	4	5	6	7	8	9
5.5	.7404	.7412	.7419	.7427	.7435	.7443	.7451	.7459	.7466	.7474
5.6	.7482	.7490	.7497	.7505	.7513	.7520	.7528	.7536	.7543	.7551
5.7	.7559	.7566	.7574	.7582	.7489	.7597	.7604	.7612	.7619	.7627
5.8	.7634	.7642	.7649	.7657	.7664	.7672	.7679	.7686	.7694	.7701
5.9	.7709	.7716	.7723	.7731	.7738	.7745	.7752	.7760	.7767	.7774
6.0	.7782	.7789	.7796	.7803	.7810	.7818	.7825	.7832	.7839	.7846
6.1	.7853	.7860	.7868	.7875	.7882	.7889	.7896	.7903	.7910	.7917
6.2	.7924	.7931	.7938	.7945	.7952	.7959	.7966	.7973	.7980	.7987
6.3	.7993	.8000	.8007	.8014	.8021	.8028	.8035	.8041	.8048	.8055
6.4	.8062	.8069	.8075	.8082	.8089	.8096	.8102	.8109	.8116	.8122
6.5	.8129	.8136	.8142	.8149	.8156	.8162	.8169	.8176	.8182	.8189
6.6	.8195	.8202	.8209	.8215	.8222	.8228	.8235	.8241	.8248	.8254
6.7	.8261	.8267	.8274	.8280	.8287	.8293	.8299	.8306	.8312	.8319
6.8	.8325	.8331	.8338	.8344	.8351	.8357	.8363	.8370	.8376	.8382
6.9	.8388	.8395	.8401	.8407	.8414	.8420	.8426	.8432	.8439	.8445
7.0	.8451	.8457	.8463	.8470	.8476	.8482	.8488	.8494	.8500	.8506
7.1	.8513	.8519	.8525	.8531	.8537	.8543	.8549	.8555	.8561	.8567
7.2	.8573	.8579	.8585	.8591	.8597	.8603	.8609	.8615	.8621	.8627
7.3	.8633	.8639	.8645	.8651	.8657	.8663	.8669	.8675	.8681	.8686
7.4	.8692	.8698	.8704	.8710	.8716	.8722	.8727	.8733	.8739	.8745
7.5	.8751	.8756	.8762	.8768	.8774	.8779	.8785	.8791	.8797	.8802
7.6	.8808	.8814	.8820	.8825	.8831	.8837	.8842	.8848	.8854	.8859
7.7	.8865	.8871	.8876	.8882	.8887	.8893	.8899	.8904	.8910	.8915
7.8	.8921	.8927	.8932	.8938	.8943	.8949	.8954	.8960	.8965	.8971
7.9	.8976	.8982	.8987	.8993	.8998	.9004	.9009	.9015	.9020	.9025
8.0	.9031	.9036	.9042	.9047	.9053	.9058	.9063	.9069	.9074	.9079
8.1	.9085	.9090	.9096	.9101	.9106	.9112	.9117	.9122	.9128	.9133
8.2	.9138	.9143	.9149	.9154	.9159	.9165	.9170	.9175	.9180	.9186
8.3	.9191	.9196	.9201	.9206	.9212	.9217	.9222	.9227	.9232	.9238
8.4	.9243	.9248	.9253	.9258	.9263	.9269	.9274	.9279	.9284	.9289
8.5	.9294	.9299	.9304	.9309	.9315	.9320	.9325	.9330	.9335	.9340
8.6	.9345	.9350	.9355	.9360	.9365	.9370	.9375	.9380	.9385	.9390
8.7	.9395	.9400	.9405	.9410	.9415	.9420	.9425	.9430	.9435	.9440
8.8	.9445	.9450	.9455	.9460	.9465	.9469	.9474	.9479	.9484	.9489
8.9	.9494	.9499	.9504	.9509	.9513	.9518	.9523	.9528	.9533	.9538
9.0	.9542	.9547	.9552	.9557	.9562	.9566	.9571	.9576	.9581	.9586
9.1	.9590	.9595	.9600	.9605	.9609	.9614	.9619	.9624	.9628	.9633
9.2	.9638	.9643	.9647	.9652	.9657	.9661	.9666	.9671	.9675	.9680
9.3	.9685	.9689	.9694	.9699	.9703	.9708	.9713	.9717	.9722	.9727
9.4	.9731	.9736	.9741	.9745	.9750	.9754	.9759	.9763	.9768	.9773
9.5	.9777	.9782	.9786	.9791	.9795	.9800	.9805	.9809	.9814	.9818
9.6	.9823	.9827	.9832	.9836	.9841	.9845	.9850	.9854	.9859	.9863
9.7	.9868	.9872	.9877	.9881	.9886	.9890	.9894	.9899	.9903	.9908
9.8	.9912	.9917	.9921	.9926	.9930	.9934	.9939	.9943	.9948	.9952
9.9	.9956	.9961	.9965	.9969	.9974	.9978	.9983	.9987	.9991	.9996

Table of Square and Cube Roots

Decimal approximations have been rounded to the nearest thousandth.

N	$\sqrt{N}$	$\sqrt[3]{N}$	N	$\sqrt{N}$	$\sqrt[3]{N}$	N	$\sqrt{N}$	$\sqrt[3]{N}$
1	1	1	36	6	3.302	71	8.426	4.141
2	1.414	1.260	37	6.083	3.332	72	8.485	4.160
3	1.732	1.442	38	6.164	3.362	73	8.544	4.179
4	2	1.587	39	6.245	3.391	74	8.602	4.198
5	2.236	1.710	40	6.325	3.420	75	8.660	4.217
6	2.449	1.817	41	6.403	3.448	76	8.718	4.236
7	2.646	1.913	42	6.481	3.476	77	8.775	4.254
8	2.828	2	43	6.557	3.503	78	8.832	4.273
9	3	2.080	44	6.633	3.530	79	8.888	4.291
10	3.162	2.154	45	6.708	3.557	80	8.944	4.309
11	3.317	2.224	46	6.782	3.583	81	9	4.327
12	3.464	2.289	47	6.856	3.609	82	9.055	4.344
13	3.606	2.351	48	6.928	3.634	83	9.110	4.362
14	3.742	2.410	49	7	3.659	84	9.165	4.380
15	3.873	2.466	50	7.071	3.684	85	9.220	4.397
16	4	2.520	51	7.141	3.708	86	9.274	4.414
17	4.123	2.571	52	7.211	3.733	87	9.327	4.431
18	4.243	2.621	53	7.280	3.756	88	9.381	4.448
19	4.359	2.668	54	7.348	3.780	89	9.434	4.465
20	4.472	2.714	55	7.416	3.803	90	9.487	4.481
21	4.583	2.759	56	7.483	3.826	91	9.539	4.498
22	4.690	2.802	57	7.550	3.849	92	9.592	4.514
23	4.796	2.844	58	7.616	3.871	93	9.644	4.531
24	4.899	2.884	59	7.681	3.893	94	9.695	4.547
25	5	2.924	60	7.746	3.915	95	9.747	4.563
26	5.099	2.962	61	7.810	3.936	96	9.798	4.579
27	5.196	3	62	7.874	3.958	97	9.849	4.595
28	5.292	3.037	63	7.937	3.979	98	9.899	4.610
29	5.385	3.072	64	8	4	99	9.950	4.626
30	5.477	3.107	65	8.062	4.021	100	10	4.642
31	5.568	3.141	66	8.124	4.041	101	10.050	4.657
32	5.657	3.175	67	8.185	4.062	102	10.100	4.672
33	5.745	3.208	68	8.246	4.082	103	10.149	4.688
34	5.831	3.240	69	8.307	4.102	104	10.198	4.703
35	5.916	3.271	70	8.367	4.121	105	10.247	4.718

Answers to Chapter 1

SECTION 1.1   (page 3)

Problem 1   **A.** -18 **B.** 5.2

Problem 2   **A.** 11 **B.** $-\dfrac{4}{5}$

Problem 3   **A.** $\dfrac{5}{6} - \dfrac{3}{8} + \dfrac{7}{9} = \dfrac{60}{72} - \dfrac{27}{72} + \dfrac{56}{72} = \dfrac{60 - 27 + 56}{72} = \dfrac{89}{72}$

B.
$$
\begin{array}{r}
12.094 \\
-\ \ 8.729 \\
\hline
3.365
\end{array}
\qquad -8.729 + 12.094 = 3.365
$$

C. $\dfrac{5}{8} \div \left(-\dfrac{15}{40}\right) = \dfrac{5}{8} \cdot \left(-\dfrac{40}{15}\right) = -\dfrac{5 \cdot 40}{8 \cdot 15} = -\dfrac{\overset{1}{\cancel{5}} \cdot \overset{1}{\cancel{2}} \cdot \overset{1}{\cancel{2}} \cdot \overset{1}{\cancel{2}} \cdot 5}{\underset{1}{\cancel{2}} \cdot \underset{1}{\cancel{2}} \cdot \underset{1}{\cancel{2}} \cdot 3 \cdot \underset{1}{\cancel{5}}} = -\dfrac{5}{3}$

Problem 4
$$
\begin{array}{r}
4.027 \\
\times\ \ \ 0.49 \\
\hline
36243 \\
16108\ \ \\
\hline
1.97323 \approx 1.97
\end{array}
\qquad -4.027(0.49) \approx -1.97
$$

Problem 5 $(-2)^4 = (-2)(-2)(-2)(-2) = 16$

$-2^4 = -(2 \cdot 2 \cdot 2 \cdot 2) = -16$

Problem 6   **A.** $-3^2 \cdot 2^2 = -(3)(3)(3) \cdot (2)(2) = -27 \cdot 4 = -108$

B. $\left(-\dfrac{2}{5}\right)^3 \cdot 5^2 = \left(-\dfrac{2}{5}\right)\left(-\dfrac{2}{5}\right)\left(-\dfrac{2}{5}\right) \cdot (5)(5) = -\dfrac{8}{5}$

Problem 7 $(3.81 - 1.41)^2 \div 0.036 - 1.89$
$(2.40)^2 \div 0.036 - 1.89$
$5.76 \div 0.036 - 1.89$
$160 - 1.89$
158.11

Problem 8
$$\frac{1}{3} + \frac{5}{8} \div \frac{15}{16} - \frac{7}{12}$$
$$\frac{1}{3} + \frac{5}{8} \cdot \frac{16}{15} - \frac{7}{12}$$
$$\frac{1}{3} + \frac{2}{3} - \frac{7}{12}$$
$$1 - \frac{7}{12}$$
$$\frac{5}{12}$$

Problem 9
$$\frac{11}{12} - \frac{\frac{5}{4}}{2 - \frac{7}{2}} \cdot \frac{3}{4}$$
$$\frac{11}{12} - \frac{\frac{5}{4}}{-\frac{3}{2}} \cdot \frac{3}{4}$$
$$\frac{11}{12} - \left[\frac{5}{4} \cdot \left(-\frac{2}{3}\right)\right] \cdot \frac{3}{4}$$
$$\frac{11}{12} - \left(-\frac{5}{6}\right) \cdot \frac{3}{4}$$
$$\frac{11}{12} - \left(-\frac{5}{8}\right)$$
$$\frac{37}{24}$$

EXERCISES   (page 13)

1. -83 **3.** 75   **5.** -9.3 **7.** 6.4   **9.** $\frac{11}{12}$ **11.** -126 **13.** 436   **15.** -16 **17.** 4.93   **19.** $-\frac{7}{8}$

21. $\frac{43}{48}$ **23.** $-\frac{67}{45}$ **25.** $-\frac{13}{36}$ **27.** $\frac{11}{24}$ **29.** $\frac{13}{24}$ **31.** $-\frac{3}{56}$ **33.** $-\frac{2}{3}$ **35.** $-\frac{11}{14}$ **37.** $-\frac{1}{24}$

39. -12.974 **41.** -6.008 **43.** 1.9215   **45.** -6.02 **47.** -6.7 **49.** -44.585 **51.** 18.90

53. -1802.87 **55.** -64 **57.** 64   **59.** 432   **61.** -1125 **63.** 512   **65.** -160 **67.** 24   **69.** -36

71. -11 **73.** $\frac{1}{4}$ **75.** $\frac{1}{4}$ **77.** 12   **79.** 25   **81.** 44   **83.** $\frac{109}{150}$ **85.** $\frac{91}{36}$ **87.** 4.4   **89.** -0.91

91. integer, rational number, real number   **93.** rational number, real number
95. rational number, real number   **97.** irrational number, real number
99. irrational number, real number   **101.** irrational number, real number   **103.** positive
105. negative   **107.** 1, -1 **109.** 3 integers; 1, 4, and 9   **111.** 1

SECTION 1.2   (page 16)

Problem 1
$(b - c)^2 \div ab$
$[2 - (-4)]^2 \div (-3)(2)$
$[6]^2 \div (-3)(2)$
$36 \div (-3)(2)$
$-12(2)$
-24

Problem 2 $(x)\left(\dfrac{1}{4}\right) = \left(\dfrac{1}{4}\right)(x)$

Problem 3   The Associative Property of Addition

Problem 4 $(2x + xy - y) - (5x - 7xy + y)$
$2x + xy - y - 5x + 7xy - y$
$-3x + 8xy - 2y$

Problem 5 $2x - 3[\,y - 3(x - 2y + 4)]$
$2x - 3[\,y - 3x + 6y - 12]$
$2x - 3[7y - 3x - 12]$
$2x - 21y + 9x + 36$
$11x - 21y + 36$

Problem 6   the first integer: n
the next consecutive integer: $n + 1$
the next consecutive integer: $n + 2$

$n + (n + 1) + (n + 2)$
$3n + 3$

Problem 7   the unknown number: n

three eighths of the number: $\dfrac{3}{8}n$

five twelfths of the number: $\dfrac{5}{12}n$

$\dfrac{3}{8}n + \dfrac{5}{12}n$

$\dfrac{9}{24}n + \dfrac{10}{24}n$

$\dfrac{19}{24}n$

EXERCISES   (page 23)

1. 10   **3.** 4   **5.** 0   **7.** $-\dfrac{1}{7}$ **9.** $\dfrac{9}{2}$ **11.** $\dfrac{1}{2}$ **13.** 2   **15.** 3   **17.** -12 **19.** 2   **21.** 6
23. -2 **25.** 6.51   **27.** -19.71 **29.** 15   **31.** 4   **33.** 0   **35.** 5   **37.** $[-(x + y)]$ **39.** x
41. ab **43.** The Inverse Property of Addition   **45.** The Commutative Property of Multiplication
47. The Associative Property of Addition   **49.** The Distributive Property
51. The Multiplication Property of Zero   **53.** The Commutative Property of Addition   **55.** $13x$
57. $-4x$ **59.** $7a + 7b$ **61.** x **63.** $-3x + 6$ **65.** $5x + 10$ **67.** $x + y$ **69.** $3x - 6y - 5$
71. $-11a + 21$ **73.** $-x + 6y$ **75.** $-30a + 140$ **77.** $30x - 10y$ **79.** $-10a + 2b$
81. $-12a + b$ **83.** $-2x - 144y - 96$ **85.** $5x - 32 + 3y$ **87.** $x + 6$
89. $n - (5 - n)$; $2n - 5$ **91.** $\dfrac{3}{8}n - \dfrac{1}{6}n$; $\dfrac{5}{24}n$ **93.** $n + \dfrac{2}{3}n$; $\dfrac{5}{3}n$ **95.** $\dfrac{1}{2}(6n + 22)$; $3n + 11$
97. $(n + 1) + (n + 2)$; $2n + 3$ **99.** $4(n + 1) + 12$; $4n + 16$ **101.** $3\left(\dfrac{2n}{6}\right)$; n
103. $2(n + 11) - 4$; $2n + 18$ **105.** $20 - (n + 4)(12)$; $-28 - 12n$ **107.** $n + (n - 12)(3)$; $4n - 36$

109. a. The Distributive Property   **b.** The Commutative Property of Addition
c. The Associative Property of Addition   **d.** The Distributive Property
111. a. The Commutative Property of Multiplication **b.** The Associative Property of Multiplication
113. a. The Distributive Property   **b.** The Associative Property of Multiplication
c. The Multiplication Property of One   **115.** $3c$, where c = the amount of cashews
117. $\frac{1}{3}x$, where x = the measure of the largest angle   **119.** $2s - 15$, where s = the age of the silver coin

SECTION 1.3   (page 28)

Problem 1 $\{2, 4, 6, 8, 10\}$

Problem 2 $\{1, 3, 5, 7, 9\}$

Problem 3   0.35 is less than 2. Yes, 0.35 is an element of the set.

Problem 4 $A \cup C = \{-5, -2, -1, 0, 1, 2, 5\}$

Problem 5   There are no integers that are both odd and even. $E \cap F = \varnothing$

Problem 6   The solution set is $\{x \mid x > -3\}$.

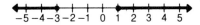

Problem 7   The solution set is $\{x \mid x \geq 1 \text{ or } x \leq -3\}$.

EXERCISES   (page 32)

1. $\{-2, -1, 0, 1, 2, 3, 4\}$ **3.** $\{2, 4, 6, 8, 10, 12\}$ **5.** $\{3\}$ **7.** $\{3, 6, 9, 12, 15, 18\}$
9. $\{x \mid x > 4, x \text{ is an integer}\}$ **11.** $\{x \mid x \geq 1, x \in \text{ real numbers}\}$ **13.** $\{x \mid -2 < x < 5, x \text{ is an integer}\}$
15. $\{x \mid 0 < x < 1, x \in \text{ real numbers}\}$ **17.** $A \cup B = \{1, 2, 4, 6, 9\}$ **19.** $A \cup B = \{2, 3, 5, 8, 9, 10\}$
21. $A \cup B = \{-4, -2, 0, 2, 4, 8\}$ **23.** $A \cup B = \{1, 2, 3, 4, 5\}$ **25.** $A \cap B = \{6\}$ **27.** $A \cap B = \{5, 10, 20\}$
29. $A \cap B = \varnothing$ **31.** $A \cap B = \{4, 6\}$
33. **35.** **37.**

39. **41.** **43.**

45. **47.** **49.**

51. **53.** $\{x \mid x > 0, x \text{ is an integer}\}$ **55.** $\{x \mid x \geq 15, x \text{ is an odd integer}\}$ **57.** a

59. b, c **61.** Closed   **63.** Closed   **65.** Closed   **67.** Not closed   **69.** Not closed

CHAPTER REVIEW (page 39)

1. 12 (Objective 1.1.1)   **2.** 2 (Objective 1.1.2)   **3.** −1.41 (Objective 1.1.2)
4. $14x + 48y$ (Objective 1.2.3)   **5.** 6 (Objective 1.1.4)   **6.** −5 (Objective 1.2.1)
7. The Distributive Property (Objective 1.2.2)   **8.** 4 (Objective 1.2.2)   **9.** −2 (Objective 1.1.1)
10. −100 (Objective 1.1.3)   **11.** −4.9 (Objective 1.1.2)   **12.** $A \cup B = \{2, 3, 4, 5, 6, 8\}$ (Objective 1.3.1)
13. $13x − y$ (Objective 1.2.3)   **14.** (Objective 1.3.2)   **15.** 10 (Objective 1.1.4)

16. −72 (Objective 1.1.3)   **17.** $A \cup B = \{-3, -1, 0, 1, 3, 5\}$ (Objective 1.3.1)   **18.** 2 (Objective 1.2.1)

19. $\frac{25}{36}$ (Objective 1.1.2)   **20.** $A \cap B = \{-1, 0, 1\}$ (Objective 1.3.1)

21.   (Objective 1.3.2)   **22.** $-\frac{4}{27}$ (Objective 1.1.2)   **23.** −270 (Objective 1.1.4)

24. −3 (Objective 1.1.1)   **25.** $A \cap B = \{-4, 0\}$ (Objective 1.3.1)   **26.** −36 (Objective 1.1.3)
27.   (Objective 1.3.2)   **28.** 1 (Objective 1.2.1)   **29.** $-5x − 2y$ (Objective 1.2.3)

30.   (Objective 1.3.2)   **31.** $13 − (n − 3)(9)$; $40 − 9n$ (Objective 1.2.4)

32. $5[n + (n + 1)]$; $10n + 5$ (Objective 1.2.4)   **33.** $8 − (4n − 12)$; $20 − 4n$ (Objective 1.2.4)
34. $6 + [n + (n + 1)]$; $2n + 7$ (Objective 1.2.4)

Note: The numbers in parentheses following the answers in the Chapter Review are a reference to the objective that corresponds with that problem. The first number of the reference indicates the chapter, and the second and third numbers indicate the objective. For example, the reference (Objective 1.2.1) stands for Chapter 1, Section 2, Objective 1. This notation will be used for all chapter and cumulative reviews throughout the text.

Answers to Chapter 2 —————————————

SECTION 2.1 (page 43)

Problem 1
$$x + 4 = -3$$
$$x + 4 + (-4) = -3 + (-4)$$
$$x = -7$$

The solution is −7.

Problem 2
$$-3x = 18$$
$$\left(-\frac{1}{3}\right)(-3x) = \left(-\frac{1}{3}\right)(18)$$
$$x = -6$$

The solution is −6.

Problem 3

$$6x - 5 - 3x = 14 - 5x$$
$$3x - 5 = 14 - 5x$$
$$3x - 5 + 5x = 14 - 5x + 5x$$
$$8x - 5 = 14$$
$$8x - 5 + 5 = 14 + 5$$
$$8x = 19$$
$$\frac{1}{8}(8x) = \frac{1}{8}(19)$$
$$x = \frac{19}{8}$$

The solution is $\frac{19}{8}$.

Problem 4

$$6(5 - x) - 12 = 2x - 3(4 + x)$$
$$30 - 6x - 12 = 2x - 12 - 3x$$
$$18 - 6x = -x - 12$$
$$18 - 5x = -12$$
$$-5x = -30$$
$$x = 6$$

The solution is 6.

Problem 5   The LCM of 3, 5, and 30 is 30.

$$\frac{2x - 7}{3} - \frac{5x + 4}{5} = \frac{-x - 4}{30}$$
$$30\left(\frac{2x - 7}{3} - \frac{5x + 4}{5}\right) = 30\left(\frac{-x - 4}{30}\right)$$
$$\frac{30(2x - 7)}{3} - \frac{30(5x + 4)}{5} = \frac{30(-x - 4)}{30}$$
$$10(2x - 7) - 6(5x + 4) = -x - 4$$
$$20x - 70 - 30x - 24 = -x - 4$$
$$-10x - 94 = -x - 4$$
$$-9x - 94 = -4$$
$$-9x = 90$$
$$x = -10$$

The solution is -10.

Problem 6   **Strategy**

▶ The first number: n
The second number: $2n$
The third number: $4n - 3$
▶ The sum of the numbers is 81.

Solution

$$n + 2n + (4n - 3) = 81$$
$$7n - 3 = 81$$
$$7n = 84$$
$$n = 12$$

$$2n = 2(12) = 24$$
$$4n - 3 = 4(12) - 3 = 48 - 3 = 45$$

The numbers are 12, 24, and 45.

Problem 7 **Strategy**

▶ The first odd integer: n
Second odd integer: $n + 2$
Third odd integer: $n + 4$
▶ Three times the sum of the first two integers is ten more than the product of the third integer and four.

Solution

$$3[n + (n + 2)] = (n + 4)4 + 10$$
$$3[2n + 2] = 4n + 16 + 10$$
$$6n + 6 = 4n + 26$$
$$2n + 6 = 26$$
$$2n = 20$$
$$n = 10$$

Since 10 is not an odd integer, there is no solution.

EXERCISES   (page 52)

1. 9 **3.** -10 **5.** -2 **7.** -8 **9.** -15 **11.** 4 **13.** $-\dfrac{2}{3}$ **15.** $\dfrac{17}{6}$ **17.** $\dfrac{1}{6}$ **19.** $\dfrac{15}{2}$ **21.** -40

23. $-\dfrac{32}{25}$ **25.** 1 **27.** $\dfrac{15}{16}$ **29.** -3.73 **31.** $\dfrac{3}{2}$ **33.** 8 **35.** 0 **37.** 6 **39.** 6 **41.** No solution

43. -3 **45.** 1 **47.** $\dfrac{7}{4}$ **49.** $\dfrac{4}{3}$ **51.** 1 **53.** $-\dfrac{4}{3}$ **55.** $\dfrac{2}{5}$ **57.** 12 **59.** $-\dfrac{3}{2}$ **61.** -3

63. No solution **65.** $\dfrac{11}{2}$ **67.** -1.25 **69.** 0.433 **71.** 6 **73.** $\dfrac{1}{2}$ **75.** -9 **77.** $\dfrac{2}{3}$ **79.** $-\dfrac{1}{3}$

81. -6 **83.** $\dfrac{5}{6}$ **85.** $-\dfrac{4}{3}$ **87.** $\dfrac{8}{9}$ **89.** $\dfrac{9}{4}$ **91.** $\dfrac{6}{7}$ **93.** $\dfrac{2}{3}$ **95.** $\dfrac{35}{12}$ **97.** -22 **99.** -1

101. -33 **103.** 4 **105.** 6 **107.** 6 **109.** $\dfrac{25}{14}$ **111.** $\dfrac{3}{4}$ **113.** $\dfrac{11}{2}$ **115.** $-\dfrac{4}{29}$ **117.** 3 **119.** -10

121. $\dfrac{5}{11}$ **123.** 1.35 **125.** 9.4 **127.** 24 **129.** 56 **131.** 3.94875 **133.** 16 and 24 **135.** 3 and 5

137. 1 and 3 **139.** 21, 44, and 58 **141.** $-30, -29,$ and -28 **143.** 8 **145.** No solution

147. 3, 5, and 7 **149.** $-\dfrac{2}{5}$ **151.** -9 **153.** -1 **155.** -21 **157.** No solution **159.** -6

161. No solution **163.** 47 **165.** 4 **167.** 50 **169.** 10 days **171.** 10 min

SECTION 2.2   (page 59)

Problem 1 $2x - 1 < 6x + 7$
$-4x - 1 < 7$
$-4x < 8$

$$-\frac{1}{4}(-4x) > -\frac{1}{4}(8)$$

$$x > -2$$

$\{x \mid x > -2\}$

Problem 2 $5x - 2 \le 4 - 3(x - 2)$
$5x - 2 \le 4 - 3x + 6$
$5x - 2 \le 10 - 3x$
$8x - 2 \le 10$
$8x \le 12$

$$\frac{1}{8}(8x) \le \frac{1}{8}(12)$$

$$x \le \frac{3}{2}$$

$$\left\{ x \mid x \le \frac{3}{2} \right\}$$

Problem 3 $-2 \le 5x + 3 \le 13$
$-2 + (-3) \le 5x + 3 + (-3) \le 13 + (-3)$
$-5 \le 5x \le 10$

$$\frac{1}{5}(-5) \le \frac{1}{5}(5x) \le \frac{1}{5}(10)$$

$$-1 \le x \le 2$$

$\{x \mid -1 \le x \le 2\}$

Problem 4 $5 - 4x > 1$ and $6 - 5x < 11$
$-4x > -4$ $-5x < 5$
$x < 1$ $x > -1$

$\{x \mid x < 1\}$ $\{x \mid x > -1\}$

$\{x \mid x < 1\} \cap \{x \mid x > -1\} = \{x \mid -1 < x < 1\}$

Problem 5 $2 - 3x > 11$ or $5 + 2x > 7$
$-3x > 9$ $2x > 2$
$x < -3$ $x > 1$

$\{x \mid x < -3\}$ $\{x \mid x > 1\}$

$\{x \mid x < -3\} \cup \{x \mid x > 1\} = \{x \mid x < -3 \text{ or } x > 1\}$

Problem 6   **Strategy**

To find the maximum height, substitute the given values in the inequality $\frac{1}{2}bh < 50$ and solve.

Solution

$$\frac{1}{2}bh < 50$$

$$\frac{1}{2}(12)(x + 2) < 50$$

$$6(x + 2) < 50$$

$$6x + 12 < 50$$

$$6x < 38$$

$$x < \frac{19}{3}$$

The largest integer less than $\frac{19}{3}$ is 6.

$$x + 2 = 6 + 2 = 8$$

The maximum height of the triangle is 8 in.

Problem 7   **Strategy**

To find the range of scores, write and solve an inequality using N to represent the score on the fifth exam.

Solution

$$80 \le \frac{72 + 94 + 83 + 70 + N}{5} \le 89$$

$$80 \le \frac{319 + N}{5} \le 89$$

$$5(80) \le 5\left(\frac{319 + N}{5}\right) \le 5(89)$$

$$400 \le 319 + N \le 445$$

$$400 - 319 \le 319 + N - 319 \le 445 - 319$$

$$81 \le N \le 126$$

Since 100 is a maximum score, the range of scores that will give the student a B for the course is $81 \le N \le 100$.

EXERCISES   (page 66)

1. $\{x \mid x < 5\}$ **3.** $\{x \mid x \le 2\}$ **5.** $\{x \mid x < -4\}$ **7.** $\{x \mid x > 3\}$ **9.** $\{x \mid x > 4\}$ **11.** $\{x \mid x \le 2\}$
13. $\{x \mid x > -2\}$ **15.** $\{x \mid x \ge 2\}$ **17.** $\{x \mid x > -2\}$ **19.** $\{x \mid x \le 3\}$ **21.** $\{x \mid x < 2\}$ **23.** $\{x \mid x < -3\}$
25. $\{x \mid x \le 5\}$ **27.** $\{x \mid x \ge 1\}$ **29.** $\{x \mid x < -5\}$ **31.** $\{x \mid x < -24\}$ **33.** $\{x \mid x < 3\}$ **35.** $\{x \mid x > -3\}$
37. $\{x \mid -1 < x < 2\}$ **39.** $\{x \mid x \ge 3 \text{ or } x \le 1\}$ **41.** $\{x \mid -2 < x < 4\}$ **43.** $\{x \mid x < -3 \text{ or } x > 0\}$
45. $\{x \mid x \ge 3\}$ **47.** $\varnothing$ **49.** $\varnothing$ **51.** $\{x \mid x < 1 \text{ or } x > 3\}$ **53.** $\{x \mid -3 < x < 4\}$ **55.** $\{x \mid 3 < x < 5\}$
57. $\left\{x \mid x > 3 \text{ or } x \le -\frac{5}{2}\right\}$ **59.** $\{x \mid x > 4\}$ **61.** $\varnothing$ **63.** The solution set is the set of real numbers.

65. $\{x \mid -4 \le x \le 2\}$ **67.** $\{x \mid -4 < x < -2\}$ **69.** $\{x \mid x < -4 \text{ or } x > 3\}$ **71.** $\{x \mid -4 \le x < 1\}$

73. $\left\{x \mid x > \dfrac{27}{2} \text{ or } x < \dfrac{5}{2}\right\}$ **75.** $\left\{x \mid 7 < x < \dfrac{15}{2}\right\}$ **77.** $\left\{x \mid -10 \le x \le \dfrac{11}{2}\right\}$ **79.** -20 **81.** 12 in.

83. less than 115 mi   **85.** less than 350 mi   **87.** \$44,000 or more   **89.** less than 50 checks

91. $58 \le N \le 100$ **93.** 10, 12, and 14; or 12, 14, and 16; or 14, 16, and 18   **95.** $\{1, 2, 3\}$ **97.** $\{1, 2\}$

99. $\{1, 2, 3, 4, 5, 6\}$ **101.** $\{1, 2, 3, 4\}$ **103.** $\{1, 2\}$ **105.** $\{1, 2, 3, 4, 5, 6\}$ **107.**

109.

   111. 20°C to 25°C   **113.** 44   **115.** 11 min

SECTION 2.3   (page 72)

Problem 1   **A.** $|x| = 25$

$$x = 25 \qquad x = -25$$

The solutions are 25 and -25.

B. $|2x - 3| = 5$

$$2x - 3 = 5 \qquad 2x - 3 = -5$$
$$2x = 8 \qquad\quad 2x = -2$$
$$x = 4 \qquad\quad x = -1$$

The solutions are 4 and -1.

C. $|x - 3| = -2$

The absolute value of a number must be nonnegative.
There is no solution to this equation.

D. $5 - |3x + 5| = 3$
$$-|3x + 5| = -2$$
$$|3x + 5| = 2$$

$$3x + 5 = 2 \qquad 3x + 5 = -2$$
$$3x = -3 \qquad\quad 3x = -7$$
$$x = -1 \qquad\quad x = -\frac{7}{3}$$

The solutions are -1 and $-\dfrac{7}{3}$.

Problem 2 $|3x + 2| < 8$

$$-8 < 3x + 2 < 8$$
$$-8 + (-2) < 3x + 2 + (-2) < 8 + (-2)$$
$$-10 < 3x < 6$$
$$\frac{1}{3}(-10) < \frac{1}{3}(3x) < \frac{1}{3}(6)$$
$$-\frac{10}{3} < x < 2$$

$$\left\{x \mid -\frac{10}{3} < x < 2\right\}$$

Problem 3 $|3x - 7| < 0$

The absolute value of a number must be nonnegative. The solution set is the empty set.

$$\varnothing$$

Problem 4 $|5x + 3| > 8$

$$5x + 3 < -8 \quad \text{or} \quad 5x + 3 > 8$$
$$5x < -11 \qquad\qquad 5x > 5$$
$$x < -\frac{11}{5} \qquad\qquad x > 1$$
$$\qquad\qquad\qquad \{x \mid x > 1\}$$

$$\left\{ x \mid x < -\frac{11}{5} \right\}$$

$$\left\{ x \mid x < -\frac{11}{5} \right\} \cup \{x \mid x > 1\} = \left\{ x \mid x < -\frac{11}{5} \text{ or } x > 1 \right\}$$

Problem 5   **Strategy**
Let b represent the diameter of the bushing, T the tolerance, and d the lower and upper limits of the diameter. Solve the absolute value inequality $|d - b| \le T$ for d.

Solution

$$|d - b| \le T$$
$$|d - 2.55| \le 0.003$$

$$-0.003 \le d - 2.55 \le 0.003$$
$$-0.003 + 2.55 \le d - 2.55 + 2.55 \le 0.003 + 2.55$$
$$2.547 \le d \le 2.553$$

The lower and upper limits of the diameter of the bushing are 2.547 in. and 2.553 in.

EXERCISES   (page 77)

1. 7 and -7 **3.** 4 and -4 **5.** 6 and -6 **7.** 7 and -7 **9.** No Solution   **11.** No Solution

13. 1 and -5 **15.** 8 and 2   **17.** 2   **19.** No Solution   **21.** $\frac{9}{2}$ and $\frac{1}{2}$ **23.** 3 and $-\frac{5}{3}$

25. $\frac{3}{2}$ and $-\frac{13}{2}$ **27.** -1 and 5   **29.** $-\frac{2}{3}$ and 2   **31.** 0 and $\frac{8}{3}$ **33.** $-\frac{3}{2}$ and 3   **35.** $\frac{3}{2}$

37. No Solution   **39.** 7 and -3 **41.** 2 and $-\frac{10}{3}$ **43.** 1 and 3   **45.** $\frac{3}{2}$ **47.** No Solution

49. $\frac{11}{6}$ and $-\frac{1}{6}$ **51.** $-\frac{1}{3}$ and -1 **53.** No Solution   **55.** 3 and 0   **57.** No Solution

59. 1 and $\frac{13}{3}$ **61.** No Solution   **63.** $\frac{7}{3}$ and $\frac{1}{3}$ **65.** $-\frac{1}{2}$ **67.** $-\frac{1}{2}$ and $-\frac{7}{2}$ **69.** $-\frac{8}{3}$ and $\frac{10}{3}$

71. No Solution   **73.** $\{x \mid x > 3 \text{ or } x < -3\}$ **75.** $\{x \mid x > 1 \text{ or } x < -3\}$ **77.** $\{x \mid 4 \le x \le 6\}$

79. $\{x \mid x \ge 5 \text{ or } x \le -1\}$ **81.** $\{x \mid -3 < x < 2\}$ **83.** $\left\{ x \mid x > 2 \text{ or } x < -\frac{14}{5} \right\}$ **85.** $\varnothing$

87. The solution set is the set of real numbers.   **89.** $\left\{ x \mid x \le -\frac{1}{3} \text{ or } x \ge 3 \right\}$ **91.** $\left\{ x \mid -2 \le x \le \frac{9}{2} \right\}$

93. 2   **95.** $\left\{ x \mid x < -2 \text{ or } x > \frac{22}{9} \right\}$ **97.** lower limit: 3.95 cc; upper limit: 4.05 cc

99. lower limit: 2.648 in.; upper limit: 2.652 in.   **101.** lower limit: $9\frac{19}{32}$ in; upper limit: $9\frac{21}{32}$ in.

103. lower limit: 28,420 ohms; upper limit: 29,580 ohms   **105.** lower limit: 23,750 ohms;

upper limit: 26,250 ohms   **107.** -8 and 13   **109.** $-\dfrac{2}{3}$ and 2   **111.** $\{x \mid x > 5 \text{ or } x < -4\}$

113. $\{x \mid -7 \le x \le 8\}$

115.   **117.**   **119.**

121.   **123.** $\{x \mid x \ge -3\}$ **125.** $\{a \mid a \ge 4\}$

127. $\{c \mid c \le -5\}$ **129.** $\{y \mid y \le 8\}$ **131.** $|ax + b| < c$

$$-c < ax + b < c$$
$$-c - b < ax < c - b$$
$$\frac{-c - b}{a} < x < \frac{c - b}{a}$$
$$\left\{ x \,\middle|\, \frac{-c - b}{a} < x < \frac{c - b}{a} \right\}$$

SECTION 2.4   (page 82)

Problem 1   **Strategy**

▶ Number of 3¢ stamps: x
Number of 10¢ stamps: $2x + 2$
Number of 15¢ stamps: $3x$

Stamps	Number	Value	Total Value
3¢	x	3	$3x$
10¢	$2x + 2$	10	$10(2x + 2)$
15¢	$3x$	15	$45x$

▶ The sum of the total values of each type of stamp equals the total value of all the stamps (156 cents).

Solution

$$3x + 10(2x + 2) + 45x = 156$$
$$3x + 20x + 20 + 45x = 156$$
$$68x + 20 = 156$$
$$68x = 136$$
$$x = 2$$

$$3x = 3(2) = 6$$

There are six 15¢ stamps in the collection.

Problem 2   **Strategy**

▶ Pounds of $3.00 hamburger: x
Pounds of $1.80 hamburger: $75 - x$

	Amount	Cost	Value
$3.00 hamburger	x	3.00	$3.00x$
$1.80 hamburger	$75 - x$	1.80	$1.80(75 - x)$
Mixture	75	2.20	$75(2.20)$

▶ The sum of the values before mixing equals the value after mixing.

Solution

$$3.00x + 1.80(75 - x) = 75(2.20)$$
$$3x + 135 - 1.80x = 165$$
$$1.2x + 135 = 165$$
$$1.2x = 30$$
$$x = 25$$

$$75 - x = 75 - 25 = 50$$

The mixture must contain 25 lb of the $3.00 hamburger and 50 lb of the $1.80 hamburger.

Problem 3   **Strategy**

▶ Rate of the second plane: r
Rate of the first plane: $r + 30$

	Rate	Time	Distance
First plane	$r + 30$	4	$4(r + 30)$
Second plane	r	4	$4r$

▶ The total distance traveled by the two planes is 1160 mi.

Solution

$$4(r + 30) + 4r = 1160$$
$$4r + 120 + 4r = 1160$$
$$8r + 120 = 1160$$
$$8r = 1040$$
$$r = 130$$

$$r + 30 = 130 + 30 = 160$$

The first plane is traveling 160 mph.
The second plane is traveling 130 mph.

Problem 4   **Strategy**

▶ Time to the resort: t
Time returning from the resort: $10 - t$

	Rate	Time	Distance
Going	60	t	$60t$
Returning	40	$10 - t$	$40(10 - t)$

▶ The distance to the resort is the same as the distance returning.

Solution

$60t = 40(10 - t)$
$60t = 400 - 40t$
$100t = 400$
$t = 4$

$d = rt = 60(4) = 240$

The distance to the mountain resort is 240 mi.

EXERCISES   (page 88)

1. 5 dimes; 17 quarters   **3.** 13 nickels   **5.** 16 dimes   **7.** 125 ten-dollar bills   **9.** 20¢ stamps: 100; 28¢ stamps: 40   **11.** 20 stamps   **13.** 80 stamps   **15.** 45 stamps   **17.** 7.5 lb of the $6 grade; 17.5 lb of the $3.50 grade   **19.** 72 adult tickets   **21.** 250 bushels of soybeans; 750 bushels of wheat   **23.** 90 oz   **25.** $6.85/oz   **27.** 15.625 kg of walnuts; 34.375 kg of cashews   **29.** 3 mph; 3.5 mph   **31.** 104 mi   **33.** 375 mi   **35.** freight train: 30 mph; passenger train: 48 mph   **37.** 1st car: 58 mph; 2nd car: 66 mph   **39.** 44 mi   **41.** $3.10/lb   **43.** $4.30   **45.** 14 stamps   **47.** 479 mi   **49.** −43

SECTION 2.5   (page 94)

Problem 1   **Strategy**

▶ Amount invested at 11.5%: x

	Principal	Rate	Interest
Amount at 13.2%	3500	0.132	0.132(3500)
Amount at 11.5%	x	0.115	$0.115x$

▶ The sum of the interest earned by the two investments equals the total annual interest earned ($1037).

Solution

$0.132(3500) + 0.115x = 1037$
$462 + 0.115x = 1037$
$0.115x = 575$
$x = 5000$

The amount invested at 11.5% is $5000.

Problem 2   Strategy

▶ Pounds of 22% hamburger: x
Pounds of 12% hamburger: $80 - x$

	Amount	Percent	Quantity
22%	x	0.22	$0.22x$
12%	$80 - x$	0.12	$0.12(80 - x)$
18%	80	0.18	$0.18(80)$

▶ The sum of the quantities before mixing is equal to the quantity after mixing.

Solution

$$0.22x + 0.12(80 - x) = 0.18(80)$$
$$0.22x + 9.6 - 0.12x = 14.4$$
$$0.10x + 9.6 = 14.4$$
$$0.10x = 4.8$$
$$x = 48$$

$$80 - x = 80 - 48 = 32$$

The butcher needs 48 lb of the hamburger that is 22% fat and 32 lb of the hamburger that is 12% fat.

EXERCISES   (page 98)
1. $3000 @ 10.5%; $5000 @ 14%   **3.** $8250   **5.** $6000 @ 11.2%; $4800 @ 14%   **7.** $5500 @ 8%;
$8800 @ 10.5%; $7700 @ 20%   **9.** $12,000   **11.** 32%   **13.** 28.8 ml   **15.** 35%   **17.** 28 lb of 21% fat;
56 lb of 15% fat   **19.** 60 g   **21.** 50 lb   **23.** 55%   **25.** 75 L of 60% solution; 25 L of 20% solution
27. 7.1%   **29.** $5000 @ 9%; $6000 @ 8%, and $9000 @ 9.5%   **31.** $4.84 lb   **33.** rate of the 1st plane:
265 mph; rate of the 2nd plane: 225 mph   **35.** 18.75 kg   **37.** $8400 or more

CHAPTER REVIEW   (page 105)

1. $\frac{5}{6}$ (Objective 2.1.1)   **2.** $-\frac{1}{5}$ (Objective 2.1.1)   **3.** $\{x \mid x > 2 \text{ or } x < -2\}$ (Objective 2.2.2)

4. $\frac{12}{7}$ (Objective 2.1.2)   **5.** -2 (Objective 2.1.1)   **6.** $\left\{x \mid x < -\frac{4}{3} \text{ or } x > 2\right\}$ (Objective 2.3.2)

7. $\frac{32}{3}$ (Objective 2.1.1)   **8.** 3 and $-\frac{9}{5}$ (Objective 2.3.1)   **9.** $\{x \mid x > -1\}$ (Objective 2.2.1)

10. -12 (Objective 2.1.1)   **11.** $-\frac{1}{8}$ (Objective 2.1.1)   **12.** $\frac{11}{13}$ (Objective 2.1.2)

13. $\{x \mid x > 2\}$ (Objective 2.2.1)   **14.** -1 (Objective 2.1.1)   **15.** $\{x \mid 1 \le x \le 4\}$ (Objective 2.3.2)
16. -24 (Objective 2.1.2)   **17.** $\{x \mid -4 \le x \le 1\}$ (Objective 2.2.2)   **18.** 4 (Objective 2.1.1)
19. 3 (Objective 2.1.2)   **20.** $\{x \mid x \le 1\}$ (Objective 2.2.1)   **21.** 7 and -2 (Objective 2.3.1)
22. $\varnothing$ (Objective 2.2.2)   **23.** 12 dimes (Objective 2.4.1)   **24.** 75 mi (Objective 2.4.3)
25. 40% (Objective 2.5.2)   **26.** 6 and 12 (Objective 2.1.3)   **27.** 42 checks (Objective 2.2.3)
28. $5000 @ 9.5%; $11,000 @ 11.2% (Objective 2.5.1)   **29.** 10 lb (Objective 2.4.2)   **30.** lower limit: 3.97 cc;
upper limit: 4.03 cc (Objective 2.3.3)

CUMULATIVE REVIEW   (page 107)

1. -108 (Objective 1.1.3)   **2.** 3 (Objective 1.1.4)   **3.** -64 (Objective 1.1.4)   **4.** -8 (Objective 1.2.1)
5. The Commutative Property of Addition (Objective 1.2.2)   **6.** $A \cap B = \{3, 9\}$ (Objective 1.3.1)
7. (Objective 1.3.2)   **8.** (Objective 1.3.2)

$-2 \quad 0$ $-2 \quad 0 \quad 1$

9. $-17x + 2$ (Objective 1.2.3)   **10.** $25y$ (Objective 1.2.3)   **11.** 2 (Objective 2.1.1)   **12.** $\frac{1}{2}$ (Objective 2.1.1)

13. 1 (Objective 2.1.1)   **14.** 24 (Objective 2.1.1)   **15.** 2 (Objective 2.1.2)   **16.** 2 (Objective 2.1.2)

17. $-\frac{13}{5}$ (Objective 2.1.2)   **18.** $\{x \mid x \le -3\}$ (Objective 2.2.1)   **19.** $\varnothing$ (Objective 2.2.2)

20. $\{x \mid x > -2\}$ (Objective 2.2.2)   **21.** -1 and 4 (Objective 2.3.1)   **22.** 7 and -4 (Objective 2.3.1)

23. $\left\{x \mid \frac{1}{3} \le x \le 3\right\}$ (Objective 2.3.2)   **24.** $\left\{x \mid x > 2 \text{ or } x < -\frac{1}{2}\right\}$ (Objective 2.3.2)   **25.** $(3n + 6) + 3n$;
$6n + 6$ (Objective 1.2.4)   **26.** 1 (Objective 2.1.3)   **27.** 11 stamps (Objective 2.4.1)   **28.** 48 adult tickets
(Objective 2.4.2)   **29.** 340 mph (Objective 2.4.3)   **30.** 3 L (Objective 2.5.2)   **31.** $6500 (Objective 2.5.1)

Answers to Chapter 3

SECTION 3.1   (page 111)

Problem 1
$$\begin{array}{r} 5x^2 + 3x - 1 \\ 2x^2 + 4x - 6 \\ + \ x^2 - 7x + 8 \\ \hline 8x^2 \qquad + 1 \end{array}$$

Problem 2 $(4x^2 + 3x - 5) + (6x^3 - 2 + x^2)$
$6x^3 + 5x^2 + 3x - 7$

Problem 3
$$\begin{array}{r} -5x^2 + 2x - 3 \\ - \ 6x^2 + 3x - 7 \\ \hline \end{array} = \begin{array}{r} -5x^2 + 2x - 3 \\ + \ -6x^2 - 3x + 7 \\ \hline -11x^2 - \quad x + 4 \end{array}$$

Problem 4 $(5x^{2n} - 3x^n - 7) - (-2x^{2n} - 5x^n + 8)$
$(5x^{2n} - 3x^n - 7) + (2x^{2n} + 5x^n - 8)$
$7x^{2n} + 2x^n - 15$

Problem 5 $(7xy^3)(-5x^2y^2)(-xy^2) = 35x^4y^7$

Problem 6   **A.** $(y^3)^6 = y^{18}$ **B.** $(x^n)^3 = x^{3n}$

Problem 7 $(-2ab^3)^4 = (-2)^4 a^{1 \cdot 4} b^{3 \cdot 4} = 16a^4 b^{12}$

Problem 8 $6a(2a)^2 + 3a(2a^2) = 6a(2^2 a^2) + 6a^3 = 6(4)a^3 + 6a^3 = 24a^3 + 6a^3 = 30a^3$

Problem 9   **A.** $\dfrac{x^2y^4}{x^3y} = \dfrac{y^{4-1}}{x^{3-2}} = \dfrac{y^3}{x}$ **B.** $\dfrac{a^{2n+1}}{a^{n+3}} = a^{(2n+1)-(n+3)} = a^{2n+1-n-3} = a^{n-2}$

Problem 10 $\left(\dfrac{4a^2}{5b^3}\right)^3\left(\dfrac{-10a^3}{2b^8}\right)^2 = \left(\dfrac{4a^2}{5b^3}\right)^3\left(\dfrac{-5a^3}{b^8}\right)^2 = \left(\dfrac{4^3a^6}{5^3b^9}\right)\left(\dfrac{(-5)^2a^6}{b^{16}}\right) =$

$\left(\dfrac{64a^6}{125b^9}\right)\left(\dfrac{25a^6}{b^{16}}\right) = \dfrac{64 \cdot 25a^{12}}{125b^{25}} = \dfrac{64a^{12}}{5b^{25}}$

Problem 11   **A.** $12a^{12}b^4 = (6a^5b^3)(2a^7b)$ **B.** $a^{4n}b^{6n} = (a^{2n}b^{3n})(a^{2n}b^{3n})$

EXERCISES   (page 118)

1. $6x^2 - 6x + 5$ **3.** $3x^2 - 3xy - 2y^2$ **5.** $-x^2 + 1$ **7.** $9x^n + 5$ **9.** $3y^3 + 2y^2 - 15y + 2$
11. $7a^2 - a + 2$ **13.** $3x^4 - 8x^2 + 2x$ **15.** $-5a^2 - 6a - 4$ **17.** $-b^{2n} + 2b^n - 7$ **19.** $2x^2 + 4x - 7$
21. $-3x^3 + 7x^2 - 5x - 10$ **23.** $7x^4 - 9x^3 - 5x^2 + 9x - 2$ **25.** $5a^3 - 8a^2b + 10ab^2 + 7b^3$ **27.** a^4b^4
29. $-18x^3y^4$ **31.** x^8y^{16} **33.** $81x^8y^{12}$ **35.** $729a^{10}b^6$ **37.** x^5y^{11} **39.** $-432a^7b^{11}$ **41.** $54a^{13}b^{17}$
43. x^{2n+1} **45.** y^{6n-2} **47.** a^{2n2-6n} **49.** x^{15n+10} **51.** $-6x^5y^5z^4$ **53.** $-12a^2b^9c^2$ **55.** $-6x^4y^4z^5$
57. $48x^{13}y^3z^{16}$ **59.** $-72x^9y^7z^{11}$ **61.** $18a^4b^3$ **63.** $15x^5y^3$ **65.** $-54a^{3n+1}b^{n+1}c^{3n+1}$ **67.** $16a^{2n+1}b^3$
69. $\dfrac{5b^5}{7}$ **71.** $-\dfrac{x^{10}}{y^5}$ **73.** $\dfrac{1}{3y^2}$ **75.** $-\dfrac{3b^7}{2a^3}$ **77.** $\dfrac{b^4}{a^4c^3}$ **79.** $\dfrac{1}{x}$ **81.** $\dfrac{2}{27xy^2}$ **83.** $-\dfrac{64}{27a^9b^{18}}$ **85.** $\dfrac{2x}{y^7}$
87. $-\dfrac{4y^5}{x^{25}}$ **89.** a^{2n} **91.** $-x^{3n}$ **93.** x^{n+2} **95.** $-\dfrac{a^{10}}{18b^3}$ **97.** $\dfrac{(a-b)^2}{4a}$ **99.** $\dfrac{a^{2n}}{b^n}$ **101.** $\dfrac{a^{n-3}}{b^{n+1}}$
103. $2a^3$ **105.** $2ab^3$ **107.** $2x^2$ **109.** $2x^3y$ **111.** $2xy^2z^2$ **113.** $4ab^3$ **115.** $-2x^2$
117. a^{2n} **119.** z **121.** No   **123.** No   **125.** Yes   **127.** Yes   **129.** $9x^2 + 6x + 3$
131. $b^3 - b^2 + 4b + 7$ **133.** $2a^4 - 3a^3 + a^2 + 5a - 8$ **135.** $8y$ **137.** $7a^2b^2c^3$ **139.** $-7x^2z^2$
141. -2 **143.** $8x^n$ **145.** $24a^2b^2$ **147.** $4x^2y^3$ **149.** $4ab$

SECTION 3.2   (page 123)

Problem 1   **A.** $-4y(y^2 - 3y + 2) = -4y(y^2) - (-4y)(3y) + (-4y)(2) = -4y^3 + 12y^2 - 8y$

B. $x^2 - 2x[x - x(4x - 5) + x^2]$
$x^2 - 2x[x - 4x^2 + 5x + x^2]$
$x^2 - 2x[6x - 3x^2]$
$x^2 - 12x^2 + 6x^3$
$6x^3 - 11x^2$

C. $y^{n+3}(y^{n-2} - 3y^2 + 2)$
$y^{n+3}(y^{n-2}) - (y^{n+3})(3y^2) + (y^{n+3})(2)$
$y^{n+3+(n-2)} - 3y^{n+3+2} + 2y^{n+3}$
$y^{2n+1} - 3y^{n+5} + 2y^{n+3}$

Problem 2

$$
\begin{array}{r}
-2b^2 + 5b - 4 \\
\times \quad\quad -3b + 2 \\
\hline
-4b^2 + 10b - 8 \\
6b^3 - 15b^2 + 12b \quad\quad\;\; \\
\hline
6b^3 - 19b^2 + 22b - 8
\end{array}
$$

Problem 3 $(5a - 3b)(2a + 7b) = 10a^2 + 35ab - 6ab - 21b^2 = 10a^2 + 29ab - 21b^2$

Problem 4   **A.** $(3x - 7)(3x + 7) = 9x^2 - 49$

B. $(3x - 4y)^2 = 9x^2 - 24xy + 16y^2$

C. $(2x^n + 3)(2x^n - 3) = 4x^{2n} - 9$

D. $(2x^n - 8)^2 = 4x^{2n} - 32x^n + 64$

Problem 5   **Strategy**

To find the area, replace the variables b and h in the equation $A = \frac{1}{2}bh$ with the given values and solve for A.

Solution

$A = \frac{1}{2}bh$

$A = \frac{1}{2}(2x + 6)(x - 4)$

$A = (x + 3)(x - 4)$

$A = x^2 - 4x + 3x - 12$

$A = x^2 - x - 12$

The area is $(x^2 - x - 12)$ ft^2.

Problem 6   **Strategy**

To find the volume, subtract the volume of the small rectangular solid from the volume of the large rectangular solid.

Large rectangular solid: Length $= L_1 = 12x$
              Width $= w_1 = 7x + 2$
              Height $= h_1 = 5x - 4$
Small rectangular solid: Length $= L_2 = 12x$
              Width $= w_2 = x$
              Height $= h_2 = 2x$

Solution

$V =$ Volume of large rectangular solid $-$ volume of small rectangular solid
$V = (L_1 \cdot w_1 \cdot h_1) - (L_2 \cdot w_2 \cdot h_2)$
$V = (12x)(7x + 2)(5x - 4) - (12x)(x)(2x)$
$V = (84x^2 + 24x)(5x - 4) - (12x^2)(2x)$
$V = (420x^3 - 336x^2 + 120x^2 - 96x) - (24x^3)$
$V = 396x^3 - 216x^2 - 96x$

The volume is $(396x^3 - 216x^2 - 96x)$ ft^3.

EXERCISES   (page 128)

1. $2x^2 - 6x$ **3.** $6x^4 - 3x^3$ **5.** $6x^2y - 9xy^2$ **7.** $x^{n+1} + x^n$ **9.** $x^{2n} + x^n y^n$ **11.** $-4b^2 + 10b$
13. $-6a^4 + 4a^3 - 6a^2$ **15.** $9b^5 - 9b^3 + 24b$ **17.** $-3y^4 - 4y^3 + 2y^2$ **19.** $-20x^5 - 15x^4 + 15x^3 - 20x^2$
21. $-2x^4y + 6x^3y^2 - 4x^2y^3$ **23.** $x^{3n} + x^{2n} + x^{n+1}$ **25.** $a^{2n+1} - 3a^{n+2} + 2a^{n+1}$ **27.** $5y^2 - 11y$
29. $6y^2 - 31y$ **31.** $24n^3 + 16n^2 - n - 12$ **33.** $y^2 + 11y + 24$ **35.** $15x^2 - 61x + 56$
37. $14x^2 - 69xy + 27y^2$ **39.** $3a^2 + 16ab - 35b^2$ **41.** $15x^2 + 37xy + 18y^2$ **43.** $30x^2 - 79xy + 45y^2$

45. $2x^2y^2 - 3xy - 35$ **47.** $x^4 - 10x^2 + 24$ **49.** $x^4 + 2x^2y^2 - 8y^4$ **51.** $x^{2n} - 9x^n + 20$
53. $10b^{2n} + 18b^n - 4$ **55.** $3x^{2n} + 7x^nb^n + 2b^{2n}$ **57.** $x^3 + 8x^2 + 7x - 24$ **59.** $a^4 - a^3 - 6a^2 + 7a + 14$
61. $6a^3 - 13a^2b - 14ab^2 - 3b^3$ **63.** $6b^4 - 6b^3 + 3b^2 + 9b - 18$ **65.** $6a^5 - 15a^4 - 6a^3 + 19a^2 - 20a + 25$
67. $x^4 - 5x^3 + 14x^2 - 23x + 7$ **69.** $3b^3 - 14b^2 + 17b - 6$ **71.** $a^{7n} - 3a^{5n} - a^{4n} + a^{3n} + 3a^{2n} - 3a^n$
73. $x^{3n} - 4x^{2n}y^n + 2x^ny^{2n} + 3y^{3n}$ **75.** $b^2 - 49$ **77.** $b^2 - 121$ **79.** $x^2 - y^2z^2$ **81.** $36 - x^2$
83. $4a^2 - 9b^2$ **85.** $x^4 - 1$ **87.** $x^{2n} - 9$ **89.** $25a^2 - 81b^2$ **91.** $4x^{2n} - 25$ **93.** $y^2 + 4y + 4$
95. $4x^2 - 4xy + y^2$ **97.** $25x^2 - 40xy + 16y^2$ **99.** $x^4 + 2x^2y^2 + y^4$ **101.** $9a^2 - 24ab + 16b^2$
103. $9x^{2n} + 12x^n + 4$ **105.** $4x^{2n} + 4x^ny^n + y^{2n}$ **107.** $x^{2n} - 2x^n + 1$ **109.** $4x^{2n} + 20x^ny^n + 25y^{2n}$
111. $\left(\frac{3}{2}x^2 - 5x - 4\right)$ ft^2 **113.** $(x^2 + 12x + 16)$ ft^2 **115.** $(3x^3 - 10x^2 - 8x)$ cm^3
117. $(4x^3 + 32x^2 + 48x)$ cm^3 **119.** $(3.14x^2 - 12.56x + 12.56)$ in.^{2} **121.** $9x^2 - 30x + 25$ **123.** $2x^2 + 4xy$
125. $5a^2 + 4$ **127.** $6x^5 - 5x^4 + 8x^3 + 13x^2$ **129.** $5x^2 - 5y^2$ **131.** $4x^4y^2 - 4x^4y + x^4$ **133.** 3
135. 1 **137.** 5 **139.** -1 **141.** a^2 **143.** $x^3 - 15x^2 + 75x - 125$ **145.** $2x^2 + 11x - 21$
147. $x^2 - 2xy + y^2$ **149.** 256 **151.** $(x^2 - y^2)^2 = (x^2 - y^2)(x^2 - y^2) = x^4 - 2x^2y^2 + y^4;\ (x^2 - y^2)^2 \neq x^4 - y^4$

SECTION 3.3   (page 133)

Problem 1   **A.** The GCF of $3x^3y - 6x^2y^2 - 3xy^3$ is $3xy$.

$$3x^3y - 6x^2y^2 - 3xy^3 = 3xy(x^2 - 2xy - y^2)$$

B. The GCF of $6t^{2n}$ and $9t^n$ is $3t^n$.

$$6t^{2n} - 9t^n = 3t^n(2t^n - 3)$$

Problem 2   **A.** $x^2 + 13x + 42 = (x + 6)(x + 7)$

B. $x^2 - x - 20 = (x + 4)(x - 5)$

C. $x^2 + 5xy + 6y^2 = (x + 2y)(x + 3y)$

Problem 3 $x^2 + 5x - 1$

There are no factors of -1 whose sum is 5.

The trinomial is nonfactorable over the integers.

Problem 4   **A.** $4x^2 + 15x - 4$

Factors of 4	Factors of −4	Trial Factors	Middle Term
1, 4	1, −4	$(4x + 1)(x - 4)$	$-16x + x = -15x$
2, 2	−1, 4	$(4x - 1)(x + 4)$	$16x - x = 15x$
	2, −2		

$$4x^2 + 15x - 4 = (4x - 1)(x + 4)$$

B. $10x^2 + 39x + 14$

Factors of 10	Factors of 14	Trial Factors	Middle Term
1, 10	1, 14	$(x + 2)(10x + 7)$	$7x + 20x = 27x$
2, 5	2, 7	$(2x + 1)(5x + 14)$	$28x + 5x = 33x$
		$(10x + 1)(x + 14)$	$140x + x = 141x$
		$(5x + 2)(2x + 7)$	$35x + 4x = 39x$

$$10x^2 + 39x + 14 = (5x + 2)(2x + 7)$$

Problem 5   **A.** $6x^2y^2 - 19xy + 10 = (3xy - 2)(2xy - 5)$

B. $3x^4 + 4x^2 - 4 = (x^2 + 2)(3x^2 - 2)$

Problem 6   **A.** $3a^3b^3 + 3a^2b^2 - 60ab = 3ab(a^2b^2 + ab - 20) = 3ab(ab + 5)(ab - 4)$

B. $40a - 10a^2 - 15a^3 = 5a(8 - 2a - 3a^2) = 5a(4 - 3a)(2 + a)$

EXERCISES   (page 141)

1. $3a(2a - 5)$ **3.** $x^2(4x - 3)$ **5.** Nonfactorable   **7.** $x(x^4 - x^2 - 1)$ **9.** $4(4x^2 - 3x + 6)$
11. $5b^2(1 - 2b + 5b^2)$ **13.** $x^n(x^n - 1)$ **15.** $x^{2n}(x^n - 1)$ **17.** $a^2(a^{2n} + 1)$ **19.** $6x^2y(2y - 3x + 4)$
21. $4a^2b^2(-4b^2 - 1 + 6a)$ **23.** $y^2(y^{2n} + y^n - 1)$ **25.** $(x - 3)(x - 5)$ **27.** $(a + 1)(a + 11)$
29. $(b + 7)(b - 5)$ **31.** $(y - 3)(y - 13)$ **33.** $(b + 8)(b - 4)$ **35.** $(a - 7)(a - 8)$
37. $(y + 1)(y + 12)$ **39.** $(x + 5)(x - 1)$ **41.** $(a + 5b)(a + 6b)$ **43.** $(x - 2y)(x - 12y)$
45. $(y + 9x)(y - 7x)$ **47.** $(7 + x)(3 - x)$ **49.** $(5 + a)(10 - a)$ **51.** Nonfactorable
53. $(2x + 5)(x - 8)$ **55.** $(4y - 3)(y - 3)$ **57.** $(2a + 1)(a + 6)$ **59.** Nonfactorable
61. $(5x + 1)(x + 5)$ **63.** $(11x - 1)(x - 11)$ **65.** $(2x + 3)(6x - 1)$ **67.** $(4a + 1)(3a - 5)$
69. $(3x + 2)(5x + 3)$ **71.** $(12x - 5)(x - 1)$ **73.** $(4y - 3)(2y - 3)$ **75.** Nonfactorable
77. Nonfactorable   **79.** $(3x + 7y)(2x - 3y)$ **81.** $(4a + 7b)(a + 9b)$ **83.** $(5x - 4y)(2x - 3y)$
85. $(3 + 2x)(8 - x)$ **87.** Nonfactorable   **89.** $(5 + 2a)(3 - 4a)$ **91.** $(3 + 4a)(4 - 7a)$
93. $(5 + 2a)(3 - 10a)$ **95.** $(xy + 3)(xy - 11)$ **97.** $(ab + 6)(ab + 4)$ **99.** $(y^2 + 2)(y^2 - 8)$
101. $(a^2 + 5)(a^2 + 9)$ **103.** $(a^2b^2 + 13)(a^2b^2 - 2)$ **105.** $(a^n + 3)(a^n - 4)$ **107.** $(5xy - 4)(xy - 11)$
109. $(ab + 1)(10ab - 7)$ **111.** $(x^2 + 4)(3x^2 + 8)$ **113.** $(2x^n + 5)(2x^n - 1)$ **115.** $2x(2x - 3)(x - 1)$
117. $2y(3y - 5)(2y + 7)$ **119.** $5(3a + 4b)(2a + 3b)$ **121.** $a^2(a + 5b)(8a - 3b)$ **123.** $x^2(10 - 3x)(5 + 4x)$
125. $2y(12 + x)(4 - x)$ **127.** $4(xy - 3)(xy - 5)$ **129.** $b^2(ab + 6)(2ab - 3)$ **131.** $3x(xy + 8)(xy - 4)$
133. Nonfactorable   **135.** $3b^2(b^2 + 2)(b^2 - 5)$ **137.** $x^2y(2x + 5)(x - 6)$ **139.** $4x^2y(4y + 5)(y + 1)$
141. $2ab(6a + b)(a - 6b)$ **143.** $x^n(x^n + 2)(x^n + 8)$ **145.** $6, 9, -6, -9$ **147.** $9, 12, 21, -9, -12, -21$
149. $5, 7, -5, -7$ **151.** $8, 16, -8, -16$ **153.** $5, 7, -5, -7$ **155.** $10, 22, -10, -22$
157. $(3y + 1)(y - 2)$ **159.** $a(3a + 1)(a + 4)$ **161.** $ab(2a + b)(a - b)$ **163.** $2y(y^2 + 4)(y^2 - 3)$
165. $p(3p + 5)$ **167.** $(3b - 4)(2b - 5)$ **169.** $y(4x - 7)(x + 1)$

SECTION 3.4   (page 145)

Problem 1 $x^2 - 36y^4 = x^2 - (6y^2)^2 = (x + 6y^2)(x - 6y^2)$

Problem 2 $9x^2 + 12x + 4 = (3x + 2)^2$

Problem 3   **A.** $8x^3 + y^3z^3 = (2x)^3 + (yz)^3 = (2x + yz)(4x^2 - 2xyz + y^2z^2)$

B. $(x - y)^3 + (x + y)^3 =$
$[(x - y) + (x + y)][(x - y)^2 - (x - y)(x + y) + (x + y)^2] =$
$2x[x^2 - 2xy + y^2 - (x^2 - y^2) + x^2 + 2xy + y^2] =$
$2x[x^2 - 2xy + y^2 - x^2 + y^2 + x^2 + 2xy + y^2] = 2x(x^2 + 3y^2)$

Problem 4 $6a(2b - 5) + 7(5 - 2b) = 6a(2b - 5) - 7(2b - 5) = (2b - 5)(6a - 7)$

Problem 5 $3rs - 2r - 3s + 2 = (3rs - 2r) + (-3s + 2) = r(3s - 2) - (3s - 2) = (3s - 2)(r - 1)$

Problem 6   **A.** $4x - 4y - x^3 + x^2y = (4x - 4y) + (-x^3 + x^2y) = 4(x - y) - x^2(x - y) =$
$(x - y)(4 - x^2) = (x - y)(2 + x)(2 - x)$

B. $x^{4n} - x^{2n}y^{2n} = x^{2n+2n} - x^{2n}y^{2n} = x^{2n}(x^{2n} - y^{2n}) = x^{2n}[(x^n)^2 - (y^n)^2] = x^{2n}(x^n + y^n)(x^n - y^n)$

EXERCISES   (page 149)

1. $(x + 4)(x - 4)$ **3.** $(2x + 1)(2x - 1)$ **5.** $(4x + 11)(4x - 11)$ **7.** $(1 + 3a)(1 - 3a)$
9. $(xy + 10)(xy - 10)$ **11.** Nonfactorable   **13.** $(5 + ab)(5 - ab)$ **15.** $(2x + y)(2x - y)$
17. $(a^n + 1)(a^n - 1)$ **19.** $(a + 2)^2$ **21.** $(x - 6)^2$ **23.** $(b - 1)^2$ **25.** $(4x - 5)^2$ **27.** Nonfactorable
29. Nonfactorable   **31.** $(x + 3y)^2$ **33.** $(5a - 4b)^2$ **35.** $(x^n + 3)^2$ **37.** $(x - 3)(x^2 + 3x + 9)$
39. $(2x - 1)(4x^2 + 2x + 1)$ **41.** $(x - y)(x^2 + xy + y^2)$ **43.** $(m + n)(m^2 - mn + n^2)$.
45. $(4x + 1)(16x^2 - 4x + 1)$ **47.** $(3x - 2y)(9x^2 + 6xy + 4y^2)$ **49.** $(xy + 4)(x^2y^2 - 4xy + 16)$
51. Nonfactorable   **53.** Nonfactorable   **55.** $(a - 2b)(a^2 - ab + b^2)$ **57.** $(x^{2n} + y^n)(x^{4n} - x^{2n}y^n + y^{2n})$
59. $(x^n + 2)(x^{2n} - 2x^n + 4)$ **61.** $(a + 2)(x - 2)$ **63.** $(x - 2)(a + b)$ **65.** $(x + 3)(x + 2)$
67. $(x + 4)(y - 2)$ **69.** $(a + b)(x - y)$ **71.** $(y - 3)(x^2 - 2)$ **73.** $(3 + y)(2 + x^2)$
75. $(2a + b)(x^2 - 2y)$ **77.** $(y - 5)(x^n + 1)$ **79.** $5(x + 1)^2$ **81.** $3x(x - 3)(x^2 + 3x + 9)$
83. $7(x + 2)(x - 2)$ **85.** $y^2(y - 3)(y - 7)$ **87.** $(x^2 + 4)(x + 2)(x - 2)$ **89.** $2x^3(2x + 7)(2x - 7)$
91. $x^3(y - 1)(y^2 + y + 1)$ **93.** $x^3y^3(xy - 1)(x^2y^2 + xy + 1)$ **95.** $x^2(x^2 + 15x - 56)$ **97.** $2x^2(2x - 5)^2$
99. $(x^2 + y^2)(x + y)(x - y)$ **101.** $(x^2 + y^2)(x^4 - x^2y^2 + y^4)$ **103.** Nonfactorable
105. $2a(2a - 1)(4a^2 + 2a + 1)$ **107.** $a^2b^2(a + 4b)(a - 12b)$ **109.** $2b^2(3a + 5b)(4a - 9b)$
111. $(x - 2)^2(x + 2)$ **113.** $4(y + 2)(x + 1)$ **115.** $(x + y)(x - y)(2x + 1)(2x - 1)$
117. $(x - 1)(x^2 + x + 1)(xy + 1)(x^2y^2 - xy + 1)$ **119.** $x(x^n + 1)^2$ **121.** $b^n(3b - 2)(b + 2)$ **123.** $18, -18$
125. $6, -6$ **127.** $112, -112$ **129.** 36 **131.** 9 **133.** $(x + 2b)(y + 1)(y - 1)$ **135.** $(2r - 3)^2$
137. $9(2b - 1)$ **139.** $(x^n + 1)(x^{2n} - x^n + 1)(x^n - 1)(x^{2n} + x^n + 1)$ **141.** $-(2a + 7)(a - 2)^2$ **143.** 10
145. $0.215x^2 \text{ cm}^3$

SECTION 3.5   (page 154)

Problem 1   **A.** $4x^2 + 11x = 3$
$4x^2 + 11x - 3 = 0$
$(4x - 1)(x + 3) = 0$

$4x - 1 = 0 \qquad x + 3 = 0$
$\qquad 4x = 1 \qquad\qquad x = -3$
$\qquad\quad x = \dfrac{1}{4}$

The solutions are $\dfrac{1}{4}$ and -3.

B. $(x - 2)(x + 5) = 8$
$x^2 + 3x - 10 = 8$
$x^2 + 3x - 18 = 0$
$(x + 6)(x - 3) = 0$

$x + 6 = 0 \qquad x - 3 = 0$
$\qquad x = -6 \qquad\qquad x = 3$

The solutions are -6 and 3.

Problem 2

$$2x^2 - x - 10 \le 0$$
$$(2x - 5)(x + 2) \le 0$$

$$\begin{array}{c} 2x-5 --- | -------- | +++ \\ x+2 --- | +++++++ | +++ \end{array}$$

$$\leftarrow \!\!\! \begin{array}{ccccccc} | & | & | & | & | & | & | \\ -3 & -2 & -1 & 0 & 1 & & 3 \end{array} \!\!\! \rightarrow$$

$$\left\{ x \mid -2 \le x \le \frac{5}{2} \right\}$$

$$\leftarrow \!\!\! \begin{array}{ccccccccc} + & + & + & \bullet & + & + & + & \bullet & + & + & + \\ -5 & -4 & -3 & -2 & -1 & 0 & 1 & 2 & 3 & 4 \end{array} \!\!\! \rightarrow$$

Problem 3

Strategy

▶ Width of the rectangle: W
Length of the rectangle: $W + 5$
▶ Use the equation $A = LW$.

Solution

$$A = LW$$
$$66 = (W + 5)(W)$$
$$66 = W^2 + 5W$$
$$0 = W^2 + 5W - 66$$
$$0 = (W + 11)(W - 6)$$

$$W + 11 = 0 \qquad\qquad W - 6 = 0$$
$$W = -11 \qquad\qquad W = 6 \qquad \text{The width cannot be a negative number.}$$

Length $= W + 5 = 6 + 5 = 11$

The length is 11 in., and the width is 6 in.

EXERCISES (page 158)

1. -4 and -6　**3.** 0 and 7　**5.** 0 and $-\dfrac{5}{2}$　**7.** $-\dfrac{3}{2}$ and 7　**9.** 7 and -7　**11.** $\dfrac{4}{3}$ and $-\dfrac{4}{3}$

13. 1 and -5　**15.** 4 and $-\dfrac{3}{2}$　**17.** 0 and 9　**19.** 0 and 4　**21.** 7 and -4　**23.** $\dfrac{2}{3}$ and -5

25. $\dfrac{2}{5}$ and 3　**27.** $\dfrac{3}{2}$ and $-\dfrac{1}{4}$　**29.** 7 and -5　**31.** 3 and 9　**33.** $-\dfrac{4}{3}$ and 2　**35.** $\dfrac{4}{3}$ and -5

37. 5 and -3　**39.** -4 and -11　**41.** 2 and 4　**43.** 1 and 2　**45.** 2 and 3

47. $\{x \mid x < -1 \text{ or } x > 3\}$ $\leftarrow \!\!\! \begin{array}{ccc} \circ & & \circ \\ -1 & 0 & 3 \end{array} \!\!\! \rightarrow$　　**49.** $\{x \mid x > -2 \text{ or } x < -3\}$ $\leftarrow \!\!\! \begin{array}{ccc} \circ & \circ & \\ -3 & -2 & 0 \end{array} \!\!\! \rightarrow$

51. $\{x \mid -5 < x < 4\}$ $\leftarrow \!\!\! \begin{array}{ccc} \circ & & \circ \\ -5 & 0 & 4 \end{array} \!\!\! \rightarrow$　　**53.** $\{x \mid -4 < x < -1 \text{ or } x > 2\}$ $\leftarrow \!\!\! \begin{array}{cccc} \circ & \circ & & \circ \\ -4 & -1 & 0 & 2 \end{array} \!\!\! \rightarrow$

55. $\{x \mid x \le -5 \text{ or } 1 \le x \le 2\}$ $\leftarrow \!\!\! \begin{array}{cccc} \bullet & & \bullet & \bullet \\ -5 & 0 & 1 & 2 \end{array} \!\!\! \rightarrow$　　**57.** $\{x \mid x \le -2 \text{ or } x \ge 2\}$

59. $\{x \mid x > -3 \text{ or } x < -3\}$　**61.** $\{x \mid x < -7 \text{ or } x > 3\}$　**63.** $\left\{ x \mid \dfrac{1}{4} < x < 2 \right\}$　**65.** $\left\{ x \mid x \le -4 \text{ or } x \ge -\dfrac{3}{2} \right\}$

67. $\{x \mid -5 < x < 2 \text{ or } x > 3\}$　**69.** $\left\{ x \mid x \le -2 \text{ or } \dfrac{1}{3} \le x \le 2 \right\}$　**71.** $\left\{ x \mid -4 < x < -\dfrac{7}{2} \text{ or } x > 2 \right\}$

73. -9 or 8　**75.** $\{n \mid -7 < n < 6\}$　**77.** 13 or -11　**79.** 10 and 12　**81.** 2 and 4　**83.** $0, 1,$ or 5

85. length: 16 ft; width: 11 ft　**87.** 20 cm　**89.** 6 s　**91.** length: 12 cm; width: 8 cm　**93.** $-7a$ and $3a$

95. $4a$ and $-4a$ **97.** $6a$ **99.** $\dfrac{2a}{3}$ and $-2a$ **101.** $8a$ and $-3a$ **103.**

105.

   107. 236 or -44 **109.** length: 22 in.; width: 10 in.

111. length: 12 m; width: 10 m   **113.** 160 cm^3

CHAPTER REVIEW   (page 165)

1. $-72x^8y^{14}$ (Objective 3.1.2)   **2.** $4a^3 - 18a^2 + 10a$ (Objective 3.2.1)   **3.** $(2x^2 - 1)(x^2 + 3)$ (Objective 3.3.4)
4. $(1 - 2a)(1 + 2a + 4a^2)$ (Objective 3.4.2)   **5.** $4a^2 - 20ab + 25b^2$ (Objective 3.2.3)
6. $-\dfrac{3a^4b^3}{4}$ (Objective 3.1.3)   **7.** 0 and $\dfrac{7}{2}$ (Objective 3.5.1)   **8.** $(x - 2)(3 - y)$ (Objective 3.4.3)
9. $9x^2 + 18xy - 16y^2$ (Objective 3.2.2)   **10.** $4xy(x^2 + 3x - y^2)$ (Objective 3.3.1)
11. $(3 + 2x)(2 - 3x)$ (Objective 3.3.3)   **12.** $(3y + 1)^2$ (Objective 3.4.1)   **13.** 4 and -6 (Objective 3.5.1)
14. $8x^3 - 2x^2 + 5x + 4$ (Objective 3.1.1)   **15.** $-6x^3 + 4x^2 - 8x$ (Objective 3.2.1)
16. $-3ab^2(4a^3b^5)$ (Objective 3.1.4)   **17.** $(a - 4)(b + 2)$ (Objective 3.4.3)
18. $2y^2(x + 2)(2x - 1)$ (Objective 3.3.5)   **19.** 4 and $-\dfrac{1}{2}$ (Objective 3.5.1)
20. $4a^5 - 2a^3 - 10a^2 - 6a - 10$ (Objective 3.2.2)   **21.** $(2x - 1)(2x - 3)$ (Objective 3.3.3)
22. $(2x^2 + 1)(x + 2)(x - 2)$ (Objective 3.4.4)   **23.** $9x^2 - 1$ (Objective 3.2.3)
24. $\dfrac{2}{3}$ and -5 (Objective 3.5.1)   **25.** $(x + 6)(x - 4)$ (Objective 3.3.2)   **26.** 8 and -5 (Objective 3.5.1)
27. $a(3x - 5)(2x + 1)$ (Objective 3.4.4)   **28.** $(x + 1)(x - 1)(y + 1)$ (Objective 3.4.4)
29. $-\dfrac{27x^2}{16y^7}$ (Objective 3.1.3)   **30.** $\{x\,|\,x \le -4 \text{ or } x \ge 1\}$ (Objective 3.5.2)
31. $(4x - 3)(2x - 5)$ (Objective 3.3.3)   **32.** $(2x - 5y)^2$ (Objective 3.4.1)   **33.** 0 and 2 (Objective 3.5.1)
34. $4a^2 - 25b^2$ (Objective 3.2.3)   **35.** $(xy - 3)(x^2y^2 + 3xy + 9)$ (Objective 3.4.2)
36. 0 and 5 (Objective 3.5.1)   **37.** $\left\{x\,\Big|\,-\dfrac{1}{2} < x < 3\right\}$ (Objective 3.5.2)   **38.** $324a^6b^8$ (Objective 3.1.2)
39. $(3x^2 + 4x - 4)$ ft^2 (Objective 3.2.4)   **40.** -6 and -4, or 4 and 6 (Objective 3.5.3)
41. 9 or -10 (Objective 3.5.3)   **42.** $(x^3 + 12x^2 + 48x + 64)$ cm^3 (Objective 3.2.4)

CUMULATIVE REVIEW   (page 167)

1. 6 (Objective 1.1.4)   **2.** $-\dfrac{5}{4}$ (Objective 1.2.1)   **3.** The Inverse Property of Addition (Objective 1.2.2)
4. $-18x + 8$ (Objective 1.2.3)   **5.** $A \cap B = \{3, 9\}$ (Objective 1.3.1)
6.

   (Objective 1.3.2)   **7.**

   (Objective 1.3.2)

8. $-\dfrac{1}{6}$ (Objective 2.1.1)   **9.** $-\dfrac{11}{4}$ (Objective 2.1.1)   **10.** $\dfrac{35}{3}$ (Objective 2.1.2)
11. $\{x\,|\,x \le 3\}$ (Objective 2.2.1)   **12.** The solution set is the set of real numbers. (Objective 2.2.2)
13. -1 and $\dfrac{7}{3}$ (Objective 2.3.1)   **14.** $\left\{x\,\Big|\,x \ge 1 \text{ or } x \le -\dfrac{7}{3}\right\}$ (Objective 2.3.2)
15. $\left\{x\,\Big|\,-\dfrac{5}{2} < x < 3\right\}$ (Objective 2.3.2)   **16.** $8x^2 - 5xy - y^2$ (Objective 3.1.1)   **17.** $144a^8b^{10}$ (Objective 3.1.2)

18. $-\dfrac{2y^3}{x^8}$ (Objective 3.1.3) **19.** $-3x^2 + 60x + 8$ (Objective 3.2.1) **20.** $4x^3 - 7x + 3$ (Objective 3.2.2)

21. $x^{2n} + 2x^n + 1$ (Objective 3.2.3) **22.** $(3x + 4)(2x - 5)$ (Objective 3.3.3)

23. $-2x(2x - 3)(x - 2)$ (Objective 3.3.5) **24.** $(9x - y)(9x + y)$ (Objective 3.4.1)

25. $(x - y)(a + b)$ (Objective 3.4.3) **26.** $(x - 2)(x + 2)(x^2 + 4)$ (Objective 3.4.4)

27. -12 and 6 (Objective 3.5.1) **28.** $-\dfrac{2}{3}$ and 3 (Objective 3.5.1) **29.** $\{x \mid -5 \le x \le 1\}$ (Objective 3.5.2)

30. $\left\{x \mid \dfrac{1}{3} < x < 5\right\}$ (Objective 3.5.2) **31.** 9 and 15 (Objective 2.1.3) **32.** 40 oz (Objective 2.4.2)

33. slower cyclist: 5 mph; faster cyclist: 7.5 mph (Objective 2.4.3) **34.** $\$4500$ (Objective 2.5.1)

35. 8, 10, and 12; or 10, 12, and 14 (Objective 2.1.3) **36.** 12 in. (Objective 3.5.3)

Answers to Chapter 4

SECTION 4.1 (page 171)

Problem 1 **A.** $\dfrac{6x^4 - 24x^3}{12x^3 - 48x^2} = \dfrac{6x^3(x - 4)}{12x^2(x - 4)} = \dfrac{\overset{1}{\cancel{6x^3}}\overset{}{(\cancel{x - 4})}}{\underset{1}{\cancel{12x^2}}\overset{}{(\cancel{x - 4})}} = \dfrac{x}{2}$

B. $\dfrac{20x - 15x^2}{15x^3 - 5x^2 - 20x} = \dfrac{5x(4 - 3x)}{5x(3x^2 - x - 4)} = \dfrac{5x(4 - 3x)}{5x(3x - 4)(x + 1)} = \dfrac{\overset{-1}{\cancel{5x(4 - 3x)}}}{\underset{1}{\cancel{5x(3x - 4)}}(x + 1)} = -\dfrac{1}{x + 1}$

C. $\dfrac{x^{2n} + x^n - 12}{x^{2n} - 3x^n} = \dfrac{(x^n + 4)(x^n - 3)}{x^n(x^n - 3)} = \dfrac{(x^n + 4)\overset{1}{\cancel{(x^n - 3)}}}{x^n\underset{1}{\cancel{(x^n - 3)}}} = \dfrac{x^n + 4}{x^n}$

Problem 2 **A.**
$$
\begin{array}{r}
5x \;-\; 1 \\
3x + 4 \overline{)\,15x^2 + 17x - 20} \\
\underline{15x^2 + 20x } \\
-3x - 20 \\
\underline{-3x - 4} \\
-16
\end{array}
$$

$$\dfrac{15x^2 + 17x - 20}{3x + 4} = 5x - 1 - \dfrac{16}{3x + 4}$$

B.
$$
\begin{array}{r}
x^2 + 3x \;-\; 1 \\
3x - 1 \overline{)\,3x^3 + 8x^2 - 6x + 2} \\
\underline{3x^3 - x^2 } \\
9x^2 - 6x \\
\underline{9x^2 - 3x } \\
-3x + 2 \\
\underline{-3x + 1} \\
1
\end{array}
$$

$$\dfrac{3x^3 + 8x^2 - 6x + 2}{3x - 1} = x^2 + 3x - 1 + \dfrac{1}{3x - 1}$$

Problem 3   **A.** $-2\rfloor$ 6   8 -5
                              -12 8
                         ———————————
                         6   -4 3

$$(6x^2 + 8x - 5) \div (x + 2) = 6x - 4 + \frac{3}{x + 2}$$

B. $3\rfloor$ 2 -3 -8 0 -2
                   6   9   3   9
              ————————————————————————
              2   3   1   3   7

$$(2x^4 - 3x^3 - 8x^2 - 2) \div (x - 3) = 2x^3 + 3x^2 + x + 3 + \frac{7}{x - 3}$$

EXERCISES   (page 176)

1. $1 - 2x$ **3.** $3x - 1$ **5.** $2x$ **7.** $-\dfrac{2}{x}$ **9.** $\dfrac{x}{2}$ **11.** The expression is in simplest form.

13. $4x^2 - 2x + 3$ **15.** $a^2 + 2a - 3$ **17.** $\dfrac{x^n - 3}{4}$ **19.** $\dfrac{x^n}{x^n - y^n}$ **21.** $\dfrac{x - 3}{x - 5}$ **23.** $\dfrac{x + y}{x - y}$

25. $-\dfrac{x + 3}{3x - 4}$ **27.** $-\dfrac{x + 7}{x - 7}$ **29.** The expression is in simplest form.   **31.** $\dfrac{a - b}{a^2 - ab + b^2}$

33. $\dfrac{4x^2 + 2xy + y^2}{2x + y}$ **35.** $\dfrac{x + y}{3x}$ **37.** $\dfrac{x - 2}{2x(x + 1)}$ **39.** $\dfrac{x + 2y}{3x + 4y}$ **41.** $-\dfrac{2x - 3}{2(2x + 3)}$ **43.** $\dfrac{x - 2}{a - b}$

45. $\dfrac{x^2 + 2}{(x - 1)(x + 1)}$ **47.** $\dfrac{xy + 7}{xy - 7}$ **49.** $\dfrac{a^n - 2}{a^n + 2}$ **51.** $\dfrac{a^n + 1}{a^n - 1}$ **53.** $-\dfrac{x - 3}{x + 3}$ **55.** $x + 8$

57. $x^2 + \dfrac{2}{x - 3}$ **59.** $3x + 5 + \dfrac{3}{2x + 1}$ **61.** $5x + 7 + \dfrac{2}{2x - 1}$ **63.** $4x^2 + 6x + 9 + \dfrac{18}{2x - 3}$

65. $3x^2 + 1 + \dfrac{1}{2x^2 - 5}$ **67.** $x^2 - 3x - 10$ **69.** $x^2 - 2x + 1 - \dfrac{1}{x - 3}$ **71.** $2x^3 - 3x^2 + x - 4$

73. $x - 4 + \dfrac{x + 3}{x^2 + 1}$ **75.** $2x - 3 + \dfrac{1}{x - 1}$ **77.** $3x + 1 + \dfrac{10x + 7}{2x^2 - 3}$ **79.** $2x - 8$ **81.** $3x - 8$

83. $3x + 3 - \dfrac{1}{x - 1}$ **85.** $x - 4 + \dfrac{7}{x + 4}$ **87.** $x - 2 + \dfrac{16}{x + 2}$ **89.** $4x - 12 + \dfrac{15}{x + 1}$

91. $2x^2 - 3x + 9$ **93.** $x^2 - 3x + 2$ **95.** $x^2 - 5x + 16 - \dfrac{41}{x + 2}$ **97.** $x^2 - x + 2 - \dfrac{4}{x + 1}$

99. $4x^2 + 8x + 15 + \dfrac{12}{x - 2}$ **101.** $2x^2 - 3x + 7 - \dfrac{8}{x + 4}$ **103.** $2x^3 - 3x^2 + x - 4$

105. $3x^3 + 2x^2 + 12x + 19 + \dfrac{33}{x - 2}$ **107.** $3x^3 - x + 4 - \dfrac{2}{x + 1}$ **109.** $2x^3 + 6x^2 + 17x + 51 + \dfrac{155}{x - 3}$

111. $x^2 - 5x + 25$ **113.** $-3, 3$ **115.** $-6, 1$ **117.** $0, 3$ **119.** $-\dfrac{3}{2}, \dfrac{1}{3}$ **121.** $-4, 3$ **123.** $3x + 2y$

125. $2a - b + \dfrac{2b^2}{3a - b}$ **127.** $3x + 2y + \dfrac{y^2}{2x + y}$ **129.** $a^3 - a^2b + ab^2 - b^3 + \dfrac{2b^4}{a + b}$

131. $x^5 + x^4y + x^3y^2 + x^2y^3 + xy^4 + y^5$ **133.** -2 **135.** 27 **137.** -2 **139.** $x + 2$

SECTION 4.2   (page 182)

Problem 1   **A.** $\dfrac{12 + 5x - 3x^2}{x^2 + 2x - 15} \cdot \dfrac{2x^2 + x - 45}{3x^2 + 4x} = \dfrac{(4 + 3x)(3 - x)}{(x + 5)(x - 3)} \cdot \dfrac{(2x - 9)(x + 5)}{x(3x + 4)} =$

$$\dfrac{(4 + 3x)(3 - x)(2x - 9)(x + 5)}{(x + 5)(x - 3) \cdot x(3x + 4)} = \dfrac{\overset{1}{(4 + 3x)}\overset{-1}{(3 - x)}(2x - 9)\overset{1}{(x + 5)}}{\underset{1}{(x + 5)}\underset{1}{(x - 3)} \cdot \underset{1}{x(3x + 4)}} = -\dfrac{2x - 9}{x}$$

B. $\dfrac{2x^2 - 13x + 20}{x^2 - 16} \cdot \dfrac{2x^2 + 9x + 4}{6x^2 - 7x - 5} = \dfrac{(2x - 5)(x - 4)}{(x - 4)(x + 4)} \cdot \dfrac{(2x + 1)(x + 4)}{(3x - 5)(2x + 1)} =$

$$\dfrac{(2x - 5)(x - 4)(2x + 1)(x + 4)}{(x - 4)(x + 4)(3x - 5)(2x + 1)} = \dfrac{(2x - 5)\overset{1}{(x - 4)}\overset{1}{(2x + 1)}\overset{1}{(x + 4)}}{\underset{1}{(x - 4)}\underset{1}{(x + 4)}(3x - 5)\underset{1}{(2x + 1)}} = \dfrac{2x - 5}{3x - 5}$$

Problem 2   **A.** $\dfrac{6x^2 - 3xy}{10ab^4} \div \dfrac{16x^2y^2 - 8xy^3}{15a^2b^2} = \dfrac{6x^2 - 3xy}{10ab^4} \cdot \dfrac{15a^2b^2}{16x^2y^2 - 8xy^3} = \dfrac{3x(2x - y)}{10ab^4} \cdot \dfrac{15a^2b^2}{8xy^2(2x - y)} =$

$$\dfrac{3x(2x - y) \cdot 15a^2b^2}{10ab^4 \cdot 8xy^2(2x - y)} = \dfrac{3x\overset{1}{(2x - y)} \cdot 15a^2b^2}{10ab^4 \cdot 8xy^2\underset{1}{(2x - y)}} = \dfrac{9a}{16b^2y^2}$$

B. $\dfrac{6x^2 - 7x + 2}{3x^2 + x - 2} \div \dfrac{4x^2 - 8x + 3}{5x^2 + x - 4} = \dfrac{6x^2 - 7x + 2}{3x^2 + x - 2} \cdot \dfrac{5x^2 + x - 4}{4x^2 - 8x + 3} =$

$$\dfrac{(2x - 1)(3x - 2)}{(x + 1)(3x - 2)} \cdot \dfrac{(x + 1)(5x - 4)}{(2x - 1)(2x - 3)} = \dfrac{(2x - 1)(3x - 2)(x + 1)(5x - 4)}{(x + 1)(3x - 2)(2x - 1)(2x - 3)} =$$

$$\dfrac{\overset{1}{(2x - 1)}\overset{1}{(3x - 2)}\overset{1}{(x + 1)}(5x - 4)}{\underset{1}{(x + 1)}\underset{1}{(3x - 2)}\underset{1}{(2x - 1)}(2x - 3)} = \dfrac{5x - 4}{2x - 3}$$

Problem 3   The LCM is $a(a - 5)(a + 5)$.

$$\dfrac{a - 3}{a^2 - 5a} - \dfrac{a - 9}{a^2 - 25} = \dfrac{a - 3}{a(a - 5)} \cdot \dfrac{a + 5}{a + 5} + \dfrac{a - 9}{(a - 5)(a + 5)} \cdot \dfrac{a}{a} =$$

$$\dfrac{a^2 + 2a - 15}{a(a - 5)(a + 5)} + \dfrac{a^2 - 9a}{a(a - 5)(a + 5)} = \dfrac{(a^2 + 2a - 15) + (a^2 - 9a)}{a(a - 5)(a + 5)} =$$

$$\dfrac{a^2 + 2a - 15 + a^2 - 9a}{a(a - 5)(a + 5)} = \dfrac{2a^2 - 7a - 15}{a(a - 5)(a + 5)} = \dfrac{(2a + 3)(a - 5)}{a(a - 5)(a + 5)} =$$

$$\dfrac{(2a + 3)\overset{1}{(a - 5)}}{a\underset{1}{(a - 5)}(a + 5)} = \dfrac{2a + 3}{a(a + 5)}$$

Problem 4   The LCM is $(x - 2)(2x - 3)$.

$$\frac{x - 1}{x - 2} - \frac{7 - 6x}{2x^2 - 7x + 6} + \frac{4}{2x - 3} = \frac{x - 1}{x - 2} \cdot \frac{2x - 3}{2x - 3} - \frac{7 - 6x}{(x - 2)(2x - 3)} + \frac{4}{2x - 3} \cdot \frac{x - 2}{x - 2} =$$

$$\frac{2x^2 - 5x + 3}{(x - 2)(2x - 3)} - \frac{7 - 6x}{(x - 2)(2x - 3)} + \frac{4x - 8}{(x - 2)(2x - 3)} =$$

$$\frac{(2x^2 - 5x + 3) - (7 - 6x) + (4x - 8)}{(x - 2)(2x - 3)} = \frac{2x^2 - 5x + 3 - 7 + 6x + 4x - 8}{(x - 2)(2x - 3)} =$$

$$\frac{2x^2 + 5x - 12}{(x - 2)(2x - 3)} = \frac{(x + 4)(2x - 3)}{(x - 2)(2x - 3)} = \frac{(x + 4)\cancel{(2x - 3)}}{(x - 2)\cancel{(2x - 3)}} = \frac{x + 4}{x - 2}$$

EXERCISES   (page 187)

1. $\dfrac{15b^4xy}{4}$ **3.** 1   **5.** $\dfrac{2x - 3}{x^2(x + 1)}$ **7.** $-\dfrac{x - 1}{x - 5}$ **9.** $\dfrac{2(x + 2)}{x - 1}$ **11.** $\dfrac{5x + 1}{5x - 1}$ **13.** $-\dfrac{x - 3}{2x + 3}$ **15.** $\dfrac{x^n - 6}{x^n - 1}$

17. $x + y$ **19.** $\dfrac{a^2y}{10x}$ **21.** $\dfrac{3}{2}$ **23.** $\dfrac{5(x - y)}{xy(x + y)}$ **25.** $\dfrac{(x - 3)^2}{(x + 3)(x - 4)}$ **27.** $\dfrac{2x + 5}{2x - 5}$ **29.** $-\dfrac{x + 2}{x + 5}$

31. $\dfrac{x^{2n}}{8}$ **33.** $\dfrac{(x - 3)(3x - 5)}{(2x + 5)(x + 4)}$ **35.** -1 **37.** 1   **39.** $-\dfrac{2(x^2 + x + 1)}{(x + 1)^2}$ **41.** $-\dfrac{13}{2xy}$ **43.** $\dfrac{1}{x - 1}$

45. $\dfrac{15 - 16xy - 9x}{10x^2y}$ **47.** $-\dfrac{5y + 21}{30xy}$ **49.** $-\dfrac{6x + 19}{36x}$ **51.** $\dfrac{7xy + 6y + 10x}{12x^2y^2}$ **53.** $-\dfrac{x^2 + x}{(x - 3)(x - 5)}$

55. -1 **57.** $\dfrac{5x - 8}{(x + 5)(x - 5)}$ **59.** $-\dfrac{3x^2 - 4x + 8}{x(x - 4)}$ **61.** $\dfrac{4x^2 - 14x + 15}{2x(2x - 3)}$ **63.** $\dfrac{2x^2 + x + 3}{(x - 1)(x + 1)^2}$

65. $\dfrac{x + 1}{x + 3}$ **67.** $\dfrac{x + 2}{x + 3}$ **69.** $-\dfrac{x^n - 2}{(x^n + 1)(x^n - 1)}$ **71.** $\dfrac{2}{x^n + 2}$ **73.** $\dfrac{12x + 13}{(2x + 3)(2x - 3)}$ **75.** $\dfrac{2x + 3}{2x - 3}$

77. $\dfrac{2x + 5}{(x + 3)(x - 2)(x + 1)}$ **79.** $\dfrac{3x^2 + 4x + 3}{(x + 4)(2x + 3)(x - 3)}$ **81.** $\dfrac{x - 2}{x - 3}$ **83.** $-\dfrac{2x^2 - 9x - 9}{(x + 6)(x - 3)}$

85. $\dfrac{3x^2 - 14x + 24}{(3x - 2)(2x + 5)}$ **87.** $\dfrac{x - 12}{2x - 5}$ **89.** $\dfrac{5x - 1}{2x - 3}$ **91.** $\dfrac{4}{(x - 2)(x^2 + 2x + 4)}$ **93.** $\dfrac{2}{x^2 + 1}$ **95.** $-x - 1$

97. $\dfrac{108a}{b}$ **99.** $\dfrac{y - 2}{x^4}$ **101.** $\dfrac{4y}{(y - 1)^2}$ **103.** $\dfrac{4}{3(a - 2)}$ **105.** $\dfrac{5(x + 4)(x - 4)}{2(x + 2)(x - 2)(x - 3)}$

107. $-\dfrac{(a + 3)(a - 2)(5a - 2)}{(3a + 1)(a + 2)(2a - 3)}$ **109.** $\dfrac{6(x - 1)}{x(2x + 1)}$ **111.** $\dfrac{3}{y} + \dfrac{6}{x}$ **113.** $\dfrac{4}{b^2} + \dfrac{3}{ab}$

115. $\dfrac{1}{3} + \dfrac{1}{5} = \dfrac{5}{15} + \dfrac{3}{15} = \dfrac{8}{15}$; $\dfrac{1}{3 + 5} = \dfrac{1}{8} \ne \dfrac{8}{15}$

SECTION 4.3   (page 194)

Problem 1   **A.** The LCM of x and x^2 is x^2.

$$\frac{3 + \dfrac{16}{x} + \dfrac{16}{x^2}}{6 + \dfrac{5}{x} - \dfrac{4}{x^2}} = \frac{3 + \dfrac{16}{x} + \dfrac{16}{x^2}}{6 + \dfrac{5}{x} - \dfrac{4}{x^2}} \cdot \frac{x^2}{x^2} = \frac{3 \cdot x^2 + \dfrac{16}{x} \cdot x^2 + \dfrac{16}{x^2} \cdot x^2}{6 \cdot x^2 + \dfrac{5}{x} \cdot x^2 - \dfrac{4}{x^2} \cdot x^2} = \frac{3x^2 + 16x + 16}{6x^2 + 5x - 4} =$$

$$\frac{(3x + 4)\,(x + 4)}{(2x - 1)\,(3x + 4)} = \frac{\overset{1}{\cancel{(3x + 4)}}\,(x + 4)}{(2x - 1)\,\underset{1}{\cancel{(3x + 4)}}} = \frac{x + 4}{2x - 1}$$

B. The LCM is $x - 3$.

$$\frac{2x + 5 + \dfrac{14}{x - 3}}{4x + 16 + \dfrac{49}{x - 3}} = \frac{2x + 5 + \dfrac{14}{x - 3}}{4x + 16 + \dfrac{49}{x - 3}} \cdot \frac{x - 3}{x - 3} = \frac{(2x + 5)\,(x - 3) + \dfrac{14}{x - 3}\,(x - 3)}{(4x + 16)\,(x - 3) + \dfrac{49}{(x - 3)}\,(x - 3)} =$$

$$\frac{2x^2 - x - 15 + 14}{4x^2 + 4x - 48 + 49} = \frac{2x^2 - x - 1}{4x^2 + 4x + 1} = \frac{(2x + 1)\,(x - 1)}{(2x + 1)\,(2x + 1)} = \frac{\overset{1}{\cancel{(2x + 1)}}\,(x - 1)}{\underset{1}{\cancel{(2x + 1)}}\,(2x + 1)} = \frac{x - 1}{2x + 1}$$

Problem 2 $3 + \dfrac{3}{3 + \dfrac{3}{y}} = 3 + \dfrac{3}{3 + \dfrac{3}{y}} \cdot \dfrac{y}{y} = 3 + \dfrac{3y}{3y + 3} = 3 + \dfrac{3y}{3(y + 1)} = 3 + \dfrac{y}{y + 1} =$

$\dfrac{3(y + 1)}{y + 1} + \dfrac{y}{y + 1} = \dfrac{3y + 3 + y}{y + 1} = \dfrac{4y + 3}{y + 1}$

EXERCISES   (page 196)

1. $\dfrac{x}{x - 1}$ 3. $-\dfrac{a}{a + 2}$ 5. $-\dfrac{a - 1}{a + 1}$ 7. $\dfrac{2}{5}$ 9. $-\dfrac{1}{2}$ 11. $\dfrac{x^2 - x - 1}{x^2 + x + 1}$ 13. $\dfrac{x + 2}{x - 1}$ 15. $-\dfrac{x + 4}{2x + 3}$

17. $\dfrac{2x + 4}{x - 4}$ 19. $\dfrac{1}{3x + 8}$ 21. $\dfrac{x - 2}{x + 2}$ 23. $\dfrac{x - 3}{x + 4}$ 25. $\dfrac{3x + 1}{x - 5}$ 27. $-\dfrac{2a + 2}{7a - 4}$ 29. $\dfrac{x + y}{x - y}$

31. $-\dfrac{2x}{x^2 + 1}$ 33. $\dfrac{6x - 12}{2x - 3}$ 35. $-\dfrac{a^2}{1 - 2a}$ 37. $\dfrac{1}{x - 1}$ 39. $3 - b$ 41. $-\dfrac{1}{x(x + h)}$ 43. $\dfrac{2c + 1}{1 - 2c}$

45. $\dfrac{x}{2 + 8x}$

SECTION 4.4   (page 199)

Problem 1   A.
$$\frac{5}{2x - 3} = \frac{-2}{x + 1}$$

$$\frac{5}{2x - 3}(x + 1)(2x - 3) = \frac{-2}{x + 1}(x + 1)(2x - 3)$$

$$5(x + 1) = -2(2x - 3)$$
$$5x + 5 = -4x + 6$$
$$9x + 5 = 6$$
$$9x = 1$$
$$x = \frac{1}{9}$$

The solution is $\frac{1}{9}$.

B.
$$\frac{x}{x + 3} = \frac{5}{x + 7}$$

$$(x + 3)(x + 7)\frac{x}{x + 3} = (x + 3)(x + 7)\frac{5}{x + 7}$$

$$(x + 7)x = (x + 3)5$$
$$x^2 + 7x = 5x + 15$$
$$x^2 + 2x - 15 = 0$$
$$(x + 5)(x - 3) = 0$$

$$x + 5 = 0 \qquad x - 3 = 0$$
$$x = -5 \qquad x = 3$$

The solutions are -5 and 3.

Problem 2   **Strategy**

To find the cost, write and solve a proportion using x to represent the cost.

Solution

$$\frac{2}{3.10} = \frac{15}{x}$$

$$\frac{2}{3.10} \cdot x(3.10) = \frac{15}{x} \cdot x(3.10)$$

$$2x = 15(3.10)$$
$$2x = 46.50$$
$$x = 23.25$$

The cost of 15 lb of cashews is $23.25.

EXERCISES   (page 201)

1. 9   **3.** $\frac{15}{2}$ **5.** 3   **7.** $\frac{10}{3}$ **9.** -2 **11.** $\frac{2}{5}$ **13.** -4 **15.** $\frac{3}{4}$ **17.** -9 and 8   **19.** 2 and 5
21. -3 and 1   **23.** $98.40   **25.** 140,000 people   **27.** 264 ft^2 **29.** 1584 ceramic tiles
31. 12 additional ounces   **33.** $3000   **35.** 2.8 additional gallons   **37.** 1200 additional pieces
39. 10 ft^3 **41.** $600   **43.** 130 hits   **45.** 175 mi   **47.** 150 ml   **49.** 120 dentists   **51.** $900,000
53. 48 boxes

SECTION 4.5   (page 205)

Problem 1
$$\frac{4x + 1}{2x - 1} = 5 + \frac{3}{2x - 1}$$

$$(2x - 1)\left(\frac{4x + 1}{2x - 1}\right) = (2x - 1)\left(5 + \frac{3}{2x - 1}\right)$$

$$4x + 1 = 10x - 5 + 3$$
$$4x + 1 = 10x - 2$$
$$-6x + 1 = -2$$
$$-6x = -3$$
$$x = \frac{1}{2}$$

$\frac{1}{2}$ does not check as a solution.

The equation has no solution.

Problem 2 $\dfrac{x}{x - 2} \le 0$

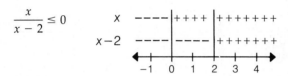

$$\{x \mid 0 \le x < 2\}$$

Problem 3   **A.**
$$\frac{1}{R_1} + \frac{1}{R_2} = \frac{1}{R}$$

$$RR_1 R_2\left(\frac{1}{R_1} + \frac{1}{R_2}\right) = RR_1 R_2\left(\frac{1}{R}\right)$$

$$RR_1 R_2\left(\frac{1}{R_1}\right) + RR_1 R_2\left(\frac{1}{R_2}\right) = R_1 R_2$$

$$RR_2 + RR_1 = R_1 R_2$$
$$R(R_2 + R_1) = R_1 R_2$$
$$R = \frac{R_1 R_2}{R_2 + R_1}$$

B.
$$\frac{r}{r + 1} = t$$

$$(r + 1) \cdot \left(\frac{r}{r + 1}\right) = t(r + 1)$$

$$r = tr + t$$
$$r - tr = t$$
$$r(1 - t) = t$$
$$r = \frac{t}{1 - t}$$

Problem 4 **Strategy**

▶ Time required for the small pipe to fill the tank: x

	Rate	Time	Part
Large pipe	$\dfrac{1}{9}$	6	$\dfrac{6}{9}$
Small pipe	$\dfrac{1}{x}$	6	$\dfrac{6}{x}$

▶ The sum of the part of the task completed by the large pipe and the part of the task completed by the small pipe is 1.

Solution

$$\frac{6}{9} + \frac{6}{x} = 1$$

$$\frac{2}{3} + \frac{6}{x} = 1$$

$$3x\left(\frac{2}{3} + \frac{6}{x}\right) = 3x \cdot 1$$

$$2x + 18 = 3x$$

$$18 = x$$

The small pipe working alone will fill the tank in 18 h.

Problem 5 **Strategy**

▶ Rate of the wind: r

	Distance	Rate	Time
With wind	700	$150 + r$	$\dfrac{700}{150 + r}$
Against wind	500	$150 - r$	$\dfrac{500}{150 - r}$

▶ The time flying with the wind equals the time flying against the wind.

Solution

$$\frac{700}{150 + r} = \frac{500}{150 - r}$$

$$(150 + r)(150 - r)\left(\frac{700}{150 + r}\right) = (150 + r)(150 - r)\left(\frac{500}{150 - r}\right)$$

$$(150 - r)700 = (150 + r)500$$

$$105{,}000 - 700r = 75{,}000 + 500r$$

$$30{,}000 = 1200r$$

$$25 = r$$

The rate of the wind is 25 mph.

EXERCISES (page 214)

1. -5 **3.** $\dfrac{5}{2}$ **5.** -1 **7.** 8   **9.** 5   **11.** No solution   **13.** 2   **15.** $\dfrac{7}{2}$ **17.** No solution

19.   **21.**   **23.**

25. $\{x \,|\, x < -1 \text{ or } x > 2\}$ **27.** $\{x \,|\, -1 < x \le 0\}$ **29.** $\{x \,|\, x > 1 \text{ or } -2 < x \le 0\}$ **31.** $\left\{x \,\middle|\, x > \dfrac{1}{2} \text{ or } x < 0\right\}$

33. $w = \dfrac{P - 2L}{2}$ **35.** $C = \dfrac{S}{1 - r}$ **37.** $R = \dfrac{PV}{nT}$ **39.** $m_2 = \dfrac{Fr^2}{Gm_1}$ **41.** $R = \dfrac{E - Ir}{I}$ **43.** $b_2 = \dfrac{2A - hb_1}{h}$

45. $R_2 = \dfrac{RR_1}{R_1 - R}$ **47.** $d = \dfrac{a_n - a_1}{n - 1}$ **49.** $h = \dfrac{S - 2wL}{2w + 2L}$ **51.** $x = \dfrac{-by - c}{a}$ **53.** $x = \dfrac{d - b}{a - c}$ **55.** $x = \dfrac{ac}{b}$

57. $x = \dfrac{ab}{a + b}$ **59.** 48 min   **61.** 24 min   **63.** 3.6 h   **65.** 8 h   **67.** 12 h   **69.** 75 min   **71.** 16 h

73. 10 min   **75.** 40 min   **77.** 30 h   **79.** 62 mph   **81.** passenger train: 60 mph; freight train: 42 mph
83. 8 mph   **85.** 12 mph   **87.** 4 mph   **89.** 70 mph   **91.** single-engine plane: 120 mph; jet: 480 mph
93. sailboat: 8 mph; cruiser: 16 mph   **95.** 5 mph   **97.** 30 mph   **99.** 3 mph   **101.** 51.43 mph

103. $2y + 5$ **105.** 0 and y **107.** $\dfrac{y}{2}$ and $5y$ **109.** 8 and 9   **111.** $\dfrac{17}{21}$

113. $\{x \,|\, x > 2, \ x \text{ is an integer}\}$ **115.** $1\dfrac{1}{3}$ h   **117.** 50 mph

CHAPTER REVIEW   (page 224)

1. $-\dfrac{2x + 3}{4(2x + 1)}$ (Objective 4.1.1)   **2.** $\dfrac{(x + 1)(x + 4)(x - 4)}{(x + 2)(x - 2)}$ (Objective 4.2.1)   **3.** $\dfrac{x + 3}{x - 3}$ (Objective 4.3.1)

4. No solution (Objective 4.5.1)   **5.** $a = \dfrac{fb}{b - f}$ (Objective 4.5.3)   **6.** $\dfrac{2}{x + 1}$ (Objective 4.1.1)

7. $\dfrac{1}{2x}$ (Objective 4.2.1)   **8.** 5 (Objective 4.4.1)   **9.** $\dfrac{25}{2}$ (Objective 4.5.1)

10. $3x^3 - 4x^2 + 4x - 2 - \dfrac{4}{x + 2}$ (Objective 4.1.3)   **11.** $\dfrac{2x - 1}{(x + 2)(x + 1)}$ (Objective 4.2.2)

12. $s = \dfrac{a}{1 + r}$ (Objective 4.5.3)   **13.** $\{x \,|\, x < -3 \text{ or } -2 \le x \le 1\}$ (Objective 4.5.2)

14. $2x^2 + 3x + 4 + \dfrac{2}{2x - 3}$ (Objective 4.1.2)   **15.** $-\dfrac{2(x + 1)}{2x - 1}$ (Objective 4.3.1)

16. $x + 3y$ (Objective 4.1.1)   **17.** $\dfrac{3xy}{x - y}$ (Objective 4.2.1)   **18.** 9 (Objective 4.4.1)

19. $\dfrac{5}{2}$ (Objective 4.5.1)   **20.** $\dfrac{4}{a - 1}$ (Objective 4.1.1)   **21.** $\dfrac{x}{2}$ (Objective 4.2.1)

22. $C = R - Pn$ (Objective 4.5.3)   **23.** $4x^2 + 10x + 10 + \dfrac{21}{x - 2}$ (Objective 4.1.3)

24. $-\dfrac{x}{(3x + 2)(x + 1)}$ (Objective 4.2.2)   **25.** 13 (Objective 4.5.1)   **26.** $\dfrac{x - 3}{x + 3}$ (Objective 4.3.1)

27. $\{x \,|\, x > 3 \text{ or } -2 < x < 2\}$ (Objective 4.5.2)   **28.** $x^2 - 3x - 4 - \dfrac{6}{3x - 4}$ (Objective 4.1.2)

29. $1500 (Objective 4.4.2)   **30.** 24 min (Objective 4.5.4)
31. freight train: 38 mph; passenger train: 50 mph (Objective 4.5.5)

CUMULATIVE REVIEW   (page 226)

1. $\dfrac{36}{5}$ (Objective 1.1.4)   **2.** -52 (Objective 1.2.1)   **3.**   (Objective 1.3.2)

4. $\dfrac{2}{3}$ (Objective 2.1.1)   **5.** $\dfrac{15}{2}$ (Objective 2.1.2)   **6.** $\{x \,|\, x \geq -16\}$ (Objective 2.2.1)   **7.** $\{x \,|\, -2 < x < 2\}$

(Objective 2.2.2)   **8.** 1 and 7 (Objective 2.3.1)   **9.** $\left\{x \,\middle|\, -2 < x < -\dfrac{2}{5}\right\}$ (Objective 2.3.2)

10. $-4a^4 + 6a^3 - 2a^2$ (Objective 3.2.1)   **11.** $(2x^n - 1)(x^n + 2)$ (Objective 3.3.4)
12. $(ab - 2)(a^2b^2 + 2ab + 4)$ (Objective 3.4.2)   **13.** $(a + b)(x + y)(x - y)$ (Objective 3.4.4)

14.   (Objective 3.5.2)   **15.** $\dfrac{1}{3}$ and -2 (Objective 3.5.1)

16. $3x^2 + 5x + 6 + \dfrac{19}{x - 3}$ (Objective 4.1.3)   **17.** $\dfrac{5ab}{a - b}$ (Objective 4.2.1)   **18.** $\dfrac{x(x - 11)}{(2x + 3)(x + 1)(x - 1)}$

(Objective 4.2.2)   **19.** $a - 1$ (Objective 4.3.1)   **20.** 18 (Objective 4.4.1)   **21.** No solution (Objective 4.5.1)

22. 7 (Objective 4.5.1)   **23.** $r = \dfrac{E - IR}{I}$ (Objective 4.5.3)   **24.** $\left\{x \,\middle|\, -4 < x \leq \dfrac{3}{2}\right\}$ (Objective 4.5.2)

25. 5 and 10 (Objective 2.1.3)
26. 15, 16, and 17; 16, 17, and 18; 17, 18, and 19; or 18, 19, and 20 (Objective 2.2.3)
27. 50 lb (Objective 2.4.2)   **28.** 75,000 people (Objective 4.4.2)   **29.** 14 min (Objective 4.5.4)
30. 60 mph (Objective 4.5.5)

Answers to Chapter 5

SECTION 5.1   (page 231)

Problem 1   **A.** $\dfrac{a^{-1}b^4}{a^{-2}b^{-2}} = ab^6$

B. $\left(\dfrac{2^{-1}x^2y^{-3}}{4x^{-2}y^{-5}}\right)^{-2} = \left(\dfrac{x^4y^2}{8}\right)^{-2} = \dfrac{x^{-8}y^{-4}}{8^{-2}} = \dfrac{8^2}{x^8y^4} = \dfrac{64}{x^8y^4}$

C. $[(a^{-1}b)^{-2}]^3 = [a^2b^{-2}]^3 = a^6b^{-6} = \dfrac{a^6}{b^6}$

Problem 2 $942,000,000 = 9.42 \times 10^8$

Problem 3 $2.7 \times 10^{-5} = 0.000027$

Problem 4 $\dfrac{5,600,000 \times 0.000000081}{900 \times 0.000000028} = \dfrac{5.6 \times 10^6 \times 8.1 \times 10^{-8}}{9 \times 10^2 \times 2.8 \times 10^{-8}} = \dfrac{(5.6)(8.1) \times 10^{6+(-8)-2-(-8)}}{(9)(2.8)} = 1.8 \times 10^4$
$= 18,000$

Problem 5   **Strategy**

To find the number of arithmetic operations:
▶ Find the reciprocal of 1×10^{-7}, which is the number of operations performed in one second.
▶ Write the number of seconds in one minute (60) in scientific notation.
▶ Multiply the number of arithmetic operations per second by the number of seconds in one minute.

Solution

$$\frac{1}{1 \times 10^{-7}} = 10^7$$
$$60 = 6 \times 10$$
$$6 \times 10 \times 10^7$$
$$6 \times 10^8$$

The computer can perform 6×10^8 operations in one minute.

EXERCISES   (page 236)

1. $\dfrac{1}{8}$ **3.** x^4 **5.** $\dfrac{2}{x^2 y^4}$ **7.** $\dfrac{y}{x^3}$ **9.** -5 **11.** $-\dfrac{1}{8}$ **13.** $\dfrac{9}{x^4}$ **15.** 1 **17.** a^2 **19.** $\dfrac{1}{x^5}$ **21.** y

23. $\dfrac{x^4}{y^8}$ **25.** $a^4 b^6$ **27.** $\dfrac{32}{x}$ **29.** $\dfrac{1}{x^3 y^9}$ **31.** $\dfrac{b^2}{a^5}$ **33.** $\dfrac{9y^8}{x^4}$ **35.** $\dfrac{2}{a^2}$ **37.** $\dfrac{4}{a^2}$ **39.** $\dfrac{1}{x^6}$ **41.** $\dfrac{b^9}{a^6}$

43. $\dfrac{x}{16y^2}$ **45.** $\dfrac{x^6}{y^4}$ **47.** $\dfrac{16y^2}{x^2}$ **49.** $\dfrac{12}{a^3 b}$ **51.** $\dfrac{4y^3}{3x}$ **53.** $x^2 y^{10}$ **55.** $a^{24} b^{12}$ **57.** $\dfrac{4x^5}{9y^4}$ **59.** $\dfrac{x^{24}}{64y^{16}}$

61. $\dfrac{a^8 b^4}{18}$ **63.** $\dfrac{1+xy}{1-xy}$ **65.** $a-b$ **67.** $\dfrac{y+x}{xy}$ **69.** $\dfrac{1+x^2 y^2}{xy}$ **71.** $\dfrac{y^{12}}{x^6}$ **73.** $\dfrac{b^2}{a^4}$ **75.** $x^{12}y^6$

77. 4.67×10^{-6} **79.** 1.7×10^{-10} **81.** 2×10^{11} **83.** 0.000000123 **85.** $8,200,000,000,000,000$

87. 0.039 **89.** $150,000$ **91.** $20,800,000$ **93.** 0.000000015 **95.** 0.0000000000178 **97.** $140,000,000$

99. $11,456,790$ **101.** 0.000008 **103.** 2.592×10^{10} km   **105.** 2×10^7 operations   **107.** 1×10^4 dollars

109. 5×10^2 s   **111.** 9.3×10^2 h   **113.** 5.785344×10^{12} mi   **115.** 5.54×10^{15} ft^2

117. 4.32×10^{10} computations   **119.** 3×10^{13} h   **121.** $-x^{18}y^{15}$ **123.** $-\dfrac{2n^3}{m^3}$ **125.** $72x^2 y^2$ **127.** $5m^4 n^2$

129. ab^2 **131.** $\dfrac{xy}{y-5x}$ **133.** $\dfrac{1}{x^2}$ **135.** 3 **137.** 1 **139.** 4 **141.** 3 **143.** 3 **145.** 4

147. $(2+3)^{-2} = 5^{-2} = \dfrac{1}{5^2} = \dfrac{1}{25}$; $2^{-2} + 3^{-2} = \dfrac{1}{2^2} + \dfrac{1}{3^2} = \dfrac{1}{4} + \dfrac{1}{9} = \dfrac{9}{36} + \dfrac{4}{36} = \dfrac{13}{36}$; $\dfrac{1}{25} \neq \dfrac{13}{36}$;
$(2+3)^{-2} \neq 2^{-2} + 3^{-2}$

149. $(2+3)^{-2} = 5^{-2} = \dfrac{1}{5^2} = \dfrac{1}{25}$; $\dfrac{1}{2^2 + 3^2} = \dfrac{1}{4+9} = \dfrac{1}{13}$; $\dfrac{1}{25} \neq \dfrac{1}{13}$; $(2+3)^{-2} \neq \dfrac{1}{2^2 + 3^2}$

SECTION 5.2   (page 241)

Problem 1   **A.** $64^{\frac{2}{3}} = (2^6)^{\frac{2}{3}} = 2^4 = 16$

B. $16^{-\frac{3}{4}} = (2^4)^{-\frac{3}{4}} = 2^{-3} = \dfrac{1}{2^3} = \dfrac{1}{8}$

C. $(-81)^{\frac{3}{4}}$

The base of the exponential expression is negative, while the denominator of the exponent is a positive even number.

Therefore, $(-81)^{\frac{3}{4}}$ is not a real number.

Problem 2 A. $\dfrac{x^{\frac{1}{2}}y^{-\frac{5}{4}}}{x^{-\frac{4}{3}}y^{\frac{1}{3}}} = \dfrac{x^{\frac{1}{2}+\frac{4}{3}}}{y^{\frac{1}{3}+\frac{5}{4}}} = \dfrac{x^{\frac{11}{6}}}{y^{\frac{19}{12}}}$

B. $(x^{\frac{3}{4}}y^{\frac{1}{2}}z^{-\frac{2}{3}})^{-\frac{4}{3}} = x^{-1}y^{-\frac{2}{3}}z^{\frac{8}{9}} = \dfrac{z^{\frac{8}{9}}}{xy^{\frac{2}{3}}}$

C. $\left(\dfrac{16a^{-2}b^{\frac{4}{3}}}{9a^{4}b^{-\frac{2}{3}}}\right)^{-\frac{1}{2}} = \left(\dfrac{2^{4}a^{-6}b^{2}}{3^{2}}\right)^{-\frac{1}{2}} = \dfrac{2^{-2}a^{3}b^{-1}}{3^{-1}} = \dfrac{3a^{3}}{2^{2}b} = \dfrac{3a^{3}}{4b}$

Problem 3 A. $(2x^{3})^{\frac{3}{4}} = \sqrt[4]{(2x^{3})^{3}} = \sqrt[4]{8x^{9}}$

B. $-5a^{\frac{5}{6}} = -5(a^{5})^{\frac{1}{6}} = -5\sqrt[6]{a^{5}}$

Problem 4 A. $\sqrt[3]{3ab} = (3ab)^{\frac{1}{3}}$

B. $\sqrt[4]{x^{4}+y^{4}} = (x^{4}+y^{4})^{\frac{1}{4}}$

Problem 5 A. $-\sqrt[4]{x^{12}} = -(x^{12})^{\frac{1}{4}} = -x^{3}$

B. $\sqrt{121x^{10}y^{4}} = \sqrt{11^{2}x^{10}y^{4}} = 11x^{5}y^{2}$

C. $\sqrt[3]{-125a^{6}b^{9}} = \sqrt[3]{(-5)^{3}a^{6}b^{9}} = -5a^{2}b^{3}$

EXERCISES   (page 247)

1. 2   3. 27   5. $\dfrac{1}{9}$ 7. 4   9. Not a real number   11. $\dfrac{343}{125}$ 13. x 15. $y^{\frac{1}{2}}$ 17. $x^{\frac{1}{12}}$ 19. $a^{\frac{7}{12}}$

21. $\dfrac{1}{a}$ 23. $\dfrac{1}{y}$ 25. $y^{\frac{3}{2}}$ 27. $\dfrac{1}{x}$ 29. $\dfrac{1}{x^{4}}$ 31. a 33. $x^{\frac{3}{10}}$ 35. a^{3} 37. $\dfrac{1}{x^{-\frac{1}{2}}}$ 39. $y^{\frac{2}{3}}$ 41. $x^{4}y$

43. $x^{6}y^{3}z^{9}$ 45. $\dfrac{x}{y^{2}}$ 47. $\dfrac{x^{\frac{3}{2}}}{y^{\frac{1}{4}}}$ 49. $x^{2}y^{8}$ 51. $\dfrac{1}{x^{\frac{11}{12}}}$ 53. $\dfrac{1}{y^{\frac{5}{2}}}$ 55. $\dfrac{1}{b^{\frac{7}{8}}}$ 57. $a^{5}b^{13}$ 59. $\dfrac{m^{2}}{4n^{\frac{3}{2}}}$ 61. $\dfrac{y^{\frac{17}{2}}}{x^{3}}$

63. $\dfrac{16b^{2}}{a^{\frac{1}{3}}}$ 65. $\dfrac{y^{2}}{x^{3}}$ 67. $a - a^{2}$ 69. $y + 1$ 71. $a - \dfrac{1}{a}$ 73. $\dfrac{1}{a^{3n}}$ 75. $a^{\frac{n}{6}}$ 77. $\dfrac{1}{b^{\frac{2m}{3}}}$ 79. $x^{10n^{2}}$

81. $x^{3n}y^{2n}$ 83. $\sqrt{5}$ 85. $\sqrt[3]{b^{4}}$ 87. $\sqrt[3]{9x^{2}}$ 89. $-3\sqrt[5]{a^{2}}$ 91. $\sqrt[4]{x^{6}y^{9}}$ 93. $\sqrt{a^{9}b^{21}}$ 95. $\sqrt[3]{3x-2}$

97. $\dfrac{1}{\sqrt[4]{b^{3}}}$ 99. $7^{\frac{1}{2}}$ 101. $y^{\frac{1}{4}}$ 103. $a^{\frac{3}{4}}$ 105. $b^{\frac{5}{4}}$ 107. $(4y^{7})^{\frac{1}{5}}$ 109. $-(4x^{5})^{\frac{1}{4}}$ 111. $2yx^{\frac{3}{2}}$

113. $(3 + y^2)^{\frac{1}{2}}$ **115.** y^7 **117.** $-a^3$ **119.** a^7b^3 **121.** $11y^6$ **123.** a^2b^4 **125.** $-a^3b^3$ **127.** $5b^5$

129. $-a^2b^3$ **131.** $5x^4y$ **133.** Not a real number **135.** $2a^7b^2$ **137.** $-3ab^5$ **139.** y^3 **141.** $3a^5$

143. $-a^4b$ **145.** ab^5 **147.** $2a^2b^5$ **149.** $-2x^3y^4$ **151.** $\dfrac{4x}{y^7}$ **153.** $\dfrac{3b}{a^3}$ **155.** $2x + 3$ **157.** $x + 1$

159. c **161.** a **163.** a **165.** x **167.** a **169.** $2x^3$ **171.** a^2 **173.** x^2y^4 **175.** $2x^6y^5$ **177.** $\dfrac{3}{4}$

179. 4 **181.** $\dfrac{4}{5}$ **183.** $\dfrac{3}{4}$ **185.** $-\dfrac{1}{3}$ **187.** $-\dfrac{1}{3}$

SECTION 5.3   (page 251)

Problem 1 $\sqrt[5]{x^7} = \sqrt[5]{x^5 \cdot x^2} = \sqrt[5]{x^5}\sqrt[5]{x^2} = x\sqrt[5]{x^2}$

Problem 2 $\sqrt{216} = \sqrt{2^3 \cdot 3^3} = \sqrt{2^2 \cdot 3^2(2 \cdot 3)} = \sqrt{2^2 \cdot 3^2}\sqrt{2 \cdot 3} = 2 \cdot 3\sqrt{6} =$
$6\sqrt{6} \approx 6(2.449) \approx 14.694$

Problem 3   **A.** $3xy\sqrt[3]{81x^5y} - \sqrt[3]{192x^8y^4} = 3xy\sqrt[3]{3^4x^5y} - \sqrt[3]{2^6 \cdot 3x^8y^4} =$
$3xy\sqrt[3]{3^3x^3}\sqrt[3]{3x^2y} - \sqrt[3]{2^6x^6y^3}\sqrt[3]{3x^2y} = 3xy \cdot 3x\sqrt[3]{3x^2y} - 2^2x^2y\sqrt[3]{3x^2y} =$
$9x^2y\sqrt[3]{3x^2y} - 4x^2y\sqrt[3]{3x^2y} = 5x^2y\sqrt[3]{3x^2y}$

   B. $4a\sqrt[3]{81a^7b^9} + a^2b\sqrt[3]{192a^4b^6} = 4a\sqrt[3]{3^4a^7b^9} + a^2b\sqrt[3]{2^6 \cdot 3a^4b^6} =$
$4a\sqrt[3]{3^3a^6b^9}\sqrt[3]{3a} + a^2b\sqrt[3]{2^6a^3b^6}\sqrt[3]{3a} = 4a \cdot 3a^2b^3\sqrt[3]{3a} + a^2b \cdot 2^2ab^2\sqrt[3]{3a} =$
$12a^3b^3\sqrt[3]{3a} + 4a^3b^3\sqrt[3]{3a} = 16a^3b^3\sqrt[3]{3a}$

Problem 4 $\sqrt{5b}(\sqrt{3b} - \sqrt{10}) = \sqrt{15b^2} - \sqrt{50b} = \sqrt{3 \cdot 5b^2} - \sqrt{2 \cdot 5^2b} =$
$\sqrt{b^2}\sqrt{3 \cdot 5} - \sqrt{5^2}\sqrt{2b} = b\sqrt{15} - 5\sqrt{2b}$

Problem 5 $(2\sqrt[3]{2x} - 3)(\sqrt[3]{2x} - 5) = 2\sqrt[3]{4x^2} - 10\sqrt[3]{2x} - 3\sqrt[3]{2x} + 15 = 2\sqrt[3]{4x^2} - 13\sqrt[3]{2x} + 15$

Problem 6   **A.** $\dfrac{y}{\sqrt{3y}} = \dfrac{y}{\sqrt{3y}} \cdot \dfrac{\sqrt{3y}}{\sqrt{3y}} = \dfrac{y\sqrt{3y}}{\sqrt{3^2y^2}} = \dfrac{y\sqrt{3y}}{3y} = \dfrac{\sqrt{3y}}{3}$

   B. $\dfrac{3}{\sqrt[3]{3x^2}} = \dfrac{3}{\sqrt[3]{3x^2}} \cdot \dfrac{\sqrt[3]{3^2x}}{\sqrt[3]{3^2x}} = \dfrac{3\sqrt[3]{9x}}{\sqrt[3]{3^3x^3}} = \dfrac{3\sqrt[3]{9x}}{3x} = \dfrac{\sqrt[3]{9x}}{x}$

   C. $\dfrac{\sqrt{x^5} + \sqrt{x}}{\sqrt{x}} = \dfrac{\sqrt{x^4}\sqrt{x} + \sqrt{x}}{\sqrt{x}} = \dfrac{x^2\sqrt{x} + \sqrt{x}}{\sqrt{x}} = \dfrac{\sqrt{x}(x^2 + 1)}{\sqrt{x}} = x^2 + 1$

Problem 7 $\dfrac{\sqrt{2} + \sqrt{x}}{\sqrt{2} - \sqrt{x}} = \dfrac{\sqrt{2} + \sqrt{x}}{\sqrt{2} - \sqrt{x}} \cdot \dfrac{\sqrt{2} + \sqrt{x}}{\sqrt{2} + \sqrt{x}} = \dfrac{(\sqrt{2})^2 + \sqrt{2x} + \sqrt{2x} + (\sqrt{x})^2}{(\sqrt{2})^2 - (\sqrt{x})^2} = \dfrac{2 + 2\sqrt{2x} + x}{2 - x}$

EXERCISES   (page 257)

1. $x^2yz^2\sqrt{yz}$ **3.** $2ab^4\sqrt{2a}$ **5.** $3xyz^2\sqrt{5yz}$ **7.** Not a real number **9.** $a^5b^2\sqrt[3]{ab^2}$ **11.** $-5y\sqrt[3]{x^2y}$

13. $abc^2\sqrt[3]{ab^2}$ **15.** $2x^2y\sqrt[4]{xy}$ **17.** 24.49 **19.** 23.219 **21.** 29.068 **23.** 774.6 **25.** $-6\sqrt{x}$

27. $-2\sqrt{2}$ **29.** $\sqrt{2x}$ **31.** $3\sqrt{3a} - 2\sqrt{2a}$ **33.** $10x\sqrt{2x}$ **35.** $2x\sqrt{3xy}$ **37.** $xy\sqrt{2y}$

39. $5a^2b^2\sqrt{ab}$ **41.** $9\sqrt[3]{2}$ **43.** $-7a\sqrt[3]{3a}$ **45.** $-5xy^2\sqrt[3]{x^2y}$ **47.** $16ab\sqrt[4]{ab}$ **49.** $6\sqrt{3} - 6\sqrt{2}$

51. $6x^3y^2\sqrt{xy}$ **53.** $-9a^2\sqrt{3ab}$ **55.** $10xy\sqrt[3]{3y}$ **57.** $2xy\sqrt[4]{3x}$ **59.** $7\sqrt{10}$ **61.** 6 **63.** $a^2b^2\sqrt{b}$

65. $5x^3y^2\sqrt{2y}$ **67.** $2ab^2\sqrt[3]{4b^2}$ **69.** $2ab\sqrt[4]{27a^3b^3}$ **71.** $10 - 5\sqrt{2}$ **73.** $y - \sqrt{5y}$ **75.** $9a\sqrt{a} - a\sqrt{3}$

77. $2x + 8\sqrt{2x} + 16$ **79.** $36x^3y^2\sqrt{6}$ **81.** $2ab^2\sqrt[3]{36ab^2}$ **83.** $-10 + \sqrt{2}$ **85.** $y - 4$ **87.** $2x - 9y$

89. $a - 5\sqrt{a} + 6$ **91.** $\sqrt[3]{a^2} + 5\sqrt[3]{a} + 6$ **93.** $6x - \sqrt{xy} - y$ **95.** $4\sqrt{x}$ **97.** $b^2\sqrt{3a}$ **99.** $\dfrac{\sqrt{5}}{5}$

101. $\dfrac{\sqrt{2x}}{2x}$ **103.** $\dfrac{\sqrt{5x}}{x}$ **105.** $\dfrac{\sqrt{5x}}{5}$ **107.** $\dfrac{3\sqrt[3]{4}}{2}$ **109.** $\dfrac{3\sqrt[3]{2x}}{2x}$ **111.** $\dfrac{\sqrt{2xy}}{2y}$ **113.** $\dfrac{2\sqrt{3ab}}{3b^2}$

115. $2\sqrt{5} - 4$ **117.** $\dfrac{3\sqrt{y} + 6}{y - 4}$ **119.** $-5 + 2\sqrt{6}$ **121.** 5 **123.** $1 - x$ **125.** $\dfrac{3\sqrt[4]{2x}}{2x}$ **127.** $\dfrac{2\sqrt[5]{2a^3}}{a}$

129. $\dfrac{a\sqrt{2b} - b\sqrt{2a}}{ab}$ **131.** $\dfrac{1 + 2\sqrt{ab} + ab}{1 - ab}$ **133.** $\dfrac{5x\sqrt{y} + 5y\sqrt{x}}{x - y}$ **135.** $24\sqrt{3}$ **137.** $30\sqrt{3} - 54$

139. $38 + 17\sqrt{5}$ **141.** $\dfrac{2\sqrt{x + 4} - 4}{x}$ **143.** $\dfrac{b - 6\sqrt{b + 9} + 18}{b}$ **145.** $\sqrt{y + 4} + \sqrt{y}$ **147.** $\sqrt[4]{a + b}$

149. $\sqrt[4]{2x(y - 2)^2}$ **151.** $\sqrt[6]{b^3(b - 1)^2}$ **153.** $(x + 3\sqrt{2})(x - 3\sqrt{2})$ **155.** $(b + 2\sqrt{6})(b - 2\sqrt{6})$

157. $(x + 3\sqrt{3})^2$

SECTION 5.4   (page 262)

Problem 1 $\sqrt{-45} = i\sqrt{45} = i\sqrt{3^2 \cdot 5} = 3i\sqrt{5}$

Problem 2 $\sqrt{98} - \sqrt{-60} = \sqrt{98} - i\sqrt{60} = \sqrt{2 \cdot 7^2} - i\sqrt{2^2 \cdot 3 \cdot 5} = 7\sqrt{2} - 2i\sqrt{15}$

Problem 3 $(-4 + 2i) - (6 - 8i) = -10 + 10i$

Problem 4 $(16 - \sqrt{-45}) - (3 + \sqrt{-20}) = (16 - i\sqrt{45}) - (3 + i\sqrt{20}) =$
$(16 - i\sqrt{3^2 \cdot 5}) - (3 + i\sqrt{2^2 \cdot 5}) = (16 - 3i\sqrt{5}) - (3 + 2i\sqrt{5}) = 13 - 5i\sqrt{5}$

Problem 5 $\sqrt{-3}(\sqrt{27} - \sqrt{-6}) = i\sqrt{3}(\sqrt{27} - i\sqrt{6}) = i\sqrt{81} - i^2\sqrt{18} =$
$i\sqrt{3^4} - (-1)\sqrt{2 \cdot 3^2} = 9i + 3\sqrt{2} = 3\sqrt{2} + 9i$

Problem 6   **A.** $(4 - 3i)(2 - i) = 8 - 4i - 6i + 3i^2 = 8 - 10i + 3i^2 = 8 - 10i + 3(-1) = 5 - 10i$

B. $(3 - i)\left(\dfrac{3}{10} + \dfrac{1}{10}i\right) = \dfrac{9}{10} + \dfrac{3}{10}i - \dfrac{3}{10}i - \dfrac{1}{10}i^2 = \dfrac{9}{10} - \dfrac{1}{10}i^2 = \dfrac{9}{10} - \dfrac{1}{10}(-1) = \dfrac{9}{10} + \dfrac{1}{10} = 1$

C. $(3 + 6i)(3 - 6i) = 3^2 + 6^2 = 9 + 36 = 45$

Problem 7 $\dfrac{2 - 3i}{4i} = \dfrac{2 - 3i}{4i} \cdot \dfrac{i}{i} = \dfrac{2i - 3i^2}{4i^2} = \dfrac{2i - 3(-1)}{4(-1)} = \dfrac{3 + 2i}{-4} = -\dfrac{3}{4} - \dfrac{1}{2}i$

Problem 8 $\dfrac{2 + 5i}{3 - 2i} = \dfrac{2 + 5i}{3 - 2i} \cdot \dfrac{3 + 2i}{3 + 2i} = \dfrac{6 + 4i + 15i + 10i^2}{3^2 + 2^2} = \dfrac{6 + 19i + 10(-1)}{13} = \dfrac{-4 + 19i}{13} = -\dfrac{4}{13} + \dfrac{19}{13}i$

EXERCISES   (page 268)

1. $2i$ **3.** $7i\sqrt{2}$ **5.** $3i\sqrt{3}$ **7.** $4 + 2i$ **9.** $2\sqrt{3} - 3i\sqrt{2}$ **11.** $4\sqrt{10} - 7i\sqrt{3}$ **13.** $2ai$ **15.** $7x^6i$

17. $12ab^2i\sqrt{ab}$ **19.** $2\sqrt{a} + 2ai\sqrt{3}$ **21.** $3b^2\sqrt{2b} - 3bi\sqrt{3b}$ **23.** $5x^3y\sqrt{y} + 5xyi\sqrt{2xy}$ **25.** $2a^2bi\sqrt{a}$

27. $2a\sqrt{3a} + 3bi\sqrt{3b}$ **29.** $10 - 7i$ **31.** $11 - 7i$ **33.** $-6 + i$ **35.** $-4 - 8i\sqrt{3}$ **37.** $-\sqrt{10} - 11i\sqrt{2}$

39. $6 - 4i$ **41.** 0 **43.** $11 + 6i$ **45.** -24 **47.** -15 **49.** $-5\sqrt{2}$ **51.** $-15 - 12i$ **53.** $3\sqrt{2} + 6i$

55. $-10i$ **57.** $29 - 2i$ **59.** 1 **61.** 1 **63.** 89 **65.** 50 **67.** $80 - 18i$ **69.** $-\dfrac{4}{5}i$

71. $-\dfrac{5}{3} + \dfrac{16}{3}i$ **73.** $\dfrac{30}{29} - \dfrac{12}{29}i$ **75.** $\dfrac{20}{17} + \dfrac{5}{17}i$ **77.** $\dfrac{11}{13} + \dfrac{29}{13}i$ **79.** $-\dfrac{1}{5} + \dfrac{\sqrt{6}}{10}i$ **81.** $-1 + 4i$

83. 1 **85.** i **87.** i **89.** 1 **91.** -1 **93.** -1 **95.** 1 **97.** $(a + 2i)(a - 2i)$

99. $(2b + 3i)(2b - 3i)$ **101.** $(6a + bi)(6a - bi)$ **103.** $(3a + 8i)(3a - 8i)$

105. a. Yes **b.** Yes   **107. a.** No **b.** No   **109. a.** Yes **b.** Yes

SECTION 5.5   (page 271)

Problem 1   **A.** $\sqrt{4x + 5} - 12 = -5$

$$\sqrt{4x + 5} = 7$$
$$(\sqrt{4x + 5})^2 = 7^2$$
$$4x + 5 = 49$$
$$4x = 44$$
$$x = 11$$

Check:

$$\sqrt{4x + 5} - 12 = -5$$

$\sqrt{4 \cdot 11 + 5} - 12$ | -5
$\sqrt{44 + 5} - 12$
$\sqrt{49} - 12$
$7 - 12$

$$-5 = -5$$

The solution is 11.

B. $\sqrt[4]{x - 8} = 3$
$(\sqrt[4]{x - 8})^4 = 3^4$
$x - 8 = 81$
$x = 89$

Check:

$\sqrt[4]{x - 8} = 3$
$\sqrt[4]{89 - 8}$ | 3
$\sqrt[4]{81}$

$$3 = 3$$

The solution is 89.

C. $x + 3\sqrt{x + 2} = 8$
$3\sqrt{x + 2} = 8 - x$
$(3\sqrt{x + 2})^2 = (8 - x)^2$
$9(x + 2) = 64 - 16x + x^2$
$9x + 18 = 64 - 16x + x^2$
$0 = x^2 - 25x + 46$
$0 = (x - 2)(x - 23)$

Check:

$x + 3\sqrt{x + 2} = 8$ | $x + 3\sqrt{x + 2} = 8$
$2 + 3\sqrt{2 + 2}$ | 8 | $23 + 3\sqrt{23 + 2}$ | 8
$2 + 3\sqrt{4}$ | | $23 + 3\sqrt{25}$
$2 + 3 \cdot 2$ | | $23 + 3 \cdot 5$
$2 + 6$ | | $23 + 15$
$8 = 8$ | | $38 \neq 8$

$x - 2 = 0$ $x - 23 = 0$
$x = 2$ $x = 23$

23 does not check as a solution.
The solution is 2.

Problem 2   **A.** $\sqrt{3x - 5} = \sqrt{x + 7}$

Check: $\dfrac{\sqrt{3x - 5} = \sqrt{x + 7}}{\begin{array}{c|c} \sqrt{3 \cdot 6 - 5} & \sqrt{6 + 7} \\ \sqrt{18 - 5} & \sqrt{13} \\ \sqrt{13} = \sqrt{13} \end{array}}$

$(\sqrt{3x - 5})^2 = (\sqrt{x + 7})^2$

$3x - 5 = x + 7$

$2x - 5 = 7$

$2x = 12$

$x = 6$

The solution is 6.

B. $\sqrt{x + 5} = 5 - \sqrt{x}$

Check: $\dfrac{\sqrt{x + 5} = 5 - \sqrt{x}}{\begin{array}{c|c} \sqrt{4 + 5} & 5 - \sqrt{4} \\ \sqrt{9} & 5 - 2 \\ 3 = 3 \end{array}}$

$(\sqrt{x + 5})^2 = (5 - \sqrt{x})^2$

$x + 5 = 25 - 10\sqrt{x} + x$

$-20 = -10\sqrt{x}$

$2 = \sqrt{x}$

$2^2 = (\sqrt{x})^2$

$4 = x$

The solution is 4.

Problem 3   **Strategy**

To find the diagonal, use the Pythagorean Theorem. One leg is the length of the rectangle. The second leg is the width of the rectangle. The hypotenuse is the diagonal of the rectangle.

Solution

$c^2 = a^2 + b^2$

$c^2 = (6)^2 + (3)^2$

$c^2 = 36 + 9$

$c^2 = 45$

$(c^2)^{\frac{1}{2}} = (45)^{\frac{1}{2}}$

$c = \sqrt{45}$

$c \approx 6.7$

The diagonal is 6.7 cm.

Problem 4   **Strategy**

To find the height above water, replace d in the equation with the given value, and solve for h.

Solution

$d = 1.4\sqrt{h}$

$5.5 = 1.4\sqrt{h}$

$\dfrac{5.5}{1.4} = \sqrt{h}$

$\left(\dfrac{5.5}{1.4}\right)^2 = (\sqrt{h})^2$

$\dfrac{30.25}{1.96} = h$

$15.434 \approx h$

The periscope must be 15.434 ft above the water.

EXERCISES   (page 276)

1. 25   **3.** 27   **5.** 48   **7.** -2 **9.** No Solution   **11.** 9   **13.** -23 **15.** -16 **17.** $\dfrac{9}{4}$ **19.** 14

21. 7   **23.** 7   **25.** $\dfrac{13}{3}$ **27.** 35   **29.** -7 **31.** 45   **33.** 23   **35.** -4 **37.** 10   **39.** 1   **41.** 3

43. 21   **45.** $\dfrac{25}{72}$ **47.** 5   **49.** -2 and 4   **51.** -2 and 2   **53.** -2 and 1   **55.** 2 and 11   **57.** 3

59. 15.2 ft   **61.** 10 ft   **63.** 6.61 ft   **65.** 2500 ft   **67.** 120 ft   **69.** 4.67 ft   **71.** 27   **73.** 243

75. 6   **77.** 0 and 1   **79.** -1 **81.** $s = \dfrac{v^2}{2a}$ **83.** $r = \dfrac{\sqrt{A\pi}}{\pi}$ **85.** $r = \dfrac{\sqrt[3]{6V\pi^2}}{2\pi}$ **87.** 7.1 cm

89. $4x + 2x\sqrt{2}$ **91.** Impossible; at least one of the integers must be even.

CHAPTER REVIEW   (page 282)

1. $81x^4y^2$ (Objective 5.1.1)   **2.** $\dfrac{16b}{a^5}$ (Objective 5.1.1)   **3.** $3x^2y^2\sqrt[3]{3y^2}$ (Objective 5.3.1)

4. $-2x^{\frac{4}{3}}$ (Objective 5.2.2)   **5.** $\dfrac{4}{3xy^5}$ (Objective 5.1.1)   **6.** $12x^2y^3$ (Objective 5.2.3)

7. $3x\sqrt{2}$ (Objective 5.3.3)   **8.** $5 - 2i$ (Objective 5.4.1)   **9.** -18 (Objective 5.5.1)

10. 0.0075 (Objective 5.1.2)   **11.** $1 - i$ (Objective 5.4.2)   **12.** $-16 + 4\sqrt{5}$ (Objective 5.3.3)

13. $2\sqrt[3]{a^2}$ (Objective 5.2.2)   **14.** $\dfrac{18}{13} + \dfrac{12}{13}i$ (Objective 5.4.4)   **15.** $\dfrac{xy\sqrt{3}}{3}$ (Objective 5.3.4)

16. 11 (Objective 5.5.1)   **17.** $-2xy\sqrt{y}$ (Objective 5.3.2)   **18.** $-6 + 43i$ (Objective 5.4.3)

19. $\dfrac{1}{a^{\frac{7}{6}}}$ (Objective 5.1.1)   **20.** $\dfrac{\sqrt{2y} + 2\sqrt{2}}{y - 4}$ (Objective 5.3.4)   **21.** $2x^2y^2$ (Objective 5.1.1)

22. $2ab^4$ (Objective 5.2.3)   **23.** $-x\sqrt{10x}$ (Objective 5.3.2)   **24.** $13 - 7\sqrt{3}$ (Objective 5.3.3)

25. $-8 - 14i$ (Objective 5.4.3)   **26.** $2x^{\frac{3}{4}}$ (Objective 5.2.2)   **27.** -20 (Objective 5.5.1)

28. $3\sqrt{2} - 5i$ (Objective 5.4.1)   **29.** 2,250,000,000 (Objective 5.1.2)   **30.** $1 - a$ (Objective 5.2.1)

31. $\dfrac{x + \sqrt{xy}}{x - y}$ (Objective 5.3.4)   **32.** $-\dfrac{1}{5} + \dfrac{3}{5}i$ (Objective 5.4.4)   **33.** $3xy^2\sqrt{6xy}$ (Objective 5.3.1)

34. $7 + i$ (Objective 5.4.2)   **35.** $-2ab^2\sqrt[3]{ab^2}$ (Objective 5.3.3)   **36.** -3 (Objective 5.5.1)

37. $\sqrt[3]{3x - 2}$ (Objective 5.2.2)   **38.** $\dfrac{x\sqrt[3]{4y^2}}{2y}$ (Objective 5.3.4)   **39.** 1.2×10 h (Objective 5.1.3)

40. 576 ft (Objective 5.5.2)

CUMULATIVE REVIEW   (page 284)

1. The Distributive Property (Objective 1.2.2)   **2.** $-x - 24$ (Objective 1.2.3)

3. $A \cap B = \varnothing$ (Objective 1.3.1)   **4.** ⟵●──●──→ (Objective 1.3.2)   **5.** $\dfrac{3}{2}$ (Objective 2.1.1)

6. $\dfrac{2}{3}$ (Objective 2.1.2)   **7.** $\{x \mid x \geq -1\}$ (Objective 2.2.1)   **8.** $\{x \mid 4 < x < 5\}$ (Objective 2.2.2)

9. $\dfrac{1}{3}$ and $\dfrac{7}{3}$ (Objective 2.3.1)   **10.** $\left\{x \mid x < 2 \text{ or } x > \dfrac{8}{3}\right\}$ (Objective 2.3.2)

11. $(8a + b)(8a - b)$ (Objective 3.4.1)   **12.** $x(x^2 + 3)(x + 1)(x - 1)$ (Objective 3.4.4)

13. $\dfrac{2}{3}$ and -5 (Objective 3.5.1)   **14.** 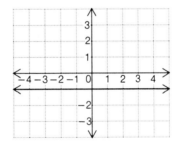 (Objective 3.5.2)   **15.** $\dfrac{4}{a - 1}$ (Objective 4.1.1)

16. $y - 2$ (Objective 4.3.1)   **17.** $R = Pn - C$ (Objective 4.5.3)   **18.** $\{x \mid x < -2 \text{ or } x \geq 5\}$ (Objective 4.5.2)

19. $\dfrac{3x^3}{y}$ (Objective 5.1.1)   **20.** $\dfrac{y^8}{x^2}$ (Objective 5.2.1)   **21.** $-x\sqrt{5x}$ (Objective 5.3.2)

22. $11 - 5\sqrt{5}$ (Objective 5.3.3)   **23.** $\dfrac{x\sqrt[3]{4y^2}}{2y}$ (Objective 5.3.4)   **24.** $22 - 7i$ (Objective 5.4.3)

25. $-\dfrac{3}{5} + \dfrac{6}{5}i$ (Objective 5.4.4)   **26.** 16 (Objective 5.5.1)   **27.** 19 stamps (Objective 2.4.1)

28. \$4000 (Objective 2.5.1)   **29.** length: 12 ft; width: 6 ft (Objective 3.5.3)

30. 250 mph (Objective 4.5.5)   **31.** 1.25 s (Objective 5.1.3)   **32.** 25 ft (Objective 5.5.2)

Answers to Chapter 6

SECTION 6.1   (page 289)

Problem 1   **A.**

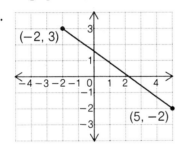

B.

$$y = -2x + 5$$

Problem 2
$$y = -2\left(\frac{1}{3}\right) + 5$$

$$y = -\frac{2}{3} + 5$$

$$y = \frac{13}{3}$$

The ordered pair solution is $\left(\dfrac{1}{3}, \dfrac{13}{3}\right)$.

Problem 3

Problem 4 $-3x + 2y = 4$
$$2y = 3x + 4$$
$$y = \frac{3}{2}x + 2$$

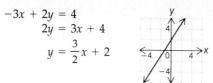

Problem 5

Problem 6   **A.** $x + 3y > 6$ **B.** $y < 2$
$$3y > -x + 6$$
$$y > -\frac{1}{3}x + 2$$

EXERCISES   (page 296)

1.   **3.**   **5.** A is $(0, 3)$, B is $(1, 1)$, C is $(3, -4)$, D is $(-4, 4)$

7. A is $(4, 0)$, B is $(-1, 1)$, C is $(-4, -2)$, D is $(1, 2)$ **9.**   **11.**   **13.** $(-3, -6)$

15. $(-6, 8)$ **17.** $(-4, -3)$ **19.** $(-2, 1)$ **21.** $\left(-2, -\frac{1}{3}\right)$ **23.**   **25.**

27.   **29.**   **31.**   **33.**   **35.**

37.   **39.**   **41.**   **43.**   **45.**

47.   **49.**   **51.** $(0, 5)$ **53.** $(-3, 5)$ **55.** $(-1, 8)$ **57.** 2   **59.** 4

61. 2   **63.** a **65.** a and b **67.** b **69. a.** $y = 3x + 13$ **b.** $(-5, -2)$ **71.** 3   **73.** 10 square units
75. $(8 + \sqrt{34})$ units

SECTION 6.2   (page 302)

Problem 1   Let $P_1 = (4, -3)$ and $P_2 = (2, 7)$.

$$m = \frac{y_2 - y_1}{x_2 - x_1} = \frac{7 - (-3)}{2 - 4} = \frac{10}{-2} = -5$$

The slope is -5.

Problem 2 x-intercept: y-intercept:

$3x - y = 2$ $3x - y = 2$
$3x - 0 = 2$ $3(0) - y = 2$
$3x = 2$ $-y = 2$
$x = \dfrac{2}{3}$ $y = -2$

$\left(\dfrac{2}{3}, 0\right)$ $(0, -2)$

Problem 3 x-intercept: y-intercept:

$$y = \frac{1}{4}x + 1 \qquad\qquad (0,b)$$

$$0 = \frac{1}{4}x + 1 \qquad\qquad b = 1$$

$$-\frac{1}{4}x = 1$$

$$x = -4$$

$$(-4, 0) \qquad\qquad\qquad (0,1)$$

Problem 4 $2x + 3y = 6$

$$3y = -2x + 6$$

$$y = -\frac{2}{3}x + 2$$

$$m = -\frac{2}{3} = \frac{-2}{3}$$

y-intercept: $(0, 2)$

Problem 5 $(x_1, y_1) = (-3, -2)$
$m = 3$

EXERCISES   (page 308)

1. -1 **3.** $\frac{1}{3}$ **5.** $-\frac{2}{3}$ **7.** $-\frac{3}{4}$ **9.** No slope   **11.** $\frac{7}{5}$ **13.** 0 **15.** $-\frac{1}{2}$ **17.** -3

19. No slope   **21.** -2 **23.** 0 **25.**   **27.**   **29.**

31.   **33.**   **35.**   **37.**   **39.**

41.   **43.**   **45.**   **47.**   **49.**

51.   **53.**   **55.**   **57.**   **59.**

61. 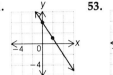   **63.** $y = \dfrac{1}{2}x - 2$ **65.** $y = -3x - 3$ **67. a.** $y = \dfrac{5}{2}x - 5$ **b.** $(2, 0)$ and $(0, -5)$

69. a. $y = 8x - 8$ **b.** $(1, 0)$ and $(0, -8)$ **71.** The slope of the line rotates clockwise.
73. The graph of the line moves down.

SECTION 6.3   (page 313)

Problem 1 $m = -\dfrac{5}{4}$ $b = 3$

$y = mx + b$

$y = -\dfrac{5}{4}x + 3$

The equation of the line is $y = -\dfrac{5}{4}x + 3$.

Problem 2 $m = -3$ $(x_1, y_1) = (4, -3)$

$y - y_1 = m(x - x_1)$
$y - (-3) = -3(x - 4)$
$y + 3 = -3x + 12$
$y = -3x + 9$

The equation of the line is $y = -3x + 9$.

Problem 3   **A.** Let $(x_1, y_1) = (4, -2)$ and $(x_2, y_2) = (-1, -7)$.

$m = \dfrac{y_2 - y_1}{x_2 - x_1} = \dfrac{-7 - (-2)}{-1 - 4} = \dfrac{-5}{-5} = 1$
$y - y_1 = m(x - x_1)$
$y - (-2) = 1(x - 4)$
$y + 2 = 1(x - 4)$
$y + 2 = x - 4$
$y = x - 6$

The equation of the line is $y = x - 6$.

B. Let $(x_1, y_1) = (2, 3)$ and $(x_2, y_2) = (-5, 3)$.

$$m = \frac{y_2 - y_1}{x_2 - x_1} = \frac{3 - 3}{-5 - 2} = \frac{0}{-7} = 0$$

The line has zero slope.
The line is a horizontal line.
All points on the line have an ordinate of 3.
The equation of the line is $y = 3$.

Problem 4 $5x + 2y = 2$ $5x + 2y = -6$

$\qquad\qquad\quad 2y = -5x + 2$ $2y = -5x - 6$

$\qquad\qquad\quad\; y = -\dfrac{5}{2}x + 1$ $y = -\dfrac{5}{2}x - 3$

$m_1 = m_2 = -\dfrac{5}{2}$

The slopes of the lines are equal. The lines are parallel.

Problem 5 $x - 4y = 3$

$\qquad\qquad\quad -4y = -x + 3$

$\qquad\qquad\qquad\; y = \dfrac{1}{4}x - \dfrac{3}{4}$

$m_1 = \dfrac{1}{4}$

$m_1 \cdot m_2 = -1$

$\dfrac{1}{4} \cdot m_2 = -1$

$\qquad m_2 = -4$

$y - y_1 = m(x - x_1)$
$y - 2 = -4[x - (-2)]$
$y - 2 = -4(x + 2)$
$y - 2 = -4x - 8$
$\qquad y = -4x - 6$

The equation of the line is $y = -4x - 6$.

EXERCISES   (page 319)

1. $y = 2x + 5$ **3.** $y = \dfrac{1}{2}x + 2$ **5.** $y = \dfrac{5}{4}x + \dfrac{21}{4}$ **7.** $y = -\dfrac{5}{3}x + 5$ **9.** $y = -3x + 9$

11. $y = -3x + 4$ **13.** $y = \dfrac{2}{3}x - \dfrac{7}{3}$ **15.** $y = \dfrac{1}{2}x$ **17.** $y = 3x - 9$ **19.** $y = -\dfrac{2}{3}x + 7$

21. $y = 0$ **23.** $y = \dfrac{3}{2}x$ **25.** $x = 0$ **27.** $y = 3x - 3$ **29.** $y = -\dfrac{5}{2}x - \dfrac{25}{2}$ **31.** $y = \dfrac{4}{3}x + 5$

33. $y = -\dfrac{2}{5}x + \dfrac{3}{5}$ **35.** $x = 3$ **37.** $y = -\dfrac{5}{4}x - \dfrac{15}{2}$ **39.** $y = -3$ **41.** $y = x + 2$

43. $y = -2x - 3$ **45.** $y = \dfrac{2}{3}x + \dfrac{5}{3}$ **47.** $y = \dfrac{1}{3}x + \dfrac{10}{3}$ **49.** $y = \dfrac{3}{2}x - \dfrac{1}{2}$ **51.** $y = -\dfrac{3}{2}x + 3$

53. $y = -1$ **55.** $y = x - 1$ **57.** $y = -2x - 1$ **59.** $x = -2$ **61.** $y = -\dfrac{3}{4}x + \dfrac{17}{4}$

63. $y = 3x - 10$ **65.** $y = -x + 1$ **67.** $y = \dfrac{3}{7}x - \dfrac{15}{7}$ **69.** $y = \dfrac{1}{2}x - 1$ **71.** $y = -4$ **73.** $y = \dfrac{3}{4}x$

75. $y = -\dfrac{4}{3}x + \dfrac{5}{3}$ **77.** $x = -2$ **79.** $y = x - 1$ **81.** Yes   **83.** No   **85.** No   **87.** Yes   **89.** Yes

91. Yes   **93.** No   **95.** Yes   **97.** Yes   **99.** $y = \dfrac{2}{3}x - \dfrac{8}{3}$ **101.** $y = \dfrac{1}{3}x - \dfrac{1}{3}$ **103.** $y = -\dfrac{5}{3}x - \dfrac{14}{3}$

105. $y = \dfrac{2}{3}x + 3$ **107.** $y = -\dfrac{5}{3}x - \dfrac{13}{3}$ **109.** $y = \dfrac{5}{2}x + 8$ **111.** $y = -\dfrac{1}{5}x - \dfrac{11}{5}$ **113.** $y = -\dfrac{2}{3}x - 5$

115. $y = -x + 4$ **117.** $y = \dfrac{1}{2}x + 1$ **119.** 0   **121.** 6   **123.** -3 **125.** -2 **127.** -3 **129.** 5

131. $y = -\dfrac{6}{5}x + \dfrac{12}{5}; \; y = \dfrac{1}{4}x - \dfrac{1}{2}; \; y = -7x - 15$

133. Let $P_1 = (1, 4)$ and $P_2 = (-3, -2)$; $m = \dfrac{-2 - 4}{-3 - 1} = \dfrac{-6}{-4} = \dfrac{3}{2}$.

Let $P_3 = (-3, -2)$ and $P_4 = (3, -6)$; $m = \dfrac{-6 - (-2)}{3 - (-3)} = \dfrac{-4}{6} = -\dfrac{2}{3}$; $m_1 \cdot m_2 = \dfrac{3}{2}\left(-\dfrac{2}{3}\right) = -1$.

135. Let $P_1 = (1, 6)$ and $P_2 = (3, 2)$; $m_1 = \dfrac{2 - 6}{3 - 1} = \dfrac{-4}{2} = -2$.

Let $P_3 = (-3, -2)$ and $P_4 = (-1, -6)$; $m_2 = \dfrac{-6 - (-2)}{-1 - (-3)} = \dfrac{-4}{2} = -2$; $m_1 = m_2$.

Let $P_5 = (-3, -2)$ and $P_6 = (1, 6)$; $m_3 = \dfrac{6 - (-2)}{1 - (-3)} = \dfrac{8}{4} = 2$.

Let $P_7 = (-1, -6)$ and $P_8 = (3, 2)$; $m_4 = \dfrac{2 - (-6)}{3 - (-1)} = \dfrac{8}{4} = 2$, $m_3 = m_4$.

SECTION 6.4   (page 325)

Problem 1   Strategy

To write the equation:
▶ Use two points on the graph to find the slope of the line.
▶ Locate the y-intercept of the line on the graph.
▶ Use the slope-intercept form of an equation to write the equation of the line.
To find the Fahrenheit temperature, substitute $40°$ for x in the equation, and solve for y.

Solution

$(x_1, y_1) = (0, 32)$ $(x_2, y_2) = (100, 212)$

$m = \dfrac{y_2 - y_1}{x_2 - x_1} = \dfrac{212 - 32}{100 - 0} = \dfrac{180}{100} = \dfrac{9}{5}$

The y-intercept is $(0, 32)$.

$y = mx + b$

$y = \dfrac{9}{5}x + 32$

The equation of the line is $y = \dfrac{9}{5}x + 32$.

$y = \dfrac{9}{5}x + 32$

$y = \dfrac{9}{5}(40) + 32 = 72 + 32 = 104$

The Fahrenheit temperature is 104°.

EXERCISES   (page 327)

1. The equation of the line is $y = 300x$. In 2.5 h, the plane traveled 750 mi.
3. The equation of the line is $y = -1200x + 12{,}000$. In 4.5 years, the value of the truck is $6600.
5. The equation of the line is $y = 25x + 2000$. The cost of manufacturing 200 bicycles is $7000.
7. The equation of the line is $y = -x + 5$. The demand for the product is 1.5 units.
9.

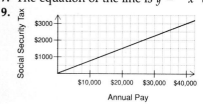

11. **a.** The slope represents the Social Security tax per $1.
b. The y-intercept represents the amount of tax paid when $0 is earned.

CHAPTER REVIEW   (page 332)

1.

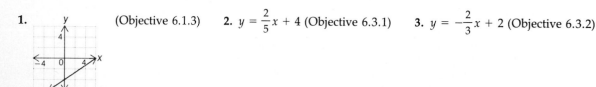

(Objective 6.1.3)   **2.** $y = \dfrac{2}{5}x + 4$ (Objective 6.3.1)   **3.** $y = -\dfrac{2}{3}x + 2$ (Objective 6.3.2)

4.   (Objective 6.1.1)   **5.** $-\dfrac{1}{6}$ (Objective 6.2.1)   **6.** $y = -\dfrac{7}{5}x + \dfrac{1}{5}$ (Objective 6.3.1)

7. $(-3, 0)$ (Objective 6.1.2)   **8.**   (Objective 6.2.2)   **9.**   (Objective 6.1.4)

10. $y = 2x + 1$ (Objective 6.3.2)   **11.** $y = -\dfrac{3}{4}x + 2$ (Objective 6.3.1)   **12.**   (Objective 6.1.3)

13.   (Objective 6.2.3)   **14.** 0 (Objective 6.2.1)   **15.** $x = 2$ (Objective 6.3.1)

16.   (Objective 6.1.4)   **17.** $(3, 0)$ (Objective 6.2.2)   **18.**   (Objective 6.1.3)

19. $y = -\dfrac{3}{2}x$ (Objective 6.3.2)   **20.** $\left(3, -\dfrac{1}{4}\right)$ (Objective 6.1.2)   **21.** No slope (Objective 6.2.1)

22. $y = -\dfrac{3}{2}x$ (Objective 6.3.2)   **23.** 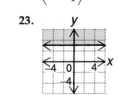   (Objective 6.1.3)   **24.** 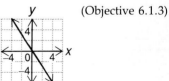   (Objective 6.1.3)

25.   (Objective 6.2.3)   **26.** $y = \dfrac{5}{2}x - 9$ (Objective 6.3.1)

27. $y = -\dfrac{10,000}{3}x + 50,000$ (Objective 6.4.1)

CUMULATIVE REVIEW   (page 334)

1. 432 (Objective 1.1.3)   **2.** 0 (Objective 1.1.4)   **3.** $-\dfrac{13}{2}$ (Objective 1.2.1)   **4.** $\dfrac{9}{2}$ (Objective 2.1.1)

5. $\dfrac{8}{9}$ (Objective 2.1.2)   **6.** $\{x \mid 1 < x < 3\}$ (Objective 2.2.2)   **7.** $\dfrac{5}{2}$ and $-\dfrac{3}{2}$ (Objective 2.3.1)

8. $\left\{x \mid -\dfrac{2}{5} < x < 2\right\}$ (Objective 2.3.2)   **9.** $\dfrac{3b^8}{8a^2}$ (Objective 3.1.3)   **10.** $a^4 - a^3 + 7a^2 - 2a$ (Objective 3.2.1)

11. $8xy(x + 2xy + 3y)$ (Objective 3.3.1)   **12.** $(2x - 3b)(3x + 2a)$ (Objective 3.4.3)

13. 5 and -1 (Objective 3.5.1)   **14.** $\{x \mid x \geq 2 \text{ or } -3 \leq x \leq 1\}$ (Objective 3.5.2)   **15.** $\dfrac{x - 8}{x - 5}$ (Objective 4.2.2)

16. $4ab\sqrt[3]{2ab}$ (Objective 5.3.2)   **17.** $14 - 27i$ (Objective 5.4.3)   **18.** 4 (Objective 5.5.1)

19. $(2, -1)$ (Objective 6.1.2)   **20.** 2 (Objective 6.2.1)   **21.** $y = -2x - 8$ (Objective 6.3.1)

22. $y = -2x$ (Objective 6.3.1)   **23.** $y = \dfrac{2}{3}x + 6$ (Objective 6.3.2)   **24.** $y = -4x - 2$ (Objective 6.3.2)

25. (Objective 6.1.3)   **26.** (Objective 6.1.4)

27. (Objective 6.2.3)   **28.** 125 oz (Objective 2.5.2)   **29.** 26 stamps (Objective 2.4.1)

30. slower plane: 200 mph; faster plane: 400 mph (Objective 2.4.3)   **31.** $2500 (Objective 2.5.1)

32. $y = 18x$ (Objective 6.4.1)

Answers to Chapter 7 ——————

SECTION 7.1   (page 339)

Problem 1 $x^2 - 3ax - 4a^2 = 0$
$(x + a)(x - 4a) = 0$

$x + a = 0 \qquad\qquad x - 4a = 0$
$\qquad x = -a \qquad\qquad\qquad x = 4a$

The solutions are $-a$ and $4a$.

Problem 2

$$(x - r_1)(x - r_2) = 0$$

$$\left[x - \left(-\frac{2}{3}\right)\right]\left(x - \frac{1}{6}\right) = 0$$

$$\left(x + \frac{2}{3}\right)\left(x - \frac{1}{6}\right) = 0$$

$$x^2 + \frac{3}{6}x - \frac{2}{18} = 0$$

$$18\left(x^2 + \frac{3}{6}x - \frac{2}{18}\right) = 0$$

$$18x^2 + 9x - 2 = 0$$

Problem 3

$$2(x + 1)^2 + 24 = 0$$

$$2(x + 1)^2 = -24$$

$$(x + 1)^2 = -12$$

$$\sqrt{(x + 1)^2} = \sqrt{-12}$$

$$x + 1 = \pm\sqrt{-12} = \pm 2i\sqrt{3}$$

$$x + 1 = 2i\sqrt{3} \qquad\qquad x + 1 = -2i\sqrt{3}$$
$$x = -1 + 2i\sqrt{3} \qquad\qquad x = -1 - 2i\sqrt{3}$$

The solutions are $-1 + 2i\sqrt{3}$ and $-1 - 2i\sqrt{3}$.

EXERCISES   (page 342)

1. 0 and 4   **3.** 5 and -5 **5.** 3 and -2 **7.** 3   **9.** 0 and 2   **11.** 5 and -2 **13.** 2 and 5

15. 6 and $-\dfrac{3}{2}$ **17.** 2 and $\dfrac{1}{4}$ **19.** $\dfrac{1}{3}$ and -4 **21.** $\dfrac{9}{2}$ and $-\dfrac{2}{3}$ **23.** $\dfrac{1}{4}$ and -4 **25.** 9 and -2

27. -2 and $-\dfrac{3}{4}$ **29.** 2 and -5 **31.** -4 and $-\dfrac{3}{2}$ **33.** $2b$ and $7b$ **35.** $7c$ and $-c$ **37.** $-\dfrac{b}{2}$ and $-b$

39. $4a$ and $\dfrac{2a}{3}$ **41.** $-\dfrac{a}{3}$ and $3a$ **43.** $-\dfrac{3y}{2}$ and $-\dfrac{y}{2}$ **45.** $-\dfrac{a}{2}$ and $-\dfrac{4a}{3}$ **47.** $x^2 - 7x + 10 = 0$

49. $x^2 + 6x + 8 = 0$ **51.** $x^2 - 5x - 6 = 0$ **53.** $x^2 - 9 = 0$ **55.** $x^2 - 8x + 16 = 0$

57. $x^2 - 5x = 0$ **59.** $x^2 - 3x = 0$ **61.** $2x^2 - 7x + 3 = 0$ **63.** $4x^2 - 5x - 6 = 0$

65. $3x^2 + 11x + 10 = 0$ **67.** $9x^2 - 4 = 0$ **69.** $6x^2 - 5x + 1 = 0$ **71.** $10x^2 - 7x - 6 = 0$

73. $8x^2 + 6x + 1 = 0$ **75.** $50x^2 - 25x - 3 = 0$ **77.** 7 and -7 **79.** $2i$ and $-2i$ **81.** 2 and -2

83. $\dfrac{9}{2}$ and $-\dfrac{9}{2}$ **85.** $7i$ and $-7i$ **87.** $4\sqrt{3}$ and $-4\sqrt{3}$ **89.** $5\sqrt{3}$ and $-5\sqrt{3}$ **91.** $3i\sqrt{2}$ and $-3i\sqrt{2}$

93. 7 and -5 **95.** 0 and -6 **97.** -7 and 3   **99.** $-5 + 5i$ and $-5 - 5i$ **101.** $4 + 2i$ and $4 - 2i$

103. $9 + 3i$ and $9 - 3i$ **105.** 1 and 0   **107.** 1 and $-\dfrac{1}{5}$ **109.** 0 and $-\dfrac{3}{2}$ **111.** $\dfrac{7}{3}$ and 1

113. $-5 + \sqrt{6}$ and $-5 - \sqrt{6}$ **115.** $2 + 2\sqrt{6}$ and $2 - 2\sqrt{6}$ **117.** $-1 + 2i\sqrt{3}$ and $-1 - 2i\sqrt{3}$

119. $3 + 3i\sqrt{5}$ and $3 - 3i\sqrt{5}$ **121.** $\dfrac{-2 - 9\sqrt{2}}{3}$ and $\dfrac{-2 + 9\sqrt{2}}{3}$ **123.** $\dfrac{3 + 12\sqrt{3}}{4}$ and $\dfrac{3 - 12\sqrt{3}}{4}$

125. $-\dfrac{1}{2} + 2i\sqrt{10}$ and $-\dfrac{1}{2} - 2i\sqrt{10}$ **127.** $\dfrac{2}{3} + \dfrac{5}{3}i$ and $\dfrac{2}{3} - \dfrac{5}{3}i$ **129.** $x^2 - 2 = 0$ **131.** $x^2 + 1 = 0$

133. $x^2 - 8 = 0$ **135.** $x^2 - 12 = 0$ **137.** $x^2 + 12 = 0$ **139.** $\dfrac{4b}{a}$ and $-\dfrac{4b}{a}$ **141.** $\dfrac{5z}{y}$ and $-\dfrac{5z}{y}$

143. $b + 1$ and $b - 1$ **145.** $-y + 3$ and $-y - 3$ **147.** $a + \dfrac{4}{b}$ and $a - \dfrac{4}{b}$ **149.** $b + \dfrac{2\sqrt{6}}{a}$ and $b - \dfrac{2\sqrt{6}}{a}$

151. $ax^2 + c = 0$

$$ax^2 = -c$$

$$x^2 = -\frac{c}{a}$$

$$\sqrt{x^2} = \sqrt{-\frac{c}{a}}$$

$$x = \pm\frac{\sqrt{ca}}{a}i$$

The solutions are $\dfrac{\sqrt{ca}}{a}i$ and $-\dfrac{\sqrt{ca}}{a}i$.

SECTION 7.2   (page 347)

Problem 1   **A.** $4x^2 - 4x - 1 = 0$

$$4x^2 - 4x = 1$$

$$\frac{1}{4}(4x^2 - 4x) = \frac{1}{4} \cdot 1$$

$$x^2 - x = \frac{1}{4}$$

Complete the square.

$$x^2 - x + \frac{1}{4} = \frac{1}{4} + \frac{1}{4}$$

$$\left(x - \frac{1}{2}\right)^2 = \frac{2}{4}$$

$$\sqrt{\left(x - \frac{1}{2}\right)^2} = \sqrt{\frac{2}{4}}$$

$$x - \frac{1}{2} = \pm\frac{\sqrt{2}}{2}$$

$$x - \frac{1}{2} = \frac{\sqrt{2}}{2} \qquad x - \frac{1}{2} = -\frac{\sqrt{2}}{2}$$

$$x = \frac{1}{2} + \frac{\sqrt{2}}{2} \qquad x = \frac{1}{2} - \frac{\sqrt{2}}{2}$$

The solutions are $\dfrac{1 + \sqrt{2}}{2}$ and $\dfrac{1 - \sqrt{2}}{2}$.

B. $2x^2 + x - 5 = 0$

$$2x^2 + x = 5$$

$$\frac{1}{2}(2x^2 + x) = \frac{1}{2} \cdot 5$$

$$x^2 + \frac{1}{2}x = \frac{5}{2}$$

Complete the square.

$$x^2 + \frac{1}{2}x + \frac{1}{16} = \frac{5}{2} + \frac{1}{16}$$

$$\left(x + \frac{1}{4}\right)^2 = \frac{41}{16}$$

$$\sqrt{\left(x + \frac{1}{4}\right)^2} = \sqrt{\frac{41}{16}}$$

$$x + \frac{1}{4} = \pm\frac{\sqrt{41}}{4}$$

$$x + \frac{1}{4} = \frac{\sqrt{41}}{4} \qquad x + \frac{1}{4} = -\frac{\sqrt{41}}{4}$$

$$x = -\frac{1}{4} + \frac{\sqrt{41}}{4} \qquad x = -\frac{1}{4} - \frac{\sqrt{41}}{4}$$

The solutions are $\dfrac{-1 + \sqrt{41}}{4}$ and $\dfrac{-1 - \sqrt{41}}{4}$.

Problem 2   **A.** $x^2 + 6x - 9 = 0$

$$a = 1, b = 6, c = -9$$

$$x = \frac{-b \pm \sqrt{b^2 - 4ac}}{2a}$$

$$= \frac{-6 \pm \sqrt{6^2 - 4(1)(-9)}}{2 \cdot 1}$$

$$= \frac{-6 \pm \sqrt{36 + 36}}{2}$$

$$= \frac{-6 \pm \sqrt{72}}{2} = \frac{-6 \pm 6\sqrt{2}}{2}$$

$$= -3 \pm 3\sqrt{2}$$

The solutions are $-3 + 3\sqrt{2}$ and $-3 - 3\sqrt{2}$.

B. $$4x^2 = 4x - 1$$
$$4x^2 - 4x + 1 = 0$$

$$a = 4, b = -4, c = 1$$

$$x = \frac{-b \pm \sqrt{b^2 - 4ac}}{2a}$$

$$= \frac{-(-4) \pm \sqrt{(-4)^2 - 4(4)(1)}}{2 \cdot 4}$$

$$= \frac{4 \pm \sqrt{16 - 16}}{8} = \frac{4 \pm \sqrt{0}}{8}$$

$$= \frac{4}{8} = \frac{1}{2}$$

The solution is $\frac{1}{2}$.

Problem 3 $3x^2 - x - 1 = 0$

$$a = 3, b = -1, c = -1$$

$$b^2 - 4ac$$
$$(-1)^2 - 4(3)(-1) = 1 + 12 = 13$$
$$13 > 0$$

Since the discriminant is greater than zero, the equation has two real number solutions.

EXERCISES   (page 354)

1. 5 and -1 **3.** -9 and 1   **5.** 3   **7.** $-2 + \sqrt{11}$ and $-2 - \sqrt{11}$ **9.** $3 + \sqrt{2}$ and $3 - \sqrt{2}$

11. $1 + i$ and $1 - i$ **13.** 8 and -3 **15.** 4 and -9 **17.** $\frac{3 + \sqrt{5}}{2}$ and $\frac{3 - \sqrt{5}}{2}$ **19.** $\frac{1 + \sqrt{5}}{2}$ and $\frac{1 - \sqrt{5}}{2}$

21. $3 + \sqrt{13}$ and $3 - \sqrt{13}$ **23.** 5 and 3   **25.** $2 + 3i$ and $2 - 3i$ **27.** $-3 + 2i$ and $-3 - 2i$

29. $1 + 3\sqrt{2}$ and $1 - 3\sqrt{2}$ **31.** $\frac{1 + \sqrt{17}}{2}$ and $\frac{1 - \sqrt{17}}{2}$ **33.** $1 + 2i\sqrt{3}$ and $1 - 2i\sqrt{3}$ **35.** -1 and $-\frac{1}{2}$

37. $\frac{1}{2}$ and $\frac{3}{2}$ **39.** $\frac{4}{3}$ and $-\frac{1}{2}$ **41.** $\frac{1}{2} + i$ and $\frac{1}{2} - i$ **43.** $\frac{1}{3} + \frac{1}{3}i$ and $\frac{1}{3} - \frac{1}{3}i$

45. $\frac{2 + \sqrt{14}}{2}$ and $\frac{2 - \sqrt{14}}{2}$ **47.** 1 and $-\frac{3}{2}$ **49.** $1 + \sqrt{5}$ and $1 - \sqrt{5}$ **51.** $\frac{1}{2}$ and 5

53. $2 + \sqrt{5}$ and $2 - \sqrt{5}$ **55.** 1.236 and -3.236 **57.** 1.707 and 0.293   **59.** 0.309 and -0.809

61. 5 and -2 **63.** 4 and -9 **65.** $4 + 2\sqrt{22}$ and $4 - 2\sqrt{22}$ **67.** 3 and -8 **69.** -3 and $\frac{1}{2}$

71. $\frac{3}{2}$ and $-\frac{1}{4}$ **73.** $1 + 2\sqrt{2}$ and $1 - 2\sqrt{2}$ **75.** 10 and -2 **77.** $6 + 2\sqrt{3}$ and $6 - 2\sqrt{3}$

79. $\frac{1 + 2\sqrt{2}}{2}$ and $\frac{1 - 2\sqrt{2}}{2}$ **81.** 1 and $\frac{1}{2}$ **83.** $\frac{-5 + \sqrt{7}}{3}$ and $\frac{-5 - \sqrt{7}}{3}$ **85.** $\frac{5}{2}$ and $\frac{2}{3}$

87. $2 + i$ and $2 - i$ **89.** $-3 + 2i$ and $-3 - 2i$ **91.** $3 + i$ and $3 - i$ **93.** $-\frac{1}{2}$ and $-\frac{3}{2}$

95. $-\frac{1}{2} + \frac{5}{2}i$ and $-\frac{1}{2} - \frac{5}{2}i$ **97.** $\frac{-3 + \sqrt{6}}{3}$ and $\frac{-3 - \sqrt{6}}{3}$ **99.** $1 + \frac{\sqrt{6}}{3}i$ and $1 - \frac{\sqrt{6}}{3}i$

101. -1 and $-\dfrac{3}{2}$ **103.** 4 and $-\dfrac{3}{2}$ **105.** Two complex   **107.** One real   **109.** Two real

111. Two real   **113.** 0.873 and -6.873 **115.** 3.236 and -1.236 **117.** 2.468 and -0.135

119. $\dfrac{\sqrt{2}}{2}$ and $-2\sqrt{2}$ **121.** $\dfrac{\sqrt{2}}{2}$ and $-3\sqrt{2}$ **123.** $\dfrac{\sqrt{3}}{2} + \dfrac{1}{2}i$ and $\dfrac{\sqrt{3}}{2} - \dfrac{1}{2}i$ **125.** $2a$ and $-a$

127. a and $-4a$ **129.** $4a$ and $2a$ **131.** $\dfrac{a}{2}$ and $-2a$ **133.** $\{p \mid p < 9,\ p \in \text{real numbers}\}$

135. $\left\{ p \mid p < \dfrac{9}{4},\ p \in \text{real numbers} \right\}$ **137.** $\{p \mid p > 1,\ p \in \text{real numbers}\}$

139. $\left\{ p \mid p > \dfrac{1}{4},\ p \in \text{real numbers} \right\}$

141. $b^2 - 4ac = b^2 - 4(1)(-1) = b^2 + 4;\ b^2 \geq 0$ for any real number $b;\ b^2 + 4 > 0$ for any real number b.

SECTION 7.3   (page 358)

Problem 1   **A.** $x - 5x^{\frac{1}{2}} + 6 = 0$

$\left(x^{\frac{1}{2}}\right)^2 - 5\left(x^{\frac{1}{2}}\right) + 6 = 0$

$u^2 - 5u + 6 = 0$

$(u - 2)(u - 3) = 0$

$u - 2 = 0 \qquad u - 3 = 0$

$u = 2 \qquad\quad u = 3$

Replace u with $x^{\frac{1}{2}}$.

$x^{\frac{1}{2}} = 2 \qquad\qquad x^{\frac{1}{2}} = 3$

$(x^{\frac{1}{2}})^2 = 2^2 \qquad (x^{\frac{1}{2}})^2 = 3^2$

$x = 4 \qquad\qquad x = 9$

4 and 9 check as solutions.
The solutions are 4 and 9.

B. $4x^4 + 35x^2 - 9 = 0$

$4(x^2)^2 + 35(x^2) - 9 = 0$

$4u^2 + 35u - 9 = 0$

$(4u - 1)(u + 9) = 0$

$4u - 1 = 0 \qquad u - 9 = 0$

$4u = 1 \qquad\quad u = -9$

$u = \dfrac{1}{4}$

Replace u with x^2.

$x^2 = \dfrac{1}{4} \qquad\qquad x^2 = -9$

$\sqrt{x^2} = \sqrt{\dfrac{1}{4}} \qquad \sqrt{x^2} = \sqrt{-9}$

$x = \pm 3i$

$x = \pm\dfrac{1}{2}$

The solutions are $\dfrac{1}{2}$, $-\dfrac{1}{2}$, $3i$ and $-3i$.

Problem 2 $\sqrt{2x + 1} + x = 7$

$\sqrt{2x + 1} = 7 - x$

$(\sqrt{2x + 1})^2 = (7 - x)^2$

$2x + 1 = 49 - 14x + x^2$

$0 = x^2 - 16x + 48$

$0 = (x - 4)(x - 12)$

$x - 4 = 0 \qquad x - 12 = 0$

$x = 4 \qquad\quad x = 12$

4 checks as a solution.
12 does not check as a solution.

The solution is 4.

Problem 3 $\sqrt{2x - 1} + \sqrt{x} = 2$

Solve for one of the radical expressions.
$$\sqrt{2x - 1} = 2 - \sqrt{x}$$
$$(\sqrt{2x - 1})^2 = (2 - \sqrt{x})^2$$
$$2x - 1 = 4 - 4\sqrt{x} + x$$
$$x - 5 = -4\sqrt{x}$$

Square each side of the equation.
$$(x - 5)^2 = (-4\sqrt{x})^2$$
$$x^2 - 10x + 25 = 16x$$
$$x^2 - 26x + 25 = 0$$
$$(x - 1)(x - 25) = 0$$

$$x - 1 = 0 \qquad\qquad x - 25 = 0$$
$$x = 1 \qquad\qquad\quad x = 25$$

1 checks as a solution.
25 does not check as a solution.

The solution is 1.

Problem 4   **A.**
$$3y + \frac{25}{3y - 2} = -8$$
$$(3y - 2)\left(3y + \frac{25}{3y - 2}\right) = (3y - 2)(-8)$$
$$(3y - 2)(3y) + (3y - 2)\left(\frac{25}{3y - 2}\right) = (3y - 2)(-8)$$
$$9y^2 - 6y + 25 = -24y + 16$$
$$9y^2 + 18y + 9 = 0$$
$$9(y^2 + 2y + 1) = 0$$
$$9(y + 1)(y + 1) = 0$$

$$y + 1 = 0 \qquad\qquad y + 1 = 0$$
$$y = -1 \qquad\qquad\quad y = -1$$

The solution is -1.

B.
$$\frac{5}{x + 2} = 2x - 5$$
$$(x + 2)\left(\frac{5}{x + 2}\right) = (x + 2)(2x - 5)$$
$$5 = 2x^2 - x - 10$$
$$0 = 2x^2 - x - 15$$
$$0 = (2x + 5)(x - 3)$$

$$2x + 5 = 0 \qquad\qquad x - 3 = 0$$
$$2x = -5 \qquad\qquad\quad x = 3$$
$$x = -\frac{5}{2}$$

The solutions are $-\dfrac{5}{2}$ and 3.

EXERCISES   (page 364)

1. $3, -3, 2, -2$ **3.** $\sqrt{2}, -\sqrt{2}, 2, -2$ **5.** 1 and 4   **7.** 16   **9.** $2i, -2i, 1, -1$ **11.** $4i, -4i, 2, -2$

13. 16   **15.** 512 and 1   **17.** 1 and 2   **19.** $1, -1, 2,$ and -2 **21.** -64 and 8   **23.** $\dfrac{1}{4}$ and 1

25. 3   **27.** 9   **29.** 2 and -1 **31.** 0 and 2   **33.** 2 and $-\dfrac{1}{2}$ **35.** -2 **37.** 1   **39.** 1

41. -3 **43.** 10 and -1 **45.** $-\dfrac{1}{2} + \dfrac{\sqrt{7}}{2}i$ and $-\dfrac{1}{2} - \dfrac{\sqrt{7}}{2}i$ **47.** 1 and -3 **49.** 0 and -1

51. $\dfrac{1}{2}$ and $-\dfrac{1}{3}$ **53.** $-\dfrac{2}{3}$ and 6   **55.** $\dfrac{4}{3}$ and 3   **57.** $-\dfrac{1}{4}$ and 3   **59.** $\dfrac{-5 + \sqrt{33}}{2}$ and $\dfrac{-5 - \sqrt{33}}{2}$

61. $\dfrac{2}{3}$ and 2   **63.** -6 and 4   **65.** $-1 + 6\sqrt{2}$ and $-1 - 6\sqrt{2}$ **67.** 4   **69.** 6   **71.** 3

73. 4 and 5   **75.** $2, -2, 1,$ and -1 **77.** $3, -3, i,$ and $-i$ **79.** $\dfrac{-2 + \sqrt{2}}{2}$ and $\dfrac{-2 - \sqrt{2}}{2}$

81. $10 + 4\sqrt{6}$ and $10 - 4\sqrt{6}$ **83.** $-\dfrac{3}{2}$ and 2   **85.** $2, -2, i\sqrt{6},$ and $-i\sqrt{6}$ **87.** $\dfrac{-3 + \sqrt{5}}{2}$ and $\dfrac{-3 - \sqrt{5}}{2}$

89. $2 + \sqrt{6}$ and $2 - \sqrt{6}$ **91.** $\sqrt{2}$ and i **93.** $6 + 4\sqrt{2}$ and $6 - 4\sqrt{2}$ **95.** $1 + \sqrt{5}$ and $1 - \sqrt{5}$

97. $4 + ab$ and $4 - ab$ **99.** $2 + ab$ and $2 - ab$

SECTION 7.4   (page 368)

Problem 1   **A.**   **B.**

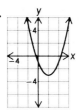
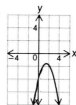

Problem 2   **A.** $y = 2x^2 - 5x + 2$
$0 = 2x^2 - 5x + 2$
$0 = (2x - 1)(x - 2)$

$2x - 1 = 0$ $x - 2 = 0$
$2x = 1$ $x = 2$

$x = \dfrac{1}{2}$

The x-intercepts are

$\left(\dfrac{1}{2}, 0\right)$ and $(2, 0)$.

B. $y = x^2 + 4x + 5$
$0 = x^2 + 4x + 5$
$a = 1, b = 4, c = 5$

$x = \dfrac{-b \pm \sqrt{b^2 - 4ac}}{2a}$

$= \dfrac{-4 \pm \sqrt{4^2 - 4(1)(5)}}{2 \cdot 1}$

$= \dfrac{-4 \pm \sqrt{16 - 20}}{2}$

$= \dfrac{-4 \pm \sqrt{-4}}{2} = \dfrac{-4 \pm 2i}{2}$

$= -2 \pm i$

The equation has no real number solutions.
There are no x-intercepts.

Problem 3 $y = x^2 - x - 6$
$a = 1, b = -1, c = -6$

$b^2 - 4ac$
$(-1)^2 - 4(1)(-6) = 1 + 24 = 25$

Since the discriminant is greater than zero, the parabola has two x-intercepts.

EXERCISES   (page 372)

1.   **3.**   **5.**   **7.**   **9.**

11.   **13.**   **15.**   **17.**   **19.**

21.   **23.**   **25.** $(2, 0)$ and $(-2, 0)$ **27.** $(0, 0)$ and $(2, 0)$

29. $(2, 0)$ and $(-1, 0)$ **31.** $\left(-\dfrac{1}{2}, 0\right)$ and $(3, 0)$ **33.** $\left(-\dfrac{2}{3}, 0\right)$ and $(7, 0)$ **35.** $\left(\dfrac{4}{3}, 0\right)$ and $(5, 0)$

37. $\left(\dfrac{2}{3}, 0\right)$ **39.** $\left(\dfrac{\sqrt{2}}{3}, 0\right)$ and $\left(-\dfrac{\sqrt{2}}{3}, 0\right)$ **41.** $\left(-\dfrac{1}{2}, 0\right)$ and $(3, 0)$ **43.** $(-2 + \sqrt{7}, 0)$ and $(-2 - \sqrt{7}, 0)$

45. No x-intercepts   **47.** $(-1 + \sqrt{2}, 0)$ and $(-1 - \sqrt{2}, 0)$ **49.** $(-2 + \sqrt{2}, 0)$ and $(-2 - \sqrt{2}, 0)$
51. $(-4 + \sqrt{2}, 0)$ and $(-4 - \sqrt{2}, 0)$ **53.** $(1 + \sqrt{5}, 0)$ and $(1 - \sqrt{5}, 0)$ **55.** No x-intercepts   **57.** Two
59. One   **61.** No x-intercepts   **63.** Two   **65.** No x-intercepts   **67.** One   **69.** Two   **71.** Two
73. No x-intercepts   **75.** No x-intercepts   **77.** No x-intercepts   **79.** Two   **81.** Two
83.   **85.**   **87.**   **89.**   **91.**

93. -2 **95.** -10 **97.** -4 **99.** 1 **101.** -4 **103.** -3
105. When the coefficient of x^2 is decreased, the parabola becomes wider.
107. When the constant term is decreased, the graph of the parabola is lower on the rectangular coordinate system.

SECTION 7.5   (page 376)

Problem 1   **Strategy**

▶ This is a geometry problem.
▶ Width of the rectangle: w
  Length of the rectangle: $w + 3$
▶ Use the equation $A = L \cdot w$.

Solution

$A = L \cdot w$
$54 = (w + 3)(w)$
$54 = w^2 + 3w$
$0 = w^2 + 3w - 54$
$0 = (w + 9)(w - 6)$

$w + 9 = 0 \qquad\quad w - 6 = 0$
$\quad\; w = -9 \qquad\qquad w = 6$

The solution -9 is not possible.

$\text{length} = w + 3 = 6 + 3 = 9$

The length is 9 m.

EXERCISES   (page 379)

1. width: 2 ft; length: 5 ft   **3.** base: 6 m; height: 4 m   **5.** 9 and 10   **7.** 2 and 4 or -2 and -4
9. -2 **11.** 3 h   **13.** larger pipe: 6 min; smaller pipe: 12 min   **15.** smaller unit: 12 h; larger unit: 4 h

17. 20 mph   **19.** 1st car: 30 mph; 2nd car: 40 mph   **21.** $\dfrac{1}{4}$ or 8   **23.** $\dfrac{2}{4}$ **25.** -7 and -6; or 6 and 7

27. 2 cm × 8 cm × 16 cm   **29.** 9 cm   **31.** 3 cm, 4 cm, 5 cm   **33.** $\dfrac{1 + \sqrt{5}}{2}$ and $\dfrac{1 - \sqrt{5}}{2}$

CHAPTER REVIEW   (page 384)

1. $\dfrac{3}{2}$ and $-\dfrac{2}{3}$ (Objective 7.1.1)   **2.** $2 + 2\sqrt{2}$ and $2 - 2\sqrt{2}$ (Objective 7.1.3)

3. $\dfrac{3 + \sqrt{15}}{3}$ and $\dfrac{3 - \sqrt{15}}{3}$ (Objective 7.2.1/7.2.2)   **4.** $-2 + 2i\sqrt{2}$ and $-2 - 2i\sqrt{2}$ (Objective 7.2.1/7.2.2)

5. Two real number solutions (Objective 7.2.2)   **6.** $2x^2 + 7x - 4 = 0$ (Objective 7.1.2)   **7.** 4 (Objective 7.3.2)
8.

(Objective 7.4.1)   **9.** $(-1 + \sqrt{5}, 0)$ and $(-1 - \sqrt{5}, 0)$ (Objective 7.4.2)

10. 1, -1, $\sqrt{3}$, and $-\sqrt{3}$ (Objective 7.3.1)   **11.** $\dfrac{2}{3}$ and -4 (Objective 7.1.1)

12. $3 + \sqrt{11}$ and $3 - \sqrt{11}$ (Objective 7.2.1/7.2.2)   **13.** 2 and -9 (Objective 7.3.3)

14. $\dfrac{1 + \sqrt{3}}{2}$ and $\dfrac{1 - \sqrt{3}}{2}$ (Objective 7.2.1/7.2.2)   **15.** Two x-intercepts (Objective 7.4.2)

16. $3x^2 + 8x - 3 = 0$ (Objective 7.1.2)   **17.** $\dfrac{1}{2}$ and -5 (Objective 7.1.1)

18. $-1 + 3\sqrt{2}$ and $-1 - 3\sqrt{2}$ (Objective 7.1.3)   **19.** $-3 + i$ and $-3 - i$ (Objective 7.2.1/7.2.2)
20.   (Objective 7.4.1)   **21.** $\sqrt{2}$, $-\sqrt{2}$, 2, and -2 (Objective 7.3.1)

22. $2 + \sqrt{10}$ and $2 - \sqrt{10}$ (Objective 7.2.1/7.2.2)   **23.** Two real number solutions (Objective 7.2.2)

24. $\left(-\dfrac{1}{2}, 0\right)$ and $(1, 0)$ (Objective 7.4.2)   **25.** 0 and 1 (Objective 7.3.2)   **26.** $\dfrac{2}{3}$ and -3 (Objective 7.1.1)

27. $\dfrac{-1 + \sqrt{7}}{2}$ and $\dfrac{-1 - \sqrt{7}}{2}$ (Objective 7.2.1/7.2.2)   **28.** -1 and $-\dfrac{3}{2}$ (Objective 7.3.3)

29. No x-intercepts (Objective 7.4.2)   **30.** $\dfrac{1 + \sqrt{7}}{3}$ and $\dfrac{1 - \sqrt{7}}{3}$ (Objective 7.2.1/7.2.2)

31. length: 10 ft; width: 4 ft (Objective 7.5.1)   **32.** 6 m (Objective 7.5.1)   **33.** 4 mph (Objective 7.5.1)
34. 6 mph (Objective 7.5.1)

CUMULATIVE REVIEW   (page 385)

1. 14 (Objective 1.2.1)   **2.**   (Objective 1.3.1)   **3.** -28 (Objective 2.1.2)

4. $\{x \mid x > -2\}$ (Objective 2.2.1)   **5.** -1 and $\dfrac{5}{2}$ (Objective 2.3.1)   **6.** $-\dfrac{3}{2}$ (Objective 6.2.1)
7. $y = x + 1$ (Objective 6.3.2)   **8.** $\{x \mid -3 < x < -2\}$ (Objective 3.5.2)
9. $-3xy(x^2 - 2xy + 3y^2)$ (Objective 3.3.1)   **10.** $(2x - 5)(3x + 4)$ (Objective 3.3.3)
11. $(x + y)(a^n - 2)$ (Objective 3.4.3)   **12.** $\{x \mid -2 < x \le 5\}$ (Objective 4.5.2)

13. $x^2 - 3x - 4 - \dfrac{6}{3x - 4}$ (Objective 4.1.2)   **14.** $\dfrac{x}{2}$ (Objective 4.2.1)   **15.** $-\dfrac{1}{2}$ (Objective 4.5.1)

16. $b = \dfrac{2S - an}{n}$ (Objective 4.5.3)   **17.** $-8 - 14i$ (Objective 5.4.3)   **18.** $1 - a$ (Objective 5.2.1)

19. $\dfrac{x\sqrt[3]{4y^2}}{2y}$ (Objective 5.3.4)   **20.**   (Objective 6.1.4)   **21.** $\dfrac{4}{3}$ and -4 (Objective 7.1.1)

22. $-2 + i\sqrt{5}$ and $-2 - i\sqrt{5}$ (Objective 7.2.1/7.2.2)   **23.** Two real number solutions (Objective 7.2.2)
24. 2, -2, $2i\sqrt{2}$, and $-2i\sqrt{2}$ (Objective 7.3.1)   **25.** 0 and 3 (Objective 7.3.2)
26. 2 and -4 (Objective 7.3.3)   **27.**   (Objective 7.4.1)

28. $\left(-\dfrac{2}{3}, 0\right)$ and $(1, 0)$ (Objective 7.4.2)   **29.** 1750 lb (Objective 2.5.2)

30. $(8x^2 - 2x - 3)$ ft^2 (Objective 3.2.4)   **31.** $66\dfrac{2}{3}$ min (Objective 4.5.4)   **32.** 60 mph (Objective 4.5.5)

Answers to Chapter 8

SECTION 8.1   (page 391)

Problem 1

$$s(t) = 2t^2 + 3t - 4$$
$$s(-3) = 2(-3)^2 + 3(-3) - 4$$
$$= 2(9) + 3(-3) - 4$$
$$= 18 + (-9) - 4$$
$$= 9 - 4$$
$$= 5$$

Problem 2

$$f(x) = x^2$$
$$f(a + 1) = (a + 1)^2$$
$$= a^2 + 2a + 1$$

Problem 3

$$f(x) - g(x) = 2x^2 - (4x - 1)$$
$$f(-1) - g(-1) = 2(-1)^2 - [4(-1) - 1]$$
$$= 2(1) - (-4 - 1)$$
$$= 2 - (-5)$$
$$= 7$$

Problem 4   The domain is $\{0, 1, 2, 3, 4\}$.
The range is $\{1, 3, 5, 7, 9\}$.

Problem 5

$$f(x) = x^2 - 2x + 1$$
$$f(-2) = (-2)^2 - 2(-2) + 1 = 4 - (-4) + 1 = 9$$
$$f(-1) = (-1)^2 - 2(-1) + 1 = 1 - (-2) + 1 = 4$$
$$f(0) = (0)^2 - 2(0) + 1 = 0 - 0 + 1 = 1$$
$$f(1) = (1)^2 - 2(1) + 1 = 1 - 2 + 1 = 0$$
$$f(2) = (2)^2 - 2(2) + 1 = 4 - 4 + 1 = 1$$

The range is $\{0, 1, 4, 9\}$.

Problem 6

$$f(x) = -4$$
$$2x - 1 = -4$$
$$2x = -3$$
$$x = -\dfrac{3}{2}$$

$$\left\{-\dfrac{3}{2}, -4\right\}$$

Problem 7 $2x - 5 < 0$

$2x < 5$

$x < \dfrac{5}{2}$

$\left\{ x \mid x < \dfrac{5}{2} \right\}$

EXERCISES   (page 396)

1. 12   **3.** 3   **5.** $3a^2$ **7.** 1   **9.** 7   **11.** $t^2 - t + 1$ **13.** -11 **15.** $4h + 5$ **17.** $4h$ **19.** 3
21. $h^2 + 2h$ **23.** $h^2 + 6h$ **25.** 6   **27.** 0   **29.** $2h^2 + 4h$ **31.** -2 **33.** $2h + 12$ **35.** -2
37. domain: $\{1, 2, 3, 4, 5\}$; range: $\{1, 4, 7, 10, 13\}$ **39.** domain: $\{0, 2, 4, 6\}$; range: $\{1, 2, 3, 4\}$
41. domain: $\{1, 3, 5, 7, 9\}$; range: $\{0\}$ **43.** domain: $\{-2, -1, 0, 1, 2\}$; range: $\{0, 1, 2\}$ **45.** $\{-3, 1, 5, 9, 13\}$
47. $\{1, 2, 3, 4, 5\}$ **49.** $\left\{ \dfrac{2}{5}, \dfrac{1}{2}, \dfrac{2}{3}, 1, 2 \right\}$ **51.** $\left\{ -1, -\dfrac{1}{2}, 1 \right\}$ **53.** -8; $(-8, -3)$ **55.** -1; $(-1, -1)$

57. 1; $(1, 1)$ **59.** ± 2; $(2, 7)$ or $(-2, 7)$ **61.** 34; $(34, 7)$ **63.** None   **65.** $0, \dfrac{1}{3}$ **67.** $-3, 2$

69. $\left\{ x \mid x < \dfrac{1}{3} \right\}$ **71.** None   **73.** $f(14) = 8$ **75.** $a = 18$ **77.** $f(3) - g(3) = 3$ **79.** $f(3\pi) = 9$
81. $\{x \mid -1 < x < 1\}$ **83.** $x = -3$ **85.** $f(2, 5) + g(2, 5) = 17$ **87.** $f(5) - f(4) = 96$ **89.** \$160

SECTION 8.2   (page 401)

Problem 1   **A.**

The domain is all real numbers.
The range is $\{y \mid y \geq 0\}$.

B.

The domain is all real numbers.
The range is all real numbers.

Problem 2   **A.** Each member of the domain is paired with only one member of the range. The relation is a function.

B. Each member of the domain is paired with only one member of the range. The relation is a function.

Problem 3   Since any vertical line would intersect the graph at no more than one point, the graph is the graph of a function.

Problem 4   **A.** Since any vertical line will intersect the graph at no more than one point, and any horizontal line will intersect the graph at no more than one point, the graph is the graph of a 1–1 function.

B. Since any vertical line will intersect the graph at no more than one point, and any horizontal line will intersect the graph at no more than one point, the graph is the graph of a 1–1 function.

EXERCISES   (page 408)

1.

D: all reals
R: all reals

3.

D: all reals
R: $y \geq -1$

5.

D: all reals
R: $y \geq 0$

7.
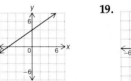
D: all reals
R: all reals

9.
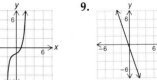
D: all reals
R: all reals

11.

D: all reals
R: $y \geq 0$

13.

D: all reals
R: $y \geq 1$

15.

D: all reals
R: all reals

17.

D: all reals
R: all reals

19.

D: all reals
R: $y \geq -5$

21.

D: all reals
R: $y \geq -1$

23.

D: all reals
R: all reals

25.

D: all reals
R: all reals

27.
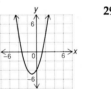
D: all reals
R: $y \geq -5$

29.

D: all reals
R: $y \leq 0$

31.

D: all reals
R: all reals

33. Yes   **35.** Yes   **37.** No   **39.** Yes   **41.** Yes   **43.** No   **45.** Yes

47. Yes   **49.** No   **51.** Yes   **53.** No   **55. a.** C **b.** d **57. a.** A **b.** r **59. a.** d **b.** t
61. a. F **b.** C   **63. a.** y **b.** x **65. a.** y **b.** x **67.** c   **69.** $f(-1) = -5$ **71.** $f(0) = -4$

73. $f(1) = 0$ **75.** $f(-1) = 5$ **77. a.** $x = 0$ **b.** $(0,0)$ **79. a.** $x = -1$ **b.** $(-1,0)$
81. $(1,0)$, $(-1,0)$ and $(-2,0)$

SECTION 8.3   (page 414)

Problem 1   **A.** $h(x) = x^2 + 1$ **B.** $h(g(x)) = h(3x - 2)$
$\quad\quad h(0) = 0 + 1 = 1$ $\quad\quad\quad\quad = (3x - 2)^2 + 1$
$\quad\quad g(x) = 3x - 2$ $\quad\quad\quad\quad = 9x^2 - 12x + 4 + 1$
$\quad\quad g(1) = 3(1) - 2 = 1$ $\quad\quad\quad\quad = 9x^2 - 12x + 5$

Problem 2 $f(x) = 4x + 2$
$\quad\quad\quad y = 4x + 2$
$\quad\quad\quad x = 4y + 2$
$\quad\quad 4y = x - 2$
$\quad\quad\quad y = \dfrac{1}{4}x - \dfrac{1}{2}$

$\quad\quad f^{-1}(x) = \dfrac{1}{4}x - \dfrac{1}{2}$

Problem 3 $h(g(x)) = 4\left(\dfrac{1}{4}x - \dfrac{1}{2}\right) + 2 = x - 2 + 2 = x$

$\quad\quad\quad g(h(x)) = \dfrac{1}{4}(4x + 2) - \dfrac{1}{2} = x + \dfrac{1}{2} - \dfrac{1}{2} = x$

The functions are inverses of each other.

EXERCISES   (page 419)

1. $f(g(0)) = -5$ **3.** $f(g(2)) = 11$ **5.** $f(g(x)) = 8x - 5$ **7.** $h(f(0)) = 8$ **9.** $h(f(2)) = 10$
11. $h(f(x)) = x + 8$ **13.** $g(h(0)) = 7$ **15.** $g(h(4)) = 7$ **17.** $g(h(x)) = x^2 - 4x + 7$ **19.** $f(h(0)) = 7$
21. $f(h(-1)) = 1$ **23.** $f(h(x)) = 9x^2 + 15x + 7$ **25.** $\{(0,1), (3,2), (8,3), (15,4)\}$ **27.** No inverse
29. $f^{-1}(x) = \dfrac{1}{4}x + 2$ **31.** No inverse   **33.** $f^{-1}(x) = x + 5$ **35.** $f^{-1}(x) = 3x - 6$

37. $f^{-1}(x) = -\dfrac{1}{3}x - 3$ **39.** $f^{-1}(x) = \dfrac{3}{2}x - 6$ **41.** $f^{-1}(x) = -3x + 3$ **43.** $f^{-1}(x) = \dfrac{1}{2}x + \dfrac{5}{2}$

45. No inverse   **47.** $f^{-1}(x) = \dfrac{1}{4}x + \dfrac{1}{2}$ **49.** $f^{-1}(x) = -\dfrac{1}{8}x + \dfrac{1}{2}$ **51.** $f^{-1}(x) = \dfrac{1}{8}x - \dfrac{3}{4}$ **53.** Yes

55. No   **57.** Yes   **59.** Yes   **61.** Range   **63.** 3   **65.** $f^{-1}(5) = 0$ **67.** $f^{-1}(9) = 2$ **69.** $f^{-1}(5) = 2$

71. $f^{-1}(2) = 2$ **73.** $f^{-1}(6) = -2$ **75.** $f^{-1}(2) = \dfrac{7}{3}$ **77.** $f(g(0)) = -3$ **79.** $f(g(4)) = 1$

81. $f(g(-4)) = 1$ **83.** $g(f(4)) = -2$ **85.**   **87.**   **89.**

91.

SECTION 8.4   (page 423)

Problem 1   **Strategy**

To find the distance:
▶ Write the basic direct variation equation, replace the variables by the given values, and solve for k.
▶ Write the direct variation equation, replacing k by its value. Substitute 5 for t, and solve for s.

Solution

$$s = kt^2$$
$$64 = k(2)^2$$
$$64 = k \cdot 4$$
$$16 = k$$

$$s = 16t^2$$
$$= 16(5)^2 = 400$$

The object will fall 400 ft in 5 s.

Problem 2   **Strategy**

To find the resistance:
▶ Write the basic inverse variation equation, replace the variables by the given values, and solve for k.
▶ Write the inverse variation equation, replacing k by its value. Substitute 0.02 for d, and solve for R.

Solution

$$R = \frac{k}{d^2}$$
$$0.5 = \frac{k}{(0.01)^2}$$
$$0.5 = \frac{k}{0.0001}$$
$$0.00005 = k$$
$$R = \frac{0.00005}{d^2}$$
$$= \frac{0.00005}{(0.02)^2} = 0.125$$

The resistance is 0.125 ohms.

Problem 3   **Strategy**

To find the strength of the beam:
▶ Write the basic combined variation equation, replace the variables by the given values, and solve for k.
▶ Write the basic combined variation equation, replacing k by its value and substituting 4 for w, and solve for s.

Solution

$$s = \frac{wk}{d^2} \qquad\qquad s = \frac{wk}{d^2}$$

$$1200 = \frac{2k}{12^2} \qquad\qquad s = \frac{4(86{,}400)}{8^2}$$

$$1200 = \frac{2k}{144} \qquad\qquad s = \frac{345{,}600}{64}$$

$$172{,}800 = 2k \qquad\qquad s = 5400$$

$$86{,}400 = k$$

The strength of the beam is 5400 lb.

EXERCISES   (page 426)

1. 8 lb   **3.** 675 bushels   **5.** 244.8 ft   **7.** 2.37 s   **9.** 30 revolutions/minute   **11.** 75 items
13. 48 lumens   **15.** 13 amps   **17.** 200 watts   **19.** 36 vibrations/s

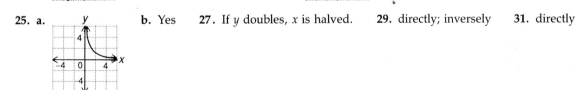

21. a.   **b.** a linear function   **23. a.**   **b.** a quadratic function

25. a.   **b.** Yes   **27.** If y doubles, x is halved.   **29.** directly; inversely   **31.** directly

CHAPTER REVIEW   (page 432)

1. $f(4) = 15$ (Objective 8.1.1)   **2.**   (Objective 8.2.1)   **3.** Yes (Objective 8.2.2)

4. $g(-2) = -12$ (Objective 8.1.1)   **5.** (Objective 8.2.1)   **6.** 0 and 8 (Objective 8.1.2)

7. $\{-11, -7, -3, 1, 5\}$ (Objective 8.1.2)   **8.** $f(g(0)) = -1$ (Objective 8.3.1)
9. $f^{-1}(x) = 2x + 2$ (Objective 8.3.2)   **10.**

The domain is all real numbers.
The range is $\{y \mid y \geq 1\}$.
(Objective (8.2.1)

11. $g(h(x)) = -3x + 14$ (Objective 8.3.1)
12. Yes (Objective 8.3.2)   **13.**   **14.** $\{3, 5, 7, 9\}$ (Objective 8.1.2)

The domain is all real numbers.
The range is $\{y \mid y \geq 0\}$.
(Objective 8.2.1)

15. $h(-1) = -1$ (Objective 8.1.1)   **16.** (Objective 8.2.1)   **17.** No (Objective 8.3.2)

18. $g(1 + h) = 1 - 2h$ (Objective 8.1.1)   **19.** $\{x \mid x < 7\}$ (Objective 8.1.2)
20. $f^{-1}(x) = \dfrac{1}{3}x - 2$ (Objective 8.3.2)   **21.** $-\dfrac{4}{3}$ (Objective 8.1.2)   **22.** $f(-2) = -\dfrac{2}{5}$ (Objective 8.1.1)
23. $g(0) + h(0) = 5$ (Objective 8.1.1)   **24.**

The domain is all real numbers.
The range is $\{y \mid y \geq -2\}$.
(Objective 8.2.1)

25. $f(-1) = -4$ (Objective 8.1.1)   **26.** $\{0, 1\}$ (Objective 8.1.2)   **27.** No (Objective 8.2.2)
28. Yes (Objective 8.2.3)   **29.** 61.2 ft (Objective 8.4.1)   **30.** 2 amps (Objective 8.4.1)

CUMULATIVE REVIEW   (page 434)

1. $-\dfrac{23}{4}$ (Objective 1.2.1)   **2.** 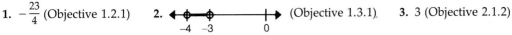 (Objective 1.3.1)   **3.** 3 (Objective 2.1.2)

4. $\{x \mid x < -2 \text{ or } x > 3\}$ (Objective 2.2.2)   **5.** The set of all real numbers (Objective 2.3.2)

6. $-\dfrac{a^{10}}{12b^4}$ (Objective 3.1.2)   **7.** $2x^3 - 4x^2 - 17x + 4$ (Objective 3.2.2)

8. $(a + 2)(a - 2)(a^2 + 2)$ (Objective 3.4.4)   **9.** $xy(x - 2y)(x + 3y)$ (Objective 3.3.5)

10. -3 and 8 (Objective 3.5.1)   **11.** $\{x \mid x < -3 \text{ or } x > 5\}$ (Objective 3.5.2)

12. $\dfrac{5}{2x - 1}$ (Objective 4.2.2)   **13.** -2 (Objective 4.5.1)   **14.** $-3 - 2i$ (Objective 5.4.4)

15. 2 (Objective 5.5.1)   **16.** (Objective 6.1.4)   **17.** $y = -2x - 2$ (Objective 6.3.1)

18. $y = -\dfrac{3}{2}x - \dfrac{7}{2}$ (Objective 6.3.2)   **19.** $\dfrac{1}{2} + \dfrac{\sqrt{3}}{6}i$ and $\dfrac{1}{2} - \dfrac{\sqrt{3}}{6}i$ (Objective 7.2.1/7.2.2)

20. 3 (Objective 7.3.2)   **21.** $f(-2) = 5$ (Objective 8.1.1)   **22.** $\{1, 2, 4, 5\}$ (Objective 8.1.2)

23. Yes (Objective 8.2.2)   **24.** 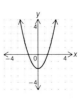 (Objective 8.2.1)   **25.** $g(h(2)) = 10$ (Objective 8.3.1)

26. $f^{-1}(x) = -\dfrac{1}{3}x + 3$ (Objective 8.3.2)   **27.** \$3.96 (Objective 2.4.2)   **28.** 25 lb (Objective 2.5.2)

29. 4.5 oz (Objective 4.4.2)   **30.** 4 min (Objective 7.5.1)   **31.** 24 in. (Objective 8.4.1)

32. 80 vibrations per minute (Objective 8.4.1)

Answers to Chapter 9

SECTION 9.1   (page 439)

Problem 1 x-coordinate: $-\dfrac{b}{2a} = -\dfrac{0}{2(1)} = 0$

The axis of symmetry is the line $x = 0$.

$$y = x^2 - 2$$
$$= 0^2 - 2$$
$$= -2$$

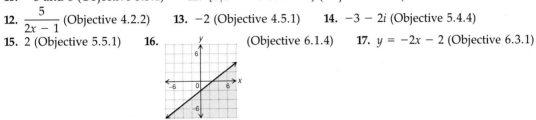

The vertex is $(0, -2)$.

Problem 2 y-coordinate: $-\dfrac{b}{2a} = -\dfrac{-4}{2(-2)} = -1$

The axis of symmetry is the line $y = -1$.

$\begin{aligned} x &= -2y^2 - 4y - 3 \\ &= -2(-1)^2 - 4(-1) - 3 \\ &= -1 \end{aligned}$

The vertex is $(-1, -1)$.

Problem 3 $x = -\dfrac{b}{2a} = -\dfrac{-3}{2(2)} = \dfrac{3}{4}$

$f(x) = 2x^2 - 3x + 1$

$f\left(\dfrac{3}{4}\right) = 2\left(\dfrac{3}{4}\right)^2 - 3\left(\dfrac{3}{4}\right) + 1 = \dfrac{9}{8} - \dfrac{9}{4} + 1 = -\dfrac{1}{8}$

Since a is positive, the function has a minimum value.

The minimum value of the function is $-\dfrac{1}{8}$.

Problem 4   **Strategy**

▶ To find the time it takes the ball to reach its maximum height, find the t-coordinate of the vertex.
▶ To find the maximum height, evaluate the function at the t-coordinate of the vertex.

Solution

$t = -\dfrac{b}{2a} = -\dfrac{64}{2(-16)} = 2$

The ball reaches its maximum height in 2 s.

$s(t) = -16t^2 + 64t$
$s(2) = -16(2)^2 + 64(2) = -64 + 128 = 64$

The maximum height is 64 ft.

Problem 5   **Strategy**

The perimeter is 44 ft.
$\quad\quad 44 = 2L + 2w$
$\quad\quad 22 = L + w$
$22 - L = w$

The area is: $L \cdot w = L(22 - L) = 22L - L^2$.
▶ To find the length, find the L-coordinate of the vertex of the function $f(L) = -L^2 + 22L$.
▶ To find the width, replace L in $22 - L$ by the L-coordinate of the vertex and evaluate.

Solution

$$L = -\frac{b}{2a} = -\frac{22}{2(-1)} = 11$$

The length is 11 ft.

$$22 - L = 22 - 11 = 11$$

The width is 11 ft.

EXERCISES   (page 444)

1.

Vertex: $(1, -5)$
Axis of symmetry:
$x = 1$

3.

Vertex: $(1, -2)$
Axis of symmetry:
$x = 1$

5.

Vertex: $(-4, -3)$
Axis of symmetry:
$y = -3$

7.

Vertex: $(1, -1)$
Axis of symmetry:
$x = 1$

9.

Vertex: $\left(\frac{5}{2}, -\frac{9}{4}\right)$
Axis of symmetry:
$x = \frac{5}{2}$

11.

Vertex: $(-6, 1)$
Axis of symmetry:
$y = 1$

13.

Vertex: $\left(-\frac{3}{2}, \frac{27}{4}\right)$
Axis of symmetry:
$x = -\frac{3}{2}$

15.

Vertex: $(4, 0)$
Axis of symmetry:
$y = 0$

17.

Vertex: $\left(\frac{1}{2}, 1\right)$
Axis of symmetry:
$y = 1$

19.

Vertex: $(-2, -8)$
Axis of symmetry:
$x = -2$

21. minimum: 2   **23.** maximum: -1 **25.** minimum: $-\frac{11}{4}$ **27.** maximum: $-\frac{2}{3}$ **29.** 114 ft
31. 5 days   **33.** \$250   **35.** 10 and 10   **37.** 12 and -12
39. domain: the real numbers; the range: $y \geq -6$ **41.** domain: the real numbers; range: $y \leq -2$
43. domain: $x \geq -14$; range: the real numbers   **45.** domain: $x \leq 7$; range: the real numbers
47. domain: the real numbers; range: $y \leq 1$ **49.** domain: $x \leq -2$; range: the real numbers
51. $y = (x - 2)^2 + 1$; vertex: $(2, 1)$ **53.** $y = (x - 3)^2 - 6$; vertex: $(3, -6)$
55. $y = \left(x + \frac{1}{2}\right)^2 + \frac{7}{4}$; vertex: $\left(-\frac{1}{2}, \frac{7}{4}\right)$

SECTION 9.2   (page 447)

Problem 1 $(x_1, y_1) = (3, -2)$ $(x_2, y_2) = (-1, -5)$

$$d = \sqrt{(x_1 - x_2)^2 + (y_1 - y_2)^2}$$
$$= \sqrt{[3 - (-1)]^2 + [(-2) - (-5)]^2}$$
$$= \sqrt{4^2 + 3^2} = \sqrt{16 + 9} = \sqrt{25} = 5$$

Problem 2

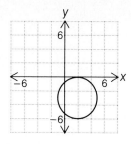

Problem 3 $(x - h)^2 + (y - k)^2 = r^2$
$(x - 2)^2 + [y - (-3)]^2 = 4^2$
$(x - 2)^2 + (y + 3)^2 = 16$

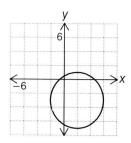

Problem 4 $x^2 + y^2 - 4x + 8y + 15 = 0$
$(x^2 - 4x) + (y^2 + 8y) = -15$
$(x^2 - 4x + 4) + (y^2 + 8y + 16) = -15 + 4 + 16$
$(x - 2)^2 + (y + 4)^2 = 5$

Center: $(2, -4)$
Radius: $\sqrt{5}$

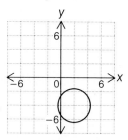

EXERCISES   (page 452)

1. $\sqrt{10}$ **3.** $\sqrt{34}$ **5.** $\sqrt{53}$ **7.** 5   **9.**   **11.**   **13.**

15.   **17.**   **19.**

21. Yes   **23.** No   **25.** $(x - 3)^2 + y^2 = 9$ **27.** $(x - 2)^2 + (y - 4)^2 = 10$ **29.** $(x + 1)^2 + (y - 1)^2 = 1$

SECTION 9.3   **(page 454)**

Problem 1   **A.** x-intercepts:
$(2, 0)$ and $(-2, 0)$

y-intercepts:
$(0, 5)$ and $(0, -5)$

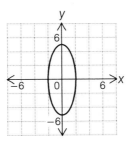

B. x-intercepts:
$(3\sqrt{2}, 0)$ and $(-3\sqrt{2}, 0)$

y-intercepts:
$(0, 3)$ and $(0, -3)$

$$\left(3\sqrt{2} \approx 4\frac{1}{4}\right)$$

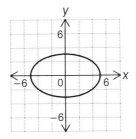

Problem 2   **A.** Axis of symmetry:
x-axis

Vertices:
$(3, 0)$ and $(-3, 0)$

Asymptotes:
$y = \dfrac{5}{3}x$ and $y = -\dfrac{5}{3}x$

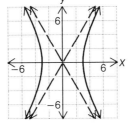

B. Axis of symmetry:
y-axis

Vertices:
$(0, 3)$ and $(0, -3)$

Asymptotes:
$y = x$ and $y = -x$

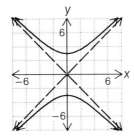

EXERCISES   (page 459)

1.

3.

5.

7.

9.

11.

13.

15.

17.

19.

21.

23.

25.

27.

SECTION 9.4   (page 461)

Problem 1   **A.** Graph the ellipse $\dfrac{x^2}{9} + \dfrac{y^2}{16} = 1$ as a solid line.

Shade the region of the plane that includes $(0, 0)$.

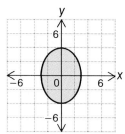

B. Graph the hyperbola $\dfrac{x^2}{9} - \dfrac{y^2}{4} = 1$ as a solid line.

Shade the region that includes $(0, 0)$.

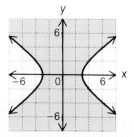

EXERCISES   (page 462)

1.   **3.**   **5.**   **7.**   **9.**

11.   **13.**

CHAPTER REVIEW   **(page 465)**

1.   (Objective 9.1.1)   **2.** $(x - 4)^2 + (y - 1)^2 = 18$ (Objective 9.2.2)   **3.** $-\dfrac{7}{2}$ (Objective 9.1.2)

4.   (Objective 9.3.1)   **5.** $x = 3$ (Objective 9.1.1)   **6.**   (Objective 9.4.1)

7. 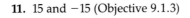 (Objective 9.3.2)   **8.** $\sqrt{58}$ (Objective 9.2.1)   **9.** $(x - 2)^2 + (y + 1)^2 = 9$ (Objective 9.2.2)

10.   (Objective 9.1.1)   **11.** 15 and -15 (Objective 9.1.3)   **12.**   (Objective 9.4.1)

13. $(x + 2)^2 + (y - 4)^2 = 9$ (Objective 9.2.2)   **14.** $(-3, 2)$ (Objective 9.1.1)

15. (Objective 9.1.1)   **16.** 100 (Objective 9.1.3)   **17.** $(x + 2)^2 + (y - 1)^2 = 32$ (Objective 9.2.2)

18. (Objective 9.3.2)   **19.** 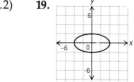 (Objective 9.3.1)

20. $(x - 2)^2 + (y + 1)^2 = 4$ (Objective 9.2.3)   **21.** 4 (Objective 9.1.2)   **22.** 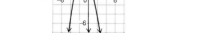 (Objective 9.1.1)

23. (Objective 9.4.1)   **24.** $5\sqrt{2}$ (Objective 9.2.1)   **25.** 12 cm × 12 cm (Objective 9.1.3)

26. $\left(\dfrac{3}{2}, \dfrac{1}{4}\right)$ (Objective 9.1.1)

CUMULATIVE REVIEW   (page 467)

1. 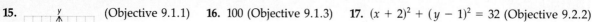 (Objective 1.3.1)   **2.** $\dfrac{38}{53}$ (Objective 2.1.2)   **3.** $\left\{x \mid -\dfrac{4}{3} < x < 0\right\}$ (Objective 2.3.2)

4. $y = -\dfrac{3}{2}x$ (Objective 6.3.1)   **5.** $y = x - 6$ (Objective 6.3.2)   **6.** $x^{4n} + 2x^{3n} - 3x^{2n+1}$ (Objective 3.2.1)

7. $(x - y - 1)(x^2 - 2x + 1 + xy - y + y^2)$ (Objective 3.4.2)   **8.** $\{x \mid -4 < x \le 3\}$ (Objective 4.5.2)

9. $\dfrac{x}{x + y}$ (Objective 4.1.1)   **10.** $\dfrac{x - 5}{3x - 2}$ (Objective 4.2.2)   **11.** $\dfrac{5}{2}$ (Objective 4.5.1)

12. (Objective 6.1.4)   **13.** $\dfrac{4a}{3b^8}$ (Objective 5.1.1)   **14.** $2x^{\frac{3}{4}}$ (Objective 5.2.2)

15. $3\sqrt{2} - 5i$ (Objective 5.4.1)   **16.** $\dfrac{-1 + \sqrt{7}}{2}$ and $\dfrac{-1 - \sqrt{7}}{2}$ (Objective 7.2.1/7.2.2)

17. 27 and −1 (Objective 7.3.1)   **18.** 6 (Objective 7.3.2)   **19.** −20 (Objective 8.1.1)

20. $f^{-1}(x) = \dfrac{1}{4}x - 2$ (Objective 8.3.2)   **21.** 0 (Objective 9.1.2)   **22.** 400 (Objective 9.1.3)

23. 5 (Objective 9.2.1)   **24.** $(x + 1)^2 + (y - 2)^2 = 17$ (Objective 9.2.2)

25.   (Objective 9.1.1)   **26.**   (Objective 9.3.1)

27.   (Objective 9.3.2)   **28.**   (Objective 9.3.2)   **29.** 82 tickets (Objective 2.4.2)

30. 60 mph (Objective 4.5.5)   **31.** 4.5 mph (Objective 7.5.1)   **32.** 18 revolutions/min (Objective 8.4.1)

Answers to Chapter 10

SECTION 10.1   **(page 473)**

Problem 1   **A.**

The solution is $(-1, 2)$.

B.

The two equations represent the same line. The equations are dependent. Any ordered pair that is a solution of one equation is also a solution of the other equation.

Problem 2   **A.** $3x - y = 3$
$6x + 3y = -4$

Solve equation (1) for y.
$3x - y = 3$
$-y = -3x + 3$
$y = 3x - 3$
Substitute into equation (2).
$6x + 3y = -4$
$6x + 3(3x - 3) = -4$
$6x + 9x - 9 = -4$
$15x - 9 = -4$
$15x = 5$
$x = \dfrac{5}{15} = \dfrac{1}{3}$
Substitute into equation (1).
$3x - y = 3$
$3\left(\dfrac{1}{3}\right) - y = 3$
$1 - y = 3$
$-y = 2$
$y = -2$

The solution is $\left(\dfrac{1}{3}, -2\right)$.

B. $6x - 3y = 6$
$2x - y = 2$

Solve equation (2) for y.
$2x - y = 2$
$-y = -2x + 2$
$y = 2x - 2$
Substitute into equation (1).
$6x - 3y = 6$
$6x - 3(2x - 2) = 6$
$6x - 6x + 6 = 6$
$6 = 6$

This is a true equation. The equations are dependent. Any ordered pair that is a solution of one equation is also a solution of the other equation.

EXERCISES   (page 476)

1.

The solution is $(3, -1)$.

3.

The solution is $(2, 4)$.

5.

The solution is $(4, 3)$.

7.

The solution is $(4, -1)$.

9.

The solution is $(3, -2)$.

11.

The lines are parallel and therefore do not intersect. The system of equations has no solution.

13.

The two equations represent the same line. Any ordered pair that is a solution of one equation is also a solution of the other equation.

15.

The solution is $(0, -3)$.

17. $(2, 1)$ **19.** $(1, 1)$ **21.** $(16, 5)$ **23.** $(2, 1)$ **25.** $(-1, -7)$ **27.** $(-2, -3)$ **29.** $(3, -4)$ **31.** $(3, 3)$

33. $(0, -1)$ **35.** $\left(\dfrac{1}{2}, 3\right)$ **37.** $(1, 2)$ **39.** $(-1, 2)$ **41.** $(2, -4)$ **43.** $(-2, 5)$ **45.** $(0, 0)$

47. $(-4, -6)$ **49.** $(1, 5)$ **51.** $\left(\dfrac{2}{3}, 3\right)$ **53.** $(5, 2)$ **55.** $(1, 4)$ **57.** $(-1, 6)$ **59.** 2 **61.** $\dfrac{1}{2}$

63. $(2, 1)$ **65.** $\left(\dfrac{13}{11}, \dfrac{13}{5}\right)$ **67.** $(-3, 1)$ **69.** 18 and 26   **71.** 56 and 72   **73.** 8 and 11

SECTION 10.2   (page 480)

Problem 1   **A.** $2x + 5y = 6$
$3x - 2y = 6x + 2$

Write equation (2) in the form
$Ax + By = C.$
$3x - 2y = 6x + 1$
$-3x - 2y = 2$

Solve the system $2x + 5y = 6$
$\qquad\qquad\qquad\qquad -3x - 2y = 2.$

Eliminate y.
$2(2x + 5y) = 2(6)$
$5(-3x - 2y) = 5(2)$

$4x + 10y = 12$
$-15x - 10y = 10$

Add the equations.
$-11x = 22$
$x = -2$

Replace x in equation (1).
$2x + 5y = 6$
$2(-2) + 5y = 6$
$-4 + 5y = 6$
$5y = 10$
$y = 2$

The solution is $(-2, 2)$.

B. $2x + y = 5$
$4x + 2y = 6$

Eliminate y.
$-2(2x + y) = -2(5)$
$4x + 2y = 6$

$-4x - 2y = -10$
$4x + 2y = 6$

Add the equations.
$0x + 0y = -4$
$0 = -4$

This is not a true equation. The system is inconsistent and therefore has no solution.

Problem 2 (1) $x - y + z = 6$
(2) $2x + 3y - z = 1$
(3) $x + 2y + 2z = 5$

Eliminate z. Add equations (1) and (2).
$x - y + z = 6$
$2x + 3y - z = 1$
$3x + 2y = 7$

Multiply equation (2) by 2 and add to equation (3).
$4x + 6y - 2z = 2$
$x + 2y + 2z = 5$
$5x + 8y = 7$

Solve the system of two equations.
(4) $3x + 2y = 7$
(5) $5x + 8y = 7$

Multiply equation (4) by -4 and add to equation (5).
$-12x - 8y = -28$
$5x + 8y = 7$
$-7x = -21$
$x = 3$

Replace x by 3 in equation (4).
$3x + 2y = 7$
$3(3) + 2y = 7$
$9 + 2y = 7$
$2y = -2$
$y = -1$

Replace x by 3 and y by -1 in equation (1).
$x - y + z = 6$
$3 - (-1) + z = 6$
$4 + z = 6$
$z = 2$

The solution is $(3, -1, 2)$.

EXERCISES (page 486)

1. $(6, 1)$ **3.** $(1, 1)$ **5.** $(2, 1)$ **7.** $(-2, 1)$ **9.** Dependent equations **11.** $\left(-\dfrac{1}{2}, 2\right)$

13. Inconsistent **15.** $(-1, -2)$ **17.** $(-5, 4)$ **19.** $(2, 5)$ **21.** $\left(\dfrac{1}{2}, \dfrac{3}{4}\right)$ **23.** $(0, 0)$ **25.** $(-1, 3)$

27. $\left(\dfrac{2}{3}, -\dfrac{2}{3}\right)$ **29.** $(1, -1)$ **31.** Dependent equations **33.** $(5, 3)$ **35.** $\left(\dfrac{1}{3}, -1\right)$ **37.** $\left(\dfrac{5}{3}, \dfrac{1}{3}\right)$

39. Inconsistent **41.** $(-1, 2, 1)$ **43.** $(6, 2, 4)$ **45.** $(4, 1, 5)$ **47.** $(3, 1, 0)$ **49.** $(-1, -2, 2)$

51. Inconsistent **53.** $(2, 1, -3)$ **55.** $(2, -1, 3)$ **57.** $(6, -2, 2)$ **59.** $(0, -2, 0)$ **61.** $(2, 3, 1)$

63. $(1, 1, 3)$ **65.** $(2, 1)$ **67.** $(1, -3)$ **69.** $(2, -1)$ **71.** $(2, 0, -1)$ **73.** $(2, -1, 0)$

75. $A = 4, B = 5$ **77.** $A = 2, B = 3, C = -3$ **79.** 16, 18, 32

SECTION 10.3   (page 489)

Problem 1   **A.** $\begin{vmatrix} -1 & -4 \\ 3 & -5 \end{vmatrix} = -1(-5) - 3(-4) = 5 + 12 = 17$

The value of the determinant is 17.

B. Expand by minors of the first column.

$$\begin{vmatrix} 1 & 4 & -2 \\ 3 & 1 & 1 \\ 0 & -2 & 2 \end{vmatrix} = 1\begin{vmatrix} 1 & 1 \\ -2 & 2 \end{vmatrix} - 3\begin{vmatrix} 4 & -2 \\ -2 & 2 \end{vmatrix} + 0$$

$$= 1(2 + 2) - 3(8 - 4)$$
$$= 4 - 12$$
$$= -8$$

The value of the determinant is -8.

Problem 2 $D = \begin{vmatrix} 6 & -6 \\ 2 & -10 \end{vmatrix} = -48$ $D_x = \begin{vmatrix} 5 & -6 \\ -1 & -10 \end{vmatrix} = -56$ $D_y = \begin{vmatrix} 6 & 5 \\ 2 & -1 \end{vmatrix} = -16$

$$x = \frac{D_x}{D} = \frac{-56}{-48} = \frac{7}{6} \qquad y = \frac{D_y}{D} = \frac{-16}{-48} = \frac{1}{3}$$

The solution is $\left(\frac{7}{6}, \frac{1}{3}\right)$.

Problem 3 $D = \begin{vmatrix} 2 & -1 & 1 \\ 3 & 2 & -1 \\ 1 & 3 & 1 \end{vmatrix} = 21$ $D_x = \begin{vmatrix} -1 & -1 & 1 \\ 3 & 2 & -1 \\ -2 & 3 & 1 \end{vmatrix} = 9$

$$D_y = \begin{vmatrix} 2 & -1 & 1 \\ 3 & 3 & -1 \\ 1 & -2 & 1 \end{vmatrix} = -3 \qquad D_z = \begin{vmatrix} 2 & -1 & -1 \\ 3 & 2 & 3 \\ 1 & 3 & -2 \end{vmatrix} = -42$$

$$x = \frac{D_x}{D} = \frac{9}{21} = \frac{3}{7} \qquad y = \frac{D_y}{D} = \frac{-3}{21} = -\frac{1}{7} \qquad z = \frac{D_z}{D} = \frac{-42}{21} = -2$$

The solution is $\left(\frac{3}{7}, -\frac{1}{7}, -2\right)$.

Problem 4 $\begin{bmatrix} 3 & -5 & -12 \\ 4 & -3 & -5 \end{bmatrix}$

$\begin{bmatrix} 1 & -\dfrac{5}{3} & -4 \\ 4 & -3 & -5 \end{bmatrix}$ • Multiply row 1 by $\dfrac{1}{3}$.

$\begin{bmatrix} 1 & -\dfrac{5}{3} & -4 \\ 0 & \dfrac{11}{3} & 11 \end{bmatrix}$ • Multiply row 1 by -4 and add to row 2.

$\begin{bmatrix} 1 & -\dfrac{5}{3} & -4 \\ 0 & 1 & -3 \end{bmatrix}$ • Multiply row 2 by $\dfrac{3}{11}$.

$$x - \frac{5}{3}y = -4 \qquad\qquad x - \frac{5}{3}(3) = -4$$
$$y = 3 \qquad\qquad\qquad x - 5 = -4$$
$$x = 1$$

The solution is $(1, 3)$.

Problem 5 $\begin{bmatrix} 3 & -2 & -3 & 5 \\ 1 & 3 & -2 & -4 \\ 2 & 6 & 3 & 6 \end{bmatrix}$

$\begin{bmatrix} 1 & 3 & -2 & -4 \\ 3 & -2 & -3 & 5 \\ 2 & 6 & 3 & 6 \end{bmatrix}$ • Interchange rows 1 and 2.

$\begin{bmatrix} 1 & 3 & -2 & -4 \\ 0 & -11 & 3 & 17 \\ 0 & 0 & 7 & 14 \end{bmatrix}$ • Multiply row 1 by -3 and add to row 2.
• Multiply row 1 by -2 and add to row 3.

$\begin{bmatrix} 1 & 3 & -2 & -4 \\ 0 & 1 & -\dfrac{3}{11} & -\dfrac{17}{11} \\ 0 & 0 & 7 & 14 \end{bmatrix}$ • Multiply row 2 by $-\dfrac{1}{11}$.

$\begin{bmatrix} 1 & 3 & -2 & -4 \\ 0 & 1 & -\dfrac{3}{11} & -\dfrac{17}{11} \\ 0 & 0 & 1 & 2 \end{bmatrix}$ • Multiply row 3 by $\dfrac{1}{7}$.

$$x + 3y - 2z = -4 \qquad y - \frac{3}{11}(2) = -\frac{17}{11} \qquad x + 3(-1) - 2(2) = -4$$
$$\qquad\qquad\qquad\qquad\qquad\qquad\qquad\qquad\qquad x - 3 - 4 = -4$$
$$y - \frac{3}{11}z = -\frac{17}{11} \qquad\quad y - \frac{6}{11} = -\frac{17}{11} \qquad\qquad x - 7 = -4$$
$$z = 2 \qquad\qquad\qquad y = -1 \qquad\qquad\qquad x = 3$$

The solution is $(3, -1, 2)$.

EXERCISES   (page 501)

1. 11   **3.** 18   **5.** 0   **7.** 15   **9.** −30   **11.** 0   **13.** (3, −4)   **15.** (4, −1)   **17.** $\left(\dfrac{11}{14}, \dfrac{17}{21}\right)$

19. $\left(\dfrac{1}{2}, 1\right)$ **21.** No solution   **23.** (−1, 0)   **25.** (1, −1, 2)   **27.** (2, −2, 3)   **29.** No solution

31. $\left(\dfrac{68}{25}, \dfrac{56}{25}, -\dfrac{8}{25}\right)$ **33.** (1, 3)   **35.** (−1, −3)   **37.** (3, 2)   **39.** No solution   **41.** (−2, 2)

43. (0, 0, −3)   **45.** (1, −1, −1)   **47.** No solution   **49.** $\left(\dfrac{1}{3}, \dfrac{1}{2}, 0\right)$ **51.** $\left(\dfrac{1}{5}, \dfrac{2}{5}, -\dfrac{3}{5}\right)$

53. $\left(\dfrac{1}{4}, 0, -\dfrac{2}{3}\right)$ **55.** 3   **57.** −14   **59.** −3   **61.** 0

SECTION 10.4   (page 503)

Problem 1   **Strategy**

▶ Rate of the rowing team in calm water: t
Rate of the current: c

	Rate	Time	Distance
With current	$t + c$	2	$2(t + c)$
Against current	$t - c$	2	$2(t - c)$

▶ The distance traveled with the current is 18 mi.
The distance traveled against the current is 10 mi.

Solution

$$2(t + c) = 18 \qquad \frac{1}{2} \cdot 2(t + c) = \frac{1}{2} \cdot 18$$

$$2(t - c) = 10 \qquad \frac{1}{2} \cdot 2(t - c) = \frac{1}{2} \cdot 10$$

$$t + c = 9$$
$$t - c = 5$$
$$2t = 14$$
$$t = 7$$

$$t - c = 9$$
$$7 + c = 9$$
$$c = 2$$

The rate of the rowing team in calm water is 7 mph.
The rate of the current is 2 mph.

Problem 2   **Strategy**

▶ Cost of an orange tree: x
Cost of a grapefruit tree: y

First purchase:

	Amount	Unit Cost	Value
Orange trees	25	x	$25x$
Grapefruit trees	20	y	$20y$

Second purchase:

	Amount	Unit Cost	Value
Orange trees	20	x	$20x$
Grapefruit trees	30	y	$30y$

▶ The total of the first purchase was $290.
The total of the second purchase was $330.

Solution

$$25x + 20y = 290 \qquad 4(25x + 20y) = 4 \cdot 290$$
$$20x + 30y = 330 \qquad -5(20x + 30y) = -5 \cdot 330$$

$$100x + 80y = 1160$$
$$-100x - 150y = -1650$$
$$-70y = -490$$
$$y = 7$$

$$25x + 20y = 290$$
$$25x + 20(7) = 290$$
$$25x + 140 = 290$$
$$25x = 150$$
$$x = 6$$

The cost of an orange tree is $6.
The cost of a grapefruit tree is $7.

EXERCISES   (page 508)

1. plane: 142.5 mph; wind: 7.5 mph   **3.** cabin cruiser: 12 mph; current: 3 mph
5. plane: 135 mph; wind: 15 mph   **7.** boat: 20 km/h; current 4 km/h   **9.** plane: 110 mph; wind: 10 mph
11. boat: 13 mph; current: 3 mph   **13.** cabin cruiser: 15 mph; current 3 mph
15. pine: $.15/ft; redwood: $.25/ft   **17.** $.08   **19.** 15 quarters   **21.** 50 color TVs
23. 1st powder: 560 mg; 2nd powder: 120 mg   **25.** 9° and 81°   **27.** 84
29. gold coin: 25 years; silver coin: 50 years   **31.** 25 nickels, 10 dimes, and 5 quarters

SECTION 10.5   (page 512)

Problem 1   **A.** $y = 2x^2 + x - 3$
$y = 2x^2 - 2x + 9$

Use the substitution method.
$$y = 2x^2 + x - 3$$
$$2x^2 - 2x + 9 = 2x^2 + x - 3$$
$$-3x + 12 = 0$$
$$-3x = -12$$
$$x = 4$$

Substitute into equation (1).
$$y = 2x^2 + x - 3$$
$$y = 2(4)^2 + 4 - 3$$
$$y = 32 + 4 - 3$$
$$y = 33$$
The solution is $(4, 33)$.

B. $x^2 - y^2 = 10$
$x^2 + y^2 = 8$

Use the addition method.
$$2x^2 = 18$$
$$x^2 = 9$$
$$x = \pm\sqrt{9} = \pm 3$$

Substitute into equation (2).

$$x^2 + y^2 = 8 \qquad\qquad x^2 + y^2 = 8$$
$$3^2 + y^2 = 8 \qquad\qquad (-3)^2 + y^2 = 8$$
$$9 + y^2 = 8 \qquad\qquad 9 + y^2 = 8$$
$$y^2 = -1 \qquad\qquad y^2 = -1$$
$$y = \pm\sqrt{-1} \qquad\qquad y = \pm\sqrt{-1}$$

y is not a real number. Therefore, the system of equations has no real number solution. The graphs do not intersect.

Problem 2   **A.** $x^2 + y^2 < 16$
$y^2 < x$

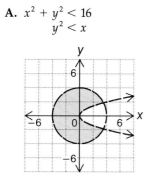

B. $y \geq x - 1$
$y < -2x$

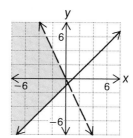

EXERCISES   (page 516)

1. $(-2, 5)$ and $(5, 19)$ **3.** $(-1, -2)$ and $(2, 1)$ **5.** $(2, -2)$ **7.** $(2, 2)$ and $\left(-\dfrac{2}{9}, -\dfrac{22}{9}\right)$

9. $(3, 2)$ and $(2, 3)$ **11.** $\left(\dfrac{\sqrt{3}}{2}, 3\right)$ and $\left(-\dfrac{\sqrt{3}}{2}, 3\right)$ **13.** No real number solution

15. $(1, 2)$, $(1, -2)$, $(-1, 2)$ and $(-1, -2)$ **17.** $(3, 2)$, $(3, -2)$, $(-3, 2)$ and $(-3, -2)$ **19.** $(2, 7)$ and $(-2, 11)$
21. $(3, \sqrt{2})$, $(3, -\sqrt{2})$, $(-3, \sqrt{2})$ and $(-3, -\sqrt{2})$ **23.** No real number solution
25. $(\sqrt{2}, 3)$, $(\sqrt{2}, -3)$, $(-\sqrt{2}, 3)$ and $(-\sqrt{2}, -3)$ **27.** $(2, -1)$ and $(8, 11)$ **29.** $(1, 0)$ **31.**

33.   **35.**   **37.**   **39.**   **41.**

43. No   **45.** 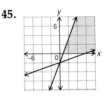   **47.**   **49.**   **51.**

CHAPTER REVIEW   (page 521)

1. (2, 1) (Objective 10.1.2)   **2.** The equations are inconsistent. (Objective 10.2.2)

3. (−1, 2, 1) (Objective 10.3.2)   **4.** (1, 1), (1, −1), (−1, 1), and (−1, −1) (Objective 10.5.1)

5.   (Objective 10.1.1)   **6.** (0, −2) (Objective 10.2.1)   **7.** 26 (Objective 10.3.1)

The solution is (−6, −2).

8. (3, −2) (Objective 10.3.3)   **9.**   (Objective 10.5.2)   **10.** (−1, −5) (Objective 10.1.2)

11. (1, −1, 4) (Objective 10.2.2)   **12.** $\left(\dfrac{8}{5}, \dfrac{7}{5}\right)$ (Objective 10.3.2)   **13.** (4, 1) and $\left(\dfrac{81}{16}, -\dfrac{9}{8}\right)$ (Objective 10.5.1)

14.   (Objective 10.1.1)   **15.** $\left(\dfrac{1}{2}, -1, \dfrac{1}{3}\right)$ (Objective 10.3.3)   **16.** 12 (Objective 10.3.1)

The solution is (2, 0).

17. (Objective 10.5.2)   **18.** $\left(\dfrac{7}{3}, -\dfrac{10}{3}\right)$ (Objective 10.2.1)   **19.** $(2, -3)$ (Objective 10.3.2)

20. $(2, -3, 1)$ (Objective 10.3.3)   **21.** rate of the boat: 12 mph; rate of the current: 4 mph (Objective 10.4.1)

22. 12.5 mph (Objective 10.4.1)   **23.** cost of the tin: \$3/lb; cost of the zinc alloy: \$8/lb (Objective 10.4.2)

24. \$3.00 (Objective 10.4.2)

CUMULATIVE REVIEW   (page 523)

1. $-\dfrac{11}{28}$ (Objective 2.1.2)   **2.** $\{x \mid -4 \le x \le 1\}$ (Objective 2.2.2)   **3.** $y = 5x - 11$ (Objective 6.3.1)

4. $(2x - 5)(3x - 2)$ (Objective 3.3.3)   **5.** $\dfrac{2x^2 - x + 6}{(x - 3)(x - 2)(x + 1)}$ (Objective 4.2.2)   **6.** -3 (Objective 4.5.1)

7. $\dfrac{1 + b}{a}$ (Objective 5.1.1)   **8.** $-2ab^2\sqrt[3]{ab^2}$ (Objective 5.3.3)   **9.** -3 (Objective 5.5.1)

10. $\dfrac{1}{2}$ and -5 (Objective 7.1.1)   **11.** $\dfrac{1 + \sqrt{7}}{3}$ and $\dfrac{1 - \sqrt{7}}{3}$ (Objective 7.2.1/7.2.2)

12. $\dfrac{1}{3}$ and $\dfrac{1}{2}$ (Objective 7.3.2)   **13.** No x-intercepts (Objective 7.4.2)   **14.** (Objective 8.2.1)

15. $f^{-1}(x) = \dfrac{3}{2}x + \dfrac{3}{2}$ (Objective 8.3.2)   **16.** $2\sqrt{10}$ (Objective 9.2.1)

17. $(x - 2)^2 + (y + 1)^2 = 9$ (Objective 9.2.3)   **18.** $(2, 2)$ (Objective 10.1.1)   **19.** $(-5, -11)$ (Objective 10.1.2)

20. $(1, 0, -1)$ (Objective 10.2.2)   **21.** 3 (Objective 10.3.1)   **22.** $\left(\dfrac{7}{6}, \dfrac{1}{2}\right)$ (Objective 10.3.2)

23. $(-1, 0, 2)$ (Objective 10.3.3)   **24.** $(1, 2)$ and $(1, -2)$ (Objective 10.5.1)

25. 16 nickels (Objective 2.4.1)   **26.** 36 in. (Objective 8.4.1)   **27.** 6 m (Objective 7.5.1)

28. 60 ml (Objective 2.5.2)   **29.** 77 or better (Objective 2.2.3)

30. rate of the rowboat: 6 mph; rate of the current: 1.5 mph (Objective 10.4.1)

Answers to Chapter 11 ——————————

SECTION 11.1   (page 527)

Problem 1 $f(x) = \left(\dfrac{2}{3}\right)^x$

$f(3) = \left(\dfrac{2}{3}\right)^3 = \dfrac{8}{27}$

$f(-2) = \left(\dfrac{2}{3}\right)^{-2} = \left(\dfrac{3}{2}\right)^2 = \dfrac{9}{4}$

Problem 2 $f(x) = 2^{2x+1}$
$f(0) = 2^{2(0)+1} = 2^1 = 2$

$f(-2) = 2^{2(-2)+1} = 2^{-3} = \dfrac{1}{2^3} = \dfrac{1}{8}$

Problem 3 $f(x) = (\sqrt{2})^x$
$f(4) = 1.414^4 = 3.9975844 \approx 3.998$

Problem 4 $f(x) = e^{-x}$
$f(-1) = 2.718^{-(-1)} = 2.718^1 = 2.718$

Problem 5   **A.**   **B.**

Problem 6 $7^3 = 343$ is equivalent to $\log_7 343 = 3$.

Problem 7 $\log_{\frac{1}{2}}\left(\dfrac{1}{8}\right) = 3$ is equivalent to $\left(\dfrac{1}{2}\right)^3 = \dfrac{1}{8}$.

Problem 8 $\log 0.1 = -1$ is equivalent to $10^{-1} = 0.1$.

Problem 9 $\ln 7.389 = 2$ is equivalent to $e^2 = 7.389$.

Problem 10 $\log_4 64 = x$
$64 = 4^x$
$4^3 = 4^x$
$3 = x$

$\log_4 64 = 3$

Problem 11

$\log_2 x = -4$

$2^{-4} = x$

$\dfrac{1}{2^4} = x$

$\dfrac{1}{16} = x$

The solution is $\dfrac{1}{16}$.

Problem 12   **A.** $f(x) = \log_2(x - 1)$

$y = \log_2(x - 1)$

$y = \log_2(x - 1)$ is equivalent to
$2^y = x - 1$.
$2^y + 1 = x$

B. $f(x) = \log_3 2x$

$y = \log_3 2x$

$y = \log_3 2x$ is equivalent to $3^y = 2x$.

$\dfrac{3^y}{2} = x$

EXERCISES   (page 535)

1. 9   **3.** $\dfrac{1}{9}$ **5.** 1   **7.** 16   **9.** $\dfrac{1}{4}$ **11.** 1   **13.** 1   **15.** 16   **17.** $\dfrac{1}{8}$ **19.** $\dfrac{1}{3}$ **21.** 9   **23.** 3

25. 2   **27.** 16   **29.** 2   **31.** 2   **33.** 1.4142   **35.** 2.718   **37.** 8.154   **39.**

41.

43.

45.

47.

49. $\log_2 32 = 5$

51. $\log_5 25 = 2$ **53.** $\log_4 \dfrac{1}{16} = -2$ **55.** $\log_{\frac{1}{2}} \dfrac{1}{4} = 2$ **57.** $\log_3 1 = 0$ **59.** $\log_a w = x$ **61.** $3^2 = 9$

63. $4^1 = 4$ **65.** $10^0 = 1$ **67.** $10^{-2} = 0.01$ **69.** $\left(\dfrac{1}{3}\right)^2 = \dfrac{1}{9}$ **71.** $b^v = u$ **73.** $10^3 = 1000$ **75.** $e^0 = 1$

77. 2   **79.** 5   **81.** 2   **83.** 3   **85.** 9   **87.** 64   **89.** $\dfrac{1}{7}$ **91.** 216   **93.**

95.   **97.**   **99.**   **101.** 2.67   **103.** 4.73   **105.** 0.38

107. 77.88   **109.** $y > 0$ **111.** $y > 0$ **113.** $x > 0$ **115.** $x > 0$ **117.**

119.   **121.**

SECTION 11.2   (page 540)

Problem 1

A. $\log_b \dfrac{x^2}{y} = \log_b x^2 - \log_b y = 2\log_b x - \log_b y$

B. $\log_b y^{\frac{1}{3}} z^3 = \log_b y^{\frac{1}{3}} + \log_b z^3 =$

$\dfrac{1}{3}\log_b y + 3\log_b z$

C. $\log_8 \sqrt[3]{xy^2} = \log_8 (xy^2)^{\frac{1}{3}} = \dfrac{1}{3}\log_8 xy^2 =$

$\dfrac{1}{3}(\log_8 x + \log_8 y^2) = \dfrac{1}{3}(\log_8 x + 2\log_8 y) =$

$\dfrac{1}{3}\log_8 x + \dfrac{2}{3}\log_8 y$

Problem 2   **A.** $2 \log_b x - 3 \log_b y - \log_b z = \log_b x^2 - \log_b y^3 - \log_b z =$

$\log_b \dfrac{x^2}{y^3} - \log_b z = \log_b \dfrac{x^2}{y^3 z}$

B. $\dfrac{1}{3}(\log_4 x - 2 \log_4 y + \log_4 z) =$

$\dfrac{1}{3}(\log_4 x - \log_4 y^2 + \log_4 z) =$

$\dfrac{1}{3}\left(\log_4 \dfrac{x}{y^2} + \log_4 z\right) = \dfrac{1}{3}\left(\log_4 \dfrac{xz}{y^2}\right) =$

$\log_4 \left(\dfrac{xz}{y^2}\right)^{\frac{1}{3}} = \log_4 \sqrt[3]{\dfrac{xz}{y^2}}$

Problem 3   Since $\log_b 1 = 0$, $\log_9 1 = 0$.

EXERCISES   (page 545)

1. $\log_8 x + \log_8 z$ **3.** $5 \log_3 x$ **5.** $\log_b r - \log_b s$ **7.** $2 \log_3 x + 6 \log_3 y$ **9.** $3 \log_7 u - 4 \log_7 v$
11. $2 \log_2 r + 2 \log_2 s$ **13.** $2 \log_9 x + \log_9 y + \log_9 z$ **15.** $\log_5 x + 2 \log_5 y - 4 \log_5 z$

17. $2 \log_8 x - \log_8 y - 2 \log_8 z$ **19.** $\dfrac{1}{2} \log_7 x + \dfrac{1}{2} \log_7 y$ **21.** $\dfrac{1}{2} \log_2 x - \dfrac{1}{2} \log_2 y$

23. $\dfrac{3}{2} \log_4 x + \dfrac{1}{2} \log_4 y$ **25.** $\dfrac{3}{2} \log_7 x - \dfrac{1}{2} \log_7 y$ **27.** $\log_b x + \dfrac{1}{2} \log_b y - \dfrac{1}{2} \log_b z$

29. $\log_3 t - \dfrac{1}{2} \log_3 x$ **31.** $\log_3 \dfrac{x^3}{y}$ **33.** $\log_8 x^4 y^2$ **35.** $\log_7 x^3$ **37.** $\log_5 x^3 y^4$ **39.** $\log_4 \dfrac{1}{x^2}$

41. $\log_3 \dfrac{x^2 z^2}{y}$ **43.** $\log_b \dfrac{x}{y^2 z}$ **45.** $\log_4 x^2 y^2$ **47.** $\log_6 \sqrt{\dfrac{x}{y}}$ **49.** $\log_4 \dfrac{s^2 r^2}{t^4}$ **51.** $\log_5 \dfrac{x}{y^2 z^2}$

53. $\log_2 \dfrac{t^3 v^2}{r^2}$ **55.** $\log_4 \sqrt{\dfrac{x^3 z}{y^2}}$ **57.** $\log_b \dfrac{\sqrt{xz}}{\sqrt[3]{y^2}}$ **59.** $\log_b \dfrac{xz^4}{\sqrt[3]{y^2}}$ **61.** 9   **63.** 0   **65.** 8   **67.** 8

69. 6   **71.** 2   **73.** 2   **75.** 3   **77.** 1.2042   **79.** 1.8063   **81.** 1.3010

SECTION 11.3   (page 548)

Problem 1   **A.** $\log 93{,}000 = \log (9.3 \times 10^4) = \log 9.3 + \log 10^4 = 0.9685 + 4 = 4.9685$

B. $\log 0.0006 = \log (6 \times 10^{-4}) = \log 6 + \log 10^{-4} = 0.7782 + (-4) = 6.7782 - 10$

Problem 2   **A.** antilog $2.3365 = 2.17 \times 10^2 = 217$

B. antilog $(9.7846 - 10) = 6.09 \times 10^{-1} = 0.609$

Problem 3 $\log 31{,}560 = \log (3.156 \times 10^4)$
$= \log 3.156 + \log 10^4$

| | Number | Mantissa | | |

```
        ┌──── 3.150   0.4983 ───┐      ┐
  0.006 │                        │  d   │
0.01 ┤  └──── 3.156   0.4983 + d ─┘     │ 0.0014
        │     3.160   0.4997            ┘
        └
```

$$\frac{0.006}{0.01} = \frac{d}{0.0014}$$

$0.0008 = d$

$\log 3.156 + \log 10^4 = (0.4983 + d) + 4 = (0.4983 + 0.0008) + 4 = 0.4991 + 4 = 4.4991$

Problem 4   antilog 1.4907

| | Mantissa | Antilog | | |

```
         ┌──── 0.4900   3.09 ───┐      ┐
  0.0007 │                       │  d   │
0.0014 ┤ └──── 0.4907   3.09 + d ─┘     │ 0.01
         │     0.4914   3.10            ┘
         └
```

$$\frac{0.0007}{0.0014} = \frac{d}{0.01}$$

$0.005 = d$

antilog $1.4907 = (3.09 + d) \times 10^1 = (3.09 + 0.005) \times 10 = 3.095 \times 10 = 30.95$

Problem 5 $N = \dfrac{125}{\sqrt{(1.1)(9.8)}}$

$\log N = \log \left[\dfrac{125}{\sqrt{(1.1)(9.8)}} \right]$

$\quad = \log (125) - \log (1.1)^{\frac{1}{2}} - \log (9.8)^{\frac{1}{2}}$

$\quad = \log (125) - \dfrac{1}{2} \log (1.1) - \dfrac{1}{2} \log (9.8)$

$\quad = 2.0969 - \dfrac{1}{2}(0.0414 + 0.9912)$

$\quad = 2.0969 - \dfrac{1}{2}(1.0326)$

$\quad = 2.0969 - 0.5163 = 1.5806$

Interpolate.

$N = $ antilog $1.5806 = 3.807 \times 10^1 = 38.07$

EXERCISES   (page 556)

1. 0.7686   **3.** 1.5378   **5.** 2.5899   **7.** 3.0128   **9.** 4.6749   **11.** 9.0569 − 10   **13.** 7.3010 − 10
15. 7.2430 − 10   **17.** 9.9930 − 10   **19.** 4.4362   **21.** 7.7566 − 10   **23.** 9.9586 − 10   **25.** 5.49

27. 34.3   **29.** 7970   **31.** 632   **33.** 0.455   **35.** 0.0245   **37.** 0.00742   **39.** 475   **41.** 0.492
43. 0.6960   **45.** 1.8286   **47.** 2.8449   **49.** 8.1623 − 10   **51.** 4.9196   **53.** 9.9918 − 10   **55.** 7.024
57. 344.8   **59.** 0.5395   **61.** 501.7   **63.** 0.03445   **65.** 47.96   **67.** 446.9   **69.** 0.01499   **71.** 1.127
73. 3.854   **75.** 6650   **77.** 2.315   **79.** 18.63   **81.** 0.8829   **83.** 0.9224   **85.** 3.14   **87.** 0.4969
89. 31.4   **91.** $N = 89{,}100$ **93.** $N = 270.0$ **95.** 0.3240   **97.** 0.0002470   **99.** 85,600   **101.** 0.4980

SECTION 11.4   (page 558)

Problem 1 $10^{3x+5} = 10^{x-3}$ Check: $\dfrac{10^{3x+5} = 10^{x-3}}{}$

$$3x + 5 = x - 3$$
$$2x + 5 = -3$$
$$2x = -8$$
$$x = -4$$

$$\frac{10^{3(-4)+5}}{10^{-12+5}} \, \bigg| \, \frac{10^{-4-3}}{10^{-7}}$$
$$10^{-7} = 10^{-7}$$

The solution is −4.

Problem 2   **A.** $4^{3x} = 25$

$$\log 4^{3x} = \log 25$$
$$3x \log 4 = \log 25$$
$$3x = \frac{\log 25}{\log 4} = \frac{1.3979}{0.6021}$$
$$3x = 2.3217$$
$$x = 0.7739$$
The solution is 0.7739.

B. $(1.06)^n = 1.5$

$$\log (1.06)^n = \log 1.5$$
$$n \log 1.06 = \log 1.5$$
$$n = \frac{\log 1.5}{\log 1.06} = \frac{0.1761}{0.0253}$$
$$n = 6.9605$$
The solution is 6.9605.

Problem 3   **A.** $\log_4(x^2 - 3x) = 1$

Rewrite in exponential form.

$$4^1 = x^2 - 3x$$
$$4 = x^2 - 3x$$
$$0 = x^2 - 3x - 4$$
$$0 = (x + 1)(x - 4)$$

$$x + 1 = 0 \qquad x - 4 = 0$$
$$x = -1 \qquad x = 4$$

The solutions are −1 and 4.

B. $\log_3 x + \log_3(x + 3) = \log_3 4$
$$\log_3[x(x + 3)] = \log_3 4$$

Use the fact that if $\log_b u = \log_b v$, then $u = v$.

$$x(x + 3) = 4$$
$$x^2 + 3x = 4$$
$$x^2 + 3x - 4 = 0$$
$$(x + 4)(x - 1) = 0$$
$$x + 4 = 0 \qquad x - 1 = 0$$
$$x = -4 \qquad x = 1$$

−4 does not check as a solution.
The solution is 1.

Problem 4   **A.** $\log_9 23 = \dfrac{\log_{10} 23}{\log_{10} 9} = \dfrac{1.3617}{0.9542} = 1.4271$

B. $\ln 3 = \dfrac{\log_{10} 3}{\log_{10} e} = \dfrac{\log_{10} 3}{\log_{10} 2.72} = \dfrac{0.4771}{0.4346} = 1.0978$

EXERCISES   (page 563)

1. 1   **3.** −3   **5.** 1.1133   **7.** 0.7211   **9.** −1.5850   **11.** 1.7093   **13.** 1.3222   **15.** −2.8076
17. 3.5854   **19.** 1.1310   **21.** 1   **23.** 6   **25.** 1.3753 and −1.3753   **27.** 1.5806   **29.** −0.6310
31. 1.6610   **33.** −1.5132   **35.** 1.4999   **37.** 8   **39.** $\dfrac{11}{2}$ **41.** 2 and −4   **43.** $\dfrac{5}{3}$ **45.** $\dfrac{1}{2}$
47. 1,000,000,000   **49.** 4   **51.** 3   **53.** 3   **55.** 2   **57.** No solution   **59.** 4   **61.** $\dfrac{5 + \sqrt{33}}{2}$
63. 1.7713   **65.** 1.5439   **67.** 0.6667   **69.** 2.4651   **71.** 0.8617   **73.** 2.1132   **75.** 1.9587
77. 0.6310   **79.** −0.6309   **81.** 0.2726   **83.** 0.1729   **85.** 3.1186   **87.** −0.6012   **89.** 2.9183
91. −2.3280   **93.** 2.3181   **95.** 4.8499   **97.** −0.8340   **99.** 1.8162   **101.** 1.1492   **103.** 1.9769
105. 1.7234   **107.** 0.8060   **109.** 5.6894   **111.** 1.7917   **113.** 1.6095   **115.** −1.2040   **117.** (3, 1)
119. (502, 498)   **121.** (−4, −9)   **123.** log (0.2) = −0.6990. In Step 2, by the Multiplication Property of
Inequality, when both sides of an inequality are multiplied by a negative number, the inequality symbol must
be reversed.

SECTION 11.5   (page 565)

Problem 1   **Strategy**

To find the pH, replace H^+ by 2.51×10^{-12} in the equation $pH = -\log (H^+)$ and solve for pH.

Solution

$$pH = -\log (H^+) = -\log (2.51 \times 10^{-12})$$
$$= -(\log 2.51 + \log 10^{-12})$$
$$= -[0.3997 + (-12)] = 11.6003$$

The pH of sodium carbonate is 11.6.

Problem 2   **Strategy**

To find how many times stronger the San Francisco earthquake was, use the Richter equation
to write a system of equations. Solve the system of equations for the ratio $\dfrac{I_1}{I_2}$.

Solution

$$7.8 = \log \frac{I_1}{I_0}$$
$$5 = \log \frac{I_2}{I_0}$$

$$7.8 = \log I_1 - \log I_0$$
$$5 = \log I_2 - \log I_0$$

$$2.8 = \log I_1 - \log I_2$$
$$2.8 = \log \frac{I_1}{I_2}$$

$$\frac{I_1}{I_2} = 10^{2.8} = 630.957$$

The San Francisco earthquake was 631 times stronger than one that can cause serious damage.

EXERCISES   (page 570)

1. $8979   **3.** 9 years   **5.** $11,827   **7.** 116 days   **9.** 24 h   **11.** 58%   **13.** 2   **15.** 1 cm
17. 10,000 times stronger   **19.** 134 decibels   **21.** 5 P.M.   **23.** 9 years   **25.** 8.5%

CHAPTER REVIEW   **(page 576)**

1. 2 (Objective 11.1.3)   **2.** $\log_3 \sqrt{\dfrac{x}{y}}$ (Objective 11.2.1)   **3.** 711.7 (Objective 11.3.3)

4. −3 (Objective 11.4.1)   **5.** 1 (Objective 11.1.1)   **6.** $\dfrac{1}{9}$ (Objective 11.1.3)

7. 7.5647 − 10 (Objective 11.3.1)   **8.** 0.06433 (Objective 11.3.4)   **9.** 6 (Objective 11.4.2)

10. (Objective 11.1.2)   **11.** $\dfrac{1}{2}\log_6 x + \dfrac{3}{2}\log_6 y$ (Objective 11.2.1)   **12.** 1.2425 (Objective 11.3.3)

13. −2 (Objective 11.4.1)   **14.** $\dfrac{1}{3}$ (Objective 11.1.1)   **15.** (Objective 11.1.4)

16. 0.0404 (Objective 11.3.2)   **17.** 0.7787 (Objective 11.3.4)   **18.** 2.6804 (Objective 11.4.3)
19. 4 (Objective 11.1.3)   **20.** 8.6911 − 10 (Objective 11.3.1)   **21.** 2.3223 (Objective 11.4.3)
22. 0.702 (Objective 11.3.3)   **23.** 3 (Objective 11.4.1)   **24.** 4 (Objective 11.1.1)

25. $\dfrac{1}{2}\log_5 x - \dfrac{1}{2}\log_5 y$ (Objective 11.2.1)   **26.** 1.9371 (Objective 11.3.3)   **27.** 125 (Objective 11.1.3)

28. 0.545 (Objective 11.3.4)   **29.** (Objective 11.1.2)

30. $\dfrac{1}{2}$ (Objective 11.4.2)   **31.** $\log_b \dfrac{x^3}{y^5}$ (Objective 11.2.1)   **32.** 4 (Objective 11.1.1)

33. (Objective 11.1.4)   **34.** 50.68 (Objective 11.3.3)   **35.** 0.719 (Objective 11.3.4)

36. 2 (Objective 11.4.1)   **37.** 33 h (Objective 11.5.1)   **38.** 0.602 cm (Objective 11.5.1)

CUMULATIVE REVIEW   (page 578)

1. 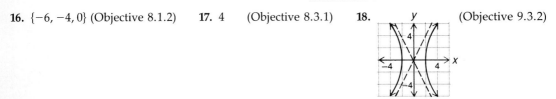 (Objective 1.3.1)   **2.** $\dfrac{8}{7}$ (Objective 2.1.2)   **3.** $\{x \mid 1 \le x \le 4\}$ (Objective 2.3.2)

4. $(4x^n + 3)(x^n + 1)$ (Objective 3.3.4)   **5.** $\{x \mid -5 \le x \le 1\}$ (Objective 3.5.2)   **6.** $\dfrac{x - 3}{x + 3}$ (Objective 4.3.1)

7. (Objective 4.5.2)   **8.** $L = \dfrac{S - 2wh}{2w + 2h}$ (Objective 4.5.3)

9. $\dfrac{x\sqrt{y} + y\sqrt{x}}{x - y}$ (Objective 5.3.4)   **10.** $-4x^2 y^3\sqrt{2x}$ (Objective 5.3.2)   **11.** $-\dfrac{1}{5} + \dfrac{2}{5}i$ (Objective 5.4.4)

12. $y = 2x - 6$ (Objective 6.3.2)   **13.** $3x^2 + 8x - 3 = 0$ (Objective 7.1.2)

14. $2 + \sqrt{10}$ and $2 - \sqrt{10}$ (Objective 7.2.1/7.2.2)   **15.**   (Objective 7.4.1)

16. $\{-6, -4, 0\}$ (Objective 8.1.2)   **17.** 4   (Objective 8.3.1)   **18.**   (Objective 9.3.2)

19. $(0, -1, 2)$ (Objective 10.2.2)   **20.** $(2,3)$ and $(-1, -3)$ (Objective 10.5.1)

21.   (Objective 11.1.2)   **22.**   (Objective 11.1.4)   **23.** 243 (Objective 11.1.1)

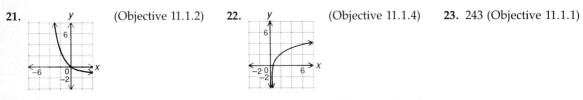

24. 64 (Objective 11.1.3)   **25.** 8 (Objective 11.4.1)   **26.** 1 (Objective 11.4.2)

27. 25% alloy: 800 lb; 50% alloy: 1200 lb (Objective 2.5.2)

28. $31,250 or more (Objective 2.2.3)   **29.** 7.5 min (Objective 4.5.4)   **30.** 500 lb/in². (Objective 8.4.1)

31. $12 (Objective 10.4.2)   **32.** $15,656.81 (Objective 11.5.1)

Answers to Chapter 12 ———————

SECTION 12.1   (page 583)

Problem 1 $a_n = n(n + 1)$

$a_1 = 1(1 + 1) = 2$ The first term is 2.

$a_2 = 2(2 + 1) = 6$ The second term is 6.

$a_3 = 3(3 + 1) = 12$ The third term is 12.

$a_4 = 4(4 + 1) = 20$ The fourth term is 20.

Problem 2 $a_n = \dfrac{1}{n(n + 2)}$

$a_6 = \dfrac{1}{6(6 + 2)} = \dfrac{1}{48}$ The sixth term is $\dfrac{1}{48}$.

$a_9 = \dfrac{1}{9(9 + 2)} = \dfrac{1}{99}$ The ninth term is $\dfrac{1}{99}$.

Problem 3

A. $\displaystyle\sum_{n=1}^{4} (7 - n) = (7 - 1) + (7 - 2) + (7 - 3) + (7 - 4) = 6 + 5 + 4 + 3 = 18$

B. $\displaystyle\sum_{i=3}^{6} (i^2 - 2) = (3^2 - 2) + (4^2 - 2) + (5^2 - 2) + (6^2 - 2)$

$= 7 + 14 + 23 + 34 = 78$

Problem 4 $\displaystyle\sum_{n=1}^{5} nx = x + 2x + 3x + 4x + 5x$

EXERCISES   (page 586)

1. $2, 3, 4, 5$ **3.** $3, 5, 7, 9$ **5.** $0, -2, -4, -6$ **7.** $2, 4, 8, 16$ **9.** $2, 5, 10, 17$ **11.** $\dfrac{1}{2}, \dfrac{2}{5}, \dfrac{3}{10}, \dfrac{4}{17}$

13. $0, \dfrac{3}{2}, \dfrac{8}{3}, \dfrac{15}{4}$ **15.** $1, -2, 3, -4$ **17.** $\dfrac{1}{2}, -\dfrac{1}{5}, \dfrac{1}{10}, -\dfrac{1}{17}$ **19.** $-2, 4, -8, 16$ **21.** $\dfrac{2}{9}, \dfrac{2}{27}, \dfrac{2}{81}, \dfrac{2}{243}$

23. 15 **25.** $\dfrac{12}{13}$ **27.** 24 **29.** $\dfrac{32}{243}$ **31.** 88 **33.** $\dfrac{1}{20}$ **35.** 38 **37.** 42 **39.** 28 **41.** 60

43. $\dfrac{25}{24}$ **45.** 28 **47.** $\dfrac{99}{20}$ **49.** $\dfrac{137}{120}$ **51.** -2 **53.** $-\dfrac{7}{12}$ **55.** $\dfrac{2}{x} + \dfrac{4}{x} + \dfrac{6}{x} + \dfrac{8}{x}$

57. $\dfrac{x}{2} + \dfrac{x^2}{3} + \dfrac{x^3}{4} + \dfrac{x^4}{5}$ **59.** $\dfrac{x^2}{3} + \dfrac{x^3}{5} + \dfrac{x^4}{7}$ **61.** $x + x^3 + x^5 + x^7$ **63.** $2x + 4x^2 + 6x^3 + 8x^4$

65. $\dfrac{1}{x} + \dfrac{1}{x^2} + \dfrac{1}{x^3} + \dfrac{1}{x^4} + \dfrac{1}{x^5}$ **67.** $a_n = 2n - 1$ **69.** $a_n = -2n + 1$ **71.** $a_n = 4n$ **73.** $a_n = 2n + 9$

75. $a_n = 13 - 10n$ **77.** $\log 8$ **79.** $14 \log x$ **81.** 3

SECTION 12.2   (page 590)

Problem 1 $9, 3, -3, -9, \ldots$

$$d = a_2 - a_1 = 3 - 9 = -6$$

$$a_n = a_1 + (n - 1)d$$
$$a_{15} = 9 + (15 - 1)(-6) = 9 + (14)(-6) = 9 - 84$$
$$a_{15} = -75$$

Problem 2 $-3, 1, 5, 9, \ldots$

$$d = a_2 - a_1 = 1 - (-3) = 4$$

$$a_n = a_1 + (n - 1)d$$
$$a_n = -3 + (n - 1)4$$
$$a_n = -3 + 4n - 4$$
$$a_n = 4n - 7$$

Problem 3 $7, 9, 11, \ldots, 59$

$$d = a_2 - a_1 = 9 - 7 = 2$$

$$a_n = a_1 + (n - 1)d$$
$$59 = 7 + (n - 1)2$$
$$59 = 7 + 2n - 2$$
$$59 = 5 + 2n$$
$$54 = 2n$$
$$27 = n$$

There are 27 terms in the sequence.

Problem 4 $-4, -2, 0, 2, 4, \ldots$

$$d = a_2 - a_1 = -2 - (-4) = 2$$

$$a_n = a_1 + (n - 1)d$$
$$a_{25} = -4 + (25 - 1)2 = -4 + 24(2) = -4 + 48$$
$$a_{25} = 44$$

$$S_n = \frac{n}{2}(a_1 + a_n)$$
$$S_{25} = \frac{25}{2}(-4 + 44) = \frac{25}{2}(40) = 25(20)$$
$$S_{25} = 500$$

Problem 5 $\displaystyle\sum_{n=1}^{18} (3n - 2)$

$a_n = 3n - 2$
$a_1 = 3(1) - 2 = 1$
$a_{18} = 3(18) - 2 = 52$

$S_n = \dfrac{n}{2}(a_1 + a_n)$

$S_{18} = \dfrac{18}{2}(1 + 52) = 9(53) = 477$

Problem 6   **Strategy**

To find the value of the 20th-place prize:
▶ Write the equation for the nth-place prize.
▶ Find the 20th term of the sequence:
To find the total amount of prize money being awarded, use the Formula for Sum of n Terms of an Arithmetic Sequence.

Solution

$10{,}000, 9700, \ldots$

$d = a_2 - a_1 = 9700 - 10{,}000 = -300$

$\begin{aligned} a_n &= a_1 + (n - 1)d \\ &= 10{,}000 + (n - 1)(-300) \\ &= 10{,}000 - 300n + 300 \\ &= -300n + 10{,}300 \end{aligned}$
$a_{20} = -300(20) + 10{,}300 = -6000 + 10{,}300 = 4300$

$S_n = \dfrac{n}{2}(a_1 + a_n)$

$S_{20} = \dfrac{20}{2}(10{,}000 + 4300) = 10(14{,}300) = 143{,}000$

The value of the 20th-place prize is $4300.

The total amount of prize money being awarded is $143,000.

EXERCISES   (page 595)

1. 141   **3.** 50   **5.** 71   **7.** $8\dfrac{1}{4}$ **9.** 17   **11.** 3.75   **13.** $a_n = n$ **15.** $a_n = -4n + 10$

17. $a_n = \dfrac{3n + 1}{2}$ **19.** $a_n = -5n - 3$ **21.** $a_n = -10n + 36$ **23.** 42   **25.** 16   **27.** 20   **29.** 20

31. 13   **33.** 20   **35.** 650   **37.** −605   **39.** $\dfrac{215}{4}$ **41.** 420   **43.** −210   **45.** −5   **47.** 11 weeks

49. 4268 seats   **51.** monthly salary for the eighth month: $945; total salary for the eight-month period: $6580
53. 5050   **55.** 8 terms   **57.** $d = 4, n = 7$ **59.** $a_1 = -3$

SECTION 12.3   (page 597)

Problem 1 $5, 2, \dfrac{4}{5}, \ldots$

$$r = \frac{a_2}{a_1} = \frac{2}{5}$$
$$a_n = a_1 r^{n-1}$$
$$a_5 = 5\left(\frac{2}{5}\right)^{5-1} = 5\left(\frac{2}{5}\right)^4 = 5\left(\frac{16}{625}\right)$$
$$a_5 = \frac{16}{125}$$

Problem 2 $3, a_2, a_3, -192, \ldots$

$$a_n = a_1 r^{n-1}$$
$$a_4 = 3r^{4-1}$$
$$-192 = 3r^{4-1}$$
$$-192 = 3r^3$$
$$-64 = r^3$$
$$-4 = r$$

$$a_n = a_1 r^{n-1}$$
$$a_3 = 3(-4)^{3-1} = 3(-4)^2 = 3(16) = 48$$

Problem 3 $1, -\dfrac{1}{3}, \dfrac{1}{9}, -\dfrac{1}{27}$

$$r = \frac{a_2}{a_1} = \frac{-\dfrac{1}{3}}{1} = -\frac{1}{3}$$

$$S_n = \frac{a_1(1 - r^n)}{1 - r}$$

$$S_4 = \frac{1\left[1 - \left(-\dfrac{1}{3}\right)^4\right]}{1 - \left(-\dfrac{1}{3}\right)} = \frac{1 - \dfrac{1}{81}}{\dfrac{4}{3}} = \frac{\dfrac{80}{81}}{\dfrac{4}{3}}$$

$$= \frac{80}{81} \cdot \frac{3}{4} = \frac{20}{27}$$

Problem 4 $\displaystyle\sum_{n=1}^{5} \left(\frac{1}{2}\right)^n$

$$a_n = \left(\frac{1}{2}\right)^n$$
$$a_1 = \left(\frac{1}{2}\right)^1 = \frac{1}{2}$$
$$a_2 = \left(\frac{1}{2}\right)^2 = \frac{1}{4}$$

$$r = \frac{a_2}{a_1} = \frac{\frac{1}{4}}{\frac{1}{2}} = \frac{1}{4} \cdot \frac{2}{1} = \frac{1}{2}$$

$$S_n = \frac{a_1(1 - r^n)}{1 - r}$$

$$S_5 = \frac{\frac{1}{2}\left[1 - \left(\frac{1}{2}\right)^5\right]}{1 - \frac{1}{2}} = \frac{\frac{1}{2}\left(1 - \frac{1}{32}\right)}{\frac{1}{2}} = \frac{\frac{1}{2}\left(\frac{31}{32}\right)}{\frac{1}{2}} = \frac{\frac{31}{64}}{\frac{1}{2}} = \frac{31}{64} \cdot \frac{2}{1} = \frac{31}{32}$$

Problem 5 $3, -2, \frac{4}{3}, -\frac{8}{9}, \ldots$

$$r = \frac{a_2}{a_1} = -\frac{2}{3}$$

$$S = \frac{a_1}{1 - r} = \frac{3}{1 - \left(-\frac{2}{3}\right)} = \frac{3}{1 + \frac{2}{3}}$$

$$= \frac{3}{\frac{5}{3}} = \frac{9}{5}$$

Problem 6 $0.36\overline{36} = 0.36 + 0.0036 + 0.000036 + \cdots$

$$S = \frac{a_1}{1 - r} = \frac{\frac{36}{100}}{1 - \frac{1}{100}} = \frac{\frac{36}{100}}{\frac{99}{100}} = \frac{36}{99} = \frac{4}{11}$$

An equivalent fraction is $\frac{4}{11}$.

Problem 7   **Strategy**

To find the total number of letters mailed, use the Formula for the Sum of n Terms of a Finite Geometric Series.

Solution

$n = 6$, $a_1 = 3$, $r = 3$

$$S_n = \frac{a_1(1 - r^n)}{1 - r}$$

$$S_6 = \frac{3(1 - 3^6)}{1 - 3} = \frac{3(1 - 729)}{1 - 3}$$

$$= \frac{3(-728)}{-2} = \frac{-2184}{-2} = 1092$$

From the first through the sixth mailings, 1092 letters will have been mailed.

EXERCISES   (page 603)

1. 131,072   **3.** $\dfrac{128}{243}$ **5.** 16   **7.** 6, 4   **9.** $-2, \dfrac{4}{3}$ **11.** 12, -48 **13.** 2186   **15.** $\dfrac{2343}{64}$ **17.** 62

19. $\dfrac{121}{243}$ **21.** 1364   **23.** 2800   **25.** $\dfrac{2343}{1024}$ **27.** $\dfrac{1360}{81}$ **29.** 9   **31.** $\dfrac{18}{5}$ **33.** $\dfrac{7}{9}$ **35.** $\dfrac{8}{9}$ **37.** $\dfrac{2}{9}$

39. $\dfrac{5}{11}$ **41.** $\dfrac{1}{6}$ **43.** 50 mg   **45.** 1.8 ft   **47.** \$101,138.92   **49.** \$411,613.50   **51.** $G, -\dfrac{1}{2}$

53. $A, 9.5$ **55.** $N, 25$ **57.** G, x^2 **59.** $A, 4 \log x$ **61.** 27   **63.** $a_1 = 80, n = 5$

SECTION 12.4   (page 606)

Problem 1 $\dfrac{12!}{7!\,5!} = \dfrac{12 \cdot 11 \cdot 10 \cdot 9 \cdot 8 \cdot 7 \cdot 6 \cdot 5 \cdot 4 \cdot 3 \cdot 2 \cdot 1}{(7 \cdot 6 \cdot 5 \cdot 4 \cdot 3 \cdot 2 \cdot 1)(5 \cdot 4 \cdot 3 \cdot 2 \cdot 1)} = 792$

Problem 2 $\dbinom{7}{0} = \dfrac{7!}{(7-0)!\,0!} = \dfrac{7!}{7!\,0!} = \dfrac{7 \cdot 6 \cdot 5 \cdot 4 \cdot 3 \cdot 2 \cdot 1}{(7 \cdot 6 \cdot 5 \cdot 4 \cdot 3 \cdot 2 \cdot 1)(1)} = 1$

Problem 3 $(3m - n)^4 =$
$\dbinom{4}{0}(3m)^4 + \dbinom{4}{1}(3m)^3(-n) + \dbinom{4}{2}(3m)^2(-n)^2 + \dbinom{4}{3}(3m)(-n)^3 + \dbinom{4}{4}(-n)^4 =$
$1(81m^4) + 4(27m^3)(-n) + 6(9m^2)(n^2) + 4(3m)(-n^3) + 1(n^4) =$
$81m^4 - 108m^3n + 54m^2n^2 - 12mn^3 + n^4$

Problem 4 $(y - 2)^{10} =$
$\dbinom{10}{0}y^{10} + \dbinom{10}{1}y^9(-2) + \dbinom{10}{2}y^8(-2)^2 + \cdots =$
$1(y^{10}) + 10y^9(-2) + 45y^8(4) + \cdots = y^{10} - 20y^9 + 180y^8 + \cdots$

Problem 5 $(r - 2s)^7$
$n = 7, a = r, b = -2s, r = 3$
$\dbinom{7}{3-1}(r)^{7-3+1}(-2s)^{3-1} = \dbinom{7}{2}(r)^5(-2s)^2 = 21r^5(4s^2) = 84r^5s^2$

EXERCISES   (page 611)

1. 6   **3.** 40,320   **5.** 1   **7.** 10   **9.** 1   **11.** 84   **13.** 21   **15.** 45   **17.** 1   **19.** 20   **21.** 11
23. 6   **25.** $x^4 + 4x^3y + 6x^2y^2 + 4xy^3 + y^4$ **27.** $x^5 - 5x^4y + 10x^3y^2 - 10x^2y^3 + 5xy^4 - y^5$
29. $16m^4 + 32m^3 + 24m^2 + 8m + 1$ **31.** $32r^5 - 240r^4 + 720r^3 - 1080r^2 + 810r - 243$
33. $a^{10} + 10a^9b + 45a^8b^2$ **35.** $a^{11} - 11a^{10}b + 55a^9b^2$ **37.** $256x^8 + 1024x^7y + 1792x^6y^2$
39. $65{,}536x^8 - 393{,}216x^7y + 1{,}032{,}192x^6y^2$ **41.** $x^7 + 7x^5 + 21x^3$ **43.** $x^{10} + 15x^8 + 90x^6$ **45.** $-560x^4$

47. $-6x^{10}y^2$ **49.** $126y^5$ **51.** $5n^3$ **53.** $\dfrac{x^5}{32}$ **55.** 1 7 21 35 35 21 7 1   **57.** 2450   **59.** n

61. $35x^3a^4$ **63.** $x^2 + 8x^{\frac{3}{2}} + 24x + 32x^{\frac{1}{2}} + 16$ **65.** $\dfrac{1}{x^3} + \dfrac{3}{x^2y} + \dfrac{3}{xy^2} + \dfrac{1}{y^3}$ **67.** $-8i$ **69.** 1.04060401

CHAPTER REVIEW   (page 615)

1. $3x + 3x^2 + 3x^3 + 3x^4$ (Objective 12.1.2)   **2.** 16 (Objective 12.2.1)   **3.** 32 (Objective 12.3.1)

4. 16 (Objective 12.3.3)   **5.** 84 (Objective 12.4.1)   **6.** $\frac{1}{2}$ (Objective 12.1.1)   **7.** 44 (Objective 12.2.1)

8. 468 (Objective 12.2.2)   **9.** -66 (Objective 12.3.2)   **10.** 70 (Objective 12.4.1)

11. $2268x^3y^6$ (Objective 12.4.1)   **12.** 34 (Objective 12.1.2)   **13.** $\frac{7}{6}$ (Objective 12.1.1)

14. $a_n = -3n + 15$ (Objective 12.2.1)   **15.** $\frac{2}{27}$ (Objective 12.3.1)   **16.** $\frac{7}{30}$ (Objective 12.3.3)

17. -115 (Objective 12.2.1)   **18.** $\frac{665}{32}$ (Objective 12.3.2)   **19.** 1575 (Objective 12.2.2)

20. $-280x^4y^3$ (Objective 12.4.1)   **21.** 21 (Objective 12.2.1)   **22.** 48 (Objective 12.3.1)
23. 30 (Objective 12.1.2)   **24.** 341 (Objective 12.3.2)   **25.** 120 (Objective 12.4.1)
26. $240x^4$ (Objective 12.4.1)   **27.** 143 (Objective 12.2.1)   **28.** -575 (Objective 12.2.2)
29. 2550 yd (Objective 12.2.3)   **30.** 20 mg (Objective 12.3.4)

CUMULATIVE REVIEW   (page 617)

1. $2(x^2 + 2)(x^4 - 2x^2 + 4)$ (Objective 3.4.4)   **2.** $\frac{x^2 + 5x - 2}{(x + 2)(x - 1)}$ (Objective 4.2.2)

3. $\{x \mid x < -2 \text{ or } x > 2\}$ (Objective 2.2.2)   **4.** $\frac{1}{x^{26}}$ (Objective 5.2.1)   **5.** $4y\sqrt{x} - y\sqrt{2}$ (Objective 5.3.3)

6. $\frac{1}{4} + \frac{\sqrt{55}}{4}i$ and $\frac{1}{4} - \frac{\sqrt{55}}{4}i$ (Objective 7.2.1/7.2.2)   **7.** 4 (Objective 5.5.1)

8.   (Objective 6.1.3)   **9.**    (Objective 6.1.4)

10. 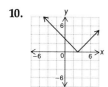   (Objective 8.2.1)   **11.** $f^{-1}(x) = \frac{1}{2}x + 2$ (Objective 8.3.2)

12. $(x + 1)^2 + (y + 1)^2 = 34$ (Objective 9.2.3)   **13.** -4 and 2 (Objective 4.5.1)

14. 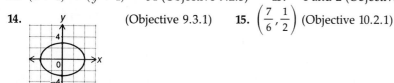   (Objective 9.3.1)   **15.** $\left(\frac{7}{6}, \frac{1}{2}\right)$ (Objective 10.2.1)

16. $(3, 4)$ (Objective 10.3.3)   **17.** -10 (Objective 10.3.1)   **18.** $\log_6 x + 3 \log_6 y + \log_6 z$ (Objective 11.2.1)

19. $\dfrac{1}{64}$ (Objective 11.1.3)   **20.** 7 (Objective 11.4.1)   **21.** 20 (Objective 12.1.1)   **22.** 6 (Objective 12.1.2)

23. -103 (Objective 12.2.1)   **24.** $\dfrac{9}{5}$ (Objective 12.3.3)   **25.** $\dfrac{7}{15}$ (Objective 12.3.3)

26. $12xy^5$ (Objective 12.4.1)   **27.** 35 mph (Objective 4.5.5)

28. rate of the plane: 105 mph; rate of the wind: 15 mph (Objective 10.4.1)   **29.** 20, 22, 24 (Objective 2.2.3)

30. 31 days (Objective 11.5.1)   **31.** 156 boxes (Objective 12.2.3)   **32.** 625 people (Objective 12.3.4)

Answers to Final Exam

1. -31 (Objective 1.1.4)   **2.** -1 (Objective 1.2.1)   **3.** $33 - 10x$ (Objective 1.2.3)   **4.** 8 (Objective 2.1.1)

5. $\dfrac{2}{3}$ (Objective 2.1.2)   **6.** $\left\{x \mid x > \dfrac{3}{2}\right\}$ (Objective 2.2.2)   **7.** 4 and $-\dfrac{2}{3}$ (Objective 2.3.1)

8. $\{x \mid -4 < x < -1\}$ (Objective 2.3.2)   **9.** $6a^3 - 5a^2 + 10a$ (Objective 3.2.1)

10. $(2 - xy)(4 + 2xy + x^2y^2)$ (Objective 3.4.2)   **11.** $(x - y)(1 + x)(1 - x)$ (Objective 3.4.4)

12. $\left\{x \mid x < -5 \text{ or } x > \dfrac{1}{2}\right\}$ (Objective 3.5.2)   **13.** $x^2 - 2x - 3 - \dfrac{5}{2x - 3}$ (Objective 4.1.2)

14. $\dfrac{x(x - 1)}{2x - 5}$ (Objective 4.2.1)   **15.** $-\dfrac{10x}{(x + 2)(x - 3)}$ (Objective 4.2.2)   **16.** $\dfrac{x + 3}{x + 1}$ (Objective 4.3.1)

17. $-\dfrac{7}{4}$ (Objective 4.5.1)   **18.** $\{x \mid -2 \le x < -1 \text{ or } x \ge 3\}$ (Objective 4.5.2)   **19.** $d = \dfrac{a_n - a_1}{n - 1}$ (Objective 4.5.3)

20. $\dfrac{y^4}{162x^3}$ (Objective 5.1.1)   **21.** $\dfrac{1}{64x^8y^5}$ (Objective 5.2.1)   **22.** $-2x^2y\sqrt{2y}$ (Objective 5.3.2)

23. $\dfrac{x^2\sqrt{2y}}{2y^2}$ (Objective 5.3.4)   **24.** $\dfrac{6}{5} - \dfrac{3}{5}i$ (Objective 5.4.4)   **25.**   (Objective 6.2.2)

26.   (Objective 6.1.4)

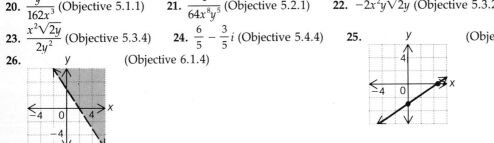

27. $y = -3x + 7$ (Objective 6.3.1)   **28.** $y = -\dfrac{2}{3}x - \dfrac{1}{3}$ (Objective 6.3.2)

29. $2x^2 - 3x - 2 = 0$ (Objective 7.1.2)   **30.** $\dfrac{3 + \sqrt{17}}{4}$ and $\dfrac{3 - \sqrt{17}}{4}$ (Objective 7.2.1/7.2.2)

31. 27 and -8 (Objective 7.3.1)   **32.** $\dfrac{3}{2}$ and -2 (Objective 7.3.3)   **33.** 11 (Objective 8.1.1)

34.   (Objective 8.2.1)   **35.** 25 (Objective 8.3.1)   **36.** $f^{-1}(x) = \dfrac{3}{2}x + 6$ (Objective 8.3.2)

37. (Objective 9.3.1)   **38.** (Objective 9.4.1)

39. $(3, 4)$ (Objective 10.2.1)   **40.** $\left(\dfrac{1}{2}, \dfrac{1}{3}\right)$ (Objective 10.3.3)   **41.** 10 (Objective 10.3.1)

42. $\left(\dfrac{5}{2}, -\dfrac{3}{2}\right)$ (Objective 10.5.1)   **43.** 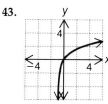 (Objective 11.1.4)

44. $\log_2 \dfrac{a^2}{b^2}$ (Objective 11.2.1)   **45.** 6 (Objective 11.4.2)   **46.** $2y + 2y^2 + 2y^3 + 2y^4 + 2y^5$ (Objective 12.1.2)

47. $\dfrac{23}{45}$ (Objective 12.3.3)   **48.** $144x^7y^2$ (Objective 12.4.1)   **49.** 7 and 17 (Objective 2.1.3)

50. $5 (Objective 2.4.2)   **51.** $4000 @ 8%; $5000 @ 6.4% (Objective 2.5.1)   **52.** 105 oz (Objective 2.5.2)
53. 80,500 people (Objective 4.4.2)   **54.** 48 h (Objective 4.5.4)   **55.** 172.8 ft (Objective 5.5.2)
56. 6 h (Objective 7.5.1)   **57.** 3.5 ft (Objective 8.4.1)   **58.** $.14 (Objective 10.4.2)
59. 79 or better (Objective 2.2.3)   **60.** $51,874.85 (Objective 12.3.4)

Index

Table of Properties

Properties of Real Numbers

Associative Property of Addition

If a, b, and c are real numbers, then $(a + b) + c = a + (b + c)$.

Associative Property of Multiplication

If a, b, and c are real numbers, then $(a \cdot b) \cdot c = a \cdot (b \cdot c)$.

Commutative Property of Addition

If a and b are real numbers, then $a + b = b + a$.

Commutative Property of Multiplication

If a and b are real numbers, then $a \cdot b = b \cdot a$.

Addition Property of Zero

If a is a real number, then $a + 0 = 0 + a = a$.

Multiplication Property of One

If a is a real number, then $a \cdot 1 = 1 \cdot a = a$.

Inverse Property of Addition

If a is a real number, then $a + (-a) = (-a) + a = 0$.

Inverse Property of Multiplication

If a is a real number and $a \neq 0$, then

$$a \cdot \frac{1}{a} = \frac{1}{a} \cdot a = 1.$$

Distributive Property

If a, b, and c are real numbers, then $a(b + c) = ab + ac$.

Properties of Equality

Addition Property of Equality

If $a = b$, then $a + c = b + c$.

Multiplication Property of Equality

If $a = b$ and $c \neq 0$, then $a \cdot c = b \cdot c$.

Properties of Exponents

If m and n are integers, then $x^m \cdot x^n = x^{m+n}$.

If m and n are integers and $x \neq 0$, then

$$\frac{x^m}{x^n} = x^{m-n}.$$

If m and n are integers, then $(x^m)^n = x^{m \cdot n}$.

If x is a real number and $x \neq 0$, then $x^0 = 1$.

If m, n, and p are integers, then $(x^m \cdot y^n)^p = x^{m \cdot p} y^{n \cdot p}$.

If n is a positive integer and $x \neq 0$, then

$$x^{-n} = \frac{1}{x^n} \text{ and } \frac{1}{x^{-n}} = x^n.$$

Principle of Zero Products

If $a \cdot b = 0$, then $a = 0$ or $b = 0$.

Properties of Radical Expressions

If a and b are positive real numbers, then $\sqrt{ab} = \sqrt{a} \cdot \sqrt{b}$.

If a and b are positive real numbers, then

$$\sqrt{\frac{a}{b}} = \frac{\sqrt{a}}{\sqrt{b}}.$$